AVIONICS NAVIGATION SYSTEMS

AVIONICS NAVIGATION SYSTEMS

SECOND EDITION

Myron Kayton and Walter R. Fried

A WILEY-INTERSCIENCE PUBLICATION

JOHN WILEY & SONS, INC.

New York • Chichester • Weinheim • Brisbane • Singapore • Toronto

Library of Congress Cataloging in Publication Data:

Avionics navigation systems / Myron Kayton, Walter Fried [editors].
     p.   cm.
   Includes bibliographical references.
   ISBN 0-471-54795-6 (cloth : alk. paper)
   ISBN 978-0-471-54795-2
   1. Avionics.   2. Aids to air navigation.   I. Kayton, Myron.

II. Fried, Walter.
TL695.A82   1996
629.135′1–dc20

96-23729

CONTENTS

PREFACE

The purpose of this book is to present a unified treatment of the principles and practices of modern navigation sensors and systems. This second edition is a total rewrite of the first edition.

During the 28 years since the first edition was published, there have been tremendous changes in the science and practice of navigation: the introduction of navigation satellites that provide, for the first time in history, global, continuous precise navigation; an enormous increase in the speed and memory of digital computers, accompanied by a sharp decrease in their size and cost; the invention of clever algorithms, based primarily on Kalman filters, that mix the outputs of several sensors to produce a best estimate of position, velocity and, sometimes, of time; and the proliferation of avionics on aircraft, interconnected by digital data buses, so that navigation is only one of several avionic subsystems.

This book was written for the navigation system engineer, whether user or designer, who is concerned with the practical application of newly developed technology, and for the technical specialist who wishes to learn about adjacent specialties. It is an engineer-oriented text that will serve a wide spectrum of readers, from the systems analyst who writes mathematical models to operations personnel who want to learn about the avionics equipment in their aircraft. This book applies to civil and military aircraft, helicopters, and unmanned aerial vehicles. It covers the speed range from hovering helicopters to hypersonic transports. For all those vehicles, it discusses the state-of-the-art and the development of new systems that are likely to be introduced in the future.

Each chapter first presents basic functions and fundamental principles. It then discusses design characteristics, equipment configurations, sources of error, and typical performance levels. It closes with a projection of future trends. Topics such as comparative performance levels, weights, and costs of equipment are covered wherever possible. Most chapters assume a knowledge of undergraduate physics and mathematics; some assume a knowledge of electronic circuits.

References are collected at the end of the book, chapter by chapter, for the interested reader to use as background reading and to pursue the subject in more depth. The index is comprehensive enough to allow readers to find topics outside their area of specialty. It includes a glossary of acronyms. Chapters 2 through 15 conclude with illustrative problems that clarify points in the text and lead the reader into new areas. These problems will be useful to university instructors who use the text as part of a course in avionics, guidance and control,

xvii

or navigation. The chapters are extensively cross-referenced for the readers' convenience; *Section x.y.z.* points to Chapter x, Section y.z.

Chapter 1 discusses the role of electronic navigation equipment in the mission of civil and military aircraft. Chapters 2 and 3 discuss navigation principles, the equations that are the basis of all navigation systems, the calculation of navigation errors, and the mechanization of multisensor systems. Chapter 2 describes the principles of terrain-matching navigation systems. The first three chapters serve as the core for the next nine, which deal with sensors.

Chapter 4 discusses radio propagation on the surface of the Earth and the method of operation of traditional ground-based radio-navigation systems. Chapter 5 treats the principles and characteristics of satellite-based radio-navigation sytems, particularly GPS and GLONASS. Chapter 6 covers integrated communication-navigation systems used on battlefields. Chapter 7 discusses inertial navigation systems that provide navigation and attitude information on civil and military aircraft. Chapter 8 describes air-data sensors and algorithms that compute airspeed, angles of attack and sideslip, and barometric altitude. Chapter 9 describes the attitude and heading sensors that continue to be used for attitude-control and dead reckoning on all aircraft. Chapter 10 covers Doppler-radar navigators, which dead-reckon aircraft and military helicopters. Chapter 10 also describes radar altimeters that are used on civil aircraft for landing and on military aircraft for terrain-following and terrain-matching. Chapter 11 covers airborne mapping and multimode radars, terrain-avoidance radars, and weather radars. Chapter 12 covers celestial navigation and high-accuracy stellar-inertial systems.

The last three chapters cover the navigation environment. Chapter 13 discusses the mechanics of landing, electronic landing aids, and naval carrier-landing systems. Chapter 14 (also chapter 1) describes the worldwide air-traffic management environment in which civil and military aircraft operate. Chapter 15 discusses the interfaces among the navigation devices, other avionic devices, displays, and electric power.

Readers may wish to consult the first edition for information on systems that are now obsolete. The first edition contained chapters on analog and digital computers and displays, which were unfamiliar to many avionics engineers in the 1970s. It had a chapter on flight control, which is now a subject in its own right that is of interest to navigation engineers because navigation-derived steering commands are executed by the flight controls. The authors regret that, this second edition could not include every subject related to aircraft navigation. Cartography is discussed only as it relates to digital-map data bases and navigation coordinates. Automatic flight control, aerodynamic stability and control, weapon control, and localization of radar emitters are omitted.

The first edition was written by a small group of authors who spoke with a single voice. Some of them have retired and some have died; many no longer work in their former fields. We want to acknowledge those who could not participate in the second edition: Richard Andeen, John Andresen, Paul Astholz, Frank Brady, Sven Dodington, Dr. R. C. Duncan, Alton Moody, Glenn Quasius,

Seymour Schoen, T. J. Thomas, Carl Wiley and Willis Wing. The Acknowledgments explain which first-edition material was re-used.

Due to increasing specialization, the second edition was written by a much larger, more diverse team, whose members are the foremost current experts in their fields. We wish to thank them for their generous contributions of time in preparing drafts, editing, and re-editing in order to give you, the reader, a coherent, unified book.

While the art and science of navigation is hundreds of years old, the last 50 years have produced exciting new sensors and systems that permit an accuracy and level of safety never before seen on moving vehicles. We hope that this second edition presents the fundamentals and enough details to stimulate innovation and the development of ever-improving systems of navigation.

MYRON KAYTON

Santa Monica, California

WALTER R. FRIED

Santa Ana, California
January 1997

ACKNOWLEDGMENTS

Dr. Kayton wishes to acknowledge Clarence Asche of Honeywell for photos of inertial instruments, Anthony Bommarito of Wilcox for information about ILS and MLS landing aids, Phil Bruner and Wayne Knitter of Litton Industries for storage of magnetic models in guidance computers, Dennis Cooper, FAA representative in Moscow, for information about Russian air traffic control. Walter Fried for his contributions to Chapters 1, 2, and 9, Professors Frank von Graas and Per Enge for information about GPS landing aids, Dr. James Huddle of Litton Guidance and Control for information about mechanization techniques, Dr. David Y. Hsu of Litton Guidance and Control for using his software to calculate CEPs, the International Civil Aviation Organization for information about worldwide airspace regulations, Jeppesen-Sanderson for information about digital aeronautical maps, Dr. Robert Kelly, formerly of Bendix Communications for the definition of airways based on required navigation performance, Bob Knutson of Honeywell for information about air data, Dan Martinec of ARINC for publications, Harold Moses of RTCA for specifications, Bill Murray and Erv Ulbrich of McDonnell-Douglas for drawings, Norman Peddie and John Quinn of USGS for information about magnetic models, Walter Schoppe for information about naval communication links, and David Scull for various government documents.

Dr. Kayton included Willis Wing as a co-author of Chapter 9 because so much of his first edition material was reused, though Mr. Wing did not directly participate in the second edition.

Mr. Walter Fried wishes to thank Gregory Soloway of GEC Marconi Electronic Systems Corporation for information on JTIDS terminals.

Dr. Edward Knobbe and Dr. Gerald Haas wish to acknowledge that most of Sections 12.4.2 and 12.4.3, including Figure 12.7 and Tables 12.2 and 12.3, were written by Glenn Quasius for the first edition.

Dr. Clyde Miller wishes to thank Gene Wong and J. C. Johns of the FAA and Capt. Colin Miller of the U.S. Air Force for providing reference materials and for reviewing various sections of Chapter 14. Chapter 14 does not necessarily represent the official views of the U.S. government.

Dr. Jack Pearson wishes to acknowledge the use of some of Carl Wiley's (deceased) material from the first edition.

Dr. Bahar Uttam wishes to acknowledge the use of Sven Dodington's Sections 4.2, 4.3, and 4.4 from the first edition.

Dr. A. J. Van Dierendonck wishes to acknowledge the contributions of Dr.

R. Grover Brown on Receiver Autonomous Integrity Monitoring, Ed Martin on the GPS spacecraft, and Jack Klobuchar on ionospheric effects on satellite navigation.

Mr. Doug Vickers wishes to thank Robert J. Bleeg of Boeing Commercial Airplane Division and Steve Osder for reviewing the flight-control sections of Chapter 13.

Dr. William Waters wishes to acknowledge the contributions of Robert Wigginton of the U.S. Naval Electronic Systems Engineering Activity in Section 13.8.

Mr. Edgar Westbrook wishes to thank Mr. Wayne Altrichter of GEC-Marconi Electronic Systems Corp. for his review of Section 6.2. He wishes to recognize Mr. Robert C. Snodgrass of the MITRE Corporation (retired) for his many contributions to the development of JTIDS RelNav.

The cover photograph is courtesy of Rockwell.

LIST OF CONTRIBUTORS

THOMSON S. ABBOTT, JR. (Co-author, Chapter 11), Hughes Aircraft Company, El Segundo, CA

DAVID H. AMOS (Co-author, Chapter 4), Senior Director, Systems Engineering, Synetics Corporation, Wakefield, MA

R. GROVER BROWN, Ph.D. (Co-author, Chapter 3), Distinguished Professor Emeritus, Iowa State University, Clear Lake, IA

HEINZ BUELL (Co-author, Chapter 10), Senior Member of Technical Staff, GEC Marconi Electronic Systems Corporation, Wayne, NJ

JOSEPH M. COVINO (Co-author, Chapter 4), Senior Engineer, Synetics Corporation, Wakefield, MA

WALTER R. FRIED, M.S. (Editor; lead author, Chapters 6 and 10), Consultant, Hughes Aircraft Company, Santa Ana, CA

GERALD N. HAAS, Ph.D. (Co-author, Chapter 12), Senior Research Engineer, Northrop-Grumman Electronic Systems, Hawthorne, CA

JAMES R. HAGER (Co-author, Chapter 10), Honeywell Military Avionics, Minneapolis, MN

JAMES R. HUDDLE, Ph.D. (Lead author, Chapter 3), Chief Scientist and Head of Advanced System Engineering, Litton Guidance and Control Division, Woodland Hills, CA

ROBERT H. JEFFERS, Ph.D. (Co-author, Chapter 11), Senior Scientist, Hughes Aircraft Company, El Segundo, CA

MYRON KAYTON, Ph.D., P.E. (Editor; author of Chapters 1, 2; co-author Chapters 7 and 13; lead author, Chapter 9), Consulting Engineer, Kayton Engineering Company, Santa Monica, CA

JAMES A. KIVETT (Co-author, Chapter 6), Hughes Aircraft Company, El Segundo, CA

EDWARD J. KNOBBE, Ph.D. (Lead author, Chapter 12), Advanced Systems Scientist (retired), Northrop-Grumman Electronic Systems, Hawthorne, CA

JOHN G. MARK, Ph.D. (Co-author, Chapter 7),   Chief Scientist, Litton Guidance and Control Division, Woodland Hills, CA

RICHARD H. MCFARLAND, Ph.D., P.E. (Co-author, Chapter 13),   Director, Emeritus, Avionics Engineering Center. Ohio University, Athens, OH

CLYDE A. MILLER, Ph.D. (Lead author, Chapter 14),   Program Director for Research, Federal Aviation Administration, Washington, DC

PETER MORRIS (Co-author, Chapter 4),   The Analytical Sciences Corporation, Reading, MA

STEPHEN S. OSDER (Author, Chapter 8),   Consultant, formerly McDonnell-Douglas Fellow, Scottsdale, AZ

JACK O. PEARSON, Ph.D. (Lead author, Chapter 11),   Vice President, Radar and Communication Systems, Hughes Aircraft Company, El Segundo, CA

JOHN A. SCARDINA, Ph.D. (Co-author, Chapter 14),   Team Leader for Air Traffic Management, Federal Aviation Administration, Washington, DC

CARY R. SPITZER (Author, Chapter 15),   President, AvioniCon, formerly, National Aeronautics and Space Administration, Williamsburg, VA

DANIEL A. TAZARTES (Lead author, Chapter 7),   Senior Member of Technical Staff, Litton Guidance and Control Division, Woodland Hills, CA

BAHAR UTTAM, Ph.D. (Lead author, Chapter 4),   President, Synetics Corporation, Wakefield, MA

A. J. VAN DIERENDONCK, Ph.D. (Author, Chapter 5),   AJ Systems, Los Altos, CA

D. B. VICKERS, M.S. (Lead author, Chapter 13),   Technical Director, Avionics Engineering Center. Ohio University, Athens, OH

WILLIAM M. WATERS, Ph.D. (Co-author, Chapter 13),   Senior Consultant, Naval Research Laboratory, Washington, DC

EDGAR A. WESTBROOK, (Co-author, Chapter 6),   Technical Staff, retired, The MITRE Corporation, Bedford, MA

WILLIS G. WING, (Co-author, Chapter 9),   Sperry Gyroscope Company, retired, Glen Head, NY

1 Introduction

1.1 DEFINITIONS

Navigation is the determination of the position and velocity of a moving vehicle. The three components of position and the three components of velocity make up a six-component *state vector* that fully describes the translational motion of the vehicle. Navigation data are usually sent to other on-board subsystems, for example, to the flight control, flight management, engine control, communication control, crew displays, and (if military) weapon-control computers.

Navigation sensors may be located in the vehicle, in another vehicle, on the ground, or in space. When the state vector is measured and calculated on board, the process is called *navigation*. When it is calculated outside the vehicle, the process is called *surveillance* or *position location*. Surveillance information is employed to prevent collisions among aircraft. The humans and computers that direct civil air traffic and most military traffic are located in Air Route Traffic Control Centers on the ground, whereas some military controllers are based in surveillance aircraft or aircraft carriers. Existing traffic control systems observe the position of aircraft using sensors outside the aircraft (e.g., surveillance radars) or reports of position from the aircraft itself. "Automatic dependent surveillance" is a term for the reporting of position, measured by sensors in an aircraft, to a traffic control center.

Traditionally, *ship navigation* included the art of pilotage: entering and leaving port, making use of wind and tides, and knowing the coasts and sea conditions. However, in modern usage, *navigation* is confined to the measurement of the state vector. The handling of the vehicle is called *guidance*; more specifically, it is called *conning* for ships, *flight control* for aircraft, and *attitude control* for spacecraft. This book is concerned only with the navigation of manned and unmanned aircraft. The calculation of the navigation state vector requires the definition of a navigation coordinate frame (as discussed in Chapter 2).

1.2 GUIDANCE VERSUS NAVIGATION

The term "guidance" has two meanings, both of which are different from "navigation":

1. Steering toward a destination of known position from the aircraft's present position. The steering equations can be derived from a plane triangle for nearby destinations or from a spherical triangle for distant destinations (Chapter 2).

2. Steering toward a destination without explicitly measuring the state vector. A guided vehicle can home on radio, infrared, or visual emissions. Guidance toward a *moving* target is usually of interest for military tactical missiles in which a steering algorithm ensures impact within the maneuver and fuel constraints of the interceptor. One of several related guidance algorithms, collectively called *proportionial navigation*, processes sensor data and steers the vehicle to impact. Guidance toward a *fixed* target involves beam-riding, as in the Instrument Landing System (Chapter 13).

1.3   CATEGORIES OF NAVIGATION

Navigation systems can be categorized as positioning or dead-reckoning. *Positioning systems* measure the state vector without regard to the path traveled by the vehicle in the past. There are three kinds of positioning systems:

1. *Radio systems (Chapters 4 to 6).* They consist of a network of transmitters (sometimes also receivers) on the ground, in satellites, or on other vehicles. The airborne navigation set detects the transmissions and computes its position relative to the known positions of the stations in the navigation coordinate frame. The aircraft's velocity is measured from the Doppler shift of the transmissions or from a sequence of position measurements.

2. *Celestial systems (Chapter 12).* They compute position by measuring the elevation and azimuth of celestial bodies relative to the navigation coordinate frame at precisely known times. Celestial navigation is used in special-purpose high-altitude aircraft in conjunction with an inertial navigator. Manual celestial navigation was practiced at sea for millennia and in aircraft from the 1930s to the 1960s.

3. *Mapping navigation systems (Section 2.6).* They observe images of the ground, profiles of altitude, or other external features.

Dead-reckoning navigation systems derive their state vector from a continuous series of measurements relative to an initial position. There are two kinds of dead-reckoning measurements:

1. Aircraft heading and either speed or acceleration. For example, heading can be measured with gyroscopes (Chapter 7) or magnetic compasses (Chapter 9), while speed can be measured with air-data sensors (Chapter

8) or Doppler radars (Chapter 10). Vector acceleration is measured with inertial sensors (Chapter 7).

2. Emissions from continuous-wave radio stations. They create ambiguous "lanes" (Chapter 4) that must be counted to keep track of coarse position. Their phase is measured for fine positioning. They must be reinitialized after any gap in radio coverage.

Dead-reckoning systems must be re-initialized as errors accumulate and if electric power is lost.

1.4   THE VEHICLE

1.4.1   Civil Aircraft

The civil aviation industry consists of air carriers and general aviation. Air carriers operate large aircraft used on trunk routes and small aircraft used in commuter service. General aviation ranges from single-place crop dusters to well-equipped four-engine corporate jets.

Most air carriers and general-aviation jet aircraft operate exclusively in developed areas where ground-based radio aids are plentiful. Others operate over oceans and undeveloped areas where, before the Global Positioning System (GPS, Chapter 5), navigation aids were nonexistent. Before the 1970s, such aircraft had astrodomes through which a human navigator took celestial fixes with a bubble-sextant (Chapter 12). From the 1970s to 2000, aircraft flying over oceans and undeveloped areas used unaided inertial systems or Omega (Chapter 4). By the year 2000, most of these aircraft will use GPS alone or in combination with inertial systems (Chapter 7). Beginning in the mid-1980s, the US-FAA allowed overwater flight with a single long-range navigation set and a separate single long-range communication set.

Simple general-aviation aircraft (including helicopters) operate over short routes, have two or fewer engines, and are flown by one or two pilots. They are used for water drops on fires, search-and-rescue, ferrying crews to offshore oil platforms, police patrols, interplant shuttles, crop dusting, and carrying logs from forests, for example. Each usage has its own navigation requirements. The simplest aircraft navigate visually or with Loran or GPS sets; the more complex aircraft use the same navigation equipment as do air carriers. In 1996 civil helicopters used VOR/DME (see Chapter 4) in developed areas. They landed visually because their approach paths were too steep for the instrument landing system (ILS, Chapter 13). Many will adopt GPS for instrument approaches.

Civil aircraft fly in a benign environment; the major electrical stresses on avionic equipment are caused by lightning and electric-power transients; the major mechanical stresses are caused by air turbulence, hard landings, and abusive handling by maintenance technicians. Figure 1.1 shows the antenna farm and avionics bays on an advanced transport that is outfitted for civil and mil-

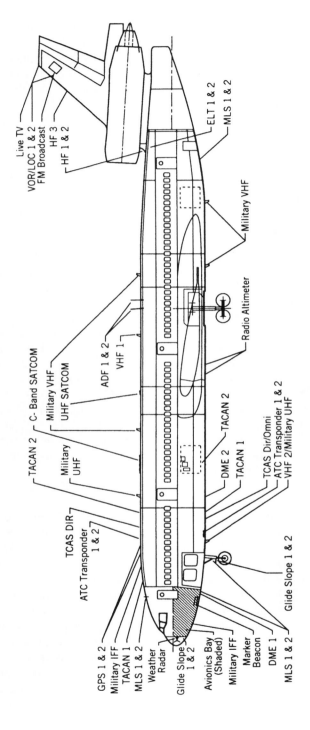

Figure 1.1 Avionics placement on multi-purpose transport (Courtesy of McDonnell Douglas, modified by author).

itary usage. The avionics bay is below the cockpit in the space between the radome and nose wheel well (in many civil aircraft, the avionics bay is aft of the nosewheel). Avionics and air-data sensors are located in the bay. Access is beneath the aircraft.

Trans-Pacific hypersonic aircraft may be developed in the twenty-first century that will navigate as does the Space Shuttle: inertial boost, GPS or celestial midcourse, and GPS or other radio approach. They will compete with electronic mail and teleconferences.

1.4.2 Military Aircraft

Fixed-wing military aircraft can be divided as follows:

1. *Interceptors and combat air patrols.* These small, high-climb-rate aircraft protect the homeland, a naval fleet, or an invasion force by seeking and destroying invading bombers, cruise missiles, and aircraft that carry contraband. Interceptors are vectored to their targets by ground-based, ship-borne, or airborne command posts. Interceptors carry on-board air-to-air radar (Chapter 11) to close on their targets. They use inertial navigation (Chapter 7) or Tacan (Chapter 4) to return to their bases. GPS may replace Tacan for returning to fixed bases leaving Tacan for returning to aircraft carriers.

2. *Close-air support.* These medium-sized aircraft deliver weapons in support of land armies. They may attack troops, tanks, convoys, or command centers. They have inertial navigators or GPS to locate the approximate position of targets and have sensors (e.g., optics and moving-target-indicating (MTI) radar) to locate the precise position of the targets. They carry communication systems that keep them in contact with local troop commanders and airborne command posts. These aircraft have relied on inertial navigation and Tacan to return to their bases. Close-air support aircraft carry inertial navigators as precise attitude references for optical sensors and as velocity references for releasing weapons.

3. *Interdiction.* These medium-sized and large aircraft strike behind enemy lines to attack strategic targets such as factories, power plants, and military installations. Nuclear strategic bombers and fighter bombers are included in this category. These aircraft carry the most precise navigation systems, based on inertial, GPS, and celestial sensors. They may obtain en-route position fixes with optical or radar image comparators or terrain matching (Section 2.6). Inertial navigators provide precise attitude and velocity references for pointing terminal-area optic sensors and for releasing weapons. Interdiction aircraft often have sensors that find tanker aircraft and allow formation flight in instrument weather conditions. Flying at treetop level to avoid enemy radars complicates the task of navigation.

4. *Cargo carriers.* These aircraft have the same navigation requirements as do civil aircraft; in addition, they drop cargos by parachute and refuel from tankers. Cargo drops require flight along a predetermined path and release at predetermined positions. Cargo aircraft are also sometimes outfitted as refueling-tankers and mobile hospitals. Tankers are equipped with radar beacons to aid in rendezvous. They may be asked to make Category III landings at third-world airports.

5. *Reconnaissance aircraft.* These aircraft collect photographic and electronic-signals data. They navigate precisely in order to annotate the data and fly close to hostile borders. They measure velocity precisely in order to compensate cameras and synthetic-aperture radars for vehicle motion.

6. *Helicopter and short-takeoff-and-landing* (STOL) *vehicles.* Military helicopters often support troops, for example, to attack tanks, to suppress artillery and small-arms fire, and to transport soldiers and casualties. They search for and destroy submarines from their bases on large ships. They measure position, so they can locate targets in the coordinate frames established by command posts. Most of them navigate by visual pilotage. Some Navy and Army helicopters dead-reckon with Doppler radar and compass (Chapters 9 and 10) using Tacan to return to their ships or land bases. They measure velocity precisely, so they can hover, handover targets, launch weapons, and transition from vertical to horizontal flight. Airspeed is difficult to measure due to the downwash from the rotors. Doppler radar can establish a coordinate frame fixed in the moving ocean surface that is useful when working with submarine-detecting sonobuoys. Search-and-rescue helicopters carry receivers that detect and direction-find emergency locator transmitters [8]. Complex helicopter weapon-platforms carry inertial navigators and optical imagers for locating targets and for landing.

7. *Unmanned air vehicles (once called "remotely piloted vehicles").* They range in size from model airplanes to ten thousand kilograms. They are used as target drones, reconnaissance vehicles, and strategic bombers. They attack high-risk targets (radiation-emitting antennas and artillery) without endangering the lives of a crew. Some have elaborate inertial, map-matching, and acoustic sensors; some carry Doppler radars. They often navigate inertially until their on-board optical or infrared sensors acquire the target, then guide themselves to impact with submeter accuracy.

Virtually every military aircraft carries an instrument-landing system (ILS) receiver (Chapter 13). In the past, some relied on a "ground-controlled approach" in which a human, watching a radar display on the ground or on an aircraft carrier, radioed steering commands to the pilot. The special problems of landing on an aircraft carrier are discussed in Chapter 13. Military aircraft often engage in high-speed, high-g, low-altitude maneuvers that challenge the mechanical design of on-board avionics. Guns and rocket launchers impose shock and vibration loads. By the year 2000, most military aircraft will carry GPS receivers.

1.5 PHASES OF FLIGHT

1.5.1 Takeoff

The takeoff phase begins upon taxiing onto the runway and ends when climb-out is established on the projected runway centerline. The aircraft is guided along the centerline by hand-flying or a coupled autopilot based on steering signals (from an ILS localizer since 1945). Two important speed measurements are made on the runway. The highest ground speed at which an aborted takeoff is possible is precomputed and compared, during the takeoff run, to the actual ground speed as displayed by the navigation system. The airspeed at which the nose is lifted ("rotation") is precalculated and compared to the actual airspeed as displayed by the air-data system. Barometric altitude rate or GPS-derived altitude rate (inertially smoothed) is measured and monitored.

1.5.2 Terminal Area

The terminal phase consists of departure and approach subphases. *Departure* begins when the aircraft maneuvers away from the projected runway centerline and ends when it leaves the terminal-control area (by which time it is established on an airway). *Approach* begins when the aircraft enters the terminal area (by which time it has left the airway) and ends when it intercepts the landing aid at an *approach fix*. In 1996, vertical navigation was based on barometric altitude, and heading vectors were assigned by traffic controllers. Major airports have standard approach and departure routes unique to each runway. In the United States, the desired terminal-area navigation accuracy is 1.7 nmi, 2-sigma per Advisory Circular 20-130. Further details are in Chapter 14.

1.5.3 En Route

The en-route phase leads from the origin to the destination and alternate destinations (an alternate destination is required of civil aircraft operating under instrument flight rules). From the 1930s to the 1990s, airways were defined by navigation aids over land and by latitude-longitude fixes over water. The width of airways and their lateral separation depended on the quality of the defining navaids and the distance between them. The introduction of inertial navigation systems and DME in the 1970s caused aviation authorities to create "area-navigation" airways (RNAV) that do not always interconnect VOR navaids [7: AC-90-45A] (see Section 2.7.4).

Beginning in the 1990s, GPS has allowed precise navigation anywhere, not just on airways. Given the extensive use of on-board collision-avoidance equipment and the trend toward reducing government budgets, "free-flight" is being introduced in controlled airspace. Each aircraft would agree on a route before takeoff and then be free to change the route, after interaircraft communication verified that the risk of collision is sufficiently low. En-route surveillance by independent ground-based radars may disappear or be replaced by position fixes and reports via com-nav satellites. Busy terminal areas and airport surfaces are likely to remain under central, positive control.

In the United States in 1996, the en-route navigation error must be less than 2.8 nmi over land and 12 nmi over oceans (2-sigma) [7: AC-20-130]. As regional maps become available in digital form with aeronautical annotation (e.g., minimum en-route altitude), aircraft in undeveloped areas will use GPS for en-route navigation and nonprecision approaches.

1.5.4   Approach

The approach phase begins at acquisition of the landing aid and continues until the airport is in sight or the aircraft is on the runway, depending on the capabilities of the landing aid (Chapter 13).

During an approach, the *decision height* (DH) is the altitude above the runway at which the approach must be aborted if the runway is not in sight. The better the landing aids, the lower the decision height. Decision heights are published for each runway at each airport (Chapter 13). The decision height for a Category III landing is 100 ft or less. By law, an approach may not even be attempted unless the horizontal visibility, measured by a runway visual range (RVR) instrument, exceeds a threshold that ranges from zero (Category IIIC, not approved anywhere in 1996) to 800 meters (Category I). A *nonprecision approach* has electronic guidance only in the horizontal direction. An aircraft executing a nonprecision approach must abort if the runway is not visible at the *minimum descent altitude*, which is typically 700 ft above the runway.

In 1996, civil aircraft outside the ex-Soviet bloc used the Instrument Landing System for low-ceiling, low-visibility approaches. A Microwave Landing System had been approved by the International Civil Aviation Organization for precision approaches and was being installed at major international airports, especially in Europe. In the United States, Loran and GPS had been approved for nonprecision approaches at many airports. (Landing aids are discussed in Chapter 13.)

1.5.5   Landing

The landing phase begins at the decision height (when the runway is in sight) and ends when the aircraft exits the runway. Navigation during flare and decrab may be visual or the navigation set's electrical output may be coupled to an autopilot. A radio altimeter measures the height of the main landing gear above the runway for guiding the flare. The rollout is guided by the landing aid (e.g., the ILS localizer). Landing navigation is described in Chapter 13.

1.5.6 Missed Approach

A missed approach is initiated at the pilot's option or at the traffic controller's request, typically because of poor visibility, poor alignment with the runway, equipment failure, or conflicting traffic. The flight path and altitude profile for a missed approach are published on the approach plates. The missed approach consists of a climb to a predetermined holding fix at which the aircraft awaits further instructions. Terminal area navigation aids are used.

1.5.7 Surface

Aircraft movement from the runway to gates, hangars, or revetments is a major limit on airport capacity in instrument meteorological conditions. Surface navigation is visual on the part of the crew, whereas the ground controllers observe aircraft visually or with a surface surveillance radar. No matter how good the surface navigation, collision avoidance among aircraft and ground vehicles requires central guidance, typically provided by a human controller with computer assistance. Position reports (e.g., via GPS) from aircraft that are concealed in radar shadows reduce the risk of collision and help keep unwanted aircraft off active runways.

1.5.8 Weather

Instrument meteorological conditions (IMC) are weather conditions in which visibility is restricted, typically less than 3 miles as defined by law. Aircraft operating in IMC are supposed to fly under *instrument flight rules* (IFR), defined by law in each country (Chapters 13 and 14).

1.6 DESIGN TRADE-OFFS

The navigation-system designer conducts trade-offs for each aircraft and mission to determine which navigation systems to use. Trade-offs consider the following attributes:

1. *Cost.* Included are the construction and maintenance of transmitter stations and the purchase of on-board hardware and software. Users are concerned only with the cost of on-board hardware and software. In the past, governments have paid to operate radio-navigation transmitters. In the future, combined com-nav aids may be operated privately and funded by user charges.

2. *Accuracy of position and velocity.* This is specified in terms of the statistical distribution of errors as observed on a large number of flights [4]. The accuracy of military systems is often characterized by circular error probable (CEP, in meters or nautical miles; Chapter 2). The maximum

allowable CEP is frequently established by the kill radius of the weapons that are released from the aircraft. For civil air carriers, the allowable en-route navigation error is based on the calculated risk of collision. In the 1990s, each subsystem was allocated a safety-related failure probability of 10^{-9} per hour [4]. The accuracy of the navigation systems is often defined as "twice the distance root mean square" (2drms), which encompasses 95% to 98% of the errors (Section 2.8.1). The allowable landing error depends on runway width, aircraft handling characteristics, and flying weather.

3. *Autonomy.* This is the extent to which the vehicle determines its own position and velocity without external aids. Autonomy is important to certain military vehicles and to civil vehicles operating in areas of inadequate radio-navigation coverage. Autonomy can be subdivided into five classes:

- Passive self-contained systems that neither receive nor transmit electromagnetic signals. They emit no radiation that would betray their presence and require no external stations. Failures are detected and corrected on board. They include dead-reckoning systems such as inertial navigators.

- Active self-contained systems that radiate but do not receive externally generated signals. Examples are radars and sonars. They do not depend on the existence of navigation stations.

- Receivers of natural radiation. These systems measure naturally emitted electromagnetic radiation. Examples are magnetic compasses, star trackers, and passive map correlators. Some unmanned military weapons guide themselves toward acoustic emissions. These systems do not announce their presence by emitting, nor do they need navigation stations.

- Receivers of artificial radiation. These systems measure electromagnetic radiation from navaids (Earth based or space based) but do not themselves transmit. Examples are Loran, Omega, VOR (Chapter 4), and GPS (Chapter 5). They require external cooperating stations but do not betray their own presence.

- Active radio navaids that exchange signals with navigation stations. These include DME, JTIDS, PLRS, and collision-avoidance systems (Chapters 4, 6, and 14). The vehicle betrays its presence by emitting and requires cooperative external stations. These are the least autonomous of navigation systems.

4. *Time delay in calculating position and velocity, caused by computational and sensor delays.* Time delay (also called *latency*) can be caused by computer-processing delays, scanning by a radar beam, or gaps in satellite coverage, for example. Forty years ago, it took five minutes to plot a fix manually on an on-board aeronautical chart. Today, navigation

calculations are completed in tens of milliseconds by a digital computer.

5. *Geographic coverage.* Terrestrial radio systems operating below approximately 100 KHz can be received beyond line of sight on Earth; those operating above approximately 100 MHz are confined to line of sight. Each satellite can cover millions of square miles of Earth, while a constellation of satellites can cover the entire Earth.

6. *Automation.* The aircraft's crew receives a direct reading of position, velocity, and equipment status, usually without human intervention, as described in Section 1.9. In years past, navigation sets were operated by skilled people, to the extent of manipulating wave forms on cathode-ray tubes.

7. *Availability.* This is the fraction of time that the system is usable for navigation. Downtime is caused by scheduled maintenance, by unscheduled outage (usually due to equipment failure), and by radio-propagation problems that cause excessive errors.

8. *System capacity.* This is the number of aircraft that the system can accommodate simultaneously. It applies to two-way ranging systems.

9. *Ambiguity.* This is the identification, by the navigation system, of two or more possible positions of the aircraft, with no indication of which is correct. Ambiguities are characteristic of ranging and hyperbolic systems when too few stations are received.

10. *Integrity.* This is the ability of the system to provide timely warnings to aircraft when its errors are excessive. For en-route navigation in 1996, an alarm must be generated within 30 seconds of the time a computed position exceeds its specified error. For a nonprecision landing aid, an alarm must be generated within ten seconds. For a precision landing aid, an alarm must be generated within two seconds. Integrity is an important issue for GPS, especially when it is used as a landing aid in the differential mode (Chapters 5 and 13). Any sensor that is the sole means of navigation must have high integrity.

1.7   EVOLUTION OF AIR NAVIGATION

The earliest aircraft were navigated visually. Pilots had an anemometer for airspeed, a barometer for altitude, and a magnetic compass for heading. Artificial horizons and turn-and-bank indicators allowed pilots to hold attitude and heading in clouds, hence motivating the installation of navigation aids. Lighted beacons were installed across the United States in the 1920s to mark airmail routes. Starting in 1929, four-course radio beacons were also added to the lighted airways to guide aircraft. Four-course beacons were installed in France, South America, and North Africa. In the 1930s, aircraft were equipped with medium-frequency and high-frequency direction finders (MF/DF and HF/DF) that measured the bearing of broadcast stations relative to the axis of the aircraft. A fix was obtained by plotting the direction toward two or more stations. Beacons near an airport allowed aircraft to fly a "nonprecision approach" to the runway (Chapter 13). Vertical beacons at 75 MHz, called *z-beacons* or *marker beacons*, were installed along the four-course airways and along approaches to runways to give a positive indication of position (Chapters 13 and 14).

Air-traffic control was procedural, following the precedent of railroad "block" clearances. Overland airways connected radio beacons; overwater airways were defined on a map, hundreds of miles apart. The airways were divided into longitudinal blocks of 20- to 30-minutes flying time. The air-traffic controller relied on the pilots' report of position, allowed only one aircraft at a time to enter a block, and kept the block free of other traffic until the pilot reported leaving. The size of the block was commensurate with the uncertainties in navigation at the time.

During World War II, meteorologists learned to route aircraft to take advantage of the cyclonic winds that circle around high- and low-pressure regions at mid- and high-latitudes. Bellamy [12] states that the transatlantic flying time was reduced an average of 10% compared to a great-circle track, with occasional savings exceeding 25%, by taking advantage of cyclonic tail winds. These *pressure-pattern* routes were plotted graphically in the 1940s–1960s but are now computed routinely in airline and military dispatch offices.

Crosswinds cause an aircraft to "drift" perpendicularly to its longitudinal axis. From the 1930s to the 1960s, drift angle was measured in flight with a downward-looking telescope that observed the direction of movement of the ground, when it was visible. From the 1940s to the 1960s, drift was estimated over oceans by observing trends in the difference, D, between the readings of the radio altimeter and pressure altimeter. Bellamy showed that in cyclonic winds, drift is proportional to the horizontal gradient of D [12]. The introduction of Doppler and inertial navigators in the 1960s and 1970s allowed drift to be observed directly. The Doppler navigator measures the direction of the ground-speed vector relative to the aircraft's centerline. The inertial navigator subtracts in-flight-measured airspeed from the measured ground velocity to calculate wind, hence lateral drift.

After World War II, VOR stations (Chapter 4) and Instrument Landing Systems (ILS, Chapter 13) were installed. VOR/DME and ILS have been the basis of navigation in western countries ever since. During the 1960s, air-traffic controllers came to rely on surveillance radar in densely populated airspace (Chapter 14). The controller identified the aircraft on his screen, hence eliminating the need for a position report from the crew. Radar surveillance of air traffic is called "positive control," which, in 1996, existed in the United States, most of Canada, western Europe, and Japan. In the late 1990s, the automatic reporting of on-board-derived position began to supplement (perhaps eventually to replace) radar surveillance.

The former Soviet republics have ICAO navigation aids and ILS at about 50 international airports and on corridors connecting them to the borders. Overflying western aircraft navigate inertially and with Omega, GPS, and nondirectional beacons. Since the late 1960s, domestic civil and military aircraft have used an L-band range-angle system known by its Russian acronym, RSBN, and not standardized by ICAO. It has 176 channels between 873 and 1000 MHz. Domestic airports guide landing aircraft with ground-based precision approach radar (PAR) using verbal commands to the crew. At international airports, PARs monitor aircraft on ILS approaches. In the 1990s, the former Soviet republics were purchasing western avionics equipment.

The People's Republic of China depended on imported Russian nondirectional beacons and PARs until the late 1970s, when it began to install western radars, ILS, VOR, and DME. In the 1990s, China installed VHF air-to-ground radio relays throughout most of the nation [13]. In 1996, western air-traffic control and navigation equipment was being installed throughout Southeast Asia and Indonesia.

Outside the developed world, major cities and some airways had VOR/DME-based procedural traffic control, so aircraft filing flight plans could be separated from each other by human controllers. Polar areas, the South Atlantic Ocean, and much of the Pacific and Indian Oceans had no navaids and no control whatsoever. Most of the rest of the world was divided into Flight Information Regions that advised crews of weather conditions and the status of airports and navigation aids but did not separate traffic. Position reports over oceans and in remote areas are mostly by HF radio but, beginning in the 1990s, were being made via satellite (e.g., North Pacific and Atlantic Oceans). In 1996, a few airlines were transmitting GPS-inertial position over digital data links via geostationary communication satellites over the Pacific Ocean, a system called Automatic Dependent Surveillance, the first step in the Future Air Navigation System (FANS, Chapter 14). Outside the United States and Canada, most aircraft pay directly for traffic control services.

Until the 1970s, precise absolute time could not economically be measured on a vehicle. Hence, radio navigation aids were built that measured the difference in time of arrival of radio signals from ground stations. The earliest (hyperbolic Loran and Decca, some military systems) date from the 1940s. As airborne clocks became more stable in the 1970s, "passive" or "one-way" ranging systems could solve for position and the absolute clock offset by processing precisely timed signals from several stations. *Direct-ranging* Loran and Omega (as distinguished from *hyperbolic* Loran and Omega, all discussed in Chapter 4), GPS and GLONASS (Chapter 5), and JTIDS (Chapter 6) are examples of such one-way ranging systems. As airborne clocks become more accurate in the twenty-first century, absolute time of arrival will be directly measurable and clock offsets will become negligible.

GPS and GLONASS are based on one-way passive range measurements to several stations, most of which are spacecraft (Figure 1.2). A few stations are ground-based *pseudolites* whose transmissions mimic those of spacecraft. Chapter 5 describes the GPS and GLONASS systems. The receiver in the airplane computes position, velocity, the offset in the airborne clock, and, in some

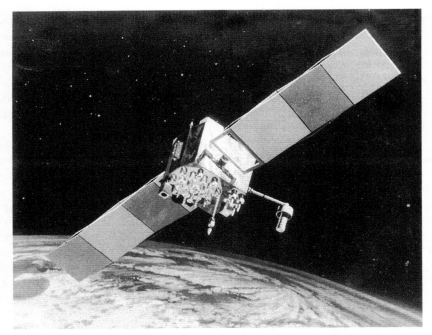

Figure 1.2   Global Positioning System Spacecraft, Block IIF (courtesy of Rockwell). L-band antenna array, S-band control antenna, and solar array are visible.

receivers, the ionospheric delay (Chapter 5). In 1996, the military modes of GPS achieved 20-meter (2drms) accuracy anywhere in the world, while the civil mode could achieve 40-meter accuracy but was intentionally degraded to 100-meters, a handicap to civil navigation that may be discontinued before the year 2000. The United States and Russia have announced that GPS and GLONASS will be available worldwide, free of charge, for a least 15 years and thereafter with 6 years' warning of the end of service. Nevertheless, worldwide civil authorities are reluctant to rely on military-controlled navigation aids that might be switched off or degraded during hostilities. The advent of continuous GPS allows the use of AHRS-quality inertial/attitude-reference systems (Chapters 7 and 9) in all but the most demanding military applications. The undetected loss of a navigation signal or the failure of a receiver could be catastrophic, especially during a landing at low decision height.

A widespread method of improving GPS accuracy and monitoring the signals is to install a ground station that receives GPS signals and transmits position errors or ranging errors and satellite failure status on a radio link to nearby aircraft. This *differential GPS* (DGPS) can achieve centimeter accuracy for fixed observers and 1- to 5-meter accuracy on aircraft that can solve at tens of iterations per second or whose velocity calculations are smoothed by an inertial navigator. The United States was experimenting with a nationwide DGPS system

(Wide Area Augmentation System, WAAS; Chapter 5) that could eventually replace the network of VORTACs.

The GPS and GLONASS systems are expensive to operate, each costing nearly a billion dollars per year for the replacement of satellites and the maintenance of the ground-control and monitoring network. The cost of collecting user charges (e.g., by selling encryption keys or taxing receivers) would exceed the revenue that could be extracted from navigation-only users. Hence, in the next generation, GPS transmitters will be installed on low-cost communication satellites as a way to augment the GPS network or as a low-cost replacement for dedicated GPS satellites. Governments would still maintain the control and monitoring stations that calculate the orbits and uplink data for rebroadcast. Taxpayer support is more likely if GPS becomes widely used in automobiles.

If present trends continue toward fee for service, taxpayer-funded navigation aids may cease to exist circa 2020, when commercial com-nav satellites will have superimposed ranging codes on their communication signals. Communication and navigation would then be available on a per-call basis, forcing aircraft again to rely on precise dead reckoning between intermittent fixes, probably from self-contained panel-mounted micro-machined inertial instruments (Chapters 7 and 9).

JTIDS and PLRS (Chapter 6) are military com-nav systems that constitute a battlefield-sized network whose terminals are in command centers and vehicles. PLRS terminals can be backpacked by soldiers.

Since airborne digital computers became available in the 1960s, algorithms have been invented and perfected that combine the measurements of diverse navigation sensors to create a "best estimate of position and velocity". They are used in "hybrid" navigation systems. From 1970 to the end of the century, various forms of Kalman filters were favored for combining data from diverse navigation sensors (Chapter 3).

1.8   INTEGRATED AVIONICS

1.8.1   All Aircraft

"Navigation" is one of several electronic subsystems, collectively called *avionics*. The other subsystems are as follows:

1. *Communication.* An airplane's communication system consists of an intercom among the crew members and one or more external two-way voice and data links.
2. *Flight control.* This consists of *stability augmentation* and *autopilot*. The former points the airframe and controls its oscillations, while the latter provides such functions as attitude-hold, heading-hold, and altitude-hold. Flaps, slats, and spoilers are often controlled electronically in addition to rudder, elevator, and ailerons.

3. *Engine control.* This is the electronic control of engine thrust, often called *throttle management.* Afterburner and thrust reversers may be controlled manually, perhaps via a thrust-by-wire control system.

4. *Flight management.* This subsystem stores the coordinates of en-route waypoints and calculates the steering signals to fly toward them. It calculates climb and descent profiles that may be followed with or without constraints on the time at which designated fixes and altitudes are crossed. Crossing fixes at predetermined times and altitudes is sometimes called *four-dimensional navigation;* it requires that the flight management subsystem control engine thrust. In 1996 all flight management subsystems stored waypoints in digital form. By the year 2000 many will store digital maps of the en-route airspace, standard approaches (called *STARs* in the United States), standard departures (called *SIDs* in the United States), approach plates, and checklists (see Chapter 14).

5. *Subsystem monitoring and control.* Faults in all subsystems are displayed, as are recommended actions to be taken. This subsystem includes wired logic and software for the automatic reconfiguration of faults in time-critical subsystems (e.g., flight control, where a fault can destroy the aircraft in less than three seconds). Quick-responses to safety-critical faults were automated in flight-control systems by the 1980s. In the 1990s, the trend was to automate the responses to slower-acting faults, thus reducing the workload in one-pilot and two-pilot aircraft. The failure-monitoring subsystem may include an on-board maintenance recorder, the radio transmission of faults to reduce repair time, and an accident recorder whose data survive a crash (required by law on many aircraft).

6. *Collision avoidance.* This subsystem predicts impending collisions with other aircraft or the ground and recommends an avoidance maneuver (Chapter 14).

7. *Weather detection.* This subsystem observes weather ahead of the aircraft so that the route of flight can be altered to avoid thunderstorms and areas of high wind-shear. The sensors are usually radars (Chapter 11) and lasers (Chapter 8).

8. *Emergency locator transmitter (ELT).* This subsystem is triggered automatically on high-g impact or manually. In 1996, ELTs emit distinctive tones on 121.5, 243, and 406 MHz [8]. These frequencies (and perhaps soon 1.6 GHz) are monitored by search-and-rescue aircraft and by SARSAT-COSPAS satellites.

1.8.2 Military Avionics

The avionics often cost 40% of the value of a military aircraft. In addition to the navigation subsystem and the subsystems described in Section 1.8.1, military avionics consist of

1. *Radar, infrared, and other target sensors.* These may have their own displays and controls or may share multipurpose devices.
2. *Weapon management*
   - Fire control. Calculates lead angle for aiming guns and unguided rockets at other aircraft and at ground targets.
   - Stores management, that initializes and launches guided weapons: missiles and bombs.
3. *Electronic countermeasures.* This subsystem detects, locates, and identifies enemy emitters of electromagnetic radiation. It may also generate jamming signals. In 1996, electronic countermeasures were often so complex that they were installed in an externally carried pod on specially equipped aircraft.
4. *Mission planning.* Pre-flight mission planning is usually done at the airbase by a computer that prepares coordinated flight plans for an entire squadron. On-board software replans routes through enemy defenses based on en-route observations. En-route replanning requires on-board digital maps of the terrain and the real-time detection of enemy radars.
5. *Formation flight.* This subsystem maintains formation flight in instrument meteorological conditions. It once consisted of beacons, transponders, and communication links but is being replaced by relative GPS.

1.8.3   Architecture

Before the 1960s, electrical and electronic systems on aircraft consisted of independent subsystems, each with its own sensors, analog computers, displays, and controls. The appearance of airborne digital computers in the 1960s created the first integrated avionic systems. The interconnectivity of airborne electronics is called *architecture.* It involves six aspects:

1. *Displays.* They present information from the avionics to the pilots (Chapters 9 and 15). The information consists of vertical and horizontal navigation data, flight-control data (e.g., speed and angle of attack), and communication data (radio frequencies). The displays show the status of all subsystems including their faults. Displays consist of dedicated gauges, dedicated glass displays, multipurpose glass displays, and the supporting symbol generators. In 1996, flat-panel vertical- and horizontal-situation displays were displacing cathode-ray tubes as "glass displays." Multipurpose displays of text and block diagrams are flat-panel matrices surrounded by buttons whose labels change as the displays change. On-board digital terrain data, used for mission planning, can be displayed on the horizontal situation display or on a head-up display.
2. *Controls.* The means of inputting information from the pilots to the avionics. The *flight controls* traditionally consist of rudder pedals and a

control-column or stick. Fly-by-wire aircraft are increasingly using either two-axis hand controllers and rudder pedals or (especially in manned spacecraft) three-axis hand controllers. The *subsystem controls* consist of panel-mounted buttons and switches. Switches are also mounted on the control column, stick, throttle, and hand-controllers; sometimes 5 buttons per hand. The buttons on the periphery of multipurpose displays control the subsystems.

3. *Computation.* The method of processing sensor data. Two extreme organizations of computation exist:

- Centralized. Data from all sensors are collected in a bank of central computers in which software from several subsystems are intermingled. The level of fault tolerance is that of the most critical subsystem, usually flight control. It has the simplest hardware and interconnections. In 1996, central computers were redundant uniprocessors or multiprocessers.

- Decentralized. Each traditional subsystem retains its integrity. Hence, navigation sensors feed a navigation computer, flight-control sensors feed a flight-control computer that drives flight-control actuators, and so on. This architecture requires complex interconnections but has the advantages that fault-tolerance provisions can differ for each subsystem according to the consequences of failure and that software is created by experts in each subsystem, executes independently of other software, and is easily modified. When designed with suitable intercomputer channels and data-reasonability tests, decentralized systems have more reliable software than do centralized systems.

Many avionic systems combine features of centralized and decentralized architectures.

4. *Data buses.* Copper or fiber-optic paths among sensors, computers, actuators, displays, and controls, as discussed in Chapter 15. Some data paths are dedicated and some are multiplexed. Complex aircraft contain parallel buses (one wire, pair of wires, or optical fiber per bit) and serial buses (bits sent sequentially on one wire-pair or fiber). A large aircraft can have a thousand pounds of signal wiring.

5. *Safety partitioning.* Commercial fly-by-wire aircraft sometimes divide the avionics into a highly redundant safety-critical flight-control system, a dually redundant mission-critical flight-management system, and a nonredundant maintenance system that collects and records data. Military aircraft sometimes partition their avionics for reasons other than safety.

6. *Environment.* Avionic equipment are subject to aircraft-generated electric-power transients, whose effects are reduced by filtering and batteries. Equipment are also subject to externally generated disturbances from radio transmitters and lightning. The effects of external disturbances (*high-intensity radiated fields*, HIRF) are reduced by shielding metal

wires and by using fiberoptic data buses. Aircraft constructed with a continuous metal skin have an added layer of Faraday shielding. Nevertheless, direct lightning strikes on antennas destroy input circuits and may damage feed cables. A nearby strike may induce enough current to do the same. Composite airframe structures can be transparent to radiation, thus exposing the avionics and power systems to external fields.

7. *Standards.* The signals in space created by navaids are standardized by the International Civil Aviation Organization (ICAO), Montreal, a United Nations agency [3]. These standards are written by committees that consist of representatives of the member governments. Interfaces among airborne subsystems, within the aircraft, are standardized by ARINC (Aeronautical Radio, Inc.), Annapolis, Maryland, a nonprofit organization owned by member airlines [1]. Other requirements are imposed on airborne equipment by two nonprofit organizations supported by member entities (mostly airframe and avionics manufacturers and government agencies). In the United States, RTCA, Inc. (Formerly Radio Technical Commission for Aeronautics), Washington D.C., defines the environmental specifications and test procedures for airborne hardware and software, and writes performance specifications for airborne equipment [5]. In Europe, EUROCAE (European Organisation for Civil Aviation Equipment), Paris, produces specifications for airborne equipment, some of which are in conjunction with RTCA [11]. Government agencies in all major nations define rules governing the usage of navigation equipment in flight, weather minimums, traffic separation, ground equipment required, pilot training requirements, and so on [6–10] (see Chapters 13, 14). Some of these rules are standardized internationally by ICAO. U.S. military organizations once issued their own standards for airborne circuit boards but have accepted civil standards since the early 1990s.

1.9 HUMAN NAVIGATOR

Large aircraft often had (and a few still had in 1996) a third crew member, the flight engineer, whose duties were to operate engines and aircraft subsystems such as air conditioning and hydraulics. Aircraft operating over oceans once carried a human navigator who used celestial fixes, whatever radio aids were available, and dead reckoning to plot the aircraft's course on a paper chart (some military aircraft still do). Those navigators were trained in celestial observatories to recognize stars, take fixes, compute position, and plot the fixes.

The navigator's crew station disappeared in civil aircraft in the 1970s, because inertial, Doppler, and radio equipment came into use that automatically selected stations, calculated position, calculated waypoint steering, and accommodated failures. Hence, instead of requiring a skilled navigator on each aircraft, a smaller number of even more skilled engineers were employed to design the automated systems. Since the 1980s, the trend has been to automate large aircraft so that subsystem management and navigation can be done by one or two pilots. Displays and controls are discussed in Chapters 9 and 15.

The key navigation skill in the twenty-first-century airplane is the operation of flight-management, inertial, satellite-navigation, and VOR equipment, each of which has different menus, inputting logic, and displays. The crew must learn to operate them in all modes, respond to failures, and enter waypoints for new routes manually. A new industry was created in the 1990s to produce computer-based trainers (called CBTs) that emulate subsystem software and include replica control panels in order to allow crews to practice scenarios without consuming expensive time on a full-mission simulator.

2 The Navigation Equations

2.1 INTRODUCTION

The navigation equations describe how the sensor outputs are processed in the on-board computer in order to calculate the position, velocity, and attitude of the aircraft. The navigation equations contain instructions and data and are part of the airborne software that also includes moding, display drivers, failure detection, and an operating system, for example. The instructions and invariant data are usually stored in a read-only memory (ROM) at the time of manufacturing. Mission-dependent data (e.g., waypoints) are either loaded from a cockpit keyboard or from a cartridge, sometimes called a *data-entry device*, into random-access memory (RAM). Waypoints are often precomputed in a ground-based *dispatch* or *mission-planning* computer and transferred to the flight computers.

Figure 2.1 is the block diagram of an aircraft navigation system. The system utilizes three types of sensor information (as explained in Chapter 1):

1. Absolute position data from radio aids, radar checkpoints, and satellites (based on range or differential range measurements).

2. Dead-reckoning data, obtained from inertial, Doppler, or air-data sensors, as a means of extrapolating present position. A heading reference is required in order to resolve the measured velocities into the computational coordinates.

3. Line-of-sight directions to stars, which measure a combination of position and attitude errors (as explained in Chapters 1 and 12).

The navigation computer combines the sensor information to obtain an estimate of the aircraft's position, velocity, and attitude. The best estimate of position is then combined with waypoint information to determine range and bearing to the destination. Bearing angle is displayed and sent to the autopilot as a steering command. Range to go is the basis of calculations, executed in a navigation or flight-management computer, that predict time of arrival at waypoints and that predict fuel consumption. Map displays, read from on-board compact discs (CD-ROM, Section 2.9), are driven by calculated position.

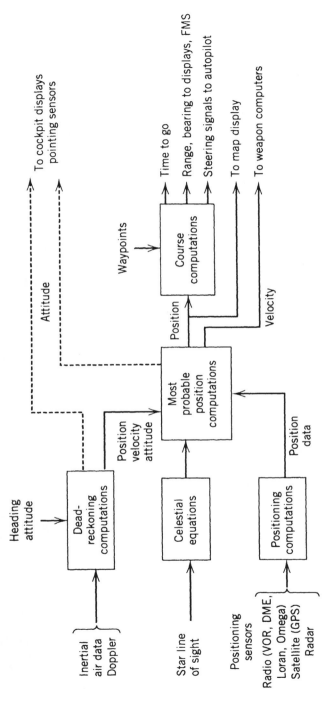

Figure 2.1 Block diagram of an aircraft navigation system.

22

2.2   GEOMETRY OF THE EARTH

The Newtonian gravitational attraction of the Earth is represented by a gravitational field **G**. Because of the rotation of the Earth, the apparent gravity field **g** is the vector sum of the gravitational and centrifugal fields (Figure 2.2):

$$\mathbf{g} = \mathbf{G} - \mathbf{\Omega} \times (\mathbf{\Omega} \times \mathbf{R}) \tag{2.1}$$

where $\mathbf{\Omega}$ is the inertial angular velocity of the Earth (15.04107 deg/hr) and **R** is the radius vector from the mass center of the Earth to a point where the field is to be computed. The direction of **g** is the "plumb bob," or "astronomic" vertical [10].

In cooling from a molten mass, the Earth has assumed a shape whose surface is a gravity equipotential and is nearly perpendicular to **g** everywhere (i.e., no horizontal stresses exist at the surface). For navigational purposes, the Earth's surface can be represented by an ellipsoid of rotation around the Earth's spin axis. The size and shape of the best-fitting ellipsoid are chosen to match the sea-level equipotential surface. Mathematically, the center of the ellipsoid is at the mass center of the Earth, and the ellipsoid is chosen so as to minimize the mean-square deviation between the direction of gravity and the normal to the ellipsoid, when integrated over the entire surface. National ellipsoids have been chosen to represent the Earth in localized areas, but they are not always good worldwide approximations. The centers of these national ellipsoids are not exactly coincident and do not exactly coincide with the mass center of the Earth [9]. In 1996, the World Geodetic System (WGS-84, [20]) was the best approximation to the geoid, based on gravimetric and satellite observations. Reference [20] contains transformation equations that convert between WGS-84 and various national ellipsoids. The differences are typically hundreds of feet, though some isolated island grids are displaced as much as a mile from WGS-84. The navigator does not ask that the Earth be mapped onto the optimum ellipsoid. Any ellipsoid is satisfactory for worldwide navigation if all points on Earth are mapped onto it.

The geometry of the ellipsoid is defined by a meridian section whose semimajor axis is the equatorial radius a and whose semiminor axis is the polar radius b, as shown in Figure 2.2. The eccentricity of the elliptic section is defined as $e = \sqrt{a^2 - b^2}/a$ and the ellipticity, or flattening, as $f = (a - b)/a$.

The radius vector **R** makes an angle F_C with the equatorial plane, where F_C is the geocentric latitude; **R** and F_C are not directly measurable, but they are sometimes used in mechanizing dead-reckoning equations.

The geodetic latitude F_T of a point is the angle between the normal to the reference ellipsoid and the equatorial plane. Geodetic latitude is our usual understanding of map latitude. The term "geographic latitude" is sometimes used synonymously with "geodetic" but should refer to geodetic latitude on a worldwide ellipsoid.

The radii of curvature of the ellipsoid are of fundamental importance to dead-

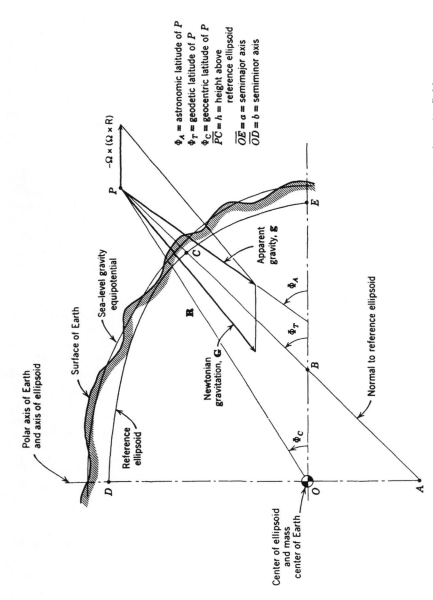

Φ_A = astronomic latitude of P
Φ_T = geodetic latitude of P
Φ_C = geocentric latitude of P
$\overline{PC}$ = h = height above
            reference ellipsoid
$\overline{OE}$ = a = semimajor axis
$\overline{OD}$ = b = semiminor axis

Figure 2.2 Meridian section of the Earth, showing the reference ellipsoid and gravity field.

24

reckoning navigation. The *meridian radius of curvature*, R_M, is the radius of the best-fitting circle to a meridian section of the ellipsoid:

$$R_M = \frac{a(1 - e^2)}{(1 - e^2 \sin^2 \mathrm{F}_T)^{3/2}} \approx a\left[1 + e^2\left(\frac{3}{2}\sin^2 \mathrm{F}_T - 1\right)\right] \qquad (2.2)$$

The *prime radius of curvature*, R_P, is the radius of the best-fitting circle to a vertical east–west section of the ellipsoid:

$$R_P = \frac{a}{(1 - e^2 \sin^2 \mathrm{F}_T)^{1/2}} \approx a\left[1 + \frac{e^2}{2}\sin^2 \mathrm{F}_T\right] \qquad (2.3)$$

The *Gaussian radius of curvature* is the radius of the best-fitting sphere to the ellipsoid at any point:

$$R_G = \sqrt{R_M R_P} \approx a\left[1 - \frac{e^2}{2}\cos 2\mathrm{F}_T\right] \qquad (2.4)$$

The radii of curvature are important, because they relate the horizontal components of velocity to angular coordinates, such as latitude and longitude; for example,

$$\dot{\mathrm{F}}_T = \frac{V_{\text{north}}}{R_M + h}$$

$$\dot{\mathrm{l}} \cos \mathrm{F}_T = \frac{V_{\text{east}}}{R_P + h} \qquad (2.5)$$

where h is the aircraft's altitude above the reference ellipsoid, measured along the normal to the ellipsoid (nearly along the direction of gravity), and l is its longitude, measured positively east.

For numerical work $a = 6378.137$ km $= 3443.918$ nmi, $f = 1/298.2572$, and $e^2 = f(2 - f)$ [18]. (One nautical mile $= 1852$ meters exactly, or 6076.11549 ft.) The angle between the gravity vector and the normal to the ellipsoid, the *deflection of the vertical*, is commonly less than 10 seconds of arc and is rarely greater than 30 seconds of arc [18]. The magnitude of gravity at sea level is

$$g = 978.049(1 + 0.00529 \sin^2 \mathrm{F}_T) \text{ cm/sec}^2 \qquad (2.6)$$

within 0.02 cm/sec^2. It decreases $10^{-6}g$ for each 10-ft increase in altitude above sea level [16].

2.3   COORDINATE FRAMES

The position, velocity, and attitude of the aircraft must be expressed in a coordinate frame. Paragraph 1 below describes the rectangular Earth-centered, Earth-fixed (ECEF) coordinate frame, y_i. Paragraph 2 describes the Earth-centered inertial (ECI) coordinate frame, x_i, which simplifies the computations for inertial and stellar sensors. Other Earth-referenced orthogonal coordinates, called z_i, can simplify navigation computations for some navaids and displays. Paragraphs 3–5 describe coordinates commonly used in inertial navigation systems. Paragraph 6 describes coordinates used in land navigators or in military aircraft that support ground troops. Paragraphs 7–9 describe coordinates that were important before powerful airborne digital computers existed.

1. *Earth-centered, Earth-fixed (ECEF)*. The basic coordinate frame for navigation near the Earth is ECEF, shown in Figure 2.3 as the y_i rectangular coordinates whose origin is at the mass center of the Earth, whose y_3-axis lies along the Earth's spin axis, whose y_1 axis lies in the Greenwich meridian, and which rotates with the Earth [10]. Satellite-based radio-navigation systems often use these ECEF coordinates to calculate satellite and aircraft positions.

2. *Earth-centered inertial (ECI)*. ECI coordinates, x_i, can have their origin at the mass-center of any freely falling body (e.g., the Earth) and are nonrotating relative to the fixed stars. For centuries, astronomers have observed

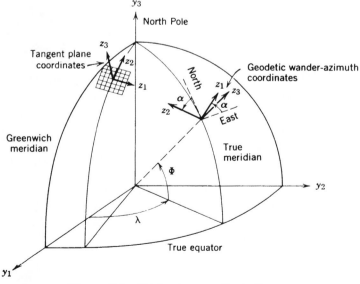

Figure 2.3   Navigation coordinate frames.

the small relative motions of stars ("proper motion") and have defined an "average" ECI reference frame [11]. To an accuracy of 10^{-5} deg/hr, an ECI frame can be chosen with its x_3-axis along the mean polar axis of the Earth and with its x_1- and x_2-axes pointing to convenient stars (as explained in Chapter 12). ECI coordinates have three navigational functions. First, Newton's laws are valid in any ECI coordinate frame. Second, the angular coordinates of stars are conventionally tabulated in ECI. Third, they are used in mechanizing inertial navigators, Section 7.5.1.

3. *Geodetic spherical coordinates.* These are the spherical coordinates of the normal to the reference ellipsoid (Figure 2.2). The symbol z_1 represents longitude l; z_2 is geodetic latitude F_T, and z_3 is altitude h above the reference ellipsoid. Geodetic coordinates are used on maps and in the mechanization of dead-reckoning and radio-navigation systems. Transformations from ECEF to geodetic spherical coordinates are given in [9] and [23].

4. *Geodetic wander azimuth.* These coordinates are locally level to the reference ellipsoid. $\hat{z}_3$ is vertically up and $\hat{z}_2$ points at an angle, a, west of true north (Figure 2.3). The wander-azimuth unit vectors, z_1 and z_2, are in the level plane but do not point east and north. Wander azimuth is the most commonly used coordinate frame for worldwide inertial navigation and is discussed below and in Section 7.5.1.

5. *Direction cosines.* The orientation of any z-coordinate frame (e.g., navigation coordinates or body axes) can be described by its direction cosines relative to ECEF y-axes. Any vector **V** can be resolved into either the y- or z-coordinate frame. The y and z components of **V** are related by the equation

$$[\mathbf{V}]_z = [C_{zy}][\mathbf{V}]_y \qquad (2.7)$$

where

$$C_{11} = -\cos a \sin l - \sin a \sin F \cos l$$
$$C_{12} = \cos a \cos l - \sin a \sin F \sin l$$
$$C_{13} = \cos F \sin a$$
$$C_{21} = \sin a \sin l - \sin F \cos a \cos l$$
$$C_{22} = -\sin a \cos l - \cos a \sin F \sin l$$
$$C_{23} = \cos F \cos a$$
$$C_{31} = \cos F \cos l$$
$$C_{32} = \cos F \sin l$$
$$C_{33} = \sin F \qquad (2.8)$$

The navigation computer calculates in terms of the C_{ij}, which are usable everywhere on Earth. The familiar geographic coordinates can be found from the relations

$$\sin F = C_{33},$$

or

$$\cos^2 F = C_{13}^2 + C_{23}^2 = C_{31}^2 + C_{32}^2$$

$$\tan l = \frac{C_{32}}{C_{31}}$$

$$\tan a = \frac{C_{13}}{C_{23}} \qquad (2.9)$$

wherever they converge. In polar regions, where a and l are not meaningful, the navigation system operates correctly on the basis of the C_{ij}. Section 7.5.1 describes an inertial mechanization in direction cosines. If the z-coordinate frame has a north-pointing axis, $a = 0$.

6. *Map-grid coordinates.* The navigation computer can calculate position in map-grid coordinates such as Lambert conformal or transverse Mercator *xy*-coordinates [13]. Grid coordinates are used in local areas (e.g., on military battlefields or in cities) but are not convenient for long-range navigation. A particular grid, Universal Transverse Mercator (UTM), is widely used by army vehicles of the western nations. The U.S. Military Grid Reference System (MGRS) consists of UTM charts worldwide except, in polar regions, polar stereographic charts [13]. The latter are projected onto a plane tangent to the Earth at the pole, from a point at the opposite pole.

7. *Geocentric spherical coordinates.* These are the spherical coordinates of the radius vector **R** (Figure 2.2). The symbol z_1 represents longitude l; z_2 is geocentric latitude F_C, and z_3 is the radius. Geocentric coordinates are sometimes mechanized in short-range dead-reckoning systems using a spherical-Earth approximation. Initialization requires knowledge of the direction toward the mass center of the Earth, a direction that is not directly observable.

8. *Transverse-pole spherical coordinates.* These coordinates are analogous to geocentric spherical coordinates except that their poles are deliberately placed on the Earth's equator. The symbol z_1 represents the transverse longitude; z_2, the transverse latitude; and z_3, the radius. They permit nonsingular operation near the north or south poles, by placing the transverse pole on the true equator. Transverse-polar coordinates involve only three z_i variables instead of nine direction cosines. However, they cannot be used for precise navigation, since the transverse equator is elliptical, com-

plicating the precise definition of transverse latitude and longitude. They are similar to but not identical to the stereographic coordinates often used in polar regions. Transverse-pole coordinates were used in inertial and Doppler navigation systems from 1955 to 1970 when primitive airborne computers required simplified computations.

9. *Tangent plane coordinates.* These coordinates are always parallel to the locally level axes at some destination point (Figure 2.3). They are locally level only at that point and are useful for flight operations within a few hundred miles of a single destination. Here z_3 lies normal to the tangent plane, and z_2 lies parallel to the meridian through the origin. Section 7.5.1 describes the mechanization of an inertial navigator in tangent-plane coordinates.

2.4 DEAD-RECKONING COMPUTATIONS

Dead reckoning (often called DR) is the technique of calculating position from measurements of velocity. It is the means of navigation in the absence of position fixes and consists in calculating the position (the z_i-coordinates) of a vehicle by extrapolating (integrating) estimated or measured ground speed. Prior to GPS, dead-reckoning computations were the heart of every automatic navigator. They gave continuous navigation information between discrete fixes. In its simplest form, neglecting wind, dead reckoning can calculate the position of a vehicle on the surface of a flat Earth from measurements of ground speed V_g and true heading w_T:

$$V_{\text{north}} = V_g \cos \mathrm{w}_T, \quad y - y_0 = \int_0^t V_{\text{north}}\, dt$$

$$V_{\text{east}} = V_g \sin \mathrm{w}_T, \quad x - x_0 = \int_0^t V_{\text{east}}\, dt \qquad (2.10)$$

where $x - x_0$ and $y - y_0$ are the east and north distances traveled during the measurement interval, respectively. Notice that a simple integration of unresolved ground speed would give curvilinear distance traveled but would be of little use for determining position.

Aircraft heading (best-available true heading) is measured using the quantities defined in Figure 2.4. With a magnetic compass, for example, the best available true heading is the algebraic sum of magnetic heading and east variation. With a gimballed inertial system, the best available true heading is platform heading (relative to the z_i computational coordinates) plus the wander angle a (Section 7.5.2). When navigating manually in polar regions, dead-reckoned velocity is resolved through best available grid heading.

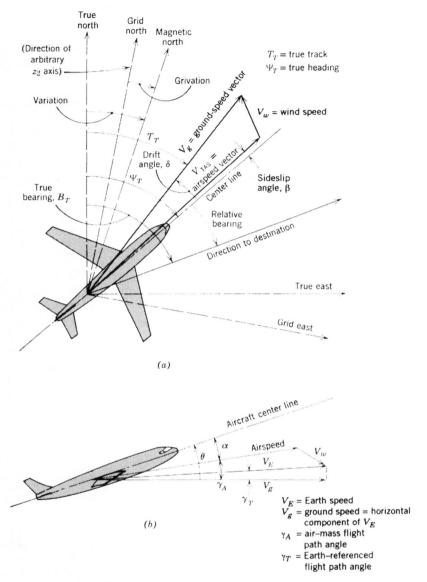

Figure 2.4   Geometry of dead reckoning.

In the presence of a crosswind the ground-speed vector does not lie along the aircraft's center line but makes an angle δ with it (Figure 2.4). The true-track angle T_T, the angle from true north to the ground-speed vector, is preferred for dead-reckoning calculations when it is available. The drift angle δ can be measured with a Doppler radar or a drift sight (a downward-pointing telescope whose reticle can be rotated by the navigator to align with the moving ground).

In a moving air mass

$$V_{\text{north}} = V_{\text{TAS}} \cos(\theta - \alpha) \cos(\psi_T + \beta) + V_{\text{wind-north}}$$

$$V_{\text{east}} = V_{\text{TAS}} \cos(\theta - \alpha) \sin(\psi_T + \beta) + V_{\text{wind-east}} \tag{2.11}$$

where θ is the pitch angle, V_{TAS} is the true airspeed and β is the sideslip angle. On a flat Earth, the north and east (or grid north and grid east) distances traveled are found by integrating the two components of velocity with respect to time. On a curved Earth, the position coordinates are not linear distances but angular coordinates. Equations 2.5 show a method for transforming linear velocities to angular coordinates. The accuracy of airspeed data is limited by errors in predicted windspeed and by errors in measuring airspeed and drift angle.

The dead-reckoning computer can process Doppler velocity. If the Doppler radar measures ground speed V_g and drift angle δ,

$$V_{\text{north}} = V_g \cos(\psi_T + \delta)$$
$$V_{\text{east}} = V_g \sin(\psi_T + \delta) \tag{2.12}$$

Equivalently, the Doppler radar can measure the components of groundspeed in body axes: V'_H along-axis and V'_D cross-axis (using the same symbols as in Chapter 10), both of which can be resolved through the three attitude angles, ψ_T, pitch and roll. Antenna misalignment relative to the heading sensor can be calibrated from a series of fixes, either in closed form or with a Kalman filter (Chapter 3).

Equations 2.5 relate velocity to λ and Ψ_T on an ellipsoidal Earth. Similar relations can be derived for the other coordinate frames discussed in Section 2.3. The velocity with respect to Earth $d\mathbf{R}/dt|_y$ can be expressed in z_i components as follows:

$$\left.\frac{d\mathbf{R}}{dt}\right|_y = \left.\frac{d\mathbf{R}}{dt}\right|_z + (\boldsymbol{\omega}_{yz} \times \mathbf{R}) \tag{2.13}$$

For example, in a spherical coordinate frame whose z_3-axis lies along the position vector $\mathbf{R}$:

$$\left.\frac{d\mathbf{R}}{dt}\right|_y = \frac{dR}{dt}\,\hat{\mathbf{R}} + (\boldsymbol{\omega}_{yz} \times \mathbf{R})$$

where the first term is the rate of change of radius, along the radius vector, and $\boldsymbol{\omega}_{yz}$ is the angular velocity of the z_i-coordinate frame relative to y_i; $\hat{\mathbf{R}}$ is the unit vector in the direction of $\mathbf{R}$. In direction-cosine mechanizations, the $\dot{C}_{ij}$ are related to the C_{ij} by Equation 7.40, where $\boldsymbol{\omega}_i - \boldsymbol{\Omega}_i$ of that equation is identical to $\boldsymbol{\omega}_{yz}$ of this one.

2.5 POSITIONING

2.5.1 Radio Fixes

There are five basic airborne radio measurements:

1. *Bearing.* The angle of arrival, relative to the airframe, of a radio signal from an external transmitter. Bearing is measured by the difference in phase or time of arrival at multiple antennas on the airframe. At each bearing, the distortion caused by the airframe may be calibrated as a function of frequency. If necessary, calibration could also include roll and pitch.

2. *Phase.* The airborne receiver measures the phase difference between continuous-wave signals emitted by two stations using a single airborne antenna. This is the method of operation of VOR azimuth and hyperbolic Omega (Chapter 4).

3. *Time difference.* The airborne receiver measures the difference in time of arrival between pulses sent from two stations. A 10^{-4} clock (one part in 10^4) is adequate to measure the short time interval if both pulses are sent simultaneously. Because Loran pulses can be transmitted 0.1 sec apart, a clock error less than 10^{-6} is needed to measure the time difference. In time-differencing and phase-measuring systems (*hyperbolic Loran*), at least two pairs of stations are required to obtain a fix.

4. *Two-way range.* The airborne receiver measures the time delay between the transmission of a pulse and its return from an external transponder at a known location. Round-trip propagation times are typically less than a millisecond, during which the clock must be stable. A 1% range error requires a clock-stability of 0.3% (3×10^{-3}), which is two microseconds at 100-km range. The calculation of range requires knowledge of the propagation speed and transponder delay. DME is a two-way ranging system (Section 4.4.6).

5. *One-way range.* The airborne receiver measures the time of arrival with respect to its own clock. If the airborne clock were synchronized with the transmitter's clock upon departure from the airfield and ran freely thereafter, a 1% range error at a distance of 100 km from a fixed station would require a clock error of one microsecond, which is 5×10^{-11} of a five-hour mission. When 25,000-km distances to GPS satellites are to be measured with a one-meter error, a short-term clock stability of one part in 3×10^8 would be needed to measure range and 10^{-4} seconds absolute time

error would be needed to calculate the satellite's position (GPS satellites are moving at 3000 ft/sec relative to Earth). Together these would require an error of one part in 10^{12} for a clock synchronized with the transmitters at the start of a ten-hour mission and allowed to run freely thereafter. Only an atomic clock had this accuracy in 1996. Therefore practical one-way ranging systems use a technique called *pseudoranging*. The transmitters contain atomic clocks with long-term stabilities of about 10^{-13}, while the airborne receiver's clocks have accuracies and stabilities of 10^{-6} to 10^{-9}. The airborne computer solves for the aircraft's clock offset (and some-times, drift rate) by making redundant range measurements. For exam-ple, measuring four pseudoranges obtains three-dimensional position and clock offset to a few nanoseconds using Equations 2.17 and 2.18. Pseu-doranging is used in GPS and GLONASS (Chapter 5) and in one-way ranging (*direct ranging*) of Loran and Omega (Chapter 4).

2.5.2   Line-of-Sight Distance Measurement

Figure 2.5 shows an aircraft near the surface of the Earth at $\mathbf{R}_0$ and a radio station that may be near the surface or in space, at $\mathbf{R}_{si}$. The slant range, $|\mathbf{R}_{si} - \mathbf{R}_0|$, from the aircraft to the station could be measured by one-way or two-way ranging. If $\hat{\boldsymbol{v}}$ is the unit local vertical vector at the aircraft, the elevation angle of the line of sight to the radio station is

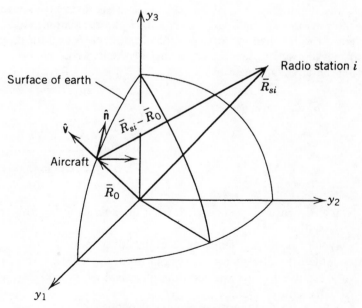

Figure 2.5   Light-of-sight distance.

$$\sin E = \frac{\hat{\mathbf{v}} \cdot (\mathbf{R}_{si} - \mathbf{R}_0)}{|\mathbf{R}_{si} - \mathbf{R}_0|} \tag{2.14}$$

If $\hat{\mathbf{n}}$ is the unit north-pointing horizontal vector at the aircraft, the azimuth angle of the line of sight is

$$\sin Az = \frac{\hat{\mathbf{v}} \times (\mathbf{R}_{si} - \mathbf{R}_0)}{|\mathbf{R}_{si} - \mathbf{R}_0| \cos E} \times \hat{\mathbf{n}} \tag{2.15}$$

These vector equations can be resolved into any coordinate frame. For example, in ECEF,

$$\hat{\mathbf{n}} = -\hat{\mathbf{x}} \sin \Phi \cos \lambda - \hat{\mathbf{y}} \sin \Phi \sin \lambda + \hat{\mathbf{z}} \cos \Phi$$

$$\hat{\mathbf{v}} = \hat{\mathbf{x}} \cos \Phi \cos \lambda + \hat{\mathbf{y}} \cos \Phi \sin \lambda + \hat{\mathbf{z}} \sin \Phi$$

$$\mathbf{R} = \hat{\mathbf{x}}(R_m + h) \cos \Phi \cos \lambda + \hat{\mathbf{y}}(R_m + h) \cos \Phi \sin \lambda$$

$$+ \hat{\mathbf{z}} \left[(R_m + h) \sin \Phi - \frac{a\epsilon^2 \sin \Phi}{\sqrt{1 - \epsilon^2 \sin^2 \Phi}} \right] \tag{2.16}$$

In one-way ranging systems, where the clock offset and range are to be calculated, the measured pseudorange vector from the ith station, $\mathbf{R}_{im}$, is $\mathbf{R}_0 - \mathbf{R}_{si}$, corrected for the unknown offset of the airborne clock and for propagation delays in the atmosphere, expressed in distance units. The magnitude of the pseudorange in any coordinate frame (e.g., ECEF) is

$$R_{im} = \eta c(\text{TOA})$$
$$R_{im(k)} = [(x_k - x_{si})^2 + (y_k - y_{si})^2 + (z_k - z_{si})^2]^{1/2} - \eta c t_k \tag{2.17}$$

where

R_{im} is the measured pseudorange from the aircraft to the ith station

$R_{im(k)}$ is the calculated pseudorange in the kth iteration.

TOA_i is the time of arrival of the signal from the ith station relative to the expected time of arrival as measured by the aircraft's clock

c is the speed of light in vacuum = 2.99792458×10^8 m/sec

η is the average index of refraction in the propagation medium; partly space, partly atmosphere

x_k, y_k, z_k is the unknown position of the aircraft (in the kth computational iteration)

x_{si}, y_{si}, z_{si} is the known position of the ith station

t_k is the computed time offset of the aircraft's clock relative to the station's clock in the kth iteration

The stations may be moving (e.g., satellites) or stationary (e.g., GPS pseudolites). Four pseudorange measurements are needed to solve for the four unknowns in Equations 2.17: aircraft position and clock offset. When more than four measurements are made, the equations are overdetermined so that a solution requires a model of the ranging errors, for example, using a Kalman filter (Chapter 3). The airborne computer usually solves for its position by assuming a position and clock offset, calculating the pseudoranges to four stations from Equation 2.17, comparing to the measured pseudoranges (with respect to its own clock), and iterating until the calculated and measured pseudoranges are close enough. The next iteration is chosen as follows: In the kth iteration, the assumed position is $x_k \; y_k \; z_k$ whose range to the ith station differs from the measured range $\mathbf{R}_{im}$ by $\Delta \mathbf{X}_k = \mathbf{R}_{im(k)} - \mathbf{R}_{si}$. The components of $\Delta \mathbf{X}_k$ in the navigation coordinates are ΔX_k, ΔY_k, and ΔZ_k. The sensitivity of R_{im} to position is

$$\Delta R_{im} = \frac{\partial R_{im}}{\partial X} \Delta X_k + \frac{\partial R_{im}}{\partial Y} \Delta Y_k + \frac{\partial R_{im}}{\partial Z} \Delta Z_k + \eta \, c \Delta t_k \qquad (2.18)$$

where $\partial R_{im}/\partial X_j$ are the direction cosines between the line of sight to the ith station and the jth coordinate axis and Δt_k is the error in estimating clock offset. If the assumed position and clock offset were correct, ΔX_k, ΔY_k, ΔZ_k, and Δt_k would be zero and $\Delta \mathbf{R}_{ik}$ would also be zero. But if the assumed position were misestimated by $\Delta \mathbf{X}_k$, the error along the line of sight would be the dot product of the unit vector along the line of sight with $\Delta \mathbf{X}_k$. Thus after computing $\mathbf{R}_{imk}$ from Equation 2.17, the next iteration is $\mathbf{R}_{im(k+1)} = \mathbf{R}_{im(k)} + \Delta \mathbf{X}_k$. Iterations cease when the difference between the calculated and measured pseudorange is within the desired accuracy. A recursive filter allows a new calculation of position and clock offset after each measurement (Chapters 3 and 5). In Equation 2.17, Earth-based line-of-sight navaids use an average index of refraction η (Chapter 4), whereas satellite-based navaids, whose signals propagate mostly in vacuum, assume that $\eta = 1$ and correct for the atmosphere with a model resident in the receiver's software (Section 5.4).

2.5.3   Ground-Wave One-Way Ranging

Loran and Omega waves propagate along the curved surface of the Earth (as explained in Chapter 4). With either sensor, an aircraft can measure the time of arrival of the navigation signal from two or more stations and compute its own position as follows:

- Assume an aircraft position.
- Calculate the exact distance and azimuth to each radio transmitter using ellipsoidal Earth equations [18].
- Calculate the predicted propagation time and time of arrival allowing for the conductivity of the intervening Earth's surface and the presence or absence of the dark/light terminator between the aircraft and the station (as described in Chapter 4).
- Measure the time of arrival using the aircraft's own clock, which is usually not synchronized to the transmitter's clock.
- Calculate the difference between the measured and predicted times of arrival to each station.
- The probable position is the assumed position, offset by the vector sum of the time differences, each in the direction of its station, converted to distance.
- Assume a new aircraft position and iterate until the residual is within the allowed error.

Three or more stations are needed if the aircraft's clock is not synchronized to the Loran or Omega transmitters. If a receiver incorporated a sufficiently stable clock, only two stations would be needed for a direct-ranging fix. Two-station fixes with a synchronized clock or three-station fixes with an asynchronous clock result in two position solutions, at the intersections of two circular, each having a vortex at a transmitter. The correct position can be found by receiving an additional station or from *a priori* knowledge.

2.5.4 Ground-Wave Time-Differencing

An aircraft can measure the difference in time of arrival of Loran or Omega signals from two or more stations (Chapter 4). To measure the time-difference within 0.1% requires a 0.03% accuracy clock, which is less expensive than the 10^{-11} clock required for uncorrected one-way ranging or the 10^{-6} to 10^{-9} clock required for pseudoranging (Section 2.5.1). As in one-way ranging, the iterative procedure is based on the precise calculation of propagation time from the aircraft to each station on an ellipsoidal Earth:

- Assume an aircraft position.
- Calculate the exact range and azimuth from the assumed position to each observed radio station using ellipsoidal Earth equations [18].
- Calculate the predicted propagation time allowing for the conductivity of

the intervening Earth's surface and the presence of the sunlight terminator between the aircraft and the station.

- Subtract the times to two stations to calculate the predicted difference in propagation time.

- Measure the difference in time of arrival of the signals from the two stations.

- Subtract the measured and predicted time differences to the two stations.

- Calculate the time-difference gradients from which is calculated the most probable position of the aircraft after the measurements (see Section 2.5.2 of the First Edition of this book and Chapter 4 of this book).

- Iterate until the residual is smaller than the allowed error.

2.6   TERRAIN-MATCHING NAVIGATION

These navigation systems obtain occasional updates when the aircraft overflies a patch of a few square miles, chosen for its unique profile [5]. A digital map of altitude above sea level, h_s, is stored for several parallel tracks; see Figure 2.6. For example, if 0.1-nmi accuracy is desired, $h_s(t)$ must be stored in 200-ft squares sampled every 0.2 sec at 600 knots.

The aircraft measures the height of the terrain above sea level as the difference between barometric altitude (Chapter 8) and radar altitude (Chapter 10); see Figure 2.7. Each pair of height measurements and the dead-reckoning position are recorded and time-tagged.

After passing over the patch, the aircraft uses its measured velocity to calculate the profile as a function of distance along track, $h_m(x)$, and calculates the cross-correlation function between the measured and stored profiles:

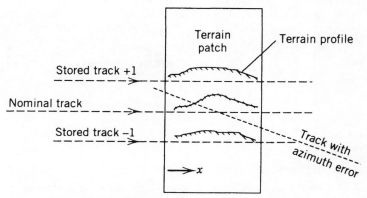

Figure 2.6   Parallel tracks through terrain patch.

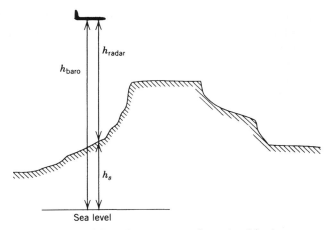

Figure 2.7   Measurement of terrain altitude.

$$\phi_{ms}(\tau) = \int_0^{nA} h_m(x)h_s(x - \tau)\, dx \qquad (2.19)$$

where the map patch has a length A. The integration is long enough ($n > 1$) to ensure that the patch is sampled, even with the expected along-track error. The computer selects the track whose cross-correlation is largest as the most probable track. The computer selects the x-shift of maximum correlation τ as the along-track correction to the dead-reckoned position. Heading drift is usually so small that correlations are not required in azimuth. The algorithm accommodates offsets in barometric altitude caused by an unknown sea-level setting. The width of the patch depends on the growth rate of azimuth errors in the dead-reckoning system. Simpler algorithms have been used ("mean absolute differences") and more complex Kalman filters have been used [5].

Terrain correlators are built under the names TERCOM, SITAN, and TERPROM. They are usually used on unmanned aircraft (cruise missiles) and can achieve errors less than 100 ft [1, 8]. The feasibility of this navigation aid depends on the existence of unique terrain patches along the flight path and on the availability of digital maps of terrain heights above sea level. The U.S. Defense Mapping Agency produces TERCOM maps for landfalls and mid-course updates in three-arcsec grids.

2.7   COURSE COMPUTATION

2.7.1   Range and Bearing Calculation

The purpose of the course computation is to calculate range and bearing from an aircraft to one or more desired waypoints, targets, airports, checkpoints, or radio beacons. The computation begins with the best-estimate of the present position of the aircraft and ends by delivering computed range and bearing to other vehicle subsystems (Figure 2.1). Waypoints may be loaded before departure or inserted en route. The navigation computer, mission computer, or flight-management computer performs the steering calculations.

Range and bearing to a destination can be calculated by using either the spherical or the plane triangle of Figure 2.8. If flat-Earth approximations are satisfactory, the xy coordinates of the aircraft are computed using the dead-reckoning Equation 2.10; x_t and y_t of the targets are loaded from a cassette or from a keyboard. Then, range D and bearing B_T to the target, measured from true north, are

$$D = [(x - x_t)^2 + (y - y_t)^2]^{1/2}$$

$$B_T = \tan^{-1} \frac{x - x_t}{y - y_t} \tag{2.20}$$

The crew will want a display of relative bearing ($B_R = B_T - \psi_T$) or relative track ($T_R = B_T - T_T$). B_T is the true bearing of the target. Relative bearing B_R is the horizontal angle from the longitudinal axis of the aircraft to the target, and relative track T_R is the horizontal angle from the ground track of the aircraft to the target (Figure 2.4).

If $\Delta\lambda$ and $\Delta\Phi$ are less than $\frac{1}{3}$ radian, the plane triangle solution exceeds the

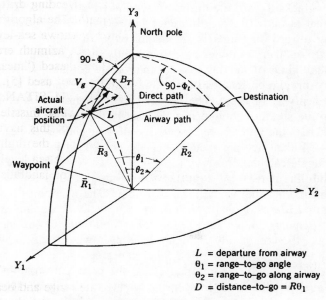

L = departure from airway
θ_1 = range–to–go angle
θ_2 = range–to–go along airway
D = distance–to–go = $R\theta_1$

Figure 2.8 Course-line calculations.

spherical triangle solution by a range ΔD:

$$\Delta D = \frac{(x - x_t)^2(y - y_t)}{2RD} \tan \Phi = \frac{D^2}{4R} \sin B_T \sin 2B_T \tan \Phi \qquad (2.21)$$

This error is 36 nmi at a range of $D = 1000$ nmi, an azimuth $B_T = 45$ deg and a latitude of $\Phi = 45$ deg. If this is not sufficiently accurate, a spherical-triangle solution may be used:

$$\cos \frac{D}{R_G} = \sin \Phi \sin \Phi_t + \cos \Phi \cos \Phi_t \cos (\lambda - \lambda_t)$$

$$\sin B_T = \frac{\cos \Phi_t}{\sin (D/R_G)} \sin (\lambda - \lambda_t) \qquad (2.22)$$

R_G is the Gaussian radius of curvature, Equation 2.4. At long range, where the absolute and percent errors are largest, they are usually least significant. Within 100 nmi of the aircraft, the Earth can be assumed flat within an error of 0.3 nmi.

Steering and range-bearing computations can be performed directly from the z_i navigation coordinates or from the direction cosines C_{ij} to prevent singularities near the north and south poles.

Knowing the measured or computed ground speed, an aircraft can be steered in such a manner that the ground speed vector—not the longitudinal axis of the aircraft—tracks toward the desired waypoint (relative track is the steering command that is nulled). The difference between heading toward the target and tracking toward the target is significant only for helicopters; the vehicle eventually arrives there in either case, by slightly different paths.

Two general kinds of steering to a destination are commonly used: (1) steering directly from the present position to the destination and (2) steering along a preplanned airway or route. The former results in area navigation (Section 2.7.4) using the shortest (though not necessarily the fastest) route to the destination, whereas the latter is representative of flying along assigned airways (Chapter 14). Either steering method may be solved by the plane-triangle or spherical-triangle calculation.

The rhumb line is used by ships and simple aircraft. It is defined by flying at a constant true heading to the local meridians. The resulting flight path is a straight line on a Mercator chart. Aircraft sometimes divide a complex route into rhumb-line segments so that each segment can be flown at constant heading.

More often, the continually changing heading toward the next waypoint is recomputed and fed to the autopilot. Since the great circle maps into a near-straight line on a Lambert conformal chart, the crew can monitor the flight path by manual plotting, if desired. In the twenty-first century, electronic map displays will show the moving aircraft on charts.

2.7.2   Direct Steering

The steering computer calculates the ground speed V_1 along the direction to the destination and V_2 normal to the line of sight to the destination. The commanded bank angle is made proportional to V_2 in order that the aircraft's heading rate $\dot{H}$ be driven to zero when flying along the desired great circle. If V_a is airspeed, $\dot{H} = (g/V_a)\tan\phi$ and the commanded bank angle is

$$\phi_c = K_1 V_2 + K_2 \dot{V}_2 \qquad (2.23)$$

K_2 provides some anticipation when approaching the correct direction of flight. The commanded bank angle ϕ_c is limited to a maximum value (e.g., 15 deg) in order to avoid violent maneuvers when the aircraft's flight direction is greatly in error.

Near the destination, the computation of lateral speed, V_2, becomes singular, and the steering signal would fluctuate erratically. To prevent this, the track angle or heading is frozen and held until the destination (computed from range-to-go and ground speed) is passed. The range at which the steering must be frozen is determined by simulation. This navigation method is sometimes called *proportional navigation*, a term derived from missile-steering techniques in which the heading rate of the vehicle is made proportional to the line-of-sight rate to the target.

The normal to the great-circle plane connecting present position $\mathbf{R}_3$ to the destination $\mathbf{R}_2$ is defined by the unit vector $\hat{\mathbf{u}}$:

$$\hat{\mathbf{u}} = \frac{\mathbf{R}_2 \times \mathbf{R}_3}{|\mathbf{R}_2 \times \mathbf{R}_3|} \qquad (2.24)$$

(Figure 2.8). The lateral speed V_2 is the magnitude of the dot product of the aircraft's velocity with this unit vector. The range to go from $\mathbf{R}_3$ to $\mathbf{R}_2$ is given by Equation (2.20) or (2.22). Time to go is calculated from the proposed velocity schedule.

The fastest route is neither the great circle nor the airway because of winds, especially because of the stratospheric jet stream whose speed often exceeds 100 knots. Thus where high-altitude aircraft are not confined to airways, they follow preplanned "pressure-pattern" routes that take advantage of cyclonic tail winds.

2.7.3   Airway Steering

The steering algorithm calculates a great circle from the takeoff point (or from a waypoint) to the destination (or to another waypoint). The aircraft is steered along this great circle by calculating the lateral deviation L (Figure 2.8) from the desired great circle and commanding a bank angle:

$$\phi_c = K_1 L + K_2 \dot{L} + K_3 \int L \, dt \qquad (2.25)$$

The integral-of-displacement term is added to give zero steady-state displacement from the airway in the presence of a constant wind and is also used in automatic landing systems (Chapter 13) in order to couple the autopilot to the localizer beam of the instrument-landing system. The bank angle is limited, to prevent excessive control commands when the aircraft is far off course. Near the destination, the track, or heading, is frozen to prevent erratic steering.

As the aircraft passes each waypoint, a new waypoint is fetched, thus selecting a new desired track. The aircraft can then fly along a series of airways connecting checkpoints or navigation stations.

The great-circle airway is defined by the waypoint vectors $\mathbf{R}_1$ and $\mathbf{R}_2$. The angle to go to waypoint 2 is:

$$\arcsin \frac{|\mathbf{R}_3 \times \mathbf{R}_2|}{|\mathbf{R}_3||\mathbf{R}_2|} \qquad (2.26)$$

Range and time to go are calculated as in Section 2.7.2. The lateral-deviation angle L/D is:

$$\arcsin \left(\frac{\mathbf{R}_3}{|\mathbf{R}_3|} \cdot \frac{\mathbf{R}_1 \times \mathbf{R}_2}{|\mathbf{R}_1 \times \mathbf{R}_2|} \right) \qquad (2.27)$$

These computations can be performed directly in terms of the navigation coordinates z_i.

2.7.4 Area Navigation

Between 1950 and approximately 1980, aircraft in developed countries flew on airways, guided by VOR bearing signals (Chapter 4). Position along the airway could be determined at discrete *intersections* (Δ in Figure 2.9) using cross-bearings to another VOR. In the 1970s DME, colocated with the VOR, allowed aircraft to determine their position along the airways continuously. Thereafter, regulating authorities allowed them to fly anywhere with proper clearances, a technique called RNAV (random navigation) or *area navigation*. RNAV uses combinations of VORs and DMEs to create artificial airways either by connecting waypoints defined by latitude/longitude or by triangulation or trilateration to VORTAC stations (as shown by the dotted lines to A_1 in Figure 2.9). The on-board flight-management or navigation computer calculates the lateral displacement L from the artificial airway and the distance D to the next waypoint A_1 along the airway [24b,c], [25].

In Figure 2.9, ρ_1 and ρ_3 are the measured distances to the DME stations at

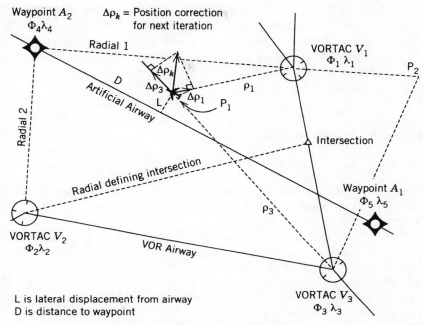

Figure 2.9   Plan view of area-navigation fix.

V_1 and V_3. The position P_1 is found from the triangle $P_1V_1V_3$. The aircraft's position must be known well enough to exclude the false solution at P_2. An artificial airway is defined by the points A_1 and A_2. D and L are usually found iteratively:

1. Assume P_1 based on prior navigation information.
2. Calculate the ranges ρ_1 and ρ_3 to the DMEs at V_1 and V_3 using the range equations of Section 2.5.2. A range-bearing solution relative to a single VOR station calculates the aircraft's position, but not as accurately as a range-range solution to two stations.
3. Correct the measured ranges for the altitudes of the aircraft and DME station.
4. Subtract the measured and calculated ranges

$$\Delta\rho_1 = \rho_{1(\text{measured})} - \rho_{1(\text{calculated})}$$
$$\Delta\rho_3 = \rho_{3(\text{measured})} - \rho_{3(\text{calculated})}$$

5. The next estimate of ρ_1 is along the vector $\Delta\rho_k$ in Figure 2.9, whose components along ρ_1 and ρ_3 are $\Delta\rho_1$ and $\Delta\rho_3$.
6. Repeat step 2 and iterate until $\Delta\rho_i$ are acceptably small.

After determining the aircraft's position P_1, the distance-to-go and lateral displacement are calculated as in Section 2.7. L is sent to the autopilot, as explained in Section 2.7.3, and D is used to calculate time-to-go.

In the 1990s, civil aircraft were being allowed the freedom to leave RNAV airways (Section 1.5.3) using GPS, inertial, Omega, and Loran navigation, none of which constrain aircraft to airways.

2.8   NAVIGATION ERRORS

2.8.1   Test Data

Navigation errors establish the width of commercial airways, the spacing of runways, and the risk of collision. Navigation errors determine the accuracy of delivering weapons and pointing sensors.

All navigation systems show a statistical dispersion in their indication of position and velocity. Test data can be taken on the navigation system as a whole and on its constituent sensors. Tests are conducted quiescently in a laboratory, in an artificial environment (e.g., rate table or thermal chamber), or in flight. As accuracies improve, the statistical dispersions, once considered mere noise, become important enough to predict (as discussed in Chapter 3).

The departure of a commercial aircraft from its desired flight path is sometimes divided into:

1. Navigation sensor errors
2. Computer errors
3. Data entry errors
4. Display error if the aircraft is flown by a pilot
5. Flight technical error, which is the departure of the pilot-flown or autopilot-flown aircraft from the computed path

Deterministic errors are added algebraically and statistical errors are root sum squared. The total is sometimes called *total system error*.

The two- or three-dimensional vector position error **r** can be defined as *indicated minus actual* position. A series of measurements taken on one navigation system or on any sample of navigation systems will yield a series of position measurements that are all different but that cluster around the actual position. If the properties of the navigation systems do not change appreciably with age, if the factory is neither improving nor degrading its quality control, and if all systems are used under the same conditions, then the statistics of the series of measurements taken on any one system are the same as those taken for a sample ("ensemble") of systems. Mathematically it is said that the statistics are *ergodic* and *stationary*.

If the position errors are plotted in two dimensions, as shown in Figure 2.10,

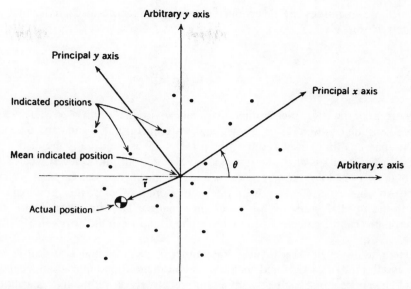

Figure 2.10 Two-dimensional navigation errors. In principal axes, the x and y statistics are independent.

it will generally be found that the average position error $\mathbf{r} = \Sigma \Delta \mathbf{r}_i / N$) is not zero ($\Delta \mathbf{r}_i$ are the individual position errors; N is the total number of measurements). Two measures of performance are the mean error and the circular error probability (CEP) (also known as the circular probable error, CPE). The CEP is usually considered to be the radius of a circle, centered at the actual position (but more properly centered at the mean position of a group of measurements) that encloses 50% of the measurements. The mean error and the CEP may be suitable as crude acceptance tests or specifications, but they yield little engineering information.

More rigorously, the horizontal position error should be considered as a bivariate (two-dimensional) distribution. The mean error $\mathbf{r}$ and the directions of the principal axes x and y, for which errors in x and y are uncorrelated, must be found. To find the principal axes, a convenient orthogonal coordinate system (x', y') is established with its origin at $\mathbf{r}$. Then the standard deviation (or rms) σ in each axis and the correlation coefficient ρ between axes are calculated:

$$\sigma_{x'}^2 = \frac{\sum (x_i')^2}{N}$$

$$\sigma_{y'}^2 = \frac{\sum (y_i')^2}{N}$$

$$\rho = \frac{\sum x_i' y_i'}{\sigma_{x'} \sigma_{y'}} \tag{2.28}$$

From these quantities are determined a new set of coordinates, xy, which are rotated θ from $x'y'$:

$$\tan 2\theta = \frac{2\sigma_{x'}\sigma_{y'}}{\sigma_{x'}^2 - \sigma_{y'}^2}\rho \qquad (2.29)$$

The origin of the new xy-coordinates coincides with the origin of $x'y'$. The components of the position errors along x and y are uncorrelated and can be considered separately. In inertial systems, the principal axes are usually the instrument axes. In Doppler systems, the principal axes are along the velocity vector and normal to it or along the aircraft axis and normal to it if the antenna is body-fixed (as was usually the case in 1996). In Loran systems, the principal axes are found by diagonalizing the covariance matrix. The rms errors in the new coordinates can be calculated anew or can be found from the errors in $x'y'$ from [7, p. 598].

The one-dimensional statistics along either of the principal axes will now be discussed. First, the mean and standard deviation are found in each independent axis. The errors along each axis are plotted separately as cumulative distribution curves, which show the fraction of errors less than x versus x. This curve allows all properties of the statistics to be determined. In many systems, experimental cumulative distributions will fit a Gaussian curve.

If the one-dimensional errors are indeed Gaussian, their statistics have the following properties:

Mean	$\bar{x}$
Standard deviation (rms)	σ
50% of the errors lie within	$\bar{x} \pm 0.675\sigma$ (probable error)
68.3% of the errors lie within	$\bar{x} \pm \sigma$
95.4% of the errors lie within	$\bar{x} \pm 2\sigma$
99.7% of the errors lie within	$\bar{x} \pm 3\sigma$

From these properties of the Gaussian distribution, it is easy to tell whether an experimentally determined cumulative distribution curve is approximately Gaussian. Reference [7] shows statistical tests for "normality" of experimental data.

Returning to the two-dimensional case, the probability that a navigation error falls within a rectangle $2m$ by $2n$, centered at the mean and aligned with the principal axes (x and y uncorrelated), is found by multiplying the tabulated error functions for m/σ_x and n/σ_y computed independently. Tables for the probability integral ([14], p. 116) can be used instead for $m/(\sigma_x\sqrt{2})$ and $n/(\sigma_y\sqrt{2})$. For example, if $\sigma_x = 0.4$ nmi and $\sigma_y = 1.0$ nmi, the probability of falling inside a rectangle 1.2 nmi (in x) by 2.4 nmi (in y) is the product of the probability integrals for $(0.6/0.4\sqrt{2}) = 1.06$ and $(1.2/1.0\sqrt{2}) = 0.85$, which is $0.866 \times 0.770 = 66\%$. Thus, the 1.2 by 2.4 nmi rectangle encloses 66% of the navigation errors.

Navigation systems, civil and military, are often specified by the fraction of navigation errors that fall within a circle of radius ϵ_P, centered on the mean. Figure 2.11 shows this probability if σ_x and σ_y are Gaussian distributed and uncorrelated. For example, if $\sigma_x = 0.4$ nmi and $\sigma_y = 1.0$ nmi, $b = \sigma_x/\sigma_y = 0.4$. If the radius of the circle is $\epsilon_P = 1.5$ nmi, then $a = \epsilon_P/\sigma_y = 1.5$, and from the graph, $P = 0.85$. Thus, 85% of the errors fall within a circle of radius 1.5 nmi.

Weapons often inflict damage in a circular pattern. Hence, military tactical navigation systems are sometimes specified by the CEP, the radius of the circle that encompasses 50% of the navigation errors (which are inherently elliptically distributed):

$$\text{CEP} = 0.59(\sigma_x + \sigma_y) \pm 3\% \qquad \text{if } \frac{\sigma_y}{3} < \sigma_x < 3\sigma_y \qquad (2.30)$$

When $\sigma_x = \sigma_y = \sigma$, the CEP $= 1.18\sigma_s$ and, from Figure 2.11, 95% of the navigation errors lie within a circle of radius 2.45σ.

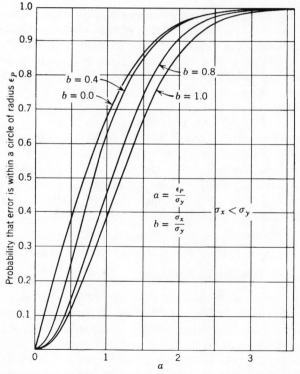

Figure 2.11 Probability of an error lying within a circle of radius ϵ_P. (σ_x and σ_y are the uncorrelated standard deviations in x and y.) From unpublished paper by Bacon and Sondberg.

Navigation errors are frequently defined in terms of a circle of radius 2drms where

$$2\text{drms} = 2\sqrt{\sigma_x^2 + \sigma_y^2} = \sqrt{(2\sigma_x)^2 + (2\sigma_y)^2} \qquad (2.31)$$

If $\sigma_x = \sigma_y = \sigma$, 2drms $= 2\sqrt{2}\sigma = 2.83\sigma$ which, from Figure 2.10, encompasses 98% of the errors. However, if $\sigma_x \neq \sigma_y$, the 2drms circle can enclose as few as 95% of the errors. Sometimes a 3drms error is specified that encompasses 99.99% of the errors; collecting enough relevant data to measure compliance with 3drms is usually impossible.

In three dimensions, the usual measure of navigation performance is the radius of a sphere, 2drms/3D:

$$2\text{drms}/3\text{D} = 2\sqrt{\sigma_x^2 + \sigma_y^2 + \sigma_z^2} = \sqrt{(2\sigma_x)^2 + (2\sigma_y)^2 + (2\sigma_z)^2} \qquad (2.32)$$

If $\sigma_x = \sigma_y = \sigma_z = \sigma$, the 2drms/3D sphere has a radius $2\sqrt{3}\sigma = 3.46\sigma$ and encloses 99% of the Gaussian-distributed navigation errors. In other words, if the single-axis standard deviations are one nautical mile, a sphere of 3.5-nmi radius encloses 99% of the errors. Military systems sometimes define a sphere whose radius is the *spherical error probable* (SEP) that encloses 50% of all three-dimensional errors. If the three variances are equal, the radius of the SEP sphere is 1.54σ.

Navigation test data are often not Gaussian distributed; they have large tails (*outliers* or *wild points*). Measures based on mean square are greatly increased by these outlying points. Hence, test specifications are often written in terms of the "95% radius," the radius of a horizontal circle, centered on the desired navigation fix, that encloses 95% of the test points. As noted earlier, if the data were Gaussian, if the two axes had equal standard deviations and if the mean were at the desired fix (no bias), then the circle radius 2.45σ would enclose 95% of the test points.

2.8.2   Geometric Dilution of Precision

Geometric dilution of precision (GDOP) relates ranging errors (e.g., to a radio beacon) to the dispersion in measured position. If three range measurements are made in orthogonal directions, the standard deviations in the aircraft's position error are the same as those of the three range sensors. However, if the range measurements are nonorthogonal or there are more than three measurements, the aircraft's position error can be slightly smaller or much larger than the error in each range.

If the variances in ranging errors to each station are equal, σ_R^2, and if the uncorrelated variances of aircraft position are σ_x^2, σ_y^2 and σ_z^2 in locally level coordinates, then, by definition, the position dilution of precision is

$$(PDOP)^2 = \frac{\sigma_x^2 + \sigma_y^2 + \sigma_z^2}{\sigma_R^2} \qquad (2.33)$$

and the horizontal dilution of precision (HDOP) is

$$(HDOP)^2 = \frac{\sigma_x^2 + \sigma_y^2}{\sigma_R^2} \qquad (2.34)$$

In pseudoranging systems, the GDOP is

$$(GDOP)^2 = (PDOP)^2 + (TDOP)^2$$

where TDOP is the time dilution of precision, the contribution of clock error to the error in pseudorange. Equations for GDOP, PDOP, and HDOP, when the standard deviations in range to each station are different, are provided in [12]. Dilution of precision plays an important role in radio-ranging computations, especially for Loran (Chapter 4) and GPS (Chapter 5). Detailed GDOP equations for Loran are in Section 4.5.1, for Omega in Section 4.5.2, and for GPS in Section 5.5.2. Receivers usually flag a PDOP or HDOP greater than approximately 6 as an indication of poor geometry of the radio stations, hence a poor fix.

2.9 DIGITAL CHARTS

Traditional aeronautical charts are printed on paper. They are of three kinds:

1. *Visual charts.* Showing terrain, airports, some navaids, restricted areas
2. *En-route instrument charts.* Showing airways, navigation aids, intersections, restricted areas, and legal boundaries of controlled airspace. Airways are annotated to show altitude restrictions; high terrain is identified.
3. *Approach plates, standard approaches (STARs), and standard departures (SIDs).* Showing horizontal and vertical profiles of preselected paths to and from the runway, beginning or ending at en-route fixes. High terrain and man-made obstacles are indicated. Missed approaches to a holding fix are described visually.

Military targeting charts show the expected location of defenses, the initial approach fix, the direction of approach, a visual picture of the target in season (e.g., snow covered) and the preplanned escape route.

Since World War II, experiments have been made with analog charts driven by automatic navigation equipment. Paper charts were unrolled or scanned onto CRTs, while an aircraft "bug" was driven by the navigation computations. The systems were limited by cost, reliability, and the need for wide swaths of chart to allow for diversion. Their use was confined to some helicopters and experimental military aircraft.

In 1996, digital maps were well-established in surveying data bases, the census, automotive navigation, and other specialized uses. Manufacturers of navigation sets created their own data base of navaids and airports or purchased one. Small digital data bases were included in the navaid's ROM whereas large data bases, especially those that included terrain, were usually delivered to customers on CD-ROM. National cartographic services in the developed world were all converting from paper maps to digital data bases. The U.S. Defense Mapping Agency (DMA) issued a standard for topographic maps on CD-ROM [19], and several other nations' cartographic agencies were doing the same. The U.S. Defense Mapping Agency produces separate data bases for terrain elevation and cultural features. They can be stored separately and superimposed on an airborne display. A U.S. National Imagery Transmission Format was created to send and store digital data. The GRASS language was widely used in the United States to manipulate DMA data [6]. Private companies were producing remarkably diverse Geographic Information System (GIS) data bases. In 1996, at least one company (Jeppesen) was producing digital approach plates on CD-ROM [4]. RTCA published a guide to aeronautical data bases [24a] as did ARINC [26a].

The technical challenges have been (1) to standardize the medium (e.g., CD-ROM), (2) to standardize the format of data stored on the medium so that any disc could be loaded onto any aircraft, just as any chart can be carried on any aircraft, and (3) to develop on-board software that displays sections of the chart across which the aircraft seems to move (moving map or moving bug, or both) and orient the chart properly (*north up* or *velocity vector up*). As the aircraft nears the edge of the chart, the software must move to a new section while avoiding hysteresis when flying near the edge of a chart. The expectation is that CD-ROM en-route and approach charts will be readily available to military users before the year 2000.

Digital chart displays have provisions for weather or terrain overlays (from airborne radar or from uplinked data), and provisions for traffic overlays (from on-board TCAS, ground uplink on Mode-S, or position broadcasts from other aircraft, Chapter 14). Civil airlines, driven by cost considerations, may gradually abandon their practice of purchasing charts and distributing them to the crews in hard copy. Instead, they may at first print charts on demand in the dispatch room from a central data base and, later, distribute portable digital charts on CD-ROM or via radio uplink to be loaded into the aircraft avionics when the crew boards.

2.10 SOFTWARE DEVELOPMENT

The sequence of activities in preparing the navigation software is as follows:

1. The vehicle requirements are decomposed into the navigation system requirements. The navigation functions are allocated to hardware and software, usually after trade-off studies.

2. A mathematical model of the vehicle and sensors is prepared. Sensors, such as radars and inertial instruments, are simulated with respect to accuracy and reliability.

3. Engineering simulations are conducted on the accuracy model to determine the scaling, calculation speed, memory size, word length, and minimum degree of complexity required to obtain acceptable accuracy. Reference trajectories are defined, that specify flight paths, speed profiles, and attitude histories. Often man-in-the-loop simulations are required using a cockpit mock-up to assess the crew's work load. Another simulation determines the system's reliability and availability, given the known reliability of the constituent sensors and computers.

4. The equations are coded for the flight computer. Prior to 1975, most navigation software was coded in assembly language to increase the execution speed. Thereafter, higher-order languages such as Fortran, C, Pascal, Jovial, and Ada have been used, though hardware drivers are often still written in assembly language. Subroutines and functional modules come from libraries of well-tested routines that are re-used. The modules are individually tested; then the complete program is gradually compiled and tested. The contents must be documented at each stage, a process called *configuration control*.

5. The code is verified by an independent agent. Mission-critical code, whose failure causes diversion of the aircraft, undergoes a simple verification, sometimes by engineers in the same company who did not participate in the development or test of the code. However, if navigation code is embedded with code that can cause loss of the aircraft (safety-critical code), the code must be further verified, usually by an independent organization using independently derived mathematical models of the aircraft and sensors. The high cost of independent verification encourages architectures in which safety-critical code is segregated. RTCA describes the certification of airborne software [24d] as does ARINC [26b].

6. The code is loaded into ROM chips (often into ultraviolet-erasable EPROMs or into electrically-erasable programmable ROMs called EE-PROMs or flash ROMs) that are installed in the computer by the manufacturer. EEPROM and flash-ROM code can be field-altered using test connectors. Sometimes, when navigation is not embedded into safety-critical software, the code is delivered to the airbase on tape or CD-ROM and

loaded into the flight computer via the on-board data bus. Revisions of flight software may be issued from time to time.

7. A copy of the flight software is usually delivered to a training facility, where it is used to check out crews in a ground simulator. The simulator may be a part-task computer-based trainer (CBT); a terminal that emulates the navigation keyboards, on-board computer, and displays; or it may be a high-fidelity emulation of the cockpit and avionics. A high-fidelity simulator may incorporate a flight computer that contains the navigation software or may rely on a scientific computer, programmed to emulate the flight computer. In a CBT, sensor inputs are simulated; in a high-fidelity simulator, they may come from real or simulated hardware. Simulator training is cheaper and often more effective than flight training.

8. The final task in the preparation of the navigation software is the evaluation of its performance during flight versus the specification. This is done by the aircraft manufacturer or operator.

2.11 FUTURE TRENDS

The increasing capability of airborne digital computers will permit more complex algorithms to be solved. Companies that specialize in aircraft navigation will continue to build libraries of proprietary algorithms that they incorporate into their products. Crew interfaces will become more graphical to reduce workload and reduce errors in loading data. Direct loads from the ground via Mode-S links and other data links (some via satellite) will be commonplace.

By the year 2000, on-board CD-ROM readers will display charts and flight-manual data on military aircraft. The civil aviation industry may prefer to print up-to-date paper charts for each flight in the dispatch rooms, downloaded from a central data base, as an alternative to procuring and distributing them.

The software verification costs assigned to each aircraft will be substantial, because they are amortized over a few hundred units, even with standard libraries of routines.

PROBLEMS

2.1. The direction cosine matrix $[C]$ transforms the Earth-centered inertial coordinates y_i into the locally-level navigation coordinates z_j. Let $\alpha = 0$ when the aircraft is on the equator, and let the initial matrix be

$$[C] = \begin{bmatrix} 0 & 1 & 0 \\ 0 & 0 & 1 \\ 1 & 0 & 0 \end{bmatrix}$$

Let $\dot{\alpha} = \dot{\lambda}$.

(a) If the aircraft flies 90° due east, show the direction cosine matrix.

$$\text{Ans. } \begin{bmatrix} 0 & 0 & 1 \\ 1 & 0 & 0 \\ 0 & 1 & 0 \end{bmatrix}$$

(b) If the aircraft flies 90° due north from its original position, show the direction cosine matrix.

Ans:

$$\begin{bmatrix} 0 & 1 & 0 \\ -1 & 0 & 0 \\ 0 & 0 & 1 \end{bmatrix}$$

(c) If the aircraft flies 30° due east on the equator at 600 knots from its original position, what are the C_{ij} and the $\dot{C}_{ij}$? Use Equation 7.40.

Ans:

$$[\dot{C}] = -\frac{5}{57.3} \begin{bmatrix} 1 & \sqrt{3} & -\sqrt{3} \\ -\sqrt{3} & 1 & 1 \\ 1 & -\sqrt{3} & 0 \end{bmatrix} \text{hr}^{-1}$$

2.2. An aircraft flies 3 hrs east then 2 hrs north at 300 knots at an altitude of 3 nmi, starting at 40° north latitude.

(a) Find its position using the flat-Earth dead-reckoning equations.

Ans. $x = 900$ nmi, $y = 600$ nmi.

(b) Find the final latitude-longitude using spherical-Earth equations with the Gaussian radius of curvature.

Ans. $\Phi = 49.98°$, $\Delta\lambda = 19.54°$.

(c) Find the final latitude-longitude using the ellipsoidal-Earth equations, (2.5).

Ans. $\Phi = 49.99°$, $\Delta\lambda = 19.50°$.

(d) Find the distance from the start to the destination of case **a** using the flat-Earth range equation, 2.20.

Ans. 1081.7 nmi.

(e) Find the distance to the destination of case **b** using the spherical-Earth approximation with the Gaussian radius of curvature.

Ans. 1022.7 nmi.

(f) Find the distance to the destination of case **b** using the flat-Earth equations plus the correction of Equation 2.21.

Ans. 1027.0 nmi.

2.3. A GPS satellite is at the ECEF coordinates:

$$x = 14367.71 \text{ nmi}, \quad y = 0, \quad z = 0$$

The observer is at latitude 45°, longitude 45°, 30,000-ft altitude. Calculate the distance from the aircraft to the satellite and the elevation of the line of sight. Use the WGS dimensions on page 25.

Ans. R = 12,986.56 nmi, θ = 43.5 deg.

2.4. Derive Equation 2.21. *Hint:* solve the spherical triangle for the cosine of the range angle and express the coordinates of the waypoint as the present position plus a small increment. Expand the sines to third order and the cosines to second order.

2.5. Verify the calculations on page 46 for the probability of falling inside a rectangle of edge 1.2 nmi (x = 0.6 nmi) by 2.4 nmi (y = 1.2 nmi). Let σ_x = 0.4 nmi and σ_y = 2.4 nmi.

3 Multisensor Navigation Systems

3.1 INTRODUCTION

Multisensor navigation is the process of estimating the navigation variables of position, velocity, and attitude from a sequence of measurements from more than one navigation sensor. There are essentially two broad categories for sensors used in avionics suites for navigation and related functions: dead-reckoning sensors and positioning sensors. The processing equations for these are given in Chapter 2.

Dead-reckoning sensors provide a measure of acceleration or velocity with respect to an Earth-referenced coordinate system, consequently requiring integration with respect to time to provide vehicle position with respect to the Earth. Examples of these types of sensors are inertial systems (Chapter 7), Doppler radars (Chapter 10), and air-data sensors (Chapter 8). The latter two require an attitude and heading reference (AHRS) or an inertial system (INS) to provide the required angular orientation with respect to the Earth.

Positioning sensors provide a position measurement that can be related to Earth-referenced coordinates. Examples of these sensors are radio systems such as the terrestrial-based Loran and the satellite-based Global Positioning System (GPS) (Chapters 4 and 5) which provide the position of the antenna in geodetic coordinates. A star-tracker can also be used for fixing position when its orientation with respect to the Earth is determinable through some means such as an AHRS or INS.

This chapter describes *filtering* algorithms whereby measurements from combinations of these sensors can be employed to satisfy the functional output variable requirements of an avionics suite. In general, avionics system users require a variety of information on the state of the air vehicle depending on the complexity of the application—aerospace plane to auto-gyro—and accordingly will include a variety of complementary sensors in their equipment suite. The information desired can include the following:

- Position and velocity in geodetic coordinates—east, north, and up which allow determination of ground speed and track angle.
- Orientation with respect to the Earth—pitch, roll, and yaw or heading angles.

- Linear and angular acceleration and rate in body coordinates for vehicle control purposes.
- Vehicle state relative to the air mass including orientation—angle of attack, sideslip, and airspeed, again for vehicle control purposes.

The discussion of navigation sensors in other chapters explains their capabilities and error characteristics. Clearly, the entire list of state variables for the vehicle cannot be provided by any one sensor at normally desired levels of accuracy and dependability. The deficiencies that exist in the individual sensors employed in a multisensor avionics suite include at least one of the following characteristics:

- Increase in error of the navigation variables as a function of time or distance traveled. This is a characteristic of all dead-reckoning sensors that eventually accumulate unbounded errors. Examples are inertial and Doppler radar navigation systems.
- High noise level or low bandwidth in any derivative variable. This is a characteristic of radio sensors that require differentiation with respect to time of the basic measured variable to obtain rate or acceleration. For example, with Loran, position must be differentiated to obtain velocity. For Doppler radar, differentiation of speed can yield acceleration, once the result is transformed to an Earth-referenced coordinate system using attitude from an AHRS or inertial system.
- Reliance on sensors requiring off-board components to accomplish the required functions. This includes radio sensors using terrestrial or satellite-based assets whose access can be denied, due to intentional (jamming) or nonintentional interference (transmission blockage or interfering transmissions) of the propagated signal from the off-board assets, or failure of the transmitters due to other causes.

To overcome individual sensor deficiencies, system designers have sought combinations of avionic sensors. These multisensor systems are designed to provide reliable, dynamically accurate measurements of the air vehicle's state for all specification-required flight conditions.

The most typical solution has been to use an inertial system in conjunction with an appropriate set of complementary sensors that arrest the random, time-increasing error in velocity, position, and orientation resulting from integration of the fundamental high-bandwidth inertial measurements of acceleration (force corrected for the a priori known effect of gravity) from the accelerometers and angular change from the gyroscopes. The following paragraphs discuss the attributes of integrating various sensors with an inertial system to obtain high-accuracy, high-bandwidth measurements of the dynamic state of the air vehicle. Other multisensor configurations are discussed in Section 3.8.

3.2   INERTIAL SYSTEM CHARACTERISTICS

Two fundamental error sources affect the error behavior of an inertial system. These are the errors in the measurements of force made by the accelerometers and the errors in the measurement of angular change in orientation with respect to inertial space made by the gyroscopes. The basic mechanization of an inertial system, which is described in detail in Chapter 7, is depicted schematically in Figure 3.1. In this figure it is seen that the force measurements made by the accelerometers are first transformed to a selected navigation coordinate frame that is typically the local geodetic coordinates of east, north, and local vertical. These measurements are then compensated for the force of gravity with a mathematical model, such that vehicle acceleration with respect to inertial space is obtained. The resulting variable, after correction for Coriolis acceleration, is then integrated once into velocity and a second time into position change with respect to the Earth. Additionally, the gyroscopic measurements of angular change with respect to inertial space are modified using the system computed velocity and the Earth's rotation rate vector to reflect the rotation of the local vertical due to earth rotation and the vehicle change in position as it travels over the surface of the Earth. In this manner the orientation of the accelerometer axes relative to the Earth at the present position of the vehicle is continuously computed. The result is that there are three sources of change in orientation error of the accelerometers with respect to an Earth-fixed reference coordinate frame: (1) integrated gyro drift rate; (2) integrated error in system computed velocity which results from error in the measurement of acceleration—due to accelerometer measurement errors, imperfect knowledge of the local force of gravity and the current error in the knowledge of orientation of the accelerometer sensing axes which causes a misresolution of any accelerometer force measurement including that of the gravity vector; and (3) error in the orientation of the navigation coordinate axes which changes as they rotate with respect to inertial space.

The result of this interaction of error effects is that the error characteristic of an inertial system for the computation of velocity, position, and instrument axes orientation is described by the sinusoidal Schuler oscillation that has a period of approximately 84.4 minutes. Due to this oscillatory characteristic, the position error response to a step of constant accelerometer measurement error is not a quadratic in time but a bounded Schuler oscillation as shown in Figure 3.2. Note that the error in position and the "tilt," which is the error in orientation of the accelerometer sensing axes with respect to the local level plane, are equal for any acceleration measurement error.

The velocity error response to a step of constant gyro drift rate error, as shown in Figure 3.3, is a bounded Schuler oscillation that has a constant error dictated by the magnitude of the gyro drift rate. The error in position is also characterized by a Schuler oscillation but diverges in time in proportion to the integrated velocity error. The tilt due to the gyro drift rate is a bounded Schuler oscillation with zero mean. This occurs because any tilt results in a counter-

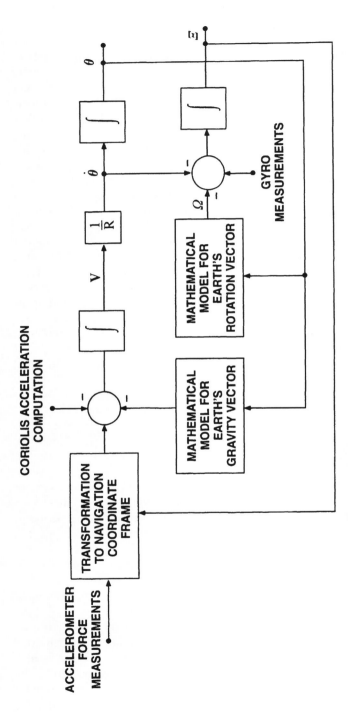

CORIOLIS ACCELERATION COMPUTATION

ACCELEROMETER FORCE MEASUREMENTS

TRANSFORMATION TO NAVIGATION COORDINATE FRAME

MATHEMATICAL MODEL FOR EARTH'S GRAVITY VECTOR

MATHEMATICAL MODEL FOR EARTH'S ROTATION VECTOR

GYRO MEASUREMENTS

$\int$

V

$\dfrac{1}{R}$

$\dot{\theta}$

$\int$

Ω

$\int$

θ

$[\Xi]$

R– LOCAL RADIUS OF EARTH'S CURVATURE
V – SYSTEM COMPUTED VELOCITY
$\dot{\theta}$ – SYSTEM COMPUTED ANGULAR RATE OF CHANGE OF POSITION (θ) OVER THE EARTH'S SURFACE
Ω – SYSTEM COMPUTED EARTH RATE VECTOR
Ξ – SYSTEM COMPUTED ACCELEROMETER SENSING AXES ORIENTATION WITH RESPECT TO THE NAVIGATION COORDINATE SYSTEM

Figure 3.1   Schematic block diagram on an inertial system mechanization.

58

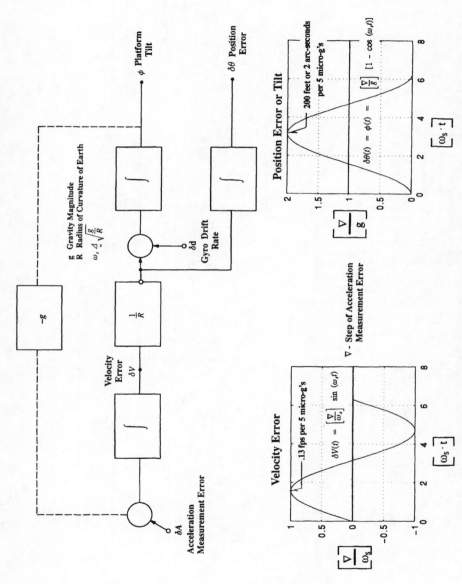

Figure 3.2 Inertial system error response to a step of acceleration measurement error. Constant acceleration measurement error induces a zero-mean Schuler oscillation in velocity error and an identical nonzero mean oscillation in tilt and position error.

59

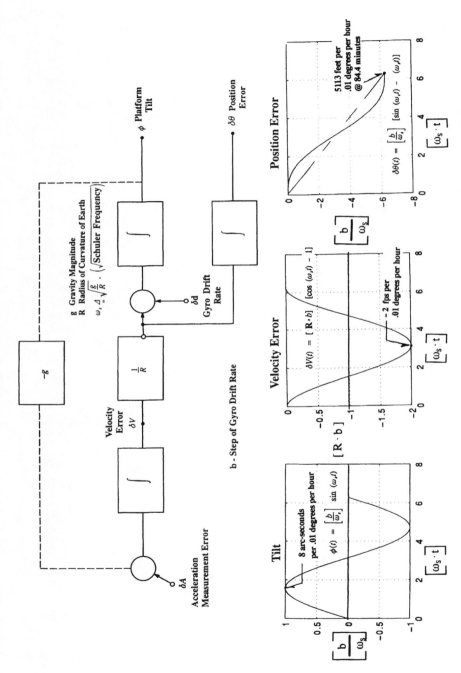

Figure 3.3 Inertial system error response to a step of gyro drift rate bias. Constant gyro drift rate induces a zero-mean Schuler oscillation in tilt, a nonzero mean oscillation in velocity error and a divergent ramp oscillation in position error.

vailing accelerometer measurement error, whose result is to produce a constant velocity error component that cancels any constant gyro drift rate at the tilt rate node.

This discussion of deterministic error responses pertains to constant error sources and initial conditions. As is well-known, the response of any oscillator to random white noise or correlated noise inputs is divergent at some fractional power of increasing time. Consequently the error response in inertial system position, velocity, and orientation due to a noise input is divergent with time. The discussion of inertial system error characteristics presented here is quite simplified in that effects due to error in the system computed Earth rate vector and error in the system computation of Coriolis acceleration lead to additional 24 hour and Foucault error oscillation characteristics as described by the error equations in Chapter 7.

Clearly inertial systems tend to "drift off" in terms of the accuracy in the system computed navigation variables of position, velocity, and orientation over an extended time of operation. Consequently other sensors are used with inertial systems to arrest this divergent error behavior for those applications that require upper bounds on system error with time. It is the usual case that an inertial system remains as the core element of an integrated aircraft navigation system because it offers the unique properties of being self-contained on the vehicle and stealthy in that it produces no emissions, cannot be jammed, or deceived because it has no reliance on signals generated by external sources; the inertial system is normally required to provide a continuous high-bandwidth reference coordinate system relative to the Earth whose use is required by other avionics sensors.

3.3 AN INTEGRATED STELLAR-INERTIAL SYSTEM

A star-tracker (Chapter 12) serves as a useful adjunct to an inertial system in that it provides the capability of calibrating the inertial system gyroscopes that are the primary source of divergence in the accuracy of a free-inertial navigation system. The star-tracker is used in a natural way with an inertial system in that the inertial system provides the reference coordinate system with respect to which the measurements of star position can be referred.

A star-tracker and inertial system work in combination as depicted schematically in Figure 3.4. Figure 3.4 shows the three basic sources of error in employing a star-tracker in combination with an inertial system all of which result in a pointing error to a selected star:

1. Orientation error of the inertial system ϕ, relative to the reference ellipsoid that results in a commensurate error in pointing to the star even if there exists no error in inertial system computed position. This situation is indicated at the true position in Figure 3.4.

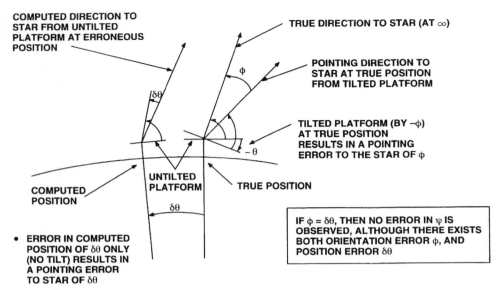

COMPUTED DIRECTION TO STAR FROM UNTILTED PLATFORM AT ERRONEOUS POSITION

TRUE DIRECTION TO STAR (AT ∞)

φ

POINTING DIRECTION TO STAR AT TRUE POSITION FROM TILTED PLATFORM

δθ

TILTED PLATFORM (BY −φ) AT TRUE POSITION RESULTS IN A POINTING ERROR TO THE STAR OF φ

−θ

UNTILTED PLATFORM

COMPUTED POSITION

TRUE POSITION

δθ

IF φ = δθ, THEN NO ERROR IN ψ IS OBSERVED, ALTHOUGH THERE EXISTS BOTH ORIENTATION ERROR φ, AND POSITION ERROR δθ

• ERROR IN COMPUTED POSITION OF δθ ONLY (NO TILT) RESULTS IN A POINTING ERROR TO STAR OF δθ

Figure 3.4   Angular position and orientation errors observed with a star-tracker with inertial stabilization. A stellar sensor observes the error ψ, the difference between the position error ($\delta\theta$) and the orientation error (ϕ) of the inertial system.

2. Error in system computed angular position on the surface of the Earth $\delta\theta$. This causes an error in pointing to the star even if the reference coordinate frame provided by the inertial system is perfectly aligned with respect to the Earth or inertial space at its actual position. This situation is indicated at the computed position in Figure 3.4.

3. Error in system time that results in an orientation error of an Earth-referenced frame with respect to the stellar background due to Earth rotation. This error is usually minimal and is not considered further here.

An interesting fact related to the stellar-inertial system combination is that an error in orientation and system-computed position can result so that the resulting pointing error to the star ψ is zero:

$$\psi = \phi - \delta\theta = 0$$

The implications of this situation is obtained by referring to the error block diagram of Figure 3.5 where the error ψ is shown. Since with an inertial system any error in the accelerometer measurement results in an equal error in both position $\delta\theta$ and orientation ϕ, then for such an error source ψ will be zero and will not be detected with a star observation. On the other hand, a constant gyro drift rate produces an oscillatory diverging error in the angular position $\delta\theta$ and a bounded oscillatory error in orientation ϕ, so the resulting error ψ is a ramp

- **LEVEL AXIS SCHULER ERROR LOOP**

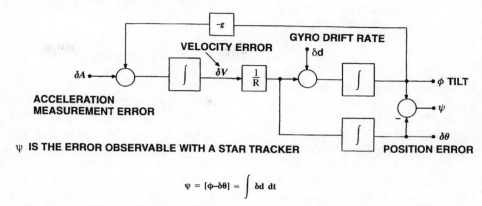

$$\psi = [\phi - \delta\theta] = \int \delta d \; dt$$

Figure 3.5   Illustration of the ψ vector observed with a star-tracker. An inertially sta-bilized stellar sensor is used to measure the effects of gyro drift rate but cannot observe the effect of acceleration measurement errors.

proportional to the gyro drift rate. Since this diverging error is detectable with the inertially stabilized star-tracker, then clearly the error in position can be corrected or "reset" when a star shot is taken.

Most stellar-inertial systems have been mechanized with inertial systems using stabilized gimbal control loops (as described in Chapter 7). This mech-anization permits a telescope and inertial instrument assembly to be manufac-tured that has only two low-bandwidth pointing loops that maintain the tele-scope line of sight to the star being observed, usually for the elevation and bearing angles to the selected star. High-bandwidth pitch, roll, and heading sta-bilization control loops employing gyro measurements of small angular dis-turbances isolate the inertial instruments and tracker assembly from vehicle angular motion. When such a gimballed mechanization is employed, the sys-tem gyros are rotated slowly with respect to inertial space to maintain a desired orientation of the instrument assembly with respect to the Earth, usually that of the local-level mechanization. Because the gyro sensing axes change slowly with respect to inertial space, the detected components of the ψ vector, which is the integral of gyro drifts projected onto inertially fixed axes, can be used to calibrate the gyros. This is possible because the detected components of the ψ vector can be correlated with specific gyro-sensing axes.

Since the component of the ψ error vector along the line of sight to an indi-vidual star cannot be observed, two star shots are generally required to com-pletely correct the system. Ideally these lines of sight are orthogonal to enhance observability. It is possible to obtain a complete calibration of the system gyro drift rate vector, when the following two conditions occur:

1. Two star shots are taken closely together in time.

2. The gyro sensing axes have not changed orientation with respect to inertial space by a great amount since the last point in time when a pair of stars was observed.

With the advent of inertial instruments suitable for mechanizing strapdown systems (discussed in Chapter 7) mechanizations of star-trackers with high-bandwidth stabilization control loops for maintaining the star line of sight axis have been considered. The dynamics relating inertial system gyro drift rate equivalent errors to the observable ψ vector error in this case clearly becomes highly complex and disallows effective calibration using conventional ad hoc fixed gain methods of the early 1960s. Fortunately the discovery of Kalman filtering theory, as elaborated later in this chapter, provides at least a theoretical approach to this problem, wherein the correlation of observed errors to the producing causes of these errors is computed in real time to provide a basis for correction of error sources regardless of the complexity of the time-variant error dynamics.

In summary, the addition of a star-tracker to an inertial system is synergistic from the point of view that the inertial system provides the stabilized reference coordinate system required by the star-tracker and the star-tracker can be used to "reset" the inertial system error growth and correct the gyro drift rates that are the principal source of position error divergence in the inertial system. On the other hand, the star-tracker does not correct for sources of error in the determination of vehicle acceleration which include the accelerometer measurement errors and any lack of knowledge about the local gravity vector deflection.[1]

Consequently, other navigation sensors are usually considered to correct for these effects. A stellar inertial system retains all the positive attributes of the inertial system—self-contained, stealthy, and virtually impossible to deceive. The use of the tracker can only be denied by adverse weather conditions.

3.4   INTEGRATED DOPPLER-INERTIAL SYSTEMS

In the early days of inertial and stellar-inertial system development, one approach to decreasing the effect of divergent navigation error was the utilization of a speed sensor such as a Doppler radar (Chapter 10 describes the theory and performance of Doppler radars). A Doppler radar, similar to a stellar-tracker, has a natural use with an inertial system in that it requires the reference coordinate system (attitude and heading) provided by the inertial system to refer its speed measurements to Earth-fixed coordinates to achieve the navigation function.

An understanding of the value in using a Doppler radar in conjunction with

---

[1] As a rule of thumb, the uncompensated effect of the deflection of the vertical results in an inertial system position error divergence of about 0.1 nmi in one hour, thereafter being in proportion to the square root of time.

an inertial system is obtained by referring to the schematic error block diagram of Figure 3.6. For simplicity in this figure the error in the Doppler radar velocity measurement is assumed to be only a constant bias, δV_R. The observation $(\delta V_R - \delta V)$ in Figure 3.6 is obtained from the difference between the reference Doppler radar velocity measurement and the inertial system computed velocity. This first requires the transformation from the Doppler radar-sensing axes to the Earth-referenced navigation coordinates using the inertial system determined orientation. Thus, both the zero and nonzero mean Schuler oscillations (Figure 3.6), due to, respectively, a step of acceleration measurement error and gyro drift rate bias, are observable by using the reference Doppler radar measurement. However, the bias in the inertial system velocity is not fully determinable, since the Doppler radar also has a bias error and only the sum of these two biases is observable.

Since the Schuler error oscillations due to inertial system error sources are observable with the Doppler radar measurement, it is possible to "damp" these Schuler error effects by feeding back the observed difference. This was done in the early days by adding a fixed feedback damping gain K_D back to the inertial velocity integrator as shown in Figure 3.7. Additionally, a fixed feed forward gain K_F was employed in more sophisticated mechanizations to provide two control parameters (damping factor and natural frequency) for this second-order mechanization.

For the older configurations of Doppler radars in which the Doppler antenna was physically stabilized using the gyro measurements, variants of the conventional mechanization described above were possible. For example, if the gyro drift rate biases significantly exceeded the Doppler radar measurement error on an equivalent basis, then a feed forward integrator could be added to compensate for the inertial system gyro drift rate error. An error in this type of compensation would of course be caused by any nonzero Doppler radar bias.

More sophisticated mechanizations for error removal are possible. For example, if the coordinate frame of the Doppler sensing axes rotates with respect to the inertial system sensing axes and the Doppler measurement error rotates accordingly, then a basis for observation and calibration of these error sources exists, as the sums of errors described above change value as a function the angle of rotation. The calibration of errors that become observable as a function of measured changes in orientation is obtained with ease through the use of Kalman filtering techniques, which capitalize on the computed real-time correlation of the modeled system error sources to the errors observed when commensurate variables from different navigation sensors are compared.

One caveat to the properties of Doppler inertial systems that is important is that changes in orientation and acceleration of the vehicle do not make the errors in the inertial system orientation directly observable. Changes in the error in both the inertial system velocity and transformed Doppler measurements due to tilt and azimuth misalignment always cancel. This is because both systems use the inertial orientation as the reference and are hence unobservable in the direct comparison of velocity measurements. Tilt can be damped out indirectly,

66

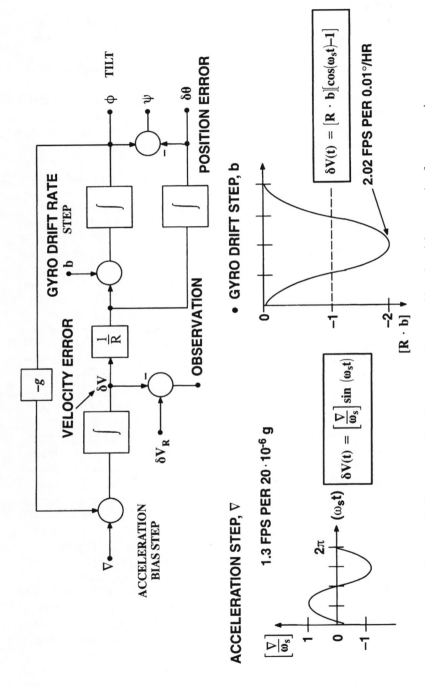

Figure 3.6 Integration of a reference speed sensor with an inertial system. A reference speed sensor is used to observe the Schuler oscillations induced by acceleration measurement error and gyro drift rate and the velocity bias error induced by constant gyro drift rate.

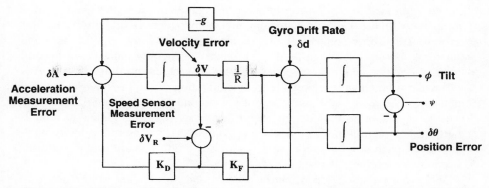

Figure 3.7   Block diagram of a fixed-gain (K_D, K_F) speed sensor damped-inertial system mechanization.

since it induces a longer-term observable Schuler oscillation. Inertial azimuth misalignment can also be corrected indirectly by the normal gyrocompassing process (as described in Chapter 7) through the Schuler oscillation it induces.

In that a Doppler radar measures the Doppler shift of a radiated signal with respect to the scattering surface, it has error characteristics to be considered in the system design. In overwater operation, the Doppler measurement is affected by major water currents as well as by the behavior of the surface due to wind. Variation in sea and terrain surface roughness also causes measurement error variation due to variation in scattering of the signal. These effects are discussed in detail in Chapter 10.1.4.

In summary, the addition of a speed sensor to an inertial system provides the capability to remove the Schuler oscillations in the errors in the inertial system computed position and velocity. Furthermore the orientation errors in tilt and azimuth can be removed in much the same manner as in normal ground alignment (as discussed in Chapter 7). Finally, a Doppler-inertial system is self-contained but not completely stealthy in that it radiates a signal to obtain the Doppler measurement. However, for all practical purposes, the Doppler radar cannot be deceived or jammed, since narrow beams at steep angles are employed.

3.5   AN AIRSPEED-DAMPED INERTIAL SYSTEM

Another way to obtain some decrease in the divergent navigation errors of an inertial system is to employ the outputs of the air-data system on the aircraft. Typically, what this sensor provides is the speed of the aircraft with respect to the air mass along with the angle of attack and sideslip angle of the vehicle (as discussed in Chapter 8). Consequently using the orientation information from the inertial system, the velocity of the aircraft with respect to the air mass can be expressed in the Earth-referenced navigation coordinate system and be compared with the inertial system computed components of velocity. The major error source in this comparison is the uncertainty in knowledge of the wind (air mass motion relative to the Earth) which is generally quite large. Several knots of error in the wind components along the east and north navigation axes will make it practically impossible to do calibration of the inertial system gyro drift rates. As a rule of thumb, 1 foot per second (fps) of reference velocity error is equivalent to 0.01 deg/hr of gyro drift rate, which is typical of a navigation grade inertial navigation system. Consequently reference velocity measurements accurate to the order of 1 fps are required to obtain useful gyro calibration of such inertial systems.

When a navigation grade inertial system and an air data sensor are integrated, the inertial system can be used to calibrate (measure) the relative wind to the order of the gyro drift rate. The calibrated vehicle speed measurement can then be used to reduce the Schuler error oscillations in the velocity components computed by the inertial system.

Since an airspeed sensor is available on any aircraft, it is worth considering combining its measurements with an inertial system to obtain the synergy described above. However, the parameters for a fixed gain mechanization or the measurement noise parameters for a Kalman filter mechanization need to be selected judiciously to reflect short-term error fluctuations in the airspeed data.

3.6   AN INTEGRATED STELLAR-INERTIAL-DOPPLER SYSTEM

When a Doppler radar (or air data sensor) is combined with a stellar-inertial system, significant benefits are obtained. Stellar measurements are used to calibrate the inertial system gyro drift rates, thereby reducing any mean error in the inertial system computed velocity components. Consequently, when the stellar-inertial system velocity components are compared with commensurate velocity components from a reference velocity sensor, the mean errors in the reference velocity sensor components can be calibrated. Effects of the errors in the measurement of acceleration (accelerometer measurement error and uncertainty in the gravity vector) can now be more effectively removed using the calibrated reference velocity sensor. Recall that the errors induced by an error in the vehicle acceleration measurement cause divergent velocity, position and orientation errors in a stellar-inertial system, not observable with the star tracker.

The most attractive attribute of a stellar-inertial system when combined with a reference velocity sensor is that the resulting errors in all the navigation variables are bounded. The residual error levels in these variables are of course a function of the accuracy of the individual sensors employed. Accuracies on the order of a hundred feet in position, a fraction of a foot per second in velocity and subarcsecond in orientation are achievable with such a system. An added attraction of such a system is that it is fully autonomous.

3.7   POSITION UPDATE OF AN INERTIAL SYSTEM

The simplest method of correcting the error drift in position computed by an inertial system is to reset the position to the coordinates of a point observable by the pilot on the surface of the Earth over which the aircraft can fly. Automatic sources of more frequent position measurement include Loran and Omega with their Earth-fixed radio transmitter stations and the Global Positioning System (GPS) using radio transmitters on the satellites.

Due to the time-variant error dynamics associated with an inertial system, principally due to acceleration effects, it has been cumbersome to design fixed gain mechanizations to feed back the observed position differences between the inertial system computation and the reference position measurements to correct inertial system computed velocity and orientation and the sources of these errors (gyro drift rate and acceleration measurement error). Fortunately, with the development of a method of recursive filtering theory by R. E. Kalman in 1960 [1] and sufficiently powerful digital computers to effect its mechanization, this problem has become mute. With Kalman filtering theory, the weighting gains between the observed position error and the other system errors are based on error correlation information computed in real time, in what is called the *error covariance matrix*. Correction of system errors through the use of such gains obtains the theoretically optimal utilization of the measurements for minimization of errors. (See Section 3.10 for Kalman filter basics.)

The most important integrated inertial and positioning measurement system in use in 1996 is that which combines GPS with an inertial sensor (Section 3.13). Not only does GPS provide position along with time but also velocity in geodetic coordinates. Mechanizations based on Kalman filtering theory are invariably employed to effect a synergistic combination of GPS and the inertial system (as discussed in Sections 3.13.3 and 3.13.4). An inertial system periodically corrected with external position measurements will have bounded errors in all the navigation variables of position, velocity, and orientation. Such a system is obviously not autonomous and is subject to intentional and unintentional denial, which is unattractive in military applications.

3.8   NONINERTIAL GPS MULTISENSOR NAVIGATION SYSTEMS

Although most integrated navigation systems are inertially based, there are exceptions. In particular, GPS receiver module cards have been developed that can be *embedded* in the chassis of other sensors such as a Doppler radar [12] or Loran [11, 14, 15]. In the case of Loran, the integration was motivated by the fact that the GPS satellites may not be sufficient in number at all times to provide positioning data with guaranteed integrity worldwide. A further attraction of Loran is that it does not suffer from the line-of-sight limitation of GPS. It employs a low-frequency signal that propagates along the surface of the Earth (the ground wave), and it is time-synchronized using atomic clock controlled transmitters. In the case of Doppler radar, the motivation for GPS integration is to significantly reduce the effect of Doppler associated error sources on navigation performance. These include speed measurement errors over water, scale factor error and bias, and the errors in a low-accuracy inertial attitude and heading reference system (AHRS). Of course, the Doppler–AHRS system provides the ability to dead-reckon position during GPS outages.

3.9 FILTERING OF MEASUREMENTS

3.9.1 Simple Sensor, Stationary Vehicle

The single example of measurement filtering is the case where a position sensor provides a sequence of measurements

$$x_i = x_a + \epsilon_{d_i} + \xi_i \tag{3.1}$$

where
  x_a is the actual position of the stationary vehicle
  ϵ_{d_i} is a deterministic error in the ith measurement
  ξ_i is a random error in the ith measurement

The deterministic error can be caused, for example, by a known sensitivity to temperature and determined a priori by calibration over the temperature range of operation. The random error is usually due to unknown causes or causes not worth the trouble to investigate. The mean value of the random error ξ is zero; otherwise, it would be compensated. For ease of analysis, we assume that these random errors have a fixed standard deviation σ_ξ over time and are independent from one measurement to the next. The error in each position measurement x_i is seen to be the random error

$$\xi_i = x_i - \epsilon_{d_i} - x_a \tag{3.2}$$

The best estimate of position in this case is the average of the measurements

$$\hat{x}_N = \frac{1}{N} \sum_{i=1}^{N} [x_i - \epsilon_{d_i}] \tag{3.3}$$

where the standard deviation of the error in the estimated position

$$\delta\hat{x}_N = \hat{x}_N - x_a \tag{3.4}$$

is the standard deviation of the random error σ_ξ divided by the square root of the number of measurements

$$\sigma_{\delta \hat{x}_N} = \frac{\sigma_\xi}{\sqrt{N}} \tag{3.5}$$

Hence the sequence of measurements can be processed to obtain an ever-improving estimate of the vehicle's position. To avoid storing all past measurements, the Nth estimate can be written in the recursive form

$$\hat{x}_N = \left[\frac{N-1}{N} \right] \hat{x}_{N-1} + \frac{x_N}{N} \tag{3.6}$$

where only the $(N-1)$th estimate and the Nth measurement are required.

3.9.2   Multiple Sensors, Stationary Vehicle

A more complex algorithm is employed if the position of a stationary vehicle is measured using sequences from multiple position sensors. In this case, assume that the optimal estimate from the Kth sensor using N measurements is denoted $\hat{x}_{N_K}$ and the standard deviation of its zero mean error is $\sigma_{\delta \hat{x}_{N_K}}$. For the case of two position sensors, the optimal minimum error estimate of position of the vehicle after the Nth measurement from each of the two independent sensors is

$$\bar{x}_N = \lambda_{1_N} \cdot \hat{x}_{N_1} + (1 - \lambda_{1_N}) \cdot \hat{x}_{N_2} \tag{3.7}$$

where the weighting factors are

$$\lambda_{1_N} = \left[\frac{\sigma_{\delta \hat{x}_{N_2}}^2}{D_N} \right]$$

$$(1 - \lambda_{1_N}) = \left[\frac{\sigma_{\delta \hat{x}_{N_1}}^2}{D_N} \right]$$

$$D_N = \sigma_{\delta \hat{x}_{N_1}}^2 + \sigma_{\delta \hat{x}_{N_2}}^2 \tag{3.8}$$

The minimum error in this least squares estimate is the weighted sum of the independent errors $\delta \hat{x}_{1_n}$ and $\delta \hat{x}_{2_N}$:

$$\delta \bar{x}_N = [\bar{x}_N - x_a] = \lambda_{1_N} \delta \hat{x}_{1_N} + (1 - \lambda_{1_N}) \delta \hat{x}_{2_N}, \tag{3.9}$$

which has variance

$$\sigma^2_{\delta\bar{x}_N} = \frac{\sigma^2_{\delta\hat{x}_{1_N}} \cdot \sigma^2_{\delta\hat{x}_{2_N}}}{D_N} \tag{3.10}$$

Alternately, this relation can be written

$$\frac{1}{\sigma^2_{\delta\bar{x}_N}} = \frac{1}{\sigma^2_{\delta\hat{x}_{1_N}}} + \frac{1}{\sigma^2_{\delta\hat{x}_{2_N}}} \tag{3.11}$$

where it is evident that the variance of the error in the combined estimate $\delta\bar{x}_N$ is less than the errors obtained from either of the two independent position sensors.

3.9.3   Multiple Sensors, Moving Vehicle

The least-squares technique described above can also be applied to obtain an instantaneous estimate of position from independent sensors for a moving vehicle. In this case, simply assume that the position measurement of the first sensor at the Nth instant of time is $\hat{x}_{1_N}$ and that of the second sensor is $\hat{x}_{2_N}$. Equation 3.7 is then the formula for a minimal error estimate of position using both sensors.

   Unfortunately, this approach is far from optimum. It does not take advantage of the capability of calibrating the errors of one sensor using measurements from another sensor. For example, in the discussion of stellar-inertial and Doppler-inertial navigation systems, the calibration in real time of gyro and Doppler bias errors was shown to be possible. Improved accuracy is obtained in the sensor outputs when such calibration can take place, which means that the errors $\delta\hat{x}_{1_N}$ and $\delta\hat{x}_{2_N}$ in the estimates are smaller than they are in Equation 3.9 above.

   Until the early 1960s ad hoc mechanizations were employed to combine the measurements from independent navigation sensors to obtain an improved overall solution. The development of Kalman filtering theory [1] provided a rigorous approach to obtrain truly optimal estimates for systems whose errors are described by linear time variant differential equations. Since the late 1960s, most integrated navigation system mechanizations have been mechanized based on this fundamentally superior method.

3.10   KALMAN FILTER BASICS

The Kalman filter requires that all error states are modelable as zero mean noise processes with known variances, power spectral densities, and time correlation parameters. Thus, the various error quantities to be estimated and the associated measurement noises are all random processes whose correlation structure is assumed to be known. The Kalman filter then obtains estimates of the states of these stochastic processes, which are described by a linear or linearized mathematical model. It accomplishes this goal by capitalizing on the known correlation structure of the various processes involved and the measurements of linear combinations of the error states. To do this, both the error propagation in time and the measurement processes are expressed in vector form. This provides a convenient way with linear matrix algebra to keep track of relatively complex relationships among all the quantities of interest. This is one of the main features that distinguishes a Kalman filter from most digital signal processing applications. The navigation system integration problem is multiple-input and multiple-output in nature, and matrix algebra is essential in keeping track of the relationships among the variables.

In the multisensor navigation system application, the behavior of the sensor error states are described by linear differential or equivalent finite difference equations. Comparison of measurements between navigation sensors are described by linear combinations of these error states. The filter accomplishes the task of error state estimation by a more complex recursive procedure, rather than forming a simple weighted sum of the individual measurements as described above.

Under the assumption of Gaussian noise distributions, the Kalman filter minimizes the mean square error in its estimates of the modeled state variables [5]. The Gaussian assumption is usually a reasonable assumption in navigation applications because the noise effects that arise in the measurements are often due to a summation of many smaller random contributions. Thus, by the central limit theorem of statistics, there is then a tendency toward the Gaussian distribution regardless of the distribution of the individual contributions. Many multisensor navigation systems have successfully implemented Kalman filters while implicitly making this assumption.

3.10.1   The Process and Measurement Models

In Kalman filtering, the equations describing the process to be estimated and the measurement relationships must first be formulated. This is sometimes referred to as the modeling part of the problem. These equations will now be described.

The first assumption is that the time propagation of the state vector to be estimated can be described by the linear finite difference equation

$$x_{k+1} = \phi_k x_k + \xi_k \tag{3.12}$$

where
x_k is the $n \times 1$ process state vector at time t_k, being errors in position, velocity, and attitude and sensor errors

ξ_k is the $n \times 1$ process white noise vector

ϕ_k is the $n \times n$ state vector transition matrix from time instant k
  to time instant $k + 1$

The process noise vector ξ_k is assumed to be a white noise vector sequence with known covariance matrix Q_k:

$$E[\xi_k \, \xi_k^T] = \boldsymbol{Q}_k \qquad (3.13)$$

where E is expectation operator, ξ_k is a column vector, and ξ_k^T is a row vector.

The comparison of commensurate measurements from the navigation sensors must satisfy the following linear relationship:

$$z_k = H_k \, x_k + \boldsymbol{v}_k \qquad (3.14)$$

where

z_k is the $m \times 1$ vector measurement at time t_k

$\boldsymbol{v}_k$ is a $m \times 1$ measurement white noise vector at time t_k

H_k is the $m \times n$ matrix giving the ideal relationship between z_k and x_k
  when no noise is present

The measurement noise covariance matrix R_k is assumed to be known:

$$E[\boldsymbol{v}_k \, \boldsymbol{v}_k^T] = \boldsymbol{R}_k \qquad (3.15)$$

Note that there are now four known matrices that describe the process and measurement models: ϕ_k, Q_k, H_k, and R_k. Thus, the design of a Kalman filter requires considerable a priori knowledge about the dynamics and statistics of the various processes to be modeled. The state transition matrix ϕ_k describes how the state vector x_k would propagate from one step to the next in the absence of a driving function. The Q_k parameter tells us something about the noise in the x_k process. If the elements of Q_k are large, this means that a large amount of randomness is inserted into the process with each step. The H_k matrix describes the linear relationship between sensor measurements z_k and the error states to be estimated x_k. Note that relatively complicated relationships can be accommodated as long as they are known and linear. Finally, R_k describes the mean-square measurement noise errors. Generally, large values in the R_k matrices means poor measurements. Note that all four of the key matrices are permitted to vary with time (i.e., with the index k).

Perhaps the most challenging aspect of applying Kalman filtering in multisensor navigation applications is that of establishing the mathematical equations that describe the physical situation at hand and casting the equations in the form dictated by Equations 3.12 and 3.14. Fortunately, this can be done with a reasonable degree of accuracy in a wide variety of integrated navigation system applications. However, one should always remember that the Kalman filter in its basic form is a model-dependent filter and not adaptive. This means that, if the model does not fit the physical situation under consideration, the filter may yield poor results.

3.10.2   The Error Covariance Matrix

The error covariance matrix defines the probable error $\tilde{x}$, in the filter estimate $\hat{x}$, of the (error) state vector x. These elements vary with time. The error covariance matrix must be computed in real time as part of the calculation of the optimum gain to be applied to measurements to update the estimates of the error state vector. Formally, the error covariance matrix is defined as

$$P = E[\tilde{x}\,\tilde{x}^T]$$ (3.16)

The diagonal elements of the error covariance matrix are the variances of the error in the filter estimate of the navigation system error state vector. The off-diagonal elements of the P matrix are covariances between different error states in the vector, and they contain important information as to the degree of correlation of one error state with another. The use of such correlation information in the gain computation is what distinguishes the Kalman filter from simpler mechanizations. In Kalman filtering, a measurement is used not only to update estimates of navigation error variables directly involved in the observation (e.g., position) but also to update estimates of error variables not directly involved (velocity, sensor errors, etc.).

3.10.3   The Recursive Filter

The derivation of the Kalman recursive filter equations can be found in references [5, 6, 7]. A schematic of equation flow is summarized in Figure 3.8. In words, the steps are as follows, beginning with the first measurement z_0 at $k = 0$. Note that an initial estimate $\hat{x}_0^-$ and its associated error covariance matrix P_0^- must be assumed to start the procedure.

1. Compute the Kalman gain K_0.
2. Update the a priori estimate $\hat{x}_0^-$ to obtain the a posteriori estimate $\hat{x}_0$. This step assimilates the measurement z_0.
3. Update the a priori error covariance matrix P_0^-, and obtain the error covariance matrix associated with the updated estimate $\hat{x}_0$, P_0.
4. Project both $\hat{x}_0$ and P_0 ahead to the next step where a measurement is available. The resulting $\hat{x}_1^-$ and P_1^- are the a priori state estimate and its error covariance matrix at $k = 1$ just prior to assimilating the measurement z_1.

Steps 1 through 4 are then repeated for $k = 1, 2$, and so on.

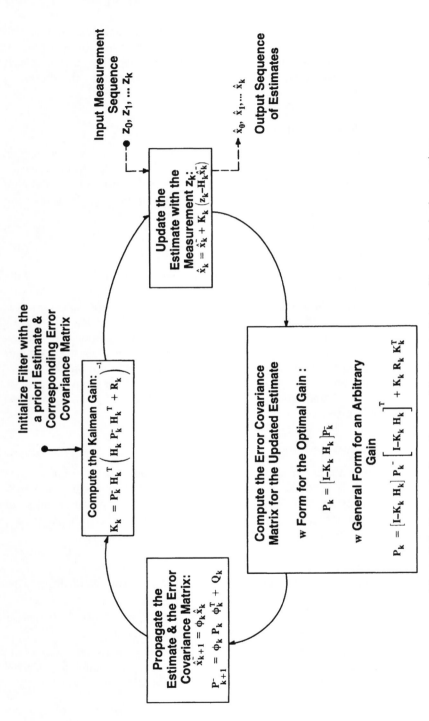

Figure 3.8 Schematic flow diagram of the discrete Kalman filter mechanization equations.

It can be seen that programming the Kalman filter equations is straightforward; it only requires a few lines of code in matrix form. Once the initial conditions and the four key matrices are defined, the resulting gain sequence is determined; that is, the sequence of gains that determine how the past measurements are weighted does not depend on the measurement sequences but rather only on the assumed model parameters and the initial error covariance matrix. The gain matrix computation at each step depends on the error covariance matrix, so the error covariance matrix must be propagated in the recursive process as a necessary adjunct for the gain matrix calculation. The error covariance matrix is useful in real time because it provides an assessment of the quality of the total navigation solution. It is of interest in off-line analysis because it can reflect the mean-square estimation error if the error model on which the Kalman filter is based is accurate. In this case, it is useful for performance analysis.

3.11   OPEN-LOOP KALMAN FILTER MECHANIZATION

The most common application of Kalman filtering in a suite of avionics navigation equipment is that of integrating the navigation data from an inertial navigation system (INS) with the navigation data from other sensors. Figure 3.9 shows a block diagram of one method for performing such an integration. The configuration shown is called the *open-loop* configuration [2] because the corrections to all the navigation sensors are made to the outputs and not fed back to internally correct these sensors. By this mechanization, the INS is used to provide the navigation variables for use in the Kalman filter transition and observation matrix equations. The filter acts on the observable deviations of the corrected inertial system outputs from the corrected aiding sensor measurements to effect system corrections. The aiding sensors are depicted schematically in one block in Figure 3.9.

This system integration scheme has been used successfully in a wide variety of applications since the mid 1960s [3, 4]. There are a number of good reasons for selecting a method that employs the corrected inertial navigation variables as the basis for the filter equations and the navigation system outputs. First of all, the method allows utilizaiton of measurements from a wide variety of aiding sensors. This is important, because the combination of aiding sensors may vary during a mission and these sensors typically provide measurements only at discrete points in time. Another reason for this integration method has to do with the restrictions placed on the Kalman filter model. Recall that the process dynamics and measurement relationship must both be linear. In general, the whole value state variables (total position, velocity, etc.) do not satisfy this requirement. The range measurement from electronic distance measuring equipment is proportional to the square root of the sum of the squares of cartesian components, which is not a linear relationship. Therefore the problem is to derive a linear representation of the situation to obtain correct application of the

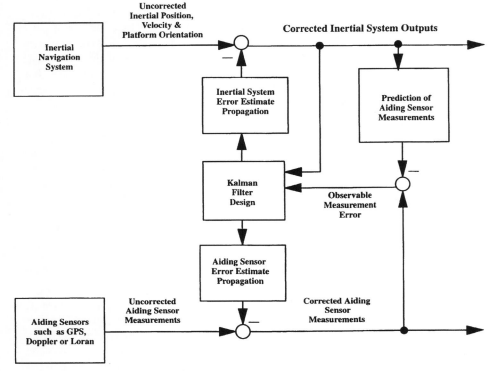

Figure 3.9 Open-loop Kalman filter architecture.

Kalman filter theory. The details as to exactly how this linearization is generally done are given in references [2, 6]. The basic idea is to choose a reference trajectory in state space that is close to the true trajectory and then write equations for the perturbation variables that represent the difference between the true and reference trajectory for use in the filter equations. The equations in terms of the perturbation variables are linear when higher-order terms are neglected. In the block diagram of Figure 3.9, the INS-corrected output is used for the reference trajectory as a matter of convenience, since it provides all the basic navigation quantities of vehicle position, velocity, acceleration, and orientation in a continuous manner. None of the other navigation sensors, when considered individually, provide the same complement of information in a continuous manner. Thus, the INS is the logical choice for the reference, even though its accuracy may be less than some of the other sensors at discrete instants in times when their measurements are available.

The final reason for choosing the INS-based integration methodology has to do with maintaining high dynamic response in the position, velocity, and attitude state variables available from the inertial system without filtering. The usual price associated with filtering is time delay or sluggish response. This lag is undesirable in most real-time navigation applications because the navigation system should follow the dynamics of the aircraft faithfully regardless of the rapidity of the change. The mechanization block diagram shown in Figure 3.9 accomplishes this and at the same time provides filtering of the measurement noise. At first glance, this may seem to be a contradiction, but note that the corrections produced by the filter are based only on the *errors* in the inertial system outputs and the aiding sensors. This approach is sometimes called *distortionless-filtering* or *dynamically exact system integration*. With this approach, the *total* dynamical quantities of interest (i.e., position, velocity, and orientation) do not need to be modeled as random processes, and the filter mechanization equations can be based on a linear model in terms of the navigation variable errors.

The implementation in Figure 3.9 is an example of the so-called *extended Kalman filter*, wherein the best estimates of the state vector are used as reference values for linearization at each filter step, rather than the true values of the trajectory which are unknown. Note that the block diagram shown in Figure 3.9 is conceptual and not literal. It is tacitly assumed that the appropriate INS outputs are converted to the aiding source frame of reference before performing the differencing operation shown. Commensurate sensor measurements must be compared in the same reference frame.

The Kalman filter might reside within the INS computer in some applications where a tightly integrated mechanization is desired. In other cases, the Kalman filter could reside within the computer associated with an aiding sensor, or it might reside in a separate mission computer. After all, the filter is just a digital computer program that accepts certain inputs to yield another set of outputs. In some system integration applications it is not possible to perform the filtering operation in a single centralized filter as shown. Equipment constraints may dictate that the outputs of filters in individual sensors be merged in a subsequent filter. This leads to a cascading of filters that is a theoretically more complicated situation discussed further in Section 3.15. The cascading of Kalman filters generally leads to some degree of suboptimality. However, system engineers sometimes have to live with this situation because of the given equipment configuration.

The estimated inertial and aiding sensor errors could be fed back to internally correct the INS and the aiding sensors as opposed to just correcting their respective outputs. This leads to the closed-loop Kalman filter mechanization which is discussed in Section 3.12.

3.12   CLOSED-LOOP KALMAN FILTER MECHANIZATION

In the mechanization of the last section, the INS is not corrected internally throughout the time span of the mission. Clearly, if the internally computed inertial system navigation variables diverge too far from their true values, the linearization assumption becomes suspect, and the associated modeling inaccuracy can lead to difficulties. In the early days of Kalman filtering, system engineers discovered that such divergence could be avoided by feeding the filter error estimates back to correct internally the inertial system at each time point where an aiding sensor measurement was available. This will then reduce the difference between the corrected real-time computed navigation variables and the true values, provided that the Kalman filter is producing good error estimates. When the filter error estimates are fed back in this manner, the mechanization is called the *closed-loop* Kalman filter mechanization [2]. This method has been used extensively and successfully in a variety of actual navigation applications. It is especially important to use the closed-loop Kalman mechanization in applications where the mission length is relatively long and the error model on which the filter equations are based are a simplification of the actual linear error model of the system. Such simplifications are made to reduce the computational burden on the real-time computer.

Figure 3.10 illustrates schematically the mechanization of the integrated INS system for the closed-loop mechanization. This diagram and that of Figure 3.9 are both conceptual in that the INS correction that takes place is usually just a matter of correcting certain numerical values in the inertial system computer. If the error modeling assumptions are valid, the performance of the open-loop and closed-loop Kalman filter mechanizations become essentially equivalent. Both mechanizations of the Kalman filter have been used extensively, so the decision as to which should be used will depend on the particular situation at hand. The closed-loop mechanization is generally used except when it is desirable to preserve the pure INS output undisturbed by internal corrections. In this case,

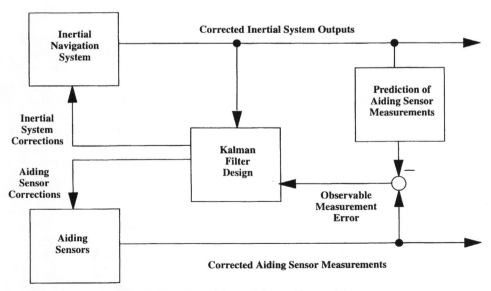

Figure 3.10   Closed-loop Kalman filter architecture.

the open-loop mechanization is employed or a separate inertial navigation solution mechanized in addition to the closed-loop corrected solution. When the filter is operating closed-loop, the inertial computer maintains the optimal estimates of the navigation variables in terms of the corrected position, velocity, and so on, rather than in terms of the uncorrected navigation variables from which the estimates of their errors are subtracted. The closed-loop configuration is more robust in that it is less sensitive to parameter variations or error-modeling simplifications.

3.13   GPS–INS MECHANIZATION

One of the most important multisensor navigation systems is that which integrates an inertial navigation system (INS) with a Global Positioning System (GPS) receiver. GPS (Section 5.5) is a very accurate, worldwide satellite navigation system using ranging that will be operating into the twenty-first century. The GPS-INS integration is nearly always done using a Kalman filter. A simplified Kalman filter error model for accomplishing such an integration is presented in Section 13.3.

3.13.1   Linearizing a Nonlinear Range Measurement

Recall that, to apply Kalman filter theory, the observation that is processed by the filter to obtain error estimate updates must have a linear relationship to the system error state vector that is being estimated. A range measurement does not satisfy this criterion. This is illustrated in Figure 3.11a with a simple two-dimensional sketch where a range measurement from the vehicle to a ground station is made using distance measuring equipment (DME). Slant range is approximated as being equal to horizontal range in this example. The aircraft is at (x, y), and the DME station is at a known location (x_1, y_1). A perfect range mea-

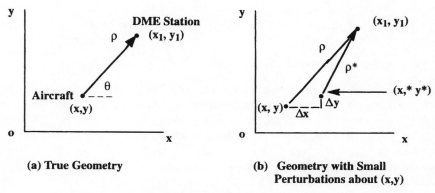

(a) True Geometry

(b)   Geometry with Small Perturbations about (x,y)

Figure 3.11   Geometry for linearized range measurement error.

surement ρ would then be related to the unknown x and y coordinates through the equation

$$\rho = \sqrt{(x - x_1)^2 + (y - y_1)^2} \qquad (3.17)$$

Clearly, ρ is a nonlinear function of x and y, which are to be estimated, and it violates the basic assumptions of the linear Kalman filter model.

Suppose that the aircraft knows its approximate position, denoted as $(x*, y*)$, which might come from an INS. Then for a small perturbation of these values by $(\Delta x, \Delta y)$ from the true aircraft position (x, y), the perturbation in the range is

$$\Delta \rho = \rho^* - \rho = \Delta x \cos(\theta) + \Delta y \sin(\theta) \qquad (3.18)$$

From Figure 3.11b, ρ is the measured range within some measurement error and $\rho*$ is the predicted range based on the aircraft's assumed reference position which is slightly incorrect. Note that there is a *linear* relationship between $\Delta \rho$ and the perturbations Δx and Δy. Thus, using $\Delta \rho$ as the observation (difference of the commensurate measurements of range from the DME and estimated range from the INS) processed by the Kalman filter and choosing the incremental quantities Δx and Δy as error state variables rather than the total position states x and y, the required linear measurement relationship for Kalman filter theory is satisfied.

Referring again to Figures 3.9 and 3.10, when the corrected INS position computation is used to estimate the range to the DME station, the differencing operation produces $\Delta \rho$ using the corrected range measurement from the aiding sensor with the addition of some measurement noise. This example accomplishes exactly what small-perturbation theory says is necessary to linearize the measurement process. Note the coefficients of Δx and Δy are just the direction cosines between the line of sight to the DME station and the respective x- and y-axis. The GPS measurement situation is similar in this regard and the direction cosines also appear in the modeling details discussed below.

3.13.2   GPS Clock Error Model

A GPS receiver obtains a measure of the difference between the time of a satellite transmission to the time of its receipt at the receiver. The time-difference is interpreted in terms of distance via a presumed value for speed of light with various corrections for propagation through the Earth's atmosphere. The local clock in the GPS receiver is usually a crystal clock whose stability is much poorer than the atomic clocks in the satellites. Thus, the drift of the local clock relative to GPS time maintained by the satellite clocks must be modeled in the Kalman filter. The time offset between GPS time and the local receiver clock

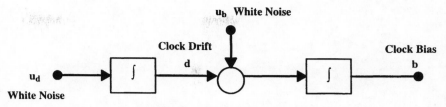

Figure 3.12   GPS clock error model.

is called *clock bias*. This bias in seconds can be scaled by the speed of light to obtain a range bias in meters. Similarly, the rate of change of clock bias is usually given in meters per second.

It is difficult to model crystal clock errors with a first-order state model; see reference [8] or Chapter 10 of reference [6]. The 2 state model shown in Figure 3.12 used in most Kalman filter implementations in GPS receivers allows for the estimation of both clock bias and drift. Numerical values for the spectral densities of the white noise forcing functions u_d and u_b depend on the quality of the crystal clock. The references show how to convert from the usual clock specifications to the Kalman filter parameters. This 2-state model is embedded in the larger 11-state error model for the whole GPS/INS system. (For additional discussion of clock characteristics, see Section 5.3.2.)

3.13.3   11-State GPS–INS Linear Error Model

The 11-state model presented here is a minimal error model of a generic GPS receiver integrated with any INS whose accelerometer force measurements are resolved into a local-level coordinate frame. The error equations presented are referred to this coordinate system. The 11-state model accounts only for platform misalignment (or equivalently the error in the knowledge of the orientation of the inertial instrument cluster), velocity and position errors, and GPS receiver clock error. It does not allow for instrument bias errors for either the accelerometers or gyros. The instrument errors are grossly simplified and are modeled only as white noise forcing functions that drive the INS error dynamics. This simple model was employed in many systems in the 1990s. There are always two parts to any Kalman filter model: the process model and the measurement model.

The error state variables in the random process model are the INS errors plus those errors in the reference source that have nontrivial time-correlation structure and need to be estimated. The block diagrams in Figures 3.13, 3.14, and 3.15 show how small errors propagate along each of the three navigation coordinate axes of the INS. The INS coordinate frame is local level and can be viewed as the x-axis pointing east, y north, and z up. The continuous differential equations for the INS errors may be obtained directly from the block diagrams. These equations are as follows:

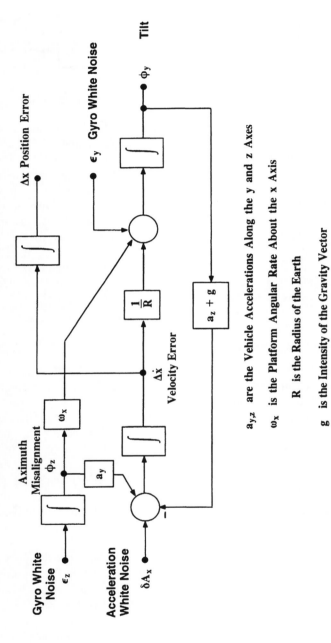

Figure 3.13 Block diagram of a simplified inertial system x-axis error model for the 11-state GPS-INS Kalman filter.

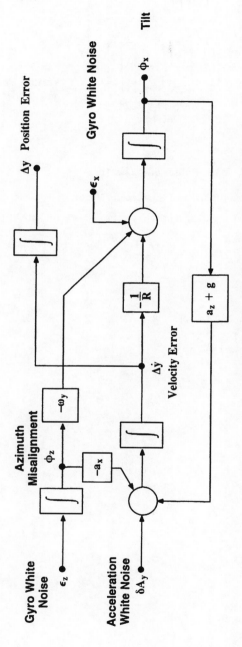

Figure 3.14 Block diagram of a simplified inertial system y-axis error model for the 11-state GPS-INS Kalman filter.

$a_{x,z}$ **are the Vehicle Accelerations Along the x and z Axes**

ω_y **is the Platform Angular Rate About the y Axis**

R **is the Radius of the Earth**

g **is the Intensity of the Gravity Vector**

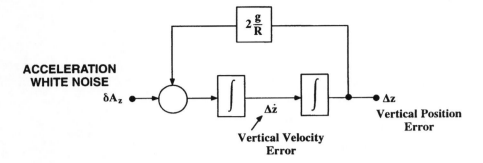

R is the Radius of the Earth

g   is the Intensity of the Gravity Vector

Figure 3.15   Block diagram of a simplified inertial system z-axis error model for the 11-state GPS-INS Kalman filter.

For east channel errors,

$$\Delta \ddot{x} = \delta A_x - (A_z + g) \cdot \phi_y + A_y \cdot \phi_z$$

$$\dot{\phi}_y = \left(\frac{1}{R} \right) \Delta \dot{x} + \omega_x \cdot \phi_z + \epsilon_y \qquad (3.19)$$

For north channel errors,

$$\Delta \ddot{y} = \delta A_y + (A_z + g) \cdot \phi_x - A_x \cdot \phi_z$$

$$\dot{\phi}_x = - \left(\frac{1}{R} \right) \Delta \dot{y} - \omega_y \cdot \phi_z + \epsilon_x \qquad (3.20)$$

For vertical channel errors,

$$\Delta \ddot{z} = \delta A_z + \left(\frac{2g}{R} \right) \Delta z \qquad (3.21)$$

For platform azimuth misalignment,

$$\dot{\phi}_x = \epsilon_z \qquad (3.22)$$

The accelerometer measurement errors δA_x, δA_y, and δA_z and the gyro drift rates ϵ_x, ϵ_y, and ϵ_z are mutually independent white noise processes with known spectral densities. Finally, the differential equations for the clock errors are from the block diagram of Figure 3.12.

For GPS clock errors,

$$\dot{b} = d + u_b$$
$$\dot{d} = u_d \tag{3.23}$$

where u_b and u_d are independent white noise processes.

The next step in the process modeling is to rewrite Equations 3.19 through 3.23 in state-space form. The 11-state variables are defined as follows:

x_1 is the east position error (meters)—Δx

x_2 is the east velocity error (meters/sec)—$\Delta \dot{x}$

x_3 is the platform tilt about y-axis (rad)—ϕ_y

x_4 is the north position error (meters)—Δy

x_5 is the north velocity error (meters/sec)—$\Delta \dot{y}$

x_6 is the platform tilt about x-axis (rad)—ϕ_x

x_7 is the vertical position error (meters)—Δz

x_8 is the vertical velocity error (meters/sec)—$\Delta \dot{z}$

x_9 is the platform azimuth error (rad)—ϕ_z

x_{10} is the user clock error in units of range (meters)—b

x_{11} is the user clock drift in units of range rate (meters/sec)—d

Once the state variables are defined, it is a routine matter to put the differential equations into state-space form:

$$\dot{x} = Fx + \xi \tag{3.24}$$

where F is the system state vector dynamics matrix describing the dynamic coupling between the (error) states. In expanded form Equation 3.24 is

$$
\begin{bmatrix} \dot{x}_1 \\ \dot{x}_2 \\ \dot{x}_3 \\ \dot{x}_4 \\ \dot{x}_5 \\ \dot{x}_6 \\ \dot{x}_7 \\ \dot{x}_8 \\ \dot{x}_9 \\ \dot{x}_{10} \\ \dot{x}_{11} \end{bmatrix}
=
\underbrace{\begin{bmatrix}
0 & 1 & 0 & 0 & 0 & 0 & 0 & 0 & 0 & 0 & 0 \\
0 & 0 & -(g+A_z) & 0 & 0 & 0 & 0 & 0 & A_y & 0 & 0 \\
0 & \dfrac{1}{R} & 0 & 0 & 0 & 0 & 0 & 0 & \omega_x & 0 & 0 \\
0 & 0 & 0 & 0 & 1 & 0 & 0 & 0 & 0 & 0 & 0 \\
0 & 0 & 0 & 0 & 0 & (g+A_z) & 0 & 0 & -A_x & 0 & 0 \\
0 & 0 & 0 & 0 & -\dfrac{1}{R} & 0 & 0 & 0 & -\omega_y & 0 & 0 \\
0 & 0 & 0 & 0 & 0 & 0 & 0 & 1 & 0 & 0 & 0 \\
0 & 0 & 0 & 0 & 0 & 0 & 2\left(\dfrac{g}{R}\right) & 0 & 0 & 0 & 0 \\
0 & 0 & 0 & 0 & 0 & 0 & 0 & 0 & 0 & 0 & 0 \\
0 & 0 & 0 & 0 & 0 & 0 & 0 & 0 & 0 & 0 & 1 \\
0 & 0 & 0 & 0 & 0 & 0 & 0 & 0 & 0 & 0 & 0
\end{bmatrix}}_{F}
\begin{bmatrix} x_1 \\ x_2 \\ x_3 \\ x_4 \\ x_5 \\ x_6 \\ x_7 \\ x_8 \\ x_9 \\ x_{10} \\ x_{11} \end{bmatrix}
$$

$$
+ \begin{bmatrix} 0 \\ \delta A_x \\ \epsilon_y \\ 0 \\ \delta A_y \\ \epsilon_x \\ 0 \\ \delta A_z \\ \epsilon_z \\ u_b \\ u_d \end{bmatrix}
\tag{3.25}
$$

The final step in determining the process model is to specify the ϕ_k and Q_k matrices. If the update interval Δt is relatively small, ϕ_k can be approximated with just the first two terms of the Taylor series expansion of $e^{F\Delta t}$:

$$
\phi_k \approx I + F\Delta t \tag{3.26}
$$

This approximation must be used with care. For example, if the aircraft experiences high dynamics, then Δt will have to be quite small or else additional terms must be included [6].

Computation of the Q_k matrix in a high-dimensional system is usually an onerous task, especially if there is nontrivial coupling among the various state variables as in this example. However, once the F matrix, the spectral densities of the forcing functions, and the Δt propagation interval are specified, then Q_k is numerical computable [6, 7].

Four pseudorange measurements (one per satellite) are sufficient for a stand-alone GPS solution. Each pseudorange measurement is of the general form

Total measured pseudorange = True range + Range bias

+ White measurement noise

To use pseudorange in the Kalman filter mechanization, it must be linearized. The observation processed by the Kalman filter is the difference between the receiver-measured pseudorange and the estimated pseudorange based on the corrected INS computed position and a corrected GPS receiver clock time. Only the incremental perturbations in the INS and clock estimate errors then appear on the right side of the measurement equation along with a white measurement noise. Direction cosines will appear as coefficients of the INS position errors, just as in the previous DME example. Clock bias enters as an additive term in units of range. There will be a linearized pseudorange measurement for each satellite, so the final form of the measurement model for the Kalman filter is

z = Actual receiver measurement

– Predicted measurement based on corrected INS position and

corrected receiver time

The resulting H matrix and observation z are

$$z = Hx + v \tag{3.27}$$

$$z = \underbrace{\begin{bmatrix} h_{11} & 0 & 0 & h_{14} & 0 & 0 & h_{17} & 0 & 0 & 1 & 0 \\ h_{21} & 0 & 0 & h_{24} & 0 & 0 & h_{27} & 0 & 0 & 1 & 0 \\ h_{31} & 0 & 0 & h_{34} & 0 & 0 & h_{37} & 0 & 0 & 1 & 0 \\ h_{41} & 0 & 0 & h_{44} & 0 & 0 & h_{47} & 0 & 0 & 1 & 0 \end{bmatrix}}_{H} \begin{bmatrix} x_1 \\ x_2 \\ x_3 \\ x_4 \\ x_5 \\ x_6 \\ x_7 \\ x_8 \\ x_9 \\ x_{10} \\ x_{11} \end{bmatrix} + \begin{bmatrix} v_1 \\ v_2 \\ v_3 \\ v_4 \end{bmatrix}$$

$$\tag{3.28}$$

where the h_{ij} are the respective direction cosines between lines of sight to the various satellites and the navigation coordinate axes. The R matrix reflects the additive white measurement noise components in which in this model would be a 4×4 matrix:

$$R = E[\nu \, \nu^T] \tag{3.29}$$

R would usually be specified as diagonal with all terms along the diagonal being equal. The numerical values of the terms are chosen to match the expected variances of the white pseudorange measurement errors. As a practical matter, these terms are usually specified to be larger than the expected error variances to compensate for the inaccuracy in modeling the measurement errors as white noise. The GPS measurement frequency (usually about 1 Hz) is very high relative to the characteristic Schuler time constant of the inertial system errors, so implementing an R matrix with large values does not significantly degrade performance due to the filtering provided by the INS.

3.13.4   Elaboration of the 11-State GPS–INS Error Model

The only way to improve the performance of the 11-state model is to add more error states. This increases the dimensionality of the error state vector and associated matrices, which in turn increases the computational burden on the system computer. Better performance is possible, but a specific application trade-off needs to be made to decide whether the added complexity will be of sufficient benefit. Additional error states might be those discussed below:

Inertial Error States   The inertial system error states that are generally the next in importance are the "biases" associated with the accelerometers and gyros. These are the random forcing functions that are calibrated at the start of a mission but then slowly wander away from their initial values during the course of the flight. It is especially important to keep calibrating these forcing functions continuously if it is anticipated that the INS might have to dead-reckon in the free-inertial mode for a significant time period without GPS measurements. If the biases are estimated in flight, the system errors grow more slowly during the free inertial period. Usually a first-order Markov or a random-walk process is used to represent an inertial instrument error source, so only one error state is added to the model for each source. Such an elaboration of the inertial system error model would add an additional three error states for three gyro drift rate errors and an additional three error states for three accelerometer measurement errors.

The primary trend is to employ low-cost, low-to-medium performance inertial components to bridge the periods of GPS outage in a cost-effective manner. These instruments will likely employ micro-machined silicon technology for the accelerometers and fiber optics for the gyros. In 1996 these technologies did not yet obtain the high-accuracy performance levels of the more expensive technologies used in the past. These instruments require more elaborate error modeling, including such additional states as scale factor error, mutual mechanical misalignments of instrument sensing axes, and bias changes as a function of measured environmental variations to obtain good calibration.

In addition to the instrument error states discussed above, the effects of unknown variations in the gravity vector should be considered as error states in some Kalman filter designs that include an inertial system. This consideration is important because the gravity disturbance vector introduces errors in the force measurements made by the accelerometers that can significantly affect some applications.

GPS Error States   Besides the receiver clock error model, filter designers have also been concerned with the error in the pseudorange measurement and the error in the Doppler or integrated Doppler (delta-range) measurement to each satellite due to residual satellite clock and orbit errors and transmission path effects. Inclusion of an error state for each of these measurements can obtain a calibration of the measurement under certain conditions. For example, precise knowledge of the vehicle location and velocity at the measurement time can provide such a condition. Further, measurements from satellites currently being tracked can be used to calibrate the measurements from a new satellite when an initial track is established.

Other refinements of the basic 11-state filter are also possible. Refinements in the error dyanmics of the model itself may be required in some applications. The example used a north-oriented coordinate frame, but a different frame can be used (e.g., the azimuth-wander frame discussed in Chapter 7). The 11-state model includes all the basic quantities that need to be estimated since the filter estimates position, velocity, instrument or platform coordinate frame orientation, and GPS receiver clock errors. All of these quantities are observable in the 11-state filter once measurements to four or more satellites have been made available. The observability implies that the usual problems of platform leveling, gyrocompassing, and damping the Schuler oscillation are all taken care of automatically by the Kalman filter without any special ad hoc procedures. In the past, the principal constraint on the elaboration of the error model has been limited computer resources. However, in the 1990s, Kalman filter designs based on models with several tens of error states were being implemented.

3.14   PRACTICAL CONSIDERATIONS

In practice, the implementation and validation of a Kalman filter design needs to address a number of topics:

- Measurement synchronization

- Measurement editing
- Tuning parameter adjustment
- Filter equation implementation

Measurements must be synchronized between an aiding sensor such as the GPS receiver and the INS. Any difference in the time between when the GPS measurement is made and when the inertial system estimate of that measurement is valid can introduce significant errors in the observable difference processed by the Kalman filter.

Measurement editing, sometimes called *reasonability testing*, is implemented in most systems to avoid spurious errors in the observable difference being inadvertently processed by the Kalman filter and consequently contaminating system performance. Typically, the observable difference is compared with some multiple of its standard deviation as computed from the error covariance matrix propagated in the real-time Kalman filter solution. Excessively large observable differences are discarded. Measurements are usually processed one at a time in any practical system rather than, for example, simultaneously processing the four GPS pseudorange measurements as defined above. With this "scalar" versus "vector" approach, Kalman corrections are made sequentially until all available satellite measurements are processed.

In the validation of any filter design, the testing with the actual hardware results in an adjustment of tuning parameters to refine performance. Such a process is necessary to compensate for the numerous error effects that are not accounted for in the filter design even when very elaborate simulation programs are employed in the design task. The parameters that are adjusted in the filter "tuning" to accommodate these errors are usually the magnitudes of the disturbance process noise power spectral densities (Q matrix) and the random observation error variances (R matrix).

Since the Kalman filter is a recursive algorithm performed on a digital computer, truncation or round-off error can lead to numerical difficulties as the number of iterations becomes large. Fortunately such problems can be nearly always avoided if proper safeguards are taken. The Kalman filter has a degree of natural stability if the system is completely observable and there is nonzero process noise driving each of the state variables at each recursive step. If a steady-state solution for the error covariance matrix exists, then this matrix will tend to converge to steady state after a small perturbation provided that the matrix positive definite.

Some techniques that have been found useful in preventing numerical problems are the following:

1. Use of high-precision arithmetic. The advance in real-time digital computer chips (to 32 and 64 bit arithmetic) facilitates this technique.
2. Symmetrize the P_k and P_k^- matrices at each step of the recursive process. Symmetry is automatically obtained if only the upper triangular portion of the error covariance matrix is employed in the implementation.

3. Avoid undriven state variables in the process model (random constants). This is the equivalent of ensuring that Q_k is positive definite, and it ensures that P_k^- will be positive definite, even if some of the measurements at the prior step are treated as perfect.

4. If the measurement data are sparse, propagate P_k through smaller time steps. Also be sure that the model parameters accurately represent the true dynamics of the processes being modeled. Otherwise, incorrect cross-correlations can be introduced in the propagation step, and they can adversely affect the gain computation.

5. At measurement time points, use the following general quadratic from [2] which is correct for an arbitrary gain matrix K to update the error covariance matrix when a measurement is processed:

$$P_k = (I - K_kH_k)P_k^-(I - K_kH_k)^T + K_kR_kK_k^T \tag{3.30}$$

This form preserves the positive definiteness of the error covariance matrix, whereas the theoretical form:

$$P_k = (I - K_kH_k)P_k^- \tag{3.31}$$

only guarantees this condition when the gain is optimal. Note that even though the equation implemented in the computer is for the optimal gain

$$K_k = P_k^-H_k^T(H_kP_kH_k^T + R_k)^{-1} \tag{3.32}$$

the *numerical* result may not realize the optimal gain.

In the past when digital computers of limited precision were available, an alternative form of the Kalman filter known as U-D factorization [6, 7, 10], a square-root formulation, was sometimes employed. The filter recursive equations are more difficult to program in U-D form than in the normal form shown in Figure 3.8. The U-D formulation has better numerical behavior than the usual equations, since the dynamic range for the equivalent error covariance information is reduced by a factor of two. As high-precision processor chips have become available, this technique has diminished in importance.

3.15 FEDERATED SYSTEM ARCHITECTURE

The architecture of avionics systems first started as decentralized, with each sensor having its own analog computer. Since the first digital computers in the

1960s were so expensive, centralized systems became universal. The development of less expensive microprocessors led back to decentralized ("federated") digital systems in the 1980s. In the 1990s, trends toward both centralized and federated systems has been noticeable.

The federated system architecture that results in a cascaded Kalman filter situation is shown in Figure 3.16. Federated system architectures have occurred because, in the past, individual "black boxes" have been procured from different specialized suppliers, each of whom has tended to implement a self-contained function. Specialists in each technology develop their own software. However, the system engineer is then faced with the task of integrating the outputs of these individual sensor subsystems into a *master* Kalman filter. Unfortunately, there is no simple optimal methodology for dealing with this problem. For an optimum solution, the system designer must process the *raw* measurements provided by each of the sensor subsystems and implement a single "centralized" Kalman filter as opposed to the "decentralized" filter approach illustrated in Figure 3.16.

Another difficulty in the decentralized filter problem is the significant time correlation structure of the estimation errors in the outputs of the sensor filters. One of the assumptions of the basic Kalman filter theory is that the measurement error sequence contains some component of random error from one measurement to the next. Two remedies to obtain this condition exist:

1. If the output data rate of a subsystem is relatively high, it can often be sampled at much lower rate for the measurement that is input to the master filter. There may be some loss of information in doing this, but often this is not severe. For example, suppose that one of the sensor subsystems provides data at 50 Hz and that the correlation time of the estimation error is about 1 sec. If the sampling rate of the master filter were reduced to 1 Hz, the measurement errors would be reasonably decorrelated, satisfying the theoretical requirement.

2. The filter parameters in the subsystem filter can be readjusted so that the output estimation errors are nearly white. This means that this filter will be operating suboptimally but optimization can be obtained by the master filter. The usual way of decorrelating the output errors of a Kalman filter is to make the Q_k matrix diagonal values artificially large and/or make those in R_k small. This gives the filter a short memory and results in light filtering of the measurement stream, which presumably contains uncorrelated errors. This method is often a viable option, because a change in parameter values is usually a minor software change.

With the advent of more centralized computer management of all sensors in future avionics system architectures with the utilization of high-speed data buses, cascaded Kalman filters should fall into disuse. The synergistic utilization of all sensor measurements should be realized more easily, including not only the optimal Kalman filtering of all sensor measurements but also the cross-use

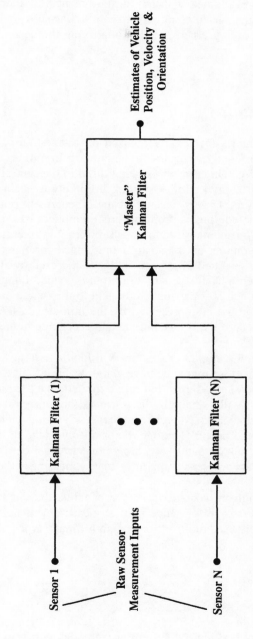

Figure 3.16 Block diagram of a federated system architecture resulting in a "cascading" of Kalman filters.

of sensor measurements between sensors for dynamic compensation. Further reinforcing this trend is the fact that the software of centralized systems was once recompiled and revalidated with any design change. In the 1990s, open software operating systems have permitted changes to one subsystem's software without affecting that of any other, greatly simplifying the design change and validation problem.

3.16 FUTURE TRENDS

The future will see significant activity focused on integration of the navigation sensors with other avionics sensors on the aircraft. In military systems, the information collected by many sensors is being fused into a central data base to ensure their optimal use. Civil aircraft systems are likely to see a slower pace of multisensor integration, largely because of the procurement practices of the airlines. They purchase equipment from different suppliers based on competitive prices and a desire to interchange "black boxes." Furthermore, reliability and safety considerations may lead to a preference for functinoal redundancy and a partially federated approach. In 1995, the civil aviation industry began to investigate centrally fused data in a program called Integrated Modular Avionics (IMA), which is somewhat equivalent to the functional integration and resource-sharing programs represented by the military ICNIA and ICNIS programs. These efforts are likely to continue and accelerate in the twenty-first century.

High-powered, low-cost digital processors will facilitate large-scale integration and enable redundant sensors to achieve a high level of fault tolerance. The self-contained black-box subsystems that drove the federated avionics systems architecture in the past will eventually disappear as the more information-rich centralized system architecture evolves. Significant savings should be realized in cost, size, weight, and power. The one major disadvantage of the federated system is that the central-computer software engineers must be expert in each sensor. Libraries of standard sensor modules may appear near the turn of the twenty-first century.

Because of its significant cost/performance benefits, the combination of a low-cost, low-to-moderate performance inertial measurement unit and a GPS receiver will be a widely used multisensor system for many types of air vehicles for years to come.

PROBLEMS

3.1. Refer to the simplified inertial system error model shown in Figure 3.14, and make the following assumptions:

$\epsilon_x = .01°/hr$ constant drift rate

$\delta A_y = 25 \ \mu g$'s constant accelerometer bias

$A_{x,z} = 0$ no acceleration

$\omega_y = \Omega \cos \Phi$ is the north component of Earth rate,

where the latitude $\Phi = 45°$

Compute the steady state values of (east) tilt ϕ_x and azimuth misalignment ϕ_z that will result from observing (north) velocity error $\Delta \dot{y}$ during the process of initial alignment of the inertial system

Hint: Also see Chapter 7.   *Ans.: $\phi_z = 200$ arcsec $\phi_x = 5$ arcsec*

3.2. Assume an inertial system has been initially aligned as in Problem 2.1 (equilibrium conditions obtained), with a variance in (east) tilt of $[5$ arcsec$]^2$ and a variance in azimuth misalignment of $[200$ arc-sec$]^2$, and, additionally, there exists an independent error in (north) position, with a variance of $[500$ feet$]^2$. What are the variances in tilt, azimuth misalignment, and position error when a star tracker is integrated with the inertial system and a star is observed directly overhead (along the local vertical), if the tracker measurement error is zero and processed by a Kalman filter?

Ans.: $\sigma_{\phi_x}^2 = [3.5 \ arcsec]^2$, $\sigma_{\Delta_y}^2 = [353 \ ft]^2$ and $\sigma_{\phi_z}^2 = [200 \ arcsec]^2$

What are these variances if a star is observed on the horizon along the local north line?

Ans.: $\sigma_{\phi_x}^2 = [5 \ arcsec]^2$, $\sigma_{\Delta_y}^2 = [500 \ ft]^2$ and $\sigma_{\phi_z}^2 = 0$

3.3. Consider an inertial system initially aligned as in Problem 3.1 but integrated with a reference speed sensor (e.g., a Doppler radar), where the vehicle accelerates in the eastward direction at $A_x = 10$ fps^2 for 10 seconds to a velocity of $V_x = 100$ fps. What are the steady-state values of the tilt and azimuth misalignment variances if Doppler speed measurements with no error are processed by a Kalman filter?

Ans.: $\sigma_{\phi_x}^2 = [5 \ arcsec]^2$ and $\sigma_{\phi_z}^2 = [200 \ arcsec]^2$

What are the steady-state values of tilt and azimuth misalignment variances if y (north) position change measurements (e.g., as obtained from a GPS receiver) with no error are processed by Kalman filter?

Ans.: $\sigma_{\phi_x}^2 = [5 \ arcsec]^2$ and $\sigma_{\phi_z}^2 = 0$

3.4. For an integrated reference speed sensor and inertial system, where the speed sensor has a constant bias error with variance of $[1 \ fps]^2$ and the inertial system has a constant gyro drift rate with variance of $[0.01°/hr]^2$, compute the steady-state variance of the system velocity error, $\sigma_{\delta V}^2$.

Hint: Refer to Figure 3.6.   *Ans.: $\sigma_{\delta V}^2 = [.707 \ fps]^2$*

What is the steady-state variance of this velocity error if, in addition, a constant step change of acceleration measurement error with variance of [25 μg]^2 is introduced?

Ans.: $\sigma^2_{\delta V} = [.707\ fps]^2$

3.5. Suppose you wanted to measure the deflection of the vertical (unknown deviation of the gravity vector from the local normal to the ellipsoid of revolution, which approximates the gravitational potential field of the earth) at a fixed point on the earth's surface for which the position is exactly known. Which two navigation sensors discussed in this chapter are necessary to perform this task, and what would the sources of residual error be in the measurements?

> *Ans.: A stationary inertial system (after an initial alignment using observations of velocity error) establishes the direction of the local gravity vector with an error primarily due to the accelerometer measurement error. A star tracker provides a measurement of the total tilt of the inertial system due to both the accelerometer measurement error and the deflection of the vertical. Error in this determination of tilt is due to the error in the tracker measurement of star position if the position on the surface of the Earth is exactly known.*

If you wanted to measure the total deflection of the vertical in an airborne vehicle moving over the surface of the Earth, which of the navigation sensors discussed in this chapter would be employed?

> *Ans.: As the vehicle moves over the surface of the earth, error in the navigated position of a stellar-inertial system will increase, degrading the ability of the system to determine deflection of the vertical as achieved above for the stationary system. Consequently, the addition of a navigation sensor that provides highly accurate positions, such as obtained with a GPS receiver, is necessary.*

4 Terrestrial Radio-Navigation Systems

4.1 INTRODUCTION

This chapter discusses the basic principles of terrestrial radio navigation systems, the radio propagation and noise characteristics and the major system performance parameters. The systems described in detail include all the important point source systems, such as direction finders, VOR, DME, and Tacan; and the hyperbolic systems, such as Loran-C, Decca, and Omega. All of these systems have been used worldwide and have provided accurate and reliable positioning and navigation in one or two dimensions for many years. In 1996 hundreds of thousands of civil and military aircraft throughout the world were equipped with these systems, and many of these will be used for years to come. Satellite radio nevigation systems (e.g., GPS) are discussed in Chapter 5 and military integrated radio communication–navigation systems are described in Chapter 6.

4.2 GENERAL PRINCIPLES

4.2.1 Radio Transmission and Reception

Figure 4.1 shows an elementary radio system. If a wire (antenna) is placed in space and excited with an alternating current of such frequency as to make the length of the wire equal to half a wavelength, almost all the applied ac power that is not dissipated in the wire will be radiated into space. A similar wire, some distance away and parallel to the first wire, will intercept some of the radiated power, and an appropriate detector connected to this receiving wire can indicate the magnitude, frequency, phase or time of arrival of the transmitted energy. This is the basis of all radio-navigation systems. Half-wavelength wires are called resonant dipole antennas.

To communicate from the transmitter to the receiver, it is necessary to *modulate* the alternating current in some manner. Early systems merely turned the alternating current on and off in accordance with the Morse code. Numerous other modulation systems have since come into use; the most important distinc-

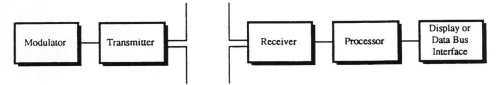

Figure 4.1   Elementary radio-navigation system.

tion, from a navigation standpoint, is whether the alternating current is left on all the time (continuous wave) or whether it is turned off most of the time and transmitted only as pulses. Within these broad categories there are many variations. For instance, the continuous wave signal may be modulated in amplitude, frequency, or phase; pulses may be modulated in amplitude, time, or arranged into various codes.

Early radio experiments were hampered by the difficulty of generating sufficiently high frequencies or building sufficiently large antennas to secure efficient transmission. They were also hampered by the low sensitivity of available detectors. However, the invention and development of the vacuum tube and solid state devices, including transistors, varactors, and many others, greatly extended the frequency spectrum that could be used, with the result that practical radio systems use frequencies from 10 kHz (30,000-meter wavelength) up to 100 GHz (3-mm wavelength), and progress is constantly being made toward the use of still higher frequencies.

By general agreement, radio frequencies have long been categorized as follows [36]:

Name	Abbreviation Frequency	Wavelength
Very low frequency	VLF	3 to 30 kHz	100 to 10 km
Low frequency	LF	30 to 300 kHz	10 to 1 km
Medium frequency	MF	300 to 3000 kHz	1 km to 100 m
High frequency	HF	3 to 30 MHz	100 to 10 m
Very high frequency	VHF	30 to 300 MHz	10 to 1 cm
Ultrahigh frequency	UHF	300 to 3000 MHz	1 m to 10 cm
Superhigh frequency	SHF	3 to 30 GHz	10 to 1 cm
Extremely high frequency	EHF	30 to 300 GHz	10 to 1 mm

At the higher frequencies, the following letter designations for certain frequency bands have been widely accepted, although they do not have official status (they are frequently related to standard wave-guide sizes) [35]:

Letter Designation	Frequency Range	Letter Designation	Frequency Range
L	0.39 to 1.55 GHz	X_b	6.25 to 6.90 GHz
L_s	0.90 to 0.95 GHz	K^1	10.90 to 36.00 GHz
S	1.55 to 5.20 GHz	K_u	15.35 to 17.25 GHz
C	3.90 to 6.20 GHz	K_a	33.00 to 36.00 GHz
X	5.20 to 10.90 GHz	Q	36.00 to 46.00 GHz

[1]Includes K_e band, which is centered at 13.3 GHz.

With progress constantly being made at still higher frequencies, other systems of nomenclature will be required. However, radio navigation, as defined in this chapter, is confined primarily to the bands lying between VLF and UHF, where the above nomenclature is likely to remain in use. Regardless of frequency, the following general rules apply in free space:

1. The propagation speed of radio waves in a vacuum is the speed of light: $299,792.5 \pm 0.3$ km/sec (usually taken as 300,000 km/sec for all but the most precise measurements).

2. The received energy is a function of the area of the receiving antenna. If transmission is omnidirectional, the received energy is proportional to the area of the receiving antenna divided by the area of a sphere of radius equal to the distance from the transmitter:

$$\frac{\text{Received power}}{\text{Transmitted power}} = \frac{\text{Receiver antenna area}}{\text{Area of a sphere } (= 4\pi R^2)} \tag{4.1}$$

where R is the range between antennas in the same units as those for the antenna area.

3. Multiple antennas may be used at both ends of the path to increase the effective antenna area. Such increases in area produce an increase in *directivity* or *gain* and result in more of the transmitted power reaching the receiver. The *gain G* of an antenna (in the direction of maximum response) is equal to its *directivity D* times its efficiency. The *maximum effective aperture* or *effective area* of an antenna is equal to $D\lambda/4\pi$. It is defined as the ratio of the power in the terminating impedance to the power density of the incident wave, when the antenna is oriented for maximum response and under conditions of maximum power transfer [19]. It is also defined as the *physical area* times the antenna *aperture efficiency* (or *absorption ratio*) [19, 36]. The *directivity* or *gain* of an antenna is usually expressed as a ratio with respect to either a hypothetical isotropic radiator or a half-wave dipole. A dipole has an *effective area* of about 0.13 times the square of the wavelength [36]. A transmitter

of power *P* and antenna gain *G* has an *effective radiated power* (ERP) of
PG along its axis of maximum gain.

For practical purposes, Equation 4.1 can be rewritten [19]

$$\frac{\text{Received power}}{\text{Transmitted power}} = \frac{A_r A_t}{R^2 \lambda^2} \qquad (4.2)$$

where

  A_r is the *effective area* of receiving antenna
  A_t is the *effective area* of transmitting antenna
  R is the range between antennas
  λ is the wavelength (the speed of light/frequency)

Thus, for fixed effective antenna areas, the power transferred from transmitter to receiver increases as the square of the frequency. However, this
is accompanied by a corresponding increase in directivity. Such directivity is of no concern in fixed point-to-point service and is of advantage in reducing external noise pickup. In many moving-platform applications, such as aircraft, a high level of directivity is a distinct disadvantage. However, when the use of tuned dipole antennas is assumed,
the power transferred *decreases* as the square of the frequency. This is
seen from Equation 4.1 where the receiver antenna area for a dipole, (i.e.,
0.13 λ^2) is substituted in the numerator.

4. The minimum power that a receiver can detect is referred to as its *sensitivity*.
   Where unlimited amplification is possible, sensitivity is limited by the noise
   existing at the input of the receiver. Such noise is of two main types:

   a. *External.* Due to other unwanted transmitters, electrical-machinery
      interference, atmospheric noise, cosmic noise, and the like.

   b. *Internal.* Depending on the state of the art and approaching, as a lower
      limit, the thermal noise across the input impedance of the receiver,
      which is given by

$$N_p = kT\Delta f \qquad (4.3)$$

where

  N_p is the noise power (in watts)
  k is the Boltzmann's constant (1.38×10^{-23} Joules/Kelvin)
  T is the temperature (in Kelvin)
  Δf is the bandwidth (in Hertz)

The factor by which a receiver fails to reach this theoretical internal-noise
limit is often expressed as a ratio, in decibels, and is known as the *noise
figure* (NF) of the receiver [37]:

$$\overline{NF} = \frac{N_0}{k T_0 \Delta f G_a} = \frac{N_0}{N_I} \qquad (4.4)$$

where N_0 is the noise power out of a practical receiver and N_I is the noise power out of an ideal receiver at standard temperature T_0, of available gain G_a, and of bandwidth Δf.

5. The minimum bandwidth occupied by the system is proportional to the information rate. For most navigational purposes, the necessary information rate is quite low. For instance, to navigate in a given direction to an accuracy of 500 ft with an aircraft that cannot change its position more than 500 ft in that direction in any one second, new information is needed only once per second. However, most practical systems have employed many times this minimum bandwidth. The reasons include (a) the need for other services, such as communications on the same channel, (b) the use of pulse techniques to aid in resolving multiple targets and to reduce the effects of multipath transmission, and (c) the use of spectrum-spreading techniques to improve signal-to-noise ratio (S/N), accuracy of range measurements, reduction of effects due to interference of site errors. (Spreading the spectrum beyond that needed by the information rate itself has the same effect as increasing the power, provided that optimum techniques are used at each end of the link [2, 27].)

In summary, to assess the free-space range of a radio system, it is necessary to have at least the following facts: transmitter power and antenna gain, receiver antenna gain and noise figure, the effective bandwidth of the system, and the effect on system performance of external or internal noise. Combining the fundamental relations in Equations 4.1 to 4.4, results in the following generalized link budget expression for the required radio transmitter power of a radio system as a function of key system parameters [35]:

$$10 \log \left(\frac{P_T}{P_N} \right) = \left(\frac{S}{N} \right)_{REQ} + L_P + NF - G_T - G_R - F_N \quad \text{(dB)} \qquad (4.5)$$

where

P_T	is the transmitter power
P_N	is the noise power in receiver
$(S/N)_{REQ}$	is the required signal-to-noise ratio in receiver
NF	is the receiver noise figure
F_N	is the noise improvement factor due to modulation method and bandwidth spreading (e.g., frequency modulation)
G_T	is the transmitter antenna gain

G_R is the receiver antenna gain

L_P is the propagation path attenuation loss

It is assumed that the polarization of the transmitting and receiving antennas are the same.

The radiation pattern from half-wave wires is a maximum along their perpendicular bisectors and a minimum along the axis of the wire, the equisignal pattern thus forming a "doughnut." (An isotropic radiator converts the volume of this doughnut into a ball, with uniform radiation in all directions. Such a radiator would have to be a point source and is theoretically impossible at coherent frequencies. However, in illumination engineering and in optics, from which radio theory borrows, such point sources are assumed to exist.)

At the lower frequencies, along the surface of the Earth, vertical polarization is universally used (the wires being vertical) with minimum signal being radiated into the ground. At frequencies where the antenna can conveniently be placed half a wavelength or more above the Earth (generally in the high-frequency band and above), either vertical or horizontal polarization is used, depending on other factors.

4.2.2 Propagation and Noise Characteristics

In free space, all radio waves, regardless of frequency, are propagated in straight lines at the speed of light. Along the surface of the earth, however, two other methods of propagation are of importance: Up to about 3 MHz, an appreciable amount of energy follows the curvature of the Earth, called the *ground wave*. Up to about 30 MHz, appreciable energy is reflected from the ionosphere; this is called the *sky wave*. The sky wave makes some types of long-range communication feasible, but it is of less value to navigation systems because its transmission path is unpredictable. At frequencies where the ground wave is useful, such ionospheric reflection actually detracts from the value of the system and requires special treatment.

Ground Waves Groundwaves are familiar as those normally received when listening to a standard AM broadcast transmitter during daylight hours. Daytime range is better at the low-frequency end of the band (550 kHz) than at the high end of the band (1650 kHz). Nighttime coverage is greater over the whole band, but depends on the sky-wave transmission. Ground-wave propagation is the primary propagation mode used in a number of modern radio navigation systems (e.g., Loran-C; Section 4.5.1).

Propagation of ground waves is dependent on several additional factors. First, ground-wave propagation is dependent on the conductivity and dielectric constant of the Earth in such a complex manner as to make the received power more nearly a function of the inverse fourth power of the distance, rather

TABLE 4.1   Typical ground-wave attenuation values (dB)

Frequency (kHz)	Over Land[b]		Over Seawater[c]	
	100 s. mi	1000 s. mi	100 s. mi	1000 s. mi
10	37	63	37	62
100	58	99	57	92
500	87	195	71	125
1000	110	245	79	145
2000	132	—	86	165

[a]Lossless isotropic antennas 30 ft above the surface, vertical polarization.
[b]Pastoral land: $\sigma = 0.005$ mho/m, $\epsilon = 15$.
[c]Seawater: $\sigma = 5$ mhos/m, $\epsilon = 80$.
Source: Reference [35].

than the inverse square, as would occur in free space. In addition, further losses, increasing with frequency, are encountered. Table 4.1 gives some typical examples. At low frequencies, ranges up to 5000 miles or more are obtainable, if sufficient power can be generated to overcome atmospheric noise, path attenuation and to compensate for low antenna efficiencies [42].

Second, at low frequencies, it is physically difficult to construct a vertical transmitting antenna large enough to be half a wavelength (or its electrical equivalent, i.e., a quarter-wave antenna above a perfectly conducting plane). Therefore, the antenna is generally much shorter than the ideal and is resonated to the operating frequency by external series inductance of the lowest possible losses. The result, despite the best engineering practice, is the radiation of considerably less power over a very narrow bandwidth than that generated by the transmitter. Nevertheless, ground-wave service is sufficiently attractive for many applications so that low efficiencies are tolerated in some lower-frequency applications.

Third, in most parts of the world and at most times of the year, atmospheric noise at low frequencies is so much greater than receiver noise that additional transmitter power must be used. This noise is generated mostly by lightning flashes. As shown in Table 4.2, at the latitude of the United States, atmospheric

TABLE 4.2   Nighttime atmospheric noise in the United States: 10 kHz bandwidth

Frequency (kHz)	Noise (μV/m)
10	2000
100	250
1000	20
10,000	0.8
100,000	Below receiver noise

noise power at the input to a receiver is typically a million times greater at 10 kHz than at 10 MHz (where it approximates typical receiver noise). At the equator, atmospheric noise power is from 10 to 25 times greater than at U.S. latitudes, whereas at the poles it is about 100 times less. The additional transmitter power required to overcome atmospheric noise levels produces one redeeming effect: Since a larger receiving-antenna area picks up more atmospheric noise along with more signal, no benefit is derived by making antennas larger than necessary. In fact, to nominally balance the effects of internal and external noises on the overall received signal, the typical antenna requires an effective area of less than one square meter.

Fourth, a characteristic of ground waves is that their propagation velocity is not entirely constant. While the variation is quite small (in percent), it is sufficient to limit the ability to obtain fixes at extreme ranges as good as the instrumentation might otherwise permit [17].

Last, despite all the handicaps listed, ground waves at low frequencies offer the only long-range radio communication means to vehicles that are not dependent on the ionosphere or airborne or satellite-borne relay stations. For this reason, there is a worldwide demand for frequencies in the VLF and LF bands. The optimum frequencies cover only approximately 100 kHz and are limited at the low end by antenna efficiency and by atmospheric noise and at the high end by poor propagation characteristics. Because of the long-range coverage, almost every single station or system in the world requires a unique frequency. In addition to the handicaps already listed, systems at these frequencies are dependent on national policies of frequency assignment.

Sky Waves In a region lying between 50 and 500 km above the Earth's surface, radiation from the sun produces a set of ionized layers called the *ionosphere* [38, 42]. The location and density of these layers depends on the time of day and, to a lesser extent, the season and the 11-year sunspot cycle. The ionosphere acts as a refractive medium; when the refractive index is high enough in relation to the frequency of a radio wave, it bends the radio wave and will, under favorable conditions, return the wave back to Earth.

Figure 4.2 shows a simplified picture of the geometry involved. At A, the radio wave strikes the refractive layer at too steep an angle and, although it is bent, is not sufficiently affected to return to Earth; it continues out into space (unless it encounters a more heavily ionized layer further out). At B, the radio wave strikes the refractive layer at a more oblique angle, is bent sufficiently to travel somewhat parallel to Earth, and is finally bent sufficiently to return to Earth. At C, the wave arrives at the refractive layer with glancing incidence and immediately returns to Earth. At D, the refractive index is too low in relation to frequency to seriously deflect the radio wave, which then travels on out to space; generally, this happens at frequencies above 30 MHz.

From this geometry it is evident that return to Earth occurs only at some minimum distance for a given frequency and degree of ionization. This is called the *maximum usable frequency* for that distance. Signals at higher frequencies,

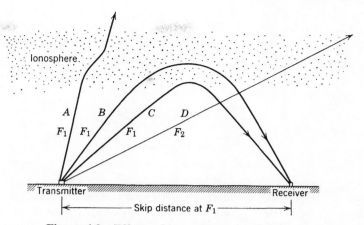

Figure 4.2   Effects of ionosphere on radio waves.

if returned at all, will be returned only at greater distances. This critical distance is known as *skip distance*; inside it there is no return to Earth at the particular operating frequency. If more than one ionizing layer are present, there may be various skip distances for the same frequency.

At those frequencies and distances where ionospheric reflection occurs, the attenuation of the radio signal is only that due to the spreading out of the power over the surface of the Earth and is, consequently, proportional to distance. Conversely, as indicated in the previous paragraphs, ground-wave attenuation is very much greater, except at the lowest frequencies. At frequencies of around 1 MHz, the signal level produced at the receiver by the two types of transmissions is likely to look like that shown in Figure 4.3.

As the frequency increases, the ground-wave curve will move to the left and the sky-wave curve to the right, leaving a gap (due to skip distance) where neither wave produces a usable received signal. In the region where the ground wave and sky wave are about equal, severe fading will occur due to the randomly varying phase of the sky-wave signal with respect to the ground-wave signal. Even when sky-wave signal strength is adequate, serious distortion of its modulation may occur due to the different paths simultaneously traveled by the signal between transmitter and receiver. These are called *multipath* effects. The differential time delay between these paths may reach several milliseconds, thus preventing faithful reproduction of modulation frequencies above a few hundred Hertz.

Therefore, sky-wave transmission is quite variable, and its efficacy is highly dependent on the distance to the receiver, the frequency used, and the time of day. For these reasons, the general practice in the 3- to 30-MHz communication band has been to use receivers and transmitters that would readily tune over the whole band and to change frequencies from hour to hour, depending on the distances required and on the condition of the ionosphere. Much work

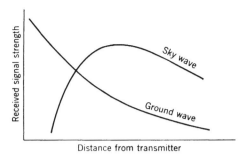

Figure 4.3   Ground-wave and sky-wave attenuation of radio signals.

has gone into the creation of charts predicting maximum usable frequencies and skip distances [42]. More recent developments include *propagation-frequency evaluators* that quickly evaluate (by frequency-scanning techniques) the best path to be used for communication between two points at a given time. By use of such techniques, ionospheric reflection has been a major long-distance communication aid. Until the advent of wideband submarine cables and communication satellites, these frequencies were the mainstay of the transoceanic telephone and radio networks. The highly variable characteristics of the ionosphere, which cause different frequencies to travel by different paths, led to the development of many ingenious steerable antenna systems for this service. The use of ionospheric reflection for navigation systems has been confined almost exclusively to ground-based direction finders.

Conversely, sky-wave transmission is considered a handicap, rather than an aid, to those navigation systems that depend on groundwaves. In such systems, the almost direct, reasonably predictable ground wave is contaminated by sky-wave energy that has arrived by a devious path. The mixture of the two often produces serious errors not only in distance but also in bearing measurements, since the effective reflecting point is not necessarily on the vertical plane joining the transmitting and receiving stations. Methods for reducing such sky-wave contamination include (1) the use of tall antenna structures for improved vertical directivity resulting in transmission fields being concentrated along the ground and less toward the sky, and (2) the use of only the leading edge of pulse transmission, since this edge arrives sooner by ground wave then by sky wave and is, therefore, uncontaminated (this usually requires greater bandwidth than that required by the information rate).

Line-of-Sight Waves   Above approximately 30 MHz, propagation follows the free-space laws listed in Section 4.2, modified by the reflecting effects of various objects on Earth. In general, the transmission path is predictable, and the wavelengths are so short as to readily permit almost any desired antenna structure; engineering for a given performance is consequently relatively straightforward. Some anomalous sky-wave effects occasionally occur up to 100 MHz,

TABLE 4.3   Attenuation (loss) versus frequency due to fog

Loss (dB/m)	Visible Distance			
	100 ft	200 ft	500 ft	1000 ft
10^{-3}	20			
10^{-4}	7	12	20	
10^{-5}		4	7	12
10^{-6}				3

Note: Table entries are frequency in GHz.

but from approximately 100 MHz to 3 GHz, the transmission path is highly predictable and is unaffected by time of day, season, precipitation, or atmospherics. Above 3 GHz, absorption and scattering by precipitation and by the atmosphere begin to be noticed, and they become limiting factors above 10 GHz. Furthermore, above that frequency, atmospheric absorption does not increase in a smooth manner but rather is characterized by narrow peak-absorption bands and by narrow "windows" of relatively reduced absorption. Tables 4.3 and 4.4 [14] show attenuation effects due to fog and rain (in addition to free-space loss).

Because of absorption above 10 GHz, transmission at such frequencies is severely limited within the Earth's atmosphere. High-flying aircraft and space vehicles of course are under no such restrictions.

In designing antenna systems for line-of-sight frequencies, it often happens that due to the relatively short wavelengths, the antenna is spaced away from a reflecting object such as the ground by a critical number of wavelengths, which has a marked effect on the overall antenna pattern. For instance, at 1 GHz, the wavelength is about 1 ft. If this practice were used at 1 GHz, a quarter-wave structure might be built with its base on the ground. Since even nearby blades of grass would seriously mar its performance (not to mention persons and vehicles moving about nearby), this would obviously be impractical. Instead, such an antenna would likely be mounted on a pole, say, 10 ft high, so as to clear the

TABLE 4.4   Attenuation (loss) versus frequency due to rain

Loss (dB/m)	Heavy (16 mm/hr)	Moderate (4 mm/hr)	Light (1 mm/hr)	Drizzle (0.25 mm/hr)
10^{-3}	15	37	100	
10^{-4}	7	12	20	43
10^{-5}	3	6	9	20
10^{-6}		3	4	8
10^{-7}				4

Note: Table entries are frequency in GHz.

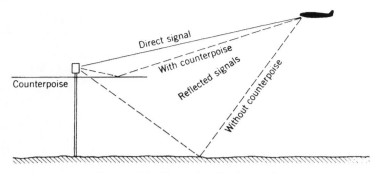

Figure 4.4 Vertical reflection paths.

immediately surrounding structures. A reflection phenomenon, lacking at lower frequencies, would now be encountered, as shown in Figure 4.4.

A receiver, at a point in space, now receives a direct ray from the transmitter and a reflected ray from the ground. Because of the short wavelength, the path difference is sufficient to cause addition or cancellation (for perfect ground reflection) as the receiver moves up and down in elevation. The resulting vertical pattern (assuming perfect ground reflection and with an exaggerated scale for clarity) is shown in Figure 4.5. Deep nulls, of virtually zero signal strength, are produced at those vertical angles at which the direct wave path and the reflected wave path differ by exactly an odd multiple of half-wavelengths. Maxima of signal strength occur where the two path lengths produce in-phase signals. The number of nulls per vertical degree of elevation increases with the height of the antenna and with frequency.

Such deep nulls, of course, occur only if the ground is smooth and perfectly reflecting. However, even when this is not the case, the null structure can produce serious loss of signal. A number of corrective steps can be used to reduce vertical nulls:

1. Raising the antenna high enough above the ground, in relation to fre-

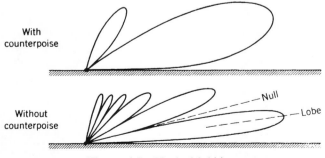

Figure 4.5 Vertical lobbing.

quency, to make the null structure so fine that the slightest irregularities in the reflecting surface will break up the null pattern. At heights of 100 wavelengths or more, the problem can usually be ignored.

2. Placing a horizontal counterpoise immediately below the antenna so as to make its effective height quite small. The counterpoise shortens the path of the reflected ray, thereby raising the angle at which cancellation occurs. To be effective, such a counterpoise must be many wavelengths in diameter.

3. Using high vertical antenna directivity (either by antenna arrays or by parabolic reflectors) and then pointing the resulting narrow antenna beam slightly above the horizon. The uptilt reduces the energy striking the ground and, therefore, reduces the reflected wave.

4. Making the null on one frequency occur at the same time that a maximum is occurring on another by using frequency diversity or wide-spectrum modulation which allows several frequencies to be used simultaneously [1].

5. Introducing vertical diversity up to an appreciable vertical angle via two antenna systems, one at half the height of the other with the null of one corresponding to the maximum of the other. However, the frequencies fed to these two antenna systems cannot be coherent; otherwise a new null pattern will appear at angles where the signals are otherwise equal. This method is, in general, limited to receiving systems where two separate receivers can be used.

6. Attempting other forms of diversity that make use of two or more paths simultaneously. The term "diversity" in radio propagation refers to this use of paths with different frequencies, polarizations, and so on in order to make reflections occur at different points on each path.

Line-of-sight systems on Earth are, of course, subject to the limitations of the horizon. The maximum range that can be obtained is illustrated in Figure 4.6. Beyond the line of sight, signal strength at these frequencies drops off almost as suddenly as does visible light when passing from day to night. Very large powers and antenna gains are, therefore, needed to produce significant performance beyond the line of sight, and such systems have not been found to be of much value in aircraft communications and navigation systems.

4.3   SYSTEM DESIGN CONSIDERATIONS

4.3.1   Radio-Navigation System Types

From the beginning use of radio, the known (nearly constant) speed of radio-wave propagation, coupled with an accurate measurement of time, has lead to the use of radio for the measurement of *distance*. Furthermore, with a mea-

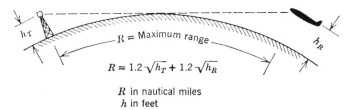

$$R = 1.2\sqrt{h_T} + 1.2\sqrt{h_R}$$

R in nautical miles
h in feet

Figure 4.6   Line-of-sight range.

surement of *differential distance* of two receiving antennas from one transmitter, the *direction* of the transmitter can be determined. In some systems, the measurement of time (and hence distance) and the angular measurement of direction are combined to determine user position, as discussed later in this section.

In Figure 4.7, two receivers (*A* and *B*) are arranged on a known baseline, which is assumed to be short with respect to the distance to the transmitter, so that the transmission paths to the two receivers can be considered parallel. A right triangle *ABC* may then be constructed and θ readily calculated if r is known. The value of r is found by comparing the outputs of the two receivers, noting the time delay between the arrival of identical parts of the signal, and dividing this time delay by the known speed of propagation. Such an arrangement is commonly called a *direction-finder* and is a widely used radio-navigation aid at all frequencies. (In many practical systems, the baseline *AB* is physically rotated until the delay is zero; the direction of arrival is then on the perpendicular bisector of *AB*.)

Some of the more common position determination methods are shown in Figure 4.8. With knowledge of distance (rho) or direction (theta) to a ground station, lines of position (LOPs) may be plotted on a map. The LOP of constant direction is a radial from the station; the LOP of constant distance is a circle centered on the station. The intersection of two LOPs provides a *fix*. It will be seen that the greatest geometrical accuracy occurs when LOPs cross at right angles.

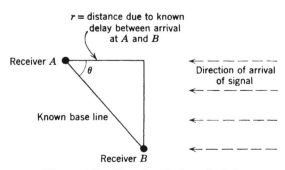

Figure 4.7   Direction-finder principle.

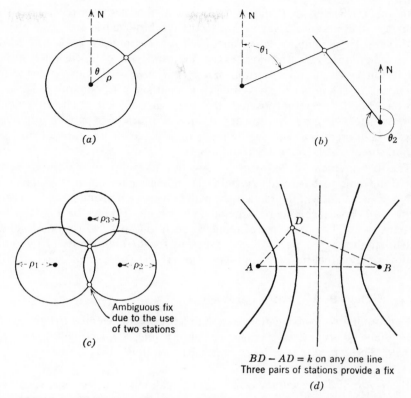

Figure 4.8 Common geometric position fixing schemes: (a) Rho-theta (ρ-θ); (b) theta-theta (θ-θ); (c) rho-rho (ρ-ρ); (d) hyperbolic.

Rho-theta systems provide a unique fix from a single station, and the LOPs always cross at right angles. *Theta-theta* systems provide a unique fix from two stations. The geometric accuracy is highest when the lines cross at right angles and is poor on a line connecting the stations. *Rho-rho* systems provide an ambiguous fix from two stations and a unique fix from three stations. Geometric accuracy is greatest within the triangle formed by the three stations and gradually decreases as the vehicle moves outside and away from the triangle.

The hyperbolic system uses LOPs that each define a constant difference in distance to two stations. Such systems operate under conditions where the determination of absolute distance to the station is impractical. Three pairs of stations are needed for a unique fix; however, for many practical applications, two pairs suffice. Geometric accuracy is very much a function of the relative station locations. Poor geometry leads to a property frequently called *geometric dilution of precision* (GDOP).

Another method, called *pseudoranging*, has been developed and is used in certain modern radio navigation systems, such as GPS (Chapter 5) and one mode of JTIDS–RelNav (Chapter 6). In this method the user receiver and the reference station(s) are assumed *not* to be synchronized in time. By measuring several (in general, at least four) such *pseudoranges* (versus *true ranges* when time synchronization does exist), the user's three-dimensional position and its time offset (from the transmitter or *system time*) can be determined (Section 2.5). Three such pseudoranges are sufficient to determine the user's two-dimensional position (provided that the user's altitude is known). Finally, the term *direct ranging* (rho-rho and rho-rho-rho) has been applied to a hyperbolic system used in a true ranging mode by achieving some form of time synchronization (see Sections 4.5.1 and 4.5.2).

The measurement of distance and direction by radio gives accurate results only if the radio path between the points being measured is direct and the propagation speed is known. In practice, the path between a transmitter and a receiver on the Earth's surface may be quite devious and may, in fact, be a combination of many paths. The problems of reducing such multipath transmissions and of recognizing the direct path are major reasons for the multiplicity of radio navigation systems that have been proposed or are in use.

4.3.2 System Performance Parameters

The utility of a radio-navigation system to a user is reflected by at least the following factors: accuracy, coverage, availability, integrity, ambiguity, and capacity [48]. These factors may be summarized as follows:

1. *Accuracy.* The accuracy of a radio-navigation system is a statistical measure of performance and is typically given as a root-mean-square (rms) measurement of its position error over some time interval. Many rms errors are expressed in two dimensions; however, some systems (e.g., GPS) can provide three-dimensional rms errors. System accuracy measures are defined in Chapter 2. The major factors affecting radio-navigation system accuracy are as follows:
   - Absolute error. The absolute or predictable error is the error in determining position relative to an Earth-referenced coordinate system. The Earth frame [5] is one such coordinate system that has its origin at the Earth's center of mass and its axes fixed in the Earth.
   - Repeatable errors. Repeatability reflects the accuracy with which a user can return to a position whose coordinates have been measured at a previous time with the same radio-navigation system.
   - Relative errors. This is the user's error with respect to a known point or to another user of the same radio-navigation system at the same time. The latter may also be expressed as a function of the distance between the two users.
   - Differential errors. These are the residual errors that remain after a user has applied the corrections broadcast by a *differential reference station*

located in the general vicinity whose position is precisely known, and the position calculated is compared using normal system signals with its known position. Depending on the distance between the user and the reference station, common errors, such as errors due to propagation effects, are eliminated or greatly reduced, resulting in a *differential* error that is much smaller than the *absolute* error. (Differential operation is widely used in such radio navigation systems as Loran-C, Omega and GPS.)

- Propagation effects. Radio-navigation system accuracy is affected by the transmission of the radio signal through the atmosphere. Error sources include reflection and refraction from the ionosphere and troposphere, variations in the conductivity of the Earth, anisotropic signal propagation, and the like. Also, the transmitter to receiver signal can simultaneously follow more than one path giving rise to *multipath* effects.

- Instrumentation errors. Instrumentation errors are the errors introduced by the radio and display equipment. These may include errors due to receiver noise, time-of-arrival (TOA) measurement circuits or angular measurement circuits, readout tolerance, and display resolution.

- Geometry effects. Geometry effects are typically expressed by a quantity called the *geometric dilution of precision* (GDOP). GDOP maps the basic (range or angle) measurements into position error and depends solely on the user-to-system geometry (see also Chapters 5 and 6, and Sections 4.5.1 and 4.5.2).

2. *Coverage*. The coverage area served by a radio navigation system is defined in terms of the specified performance of the system. In general, the coverage limit is defined by a requirement that a navigation receiver be able to acquire the radio navigation signal as well as use it and that the navigation solution meets a specific accuracy at a specified signal-to-noise ratio (S/N) value.

3. *Availability*. Radio-navigation system availability is the probability that the system is available for navigation by a user. In the United States, navigation system availabilities below 99.7% will not meet the requirements of the Federal Radio Navigation Plan [48] for safety of navigation purposes.

4. *Integrity*. Integrity in a radio-navigation system is the ability of the system to provide the user with warnings when it should not be used for navigation. For example, VOR and Tacan perform integrity monitoring using an independent receiver. When an out-of-tolerance condition is detected by the integrity monitors, the VOR and Tacan receivers remove their signal from use within ten seconds of this detection. Systems like GPS use of a variety of integrity monitoring methods (see Chapter 5).

5. *Ambiguity*. System ambiguity exists when the radio-navigation system

identifies two or more possible positions of the vehicle, with the same set of measurements, with no indication of which is the most nearly correct position. (This is not a problem with Loran-C, since the ambiguous fix is a great distance from the desired fix. Ambiguous lines of position (LOP) occur in the Omega system, since there is no means to identify particular points of contact phase (lanes) that recur throughout the coverage area. Because of this ambiguity, Omega receivers must be initialized to a known position, and the lanes counted as they are crossed.)

6. *Capacity.* Capacity is the number of users that the radio-navigation system can accommodate simultaneously. For example, there is no restriction on the number of receivers that may use Loran-C, Omega, or a VOR station simultaneously; on the other hand, DME and Tacan are currently limited to about 110 users for traffic handling.

Therefore, in considering overall accuracy, most systems must be compared on the probability of a certain accuracy being obtained under specified conditions of use. In the final analysis, such a probability must be a mixture of many probabilities, including the probability that the user will read the instrument correctly (typically called *flight technical error*) and the probability that the entire radio equipment is functioning properly. In this latter connection the fail-safe concept is generally strived for, on the theory that no information is preferable to false information. Alternatively, redundancy may be implemented, with the outputs of several systems compared and the most likely selected either by human or automatic means.

4.4 POINT SOURCE SYSTEMS [5, 15, 16, 17]

4.4.1 Direction-Finding

Direction-finding represents the earliest use of radio for navigational purposes; it continues to perform a useful function, particularly in those parts of the world that have not yet adopted the more specialized navigation aids. Its chief attraction lies in the fact that, with the proper receiving equipment, the direction of a transmitter can be found. Such transmitters do not necessarily have to be specially designed for direction-finding; they can be broadcast stations, communication stations, navigation stations, or any other kind of radiating system.

The chief drawback of direction-finding is that quite elaborate receiving equipment must be used if the best accuracy is to be obtained. Most aircraft are unable to accommodate such equipment. Direction-finders for aircraft navigation may, therefore, be grouped into two broad classes:

1. *Ground-based direction-finders.* These take bearings on airborne transmitters and then advise the aircraft of its bearing from the ground station. Such stations can afford the necessary complex equipment, but the

operation is cumbersome and time-consuming, and requires an airborne transmitter and communication link.

2. *Airborne direction-finders and homing adapters.* These take bearings on ground transmitters. These direction-finders typically can afford only the simplest of systems and must, therefore, tolerate large errors. However, even the largest bearing errors will not prevent an aircraft from homing in on the source of that bearing, though not necessarily by the most direct route. Direction-finding therefore continues to be used as a backup aid to more accurate systems.

With simple antenna systems, reliable directional information can be obtained only from ground waves or from line-of-sight waves. In the low- and medium-frequency bands, ground waves are useful to hundreds of miles; however, they are subject to sky-wave contamination (especially at night) at much shorter distances. This *night effect* is now recognized by users of low- and medium-frequency direction-finders. Reliance is placed on the readings only when the direction-finder is close enough to the station to be within good ground-wave coverage (typically 200 mi at 200 kHz and 50 mi at 1600 kHz). During thunderstorms, these distances are further reduced.

The state-of-the-art has progressed through the following stages:

1. *Fixed loop.* Intended for flying radial courses to and from the ground station by orienting the aircraft for minimum signal

2. *Rotatable loop.* Hand-operated systems that were abandoned because of the work load they imposed on the pilot

3. *Rotating loop.* Driven by a motor and forming part of a servo system that automatically rotates the loop until a null is found and then stops, sometimes referred to as a *radio compass.* Early loops were about nine inches in diameter and were housed in teardrop-shaped plastic enclosures about one foot away from the aircraft skin.

4. *Fixed, crossed loops, with a motor-driven goniometer.* Forming part of a servo system that automatically displays bearing in the cockpit. The prime advantage of this system over those using the physically rotating loop is that all moving parts (except the indicator) are in the radio-receiver box. Antenna projection from the aircraft with such a system in as low as one inch, with horizontal dimensions of about one foot. Typical airline-type equipment weigh less than 20 lb.

Loop Antenna Direction-Finder Principles.   This type of receiver is no longer in production, but its basic principles still apply to the current generation of equipment. The basic principle of direction finding is the measurement of differential distance to a transmitter from two or more known points. To reduce instrument errors, it is desirable to use common circuitry at both of the measur-

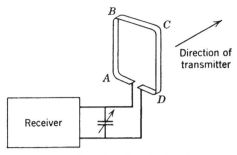

Figure 4.9   Direction-finding loop.

ing points. Furthermore, for operational convenience, it is desirable to have the two points close together. The loop antenna fulfills both of these requirements admirably.

The loop antenna in Figure 4.9 is a rectangular loop of wire whose inductance is resonated by a variable capacitor to the frequency to be received. The signal is assumed to be vertically polarized, and, consequently, it induces voltage in the arms AB and CD of the loop. If the loop were constructed accurately, these currents would be equal in amplitude and phase when the plane of the loop is exactly 90° to the direction of arrival of the signal. This is referred to as the null position of the loop (zero loop current). Physically rotating the loop to the null position indicates the direction to the transmitting station (i.e., the station is 90° from the plane of the loop).

If the correct amplitudes and phases are maintained, the horizontal antenna pattern is a figure of eight as shown in Figure 4.10 by circles A and B. This pattern has two null positions 180° apart. This ambiguity will cause the antenna system to give the same indication whether it is pointing toward a station or

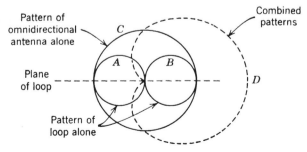

Figure 4.10   Loop and sense antenna patterns: A = left-hand loop pattern; B = right-hand loop pattern, 180° out of phase with A; C = omnidirectional pattern, 180° out of phase with A; $D = C + B - A$.

away from it. A *sense antenna* can be added when the signal ambiguity must be resolved. The sense antenna adds an additional 90° phase shift. As the loop changes direction, its phase will vary with respect to the constant sense antenna voltage resulting in the cardioid pattern shown in Figure 4.10. The combined pattern has only one null position. Since the omnidirectional antenna, and its phase and amplitude relation to the loop, are less precisely definable than the loop itself, it is customary to use the loop alone for precise directional measurement. The sense of the bearing can then be determined by coupling the vertical or sense antenna to the loop and rotating the loop 90° in a specified direction, noting whether the signal increases or decreases as the loop antenna is rotated.

Goniometer Direction-Finder Principles   It is not strictly necessary to rotate the loop mechanically in order to find the null. Instead, two fixed loops may be placed at right angles to each other and their electrical outputs combined in a goniometer. The goniometer has two sets of fixed windings at right angles to each other, each set connected to one loop. In the magnetic field of these windings is a rotor, with a winding connected to the receiver. The goniometer, in effect, translates the received radio field at the loops into a miniature magnetic field in which the rotor can operate. The nonrotating antenna can be attached to the skin of the aircraft, thereby reducing drag and increasing reliability. The accuracy of these systems is of the order of 2°, exclusive of errors induced by aircraft structure. These errors are of considerable magnitude, except in the fore-and-aft directions, where aircraft symmetry helps to minimize them. Some of these errors can be calibrated out for a given airframe, but this calibration is correct only for one frequency and/or condition of pitch and roll. As a result, low- and medium-frequency airborne direction-finders that use ground waves cannot produce reliable results better than ±5°. Sky-wave contamination can raise this figure to 30° or more.

Airborne VHF/UHF Direction-Finder Systems   Civil-aviation communication over land is conducted chiefly in the 118- to 156-MHz band, whereas the military use the 225- to 400-MHz band. Since virtually every aircraft carries a receiver for one or the other of these bands, a direction-finding "attachment" for these frequencies could be of interest to many aircraft operators. Considerable work was done during World War II in both of these bands, much of it involving variations on the Adcock principle. However, it was apparent that, in order to avoid prohibitively high site errors, the antennas would have to be of large aperture and project into the slipstream, thus generating drag. All that survives from these efforts is some VHF equipment used by the Coast Guard for air-sea rescue on the 121.5 MHz distress frequency and a military direction-finding attachment in the 225- to 400-MHz communication band on the distress frequency of 343 MHz. This equipment is of two possible types, depending on whether it is designed strictly for homing or for both homing and direction-finding. Equipment designed only for homing may use a fixed-antenna system that generates two sequentially switched cardioid patterns whose equisignal crossover direction is found by turning the aircraft toward the transmitting station. Equipment designed for both direction finding and homing uses a rotating antenna that generates a similar pair of cardioid patterns, whose equisignal crossover direction is found. Accuracy is about 5° along the axis of the aircraft but reaches 30° broadside. The direction-finding attachment is carried by many U.S. military aircraft and is useful for air-to-air direction finding and homing during rendezvous and refueling. It is also of value in locating downed flyers who carry small UHF rescue beacons.

4.4.2   Nondirectional Beacons

The universal use of low- and medium-frequency airborne direction-finders in commercial aircraft has prompted the installation of special transmitters whose sole function is to act as omnidirectional transmitters for such direction-finders. Aircraft use radio beacons to aid in finding the initial approach point of an instrument landing system as well as for nonprecision approaches at low-traffic airports without convenient nonprecision or precision approach systems (Chapter 13). Operating in the 200- to 1600-kHz bands, they have output power ranging from as low as 20 watts up to several kilowatts. Modern designs are 100% solid state.

Nondirectional beacons are connected to a single vertical antenna and produce a vertical pattern, as shown in Figure 4.11. In addition to the directional information given to direction-finders some distance away, such beacons have another useful property; namely, there is a sharp reduction in signal strength as the aircraft flies directly over the beacon, thereby providing a specifically defined fix. The accuracy of the fix produced by this "cone of silence" is somewhat dependent on the airborne antenna. It is improved if the airborne-antenna pattern contains a null in the downward direction and is degraded if the airborne-antenna null is off to the side.

Generally, purity of signal is obtained only from ground waves uncontaminated by sky waves; even in the absence of skywaves, considerable trouble has been experienced with the ground wave in terrain of nonuniform character, particularly near mountains. These two drawbacks (night effect and mountain effect) limit the usefulness of nondirectional beacons. They have retained their

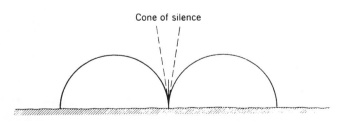

Cone of silence

Figure 4.11   Nondirectional beacon, vertical pattern.

popularity because (1) they are inexpensive, (2) they are omnidirectional, and (3) they place responsibility for accuracy entirely on the airborne receiver.

Nondirectional beacons are probably the least expensive way by which a government can claim that it has equipped its airways with "radio aids to navigation." In 1996, many thousands were in service around the world, and the United States maintains approximately 177,000 nondirectional beacons for civil aviation use. This number is expected to increase by about 7000 a year for the next ten years [48].

4.4.3 Marker Beacons

Aside from the null measured by flying directly over the station, all the facilities so far described provide only *directional* information to the aircraft. To provide better fixes along the airways, so-called marker beacons were developed. These marker beacons all operate at 75 MHz and radiate a narrow pattern upward from the ground, with little horizontal strength, so that interference between marker beacons is negligible. Each beacon generates a fan-shaped pattern, the axis of the fan being at right angles to the airway, as shown in Figure 4.12. The antenna pattern can be generated by an array of the type shown in Figure 4.13.

Four horizontal half-wave radiators are arranged in line with the airway and are fed so that their radio-frequency currents are all in phase. At right angles to their own axis, they generate a narrow vertical beam. A wire-mesh counterpoise below this array reinforces the upward beam. By placing the counterpoise a few feet above ground, the effects of vegetation and snow are reduced.

The transmitter is crystal-controlled and delivers up to 100 w. It is tone modulated, with its identity in Morse code indicated by gaps in the tone. The airborne equipment is a crystal-controlled superheterodyne receiver, with its output supplying the cockpit audio system and an indicator lamp. Automatic gain control is required to prevent saturation when the aircraft is passing directly over the marker station at low altitude. Some receivers also provide a high-low switch to increase the receiver sensitivity at higher altitudes [6].

The airborne antenna comprises a quarter-wave element on the bottom of the airplane, parallel to the axis of flight and as far aft as possible. Its own pattern consequently has directivity downward, increasing the directivity of the fan marker when crossing it. On high-speed aircraft, the antenna is foreshort

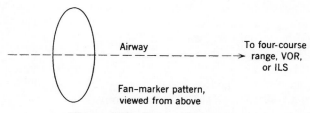

Figure 4.12 Fan-marker pattern.

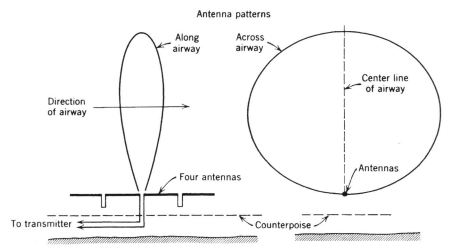

Figure 4.13   Fan-marker antenna.

ened by capacitive loading, recessed into the aircraft, and covered by a dielectric sheet. Streamlined antenna packages are $3\frac{1}{2}$ by 6 in. weighing 18 oz. The marker beacons are gradually being phased out as an en-route aid in view of the implementation of area-coverage systems, such as VOR/DME RNAV, Loran-C, and GPS. However, along instrument landing approaches, the 75 MHz marker remains a standard piece of equipment (Chapter 13).

4.4.4   VHF Omnidirectional Range (VOR)

Since the VHF band was being adopted for voice communication, it was only natural to consider the combination of communication and navigation within one band. Some early schemes involved VHF two- and four-course ranges. However, by 1946, the VHF Omnidirectional Range (VOR) had become the U.S. standard and was later was adopted by the International Civil Aviation Organization (ICAO) as an international standard.

The VOR [5, 17] operates in the 108- to 118-MHz band, with channels spaced 100 kHz apart. As soon as present low-selectivity airborne receivers (mostly used by small aircraft) can be served by better channel arrangement, the number of available channels will be doubled by spacing them 50 kHz apart.

The principle of operation is simple and straightforward. The ground station radiates a cardioid pattern that rotates at 30 rps, generating a 30-Hz sine wave at the airborne receiver. The ground station also radiates an omnidirectional signal, which is frequency modulated with a fixed 30-Hz reference tone. The phase difference between the two 30-Hz tones varies directly with the bearing of the aircraft. Since there is no sky-wave contamination at very high frequencies and no interference from stations beyond the horizon, performance is relatively consistent and is limited by only two major factors: (1) propagation effects, including vertical pattern effects, and site and terrain errors, and (2) instrument errors in reading 30-Hz phase differences in the airborne equipment.

Transmitter Characteristics VOR adopted horizontal polarization, even though aircraft VHF communication uses vertical polarization. Each radiator in the ground station transmitter is an Alford loop. The Alford loop generates a horizontally polarized signal having the same field pattern as a vertical dipole [2] and is shown in Figure 4.14.

Four radiators are arranged in a square, whose plane is horizontal. Each radiator is less than a quarter-wave long and is end loaded with capacity so as to place the maximum current at the center of the radiators. At any one instant, the currents in all the radiators are equal and alternately move clockwise and counterclockwise. The result is a doughnut-shaped field pattern, with zero radiation in the upward and downward directions, exactly like a vertical dipole (but horizontally polarized). Four such loops generate a rotating figure eight.

The omnidirectional signal comprises a continuous wave that can be voice modulated and carries Morse-code identity keying a 1020-Hz tone. Present at all times is another tone of 9960 Hz, which is sinusoidally varied ($\pm$ 480 Hz) at a rate of 30 Hz. This is the bearing reference frequency. A simplified block diagram is shown in Figure 4.15.

The transmitter is crystal-controlled, with a power output of 200 w. It is amplitude modulated to a depth of 30% by the output of a mechanically driven tone wheel, which has 332 teeth and is driven by an 1800-rpm motor. The teeth are slightly staggered so as to impart a cyclical variation from 9480 to 10,440 Hz with the rotation of the motor. On the same shaft is a capacitive goniometer, fed by the same transmitter via a modulation eliminator that strips off the amplitude modulation by means of diodes. About one-quarter of the applied power appears at the output of the modulation eliminator.

The goniometer feeds the unmodulated transmitter power first to the

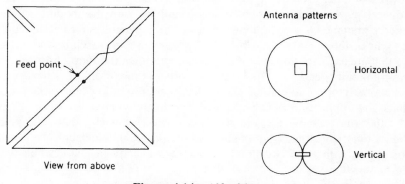

Figure 4.14 Alford loop.

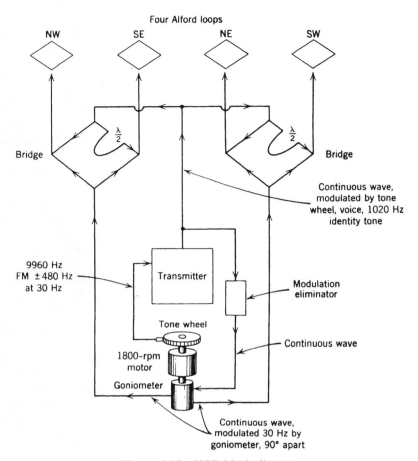

Figure 4.15   VOR block diagram.

northwest-southeast pair of Alford loops and then, 90° later, to the northeast-southwest pair of Alford loops. When combined with the modulated energy applied simultaneously to all loops, this variation generates a rotating cardioid. Each pair of loops is fed via a balanced bridge network. Each bridge has three arms that are each about one-quarter wavelength long, the fourth arm being half a wavelength longer. Energy fed into one corner of the bridge does not appear at the diagonally opposite corner. The bridge, therefore, allows the mixing of two signals and application of the result of two loads without the loads affecting each other and without the signal sources affecting each other. The phasing between tone wheel and goniometer and the physical placement of the Alford loops are such that the two 30-Hz signals are exactly in phase when viewed from magnetic north.

This seemingly elaborate arrangement serves two main purposes:

1. The division of power between the antennas is a function of passive elements only. The power of the transmitter can go up or down without affecting bearing information.

2. The two 30-Hz signals are rigidly locked together by being derived from a common rotating shaft. Motor-speed variation can alter their frequency slightly, but their phase relationship will not change.

The four Alford loops are arranged in a tight square and then placed half a wavelength above a metal-mesh counterpoise about 39 ft in diameter. This counterpoise also acts as the roof of the transmitter house. The loops are protected from the weather by a plastic randome, often hemispherical in shape. If a Tacan antenna is collocated with the VOR, the randome is conical in shape, somewhat resembling an Indian tepee.

Receiver Characteristics The airborne equipment comprises a horizontally polarized receiving antenna and a receiver. This receiver detects the 30-Hz amplitude modulation produced by the rotating pattern and compares it with the 30-Hz frequency-modulated reference. The basic receiver functions are shown in Figure 4.16.

At the output of the receiver is an amplitude-modulation detector. Its output comprises (1) a 30-Hz tone produced by the rotating cardioid, (2) voice modulation (if used at the transmitter), (3) a morse-code-modulated 1020-Hz identity tone, and (4) a 9960-Hz tone, frequency modulated (±480 Hz) by the 30-Hz reference tone. The voice frequencies and the identity tone are fed to the aircraft's audio-distribution system. The 30-Hz information is filtered to remove the other components and fed to the phase-comparison circuitry. The 9960-Hz information is filtered out, limited (to remove the 30-Hz amplitude modulation), and then applied to a frequency-modulation detector whose output is the 30-Hz

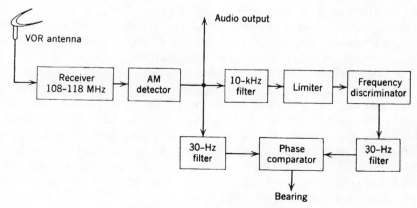

Figure 4.16 VOR receiver.

reference frequency. After filtering, this is compared with the variable phase. Several grades of receivers are currently in use.

The airline type of equipment uses a remotely tuned crystal-controlled super-heterodyne receiver and has at least two types of display. One display compares one 30-Hz sine wave with the other 30-Hz sine waves, the two signals being brought into phase by a motor-driven phase shifter forming part of a servo loop. The shaft position of this motor, therefore, displays bearing directly and may be remoted by selsyns to other parts of the aircraft and to auto-pilots. Another display shows (on a vertical left–right needle) the phase difference between one 30-Hz signal and a manually phase-shifted 30-Hz signal representing the desired bearing. The sensitivity of the vertical needle is usually arranged for a full-scale deflection of ±10° around the manually selected bearing and thus shows angular deviation from the desired track.

The simplest types of receivers use manual tuning and only the left–right type of display around a manually selected bearing. Both types of receivers are commonly arranged to also receive the 108- to 112-MHz instrument-landing-system localizer signals. Typical receivers weigh 20 lb for the airline type and 5 lb for the simplest type, exclusive of antenna. Over 200,000 airborne sets have been installed, about half of them for light aircraft.

It was previously mentioned that one of the problems of the VOR is the difficulty of accurately measuring phase shifts at 30 Hz. Much circuit refinement has taken place for the better grades of receiver. This includes, for instance, the use of identical circuits for both 30-Hz signal paths wherever possible so that temperature effects will be common to both. The result is that instrument accuracy of better than 1° is achieved in airline-type equipment.

4.4.5 Doppler VOR

Doppler VOR [3] applies the principles of wide antenna aperture to the reduction of site error. The solution used in the United States by the Federal Aviation Agency involves a 44-ft diameter circle of 52 Alford loops, together with a single Alford loop in the center. In the Doppler VOR, the role of the central radiator and the role of the array are reversed; however, the phase relations remain the same, allowing a standard airborne receiver to operate without any modification. A simplified description is given below.

The central Alford loop radiates an omnidirectional continuous wave that is amplitude modulated at 30 Hz by any conventional means; this forms the reference phase. The circle of 52 Alford loops is fed by a capacitive commutator so as to simulate the rotation of a single antenna at a radius of 22 ft. Rotation is at 30 rps, and a carrier frequency 9960 Hz higher than that in the central antenna is fed to the commutator. This 9960-Hz higher frequency is frequency modulated by the simulated rotation of the antenna, increasing in frequency as the antenna appears to move toward the receiver and decreasing in frequency as it recedes from the receiver. With a 44-ft diameter and a rotation speed of 30 rps, the peripheral speed is on the order of 1400 meters per second, or about

480 wavelengths per second at VOR radio frequencies. The 9960-Hz frequency difference is consequently varied by ±480 Hz at a 30-Hz rate, with a phase dependent on the bearing of the receiver.

In the receiver, the output of the amplitude-modulation detector contains all the signals present with the conventional VOR. Phase comparison between the two 30-Hz sine waves is performed as before, the only difference being that the 30-Hz amplitude-modulated signal is the reference and the 30-Hz frequency-modulated signal is the variable. Since the instrumentation is concerned only with the difference between the two, normal operation results with a standard VOR receiver.

However, since the aperture of the ground antenna is approximately five wavelengths, as compared with less than half a wavelength with the four Alford loops in a standard VOR ground station, a tenfold reduction in site error is theoretically possible. Actual measurements at formerly "impossible" sites verify this. At a good site, maximum deviations measured during a 20-mi orbital flight were reduced from 2.8° with a standard VOR to 0.4° with a Doppler VOR [3]. Residual errors can probably be reduced to 0.1°.

The importance of the Doppler VOR lies in the improvement it provides without any change being made to the airborne equipment. Every airborne set can benefit from it.

4.4.6   Distance-Measuring Equipment (DME)

DME is an internationally standardized pulse-ranging system for aircraft, operating in the 960- to 1215-MHz band. When the ground station is collocated with a VOR station, the resulting combination forms the standard ICAO rho-theta short-range navigation system [21]. In the United States in 1996, there are over 4600 sets in use by scheduled airlines and about 90,000 sets by general aviation.

The operation of DME can be described by means of Figure 4.17 where the aircraft interrogator transmits pulses on one of 126 frequencies, spaced 1 MHz apart, in the 1025- to 1150-MHz band. The pulses are in pairs, 12 μsec apart, each pulse lasting 3.5 μsec, with the pulse-pair-repetition rate ranging between 5 pulse-pairs per sec up to a maximum of 150 pulse-pairs per sec. The peak pulse power is on the order of 50 w to 2 kw. Paired pulses are used in order to reduce interference from other pulse systems.

The ground beacon (or transponder) receives these pulses and, after a 50-μsec fixed delay, retransmits them back to the aircraft on a frequency 63 MHz below or above the airborne transmitting frequency. The peak power of this beacon is in the range of 1 to 20 kw. The airborne interrogator automatically compares the elapsed time between transmission and reception, subtracts out the fixed 50-μsec delay, and displays the result on a meter calibrated in nautical miles, each nautical mile representing about 12 μsec of elapsed round-trip time.

Each beacon is designed to handle at least 50 aircraft simultaneously, with

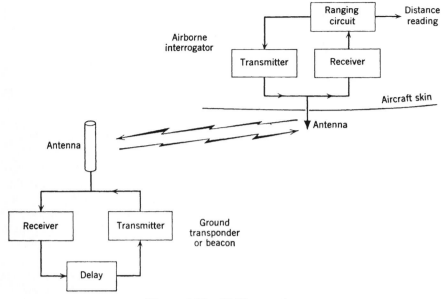

Figure 4.17 DME operation.

100 being a more typical number. The pulse-repetition rate of the interrogators is deliberately made somewhat unstable, and the interrogator is designed to recognize only those replies whose pulse-repetition rate and phase are exactly the same as its own.

In any line-of-sight geographical area, there is the possibility of providing 136 beacons, each handling 100 or more aircraft. Since each beacon's duty cycle is only 2% under these conditions, room exists to expand the system to handle heavier traffic. Modern techniques permit the airborne interrogation rate to be decreased substantially.

Receiver Characteristics Since the airborne receiver always operates 63 MHz away from its own transmitter frequency, a common crystal-controlled local-oscillator chain may be used for the local oscillator during reception and for transmission, provided that the receiver intermediate frequency is 63 MHz. A typical block diagram is shown in Figure 4.18. A photograph of a DME interrogator/receiver for commercial aviation applications is depicted in Figure 4.19.

Frequency generation commences with a crystal oscillator at about 45 MHz. This may be a single oscillator with 126 crystals or a mixture of two oscillators, one with 13 and the other with 10 crystals. This frequency is multiplied by 24 and furnishes about one milliwatt of energy to the receiver mixer crystal. It is also amplified in a chain of triodes that are pulsed during transmission to a level of 50 watts in the simplest equipment (intended for 100-mi range) or to one

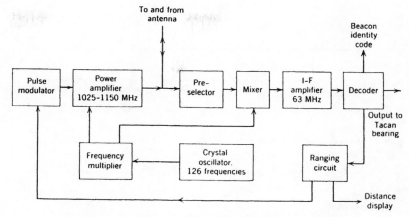

Figure 4.18   Airborne DME functional block diagram.

kilowatt in airline equipment (intended for 300-mi range). A common antenna is used for transmission and reception; mixer-crystal overload is prevented by the preselector, which is tuned 63 MHz away from the transmitter.

Amplification occurs in a 63-MHz intermediate-frequency amplifier with automatic gain control. The reply is then compared with the transmitted signal in the ranging circuit. This ranging circuit also sees all other pulses transmitted from the ground beacon (about 3000 pps), and it therefore must perform at least two major functions: (1) Recognize its own replies and reject all others (this is called *searching*) and (2) convert these into a meaningful display (this is called *tracking*).

Figure 4.19   Typical airborne DME interrogator/receiver.

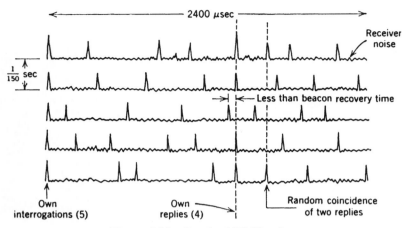

Figure 4.20   Received DME pulses.

Many different forms of circuit have been devised for these functions. They all depend on the sequence of wave forms shown in Figure 4.20. This figure shows five consecutive snapshots of an imaginary oscilloscope whose sweep is started by the airborne interrogation from a single aircraft and whose deflection circuit is connected to the output of the receiver in that aircraft.

In this instance, if one assumes a maximum desired range of 200 nm, the sweep is 2400 μsec long. Since the ground beacon is transmitting an average of 3000 pulse-pairs per sec, each sweep will display, on the average, about seven pulses. These will be quite randomly spaced, except those generated in response to our own interrogation. At an interrogation rate of 30 per sec, even the fastest aircraft does not move by as much as a pulse width from one interrogation to the next. The desired replies therefore occupy an almost fixed position on the oscilloscope display, whereas those intended for other aircraft move in a random manner. The dotted line shows the fixed (or slowly changing) position of the desired reply. On scan 3, the desired reply is missing; this is because the beacon has just replied to another aircraft and has not yet "recovered." Recovery time is typically on the order of 100 μsec. Desired replies may also be missing because of other random effects. However, all airborne DME ranging circuits are based on the principle that, within a given time slot, many more desired replies will be received than undesired replies.

The basic objective of all DME ranging circuits is to locate the time slot in which the desired replies are actually occurring. This is the search process, and it is usually conducted at the highest permissible pulse-repetition rate (150 pulse-pairs per second) in order to save time, which, depending on the technique, may vary from 1 to 20 sec. Once this time slot has been found, the track mode commences and can be conducted at a much lower pulse-repetition rate, usually between 5 and 25 pulse-pairs per second.

Search is typically performed as follows: A gate is generated 10 μsec wide, at the transmitter-interrogation rate. This gate is slowly moved outward, from a 0- to 300-mi delay with respect to the transmitted pulse. This gate strobes the received pulses; only when there is coincidence between the gate and a received pulse is the received pulse passed to an integrating circuit. During the search period, the interrogation rate is held at 150 pulse-pairs per sec. The gate is therefore opened 150 times per sec, for 10 μsec at a time, a total of 3000 μsec/sec, or 1/333 of the total time. If the beacon is trasmitting 3000 random pps, 9 random pps pass through the gate, on the average. However, if the gate is moved at about 10 mi per sec, almost 30 desired pulses will pass through the gate per second when the gate delay corresponds to the round-trip delay time. This ratio of 9 to 30 allows ample margin to switch the operation automatically from search to track.

In this example, a gate movement of 10 mi per sec would mean that 30 sec would be needed to search over the 3600 μsec period corresponding to 300 mi. Other circuits have been built in which gate movement is very fast until a pulse is received; with such circuits, 300 mi may be searched in less than a second.

In the track mode, the same 20-μsec gate follows the desired reply, controlled by the integrator. If the reply falls in the early part of the gate, the gate advances; if it falls in the late part of the gate, the gate is delayed. Since the possible change in aircraft position is quite small from one pulse to the next, the interrogation rate can be safely reduced during track. Typically, this is about 25 pps; however, with newer equipment, rates as low as 2 pps have been used, even at tracking rates of 3000 knots. The gate is usually arranged to have some memory so that it does not immediately go into search upon loss of signal. For 10 sec or so, it is arranged to say at its last position (called *static memory*) or to move at its last rate (called *velocity memory*). The position of the gate is determined in the simplest equipment by an analog voltage that operates a voltmeter, calibrated in miles. Such equipment has accuracies of about 3% of full scale and often uses two ranges, such as 0 to 20 and 0 to 100 mi. In more elaborate equipment, the position of the gate is a function of an analog shift rotation involving fine and coarse time delays; these have accuracies of 0.1 mi, independent of distance. Still other equipment uses digital counting techniques, with accuracies of 0.01 mi. In the latter equipment, overall accuracy is limited by system accuracy, which includes stability of beacon delay, accuracy of pulse rise times, and so on.

In 1996 typical high-performance airborne equipment had a weight of less than 13 lb. All circuits were solid state, and used digital microprocessors with the exception of the pulsed transmitter-amplifier chain. Airborne antennas are typically quarter-wave stubs (about 3 in. long) projecting from the bottom of the aircraft. Vertical polarization is used.

Transmitter Characteristics   Whereas the airborne equipment must operate on 126 channels, the ground beacon usually stays on one channel for long periods of time. The beacon consequently may, for a given state of the art, use a more sensitive receiver and a more powerful transmitter. One typical set provides 3-kw peak output, together with an antenna gain of 9 dB. Otherwise, the ground circuits follow the same principles as the airborne ones, with a 63-MHz intermediate-frequency receiver amplifier being used. The number of aircraft that a beacon can handle is usually based on the assumption that 95% of the aircraft will be in the track mode at not over 25 interrogations per sec; 5% are in the search mode, at not over 150 interrogations per sec. For 100 aircraft, this means about 3000 pulse-pairs per sec.

The duty cycle of the ground transmitter is therefore much greater than that of the airborne equipment, and the average power consumption is also greater. Most beacons are operated on the constant-duty-cycle principle, whereby receiver gain is increased until 3000 pps appear at the output of the receiver. In the absence of interrogation, these pulses will all be due to receiver noise; with interrogation from less than 100 aircraft, they are a mixture of noise and inter-rogations; with interrogation from more than 100 aircraft, they are the inter-rogations from the 100 nearest aircraft. After the 3000-pulse limit is passed, the gain is automatically reduced. This constant-duty cycle has the following advantages:

1. The beacon is automatically maintained in its most sensitive condition.
2. The transmitter duty cycle is maintained within safe limits.
3. The airborne automatic gain control circuit always has a constant number of pulses to work on, thereby simplifying its design.
4. In case of interrogation by too many aircraft, the nearest aircraft are the last to be deprived of service.

For the simplest, most reliable circuitry, the beacon is arranged not to receive while transmitting (self-oscillation could otherwise result); furthermore, to reduce interrogation by multipath echoes of strong interrogation pulses, it is desirable to reduce receiver gain for a short while after each genuine interroga-tion. Some interrogations are consequently lost; the amount of this countdown is typically on the order of 20%. Thus, an airborne equipment interrogating at 25 pps receives only 20 replies. Airborne tracking circuits are, however, designed to operate at this reduced rate.

The delay between transmission and reception is nominally 50 μsec. For greatest accuracy, this must be maintained constant; considerable circuit refine-ment is used to retain this value, independent of interrogation strength and envi-ronmental effects. Typical en-route-type beacons exhibit a total variation for ± 0.5 μsec, corresponding to a distance error of ± 0.04 mi [30]. Beacons asso-ciated with instrument-landing systems may be designed to be more accurate, due to the smaller spread of interrogation-signal levels. The ICAO requires an overall system accuracy of 0.5 mi or 3%, whichever is greater.

Under the control of an external keyer, usually common to the associated VOR, the beacon transmits an identify signal. Typically, this occurs for about 3 sec every 37 sec. During this time the random pulses are replaced by regularly spaced pulses at 1350 pulse-pairs per sec. These activate a 1350-Hz tuned circuit in the aircraft and are keyed with a three-letter Morse code, $\frac{1}{8}$ sec per dot and $\frac{3}{8}$ sec per dash. During this time, the airborne ranging circuit is in the memory condition.

Since the DME system, unlike the VOR system, is not a passive system, it has an inherent capacity limitation. The value generally quoted is 110 aircraft per beacon.

4.4.7   Tactical Air Navigation (Tacan)

Tacan [7] is a military omnibearing and distance measurement system using the same pulses and frequencies for the distance measurement function as the standard DME system. A Tacan beacon comprises a constant-duty-cycle DME beacon and antenna, to which the following additions are made:

1. A parasitic element rotating around the antenna at 900 rpm, generating an amplitude-modulated pattern at 15 Hz, with phase proportional to the bearing of the receiver.

2. Nine other parasitic elements, also rotating at 900 rpm, generating a multilobe pattern at 135 Hz, to improve the bearing accuracy.

3. Reference pulses at 15 and 135 Hz to which the above variable phases are compared in the aircraft, to establish its bearing.

The Tacan airborne equipment comprises a DME interrogator to which Tacan bearing circuits have been added. All Tacan beacons provide full service to all DME interrogators, and all DME beacons provide distance readings to all airborne Tacan sets. The principal advantages of Tacan over VOR/DME are the following:

1. Because of its higher frequency (960 to 1215 MHz versus 108 to 118 MHz), the Tacan beacon antenna can be smaller, it is therefore more suitable for shipboard and mobile use.

2. The multilobe principle, to enhance bearing accuracy, is built into all equipment, ground based and airborne.

3. Both distance and bearing are obtained via the same radio-frequency channel, providing certain equipment economies.

The system is in general use by the U.S. Navy and Air Force, and by NATO military forces. In 1996, over 800 facilities were maintained for the U.S. DoD with a DoD user population of 13,000 [48].

Transmitter Characteristics   A diagram of the Tacan ground beacon is given in Figure 4.21. The antenna comprises a central radiator, broadbanded to cover

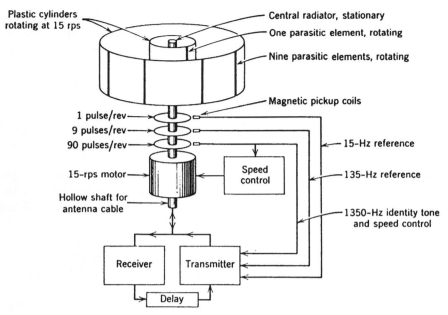

Figure 4.21 Tacan ground beacon.

the 960- to 1215-MHz range. Equipment has been built with from 1 to 11 vertical elements, depending on the kind of site for which the set is intended. All transmission and reception is by this central radiator. At a radius of about 3 in. and usually mounted on a plastic cylinder is the 15-Hz parasitic rotating element. At a radius of about 18 in. is another plastic cylinder on which are mounted nine parasitic elements, 40° apart. These superimpose a 135-Hz amplitude modulation on the transmitted signal. Depth of modulation is about 20% for each of these signals. On the same shaft that rotates the parasitic elements are three reference-pulse disks. These generate 1, 9, and 90 low-level pulses per revolution, respectively, by varying the magnetic inductance of a solenoid. These pulses are fed down to the transponder. The motor that rotates this whole assembly is usually of ac type, its speed controlled to better than 1% by a servo system in which the reference-pulse frequency is compared to a frequency standard, such as a tuning fork.

When installed aboard a ship, the Tacan antenna is stabilized in two planes. In the horizontal plane, compensation is provided to ensure that the reference pulses do not shift with the heading of the ship but remain oriented to north. In the vertical plane, compensation is provided for the roll of the ship. (Early systems also provided for pitch compensation, but this was subsequently found to be unnecessary.)

The transponder is a constant-duty-cycle DME beacon to which the bearing-reference pulses have been added. Once per revolution, coincident with the maximum of the antenna pattern pointing east, a so-called north reference pulse

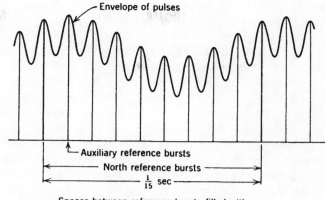

Figure 4.22   Transmitted Tacan signal.

code is emitted. This comprises 24 pulses, the spacing between pulses being alternately 12 and 18 μsec. When these pulses are decoded in the airborne equipment, they become 12 pulses, spaced 30 μsec apart. This pulse train is initiated by the one-per-revolution reference from the antenna.

Eight times per revolution, the 135 Hz reference pulse group is emitted. (The ninth group coincides with the north pulse and is intentionally omitted.) This comprises 12 pulses spaced 12 μsec apart. The circuitry of the transponder is arranged in such a way that the reference pulse groups take priority over the normal constant-duty-cycle pulses. The overall transmitted pulse envelope is shown in Figure 4.22.

The 1350-Hz identity tone, transmitted every 30 sec, is derived from the 90 pulses-per-revolution disk on the antenna shaft, thus producing phase coherence between identity and reference pulses and allowing each to be received without interference from the other. The identity code comprises 1350 groups per sec, each composed of four pulses spaced 12, 100, and 12 μsec, respectively. The reason for the 100 μsec spacing between the 12 μsec pairs is that this combination produces the least bearing error during identity transmissions, reducing the necessity for bearing memory circuits in the airborne equipment.

The DME interrogations are amplitude modulated by the rotating antenna, reducing the effective sensitivity of the Tacan beacon about 3 dB below that of an ordinary DME beacon. Although the use of a separate, nonmodulated receiving antenna would avoid this loss, such an arrangement has not been found necessary in actual practice.

Airborne Receiver Characteristics   The airborne receiver comprises a DME interrogator to which the Tacan bearing circuitry has been added. The DME interrogator must have an effective automatic gain control, so as to preserve the amplitude modulation of the pulses over the required range of expected signal strengths. This is usually taken to vary from minimum usable signal up to about 1 mw of signal at the receiving antenna.

Figure 4.23 is a generic block diagram of the airborne Tacan bearing circuit. Following decoding, the amplitude-modulated signal is filtered into two sine waves, one at 15 and one at 135 Hz. The "north" pulse activates a 33.3 kHz ringing circuit, whereas the 135 Hz reference pulse group activates an 83.3 kHz ringing circuit. These reference pulses are continually compared with the two sine waves and actuate two motor-driven servo systems, geared together $9:1$. Whenever the 135-Hz signal is present and the 15-Hz signal is within $\pm 20°$ of its correct position, the 135-Hz signal controls the servo. In effect, the bearing accuracy is determined by the nine-lobe antenna pattern of the ground beacon, with the one-lobe pattern used to resolve ambiguity, which otherwise would occur every $40°$. As with DME, both static and velocity memories have been applied to airborne bearing circuits to carry them through short-term signal dropouts. Solid-state airborne equipment typically weights 20 lb and occupies about $\frac{1}{2}$ft^3. Modern receivers incorporate digital implementations of some of the receiver functions depicted in Figure 4.23.

System Considerations   Tacan was the first operational rho-theta system to exploit the multilobe bearing principle. At "perfect" sites, bearing errors measured under carefully controlled conditions were on the order of $\pm 0.1°$ for 77% of the readings and $\pm 0.2°$ for 93% of readings. Compared with previous systems, these results were sensational and led many users to plan applications for which the system was not designed. The chief misconception concerning the performance of Tacan stemmed from the basic fact that the nine-lobe system gives an improvement only if the one-lobe system is functioning properly. Since the airborne equipment is controlled most of the time by the nine-lobe system, there was little opportunity to evaluate the performance of the one-lobe system by itself.

The one-lobe system suffers from many of the siting problems common to other point-source bearing measuring systems. Chief among these are the effects of reflecting objects near the transmitting antenna. This problem is greatly increased if the antenna is mounted relatively close to the ground and has little or no vertical directivity. The resulting strong vertical lobe-and-null structure may then create the condition where the aircraft is in the direction of a null, whereas an unwanted reflection is in the direction of a lobe. High vertical directivity, with uptilt, has been the most effective means to reduce this problem. Being a pulse system, Tacan is somewhat less susceptible to site errors than continuous-wave systems, since reflections from objects farther away than the pulse duration (about $\frac{1}{2}$ mile) are of much less importance. The U.S. 1994 Federal Radio Navigation Plan [48] cites the signal-in-space Tacan 2σ (95% probability) azimuth accuracy to be $\pm 1\%$ (± 63 m at 3.75 km) and the distance accuracy to be 185 m (± 0.1 nmi). The capacity for distance measurement is 110 aircraft, and it is unlimited for azimuth measurement.

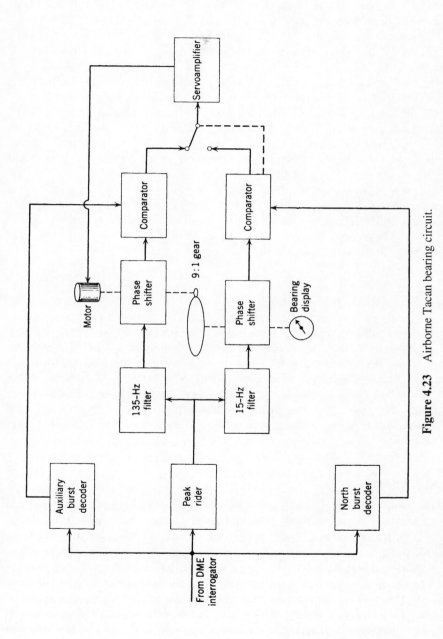

Figure 4.23 Airborne Tacan bearing circuit.

137

4.4.8 VORTAC

Since Tacan beacons can be more readily installed on ships and at tactical sites than VOR beacons, large numbers of military aircraft are equipped with Tacan. To save these aircraft the cost of carrying additional equipment for navigating the ICAO VOR/DME airways, several countries, including the United States, use the VORTAC system. In this system each VOR station, instead of being collocated with a DME, is collocated with a Tacan beacon (which also provides DME service) to provide rho-theta navigation to both civil and military aircraft. Civil aircraft read distance from the Tacan beacon and bearing from the VOR beacon. Military aircraft read both distance and bearing from the Tacan beacon. Thus each type of aircraft fits into the same air-traffic management system, regardless of which type of airborne equipment it carries. In 1996, it is estimated that there are more than 200,000 users in the United States alone.

At the ground station, the VOR central antenna is housed in a plastic cone that supports the Tacan antenna. Leads to the Tacan antenna pass through the middle of the VOR antenna, along its line of minimum radiation, and do not disturb the VOR pattern. In the case of Doppler VOR (Section 4.4.5), the antennas are arranged in a circle outside the cone.

4.5 HYPERBOLIC SYSTEMS

Hyperbolic navigation systems are so designated because of the hyperbolic lines of position (LOP) that they produce rather than the circles and radial lines associated with the systems that measure distance and bearing (see Figure 4.8*d*) [4]. The Loran-C, Omega, Decca, and Chayka systems are described in this section. They differ in that Loran-C and Chayka measure the time-difference between the signal from two or more transmitting stations, while Omega and Decca measure the phase-differences between the signals transmitted from pairs of stations.

4.5.1 Loran

Loran (long-range navigation) is a hyperbolic radio-navigation system that has evolved over a period of years, beginning just before the outbreak of World War II in Europe. The Loran-C system [8, 11, 44] has benefited greatly from analysis of the shortcomings of previous systems. It uses ground waves at low frequencies, thereby securing an operating range of over 1000 mi, independent of line of sight. Second, it uses pulse techniques to avoid sky-wave contamination. Third, being a hyperbolic system, it is not subject to the site errors of point-source systems. Fourth, it uses a form of cycle (phase) measurement to improve precision. It inherently provides a fine–coarse readout of low inherent ambiguity. All modern Loran systems are of the Loran-C variety. (Loran-A and Loran-D configurations no longer exist.)

Loran-C users fall into the two general categories: navigation users and precise time and time interval (PTTI) users. By far the larger population of direct users is in the navigation category. An even larger group of indirect users benefits from a PTTI application of Loran-C, in which digital switching, signaling, and timing of the nation's telephone system is accomplished using Loran-C. Every telephone subscriber in the United States is an indirect beneficiary of the Loran-C system.

Principles and System Configuration   Loran-C is a low-frequency radio-navigation aid operating in the radio spectrum of 90 to 110 kHz. It consists of transmitting stations in groups forming chains. At least three transmitter stations make up a chain. One station is designated as master, while others are called secondaries. The chain coverage area is determined by the transmitted power from each station, the geometry of the stations, including the distance between them and their orientation. Within the coverage area, propagation of the Loran-C signal is affected by physical conditions of the Earth's surface and atmosphere, which must be considered when using the system. Natural and man-made noise is added to the signal and must be taken into account. Receivers determine the applied coverage area by their signal-processing techniques and can derive position, velocity, and time information from the *time difference* (TD) between the *time of arrival* (TOA) of a radio wave from a secondary minus the TOA of a radio wave from the master station. Methods of application provide for conversion of basic signal time of arrival to geographic coordinates, bearing and distance, along track distance and cross-track error, velocity, and time and frequency reference.

Each of the stations in all Loran-C chains transmit pulses that have standard characteristics. The pulse consists of a 100-kHz carrier that rapidly increases in amplitude is a carefully controlled manner and then decays at a specified rate forming an envelope of the signal. Each station in a chain repetitively transmits a series of closely spaced pulses called a *pulse group* at the *group repetition interval* (GRI) of the chain. The GRI uniquely identifies the chain. When the chain is synchronized to universal time (UT) the master station also sets the time reference for the chain. Other stations of the chain are secondaries and transmit in turn after the master. Each secondary pulse transmission is delayed in time so that nowhere in the coverage area will signals from one station overlap another.

The number of pulses in a group, pulse spacing in a group, carrier phase code of each pulse, time of transmission, the time between repetition of pulse groups from a station, and the delay of secondary station pulse groups with respect to the master signals constitute the signal format. Each station in a chain is assigned a signal format based on its function.

The signal format is modified by *blinking* certain pulses to notify the user of faulty signal transmission. The signal format is also modified to accommodate a single transmitter station in two chains. This is accomplished by permitting transmission for one of the chains to take precedence over the other when the signal format calls for simultaneous transmission in both chains. This function is called *blanking*.

Wave Form and Signals in Space   Each station transmits signals that have standard pulse leading-edge characteristics. Each pulse consists of a 100-kHz carrier that rapidly increases in amplitude in a prescribed manner and then decays at a rate that depends upon the particular transmitter and transmitting antenna characteristics. The leading edge of the standard Loran-C pulse antenna current wave form, against which the actual antenna current wave form is compared, is defined as $i(t)$:

$$i(t) = A(t-\tau)^2 \exp \left\{ \frac{-2(t-\tau)}{65} \right\} \sin(0.2\pi t + \phi) \qquad \text{for} \quad \tau < t < 65+\tau, \qquad (4.6)$$

where

$$i(t) = 0 \qquad \text{for} \quad t < \tau,$$

and A is a normalization constant related to the magnitude of the peak antenna current in amperes, t is time in μsec, τ is the *envelope-to-cycle difference* (ECD) in μsec, ϕ is the phase code parameter (in radians) which is 0 for positive phase code and π for negative phase code. ECD is the displacement between the start of the Loran-C pulse envelope and the third zero crossing of the Loran-C carrier (phase). ECD arises because the phase velocity and the group velocity of the signal differ. As the signal propagates over seawater, the ECD decreases because the phase velocity exceeds the group velocity.

The pulse trailing edge is defined as that portion of the Loran-C pulse following the peak of the pulse or 65 μsec after the pulse is initiated. The pulse trailing edge is controlled in order to maintain spectrum requirements. At different transmitting sites, or with different transmitting equipment, the pulse trailing edge may differ significantly in appearance and characteristics. Regardless of these differences, for each pulse the antenna current $i(t)$ will be less than or equal to 1.4 mamps or 16 mamps depending on transmitter type.

To prevent contamination of the rising edge of a Loran-C pulse by the tail of the previous pulse, ideally, the amplitude of the tail should be well attenuated before the next pulse starts. Because of the sky-wave effects, a Loran-C pulse should be attenuated as fast as possible after attaining its peak amplitude. Unfortunately, a serious constraint in the form of the frequency spectrum bound must be considered. A compromise between these two requirements lead to a pulse length of 500 μsec. By requiring the amplitude of the pulse at 500 μsec to be 0.001A (-60 dB) where A is the peak amplitude of the pulse, the spectrum specification can be met and the pulse tail/sky-wave contamination problem can, in most cases, be avoided.

Figure 4.24 shows a Loran-C pulse [44]. Zero crossing stability is important

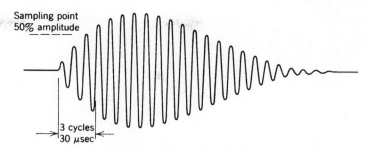

Figure 4.24   Single Loran-C pulse.

because it affects the system phase accuracy. The standard sampling point for a Loran-C receiver is the positive going zero crossing of the phase decoded pulse on its third cycle (approximately 30 μsec) after the arrival of the ground wave. This tracking is accomplished by a phase locked loop. In addition, it affects the apparent signal-to-noise ratio as seen by the receiver and, therefore, the available receiver accuracy at a given averaging time. Amplitude stability is important, because it affects the ECD of a transmitted Loran-C pulse and thereby affects the ability of a receiver to lock-on and track the correct cycle.

Propagation Effects   The low-frequency propagation of 100 kHz is influenced by the properties of the Earth's surface as well as the ionosphere. Because of the ionospheric changes, the portion of the propagated wave that is reflected from the ionosphere (sky wave) is not very stable. To make the received 100 kHz signal more stable and reliable within a given coverage area, the Loran-C radio navigation system is designed as a pulse system, which enables it to separate the ground wave from the sky wave. The ground wave consists of a space wave and a surface wave. The space wave is made up of the direct wave (a signal traveling between the transmitter and the receiver via the direct path) and the ground-reflected wave (a signal that arrives at the receiver after being reflected from the surface of the Earth). The space wave also includes the diffracted waves around the Earth and the refracted waves in the upper atmosphere. The surface wave is guided along the surface of the Earth, which absorbs energy from the wave causing its attenuation. When both antennas (transmitter and receiver) are located very near the surface of the Earth, the direct and ground reflected terms in the space wave cancel each other, and the transmission of Loran-C signal takes place entirely by means of surface wave (assuming that no sky wave exists). Figure 4.25 shows ground-wave and sky-wave modes of propagation [34].

Propagation-induced errors arise from variations in the ground-wave signal propagation velocity, which are caused by the Earth's ground conductivity along the signal path and to a lesser extent by atmospheric effects. These propagation anomaly errors represent a major contributor to the total Loran-C error budget.

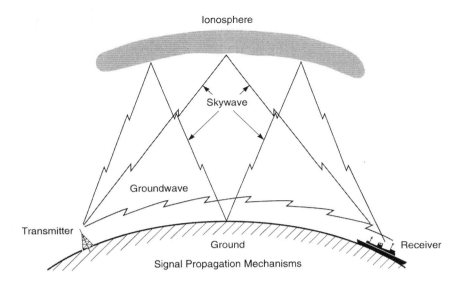

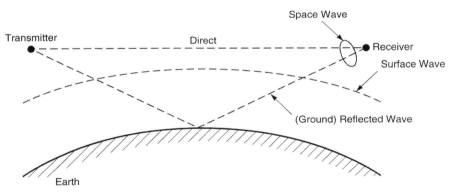

Figure 4.25   Ground-wave and sky-wave modes of propagation.

Over the chain coverage area, the propagation anomalies exhibit both spatial and temporal variations.

The temporal variations fall into two primary categories: *diurnal* and *seasonal*. The diurnal variations are short-term propagation effects caused primarily by local weather changes and day/night transitions along the signal path. Variations in the refractive index of the atmosphere versus height from the ground (vertical lapse rate) contribute to the short-term propagation errors. The diurnal time-difference (TD) variations tend to be relatively small, on the order of tens of nanoseconds.

The larger category of temporal variations are the seasonal effects, which are most pronounced over land paths. These long-term errors tend to be periodic with an annual cycle and can result in peak-to-peak TD excursions of up to several hundred nanoseconds. The seasonal variations are caused primarily by changes in ground conductivity and vertical lapse rate between winter and summer conditions.

The other major category of propagation anomaly is the *spatial* variation. The spatial portion of the Loran-C propagation error is caused by fixed topographical and surface conductivity properties along the signal path. The most uniform propagation occurs over all seawater paths. Using the known, constant speed of Loran-C signals over seawater, the predicted Loran-C signal phase variation can be computed as a function of path length, commonly called the *secondary phase factor* (SF). Signals over land travel more slowly, by an amount that varies with the conductivity and terrain irregularity. Differences in conductivity vary from 0.5×10^{-3} mhos/meter for snow covered mountains to 5.0 mhos/meter for seawater. The additional phase error effect due to the land mass is commonly referred to as the *additional secondary phase factor* (ASF). As in the case of the temporal errors, the ASF errors tend to have long correlation distances, on the order of 90 to 100 nm.

The previously discussed characteristics of signal phase were related to ground-ground-type measurements. If the receiver is raised to any altitude, as in the airborne application, then the altitude effect must be considered. Theoretical predictions yield altitude, conductivity, and distance from transmitter as the important parameters. Also, beyond a distance of 250 mi from the transmitter, the correction is essentially constant. In this case, if one is operating at this distance or greater from two transmitters and trying to form hyperbolic LOPs, then the altitude effect on both paths will cancel. If one of the ranges is much shorter than the other, errors of a few tenths of a microsecond can easily result if this height gain function is not considered. A variety of formulations for predicted Loran time difference as a function of propagation effects have been used in receivers. Examples of these are given in Chapter 2, Section 2.5.2 of the first edition.

Absolute Accuracy Performance   Within a published coverage area, Loran-C will provide the user who employs an adequate receiver with predictable accuracy of 0.25 nm 2drms or better. The repeatable and relative accuracy of Loran-C is usually between 18 and 90 meters. The total accuracy is dependent upon the *geometric dilution of precision* (GDOP) factor at the user's location within the coverage area.

GDOP is a dimensionless factor that expresses the sensitivity of position fix accuracy to errors in TD measurement. As GDOP increases in a given area, the impact of atmospheric noise, interference, and propagation vagaries inherently increases. GDOP is a function of the gradient of each LOP and the angle at which LOPs cross. Lines of constant GDOP are lines on which fix accuracy is expected to be equal.

For a triad, GDOP is defined as

$$\text{GDOP} = \frac{1}{\sin(\phi_1 + \phi_2)} \sqrt{\frac{1}{\sin^2 \phi_1} + \frac{1}{\sin^2 \phi_2} + \frac{2r\cos(\phi_1 + \phi_2)}{\sin \phi_1 \sin \phi_2}} \qquad (4.7)$$

where ϕ_i is the half-angle subtended by station i and r is the correlation coefficient between two LOPs, which is taken to be 0.5. Note that $\phi_1 + \phi_2$ is equal to the crossing angle of the LOPs (θ). The relationship between 2drms and GDOP is

$$2\text{drms} = 2\sigma \frac{K_0}{\sin \theta} \sqrt{\frac{1}{\sin^2 \phi_1} + \frac{1}{\sin^2 \phi_2} + \frac{2r\cos \theta}{\sin \phi_1 \sin \phi_2}} \qquad (4.8)$$

where σ is the timing error and K_0 is the constant 500 ft/μsec.

Availability   Although individual Loran-C transmitting equipment is very reliable, redundant equipment is used to reduce system downtime. Loran-C transmitting station signal availability is therefore greater than 99.9%.

Reliability   Reliability is a measure applied to system equipment such as receivers, timers, and transmitters. The weakest link in Loran-C system reliability is the highest-stressed component, the ground transmitter. Redundant equipment at tube-type transmitting stations, and a "graceful degradation" capability at solid-state transmitter stations, keep the system in an almost fail-safe mode. The only significant failures in service have occurred when transmitting antennas have collapsed or a severe lightning strike has completely destroyed the output modules in a solid-state transmitter.

Repeatability   The Loran-C system repeatability is excellent in terms of days to weeks or longer. This means that, once the location of a reference point or waypoint is known in the Loran-C frame of reference (grid), a navigator can return to that point with very high accuracy. The frame of reference can be either in Loran-C TDs or latitude/longitude coordinates. Repeatability is a particular strength of the Loran-C system for the majority of uses. Repeatability declines as the period of time between measurement of the reference point location and return to that point increases, due to seasonal effects on Loran-C signal propagation. If a plot of time difference or Loran-C latitude/longitude is made for a fixed user location over a period of several years, a definite periodicity in the data is clearly seen. The data have a sinusoidal pattern in a period of one year and are generally repeatable year to year. Figure 4.26 is an example of the seasonal variation in TD data and clearly shows why repeatability of the Loran-C system is good over the days-to-weeks time period but is poorer over a period of months, reaching its worst over the winter-to-summer time interval. Over one-year intervals or over the spring-to-fall interval, repeatabil-

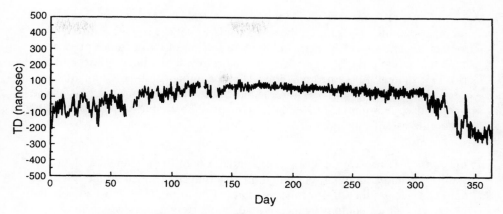

Figure 4.26 Seasonal variation of repeatable accuracy.

ity is very good. The 1994 Federal Radio Navigation Plan [48] cites a Loran-C repeatability error range at 60–300 ft 2drms.

Integrity   Loran-C stations are constantly monitored to detect signal abnormalities that would render the system unusable for navigation purposes. The secondary stations *blink* to notify the user that a master-secondary pair is unusable. Blink is a repetitive on–off pattern (approximately 0.25 sec on, 3.75 sec off) of the first two pulses of the secondary signal, which indicates that the baseline is unusable for one of the following reasons:

- TD out of tolerance
- ECD out of tolerance
- Improper phase code or GRI
- Master or secondary station output power or master station off-air

When a secondary station is blinking it continues to transmit its normal eight pulses at the normal GRI. However, it is only during a 0.25-sec period that all eight pulses are present. During the next 3.75 sec only the last six pulses of the eight pulses are present, the first two having been turned off. The on–off cycle is repeated until normal operations are resumed. This blink period should be sufficient to permit automatic blink detection circuits in receivers to activate and warn users that the baseline is unusable. In 1996, the USCG and the FAA are pursuing an "aviation blink," based on factors consistent with aviation use.

Direct Ranging   There are some Loran-C users who do not employ Loran-C in the hyperbolic mode but rather in the direct range rho-rho-rho mode or the rho-rho mode. The rho-rho-rho process involves a minimum of three transmitters and use of an iterative computation to obtain a fix. Direct range rho-rho mode requires two stations as a minimum, a highly stable user frequency standard, and precise knowledge of the time of transmission of the signal. A direct range from each station is developed, and the intersection of these circles produces a fix (actually two possible fixes that must be resolved by either a third range or prior knowledge of assumed position). Direct ranging can be used in situations where the user is within reception range of individual stations but beyond the hyperbolic coverage area. In 1996, the high cost of the stable frequency standard limited the use of this mode.

Differential Loran-C   In using differential Loran-C, a reference station is established, and a nominal set of TDs are determined. Thereafter, the reference station broadcasts the offset of the measured TDs from the established nominal values. The user's equipment incorporates the differential values to produce a highly stable and long-term high-precision position accuracy in the vicinity of the reference station. These corrections are generally valid for the "correlation distance" of approximately 100 nmi from the reference station. Real-time corrections to remove both seasonal and diurnal errors can be broadcast. For most users, correction of the seasonal variation alone would be sufficient. Diurnal variations tend to be small enough so that, within areas of good GDOP, sufficient accuracy is obtained to meet most requirements. Studies have shown that publishing the previous day's correction to the baseline TDs is entirely satisfactory. This approach reduces the electronic equipment requirements and complexity for both the user and the provider of the service and is a process that may fit within the envelope of aviation flight planning.

Grid Calibration Techniques   A variety of theoretical methods can be used to calibrate the spatial propagation variations over the Loran-C coverage area. One such technique, known as Millington's method [34], breaks the signal path between the transmitter and receiver into finite segments of different conductivity levels, based on conductivity maps. The incremental phase delay is then computed as a function of range and conductivity for each path segment in both the forward and backward propagation directions, summed, and averaged to provide an estimate of the ASF.

While purely theoretical models of ASF calibration can substantially improve the overall accuracy of the Loran-C solution, combining theoretical models with empirical data can improve accuracy further. This semiempirical grid calibration approach has been used successfully to calibrate coefficients of a Loran-C signal propagation error model using surveyed data points of empirically measured TD errors. Because of the long correlation distances of the spatial propagation errors, only a few widely separated survey data points are sufficient to provide a reasonably good grid calibration using this semiempirical modeling approach.

Modern Loran-C receivers often include some type of grid calibration algorithm or table lookup process to correct for estimated ASF errors. Such calibration can substantially minimize the effects of spatial propagation anomalies on accuracy.

Transmitter Characteristics The Loran-C transmitting system has evolved over the years from the original tube-type transmitter, some of which are still operating, to higher-powered tube transmitters and eventually to solid-state transmitters. Highly precise cesium frequency standards are employed at each transmitting station to time the transmitted signals.

The modern solid-state transmitters are described in this section. Each transmitter station is physically divided into two groups of units to provide system redundancy. At the appropriate interfaces switching units are provided between them. In Figure 4.27 a simplified block diagram depicts one set of the redundant equipment.

The timer provides all timing signals to the transmitter, including 5-MHz clock signals. Dual redundant pulse amplitude and timing controllers (PATCO) accept timing signals from the timer and derive from this all the signals needed by the transmitter. Signals generated by the PATCO include start triggers, charging triggers, digital amplitude reference signals, amplitude compensation signals, and megatron reference trigger. Each PATCO contains an ECD module to make small changes in the amplitude of each pulse group and allow fine adjustment to be made to the pulse shape. The transmitter power level is also monitored in the PATCO to identify problems with the half-cycle generators (HCG).

The HCGs are the basic building blocks of the solid-state transmitter. Each HCG contributes a portion of the power contained in the Loran-C pulse. Thirty-two HCGs comprise the standard set. The basic set can be expanded in multiples of eight HCGs, with associated ancillary equipment, to transmitting stations of 40, 48, 56, and 64 HCGs. The power output by the transmitting station depends on the HCG configuration and the type of transmitting antenna used. When used with standard transmitting antennas of 625 to 1000 ft, power outputs can range between 400 kw and 1000 kw for the 32 through 64 HCG configurations.

Each HCG takes the PATCO signal and generates a 5 μsec pulse in the megatron of 4000 amps peak that is shaped like a half-cycle of a 100-kHz sine function. These pulses are sent to the coupling and output networks during each Loran-C pulse. The coupling network unit is a passive pulse shaping network that contains coupling capacitors, coupling inductors, and tailbiter modules. The coupling network receives four 5-μsec pulses and transforms them into the required Loran-C pulse with a peak at 65 μsec. After the peak of the pulse, the tailbiter module controls the shape of the pulse so that it exponentially decays to zero. The output network presents the correct impedance to the coupling network for the transmitting antenna and provides isolation between the coupling network and the antenna ground system.

The transmitter operational control (TOPCO) and display units perform several primary functions. They permit selection between redundant PATCOs, cou-

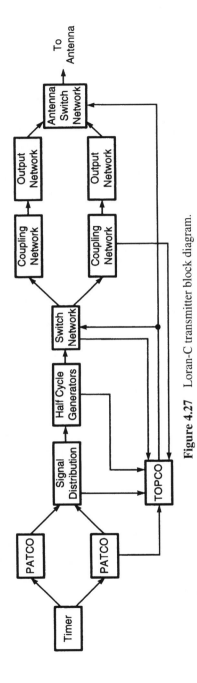

Figure 4.27 Loran-C transmitter block diagram.

pling networks, and output networks. The TOPCO and display units contain built-in monitoring and fault detection circuitry, and if a fault is detected the TOPCO automatically switches coupling and output networks. The TOPCO and display units serve as centralized alarm panels and status displays for the transmitting station.

Receiver Characteristics   In the past, navigation users employed receivers that read out Loran-C TD coordinates, but by the 1970s Loran-C receiver designers had automated much of this process to the extent of selecting the best triads, ensuring correct signal lock-on and computing a latitude and longitude from the time-differences. Many receivers contain correction tables in memory to make the navigation solution as accurate as possible with relation to the geodetic grid. By 1996, computational and processor power had resulted in user equipment with the capability to store multiple waypoints and indicate data to the user that include course and distance to go to the waypoint, and course and speed made good. Very few users operate in a time-difference output environment today.

An airborne Loran-C receiver block diagram is shown in Figure 4.28; a photograph of a Loran-C receiver for commercial aviation applications is depicted in Figure 4.29.

Loran-C receivers are commonly referenced by the rate (number of chains tracked), the source of the time reference, number of stations tracked, and the measurement type. For example, a single-rate, master-referenced, two-pair, time-difference receiver tracks a single chain selected by the user; time initialization is obtained from the master station; and two stations are tracked to obtain a TD measurement. Dual-rated receivers track two chains to produce a single position solution, while cross-chain receivers use stations from two chains to define LOPs. In 1996, master-independent cross-chain receivers, which use a priori information to define LOPs between secondary stations in different chains, were being investigated.

All Loran-C receivers used for navigation enter four functional states: initialization, acquisition, pulse group time reference (PGTR) identification, and tracking. Initialization is the process of providing to the set all the a priori knowledge of the signals to be tracked and adjusting the set to minimize the effects of interference. Initialization may include GRI selection, estimates of secondary time difference, adjustment of interference filters, setting of clipping levels, and the determination to search for the strongest path.

Acquisition is the process of searching for and locating the signals identified during initialization. Generally, a receiver will locate the signals from each station in time slots, called *intervals*, that repeat at the GRI. The search mechanization is dependent on the extent of automation (manual or automatic search) and displays built into the set. There are two general forms of manual search: pulse alignment and phase code. Automatic search invariably operates on the phase code of the signals and cross-correlates an approximation of the known signals with the received signals. Generally, automatic search will operate at lower S/N ratios than manual search and can be designed to search faster, using multipoint

150

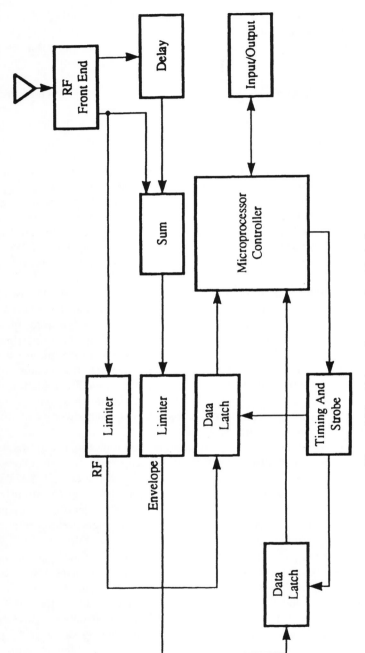

Figure 4.28   Airborne Loran receiver block diagram.

Figure 4.29   Airborne Loran receiver.

search for master and limited range search for secondaries. When the correlation reaches a certain threshold, after ensuring no false locks have occurred on secondary signals, acquisition is complete.

PGTR identification is the process of ensuring that the receiver is operating on the ground wave of the signals. Ensuring operation of the ground wave, sometimes called *guard sampling* or *ground-wave location*, operates on the principle that the ground-wave signal from a station always arrives at a receiver before the sky wave because of the longer sky-wave path. It is necessary to find the ground wave because its timing, and hence its position-locating qualities, is stable, while the sky wave is not. Typically, the acquisition process locates the sky wave because of its much larger amplitude. Ground-wave location proceeds by using signal detection algorithms at the signal 30 to 60 μsec ahead of the receiver's reference time. If signals are found, the receiver timing is advanced and the process repeated. This continues until no signals are found at two or more successive locations. Often, multiple independent tests are made after no signal is detected to account for the possibility of the ground-wave and sky-wave signals summing out of phase and creating a null that might otherwise be presumed to be the start of the ground wave.

Tracking is the process of maintaining a constant timing relationship between the receiver's time reference and the PGTR for each signal being tracked. In an automatic tracking receiver, circuits within the equipment automatically adjust the time reference and update the display to provide continuous readings. These receivers also provide alarms or warnings advising the operator of undesirable signal conditions or transmitter blinking.

Loran-C signal reception can be impaired by interference from other signals broadcast on slightly different frequencies (generally low-frequency communications). To avoid the degradation in S/N associated with these interfering sources, Loran-C sets are equipped with *notch filters* that can be used to attenuate the interfering signal. Some receivers are equipped with preset notch filters, others with adjustable notch filters, and yet others that automatically search for interfering signals near the Loran-C band and dynamically notch out any interference.

TABLE 4.5   Existing continental U.S. Loran-C chains (excluding Aslaka)

Chain Name	GRI (μ sec)
Canadian east coast (5930)	59,300
North-east United States (9960)	99,600
South-east United States (7980)	79,800
U.S. west coast (9940)	99,400
Canadian west coast (5990)	59,900
Great Lakes (8970)	89,700
North central United States (8920)	89,200
South central United States (9610)	96,100

Global Coverage   Loran-C was used worldwide in 1996; it covered maritime Canada, the North Atlantic, Norwegian Sea, Mediterranean Sea, the Bay of Biscay, an area of Russia, China, and India, South Korea and the Sea of Japan, the Northwest Pacific, the Gulf of Alaska, and the entire continental United States. New system developments are underway or in the planning stages for northwest Europe, Argentina, Brazil, and inland Canada. A Loran-C chain consists of a master station and a number of secondaries, usually no more than four. Each chain is uniquely identified by its GRI, which represents the number of microseconds between subsequent transmissions of the master station signal. Table 4.5 lists the Continental U.S. chains and their GRIs.

Charts are published by the U.S. National Oceanic and Atmospheric Administration (NOAA) that depict the geographic coverage area served by the Loran-C system. These depicted coverage contours define the geographic limits at which a receiver with a 20-kHz bandwidth will acquire and track a master and two secondary stations, each providing a signal-to-noise ratio (S/N) better than -10 dB and a fix accuracy of better than 0.25 nm 2drms (95% of the time).

The difference between acquisition of the signal by the receiver and tracking of the signal is important. Since a receiver should always be able to track a signal it has previously acquired (under the same S/N environment) and also to continue to track a signal at a much lower S/N than that which it experienced when it acquired the signal, acquisition is the more difficult process for the receiver and is the limiting factor in receiver performance. Coverage area must therefore be defined in terms of acquisition, since that process defines the operational limits at which a navigation solution can be initiated. S/N is the major factor in determining a receiver's ability to acquire the signal. This factor is the ratio of the field strength of the Loran-C signal, attenuated over the propagation path from transmitter to user, to the field strength of noise in the receiver's bandwidth at the user location. Noise in this context is generally assumed to be atmospheric noise.

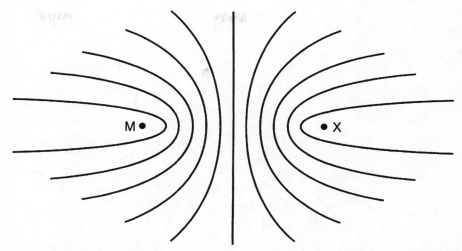

Figure 4.30   Hyperbolic lines of constant TD for a typical master-secondary pair.

Chain Geometry   Figure 4.30 shows schematically a set of hyperbolic LOPs for a typical master-secondary Loran-C transmitter pair. Each hyperbolic line contains all points having the same time difference between arrival of signals from the master and secondary. Along the baseline itself, the distance between lines of equal TD is smallest, and increases to each side of the baseline. The term applied to this parameter is gradient, and has units of feet per microsecond. The gradient is at a minimum along the baseline and deteriorates as the user moves away from it. Since the LOPs are much closer together along the baseline than they are at large distances away from it, a 100 ns standard deviation of the TD estimate represents much less error in position near the baseline than in the extremities of the coverage area. Refer to Figure 4.31 which shows two secondaries and a common master, along with hyperbolic LOPs for each master-secondary pair. If the location of each LOP is assumed to be normally distributed with a standard deviation of 100 ns, then a minimum area will be covered at the intersection of two LOPs when they cross at right angles. This area (area of fix uncertainty) will increase as the crossing angle decreases (see Figure 4.32 and 4.33). The combined effect of crossing angle and gradient is called GDOP. (See previous discussion of GDOP expression.)

Simple analysis reveals that along baseline extensions, a singularity exists. The measured TD does not change (being equal to the time from master to secondary), and hence no solution can be found.

The U.S. Coast Guard publishes predictions of the Loran-C ground-wave coverage and geometric fix accuracy limits. These coverage diagrams describe the area over which the signal-to-noise ratio of the master and two secondary stations is -10 dB or better for a nominal expected atmospheric noise level in that geographic region, and the position accuracy is 0.25 nm 2drms (95% of the time).

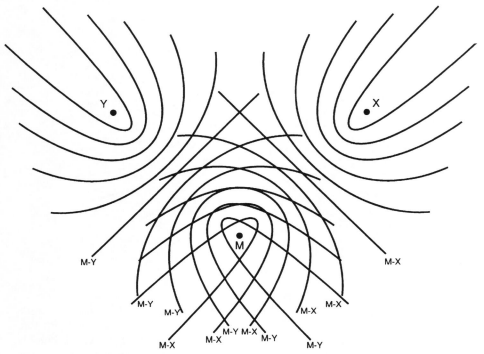

Figure 4.31   Hyperbolic lines of constant TD for a typical triad.

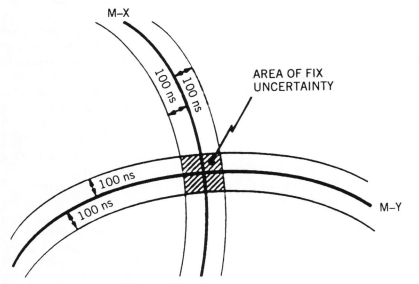

Figure 4.32   90° crossing angle.

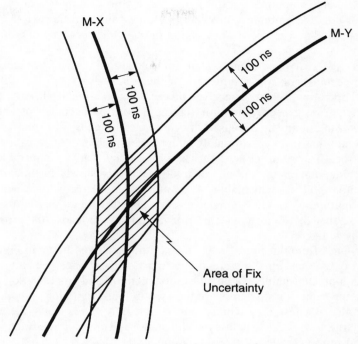

Figure 4.33   Shallow crossing angle.

4.5.2   Omega

Principles and System Configuration   In 1996, the Omega VLF radio-navigation system comprised eight transmitting stations located throughout the world. At each station, continuous-wave (CW) signals are transmitted on four common frequencies and one station-unique frequency. The signal frequencies are time-shared among the stations so that a given frequency is transmitted by only one station at any given time.

To support medium accuracy navigation, the signal transmissions from all stations are phase-synchronized to about 1 μsec. For purposes of time transfer and to facilitate the systemwide synchronization procedure, the signal timing is maintained to within an accuracy of about 0.5 μsec with respect to coordinated universal time (UTC).

Omega signals are subionospheric; that is, they are propagated between the Earth's surface and the D-region of the ionosphere. Because VLF signal attenuation is low, the signals are propagated to great ranges, typically 5000 to 15,000 nmi. Signals with amplitudes as low as 10 μV/meter can often be detected and used for navigation. Of primary interest to navigation users is the signal phase which provides a measure of transmitter-receiver distance. The fractional part of a cycle (or lane, which is the equivalent distance measure) is generally the only measurable component of the signal phase, thus leading to lane ambiguity. However, the lane ambiguity problem is reduced through the use of multiple frequencies and is resolved for navigation through a process of continuous lane count.

When used as a stand-alone system for navigation, an Omega receiver provides an accuracy of 2 to 4 nmi 95% of the time [43, 48]. In the differential mode of operation, where a receiver utilizes Omega signal phase corrections transmitted from a nearby monitor station, a position accuracy of about 500 meters can the attained. Because Omega is a continuous VLF phase-measuring system, it has been appropriately integrated with noncontinuous, high-accuracy sensors. The resulting system has an accuracy that is comparable to the high-accuracy navigation aid and degrades relatively slowly in time when the high-accuracy aid is unavailable. As commonly used in overocean civil airline configurations, an Omega receiver is combined with an inertial navigation system, so that the Omega system error effectively "bounds" the error of the inertial system.

The signals from the eight Omega transmitting stations shown in Figure 4.34 provide continuous signal coverage over most of the globe. The suite of electronics equipment (mainly signal generation, control, and amplification units) is virtually the same for all stations in the system, but the station antennas differ substantially. Because they radiate long-wavelength VLF signals, the antennas are the largest physical structures at the stations. Three types of antennas are employed in the Omega system: (1) grounded tower, (2) insulated tower, and (3) valley-span. Each has an associated signal monitoring facility about 20 to 50 km from the effective phase center of the antenna. These unmanned facilities perform several functions, including monitoring the performance of the asso-

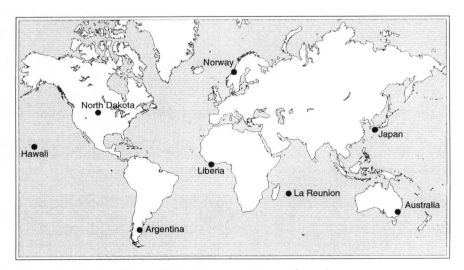

Figure 4.34   Omega station configuration.

ciated station, providing data necessary to phase-synchronize the stations, and detecting solar-terrestrial events that cause anomalous shifts of the propagated signal phase.

The Omega signal transmission format is illustrated in Figure 4.35. Across each of the eight rows in the figure is a 10-sec sample of the signal frequencies transmitted by a particular station. Important features of this time/frequency multiplex format include these four:

1. Four common transmitted signal frequencies: 10.2, $11\frac{1}{3}$, 13.6, and 11.05 kHz.

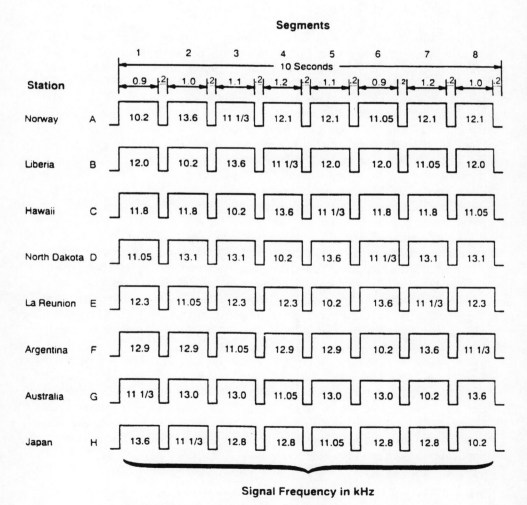

Figure 4.35   Omega system signal transmission format.

2. One unique signal frequency transmitted by each station.

3. A separation interval of 0.2 sec between each of the eight transmissions.

4. Variable-length transmission periods.

The fourth feature makes it possible to synchronize an Omega receiver to the signal format with no additional external information. For example, if a user determines that a 10.2-kHz transmission segment (repeated every 10 sec) is 1.2 sec in duration, then according to Figure 4.35, the transmitting station could be either station D (North Dakota) or station G (Australia). However, a measurement of the duration of the succeeding transmission segment at a frequency of 13.6 kHz would discriminate between station D (1.1 sec) and station G (1.0 sec).

Navigational use of Omega is based almost entirely on measurement of the signal phase transmitted from one or more stations. This is because the total signal phase is closely related to the distance from a transmitting station with known coordinates. The total phase, or cumulative phase, developed by a signal between a station and a receiver is composed of a whole number of cycles and a fraction of a cycle. In virtually all cases, the whole number component of the cumulative phase is not measurable. Since only the fractional part of the phase is measured, an ambiguity (termed *lane ambiguity*) exists with regard to the whole-cycle count. In terms of distance–unit quantities, the equivalent of the phase cycle is the signal wavelength. If a marker is made at the station and at each wavelength on the station-receiver path, the intervals between markers serve to define lanes, which are one wavelength in width and over which the phase varies from 0 to 2π rad.

The existence of lane ambiguity is not generally a problem for a user if Omega is used in a navigation mode from a known initial position. Successive positions are computed from corresponding incremental phase change measurements. Also, the presence of multiple signals reduces the chances of incorrect lane identification (or count). However, if lane count is temporarily lost (e.g., because of a receiver outage or propagation anomaly), the Omega format is designed to help resolve the ambiguity by using difference frequency techniques.

To determine a receiver's position based on the measurements of phase from the multiple station signals/frequencies, two basic methods are employed: hyperbolic and direct ranging. Hyperbolic methods use phase difference as the unit of measurement, while ranging techniques utilize phase measurements. Since about 1980, the ranging method has been employed almost exclusively in airborne receivers, while the hyperbolic method is reserved for specialized applications, such as submarine navigation.

Wave Form and Signal in Space   The principles of Omega navigation usage depend almost entirely on the assumed relationship between the signal phase received from a transmitting station and the station-receiver distance. Ideally, changes in Omega signal phase, as measured on a moving platform, bear a fixed linear relationship to corresponding changes in the position of the platform over the surface of the Earth. With this idealization, navigation and positioning become relatively simple procedures requiring only the station locations as external knowledge, in addition to measured quantities (internal to the receiver) such as the signal frequency and phase. However, since Omega signals propagate to very long ranges, they are substantially influenced by electromagnetic and geophysical variations in the Earth and ionosphere. These effects on the signal phase lead to a marked departure from a linear dependence on distance, thus complicating direct use of the signals for navigation.

Several methods have been developed for eliminating or reducing the complex signal propagation effects on navigation and positioning. One method is simply to subtract the signal phases at two of the frequencies (e.g., 10.2 and 13.6 kHz) transmitted from the same station. To eliminate propagation effects, this procedure (which is similar to the method for resolving ambiguities) relies on the assumption that propagation effects at the two frequencies are completely correlated. In reality, the propagation effects at the two frequencies are only partially correlated, so the complexity of the resulting signal is lessened but not eliminated. A related, but improved method is to take an appropriate linear combination of the signal phases at the two frequencies that minimize the variation over 24 hours (diurnal variation). This technique, known as composite Omega [29], reduces the diurnal variation but does little to reduce the wide variation in phase behavior exhibited by paths of equal length over substantially different electromagnetic/geophysical environments. By far the most common method in use is the application of *propagation corrections* (PPCs) to the measured signal phase. In contrast to PPCs (which are *predicted* variations from the nominal phase based on semiempirical models of geophysical effects and the "normal" ionosphere), *real-time* corrections are provided by differential Omega systems in local operating areas.

Omega PPCs are those predicted phase values that, when applied to the received Omega signal phase measurements, provide an idealized phase function (at spatially distinct points) which depends linearly on distance. Thus, for a receiver on a moving vehicle, two successive phase measurements are proportional to the distance traveled: $\Delta\phi = k\Delta r$, where $\Delta\phi$ is the phase difference corrected by the PPC and Δr is the corresponding distance difference. The proportionality constant k is called the *wave number*. The reduction of the phase measurement to a linear function of distance can be traced back to when navigational charts were used for manual plotting of Omega lines of position (LOPs). It is much easier to plot these LOPs if the phase (difference) is linearly related to the distance (difference) from the transmitting station(s). The particular wave number used to construct the charts is known as the "nominal" wave number, which is simply the ratio of the cumulative "idealized" phase developed by a signal to the distance over which the signal is propagated.

In free space, the wave number is given by[1] $k_0 = f/c$, where f is the frequency of the signal and c is the speed of light. The nominal wave number is given by $k_{nom} = 0.9974k_0$, which is chosen as an intermediate value between observed night and day wave numbers on seawater paths. The exact value of the nominal wave number is not critical; it is only important that the value be near the average over all time and space conditions.

The Omega PPC may be thought of as the variation of the "true" Omega signal phase (ϕ) from the nominal phase (ϕ_{nom}):

$$\text{PPC} = \phi_{nom} - \phi \qquad (4.9)$$

where $\phi_{nom} = k_{nom}r$, and where r is the distance between a transmitter and a receiver over the surface of the Earth. The assumption of a fixed wave number that relates nominal phase and distance over the surface of the Earth is the basis for the so-called nominal model of Omega signal phase/distance relationships. The calculation of the PPC is based on a semiempirical model of phase variation as a function of the electromagnetic characteristics of a signal path from transmitter to receiver [22–23, 24–25].

Thus, by Equation (4.9), the PPC and the nominal model together determine the predicted phase for a given station, signal frequency, position, and time. In this relation the nominal phase is the "dominant" term in the sense that it accounts for approximately 99% of the cumulative phase from the signal source, that is, the distance between the transmitting station and the receiver in units of wavelength. Measured in cycles of nominal signal wavelength (somewhat larger than a free-space wavelength) at 10.2 kHz, the nominal phase is 100 to 500 for typical paths, whereas the PPC is usually between −3.00 and +3.00 cycles, with a resolution of 0.01 cycle (a unit referred to as a centicycle). The predicted phase has a typical diurnal variation of 0.5 to 2 cycles, amounting to about 0.2 to 2% of the nominal phase.

Figure 4.36 shows a typical diurnal observed phase profile (measured with respect to a precise time or frequency standard) in which the path illumination conditions, nominal phase, and two sample PPC values are identified. The figure illustrates the higher (retarded) phase during path night and the lower (advanced) phase in path day, with a total diurnal shift of about 0.65 cycle. Since the phase is a function of effective ionospheric height which varies with the relative sun angle (solar zenith angle), the observed phase exhibits a "bowl-shaped" profile during the day with less variation at night. The phase profile changes from day to night behavior during path transition when the sunset terminator cuts the path. The figure illustrates the time-independence of the nominal phase and the consequent time-dependence of the PPC values.

Propagation Effects   In the wave-guide model of VLF wave propagation [40], the region in which the Omega signals are confined is known as the *Earth-*

---

[1] An alternative definition of the free-space wave number used in many texts is $k_0 = 2\pi f/c$.

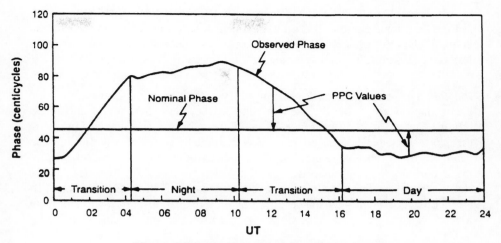

Figure 4.36   Typical diurnal phase behavior.

ionosphere (EI) wave guide. Propagation of Omega signals is mostly confined in the EI wave guide for three principal reasons:

1. The lower boundary (the Earth's surface) has a relatively high conductivity (greater than 10^{-3} mho/m over most areas of the Earth), so waves do not readily penetrate the surface.
2. The Earth's atmosphere (at altitudes between 0 and 70 km) has an extremely low concentration of charged particles and thus acts as a vacuum to VLF waves.
3. The D and E regions of the ionosphere (70 to 110 km) have low average conductivity (about 10^{-5} mho/m) but have a steep conductivity gradient between 70 and 100 km which serves to reflect VLF waves.

The above conditions also lead to low attenuation of the propagated signal; for example, over a range of about 1000 km the signal amplitude is reduced (on average) by a factor of two.

Factors effecting Omega signal propagation include the action of the Earth's magnetic field, the structure of the ionosphere, solar control and the effects of the 11-year sunspot cycle, and the presence of two or more propagation modes.

The Earth's magnetic field introduces an anisotropy into the behavior of VLF waves interacting with the ionosphere. That is to say, signal propagation depends upon the direction of propagation. This anisotropy is strongest on paths perpendicular to the geomagnetic field (east–west paths). The presence of two or more signal propagation modes with comparable amplitude will cause the phase to become a strongly oscillatory function of distance, thus rendering the signals unusable for navigation/positioning.

The ionosphere is quite sensitive to the net incident solar illumination. During the day, solar photoionization maintains a small, but stable ionized component which is not present in nighttime regions. Solar control of the ionosphere introduces a strong diurnal dependence on Omega signal propagation.

Position Determination and Accuracy   Two methods of successive position determination are currently used in Omega receivers: hyperbolic and direct ranging. In the hyperbolic mode of position determination, the difference in the phase of signals received from two distinct transmitting stations is measured. In this mode navigation equipment is assumed to know the identity of the hyperbolic lane in which it is located, either from initialization at a known starting point or from the results of continuous navigation. Since phase difference is equivalent to distance difference, the phase-difference measurement locates the receiver on a hyperbolic curve within the known lane (see Figure 4.37).

A hyperbolic lane has a width that is one-half the signal wavelength on the baseline between the two stations located at the foci of the hyperbolas. This can be seen by first noting that a lane is defined as those locations for which the phase-difference measurement $(\phi_A - \phi_B)$ with respect to two stations (A and B) varies between 0 and 1 cycle. Let the distance corresponding to this phase-difference change be Δr, and assume that the wave number k is constant. Without loss of generality, it can be assumed that the initial phase difference

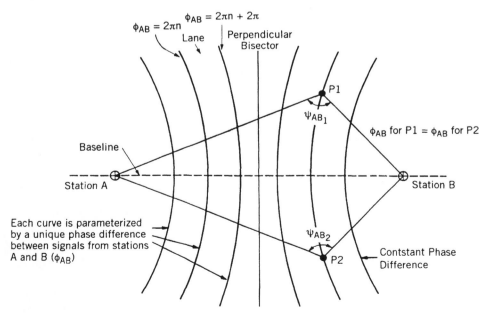

Figure 4.37   Hyperbolic geometry for phase-difference measurements.

(before moving distance Δr) is zero. However, after moving distance Δr toward the station B along the baseline, the lane phase difference is given by

$$\text{Lane phase difference} = 1 \text{ (cycle)}$$
$$= (\phi_A + k\Delta r) - (\phi_B - k\Delta r)$$
$$= \phi_A - \phi_B + 2k\Delta r$$
$$= 2k\Delta r \qquad (4.10)$$

Since the wave number is given by $k = 1/\lambda$, where λ is the signal wavelength, it follows from the above that $\Delta r = 1/(2k) = \lambda/2$. At points away from the baseline, the lane width increases with the diverging hyperbolic curves as $\lambda/(2 \sin (\psi_{AB}/2))$, where ψ_{AB} is the angle subtended by the two stations at the receiver location.

A hyperbolic lane is actually the family of all hyperbolic curves with phase differences between 0 and 1 cycle that lie within the lane boundaries. A hyperbolic curve normally has two branches, corresponding to the positive and negative values of the range difference. In the case of Omega, the sign of the phase difference is known from measurement, which limits the receiver location to a single branch. In the vicinity of the receiver (near the baseline), the hyperbolic curve resembles a planar hyperbola with foci located at the two associated transmitting stations. On scales of 1 to 10 nmi, these hyperbolic curves are well-approximated by straight lines. At points well away from the baseline, however, the spherical shape of the Earth causes the hyperbolic curve to close on itself in a quasi-elliptical shape, with one of the stations at one focus and the antipode of the other station at the other focus.

If the appropriate lane is shown for a second pair of stations (which may include a station common to the first pair), then a phase-difference measurement with respect to these two stations establishes a second hyperbolic curve, whose intersection with the first curve determines the receiver position (see Figure 4.38). It is possible that these two curves could intersect in two locations, but the correct intersection is easily resolved for one or more of the following reasons:

1. One of the two intersections is usually relatively far from the known approximate location of the receiver.

2. An independent third pair of stations (if available) provides another hyperbolic curve that passes near one of the two intersections.

3. A moving receiver shows successive fixes consistent with vehicle speed for one of the intersections and inconsistent for the other.

In cases where more than two independent hyperbolic curves are available (i.e., more than four usable Omega station signals are accessible), the multiple curves do not, in general, intersect at a single point due to the effects of noise

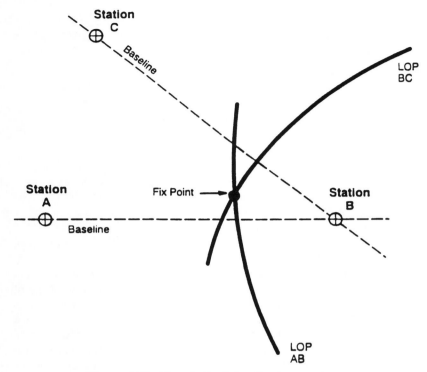

Figure 4.38   Hyperbolic fixing for three stations.

and station signal-dependent prediction errors. In these cases, a least-squares technique is often used to obtain a good estimate of the true position.

Direct ranging techniques for Omega position determination are divided into two principal types; rho-rho and rho-rho-rho. These types are differentiated because of the different receiver equipment required for each. In the ranging mode, individual phase measurements on each station signal are made so that the lane width is a full signal wavelength at all locations.

Rho-Rho Method   The rho-rho technique requires only two range measurements for a fix. As in the hyperbolic case, it is assumed that the correct lane is initially known and successive measurements are processed over small enough distance/time intervals so that lane changes are readily tracked. To obtain an accurate estimate of the distance traveled based on successive phase measurements, the processor must have access to a frequency/time reference (clock) of sufficient stability so that the reference is effectively synchronized to the Omega station during the period between precision updates.

The change in station-receiver distance, obtained from two successive phase measurements of the station signal, places the receiver's new position on a circular curve (centered on the station location) within the appropriate lane. Since the receiver's previous position is assumed known, some points on the new circle are more likely candidates than others for the new position, based on platform velocity and maneuvering limits. However, the new position is accurately determined only when a distance change to a second station is obtained from successive phase measurements. In this case, a second circle is established that intersects the first at the new receiver location. Although two intersections are possible, the correct intersection can be resolved using methods listed above for the hyperbolic case. This method corresponds to a system of two equations and two unknowns. In cases in which more than two usable station signals are available, a least-squares technique can be used to resolve the multiple intersections that arise as a result of measurement noise or phase prediction error.

Rho-Rho-Rho Method   The rho-rho-rho type of ranging is similar in principle to the rho-rho method except that it requires less precision in the onboard clock, essentially permitting oscillator (clock) self-calibration. In particular, the method assumes that the clock has a fixed *frequency offset* from its "correct" value during the period between precision updates. A fixed offset in frequency implies that the phase reference supplied by the clock drifts linearly in time away from the correct phase reference with a slope $\Delta\phi/\Delta t$ proportional to the frequency offset. The unknown slope of this clock drift introduces a third variable into the problem of determining the two components (e.g., north and east) of the incremental position change. For this reason, three phase change measurements (corresponding to three independent equations) are required during successive updates. Though nearly as accurate as rho-rho ranging, rho-rho-rho has the disadvantage that usable signals from three stations are required, instead of two. When more than three usable station signals are available, the redundant information is best handled by a using a least-squares method.

Position Accuracy   Total Omega position error can be traced to a variety of sources, including station synchronization offset, receiver processing (e.g., lane slip/jump), operator mistakes (e.g., initializing with coordinate insertion error), and temporal anomalies (e.g., a polar cap disturbance (PCD)). The predominant error source, however, is the PPC.

The PPCs are obtained from a semiempirical model of Omega signal phase behavior which is calibrated largely from phase measurements at globally distributed fixed Omega monitor sites. Analysis of these measurements reveals important features of Omega phase behavior as well as provide insights into Omega PPC error. A basic property indicated by these measurements is that, at a fixed site, Omega phase (and phase error) generally varies more over 24 consecutive hours (see Figure 4.36) than over a year at a fixed hour. Because the observed phase measurements show little systematic change over a month at a fixed hour, the average observed phase over 15 to 30 consecutive days is a robust aggregate measure of the phase for a given hour and specific month. The predicted phase (obtained from the PPCs) over the same time period is nearly constant but often differs significantly from the average observed phase. This difference is referred to as the PPC bias error which varies in magnitude from 0 to 30 centicycles (cecs). Also occurring in this 15–30 day period at a fixed hour are random (nonsystematic) day-to-day variations in the observed phase on the order of 1 to 5 cecs. Since these random variations (which are due to ionospheric fluctuations) are not reflected in the PPCs, they make up the random component of PPC error.

When converting phase measurements to position, the bias and random components of phase error produce corresponding bias and random components of position error. The transformation of the phase error components to position error components depends upon the individual phase errors of all signals received and the geometrical configuration of the receiver and stations corresponding to the received signals. If the magnitude of the random phase errors is assumed to be the same for all signals received and the bias error is assumed to be zero, then the radial position error standard deviation (σ_r) can be obtained by multiplying the phase error standard deviation (σ_ϕ) by a scalar factor known as the geometric dilution of precision (GDOP). For a least-squares method of position determination, used when multiple redundant signals are present, the following form[2] of GDOP [24, App. B] is obtained:

$$
\text{GDOP} = \frac{1}{2}
\left[
\frac{\displaystyle\sum_{i=1}^{q-1} \sum_{j=i+1}^{q} \sin^2\left((\beta_i - \beta_j)/2\right)}
{\displaystyle\sum_{i=1}^{q-2} \sum_{j=i+1}^{q-1} \sum_{k=j+1}^{q} \sin^2\left((\beta_k - \beta_j)/2\right) \sin^2\left((\beta_i - \beta_k)/2\right) \sin^2\left((\beta_j - \beta_i)/2\right)}
\right]^{1/2}
$$

$$(4.11)$$

where q is the number of usable signals received and β_i is the bearing to the ith station (corresponding to the ith usable signal). The GDOP becomes very large whenever at least $q - 1$ stations have bearings which are nearly equal. Another property of the GDOP is that the GDOP for q station signals is never greater than the GDOP for any subset (> 3 stations) of q. This means that for least-squares position processing, the use of additional (usable) signals does not degrade, and typically improves, the resulting position accuracy.

For moving vehicles performing navigation, the bias error is effectively removed at initialization, leaving only phase error due to noise (typically less than 1 cec). However, the paths from the station to the receiver eventually change (both in space and time) enough so as to become decorrelated with the original configuration of station signal paths to the receiver and the initial

---

[2]GDOP = $\sigma_r/(\lambda\sigma_\phi)$ where λ = signal wavelength.

correction no longer applies. From this decorrelation time until the next precision update, the Omega receiver is subject to PPC bias and random errors and the effect of GDOP. Omega-only accuracies have been reported for aircraft of 2.7 to 3.3 nmi 95% of the time [32, 33].

Differential Omega Like the differential techniques associated with many other radio-navigation systems, differential Omega systems provide a way of enhancing the position accuracy in a local region through the transmission of local corrections. The corrections are obtained from a central monitoring facility that compares observed signal phase readings from each of the Omega stations with the "correct" phase (using a nominal model) based on the location of the monitor and the signal frequency. The accuracy of the correction depends on the spatial correlation of the Omega signal phase between the position of the monitor and the user's position. Within a radius of about 50 km from the monitor, the correlation peak is within about 1 cec (for typical time constant receivers); for greater distances, the degree of correlation gradually degrades.

Operational differential Omega systems in place in 1993 were tailored primarily to marine users, although a number of experimental differential systems for aircraft have been tested. The correction information for marine use is normally broadcast to all users in the local area (having a typical radius of 200 to 500 nm) using a 20-Hz modulation of LF beacon signals with frequencies between 285 and 415 kHz. Measured position errors vary from 0.3 nm (100 nm from the monitor station) to about 1 nm (500 nm from the monitor station) 95% of the time [26]. As of 1996, 30 differential Omega systems were in operation throughout the world, including the Atlantic coasts of Europe and Africa, the Mediterranean Sea, the Caribbean, Eastern Canada, India, and Indonesia [6].

Omega/VLF Operation The most common external radio navigation source integrated with Omega in aircraft receiver systems arise from the network of VLF communication stations. Unlike the Omega stations, the VLF communication stations are not synchronized, so only phase changes from each station can be processed in a navigation mode. This means that VLF signal processing is used to supplement Omega navigation rather than act as a substitute. Moreover, these communication signals are broadcast for national/international security purposes, so stations can switch frequency, change modulation, or temporarily cease operation with no advance warning. Thus, although VLF signals serve a very useful supplementary function in many airborne modern Omega receivers, they do not play a primary navigational role, because the VLF communication signals are not intended for navigation.

One important feature of Omega/VLF receivers is the difference in the algorithms for processing of Omega and VLF signals. Some of these distinctions arise from inherent differences in the two transmitting systems. For example, since the stations in the VLF network are not synchronized (although the carrier signals are synthesized from precise standards), no receiver acquisition of a time-frequency pattern is required as for Omega signals. This also means that signal phase from different stations cannot be compared (in an absolute sense) to determine position. Because the received VLF signal is generally stable in time, VLF navigation requires an initial calibration to permit subsequent phase tracking of the signals from selected VLF stations. Accurate phase tracking requires an on-board precise frequency standard or a correction based on an estimate of the frequency/time offset of the receiver's internal clock. This estimate is usually obtained from Omega signal processing in the rho-rho-rho mode.

In addition to internal differences in signal processing, signals from the two systems are processed differently regarding external information. For example, all known Omega/VLF receivers use externally supplied PPCs to correct the measured Omega phase prior to navigation use, whereas few, if any, currently operational receivers correct VLF signal phase measurements. This means that, for most receivers, the received VLF signal phase is not accurately related to distance over the Earth's surface, a problem that is not necessarily ameliorated by redundant measurements. External deselection data regarding modal and long-path signals are available for Omega but not for VLF. Failure to deselect VLF modal signals is potentially a more serious problem for navigation than the lack of VLF PPCs, since modal phase excursions can be large and sudden, often resulting in cycle slips or advances.

As a result of the signal-processing differences, due to the internal and external information bases, receiver-processing algorithms treat Omega and VLF signals differently. Once acquired and initialized, Omega signal-processing alone is robust and will fail only under unusual circumstances (e.g., cycle shifts or fewer than three signals above the minimum S/N). VLF signal-processing schemes generally rely on the presence of Omega signals and other aids in the receiver's navigation filter. In most receivers VLF signals are closely monitored with frequent cross-consistency checks. Normally, Omega/VLF receivers are programmed to exclude initialization with VLF signals alone, since this represents a "degraded" mode. Current FAA certification procedures require than an Omega/VLF receiver system operate satisfactorily with Omega signals alone.

Transmitting Station Characteristics   The transmitting equipment at each station is generally described as belonging to one of three functional groups, or subsystems:

1. Timing and control subsystem
2. Transmitter subsystem
3. Antenna-tuning subsystem

The principal functions of the timing and control subsystem are signal generation and phase control. The signal source is a precision cesium beam frequency standard of 9.193 GHz with a stability of 5 parts in 10^{12}. Three cesium standards are used for frequency drift comparison and control, and are maintained as reserves in the event of failure of the on-line standard. Phase control is maintained by comparing the RF signal phase to the phase of the antenna current reference signal fed back from the antenna tuning subsystem. The signal phase is advanced or retarded to insure that its phasing at the antenna coincides with the appropriate UTC epoch.

The transmitter subsystem consists of those devices that amplify the signal generated in the timing and control subsystem. The RF signal from the timing and control subsystem is first raised to a level of 160-V RMS by the input amplifier. The driver amplifier further raises the signal level to a nominal 520-v RMS and the final amplification is performed by the power amplifier that boosts the signal voltage and current to a peak power of 150 kw. Following this final amplification stage, the signal is fed to the antenna tuning subsystem.

The antenna-tuning subsystem is designed to tune the antenna at the RF signal frequency by impedance matching the antenna to the input circuit. This ensures the maximum effective radiated power at the antenna for a given input signal power. Based on the long keying pulses from the timing and control subsystem and the current samples received from the current transformer, the antenna-tuning control first implements fine inductive tuning through the variometers. The antenna-tuning control signal activates a mechanical drive that moves the variometer coil to the appropriate position for matching impedance. The long keying pulses activate antenna relays that connect the appropriate variometer into the main antenna circuit. The RF signal is then transferred to the "helix," a large helical coil that acts as a coarse tuning device for the antenna. The helix is equipped with separate taps for each signal frequency transmitted. Finally, the RF signal is conducted to the antenna structure itself from which the signal is radiated. The structural feature which principally differentiates Omega stations is the antenna structure. Two basic designs are utilized: tower and valley span. The tower antennas are further classified as either the grounded or insulated type.

Receiver Characteristics   Narrow-band Omega signals and noise from all sources (including harmonic interference) are received at the E-field (probe) or H-field (loop) antenna having a bandwidth of about 4 kHz. The electromagnetic energy is passed to the detector stage of the Omega receiver for conditioning. Filtering is performed at the front end of the detector, reducing the bandwidth to about 100 Hz. The signal is also amplitude limited at this stage to prevent swamping from large impulsive noise spikes. In successive stages of the detection process, the signal is either processed at its original frequency (tuned RF) or mixed with a reference signal to produce a lower-frequency signal (heterodyning). In either case, the signal is compared with a reference signal having sufficient stability over one or more 10-sec Omega format periods.

In most modern Omega receivers, the signal is acquired and tracked by means of digital techniques. With these techniques, phase is usually measured as the interval between a reference clock pulse and the next zero crossing of the input signal in units of clock cycles.[3] A digital phase tracking technique used in many Omega receivers is the phase lock loop in which the reference phase is shifted by an amount that depends on the previous phase measurement and the time-averaged phase computed at the previous measurement time. The time-averaging refers to a moving average that differs from the average at the previous loop cycle by a weighted value of the previous measurement. A second-order phase lock loop is designed to track the time rate of change of phase in a manner similar to that of the first-order loop.

Like most signal-tracking circuitry, the basic function of the phase lock loop is to reduce the effective bandwidth (inversely proportional to the effective time constant) so as to best reproduce the desired signal. For aircraft receivers, time constants typically range from 100 to 200 sec. Shorter time constants do not provide sufficient averaging or noise rejection, and longer time constants may exceed the time required for aircraft maneuvers, such as sharp turns. Since the duty cycle for each of the common frequency Omega signals is 10%, the *effective* phase measurement time constant is 10 to 20 sec. Using standard assumptions [24], these time constants correspond to noise equivalent bandwidths of 0.025 to 0.013 Hz. When compared to the input bandwidth of 100 Hz, these narrow output bandwidths correspond to gains of better than 35 dB. Thus, signals with S/N as low as -20 to -30 dB in the 100 Hz receiver input bandwidth can be effectively utilized in aircraft Omega receivers.

After the signal phase measurements are made, PPCs are computed using an appropriate model/algorithm and added to the measured phase to produce an "idealized" phase value that can be readily used in the subsequent positioning calculations. Although the PPCs require receiver position as an input, the PPCs are not sensitive to precise position since they vary less than 0.05 cycle over ranges of 50 to 100 km. Thus, the PPCs can be accurately computed from only approximate knowledge of position.

Before determining position, the (idealized) signal phases are usually weighted based on the expected relative accuracy of the phase measurement. This accuracy is most commonly determined by the estimated S/N, which, for phase lock loop receivers, is closely related to the rms loop error. If the estimated S/N is below a preset threshold, such as -30 dB in a 100-Hz bandwidth, the signal phase is usually excluded (given zero weight) in the position solution. In addition to these weighting and exclusion procedures based on internally derived S/N data, the signal phases are edited by invoking external information concerning the signals. External information usually refers to signal deselection data that are generally extracted from known coverage information, including modal "maps" and data on the occurrence of long-path signals. The resulting signals that are not deselected or excluded are further screened for acceptable geometry. In some receiver mechanizations, all common frequency signals from a station must be acceptable to be used in the position fix; in others, only a sin-

---

[3]The reference clock/oscillator commonly has a frequency of 1 to 5 MHz but may be converted to a lower frequency.

gle acceptable signal frequency from a station is necessary for inclusion in the fix algorithm. Position change estimates are then formed from the weighted and edited Omega signal phase data at the common Omega signal frequencies. The estimates are computed by means of a least-squares or Kalman estimation technique (see Chapter 3). The Omega-based calculation of position change is frequently combined with the aircraft-supplied true airspeed and heading or inertial system information to furnish the best position estimate.

An airborne Omega receiver block diagram is shown in Figure 4.39; a photograph of an Omega/VLF receiver for commercial aviation applications is depicted in Figure 4.40.

4.5.3   Decca

Hyperbolic systems other than Loran and Omega exist and are used for navigation. One such example is the Decca system [31] developed by the British and used extensively during the later stages of World War II. In 1996, its major area of implementation is in northwestern Europe where it is primarily used by shipping companies.

Decca is based on the measurement of differential arrival times (at the vehicular receiver) of transmissions from two or more synchronized stations (typically 70 mi apart). As an illustration, consider two stations (A and B) 10 mi apart and each radiating synchronized radio-frequency carriers of 100 kHz. Assume that there is some way by which each station can be identified. The wavelength at this frequency is 3000 meters, or about 2 mi. On a line between the stations, the movement of a vehicle D one mile toward one station and one mile away from the other station will cause the vehicle to traverse one cycle of differential radio-frequency phase. There will, therefore, be 10 places along the line AB where the signals from the two stations will be in phase. As the vehicle moves laterally away from this line, isophase LOPs can be formed (each line being a hyperbola) with the stations as foci and $BD - AD$ as a constant for each LOP.

Site error virtually vanishes in such a system, and the accuracy depends entirely on the constancy of propagation between the stations and the vehicle. In an effort to avoid line-of-sight limitations, Decca uses a low frequency (70 to 130 kHz), which is subject to sky-wave contamination, and uses continuous waves, which preclude the separation of ground waves from sky waves. Thus, despite the low frequency (whose ground-wave range is on the order of 1000 mi), practical Decca coverage is limited to areas where sky-wave strength does not exceed about 50% of ground-wave strength. This is typically 200 mi.

A typical Decca chain consists of a master station and three slave stations. A typical station has a 2-kw crystal-controlled transmitter feeding a 300-ft antenna. The slave stations are referred to by the color of the phase meter associated with each at the receiver. Each station transmits a stable continuous wave frequency that bears a fixed relationship to the frequencies of the other three stations. Phase comparison therefore produces a family of hyperbolic LOPs of

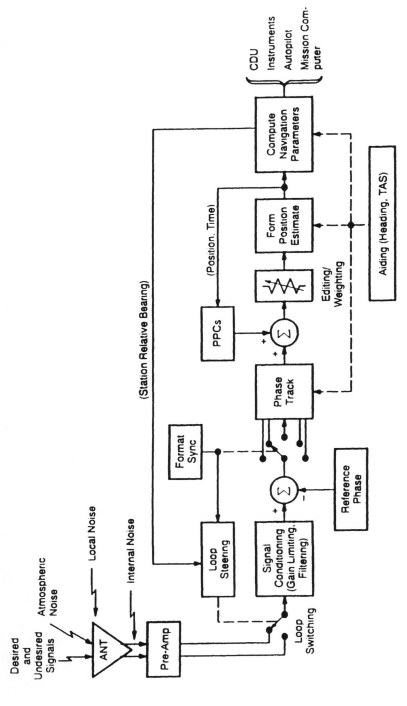

Figure 4.39　Airborne Omega receiver block diagram.

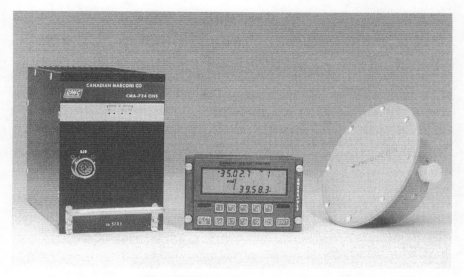

Figure 4.40 Typical airborne Omega/VLF receiver.

constant phase. The spaces between these lines are called *lanes*. The intersection of two LOPs provide a position fix.

4.5.4 Chayka

Chayka (meaning "sea gull") is a pulse-phase radio-navigation system similar to the Loran-C system. It is used in Russia and surrounding territories and seas. By using ground waves at low frequencies, the operating range is over 1000 mi; by using pulse techniques, sky-wave contamination can be avoided. The system is designed to provide both a means of determining an accurate user position and a source of high-accuracy time signals. The system can support an unlimited number of users since the computations are performed at the user receiver and position determination is possible at any time of day or year, regardless of meteorological conditions.

Principles and System Configuration   Chayka is analogous to the Loran-C system employed in the United States and other parts of the world (see Section 4.5.1). Chayka is a low-frequency pulse-phase radio-navigation system with a pulse-modulated frequency of 100 kHz. Contrary to the multiple chains of Loran-C, the Chayka system consists of only two "networks" of stations, each with a master and four slave stations. Receivers measure the time difference between the arrivals of a given wave form from the master and any particular slave station. This time-difference information can then be converted into position, velocity, and time and frequency reference information. Additional processing can produce bearing, distance, and along-track and cross-track errors.

Each of the stations in the Chayka networks transmit pulses with standard characteristics. The pulse consists of a 100-kHz carrier wave that increases from zero to a maximum and then decays at a specific rate to form the envelope of the signal. All slave stations transmit signals in packets of eight pulses; the masters emit a ninth pulse for identification. The interval between pulse onsets is 1.0 ± 0.05 μsec. In addition, in order to provide the possibility for automatic detection and identification of signals and to reduce the influence of multiply reflected signals, the signals are phase-coded with the slaves all having the same phase (i.e., $0°$ or $180°$) and the master phase differing by exactly $180°$.

The repetition periods of the radio signals are selected based on a trade-off: maximizing the average power of the signals at the receivers while preventing any signal overlap within the network operating region. Since all slave stations transmit signals with identical phase codes, each slave station transmits with its own specific code delay relative to the master signal. The magnitude of the code delays are selected such that the order of reception of slave station signals is identical everywhere within the network operating region.

Wave Form and Signals in Space Each station transmits signals with standard pulse modulation characteristics. Each pulse consists of a 100-kHz carrier wave modulated by an envelope that depends on the specific transmitting station equipment. Two types of radio transmitter (RT) stations are currently in use: those with vacuum tubes and those using impact excitation of the linear output circuit. The envelope of the vacuum tube RTs can be approximated by

$$U(t) = U_m \left[\frac{t}{t_m} e^{(1 - (t/t_m))} \right]^2 \qquad (4.12)$$

where U_m is the pulse amplitude and t_m is the time interval from the onset to the peak of the pulse.

The envelope of the impact excitation RTs can be approximated by

$$U(t) = U_{\max} \left[\frac{\sin \beta t}{\beta} e^{-\alpha t} \right]^m \qquad (4.13)$$

where $m = 1$ or 2 for two or three coupled circuits, respectively, including the circuit of the transmitting antenna, and α and β are approximation parameters chosen to provide a given steepness of growth and speed of attenuation of the radio wave envelope. Existing impact excitation RTs use linear circuits consisting of three coupled circuits (i.e., $m = 2$).

Coverage In 1996 the Chayka system consisted of only two networks, European and Eastern. Coverage from the European network is centered near Moscow and includes most of the area between $5°$ and $50°$ East longitude and

40° and 65° North latitude (e.g., Eastern Europe, western portion of the former Soviet Union, the Black Sea, and part of the Caspian Sea). Coverage from the Eastern network includes most of the area between 135° and 160° East longitude, and 35° and 65° North latitude (e.g., eastern shoreline of the former USSR, portions of Japan, and the surrounding areas of the Pacific Ocean).

4.6 FUTURE TRENDS

Terrestrial radio-navigation systems will continue to play a major role for aircraft navigation throughout the world for many years. Since the U.S. satellite-based GPS had achieved full operational capability (FOC) in 1995, followed in 1996 by the Russian GLONASS, there had been expectations that these satellite systems would quickly replace the terrestrial systems such as VOR, VOR/DME, Loran, and Omega. However, this was not the case and is not likely to occur in the near future. The reasons for this include (1) the widespread implementation of equipment by aircraft owners and the cost of replacement by satellite receivers, (2) the lack of available air-traffic management operational procedures compatible with satellite-based systems, (3) the absence of full sole-means navigation system status of GPS, and (4) the fact that issues involving system accuracy, integrity, availability, and continuity of service of the satellite systems had not been fully resolved (Chapters 5, 13, and 14). Therefore, the terrestrial radio-navigation systems will continue to be used for many years on a global basis. In the more distant future, some of these systems will be decommissioned when their utility will have been fully replaced by that of the satellite systems.

By 1993, the U.S. Coast Guard had implemented full coverage of the continental United States by Loran-C chains and the FAA had authorized Loran-C for supplemental navigation for en-route and nonprecision approaches. At least ten U.S. airports had received approval for Loran-C approaches. As a result, there was extensive use of airborne Loran-C receivers on U.S. General Aviation aircraft and that usage is likely to continue for some time until GPS receivers are widely implemented on General Aviation aircraft.

Since 1990, a number of major studies have been conducted and published that show the advantages and discuss techniques of combining data from Loran and GPS for aircraft navigation [45, 46, 47]. Among the major advantages are the mitigation of the effects of GPS coverage outages caused by satellite shutdowns or poor geometry and, conversely, that of Loran coverage outages due to ground station shutdowns, high atmospheric noise levels, or precipitation static. In addition Loran data could provide on-board fault detection and isolation of GPS satellites, in connection with GPS Receiver Autonomous Integrity Monitoring (RAIM, Section 5.7.2). The combining of GPS and Loran data (e.g., with a Kalman filter) can be at the pseudorange level and mutual time synchronization can also be included [46]. Therefore, research and development on the integration of Loran and GPS is likely to continue in the future.

In 1995, a considerable number of countries had indicated that they would continue or increase use of Loran-C. These included countries in Northern Europe, the Mediterranean and the Far East, including Russia, France, Saudi Arabia, the People's Republic of China, and India. Also, Russia promoted the use of their Chayka and their VLF Radio-Navigation System (called Alpha) for civil use.

Cooperative efforts between the United States and Russia resulted in implementing a Loran-C/Chayka chain to provide aviation and marine coverage over the five-hundred-mile coverage gap that existed in the North Pacific, between the North Pacific Loran chain and the Northwest Pacific and Eastern Russian chains [41]. These cooperative efforts are likely to continue in the future.

Several options were being considered to integrate the Russian Alpha and the Omega system [28], and this effort may also continue for some time. However, direct participation by the United States in Omega may terminate in the not too distant future, possibly as early as September 1997.

PROBLEMS

4.1. A 1000-MHz DME transponder on the ground is triggered by a signal 10 dB above its receiver-noise level. The receiver-noise figure is 10 dB. What transmitter power is needed on an aircraft to produce triggering from a distance of 100 nm? Assume simple dipoles at each end of the link, no transmission-line losses, a transponder-receiver bandwidth of 1 MHz, and a temperature of 293° Kelvin.

Ans.: 8 w.

4.2. What is the principal advantage of an omnidirectional range on the ground versus a direction-finder in an aircraft?

4.3. What are some techniques used to reduce site errors in directional-antenna systems?

4.4. What magnetic variation must be used to plot the true course of a VOR radial?

   (a) Variation at the aircraft location

   (b) Variation shown on the chart for the VOR station location

   (c) Variation used to calibrate the VOR station by maintenance personnel.

Ans.: c.

4.5. How does magnetic variation affect the accuracy of a Loran-C position?

Ans.: It has no effect.

4.6. What variation should be used to plot an initial course from an aircraft's Loran-C position to an airport 200 mi away.

Ans.: The magnetic variation shown on the chart for the aircraft's position.

4.7. What are the factors that impact the accuracy of a Loran-C fix?

> *Ans.: Signal-to-noise ratio at the receiver, crossing angle of the Loran-C lines of position, calibration of the Loran-C time difference to latitude/longitude coordinate converter.*

4.8. What are the two categories of Loran-C system accuracy? What do they mean?

> *Ans.: Repeatable accuracy to which one can return to a point visited before; absolute or predictable accuracy of the fix against some external reference grid such as latitude and longitude.*

4.9. Suppose that an Omega receiver processes four 10.2 kHz signals from stations with geographic bearing angles (at the receiver) of 31°, 121°, 211°, and 301° using a least-squares algorithm to estimate position change.

(a) What is the GDOP?

> *Ans.:* $3/2\sqrt{2}$.

(b) If the signal phase error associated with each of the four stations is 4 cecs, what is the corresponding position error in kilometers.

> *Ans.:* $\sigma_r = 1.25$ Km.

(c) If one of the station signals becomes unusable (e.g., due to modal interference as the path becomes dark), by what factor is the position accuracy degraded?

> *Ans.:* σ_r becomes larger by a factor of 4/3.

5 Satellite Radio Navigation

5.1 INTRODUCTION

Since the 1960s, the use of satellites was established as an important means of navigation on Earth. The earliest systems were designed primarily for position updates of ships, but were also found useful for the navigation of land vehicles. Beginning in the early 1970s, satellite-navigation systems for aircraft (as well as other platforms) were under intense development. Those efforts benefited from the techniques used and the experience gained with the earlier systems. In the 1980s, systems suitable for aircraft became mature and by 1996 their use for aircraft navigation was increasing at a widespread and rapid pace.

The satellite-navigation systems described in this chapter are comprised of a system of satellites that transmit radio signals. Appropriately equipped aircraft receiving these transmitted signals can derive their three-dimensional position and velocity and time. Two systems are described in detail, namely the U.S. Department of Defense's NAVSTAR Global Positioning System (GPS) and the Russian Federation's Global Orbiting Navigation Satellite System (GLONASS). The International Civil Aviation Organization (ICAO) and RTCA, Inc. have defined a more global system that includes these two systems, geostationary overlay satellites, along with any future satellite navigation systems, in what has been named the Global Navigation Satellite System (GNSS) [1, 2]. A third major system, the United States Navy's Transit System, also called the Navy Navigation Satellite System (NNSS), is a low-altitude Doppler satellite radio navigation system. In Russia, a similar system was developed, called Tsikada. Since GPS was fully operational, after 32 years the U.S. Navy ceased operations of Transit on December 31, 1996 [124]. It will not be discussed here further. (Design details are given in references [3] and [4].)

The systems described in this chapter provide users with a passive means of navigation; that is, there is no requirement for their equipment to transmit, only to receive. Both GPS and GLONASS are ranging systems. They provide both range and range rate (or change in range) measurements. Once initialized, they provide an instantaneous and continuous navigation solution in a dynamic environment. Details of these solutions are described later in this chapter.

The advantage of satellite navigation systems is that they provide an accurate all-weather worldwide navigation capability. The major disadvantages are that they can be vulnerable to intentional or unintentional interference and temporary unavailability due to signal masking or lack of visibility coverage. In some critical applications, external augmentation is required. Various means of augmentation are also described later in this chapter.

5.1.1 System Configuration

The overall system configuration for these two systems are in common. They all consist of three system segments—a space segment, a control segment, and a user segment (Figure 5.1). The space segment is comprised of the satellite constellation made up of multiple satellites. The satellites provide the basic navigation frame of reference and transmit the radio signals from which the user can collect measurements required for his navigation solution. Knowledge of the satellites' position and time history (ephemeris and time) is also required for the user's solutions. The satellites also transmit that information via data modulation of the signals.

The control segment consists of three major elements:

1. Monitor stations that track the satellites' transmitted signals and collect measurements similar to those that the users collect for their navigation.
2. A master control station that uses these measurements to determine and predict the satellites' ephemeris and time history and subsequently to upload parameters that the satellites modulate on the transmitted signals.

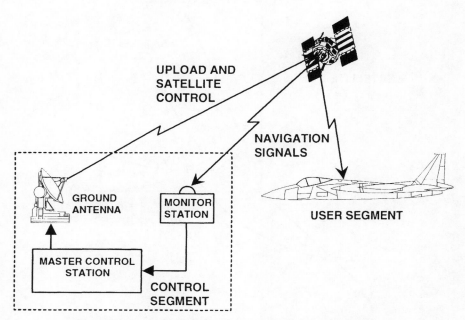

Figure 5.1 Satellite radio-navigation system configuration.

3. Ground antennas that perform the upload and general control of the satellites.

A secondary, but very important, purpose of these elements is to monitor the satellites' health and to control their operation. These elements are connected via appropriate communication links, which themselves may be via communication satellites. Normally, orbit injection of the satellites is controlled via an independent system.

The user segment is comprised of the receiving equipment and processors that perform the navigation solution. These equipments come in a variety of forms and functions, depending upon the navigation application. Details of the user equipment appear later in this chapter.

5.2 THE BASICS OF SATELLITE RADIO NAVIGATION

The concept of satellite radio navigation is illustrated in Figure 5.2. Although types of user equipment may differ, they all solve a basic set of equations for their solutions, using the ranging and/or range rate (or change in range) measurements as inputs to a least-squares, a sequential least-squares, or a Kalman

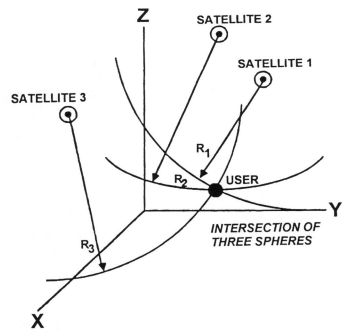

Figure 5.2   Ranging satellite radio-navigation solution.

filter algorithm. The measurements are *not* range and range rate (or change in range), but quantities described as pseudorange and pseudorange rate (or change in pseudorange). This is because they consist of errors, dominated by timing errors, that are part of the solution. For example, if only ranging type measurements are made, the actual measurement is of the form

$$PR_i = R_i + c\Delta t_{si} - c\Delta t_u + \epsilon_{PR_i} \qquad (5.1)$$

where PR_i is the measured pseudorange from satellite i, R_i is the geometric range to that satellite, Δt_{si} is the clock error in satellite i, Δt_u is the user's clock error, c is the speed of light and ϵ_{PR_i} is the sum of various correctable or uncorrectable measurement errors. These errors and/or corrections are comprised of atmospheric delays, the Earth's rotation correction, and multipath and receiver noise. These errors will be discussed in more detail later in this chapter. Equations for pseudorange rate or change in pseudorange are based on Equation 5.1 by differentiating or differencing.

Neglecting for the moment the clock and other measurement errors, the range to satellite i is given as

$$R_i = \sqrt{(X_{si} - X_u)^2 + (Y_{si} - Y_u)^2 + (Z_{si} - Z_u)^2} \qquad (5.2)$$

where X_{si}, Y_{si}, and Z_{si} are the Earth-centered, Earth-fixed (ECEF) position components of the satellite at the time of transmission and X_u, Y_u, and Z_u are the ECEF user position components at that time. For the three satellites in Figure 5.2, Equation 5.2 represents the equations for spheres whose centers are located at the satellites. The user position is the reasonable intersection of the three spheres. (There is another solution, but not near the Earth.)

5.2.1   Ranging Equations

Equation 5.2 is obviously nonlinear. The standard solution is a linearization of that equation, resulting in

$$\delta R_i = R_{im} - \hat{R}_i$$

$$= \left.\frac{\partial R_i}{\partial X_u}\right|_{\overline{X}_u = \hat{\overline{X}}_u} (X_u - \hat{X}_u) + \left.\frac{\partial R_i}{\partial Y_u}\right|_{\overline{X}_u = \hat{\overline{X}}_u} (Y_u - \hat{Y}_u) + \left.\frac{\partial R_i}{\partial Z_u}\right|_{\overline{X}_u = \hat{\overline{X}}_u} (Z_u - \hat{Z}_u) =$$

$$= \frac{X_{si} - \hat{X}_u}{\hat{R}_i} \, \delta X_u + \frac{Y_{si} - \hat{Y}_u}{\hat{R}_i} \, \delta Y_u + \frac{Z_{si} - \hat{Z}_u}{\hat{R}_i} \, \delta Z_u$$

$$= 1_{xi} \cdot \delta X_u + 1_{yi} \cdot \delta Y_u + 1_{zi} \cdot \delta Z_u$$

$$= \overline{1}_i \cdot \delta \overline{\mathbf{X}}_u \tag{5.3}$$

where δR_i is the range measurement residual, R_{im} is the range measurement to satellite i, $\hat{R}_i$ is the estimated range to satellite i, 1_{xi}, 1_{yi}, and 1_{zi} are the components (directional cosines) of the estimated line-of-site (LOS) unit vector $\overline{1}_i$ between the user and satellite i and δX_u, δY_u, and δZ_u are the components of the vector $\delta \overline{\mathbf{X}}_u$ of differences between the position solution $\overline{\mathbf{X}}_u$ and the estimated position $\overline{\mathbf{X}}_u$. This vector represents an offset from the intersection of the three spheres. Figure 5.3 illustrates this linearization in two dimensions, where the inner circles represent the measured ranges and the outer circles represent the "computed" ranges based on the estimated position. The shaded areas represent the range measurement residuals.

Solving three equations representing range measurement residuals from three satellites give a solution for the position correction vector, provided that the geometry is sufficient (i.e., the solution exists). If the differences are large, as in the exaggerated example of Figure 5.3, so that they exceed the range of the linearization, an iterative solution is generally required. This can be accomplished by using either the same set of measurements or subsequent measurements,

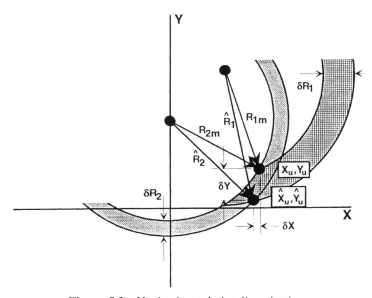

Figure 5.3 Navigation solution linearization.

where the user position in the computation of new LOS vectors is propagated from the previous solution.

5.2.2   Range-Rate (Change-in-Range) Equations

In a ranging system, the Doppler (range-rate) measurement can be used as a measure of user velocity. The range-rate measurements can be derived from Equation 5.1 in three ways: by differentiating Equation 5.2 and linearizing, or by differentiating or differencing Equation 5.3. Differentiating Equation 5.2 yields

$$\dot{R}_i = \frac{\overline{\mathbf{X}}_i - \overline{\mathbf{X}}_u}{R_i} \cdot [\overline{\mathbf{V}}_i - \overline{\mathbf{V}}_u]$$
$$= \overline{\mathbf{1}}_i \cdot [\overline{\mathbf{V}}_i - \overline{\mathbf{V}}_u] \tag{5.4}$$

where

$\overline{\mathbf{V}}_i = [\dot{X}_i \quad \dot{Y}_i \quad \dot{Z}_i]^{\mathrm{T}}$ is the known satellite i velocity vector
$\overline{\mathbf{V}}_u = [\dot{X}_u \quad \dot{Y}_u \quad \dot{Z}_u]^{\mathrm{T}}$ is the unknown user velocity vector
$\overline{\mathbf{X}}_i = [X_i \quad Y_i \quad Z_i]^{\mathrm{T}}$ is the known satellite i position vector
$\overline{\mathbf{X}}_u = [X_u \quad Y_u \quad Z_u]^{\mathrm{T}}$ is the unknown user's position vector

Note that Equation 5.4 is also nonlinear because the LOS vector is also a function of the user's unknown position. However, linearizing about an estimate of position and velocity yields

$$\delta\dot{R}_i = \overline{\mathbf{1}}_i \cdot \delta\overline{\mathbf{V}}_u \tag{5.5}$$

where $\delta\overline{\mathbf{V}}_u$ is the perturbation of user velocity.

Differentiating Equation 5.3 yields

$$\delta\dot{R}_i = \overline{\mathbf{1}}_i \cdot \delta\dot{\overline{\mathbf{X}}}_u + \dot{\overline{\mathbf{1}}}_i \cdot \delta\overline{\mathbf{X}}_u$$
$$\approx \overline{\mathbf{1}}_i \cdot \delta\overline{\mathbf{V}}_u \tag{5.6}$$

The second term of the equation can be neglected under normal circumstances because the rate of change of the LOS vector $\overline{\mathbf{1}}_i$ is small, which is due to the large distance to the satellite.

In some precision landing applications, the measurements may be Doppler count measurements. Then, change in range can be computed by differencing Equation 5.3 over a time interval $[t_{j-1}, t_j]$, resulting in the equation for a change in range as

$$\Delta \delta R_{ij} = \delta R_{ij} - \delta R_{i,j-1} = \overline{\mathbf{1}}_{i,j} \cdot \delta \overline{\mathbf{X}}_{uj} - \overline{\mathbf{1}}_{i,j-1} \cdot \delta \overline{\mathbf{X}}_{u,j-1} \qquad (5.7)$$

where the subscript j indicates measurements taken at time t_j. Note that over short time intervals, this change in range measurement can be used to estimate range rate by dividing by $t_j - t_{j-1}$.

5.2.3 Clock Errors

Equations 5.2 through 5.6 are based on the assumption that epoch times in the satellites and the user equipment are known. The epoch times in the satellites are generally known to the users to the required accuracy, because the control segment determines their offsets in time and frequency. These values are subsequently loaded into the satellites and broadcast to the users as part of the modulated data stream. The user's time error, however, is usually not known to the accuracy required for a good navigation solution. Thus, if it is included in Equations 5.2 through 5.6, the term $-c\Delta t_u$ is added to Equations 5.2 and 5.3, and the term

$$-c(\Delta t_{uj} - \Delta t_{u,j-1}) = -c \ \frac{\Delta f_u}{f_0} \ (t_j - t_{j-1}) \qquad (5.8)$$

is added to Equation 5.7, where $\Delta f_u / f_0$ is the fractional frequency offset of the user equipment oscillator with respect to system time. In the case of the pseudorange rate measurements of Equations 5.5 and 5.6, the added term due to clock drift is simply Equation 5.8 divided by the short time interval $t_j - t_{j-1}$, resulting in a solution for the oscillator fractional frequency offset.

The result of these time and frequency errors is that one more linearly independent measurement is required for a solution for the one more unknown. The effect of not doing this is that the spheres of Figure 5.2 or circles of Figure 5.3 no longer intersect at the user, unless the clock error correction is made. A fourth satellite (or fourth set of measurements) provides the ability to make this correction.

5.3 ORBITAL MECHANICS AND CLOCK CHARACTERISTICS

5.3.1 Orbital Mechanics

All of the equations in Section 5.2 have the satellite's position and velocity in ECEF coordinates as variables, either directly or as part of the LOS vector. The linear independence of the equations, which dictate the observability of the navigation solution, is a function of the relative position of the satellites in orbit. Thus, the placement of these orbits in a constellation and the evaluation of the satellites' positions and velocities is of primary importance, and orbit mechanics is fundamental to the navigation problem.

Orbital Elements   The orbit of an Earth satellite is nominally a plane (Figure 5.4) [5]. The satellite moves in a nearly elliptical orbit in the plane, with perturbations, in inertial space (Figure 5.5). In the satellite systems described in this chapter, the orbits are nominally circular, which is simply a special case of an elliptical orbit. However, to maintain the accuracy of the description of the orbits, they must be represented as elliptical orbits.

The line of nodes in Figure 5.4 is the intersection of the orbit plane with the Earth's equatorial plane. The ascending node is the point where the satellite crosses the equatorial plane from the southerly latitude to northerly. The inclination of the orbit is the angle between the orbital plane and the Earth's equatorial plane.

Six independent constants are needed to specify the nominal orbit. These can be the three components of position and velocity at any instant of time, as used in the equations of Section 5.2, or the classical orbital elements. The advantage of the latter is that they represent the total orbit, rather than one point in the orbit, from which the position and velocity at any time can be derived. Three of these orbital elements are the _orientation parameters_ shown in Figures 5.4 and 5.5:

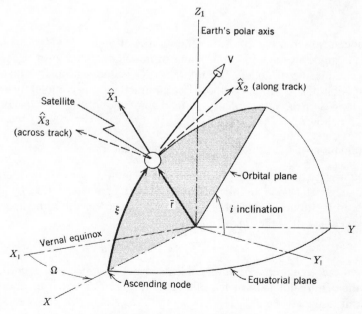

Figure 5.4   The orbital plane.

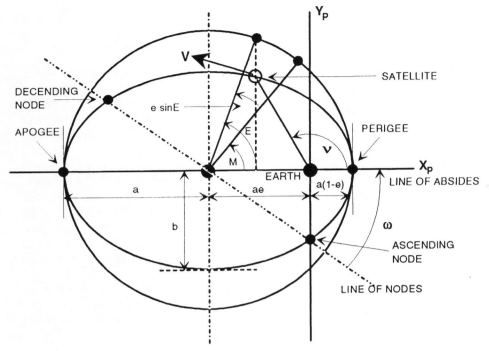

Figure 5.5 The elliptical orbit.

1. Geocentric longitude of the ascending line of nodes, Ω (Figure 5.4). This is the angle in the equatorial plane from some arbitrary reference to the ascending line of nodes. The reference X_1 is conventionally taken as the direction of the vernal equinox, but, in practice, it is completely arbitrary as long as it is appropriately defined for the application. (See Equation 5.17 below.)

2. Inclination of the orbital plane with the equatorial plane, i (Figure 5.4).

3. The argument of perigee ω, which is the angle between the direction of the ascending node and the direction of perigee measured in the plane of the orbit (Figure 5.5). The axis of the ellipse connecting the apogee and perigee is called the *line of apsides*.

Two of the remaining orbital elements are the *dimensional parameters*; namely the semimajor axis of the ellipse, a (Figure 5.5), and the eccentricity of the orbit, e, where

$$e = \sqrt{1 - \left(\frac{b}{a} \right)^2} \qquad (5.9)$$

where b is the semiminor axis. Note that for a circular orbit, where the two axis are equal, the eccentricity is zero. The sixth orbital element is the time of perigee passage, t_p, measured with respect to some arbitrary time scale. t_p establishes the phase of the satellite along the geometric path defined by the other elements.

Useful Orbital Parameters and Equations   The six orbital elements describe the path of the satellite in its unperturbed orbit. To perform a navigation solution, however, it is usually better to describe the satellite's position and velocity in ECEF coordinates. The relationship between the six coordinates and satellite position is as follows [6]:

The mean motion is the average angular rate of the satellite radius vector $\bar{r}$. It is defined as

$$n = \sqrt{\frac{\mu}{a^3}} \qquad (5.10)$$

in radians per second, where $\mu = 3.98605 \times 10^{14}$ meters^3/sec^2 is the WGS 84 value of the Earth's universal gravitational parameter [7, 8]. The period of the orbit is then

$$T_p = \frac{2\pi}{n} \qquad (5.11)$$

in seconds. The mean motion is used to compute the mean anomaly

$$M = n(t - t_p) \qquad (5.12)$$

in radians, which is the basis for the solution of Kepler's equation [6]. Kepler's equation defines the relationship for the eccentric anomaly, where

$$E - e \sin E = M \qquad (5.13)$$

in radians. This relationship is illustrated in Figure 5.5, where E is the angle between a line from the center of the ellipse to a point projected from the position of the satellite to a circumscribing circle perpendicular to the major axis and the line of absides. The eccentric anomaly can be used to compute the instantaneous radius of the orbit from the center of the Earth as

$$r = a(1 - e \cos E) = \frac{a(1 - e^2)}{1 + e \cos \nu} \qquad (5.14)$$

where ν is the true anomaly in radians. Note from Figure 5.5 that to compute the cartesian components of the satellite's position (X_p, Y_p) in the orbit plane, the cosine and sine of the true anomaly is required. They can be computed as

$$\sin \nu = \frac{\sqrt{1 - e^2} \sin E}{1 - e \cos E} \qquad (5.15)$$

$$\cos \nu = \frac{\cos E - e}{1 - e \cos E} \qquad (5.16)$$

This position in the orbit plane can then be transformed into ECEF coordinates by performing a Euler transformation through the *orientation parameters* ω, i, and Ω, in that order. Note, however, that Ω is not constant because the Earth is rotating. It varies from some predefined epoch value Ω_0 (right ascension) at some time t_0 as

$$\Omega = \Omega_0 - \dot{\Omega}_E(t - t_0) \qquad (5.17)$$

in radians, where $\dot{\Omega}_E$ is the earth's rotational rate, which has a WGS 84 value of $7.2921151467 \times 10^{-5}$ rad/sec [7, 8].

The Perturbed Orbit   If the Earth were a spherically symmetrical body in empty space, the parameters described above would stay fixed. However, other forces, called *perturbations*, cause the orbital plane to rotate and oscillate and the satellite's path to vary from its elliptical path. These forces include (1) spherically asymmetrical components of the Earth's gravitational field, (2) luni-solar perturbations, (3) air drag, and (4) magnetic and static-electric forces. Other perturbations are due to the variations in the Earth's rotation rate and polar wander. Such perturbations do not affect the orbit in inertial space but rather the *orientation parameters* as they relate to ECEF coordinates.

The effects of these forces vary, depending upon the altitude of the orbit. For example, higher orbits are less affected by the Earth's gravitational field variations and air drag than the lower orbits are, but are more affected by forces such as lunar and solar gravity and solar radiation pressure. In any case, since the reference coordinate system is ECEF, the Earth's gravitational field variations are the most significant, of which the second zonal harmonic is by far the most dominant. If a truly inertial coordinate system were used, the solar gravity would be the dominant perturbation force. The gravity potential of the Earth developed in zonal harmonics is given by [9]

$$U(r,\lambda,\phi) = \frac{\mu}{r}\left[1 + \sum_{n=2}^{\infty}\left(\frac{A_E}{r}\right)^n \sum_{m=0}^{n}(C_{n,m}\cos m\lambda + S_{n,m}\sin m\lambda)P_{n,m}(\sin\phi)\right]$$

$$(5.18)$$

where

A_E	is the WGS-84 semimajor axis of the Earth's ellipsoid = 6378.137km [9]
n, m	are degree and order
ϕ, λ	are geocentric latitude and longitude
$P_{n,m}(\sin\phi)$	are Legendre polynomials
$C_{n,m}, S_{n,m}$	are geopotential coefficients

Neglecting the effects of longitude ($m = 0$), which are relatively small compared to the effects of latitude, the gravity potential due to the second zonal harmonic is [6]

$$U_2(r,\phi) = \frac{\mu A_E^2 C_{2,0}}{r^3}\left(\frac{3}{2}\sin^2\phi - \frac{1}{2}\right) = \frac{\mu A_E^2 C_{2,0}}{4r^3}(1 - 3\cos 2\phi) \quad (5.19)$$

where the WGS-84 value for $C_{2,0}$ is -1.08263×10^{-2} [9].

The radial and meridional gravity force of the second zonal harmonic are then

$$g_{2r} = \frac{\partial U_2}{\partial r} = \frac{3\mu A_E^2 C_{2,0}}{4r^4}(3\cos 2\phi - 1) \quad (5.20)$$

$$g_{2\phi} = \frac{1}{r}\frac{\partial U_2}{\partial \phi} = \frac{3A_E^2 C_{2,0}\sin 2\phi}{2r^4} \quad (5.21)$$

Note that the radial force has two components—a constant that adds to the nominal gravitational force and one that oscillates as function of the satellite's latitude ϕ, where

$$\phi = \sin^{-1}[\sin(\nu + \omega)\sin i] \quad (5.22)$$

based on the argument of latitude, the argument of perigee, and the inclination angle of the satellites orbit. Note that the period of the oscillating force is one-half the orbit's period. The force in the direction of latitude also oscillates with latitude.

These oscillating perturbations are important for defining parametes to represent the satellites' orbits for the users.

5.3.2 Clock Characteristics

As indicated in Section 5.2, the candidate measurements used in the navigation solution include terms representing the satellite and user clock time offsets and/or clock drift. Thus, the characteristics of these clocks are important as they are a potential navigation error source in two ways. First of all, the navigation message from a satellite includes parameters describing the satellite's clock offset and drift, which are predicted by the control segment. Any instability in the satellite's clock causes this prediction to be in error, thus resulting in range and range rate errors that degrade the user's navigation solution. Second, the user's own clock may drift between, or during, navigation solution updates, which also results in a solution degradation. This degradation is due to an estimation of a "moving target," preventing averaging of the clock solution.

Clock errors are characterized in terms of a time offset, a frequency offset, aging (frequency drift), and measures of clock instability. Time offset, frequency offset, and sometimes frequency drift are part of the solution, whether it be the control segment's solution and prediction of the satellite clocks or the user's solution for his own time and frequency offset. Clock instability, on the other hand, hampers the capability to perform these functions. The most common measure of clock instability is in the form of the square root of an Allan variance [10], which is defined as

$$\sigma_y(\tau) = \sqrt{\frac{1}{2(M-1)} \sum_{k=1}^{M-1} (y_{k+1} - y_k)^2} \tag{5.23}$$

where

$$y_k = \frac{\phi(t_k + \tau) - \phi(t_k)}{2\pi f_0 \tau} = \frac{\Delta\phi(t_k)}{2\pi f_0 \tau} \tag{5.24}$$

is the fractional frequency offset averaged over τ seconds after the systematic frequency offset and frequency drift have been removed. $\Delta\phi(t_k)$ is the change in clock phase, measured in radians over τ seconds as illustrated in Figure 5.6. f_0 is the nominal frequency of the oscillator in Hertz and M is the number of samples used in the computation.

A typical square root of Allan variance for a good quality crystal oscillator is shown in Figure 5.7. Three typical stability characteristics are shown, depending upon averaging time: white frequency noise, flicker frequency noise, and random walk frequency noise. They are defined using the coefficients h_α of

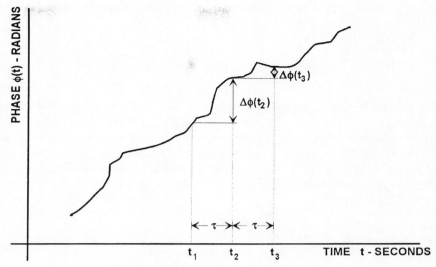

Figure 5.6 Measuring clock stability.

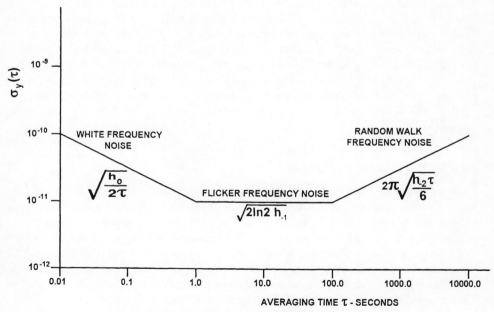

Figure 5.7 Typical square root of Allan variance.

a single-sided spectral density of fractional frequency fluctuations of the form [11]

$$S_y(f) = h_2 f^2 + h_1 f + h_0 + \frac{h_{-1}}{f} + \frac{h_{-2}}{f^2} \qquad (5.25)$$

The first two terms define high-frequency phase noise, which can affect signal tracking, but they are not considered stability terms. The square root of the Allan variance is a measure of frequency stability. An estimate of clock phase (or time) stability, in seconds, can be obtained by multiplying the ordinate axis of Figure 5.7 times the abscissa. However, a more accurate representation in terms of the coefficients h_α is given as a function of time since last measured or estimated as [12]

$$\sigma_{\Delta t}(t) = \sqrt{\frac{h_0}{2} t + 2h_{-1} t^2 + \frac{2}{3} \pi^2 h_{-2} t^3} \quad \text{sec} \qquad (5.26)$$

which can be used to determine the ability to predict time.

5.4 ATMOSPHERIC EFFECTS ON SATELLITE SIGNALS

A major source of errors in satellite navigation signal measurements (components of the ϵ_{PR_i} of Equation 5.1) is due to signal refraction through the atmosphere, which includes the ionosphere and the troposphere. In the ionosphere, the larger one of these two sources of error, the signal interacts with free electrons and ions, causing an increase in its phase velocity with a corresponding decrease in its group velocity. The troposphere causes the propagation velocity of the signal to be slowed, compressing the signal wavelength. In this section, the effects of these phenomena on measurement accuracy and quality will be examined.

5.4.1 Ionospheric Refraction

The ionosphere is a shell of electrons and electrically charged atoms and molecules that surrounds the Earth, stretching from a height of about 50 km to more than 1000 km [13]. It owes its existence primarily to ultraviolet radiation from the sun. The photons making up the radiation possess a certain amount of energy. When the photons impinge on the atoms and molecules in the upper atmosphere, the photons' energy breaks some of the bonds that hold electrons to their parent atoms. The result is a large number of free, negatively charged, electrons and positively charged atoms and molecules called *ions*.

The free electrons in the ionosphere affect the propagation of radio waves.

At frequencies below about 30 MHz the ionosphere acts almost like a mirror, bending the path traveled by a radio wave back toward the Earth, thereby allowing long-distance communication. At higher frequencies, such as those used in satellite radio navigation, radio waves pass through the ionosphere. They are, nevertheless, affected by it.

The Refractive Index and Phase and Group Velocity   The velocity of propagation of a radio wave at some point in the ionosphere is determined by the density of electrons there. The velocity of a carrier, the pure sinusoidal radio wave conveying the signal, is actually increased by the presence of the electrons. The greater the density of electrons, the greater the velocity. The net effect on a radio wave is obtained by integrating the electron density along the whole path that the signal travels from the satellite to a receiver. The result is that a particular phase of the carrier arrives at the receiver earlier than it would have had the signal traveled in complete vacuum. The early arrival is termed a *phase advance*.

The increased phase velocity of propagation is related to what is called the *refractive index n* by the expression [14]

$$v_\phi = \frac{c}{n} \tag{5.27}$$

where c is the velocity of light. The refractive index is given as [5]

$$n = \sqrt{1 - \left(\frac{f_c}{f}\right)^2} \tag{5.28}$$

where the critical frequency f_c, or plasma frequency, below which complete reflection occurs, is, in Hertz [14],

$$f_c = \frac{1}{2\pi}\sqrt{\frac{N_e e^2}{m\epsilon_0}} \approx 9\sqrt{N_e} \tag{5.29}$$

where N_e is the electron density in electrons per cubic meter, e is the electron charge, m is the mass of an electron, ϵ_0 is the permittivity in free space, and f is the signal's carrier frequency. Note that the higher this carrier frequency is, the closer n is to 1, and thus the less the ionosphere affects signal propagation. Note also that the higher the electron density is, the higher the plasma frequency is, and thus the more the ionosphere affects signal propagation.

On the other hand, the signal that is modulating the carrier (e.g., pseudorandom noise codes and navigation data) is delayed by the ionosphere. Since the composite signal can be thought of as being formed by the superposition of a large group of pure sinusoids of slightly different frequencies, the delay of the modulation is called the *group delay*. The magnitude of this group delay is identical to the phase advance. The group refractive index is given as [14]

$$n_g = \frac{d}{d\omega}(n\omega) = \frac{1}{n} \tag{5.30}$$

Thus, the relationship of the phase velocity and group velocity (the rate of energy propagation) satisfies that for a signal passed through a wave guide, which is

$$v_g v_\phi = c^2 \tag{5.31}$$

Electron Density and Total Electron Content   The electron density is quantified by counting the number of electrons in a vertical column with a cross section of one square meter [13]. This number is called the *total electron content* (TEC). The TEC is a function of the amount of incident solar radiation. On the night side of the Earth, the free electrons have a tendency to recombine with the ions, thereby reducing the TEC. As a consequence, the TEC above a particular spot on the Earth has a strong diurnal variation.

Changes in TEC can also occur on much shorter time scales. One of the phenomena responsible for such changes is the *traveling ionospheric disturbance* (TID). TIDs, which have characteristic periods on the order of 10 minues, are manifestations of waves in the upper atmosphere believed to be caused in part by severe weather fronts and volcanic eruptions. There are also seasonal variations in TEC and variations that follow the sun's 27-day rotational period and the roughly 11-year cycle of solar activity.

Dual Frequency Corrections   All these changes in TEC make it difficult to consistently predict the phase advance and group delay accurately using models. Thus, all of the satellites of the satellite radio navigation systems described herein transmit signals at two frequencies, allowing some users to measure and correct for these quantities. Note that the refraction index defined in Equation 5.28 is a function of the inverse of the carrier frequency. A Taylor series expansion of that equation yields

$$n \approx 1 - \frac{f_c^2}{2f^2} + \frac{f_c^4}{4f^4} \approx 1 - \frac{40.5 N_e}{f^2} + \frac{1640.25 N_e^2}{f^4} \tag{5.32}$$

If the carrier frequency is chosen to be high enough so that the fourth-order term is negligible, then the deviation of n from a free space value of one is inversely proportional to f^2. Thus, by making measurements on two widely spaced frequencies and combining them, the electron density N_e can be determined, and almost all of the ionospheric effect can be removed. This is true whether the measurements are pseudorange, Doppler, or integrated Doppler measurements, since they all have an error component that is a function of the refraction index. However, the correction is applied with a different sign, depending upon whether the measurement is obtained from the carrier or the modulated signal.

Applications of ionospheric corrections using models or dual frequency measurements are peculiar to satellite radio navigation system, and thus will be described later in this chapter.

Ionospheric Scintillation Effects   Irregularities in the Earth's ionosphere can produce both diffraction and refraction effects causing short-term signal fading that can severely stress the tracking capabilities of a satellite-navigation receiver [13]. The fading can be so severe that the signal level will drop completely below the receiver's lock threshold. The geographic regions where scintillation effects normally occur are in the equatorial region ($\pm 30°$ either side of the geomagnetic equator) and the polar regions. Scintillation is the strongest in the equatorial regions. Strong fading can also be accompanied by rapid carrier phase changes. Fortunately, strong scintillation effects are rare or localized at certain times of the night, and usually only during periods of high solar activity. Also, by design, a satellite navigation receiver can be made to track through the severe fading, although the collection of navigation data can be disrupted with parity errors. Fortunately these data are periodically repeated, so the disruption is only temporary.

5.4.2   Tropospheric Refraction

Refractivity   Unlike the refractivity of the ionosphere, the refractivity N of the troposphere is not a function of carrier frequency. At a given altitude it is commonly determined from the following equation [14, 15]:

$$N = 10^6(n - 1) = \frac{77.6}{T} \left(P + \frac{4810e}{T} \right) \tag{5.33}$$

where P is total pressure in millibars, T is absolute temperature in K, and e is partial pressure of water vapor in millibars, where one definition of e is given as [15]

$$e = 6.1 \frac{RH}{100} \, 10^{7.4475T_C/(234.7 + T_C)} \tag{5.34}$$

where RH is relative humidity in % and T_C is temperature in °C. The first term in Equation 5.33, which does not depend on relative humidity, is called the *dry term*, while the second is called the *wet term*. Equation 5.33 provides the refractivity at any altitude given the variables measured at that altitude.

Different scientists have modeled refractivity as a function of altitude h but have done so differently on the dry and wet terms, yielding the relationship

$$N(h) = \frac{77.6}{T} P f_d(h) + \frac{373256e}{T^2} f_w(h) \qquad (5.35)$$

Propagation Delay Through the Troposphere   The propagation times for waves traveling through the troposphere between a satellite and the user are longer than that for free space for two reasons [14]:

1. The path does not follow a straight line. The consequence of this is small and can be neglected except for very small elevation angles.
2. The wave velocity is slightly lower than it is in a vacuum, producing an apparent increase in the length of the path given as

$$\Delta L = \int_0^R (n - 1)\, ds \qquad (5.36)$$

where s is the curved abscissa on the path and R is the distance to the satellite, which can be treated as infinite.

Since the real path does not deviate much from a straight line for all but very small elevation angles, Equation 5.36 can be made a function of elevation angle and integrated with respect to altitude. The result is a correction model for ranging measurements.

In general, the dry and wet terms need to be integrated separately because they include different functions of altitude and elevation angle. This is the case when actual surface measurements are used and the ultimate accuracy is desired, such as in the control segment of the radio-navigation system. However, in the case of avionics applications, those measurements are not generally available, and a *standard day* is used to define the correction model. Then, commensurate with the accuracy of that standard day, Equation 5.36 becomes

$$\Delta L = \int_{h_0}^\infty \frac{(n - 1)}{\sin \phi_0}\, dh \qquad (5.37)$$

where h_0 is the altitude of the user and ϕ_0 is the elevation angle of the signal path to the satellite. An approximation of ΔL can be obtained by assuming the atmosphere is exponential. That is, let

$$n - 1 = (n_0 - 1)e^{-bh} \qquad (5.38)$$

so that

$$\Delta L = \frac{n_0 - 1}{b \sin \phi_0} e^{-bh_0} \tag{5.39}$$

For a typical standard day model, n_0 is 1.00032 and b is 0.000145 meters^{-1}, resulting in a zenith delay equivalent to 2.208 meters at sea level. At 5° elevation angle, the equivalent delay becomes 25.33 meters.

5.5   NAVSTAR GLOBAL POSITIONING SYSTEM

The NAVSTAR Global Positioning System (GPS) was conceived as a U.S. Department of Defense (DoD) multi-service program in 1973, bearing some resemblance to and consisting of the best elements of two predecessor development programs: the U.S. Navy's TIMATION program and the U.S. Air Force's Program 621B [16, 121 (Chapter 1)]. The success of Transit had stimulated both of these programs. The Air Force, as the executive service, manages the overall program at the GPS Joint Program Office (JPO) located at the Space Division Headquarters in Los Angeles, CA. The other U.S. military services, as well as representatives from the Defense Mapping Agency, Department of Transportation, NATO, and Australia maintain active participation at the JPO. The result of this development program is an all-weather global radio-navigation system. GPS is a passive, survivable, continuous, space-based system that provides any suitably equipped user with highly accurate three-dimensional position, velocity, and time information anywhere on or near the Earth.

5.5.1   Principles of GPS and System Operation

GPS is basically a ranging system, although precise Doppler measurements are also available. To provide accurate ranging measurements, which are time-of-arrival measurements, very accurate timing is required in the satellites. Thus, the GPS satellites contain redundant atomic frequency standards [17]. Second, to provide continuous three-dimensional navigation solutions to dynamic users, a sufficient number of satellites are required to provide geometrically spaced simultaneous measurements. Third, to provide those geometrically spaced simultaneous measurements on a worldwide continuous basis, relatively high-altitude satellite orbits are required. These three capabilities are all related, since they all are necessary to provide the high-dynamic user navigation capability in three dimensions.

General System Characteristics   The GPS satellites are in approximately 12-hour orbits (11 hours, 57 minutes, and 57.27 seconds) at an altitude of approximately 11,000 nmi. The total number of satellites in the constellation has changed over the years, the number being tied to budget constraints. The intent, however, is to provide coverage at all locations on the Earth as nearly to 100% of the time as possible. Each satellite transmits signals at two frequencies at L-Band [1575.42 (L1) and 1227.6 (L2) MHz] to permit ionospheric refraction corrections by properly equipped users [18]. These signals are modulated with synchronized, satellite-unique, pseudorandom noise (PRN) codes that provide the instantaneous ranging capability. Those codes are modulated with satellite position, clock, and other information, in order to provide the user with that information. Details on the constellation and signal structure appear in later sections of this chapter.

All equations of Section 5.2 apply to the GPS navigation solution in that all three measurement capabilities—ranging, Doppler, and integrated Doppler—exist, and, in general, the solution for the user's clock and clock drift are required. It is not uncommon for a specific user equipment to use at least two of the three measurement capabilities to simultaneously solve for position, velocity, clock offset, and clock drift, in some cases using all satellites in view [19, 20]. That number can be as high as 12.

System Accuracy   GPS provides two positioning services, the Precise Positioning Service (PPS) and the Standard Positioning Service (SPS) [21]. The PPS can be denied to unauthorized users, but the SPS is available free of change to any user worldwide. Users that are *crypto capable* are authorized to use crypto keys to always have access to the PPS. These users are normally military users, including NATO and other friendly countries. These keys allow the authorized user to acquire and track the encrypted precise (P) code on both frequencies and to correct for intentional degradation of the signal.

Encryption of the precise code provides GPS with an anti-spoofing (A-S) capability. A-S is not meant to deny the P code to unauthorized users but to prevent the spoofing of the precise code by an unfriendly force. Unfortunately, A-S denies the P code to unauthorized users. Thus, A-S prevents these users from correcting for ionospheric refraction, since the L2 signal only carries the P code, although there are "codeless cross-correlation" techniques that do allow this measurement [22, 23]. A-S does not prevent the use of the coarse/acquisition (C/A) code, which is only carried on the L1 signal.

The intentional degradation, on the other hand, is meant to deny accuracy to an unfriendly force. It is called *selective availability* (SA). Unfortunately, SA also denies accuracy to unauthorized users that are friendly, which is the entire civil community. The peace-time policy of the DoD is to provide an unauthorized accuracy of 100 meters, 2drms (horizontal accuracy) [24].

Either A-S or SA, or both, may be turned on. If neither is turned on, SPS accuracy is the same as PPS. In 1996, the U.S. stated that it is its intention to turn off SA within a decade [135]. More details on GPS system accuracy are provided later in this chapter.

The GPS Segments   GPS has the basic system configuration illustrated in Figure 5.1. The monitoring and satellite control sites are dispersed around the

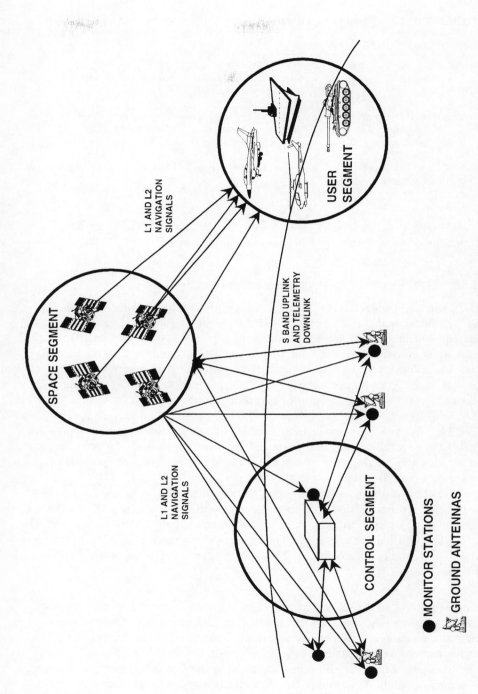

SPACE SEGMENT

USER SEGMENT

L1 AND L2 NAVIGATION SIGNALS

S BAND UPLINK AND TELEMETRY DOWNLINK

L1 AND L2 NAVIGATION SIGNALS

CONTROL SEGMENT

● MONITOR STATIONS

GROUND ANTENNAS

Figure 5.8 GPS system configuration.

TABLE 5.1   GPS segment functions

Segment	Input	Function	Product
Space	Satellite commands Navigation messages	Provide atomic time scale Generate PRN RF signals Store and forward navigation message	PRN RF signals Navigation message Telemetry
Control	PRN RF signals Telemetry Universal coordinated Time (UTC)	Estimate time and ephemeris Predict time and ephemeris Manage space assets	Navigation message Satellite commands
User	PRN RF signals Navigation messages	Solve navigation equations	Position, velocity, and time

world. GPS is comprised of three segments as illustrated in Figure 5.8. The functions of these three segments are summarized in Table 5.1.

5.5.2   GPS Satellite Constellation and Coverage

GPS Satellite Constellation   The fully operational GPS satellite constellation, as described in the 1992 Federal Radio Navigation Plan [24], comprises 24 satellites, four each in six 55°-inclined orbit planes spaced 60° apart in longitude. The nominal GPS 24 satellite constellation is given as orbit parameters for an epoch of July 1, 1993, at 0000 GMT [25] as follows: semimajor axis $A = 26,559.8$ km; eccentricity $e = 0$; inclination $i = 55°$; argument of perigee $\omega = 0°$; right ascension of ascending nodes $\Omega = 272.847°$ (plane A), $332.847°$ (plane B), $32.847°$ (plane C), $92.847°$ (plane D), $152.847°$ (plane E), $212.847°$ (plane F); and mean anomalies M_0 that are nonuniform but nominally 90° between satellites within a plane and 15° between planes. The actual mean anomalies vary significantly from those nominal values in order to provide optimum coverage over regions of interest.

Because of the altitude of the orbits, the satellite paths, as projected onto the surface of the Earth, along a line to the center of the Earth, repeat every day once a day. However, the time at any point along the path occurs three minutes and 56 seconds earlier every day. This is known as a *sidereal day*, even though the orbit period is approximately 12 hours, because of the daily rotation of the Earth.

Coverage   As stated earlier, the coverage, defined as providing a satisfactory instantaneous navigation solution, is not 100%. In turn, a satisfactory instantaneous navigation solution is one where there are at least four satellites visible to the user above a desired elevation angle, usually taken to be 5°, and line-of-sight vectors to those satellites provide an adequate geometric solution. Simply having four satellites visible is not sufficient. There are even times when more than four satellites are visible and the geometry is not adequate. To measure this adequacy, we introduce the concept of *geometric dilution of precision* (GDOP) [26]. GDOP is defined from the linearized navigation/time solution derived from a vector of N ($N \geq 4$) pseudorange residuals:

$$\delta \overline{\mathbf{PR}} = \overline{\mathbf{H}}[\delta \overline{\mathbf{X}}_u^{\mathrm{T}} \quad -\delta c \Delta t_u]^{\mathrm{T}} \tag{5.40}$$

where the ith row (for satellite i) of the measurement matrix $\mathbf{H}$ ($\mathbf{h}_i$) is given as

$$\mathbf{h}_i = [1_{xi} \quad 1_{yi} \quad 1_{zi} \quad -1] \tag{5.41}$$

in terms of Equation 5.3. (The first three elements of Equation 5.41 are the directional cosines from the user position to the satellite position.) GDOP is defined as

$$\mathrm{GDOP} = \sqrt{\mathrm{trace}[\mathbf{H}^{\mathrm{T}}\mathbf{H}]^{-1}} \tag{5.42}$$

where trace[·] indicates the sum of the diagonal elements of [·]. If the vector of pseudorange residuals defined in Equation 5.40 are statistically uncorrelated with equal 1-sigma errors of σ_{PR}, then the 1-sigma position/time error is

$$\sigma_{X,t} = \mathrm{GDOP} \cdot \sigma_{PR} = \sqrt{\sigma_x^2 + \sigma_y^2 + \sigma_z^2 + c^2 \sigma_{\Delta t}^2} \tag{5.43}$$

Position dilution of precision (PDOP) is computed by deleting the fourth diagonal element in the trace of Equation 5.42 (of which the square root represents *time dilution of precision*, TDOP).

To compute *horizontal* and *vertical dilution of precision* (HDOP and VDOP), the ECEF coordinate residuals of Equation 5.40 must first be transformed to *local tangent plane* (LTP) residuals, defined as north, east, and up residuals at the estimated location. This results in the new measurement matrix

$$\mathbf{H}_{\mathrm{LTP}} = \mathbf{HT}_{\mathrm{ECEF} \to \mathrm{LTP}} = \begin{bmatrix} \cos \mathrm{El}_1 \cos \mathrm{Az}_1 & \cos \mathrm{El}_1 \sin \mathrm{Az}_1 & \sin \mathrm{El}_1 & -1 \\ \cos \mathrm{El}_2 \cos \mathrm{Az}_2 & \cos \mathrm{El}_2 \sin \mathrm{Az}_2 & \sin \mathrm{El}_2 & -1 \\ \vdots & \vdots & \vdots & \vdots \\ \cos \mathrm{El}_N \cos \mathrm{Az}_N & \cos \mathrm{El}_N \sin \mathrm{Az}_N & \sin \mathrm{El}_N & -1 \end{bmatrix}$$
$$\tag{5.44}$$

where El_i and Az_i are the elevation and azimuth angles from the user to satellite i and where

$$
\mathbf{T}_{\text{ECEF}\to\text{LTP}}^{\text{T}} =
\begin{bmatrix}
-\cos\lambda_u \sin\phi_u & -\sin\lambda_u \sin\phi_u & \cos\phi_u & 0 \\
-\sin\lambda_u & \cos\lambda_u & 0 & 0 \\
\cos\lambda_u \cos\phi_u & \sin\lambda_u \cos\phi_u & \sin\phi_u & 0 \\
0 & 0 & 0 & 1
\end{bmatrix}
\tag{5.45}
$$

where ϕ_u and λ_u are the estimated latitude and longitude of the user. This new **H** matrix then replaces the one in Equation 5.42, and σ_x, σ_y, and σ_z are replaced with σ_N, σ_E, and σ_h, respectively. Then, the sum of only the first two diagonal elements in the trace of Equation 5.42 are included for computing HDOP, and only the third diagonal element is included for computing VDOP. The linear transformation of Equations 5.44 and 5.45 is only appropriate for the transformation of location residuals and is not appropriate for total state ECEF coordinates to latitude, longitude, and altitude, which requires a nonlinear transformation (see References [9] and [23] of Chapter 2).

Adequate coverage is usually defined by the U.S. DoD when PDOP is less than 6 for elevation angles greater than 5°. However, in some applications of GPS, such as civil and commercial aviation, this is not adequate coverage. In these applications, augmentation using pseudolites or geostationary satellites is necessary. These augmentations are discussed later in this chapter. Because of these advanced applications, the DOP concept has been replaced with the concept of availability of accuracy, accounting for satellite failures. This availability is based upon analytical procedures using the mean-time-between-failures (MTBF) and mean-time-to-repair (MTTR) characteristics of satellites (and appropriate augmentations), a concept originally developed by the French space agency CNES [27, 28] and later extended [29, 30]. With these extensions, coverage is more appropriately defined in terms of probabilities (or availability).

An Example of Coverage in Terms of Availability of DOP Using all nonfailed satellites in view (above 5°) of the nominal GPS satellite constellation [25], the average and worst-case location availability HDOP and VDOP over the continental United States (CONUS) is illustrated in Figure 5.9. This availability was computed with a 5-minute time and a 2° grid granularity. Note that HDOP is at least twice as good as VDOP, especially at the worst-case location. Since the HDOP and VDOP values plotted in Figure 5.9 are probabilistic and do not necessarily occur at the same location, they cannot be root-sum-squared to obtain PDOP. The inset in Figure 5.9 is a plot of the underlying joint probability of the unavailability of exactly N satellites in the constellation of 24 satellites for $0 \leq N \leq 24$.

As described later in Section 5.7.4, availability of 99.999% of at least a finite HDOP is required for civil and commercial aviation. In addition, availability of 99.9% of good VDOP is required for precision approach applications. The 99.999% available HDOP cannot be observed in Figure 5.9, and, in fact, a finite HDOP is *not* available with that probability. The 99.9% available VDOP is 4.29 averaged over CONUS and 5.62 at the worst-case location. None of

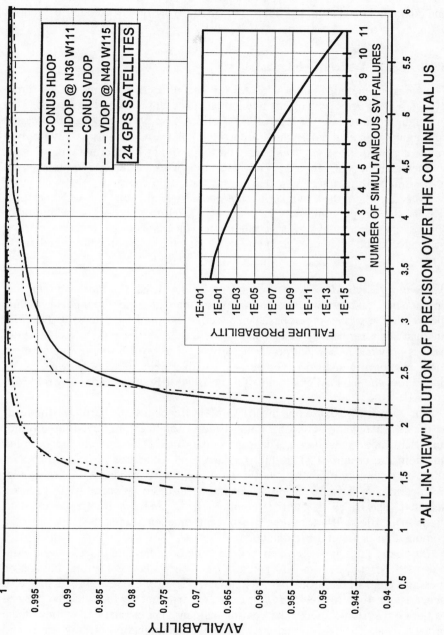

Figure 5.9 Availability of HDOP and VDOP over the continental United States (CONUS).

203

these values is acceptable, especially since real-time integrity of the signals is not guaranteed. Thus, some type of augmentation to GPS is required for civil and commercial aviation. Possible augmentations are described later in Sections 5.7.3, 5.7.4, and 5.8.

5.5.3 Space Vehicle Configuration

The various generations of NAVSTAR GPS satellites that are in orbit are three-axis stabilized vehicles with the navigation subsystem L-band antenna pointing in the direction toward the Earth. The satellites have been designed to track the sun about their yaw axis, which allows the solar array panels to have only a single degree of freedom. This concept also simplifies the thermal control environment for the satellite and its navigation payload, since one side is exposed to the sun and the other two sides always face deep space.

In addition to the navigation subsystem, which from a user's perspective is the primary mission payload, the satellite consists of seven other subsystems—electrical power; attitude and velocity control; reaction control; thermal control; telemetry, tracking and command; orbit injection; and structure.

Navigation Payload The navigation payload consists of the following key components and assemblies: transmitting antenna array, redundant atomic clocks, digital processing or baseband assemblies, and RF equipment. The GPS antenna array formed by inner quad helices encircled by a ring of eight outer helices provides near equal power density to all terrestrial users. This arrangement results in a shaped coverage beam with a 28.6° field of view. A block diagram of the payload electronics for the Block IIR satellites is shown in Figure 5.10.

The atomic clocks used as the 10.23-MHz frequency standards are the heart of the GPS navigation concept. Initial Block I satellites employed three rubidium (Rb) atomic standards. The subsequent Block IA versions incorporated an additional cesium (Cs) atomic standard. Both standards provide equivalent short-term stability of about 1 part in 10^{13}, but the Rb standards tend to wander over periods of days, requiring a drift rate term to be uploaded by the control segment. The Block II satellites employ two Cs and two Rb frequency standards. The Block IIR satellites use three frequency standards made up of a combination of Cs and Rb standards.

Baseband GPS processing to generate the specific satellite pseudorandom noise (PRN) codes (P and C/A) and the 50 bit per second digital navigation message data are controlled by redundant processors. These processors also generate the time-of-week (TOW) count (or Z-count), which can be aligned to the GPS system time by the control segment. The primary functions of the processors are to store, format, and modulate the navigation message data onto the signals transmitted via the L-band subsystem.

The L-band subsystem consists of carrier frequency synthesizers, modula-

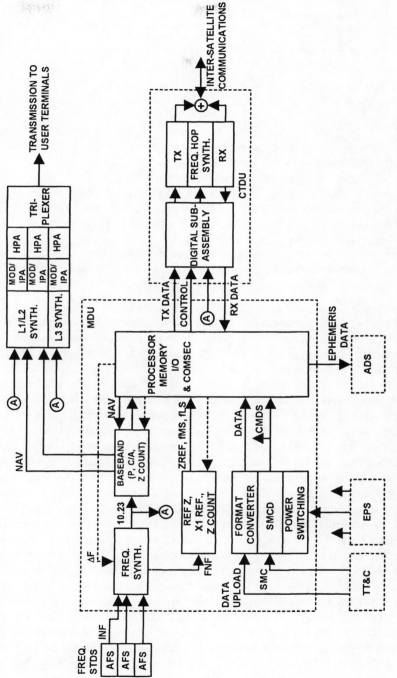

Figure 5.10 Block diagram of the Block IIR GPS satellite payload electronics (courtesy, ITT Aerospace/Communications Division, ITT Defense).

TABLE 5.2   Received RF signal levels (minimum level received) [7]

Frequency	P-Code	C/A-Code
L1	−163 dBw	−160 dBw
L2	−166 dBw	N/A

tors, and intermediate-power and high-power amplifiers. These RF equipments are totally redundant and switchable. The two L-band carriers at 1575.42 and 1227.6 MHz are combined in a triplexer/filter, which then feeds energy into the 12-element helix antenna array. (A third L3 signal is also generated, when needed, for another non-navigation satellite function.) The transmitted power is such that the received power level for a user, defined at the output of a 3 dBi linearly polarized antenna, is given in Table 5.2.

In Figure 5.10 an intersatellite communications function is also shown. On the Block IIR satellites, this function includes an intersatellite ranging capability that will be used for a future autonomous navigation capability.

Block Characterization Summary   There are to date six generations of GPS satellites—Block I, IA, II, IIA, and IIF. To a great extent, the variations in these vehicles are transparent to the navigation users. The differences (especially in satellite weight, power, and complexity) simply reflect the addition and evolution of military requirements for enhanced survivability and auxiliary payload growth. As an example, the Block I satellites consisted of approximately 33,000 individual parts with limited radiation hardening requirements. The Block II production satellites, on the other hand, contain almost 65,000 parts, have a much larger form factor, and are planned to achieve a 7.5-year design life. With the Block IIR replenishment launches, the navigation user should be assured of continued and unchanging navigation coverage and performance well into the next century. In 1996, a contract was awarded for Block IIF (follow-on) satellites with initial deliveries scheduled for the year 2001 [125]. Two of the important new features of the Block IIF satellites are a design life of 11.5 years and the capability for inclusion of a new, second, civil-transmitted frequency (L5). That frequency would permit civil users to perform automatic ionospheric delay error correction that is normally only done by authorized (military) users (Sections 5.4.1 and 5.5.7). The frequencies under consideration for L5 are in the ranges of 1207 to 1217 MHz and 1309 MHz to 1319 MHz. The frequency should be at least 200 MHz away from L1 for optimum ionospheric corrections and it must not be subject to interference by other systems operating in the same band, such as JTIDS, DME, and L-band radars [134].

5.5.4   The GPS Control Segment [31]

The principal product of the GPS control segment (CS) is the GPS navigation message data representing the predicted state of each GPS satellite, which is put into a standardized format and periodically uploaded to each satellite memory for continuous retransmission to the user. The CS function of controlling and maintaining the status, health, and configuration of the space segment (SS) assets is of equal importance to total GPS integrity. Scheduling the contacts required for the full constellation while not compromising the basic navigation service is a CS challenge second only to ensuring the navigation service integrity. The third CS function of continuously monitoring the navigation service availability and accuracy as available on L-band is essential for quality control. The fourth CS function is to monitor and manage ground-based assets so as to provide uninterrupted GPS support.

CS Performance Requirements   The following are the basic CS performance requirements:

1. *Navigation range service.* A six-meter RMS user range error (URE) is required at the GPS satellite to support the 15-meter spherical error probable (SEP) navigation service to full capability users. This requirement is derived from a constellation geometry criterion that the PDOP be no greater than six. The responsibility of the URE budget component is shared by the CS and SS, since both satellite process predictability and CS process state estimate fidelity are significant performance factors.

2. *Time transfer service.* A 90-nsec (1-sigma) calibration of the GPS time scale relative to universal coordinated time (UTC) (USNO) is required to meet a 97-nsec (1-sigma) apparent uncertainty at the satellite. GPS time (modulo 1 sec) is aligned to UTC within 1 μsec for user convenience.

3. *Constellation accommodation.* The CS must accommodate up to 24 operational satellites providing full navigation service. In addition, up to six spare satellites must be maintained within the S-band command and telemetry processing resource capacities.

4. *Orbital operations.* The CS must support pre-launch compatibility validation tests prior to launch of each satellite. Full CS responsibility commences after the satellite is three-axis stabilized on-orbit with an active L-band capability.

5. *Space vehicle communications.* The CS must provide full command, telemetry, and GPS navigation message upload support to satellites of any Block design. This includes memory-map management and provisions for telemetry data analysis and archiving. Assets to support three navigation message uploads per day for each satellite are required to achieve the

6-meter URE performance requirement with specification satellite atomic clocks (Allan standard deviation not exceeding two parts in 10^{-13} beyond one day).

CS Configuration The CS consists of a centralized master control station (MCS) and remote RF facilities coupled by dedicated communication services. Establishing locations for these RF facilities was a compromise between utility and the availability of adequate facilities to establish global satellite visibility so as to strengthen tracking geometry, to provide maximum monitoring opportunity, and to preserve operational flexibility.

Monitor stations (MSs) are the L-band facilities that receive the same signal as the user. These passive stations measure pseudoranges and collect the navigation messages for two purposes. First, the pseudorange histories are required to estimate the satellite trajectories and clock calibrations. Second, both are required to faithfully monitor the navigation service as provided to the user community. Colorado Springs, Ascension Island, Diego Garcia, Kwajalein, and Hawaii comprise the CS MS sites. The resulting track coverage is provided by these sites for satellites above 5° elevation angle. The ground projection latitude of the satellites will never exceed the 55° orbit inclination angle, and thus better than 90% average constellation coverage is provided. Contact with each of the six ground tracks is different, varying from 90% to 100%. Segments of uncovered ground tracks occur west of South America. Regions of coverage overlap are important, since they result in the common MS viewing of satellites which enable the direct MS- to MS-time transfers with the estimation process that stabilize the relative MS-time scales.

The MCS is the operations center for GPS. It is composed of the personnel and the facilities to provide overall system management. Here the processing to support the navigation mission and to maintain the satellite constellation are performed. Both equipment configuration and procedures were designed to assure continuous integrity of the total system. The Consolidated Space Operations Center (CSOC) is the permanent MCS site.

Ground antennas (GAs) are the S-band facilities that provide duplex communication with the multiple satellites by receiving telemetry and transmitting both commands and upload data.

Contact opportunities are provided by the GA sites of Ascension Island, Diego Garcia, and Kwajalein. Regions of coverage overlap provide scheduling flexibility to best meet the MCS contact requirements. A fourth GA without redundancy is located at the Eastern Launch Site to support segment compatibility verification but is of limited operational utility because of range transmission restrictions.

Figure 5.11 illustrates the total CS configuration. External operational interfaces to the Air Force Satellite Control Facility (AFSCF), United States Naval Observatory (USNO), and the Defense Mapping Agency (DMA) are implemented to permit initial satellite handover, to provide UTC time coordination, and to provide Earth orientation data, respectively. Less formal interfaces also exist which import Jet Propulsion Laboratory (JPL) sun/moon data and

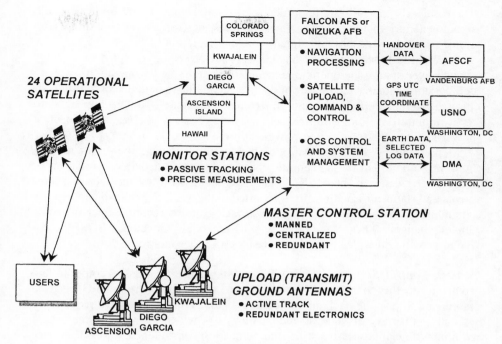

Figure 5.11   GPS control segment configuration.

exchange pertinent information with other GPS segments. The following is a brief description of the various components of the CS.

Monitor Stations   Navigation visibility is provided to the CS by the globally distributed MSs that are unmanned. The MSs are comprised of antenna electronics, a multichannel dual frequency receiver, a cesium frequency standard, a test and calibration signal generator, meteorological sensors, a communications subsystem, a central processor, and backup power supplies. The antenna electronics consists of a single beam pattern antenna that receives both the L1 and L2 frequencies, a conical ground plane with annular chokes for multipath signal rejection, and a low-noise amplifier.

The MS receiver has 12 dual frequency channels (24 demodulators) that acquire and track up to 12 GPS satellites simultaneously. These channels each provide accurate pseudorange and accumulated delta-range (phase) measurements on both the L1 and L2 frequencies every 1.5 seconds. They also demodulate the GPS navigation message and measure carrier-to-noise density as signal quality measurements. All of these quantities are set into a message format for transfer to the central processor for subsequent transfer to the MCS. The receiver uses as its frequency reference the output of one of two cesium frequency standards, which provide the basis for the highly stable GPS time.

The receiver's channels are periodically calibrated using a dual frequency GPS signal generator that is slaved to the same frequency reference. This signal generator is also used to provide signals for fault isolation.

One of the two frequency standards is on hot standby to provide a quick change-over in case the operational standard exhibits faulty operation. A phase comparator compares the phases of the two standards to help detect faulty operation, as well as to provide an initial frequency estimate of the backup standard for a smooth transition. Both frequency standards are kept alive with a backup power supply in case of prime power failure.

Meteorological sensors (barometric pressure, temperature, and relative humidity) are polled by the central processor to permit accurate correction for tropospheric delay at the MCS. These data, along with the receiver measurements and status information, are forwarded to the MCS over a dedicated secure communication channel in response to tracking orders received over the same duplex channels. This channel utilizes a commercially developed SDLC protocol to provide error detection and data block re-transmission.

Master Control Station The MCS consists of the processing complex and controller facility to completely manage and control the operational GPS space assets and to provide navigation messages. The navigation mission requires an upload availability of 98%, so redundancy is provided for all mission-critical equipment. Dual processors with communication controllers and the customary compliment of peripherals are configured to permit processing of the on-line navigation processing and satellite control functions with either unit. Personnel at the MCS control all navigation processing, constellation, and CS assets and are responsible for the system integrity. This requires established procedures and efficient access to critical mission data.

The navigation process is illustrated in Figure 5.12. It is based upon a linear expansion about a reference trajectory, which is obtained by integrating the equations of motion forward in time from an initial position and velocity state. The satellite force model used to accomplish this integration includes mathematical expressions for the following effects: (1) WGS-84 geopotential expansion, (2) sun gravitational attraction, (3) moon gravitational attraction, (4) Earth gravitational tides, (5) solar flux reaction (including eclipse), and (6) satellite y-axis acceleration.

Partial derivatives with respect to the epoch states of inertial position and velocity are generated to reduce the estimation and prediction process to linear mathematical relationships. Given values for estimated state residuals, evaluation of the first-order expansion provides the position trajectory used to process measurements or to generate the navigation message. The inertial-to-ECEF coordinate transformation matrix is generated to be consistent with externally supplied Earth orientation data.

The measurement update process consists of data editing, smoothing, measurement model transformation, and estimation steps. The raw MS data are examined and correlated with receiver fault indicators to ensure track conti-

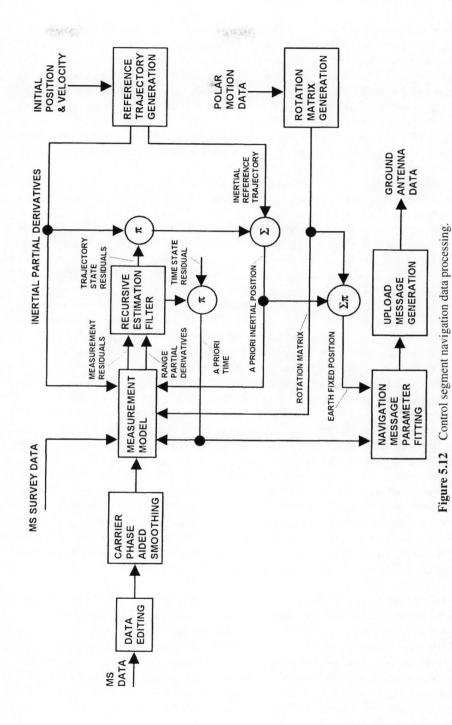

Figure 5.12 Control segment navigation data processing.

nuity. Both first- and second-difference histories are compared with threshold values to detect any data inconsistencies. Smoothing is applied to refine the MS measurements, combining the 1.5-sec measurements into one low-noise data point every 15 min. Corrections for the ionosphere and troposphere induced delays and interchannel receiver delays are made. Then the measurement model accounts for the Earth's rotation during the signal propagation time, the effects due to crustal tides, relativistic time distortions and the a priori knowledge of estimated states to form measurement residuals. Measurement partials with respect to the estimated states are also evaluated to support the estimation process.

A Kalman filter recursive estimator (Chapter 3) is implemented to estimate the following states: (1) satellite position at epoch time, (2) satellite velocity at epoch time, (3) satellite clock phase at epoch time, (4) satellite clock frequency at epoch time, (5) satellite clock aging (rubidium frequency standards only), (6) solar flux, (7) satellite y-axis acceleration bias, (8) MS clock phase at epoch time, (9) MS clock frequency at epoch time, and (10) MS wet troposphere height. The estimation process is partitioned to conserve computational resources. A common time scale ensemble of MS clock states is determined and this common time scale holds the partitions together. The partitioned estimation performance penalty is insignificant when an adequate number of satellites exist in each partition to maintain a solid MS-time transfer network.

Every eight hours the state of each satellite is used to generate a prediction of the time-scale correction and trajectory in ECEF coordinates. The same linear expansion and coordinate transformations are used. Navigation message parameters are obtained as a weighted least-squares fit to these data in accordance with ICD-GPS-200B-PR and the GPS SPS signal specification [7, 8], and the generated user information is put into a standard format and forwarded to the GAs for upload. This process is time-phased across satellites to evenly distribute the MCS work load.

Navigation service integrity is monitored using MS data for four-satellite position solutions and by comparing the pseudorange and received message data with MCS expectations. Measurement residuals are evaluated by the MCS whenever tracking data exist, and performance statistics are maintained as information available to system operator personnel and the user community. Each bit of the received navigation message is compared with the corresponding upload data-base value to verify proper dissemination.

Ground Antennas   Communications with the satellite constellation is provided to the CS by the globally distributed, unmanned GA facilities. These installations consist of a 10-meter S-band antenna and extensive dual-string electronics consisting of an RF exiter, high-power transmitter, receiver, servo equipment, communication channels, and a processor. Command, telemetry, and navigation upload traffic is handled by these GA installations. Tracking orders, equipment configuration commands, and data are received over secure dedicated duplex communication channels, using SDLC for the error detect and retransmission features. Data from the MCS are buffered on disk prior to satellite contact as a telemetry data from the satellite prior to transmission to the MCS.

5.5.5   GPS Signal Structure

The GPS satellites broadcast two signals: Link 1, L1, at a center frequency of 1575.42 MHz and Link 2, L2, at a center frequency of 1227.6 MHz [6, 17]. Each of these frequencies is an integer multiple of the 10.23 MHz clock, 154 for L1 and 120 for L2. The purpose of broadcasting signals at two frequencies is to allow the appropriately equipped user to correct for the ionospheric refraction described in Section 5.4.1.

The L1 signal consists of two carrier components: One carries a precise (P) pseudorandom noise (PRN) code, while the other, transmitted in quadrature, carries a coarse/acquisition (C/A) PRN code. The L2 signal consists of only one carrier that carries only the P code. Both codes are modulated with a 50-bit-per-second (bps) data message. The format and contents of that message are described in Section 5.5.6.

Signal Modulation   The PRN codes and data are modulated onto the carriers using binary phase shift keying (BPSK). This modulation shifts the phase of the carrier 180° each time there is a change in state of the digitally defined code and data. First of all, the PRN code is a sequence of 1's and 0's, as are the message data. The message data sequence is modulo-2 added to the code sequence, which is nothing more than an *exclusive or* of a code bit and a data bit, although the data bits transition at a much slower rate than do the codes. The resulting sequence of 1's and 0's are converted to 180° and 0° phase shifts of the carrier, respectively. Since the 180° phase shifts simply change the sign of the carrier, an equivalent representation is simply an amplitude modulation of ±1's. The result is a mathematical representation of the signals as follows:

$$s_{L1}(t) = AP(t)D(t)\cos(2\pi f_1 t + \phi_{01}) + \sqrt{2}AC(t)D(t)\sin(2\pi f_1 t + \phi_{01}) \quad (5.46)$$

$$s_{L2}(t) = \frac{A}{\sqrt{2}}P(t)D(t)\cos(2\pi f_2 t + \phi_{02}) \quad\quad\quad (5.47)$$

where A is the L1 P signal amplitude, $P(t)$ and $C(t)$ are the ±1 P and C/A code PRN sequences, $D(t)$ is the ±1 message data bit sequence, f_1 and f_2 are the L1 and L2 carrier frequencies, and ϕ_{01} and ϕ_{02} are the ambiguous L1 and L2 carrier phases. Note that the two signals are not phase coherent, even though they are derived from the same frequency reference. There are also slight group (code) delay differences between the P code and C/A code modulations and between the P code modulations at the two frequencies [7].

PRN Code Properties   The PRN codes are generated as products (modulo-2 sums, if expressed as 1's and 0's) of two other codes clocked at the same chipping rate, where

$$C_i(t) = G1(t)G2(t + n_i T_c) \tag{5.48}$$
$$P_i(t) = X1(t)X2[t + (i - 1)T_p] \tag{5.49}$$

for satellite i, where T_c is the C/A code chip width, or the inverse of the C/A code chipping rate of 1.023 MHz, T_p is the P code chip width, or the inverse of the P code chipping rate of 10.23 MHz, and n_i is an integer assigned to satellite i for the C/A code. In case of the P code, n_i takes on a value between 1 and 37 for 32 satellites and 5 reserved for ground transmitters (GTs) [7, 8]. In the case of the C/A code, the n_i are selected values between 1 and 1023 for codes exhibiting desirable properties [32].

These PRN codes provide the desirable code-division, multiple-access (CDMA) property that, to an extent, the codes received from the various satellites do not correlate with each other, nor do they correlate with a reference code in the user's receiver unless the state of the received code matches that of the reference code. Thus, all satellite signals can be received at the same frequency (except for Doppler differences) and selectively acquired and tracked, depending upon the selection of the code in the reference code generator. These PRN codes also exhibit the property that, to an extent, they *spread* interference signals over the signal's bandwidth, providing a degree of interference rejection. These *spread spectrum* properties of the GPS PRN codes will be discussed later in this chapter. Here we will discuss the correlation properties of the codes.

1. *C/A codes.* The C/A codes are Gold codes [18], where the G1 and G2 codes are generated in maximal-length, 10-stage linear feedback shift registers, each of which generate repeating maximal-length codes of length 1023 chips. Figure 5.13 represents an implementation of a C/A coder [32]. In this implementation the initial state of the G2 register represents the delayed state of that maximal-length code (the n_i in Equation 5.48) from its initial state of all 1's.

The resulting C/A (Gold) code is not a maximal-length code. A maximal-length code $x(t)$ has the autocorrelation property that [33]

$$\sum_{j=1}^{2^M - 1} x(t_j)x(t_j + kT_c) = -1, \qquad k \neq 0$$
$$= 2^M - 1, \qquad k = 0 \tag{5.50}$$

where k is an integer number and M is the number of stages in the shift register. This is partly true for the C/A codes, but they do not have perfect correlation properties. On the average, a C/A code has the autocorrelation property that, for $k \neq 0$ [17],

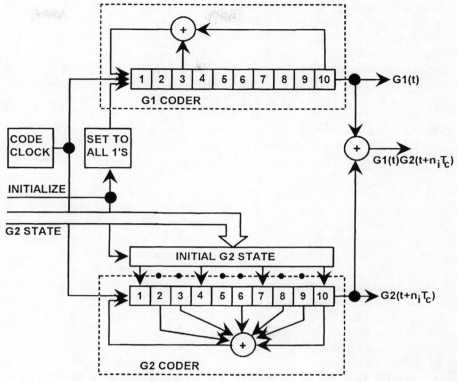

Figure 5.13   C/A coder implemented with initial G2 state.

$$\sum_{j=1}^{1023} C(j)C(j + kT_c)$$

$$= -1 \qquad \text{with probability } 0.75$$
$$= 2^{(10+2)/2} - 1 = 63 \qquad \text{with probability } 0.125$$
$$= -2^{(10+2)/2} - 1 = -65 \qquad \text{with probability } 0.125 \qquad (5.51)$$

although the probabilities vary somewhat, depending upon which code is evaluated [32]. Also, whereas the maximal-length codes are always balanced, 256 of the 1023 C/A codes are not. That is [32],

$$\sum_{1}^{1023} C(t_j) \neq -1 \qquad (5.52)$$

All of the 32 codes selected for the GPS satellites and the four codes selected for the ground transmitters are balanced.

The autocorrelation property described above carries over to the cross-correlation between the different C/A codes. This means that the C/A codes are not quite orthogonal, and care must be taken during acquisition of the satellite signals to prevent false acquisitions and false alarms. At zero Doppler difference between satellite signals, Equation 5.51 above represents a separation of 23.9 dB between signals of equal power. There is also a degree of cross-correlation between signals at other Doppler differences. This property will be described later when we discuss the spectral characteristics of the PRN codes.

2. *P codes.* Whereas the C/A codes are linear codes, the P codes are nonlinear. That is, the C/A codes are made up of two maximal-length codes that are clocked synchronously and are allowed to proceed through all of their 1023 states. This is not true for the P codes. The underlying linear codes of the P code are short-cycled before creating the product of the codes. This has the effect of creating an extremely long code. In fact the 37 individual P codes are simply a one-week piece of a long code that is approximately 38 weeks long.

A typical P code implementation is presented in Figure 5.14 [7]. Note that there are four shift registers, two each for the X1 and X2 code generators. Each of these 12-stage shift registers that have $2^{12} - 1 = 4095$ possible states are short-cycled, either at 4092 or 4093 states, and reset. Both of the X1 shift registers are reset on X1 epochs (every 1.5 seconds), while the X2 shift registers are reset every 1.5 seconds plus 37 chip clock cycles. All shift registers are reset at the end of the week. Although the X1 and X2 coders repeat every 1.5 and 1.5+ sec, the fact that they are running asynchronously at 10.23 MHz, their modulo-2 addition generates an extremely long code. Delaying the X2 code with respect to the X1 code an additional $i - 1$ chips for the ith satellite provides codes for each satellite. The count of the X1 epochs provides a Z-count that is used as basic timing for the system to which the data message and the C/A coder are synchronized. If the coder were not set at the end of week, the code would eventually run into the code of the other satellites and return to the beginning almost 38 weeks later. Unlike the C/A codes, the P codes have excellent cross-correlation properties.

3. *Spectral characteristics of the PRN codes.* In the frequency domain the spectral density of the signal is the spectral density of the PRN codes centered at $\pm f_i$, $i = 1, 2$. At baseband, the spectral density of the P code is

$$S_{spi}(f) = P_{pi} T_c \frac{\sin^2(\pi f T_c)}{(\pi f T_c)^2}, \qquad -\infty < f < \infty \qquad (5.53)$$

where T_c is the P code chip width, or the inverse of the P code chipping rate of 10.23 MHz, and P_{pi} is the appropriate P code carrier power. Actually, this spectral density is bandlimited in the GPS satellites in order to protect radio astronomers, so its range is not infinite. This could prevent a user's receiver

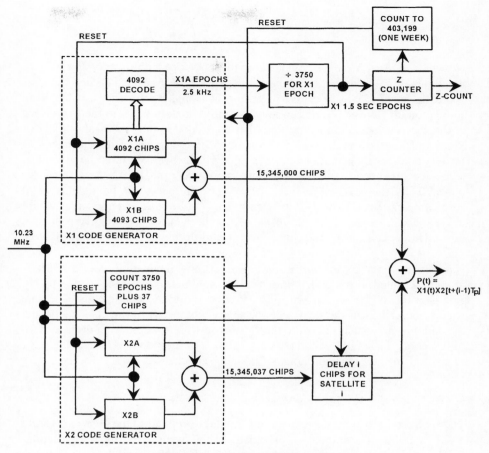

Figure 5.14 Typical P coder implementation.

from achieving full correlation. However, all P code receivers also are band limited. The resulting effect is known as *correlation loss* due to filtering.

The C/A code for satellite i is a short one millisecond repeating code whose spectral density is a line spectrum with components c_{ji}, $j = -\infty$ to $+\infty$, where j represents spectral lines 1 kHz apart and

$$\sum_{j=-\infty}^{j=+\infty} c_{ji} = P_{ci} \tag{5.54}$$

where P_{ci} is the C/A code carrier power. The envelope of this line spectrum takes on the form

$$S_{ci}(f) = 1000 P_{ci} T_c \, \frac{\sin^2(\pi f T_c)}{(\pi f T_c)^2} \tag{5.55}$$

However, the spectral lines deviate from this envelope significantly. Figure 5.15 illustrates this for the first two lobes of the spectrum. As stated earlier, the C/A codes do have a level of cross-correlation at Doppler differences. It has been shown that this cross-correlation level is approximately equal to the magnitude of the spectral line component of another C/A code in the family of 1023 [18]. Since the line spectrum shown in Figure 5.15 is typical of all the codes as far as variation about the envelope is concerned, that spectrum shows typical levels for this type of cross-correlation—on the order of 21 dB below the carrier power. Thus, the cross-correlation at Doppler differences can be higher than at zero Doppler offset. However, its occurrence is quite rare, and a receiver would never track it because it would disappear as the relative codes move past each other. These occurrences can cause false alarms during initial signal acquisition.

Navigation Data Synchronization   The 50-bps navigation data stream is modulo-2 added to the codes prior to their modulation onto the carrier [7, 8]. The data bits are synchronized to the P code and C/A code epochs. The data frames, described in the next section, are synchronized to the X1 epochs and the data bits are synchronized to the C/A code 1 ms epochs. Each data bit is 20 milliseconds in length, encompassing 20 C/A code epochs. The leading edge of every 75th bit is synchronous with an X1 epoch.

5.5.6   The GPS Navigation Message

The GPS navigation message is the information supplied to the GPS users from a GPS satellite. These data are provided via the 50-bps data bit stream modulated on the PRN codes described in Section 5.5.5, providing the user with the information needed to navigate [34]. Among other data, the user is provided with information from which can be computed the position and velocity of the satellite and the time and frequency offset of its clock, as well as information to resolve ambiguities in the received C/A code. The other information includes almanacs for determining the position, velocity, and clock offsets of the other satellites, an ionosphere model and a description of the time offset between GPS system time and universal coordinated time (UTC).

Frames, Subframes, and TLM and HOW Words   The GPS navigation message consists of a frame of five 300-bit subframes spanning 30 seconds of time as illustrated in Figure 5.16 [7, 8]. Each six-second subframe consists of ten 30-bit words, the first two of which repeat in each subframe. These words, also illustrated in Figure 5.16, are the Telemetry Word (TLM Word) and the Hand-Over Word (HOW Word). These two words are generated by the satellite, while the other eight words are generated by the CS and uploaded to the satellite as

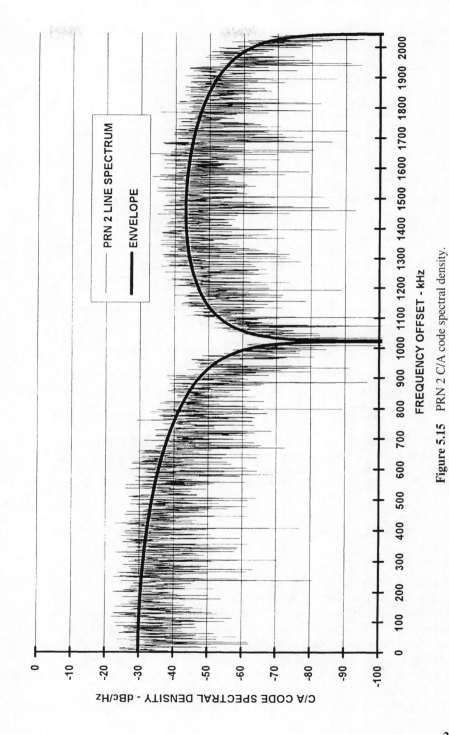

Figure 5.15 PRN 2 C/A code spectral density.

219

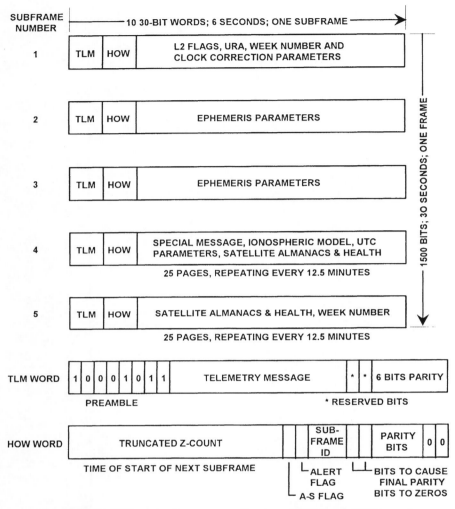

Figure 5.16 GPS navigation message frame and TLM and HOW words.

described in Section 5.5.4. The TLM Word consists of an 8-bit preamble and a satellite telemetry message. The preamble (10001011) allows the user's receiver to synchronize with the subframes to establish time of reception and subsequent resolution of the C/A code ambiguity. This ambiguity exists because the signal transit time from the satellite is on the order of 80 msec, while the length of the repeating C/A code is only 1 msec. For the most part, the telemetry message is of no use to the user; it primarily provides real-time telemetry information from the satellite to the CS.

The HOW Word contains a 17-bit truncated Z-count indicating the time of the start of the next subframe in units of six seconds, or four X_1 epochs. It also contains two flags, an A-S on/off flag and an *alert* flag, and a subframe identification (ID) (1 to 5 for subframes 1–5). The Z-count is used by the user's receiver to establish GPS time. For P code users the Z-count also provides the time information required to hand over to P code, and thus the reason for the name of the word.

The A-S on/off flag alerts non-PPS users as to whether or not they are able to acquire the P code, and it alerts PPS users as to whether or not they should encrypt their reference P code [7]. The *alert* flag indicates to unauthorized users that the satellite user range accuracy (URA) may be worse than indicated in subframe 1 and that they should use the satellite at their own risk [7].

Parity   Both the TLM and HOW words contain satellite-generated parity, six bits per word. The other eight words of each subframe contain MCS-generated parity. The HOW Word contains two noninformation bearing bits so that the last two parity bits can be forced to 0's, since the parity algorithm overlaps word boundaries.

The parity algorithm is a (32, 26) Hamming error detection algorithm. It will detect 1-, 2-, or 3-bit errors. As stated above, the parity overlaps the 30-bit words, since it is based upon 32 bits, always including the last two bits of the previous word. Thus, since the parity in the TLM and HOW words are generated in the satellite, and the parity on the other words are generated in the MCS, two bits in the HOW Word and Word 10 of the MCS generated words are wasted. This is so that both the satellite and the MCS know the state of the last two bits of the previous word (set to 0) when generating the TLM Word (satellite) and Word 3 (MCS). The Hamming error detection algorithm used is specified in [7, 8].

Words 3 through 10 of each subframe contain the message data. Subframes 1, 2, and 3 repeat every 30 seconds consisting of the same data for nominally one hour but can change more often or less frequently. Subframes 4 and 5 subcommutate 25 times each, so that a complete data message requires the transmission of 25 full 1500-bit frames, or 12.5 minutes. Thus every 30 seconds one *page* of subframes 4 and 5 are transmitted. These pages consist of less timely data that change whenever the satellite is uploaded, while subframes 1, 2, and 3 consist of more timely data that change periodically. This periodicity is either once per hour, on the hour, once per 4, 6, 12, or 24 (or more) hours on 4-, 6-, 12-, or 24-hour (or more) boundaries. Any period greater than an hour represents a case when a satellite could not be uploaded within a day of a previous upload, which would be rare. In addition, the data could change at any time with a quick upload. Again, this is a rare condition required to change substandard or erroneous data. One-hour data sets are valid for three additional hours. All other data sets are valid for two hours past their period of transmission.

Subframes 1 and 2 Data Content   The primary content of subframe 1 consist of clock parameters describing satellite time during the valid time interval.

TABLE 5.3   Subframe 1 parameters

Parameter	Description
Code on L2 flag	C/A or P or reserved; nominally P
Week number	GPS week since 1, January 1980, mod 1024
L2 P data flag	Navigation data on or off; nominally on
URA parameter	Exponent parameter for URA evaluation
Satellite health	Discretes describing the satellites health
T_{GD}	Additional clock correction for L1-only users
IODC	Issue of data—clock
t_{0c}	Reference time for clock corrections
a_{f0}	Satellite clock offset relative to GPS time
a_{f1}	Satellite clock fractional frequency offset
a_{f2}	Satellite clock drift rate

Other contents include the GPS week number, URA, health of satellite, various flags, and data reserved for authorized users. Content details are given in Table 5.3.

1. *Satellite clock corrections.* The satellite clock parameters describe a polynomial that allows the user to determine the effective satellite PRN code phase offset and carrier frequency offset. Δt_s is referenced to the phase center of the satellite antennas and with respect to GPS system time t. The equation for correcting the time t_s received from the satellite in seconds is

$$t = t_s - \Delta t_s \tag{5.56}$$

where

$$\Delta t_s = a_{f0} + a_{f1}(t_s - t_{0c}) + a_{f2}(t_s - t_{0c})^2 + \Delta t_r \tag{5.57}$$

where a_{f0}, a_{f1}, and a_{f2} are the polynomial coefficients transmitted in subframe 1, t_{0c} is the clock data reference time in seconds, and Δt_r is the relativistic term in seconds computed as

$$\Delta t_r = \frac{2\overline{\mathbf{X}}_s \cdot \overline{\mathbf{V}}_s}{c^2} = Fe\sqrt{A}\sin E_k \tag{5.58}$$

where the orbit parameters e, A, and E are derived from the data contained in subframes 2 and 3, and

$$F = \frac{-2\sqrt{\mu}}{c^2} = -4.442807633 \times 10^{-10} \quad \text{sec}/\sqrt{\text{meter}} \tag{5.59}$$

where

$$\mu = 3.986005 \times 10^{14} \quad \text{meters}^3/\text{sec}^2 \tag{5.60}$$

is the value of the Earth's universal gravitational parameter, and

$$c = 2.99792458 \times 10^8 \quad \text{meters}/\text{sec} \tag{5.61}$$

is the speed of light. These relativistic effects on the GPS satellite clocks are described in [35]. It suffices to use the *uncorrected* value of t_s in the evaluation of the above equations, since, prior to correction, it is within 1 msec of the corrected value.

Since the clock correction parameters estimated by the MCS are based upon L1 and L2 measurements combined for correction of the delay through the ionosphere, and since there is a potential L1/L2 differential bias in the satellite, the L1-only user must revise his computations of the satellite's clock offset with [7, 8, 36]

$$(\Delta t_s)_{\text{L1}} = \Delta t_s - T_{\text{GD}} \tag{5.62}$$

By specification [7, 8], this correction can be as large as 23.19 nsec.

The issue of data clock (IODC) indicates changes in the subframe 1 data, which occur nominally once per hour. It alerts the user if new data are being transmitted but does not indicate the age of the data. It should also be used by the user's receiver to match the issue of the ephemeris data provided in subframes 2 and 3.

2. *User range accuracy.* URA is the MCS's prediction of the pseudorange accuracy provided to the user for the satellite, indicating *no better than X meters*, where

$$\begin{aligned} X &= 2^{(1+N/2)} & N \le 6 \\ &= 2^{(N-2)} & N > 6 \end{aligned} \tag{5.63}$$

where $0 \le N \le 14$ is the parameter provided in subframe 1. $N = 15$ indicates the absence of an accuracy prediction. This accuracy does not include the effects of ionospheric delay modeling performed by a single frequency user.

3. *Satellite position computations.* Subframes 2 and 3 contain the ephemeris parameters required to compute the position $\overline{\mathbf{X}}_s$ and velocity $\overline{\mathbf{V}}_s$ of the satellite in ECEF coordinates. Table 5.4 lists these parameters. These parameters are

TABLE 5.4   Parameters of subframes 2 and 3

Parameter	Description
M_0	Mean anomaly at reference time
Δn	Mean motion differences from Computed Value
e	Eccentricity
$\sqrt{A}$	Square root of the semimajor axis
Ω_0	Longitude of ascending node of orbit plane at weekly epoch
i_0	Inclination angle at reference time
ω	Argument of perigee
$\dot{\Omega}$	Rate of right ascension
di/dt	Rate of inclination angle
C_{uc}	Amplitude of the cosine harmonic correction term to the argument of latitude
C_{us}	Amplitude of the sine harmonic correction term to the argument of latitude
C_{rc}	Amplitude of the cosine harmonic correction term to the orbit radius
C_{rs}	Amplitude of the sine harmonic correction term to the orbit radius
C_{ic}	Amplitude of the cosine harmonic correction term to the angle of inclination
C_{is}	Amplitude of the sine harmonic correction term to the angle of inclination
t_{0e}	Reference time for ephemeris computations
IODE	Issue of data—ephemeris

that of a curve fit over the time of validity (fit interval) and *not* necessarily an entire orbit. A *fit interval* flag in subframe 2 indicates whether the interval is four hours or greater than four hours. IODE has similar meaning for the ephemeris parameters as IODC has for the subframe 1 parameters.

The user receiver applies the parameters to a variation on the equations given in Section 5.3. However, not all the parameters are indicated in that section. These additional parameters reflect orbit drift and other orbit perturbations such as the sun and moon gravitational forces and solar radiation pressure. Typical equations for the computation of the satellite's position in ECEF coordinates are given in Table 5.5 [7, 8].

The accuracy of curve fit used in generating these parameters and subsequent truncation is quite good. For the four-hour fit, the user range error (URE) based on a projection of the curve fit error onto the user range is less than 0.4 meters, 1-sigma. For a six-hour fit, the URE degrades to 1.6 meters, 1-sigma. The equations provide the satellite's antenna phase center position in the WGS-84 ECEF reference frame defined in [7, 8].

Subframe 4 and 5 Data Content.   All of the data contained in subframes 4 and 5 are of the almanac variety; that is, the data describe long-term parameters associated with all GPS satellites and the GPS system itself. These subframes consist of 50 pages, not all of which are in use. Table 5.6 presents a summary of the data contained by page number.

1. *Satellite almanacs.* The ephemeris and clock almanac data are a subset

TABLE 5.5   Typical satellite position computations

Equation	Computation
$t_k = t - t_{0e}$	Time from ephemeris reference spoch t_{0e}
$n = n_0 + \Delta n$	Corrected mean motion
$M_k = M_0 + nt_k$	Mean anomaly at time t_k
$E_k = M_k + e \sin E_k$	Eccentric anomaly at time t_k
$\nu_k = \tan^{-1}\left(\dfrac{\sqrt{1 - e^2} \sin E_k}{\cos E_k - e} \right)$	True anomaly at time t_k
$\Phi_k = \nu_k + \omega$	Argument latitude at time t_k
$\delta u_k = C_{us} \sin 2\Phi_k + C_{uc} \cos 2\Phi_k$	Second harmonic perturbation to argument of latitude at time t_k
$\delta r_k = C_{rs} \sin 2\Phi_k + C_{rc} \cos 2\Phi_k$	Second harmonic perturbation to orbit radius at time t_k
$\delta i_k = C_{is} \sin 2\Phi_k + C_{ic} \cos 2\Phi_k$	Second harmonic perturbation to inclination angle at time t_k
$u_k = \Phi_k + \delta u_k$	Corrected argument of latitude at time t_k
$r_k = A(1 - e \cos E_k) + \delta r_k$	Corrected orbit radius at time t_k
$i_k = i_0 + \delta i_k + di/dt \cdot t_k$	Corrected inclination angle at time t_k
$x'_k = r_k \cos u_k$	Position in orbital plane at time t_k
$y'_k = r_k \sin u_k$	
$\Omega_k = \Omega_0 + (\dot{\Omega} - \dot{\Omega}_E)t_k - \dot{\Omega}_E t_{0e}$	Corrected longitude of ascending node at time t_k accounting for earth's rotation rate $\dot{\Omega}_e$
$x_k = x'_k \cos \Omega_k - y'_k \sin \Omega_k \cos i_k$	ECEF coordinates at time t_k
$y_k = x'_k \sin \Omega_k + y'_k \cos \Omega_k \cos i_k$	
$z_k = y'_k \sin i_k$	

of that provided in subframes 1 and 2, describing the complete orbits of all satellites in the constellation. The ephemeris parameters effectively consists of only the six basic Keplerian parameters plus the rate of right ascension. The clock parameters consist of only the clock offset and fractional frequency offset. All of the parameters are referenced to a common time 3.5 days in advance of the start of transmission. These parameters can be used for an extended period of time in the future provided that the number of weeks (since the almanac reference week number transmitted on page 25 of subframe 25) are accounted for in projection of the parameters to that time.

2. *Ionospheric delay model.* The ionospheric delay model is intended for the L1-only users. Eight model parameters are provided on page 18 of subframe 4 that serve as polynomial coefficients describing the maximum zenith amplitude and the time dependency of the model as a function of geomagnetic latitude of the Earth's projection of the ionospheric intersection point (IIP), assuming a mean ionospheric height of 350 km. This latitude is computed as a function of the user's geodetic latitude and lon-

TABLE 5.6   Page content of subframes 4 and 5

Pages	Content
Subframe 4, pages 1, 6, 11, 12, 16, 19, 20, 21, 22, 23 and 24.	Reserved for authorized users
Subgrame 4, pages 2, 3, 4, 5, 7, 8, 9 and 10	Ephemeris and clock data for satellites with PRNs 25 through 32
Subframe 4, pages 13, 14 and 15	Spare pages
Subframe 4, page 17	Special messages
Subframe 4, page 18	L1-only user's ionospheric delay model and GPS time/UTC conversion parameters
Subframe 4, page 25	A-S flags and satellite configurations for 32 satellites; health data for satellites 25 through 32
Subframe 5, pages 1 through 24	Ephemeris and clock data for satellites with PRNs 1 through 24
Subframe 5, page 25	Almanac reference week number and health data for satellites 1 through 24

gitude. Details of the algorithm are given in [7, 8] and [36]. The algorithm reduces the delay error for the single-frequency users on the order of 50% to 60% [37].

3. *UTC parameters.* Page 18 of subframe 4 also provides the parameters needed to relate GPS time to UTC and notices to the users regarding the scheduled future or recent past changes due to leap seconds. The parameters include the number of integer seconds between UTC and GPS time, plus first-order polynomial coefficients describing the drift between GPS time and UTC. However, this drift is kept to a minimum by the MCS by *steering* GPS time toward UTC modulo 1 second. Details of this relationship are described in [7, 8].

5.5.7 GPS Measurements and the Navigation Solution

Measurements   The basic GPS measurements are pseudoranges from N satellites as described in Equation 5.1, where $N \geq 4$, unless the solution is augmented with another type of measurement such as barometric altitude. Two other types of measurements are available to the avionics user—pseudorange rate (Doppler) and delta-pseudorange (integrated Doppler). These measurements are used to strengthen the determination of velocity and user clock drift, although the integrated Doppler measurement is sometimes used as a change in pseudorange, and thus change of position, in precise kinematic modes of operation being considered for precision landing applications [2].

The pseudorange rate is simply the derivative of Equation 5.1, where

$$\frac{dPR_i}{dt} = \dot{R}_i + c\Delta\dot{t}_{si} - c\Delta\dot{t}_u + \dot{\epsilon}_{PR_i} \tag{5.64}$$

In a linearized sense, the perturbation of the pseudorange rate is

$$\delta\,\frac{dPR_i}{dt} = \overline{\mathbf{1}}_i \cdot \delta\dot{\overline{\mathbf{X}}}_u - c\Delta\dot{t}_u + \dot{\epsilon}_{PR_i}$$

$$= \mathbf{h}_i[\delta\dot{\overline{\mathbf{X}}}_u \quad c\Delta\dot{t}_u]^T + \dot{\epsilon}_{PR_i} \tag{5.65}$$

where $\mathbf{h}_i$ is as defined in Equation 5.41.

Delta-pseudorange is usually used by an avionics receiver to approximate a pseudorange rate measurement using a relatively short integration time [38] (0.1 to 1 second, depending upon dynamics), where, in a linearized sense, the perturbation equation is simply Equation 5.65 multiplied by $\Delta t = t_j - t_{j-1}$, or multiplying the components of $\mathbf{h}_i$ and the error term by that quantity. In the case of the kinematic mode of operation, Equations 5.7 and 5.8 hold.

Measurement Errors   The error term is handled with a combination of corrections and modeled uncorrected error sources. The corrections for pseudorange measurements are as follows:

1. *Satellite clock relativity correction.* This correction is described by Equation 5.58. It is caused by the conservation of energy due to the difference in kinetic and potential energy of the satellite and the user. The correction is based on the theory of general relativity [35].
2. *Ionospheric refraction correction.* This correction is either made by measuring the difference in pseudorange $(PR_{1i} - PR_{2i})$ at two different frequencies and applying it to a dual frequency correction, or applying the single-frequency model defined in Section 5.5.6 and in reference [36]. The dual frequency correction is based on Equation 5.32 [18]. The correction for an L1 pseudorange is

$$\Delta PR_{1i} = \left(\frac{f_2^2}{f_1^2 - f_2^2} \right) (PR_{1i} - PR_{2i})$$

$$= 1.546(PR_{1i} - PR_{2i}) \tag{5.66}$$

Since the L2 delay is greater than the L1 delay, the correction is negative, decreasing the pseudorange.

3. *Tropospheric refraction correction.* This correction is made by subtracting the effects of Equation 5.36. There are many variations to this correction, depending upon the desired accuracy or complexity.

4. *Earth's rotation correction.* The error in pseudorange caused by the Earth's rotation is due to the fact that the Earth is rotating while the signal is propagating from the satellite to the user. The pseudorange error can be as large as 30 meters with a sign dependent on the user's position with respect to the satellite. The correction can be done in one of two ways [6]. The first is a correction in pseudorange, which is

$$\Delta PR_{\dot{\Omega}_{E}i} = -\frac{\dot{\Omega}_{E}}{c} \, (X_{si}Y_u - Y_{si}X_u) \tag{5.67}$$

The second correction is realized by rotating the satellite back in time about the Earth's axis to compute its position for a *slant* range rather than a *geometric* range. This is accomplished by modifying the equation for the corrected longitude of ascending node described in Table 5.5 by the value

$$\Delta\Omega_k = -\frac{\hat{R}_i}{c} \, \dot{\Omega}_E \tag{5.68}$$

where $\hat{R}_i$ is the estimated range to the satellite.

5. *Selective availability errors.* The authorized user (PPS mode) is also allowed to correct for system-induced SA errors.

Uncorrected Error Sources   After all the corrections stated above, residual error sources remain. Among the most predominant are SA errors for the unauthorized user (SPS mode). The next most predominant error source is the residual ionospheric refraction error for the L1-only users. Next is the standard system errors in the satellite clock parameters, followed by multipath errors, the tropospheric refraction, the satellite ephemeris, and finally receiver noise. New receiver technology has reduced the effects of multipath and receiver noise, with the exception of operating in a jamming environment.

The Navigation Solution   The standard methods for producing a navigation solution (position and velocity) are using Kalman filtering and sequential least-squares methods (see Chapter 3). Military applications generally use Kalman filtering, while commercial applications generally use least squares. A general least-squares method is illustrated here.

The basic assumption is that an initial (or previous) solution estimate is known (which could initially simply be the center of the Earth). Then residuals are computed with respect to that initial (or previous) estimate and sequentially updated based upon the latest projected solution estimate. The resulting measurement residuals are then defined as

$$\delta \overline{\mathbf{M}} = \left[\delta \overline{\mathbf{PR}}^{\mathrm{T}} \left(\frac{\mathrm{d} \delta \overline{\mathbf{PR}}}{\mathrm{d} t} \right)^{\mathrm{T}} \right]^{\mathrm{T}}$$

$$= \mathbf{H}_T [\delta \overline{\mathbf{X}}_u^{\mathrm{T}} \quad \delta c \Delta t_u \quad \delta \dot{\overline{\mathbf{X}}}_u^{\mathrm{T}} \quad \delta c \Delta \dot{t}_u]^{\mathrm{T}}$$

$$= \mathbf{H}_T \delta \overline{\mathbf{PV}}_u \tag{5.69}$$

where $\delta \overline{\mathbf{M}}$ is a $2N$ vector of pseudorange and pseudorange rate (or delta-pseudorange) residuals (difference between estimated and measured) for N satellites, $\delta \overline{\mathbf{PV}}_u$ is a 8×1 state vector made up of position, velocity, and time state residuals, and

$$\mathbf{H}_T = \begin{bmatrix} \mathbf{H} & \mathbf{0} \\ \mathbf{0} & \mathbf{H} \end{bmatrix} \tag{5.70}$$

is a $2N \times 8$ matrix mapping the position, velocity, and time state residuals into the $2N$ measurement residuals. The $\mathbf{H}$ matrix could be the $\mathbf{H}_{\mathrm{LTP}}$ matrix of Equation 5.44 if the solution is in the north, east, and up domain. Since the $\mathbf{H}_T$ is general not a square matrix, the solution is overly determined. In that case, the solution for $\delta \overline{\mathbf{PV}}_u$ in a weighted least-squares sense is

$$\delta \overline{\mathbf{PV}}_u = [\mathbf{H}_T^{\mathrm{T}} \mathbf{W} \mathbf{H}_T]^{-1} \mathbf{H}_T^{\mathrm{T}} \mathbf{W} \delta \overline{\mathbf{M}} \tag{5.71}$$

where $\mathbf{W}$ is a matrix weighting the measurements from the various satellites according to their accuracy or other weighting criteria. This perturbation solution is then added to the projected estimate previously used to compute the residuals. If the weights are inversely proportional to the variance of the measurement errors, the position, velocity, and time estimation error covariance matrix is then

$$\mathbf{C}_{\overline{\mathbf{PV}}_u} = [\mathbf{H}_T^{\mathrm{T}} \mathbf{W} \mathbf{H}_T]^{-1} \tag{5.72}$$

5.5.8   Aviation Receiver Characteristics [121 (Chapter 8)]

GPS avionics receiver technology has evolved significantly since the start of development in 1974 and is continuing to evolve for military avionics applications and especially for commercial avionics applications. The GPS avionics receiver characteristics vary considerably, from the low-end general aviation receiver to the high-performance military aircraft and missile avionics receivers. Only the high-end commercial and military avionics receiver characteristics are described in this chapter.

General Characteristics   The architecture of commercial and military avionics receivers is generally the same. They only differ in design parameters related to the environmental, dynamic capability, anti-jam (AJ), integrity, and security requirements. Except for the integrity requirements (see Sections 4.3.1, 5.7.1, and 5.7.3), the military avionics receiver dynamic performance requirements are usually more stringent, while commercial avionics applications are more concerned with safety-of-flight integrity requirements. Prior to describing the differences related to these requirements, a summary description of a generic GPS avionics receiver is given. More details follow in the description of commercial and military avionics receivers, respectively.

A system level functional block diagram of a generic GPS avionics receiver is shown in Figure 5.17. The receiver consists of the following major functions: (1) antenna, (2) preamplifier, (3) reference oscillator, (4) frequency synthesizer, (5) downconverter, (6) an intermediate frequency (IF) section, (7) signal processing, and (8) navigation processing.

The antenna may consist of one or more elements and associated control electronics, and it may be passive or active, depending upon its performance requirements. Its function is to receive the GPS satellite signals while rejecting multipath and, in some military applications, to reject interference signals. The passive antenna usually has a hemispheric coverage pattern.

The parameters that dictate the antenna requirements are as follows: gain versus azimuth and elevation, multipath rejection, interference rejection, profile, size and environmental conditions. The gain requirements are a function of satellite visibility requirements and are closely related to multipath rejection and somewhat related to interference rejection. The goal is to have near-uniform gain toward all satellites above a specified elevation angle while rejecting multipath signals and interference typically present at low-elevation angles. These are usually conflicting requirements. Some multipath rejection can also be achieved by reducing the left-hand circularly polarized (LHCP) gain of the antenna without reducing the right-hand circularly polarized (RHCP) gain. This is because the satellite signals are RHCP signals, whereas reflected multipath signals usually tend to be either linearly polarized (LP) or LHCP, depending upon the dielectric constant of the reflecting surface.

Interference rejection can also be achieved using a phased-array antenna, where the relative phase received from each antenna is controlled to "null" out the interference in the combined reception. This type of an antenna is called a *controlled reception pattern antenna* (CRPA), and it is usually reserved for military applications. A low antenna profile is important in the avionics application to minimize drag, but this is traded off against a desired gain pattern.

Environmental conditions dictate the type of material used for the antenna and whether a radome is required. Some materials change their dielectric properties as a function of temperature.

The minimum operational performance standards for commercial avionics antennas are given in [122b].

The preamplifier generally consists of burnout protection, filtering, and a

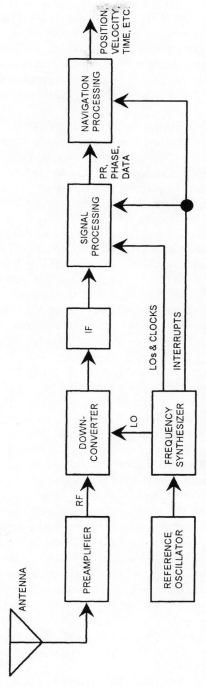

Figure 5.17 Generic GPS avionics receiver functional block diagram.

231

low-noise amplifier (LNA). Its primary function is to set the receiver's noise figure (see Section 4.2.1) and to reject out-of-band interference. The parameters that dictate the preamplifier requirements are the unwanted RF environment as received through the antenna and losses that precede and follow the preamplifier and desired system noise figure (or noise temperature) as derived from overall receiver performance requirements. The gain of the preamplifier is not a system level requirement per se, but a derived requirement that satisfies the system level requirement.

The unwanted RF environment as received through the antenna affects the preamplifier in two ways. It can cause damage to the preamplifier electronics, or it can cause saturation of the preamplifier and circuitry that follows. Normally, except for damage prevention, one can do nothing to suppress the RF environment, as passed by the antenna, at frequencies that are in the bandwidth of the desired GPS signal. That environment is considered to be either jamming or unintentional interference. There are, however, more advanced temporal interference suppression techniques that can be used to suppress narrowband interference [121, (Chapter 20)]. Suppression of the RF environment out of the desired GPS signal band can be accomplished by filtering before, during, and/or after amplification. When it is accomplished it is based upon a trade-off between system noise figure requirements, filter insertion loss, and bandwidth efficiency. Suppression of in-band and out-of-band damaging interference is usually accomplished with diodes that provide a ground path for strong signals. In the case of lightning protection, more complex lightning arrestors may be used.

The system noise figure is set using a low-noise amplifier (LNA) that provides enough gain to cause any losses inserted after the LNA to have a negligible effect. An LNA cannot account for losses inserted before its operation or for its own noise floor.

The reference oscillator provides the time and frequency reference for the receiver. Since GPS receiver measurements are based on the time of arrival of PRN code phase and received carrier phase and frequency information, the reference oscillator is a key function of the receiver. Its output is used by the frequency synthesizer, which converts the oscillator output to local oscillators (LOs) and clocks used by the receiver. One or more of those LOs are used by the downconverter to convert the radio frequency (RF) inputs to IF frequencies. The signals are easier to process in the IF section of the receiver.

The requirements on reference oscillators for avionics receivers vary depending upon the avionics application. A high-quality oscillator can be the most significant cost item of a modern receiver. Thus, there are compromises made on oscillator performance. There are some commercial and military applications where refrence oscillator performance is critical.

Typical requirements parameters applied to reference oscillators are as follows:

- Size. Stable oven-controlled crystal oscillators (OCXOs) and rubidium oscillators can be quite large. Temperature compensated crystal oscillators (TCXOs) are relatively small. Larger oscillators have more temperature inertia.

- Power. OXCOs and rubidium oscillators consume significant power.
- Short-term stability (less than 10 sec) due to temperature, power supply, and natural characteristics. Short-term stability affects the ability to estimate and predict time and frequency in the receiver.
- Long-term stability (greater than and 10 sec, up to hours and days) due to natural characteristics, including crystal aging.
- Sensitivity to acceleration—g force and vibration sensitivity. Vibration causes phase noise and dynamic g forces affect the ability to estimate time and frequency in the receiver.
- Phase noise, High-frequency stability (frequencies above 1 Hz offsets from the nominal frequency).

Mostly, the requirements placed on the frequency synthesizer are derived requirements and are the receiver designer's choice. Its design is based on the designer's *frequency plan* that defines the receiver's IF frequencies, sampling clocks, signal processing clocks, and so on. The *frequency plan* requires careful analysis to insure adequate rejection of mixer harmonics, LO feedthrough, unwanted sidebands, and images. A key design parameter for the synthesizer is the minimization of phase noise generated in the synthesizer. The frequency synthesizer is also required to generate local clocks for signal processing and interrupts for the navigation processing. These local clocks comprise the receiver's time base.

The downconverter mixes LOs generated by the frequency synthesizer with the amplified RF input to an IF frequency and, if so designed, the IF frequency to lower IF frequencies. This process implements the *frequency plan*, which again is the receiver designer's choice.

The purpose of the IF section is to provide further filtering of out-of-band noise and interference and to increase the amplitude of the signal-plus-noise to a workable signal-processing level. The IF section may also contain automatic gain control (AGC) circuits to increase the dynamic range of the receiver and to suppress pulse-type interference.

The requirements on the IF section are as follows:

1. Final rejection of out-of-band interference, unwanted sidebands, LO feedthrough, and harmonics. The bandwidth of this rejection is a trade-off against correlation loss due to filtering. In addition, rejection of wideband noise is required to minimize aliasing in the signal-processing sampling process.
2. Increase the amplitude of the signal-plus-noise to workable levels for signal processing and control that amplitude, as required, for signal processing (AGC).

3. Suppress pulse-type interference.

4. Converts the IF signal to a baseband signal comprised of in-phase (*I*) and quadraphase (*Q*) signals.

The signal-processing function of the receiver is the core of a GPS receiver that performs the following functions: (1) splits the signal-plus-noise into multiple signal-processing channels for processing of multiple satellite signals simultaneously, (2) generates the reference PRN codes of the signals, (3) acquires the satellite signals, (4) tracks the code and the carrier of the satellite signals, (5) demodulates the navigation message data from the satellite signals, (6) extracts code phase (pseudorange) measurements from the PRN code of the satellite signals, (7) extracts carrier frequency (pseudorange-rate) and carrier phase (delta-pseudorange) measurements from the carrier of the satellite signals, (8) extracts signal-to-noise ratio information from the satellite signals, and (9) maintains a relationship to GPS system time.

The prime requirement of the signal processing is to provide the GPS measurements and navigation message data from selected satellites required to perform the navigation-processing function. The outputs of the signal processing function are pseudoranges, pseudorange rates and/or delta-pseudoranges, signal-to-noise ratios, local receiver time-tags, and GPS system data for each of the GPS satellites being tracked.

The navigation-processing function controls the signal-processing function and uses its outputs to satisfy the navigation requirements. It does this by performing some or all of the following functions:

1. Selects satellites to be acquired and tracked by the signal-processing function.

2. Computes signal acquisition and tracking aiding information for the signal-processing function.

3. Reinitializes the signal-processing function in case of loss of lock.

4. Collects measurements and navigation message data from the signal-processing function and maintains a system data base.

5. Computes the satellites' positions, velocities, and time corrections as described in Section 5.5.6.

6. Corrects the measurements as described in Section 5.5.7.

7. Accepts and processes external navigation data to aid the navigation processing.

8. Solves for position, velocity, and time as described in Section 5.5.7.

9. Determines the integrity of the position and velocity solutions.

10. Performs area navigation as described in Chapters 2 and 14.

11. Performs input/output processing. In the case of the commercial avionics receivers, this processing is usually in accordance with ARINC 429 standards. In the case of the military avionics receivers, the usual inter-

face is in accordance with MIL-STD-1553 bus standards exercising GPS 1553 protocols.

12. Accepts appropriate crypto keys in military avionics receivers and performs computations required for SA and A-S.

Commercial GPS Aviation Receivers   In 1996, GPS receivers for commercial avionics are still in the developmental stage, awaiting the definition of required navigation performance (RNP) for sole means navigation. However, a properly designed and tested receiver can be used for supplemental means navigation. Requirements for a supplemental means GPS receiver are defined in the Federal Aviation Administrations (FAA) Technical Standard Order TSO-C129 [39] for all phases of flight except precision approaches, which is based upon RTCA's *Minimum Operational Performance Standards (MOPS) for Airborne Supplemental Navigation Equipment Using Global Positioning System (GPS)*—RTCA/DO-208 [40]. The European minimum operational performance specifications for airborne GPS receiving equipment are given in [123]. Similar standards are being developed for sole means navigation, including those for precision approach [41]. Precision approach requires the concept of differential GPS (DGPS), which is described in Section 5.5.9.

The characteristics for a GPS avionics receiver (GPS sensor) for commercial transport applications are specified in *ARINC Characteristic 743A-1 GNSS Sensor* [42]. These characteristics are specified for two configurations, one where the receiver and LNA are packaged separately (2 MCU configuration) and one where the receiver and LNA are packaged together for installation near the antenna (alternate configuration).

Commercial GPS avionics receivers are specified in ARINC 743A-1 to use the C/A code only, which limits them to the L1 frequency. Since some of these receivers are being developed with precision landing capabilities in mind, high-accuracy C/A code tracking is required. Thus, in this section, the receiver used as a model for the description of commercial GPS avionics receivers is one that uses state-of-the-art technology for this high accuracy, namely the 12-channel NovAtel GPSCard™. This receiver uses the *narrow correlator spacing* technology that reduces the effects of ambient noise and multipath [43, 44]. An OEM card version is illustrated in Figure 5.18. This card is being upgraded for the commercial aviation environment. A functional block diagram of this receiver card is illustrated in Figure 5.19. The following is a description of those functions:

1. *Antenna/LNA*. The antenna shown in Figure 5.20 is not peculiar to the described receiver card. It was developed and is manufactured by Sensor Systems (P/N S67-1575-39). It is an FAA TSO-C115a certified airborne radome-protected antenna that includes a built-in 26 dB LNA, bandpass filtering and lightning protection. The LNA is powered through the coaxial cable with a dc voltage between 4 and 24 v at a maximum of

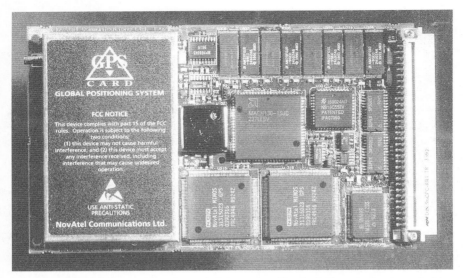

Figure 5.18 OEM GPS receiver card (courtesy, NovAtel Communications, Ltd.)

100 mw. The antenna is 3.5 in. in diameter and weighs 5 oz. It meets the requirements specified in references [42] and [122].

2. *Reference oscillator.* Like most commercial receivers, this receiver uses a small TCXO as its reference oscillator (on the center of the board shown in Figure 5.18). These small TCXOs have marginal stability and phase noise characteristics for some GPS applications, but they are satisfactory when used in conjunction with a multiple-parallel-channel receiver with tracking loop bandwidths commensurate with dynamic applications. Note that the reference oscillator output (20.473 MHz) is used directly as clocks for sampling and signal processing.

3. *Synthesizer/downconverter.* A block diagram of the synthesizer and downconverter is shown in Figure 5.21 [43]. A commercial synthesizer chip with a programmable divider (prescaler) is used to phase lock a voltage-controlled oscillator (VCO) to the reference oscillator. The VCO output frequency is doubled to provide the downconverter LO.

4. *Filtering.* IF filtering is realized with a 8 MHz bandwidth surface acoustic wave (SAW) filter centered at the IF frequency of 35 MHz [43]. Although this SAW filter has a significant group delay, the delay is stable, and the filter provides excellent rejection of out-of-band noise and interference.

5. *IF sampling and A/D conversion.* The IF sampling and A/D conversion process is illustrated in Figure 5.22. This processing includes a precorrelation automatic gain control (AGC) [43] that controls the level

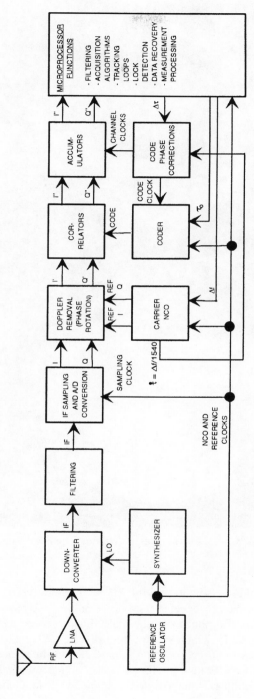

Figure 5.19 Functional block diagram of commercial receiver (courtesy, NovAtel communications, Ltd.)

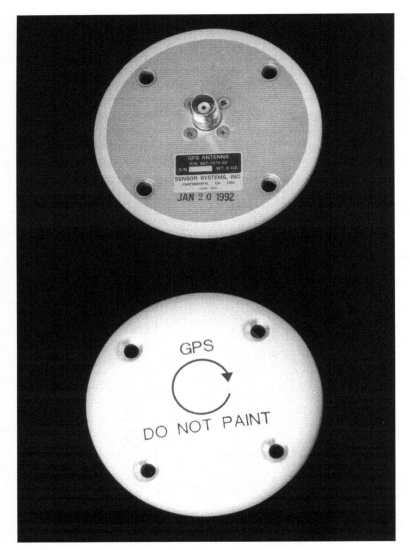

Figure 5.20   Civil aviation antenna (courtesy, Sensor Systems)

of the signal-plus-noise entering the 2.5-bit A/D converter for optimum quantization in the presence of both noise and narrowband interference. Typical lower-performance commercial GPS receivers use 1-bit quantization with no AGC, which suffer 2 to 3 dB loss of signal-to-noise ratio, depending upon the precorrelation bandwidth, in the presence of noise, and up to 6 to 8 dB loss in the presence of narrowband interference [45–50]. The multi-bit quantization coupled with AGC is important in

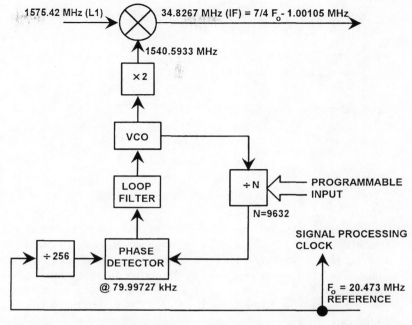

Figure 5.21 Synthesizer/downconverter block diagram (courtesy, NovAtel Communications, Ltd.)

aviation applications in which good receiver sensitivity and interference rejection is a characteristics [42]. The IF sampling rate of $4f_{IF}/N_s$, where N_s is an odd number (designer's choice), is a method of converting in-phase (I) and quadraphase (Q) samples directly, without first converting the IF signal to baseband. The result is a sequence of samples of the signal components [43]:

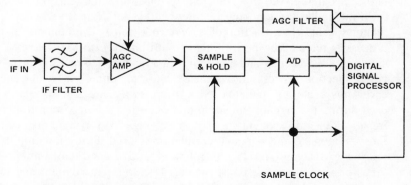

Figure 5.22 Typical IF sampling and A/D conversion process.

$$I_{sk}, Q_{sk}, -I_{sk}, -Q_{sk}, I_{sk}, -I_{sk}, -Q_{sk}, \cdots \qquad (5.73)$$

or

$$I_{sk}, -Q_{sk}, -I_{sk}, Q_{sk}, I_{sk}, -Q_{sk}, -I_{sk}, Q_{sk}, \cdots \qquad (5.74)$$

depending upon the value of N_s, where

$$I_{sk} = AC_k D_k \cos \phi_k \qquad (5.75)$$
$$Q_{sk} = AC_k D_k \sin \phi_k \qquad (5.76)$$

where C_k, D_k, and ϕ_k are the code, data bit, and signal phase at sample time t_k.

6. *Doppler removal (phase rotation).* The Doppler removal process of Figure 5.19 is part of the signal-phase or frequency-tracking function, which is a complex multiplication between the signal I and Q samples and reference I and Q samples generated by the carrier number-controlled oscillator (NCO). This NCO is controlled by the microprocessor's portion of the carrier-tracking loop. Since the C/A PRN code Doppler is related to the carrier Doppler by a factor of 1540, the carrier NCO also outputs the basic code clock Doppler correction, which is further corrected by the microprocessor's code tracking loop function with code phase corrections. In some receiver implementations a completely separate NCO is used to derive the code Doppler and phase.

7. *Coder.* The implementation of a typical C/A coder is shown in Figure 5.13. This implementation, in which the G2 state is initialized, allows the generation of *all* 1023 codes in the C/A code family, which is important for future implementations [32].

8. *Correlators.* The correlation process for narrow correlator processing is illustrated in Figure 5.23 [43, 44]. In this process early, punctual, and late codes are derived in a shift register that shifts the (early) C/A code from the coder at a clocking rate defined by the desired early/late correlator spacing (in a fraction of a C/A code chip). The dual correlation process is realized by performing a multi-bit *exclusive or* between the single-bit PRN codes and the multi-bit I and Q samples. A discriminator selection process allows the selection of either early and late or early-minus-late and punctual correlation. Early and late correlation is used during the signal acquisition process using the maximum (approximately) one chip spacing ($N = 10$ in Figure 5.23) for rapid acquisition. Early-minus-late and punctual correlation is used during tracking using a 0.1 chip spacing ($N = 1$) for optimum parallel code (early-minus-late times punctual) and carrier (punctual) tracking. This dynamic spacing concept provides fast acquisition and C/A code-tracking performance

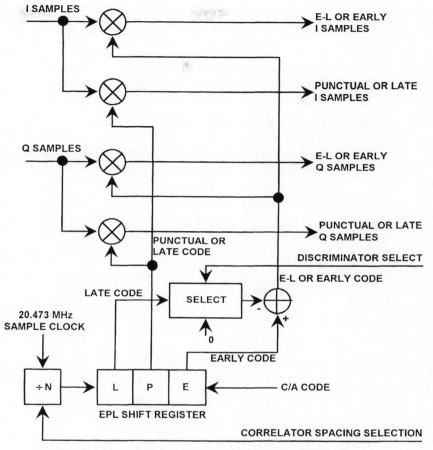

Figure 5.23   Dynamic correlation spacing process [43].

that approaches that of conventional P code-tracking performance. This is because the code-tracking error variance as a function of correlator spacing d and signal-to-noise density S/N_0 is given as [44]

$$\sigma_\tau^2 = \frac{B_L d}{S/N_0} \left[1 + \frac{B_{IF}}{S/N_0} \right] \quad \text{chips}^2 \tag{5.77}$$

where B_L is the single-sided tracking loop bandwidth and B_{IF} is the two-sided predetection bandwidth. The d for $N = 1$ is one-tenth that for the conventional $N = 10$, while the signal-to-noise density for C/A code tracking is twice that for P code tracking.

9. *Postcorrelation filtering (accumulators).* After correlation, the two sets

of I and Q samples are accumulated and dumped to the microprocessor (Figure 5.19) at the C/A code epoch rate to provide 1-kHz predetection bandwidth I and Q samples for further processing at that rate. That rate is used directly for wideband signal acquisition and bit synchronization after acquisition. Bit synchronization determines bit timing, after which the predetection bandwidth is reduced to the navigation message data rate of 100 Hz for optimum signal processing.

10. *Microprocessor signal processing.* The receiver in Figure 5.18 uses a 32-bit microprocessor with a built-in math coprocessor to perform both the signal processing and the navigation processing [43]. The different signal-processing functions are indicated in Figure 5.19. In summary, these are as follows:

- Signal acquisition is accomplished by computing signal-plus-noise power $\sum_{k=1}^{K} (I_k^2 + Q_k^2)$ at one-half C/A code chip increments until it exceeds a threshold based upon an estimate of noise power.

- Carrier phase tracking is accomplished by minimizing $\tan^{-1}(Q_k/I_k)$.

- Code tracking is accomplished by minimizing either a dot product discriminator

$$d\tau = I_{E-L}I_P + Q_{E-L}Q_P \tag{5.78}$$

or an early-minus-late power discriminator

$$d\tau = I_E^2 + Q_E^2 - I_L^2 - Q_L^2 \tag{5.79}$$

both of which minimize early-minus-late correlation amplitude, where the I and Q options and early/punctual/late correlator spacing are controlled as indicated in Figure 5.23.

- Data demodulation is accomplished by sampling the sign of I_k while tracking the carrier phase.

- Bit synchronization is accomplished by sensing sign changes in 1-kHz data samples.

- Frame synchronization is accomplished by correlating with the navigation message preamble at the beginning of the TLM Word (see Section 5.5.6).

- Phase lock is verified with the computation of the correct data parity or through the computation of $\sum_{k=1}^{K} (I_k^2 - Q_k^2)$, which is an estimate of the cosine of carrier phase.

- Signal-to-noise density computations.

- Formulation of pseudorange and delta-pseudorange measurements.

In the receiver of Figure 5.18, signal processing is optimized to accommodate up to 6 g of acceleration.

Military GPS Aviation Receivers   In 1996, the standard GPS receiver for military avionics was the miniature airborne GPS receiver (MAGR) produced by Rockwell International, which also produces another variety intended for embedded applications, where the entire receiver (the miniature GPS receiver, MGR), less antenna, is housed inside another avionics assembly such as an inertial navigation system [51]. The requirements for the MAGR are defined in [52]. Guidelines for embedded military receivers are specified in [53] and described in [54]. The MAGR receiver is illustrated in Figure 5.24.

Figure 5.24   MAGR receiver (courtesy, Rockwell International).

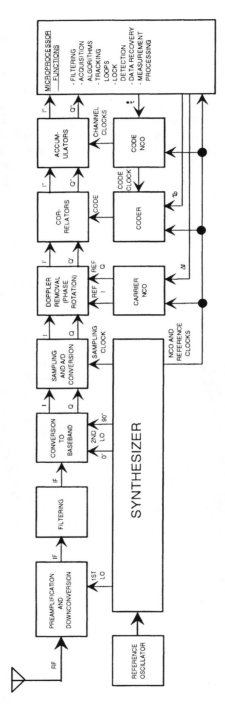

Figure 5.25 Functional block diagram of MAGR receiver functions [55].

The MAGR is a stand-alone five-channel PPS L1/L2 receiver, although there are two configurations (Air Force and Navy). The Air Force configuration does not include the Antenna Electronics (AE), which consists of the LNA and down-converter, while the Navy configuration does [52]. In the Air Force configuration, the AE is remote at the antenna. The Navy configuration is intended to be located near the antenna.

The functional description to follow will be for the Navy MAGR configuration only. Its functional block diagram is given in Figure 5.25 [55]. Note that it is functionally similar to the commercial receiver illustrated in Figure 5.19. The following is a description of those functions:

1. *Antenna.* The antenna shown in Figure 5.26, a version of the military Fixed Radiation Pattern Antenna (FRPA-3), is the Dorne & Margolin DM C146-10-2 L1/L2 antenna being used on the F-18, the AV-8-B Harrier and the TR-1, the new version of the Lockheed U-2 [56]. It is a low-profile antenna (1.5 in. high) with a diameter of less than 5 in. It weighs 0.5 lb.

2. *Preamplification, downconversion, reference oscillator and synthesizer.* The pre-select filtering, burnout protection, and LNA are made up of conventional discrete components [50]. The downconversion and synthesizer comprise two custom silicon bipolar integrated circuit chips as shown in

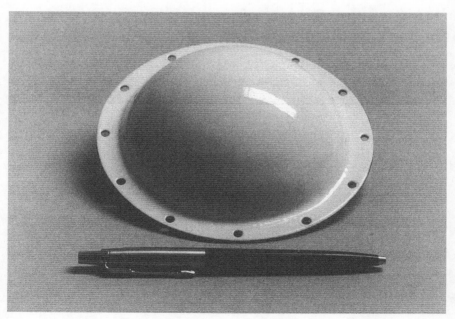

Figure 5.26   FRPA antenna (courtesy, Dorne & Margolin).

Figure 5.27. Dual downconversions are accomplished, one for L1 and one for L2. The LO for these downconversions are common, converting both RF frequencies to identical IF frequencies. This LO is derived in the synthesizer, which is driven with the output of an ovenized reference oscillator at a frequency of approximately 10.95 MHz. The synthesizer also generates in-phase and quadraphase LOs for later conversion to baseband and clocks for the signal-processing function.

3. *AGC, conversion to baseband and A/D conversion.* This process is applied to both the L1 and the L2 IF signals through identical wideband IF silicon bipolar chips, whose block diagram is shown in Figure 5.28

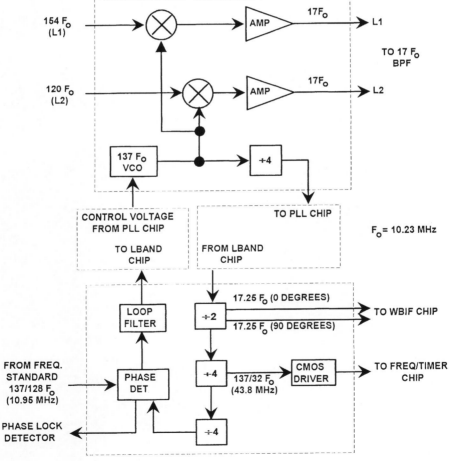

Figure 5.27   MAGR downconversion/synthesizer silicon bipolar chips [55].

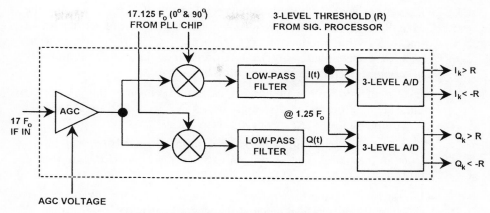

Figure 5.28   MAGR wideband IF chip [55].

[55]. Conversion to baseband is realized through mixing the L1 and L2 IF signals with in-phase and quadraphase LOs from the synthesizer. Filtering follows to limit the bandwidth to 20 MHz, the bandwidth of the P code, to prevent aliasing in the sampling process. The AGC is a rapid wideband AGC, whose purpose is to suppress pulsing interference as well as to provide a large dynamic range (70 dB) to accommodate high levels of jamming. The sampling process, at double code clock rate of 20.46 MHz, provides 1.5 bit I and Q samples to the signal-processing function. The thresholds (R and $-R$) are controlled by the signal processing for optimum performance in the presence of noise and narrowband interference.

4. *Doppler removal, coder, correlators, and postcorrelation filtering.* These functions are performed in the MAGR in a similar manner to that of the receiver shown in Figure 5.19, with the following exceptions:
   - The code Doppler removal is realized with an independent code NCO.
   - The coder function consists of both the C/A and P coders as well as the P code encryption function to produce the Y code.
   - The early-minus-late correlator spacing is fixed at one C/A or one P chip depending upon which coder is being used.
   - Rather than two sets of I and Q correlators and accumulators in each of twelve channels, the MAGR has ten sets of I and Q correlators and accumulators in each of five channels. This configuration provides fast signal acquisition and reacquisition in the presence of jamming.
5. *Microprocessor signal processing.* The MAGR uses Rockwell's AAMP-2 microprocessor to perform both the signal processing and the navigation processing [57]. The signal-processing functions are indicated in Figure 5.25. This includes variations on all the functions depicted in Figure 5.19 for the commercial receiver, with the following exceptions [58]:

- Because of the very high dynamic requirements (9g, 10g/sec [59]), the MAGR performs carrier frequency tracking, instead of carrier phase tracking, by minimizing $\tan^{-1}(Q_k/I_k) - \tan^{-1}(Q_{k-1}/I_{k-1})$ (time difference of carrier phase error). Data are demodulated differentially by observing the changes in the sign of I_k.

- Instead of verification of phase lock, measurement validity is determined comparing signal-plus-noise power $\sum_{k=1}^{K} (I_k^2 + Q_k^2)$ to a threshold based upon an estimate of noise power.

- Code-tracking loop aiding processed from corrected external inputs from an inertial navigation system to achieve a high AJ tracking capability [52, 60].

- L2 tracking for ionospheric delay corrections is normally performed sequentially on the fifth channel. However, if it is determined that L1 is jammed, L2 tracking can be performed on some or all channels [52, 60].

6. *A-S and SA processing.* The MAGR incorporates a precise positioning service-security module (PPS-SM) and auxilliary output chip (AOC) devices to perform A-S and SA processing [52, 59]. The AOC devices are provided for each channel to allow tracking of the encrypted code when the receiver is properly authorized. The PPS-SM performs crypto-key processing and management for the A-S and SA processing. It operates on battery power so that keys may be loaded or zeroized without receiver prime power. A dedicated data path from the PPS-SM to the AOC devices prevents sensitive data from being handled by the other processors.

5.5.9 Differential GPS

The concept of differential GPS (DGPS) is illustrated in Figure 5.29 [61, 62, 21]. DGPS requires a reference station at a known location that receives the same GPS signals as does the avionics user. This reference station processes its GPS measurements, deriving pseudorange, delta-pseudorange, and pseudorange-rate errors with respect to its accurately known location and then transmits these corrections to participating users in the area. The avionics user then applies these corrections to his measurements, thus canceling all common errors. Sub-meter accuracies to accuracies of 10 meters have been experienced using DGPS depending upon techniques used and distance from the reference station.

This differential technique works if the common errors are bias errors due to causes outside of the receiver. The major sources of common errors are the following [63, 64]:

1. *Selective availability errors.* Although these errors are not biases, they have correlation times that are long enough to be eliminated if the cor-

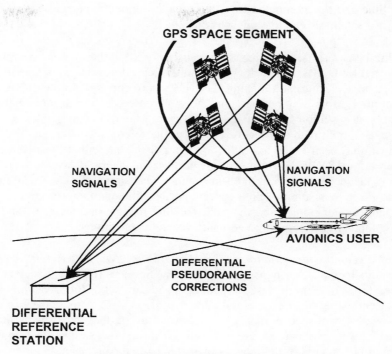

Figure 5.29 Differential GPS concept.

rection update rate is high enough. Typical pseudorange errors are about 30 meters, 1-sigma, but they have the potential to be higher.

2. *Ionospheric delays.* These propagation group delay errors can be as high as 20 to 30 meters during the afternoon hours to 1 to 6 meters at night, if not removed using two frequency corrections. The single frequency model will reduce this by approximately 50%. These errors are slowly varying biases but spatially decorrelate over larger distances.

3. *Tropospheric delays.* These propagation delays can be as much as 30 meters to a low-elevation satellite but are quite consistent and can be modeled. However, variations in the index of refraction can cause differences between the reference station and the user of 1 to 3 meters for low-elevation satellites. They are also slowly varying biases, and they spatially decorrelate over larger distances.

4. *Ephemeris errors.* Normally, the difference between the actual satellite location and the location computed from the broadcast ephemeris is small, less than 1 to 3 meters, but this error can be increased significantly with selective availability. Ephemeris errors are very slowly varying biases but can spatially decorrelate over large distances.

5. *Satellite clock errors.* The differences between the actual satellite clock

time and that computed from the broadcst corrections can become large if a satellite's clock is misbehaving.

Satellite clock errors, including those caused by SA dithering, are completely eliminated by DGPS, except for the SA dithering effects due to delays in estimating, broadcasting, and making the DGPS corrections. As noted above, all of the other errors may not be completely eliminated as the distance between the reference station and the user increases. The following errors are not eliminated—multipath and receiver noise at the reference station and the user. Multipath has been found to be the dominant error source that limits the accuracy in a local DGPS environment, while the ionospheric delay is usually the limiting factor for achieving the best accuracy over large distances. Care must also be taken to ensure that both the reference station and the user are performing their computations using the same satellite navigation messages, and that their computations are performed using accurate algorithms.

RTCM Recommended Standards for Differential NAVSTAR GPS Service
The Radio Technical Commission for Maritime Services (RTCM) Special Committee SC104 took on the task of defining a standard set of broadcast messages for disseminating GPS differential corrections [63, 64]. Although these standards were established primarily for maritime users, they have been applied successfully to aeronautical uses as well. However, the aeronautical community is in the process of defining its own standards for precision landing applications [41]. The RTCM SC104 standard has, however, provided the framework for other applications, including differential techniques similar to that used by the surveyors for kinematic surveying [65]. Data links and data link protocols have not been standardized.

The key standardized message types that have been fixed include the following [63, 64]:

1. Differential GPS corrections made up of pseudorange and pseudorange rate corrections for all satellites in view of the reference station. A user differential range error (UDRE) indicating the accuracy of each correction, plus the issue of satellite navigation data (IOD) are also included. The corrected pseudorange for satellite i is then

$$PR_{ic}(t) = PR_i(t) + PRC_i(t_0) + \frac{dPRC_i(t_0)}{dt}(t - t_0) \qquad (5.80)$$

where $PRC_i(t_0)$ and $dPRC_i(t_0)/dt$ are the broadcast corrections for satellite i at their time of applicability t_0.

2. Delta-differential GPS corrections made up of corrections to the broadcast corrections applicable to the previous issue of satellite navigation data (IOD) for a period of time after an IOD change. These delta cor-

rections for the pseudorange and pseudorange corrections for satellite i, respectively, are

$$\Delta PRC_i(t_0) = PRC_i(t_0, \text{IOD}_{\text{old}}) - PRC_i(t_0, \text{IOD}_{\text{new}}) \tag{5.81}$$

$$\Delta \frac{dPRC_i(t_0)}{dt} = \frac{dPRC_i(t_0, \text{IOD}_{\text{old}})}{dt} - \frac{dPRC_i(t_0, \text{IOD}_{\text{new}})}{dt} \tag{5.82}$$

3. Reference station parameters made up of the WGS 84 ECEF coordinates of the reference station with a resolution accuracy of 0.01 meters.
4. High-rate differential GPS corrections made up of the same contents as the differential GPS corrections, but for only those satellites having high rates of change of differential corrections. This message is used in the case when the normal broadcast rate of transmission of differential GPS corrections is not high enough to maintain the desired accuracy for those satellites. The cause of this would be excessively high-bandwidth SA errors, coupled with a low-bandwidth differential broadcast link.

Although DGPS provides a significant increase in accuracy over standard GPS, especially when SA is invoked, the user time solution is no longer a solution with respect to GPS time unless the reference station's time is synchronized to GPS time. The time solution is now with respect to the time base of the reference station. Normally, the practice is to maintain a time solution at the station that drives the average of the corrections of all satellites in view to zero, under the assumption that the mean of all errors are zero. Depending upon how this average is filtered and the quality of the frequency standard or oscillator used in the reference station, the effective time base will vary with respect to GPS time. Furthermore, any biases present in the reference station's receiver will be an offset with respect to GPS time. Receiver calibration can minimize this offset.

Special Category I Precision Approach Operations Using DGPS   At the request of the FAA, RTCA, Inc.'s Special Committee SC159 established a special ad hoc development team to prepare Minimum Aviation System Performance Standards (MASPS) for Special Category I Precision Approach Operations Using DGPS [41]. This special capability is intended for designated aircraft at special use airports and later at public use airports to provide standards for the operational use and evaluation of DGPS techniques in actual precision approach and landing conditions. Such a capability has been demonstrated using the high-performance receiver described in Section 5.5.8, Figure 5.18, for both the reference station and the airborne receiver using the RTCM DGPS messages described above with a three second update rate. Ninety-five percentile accuracies of 1 meter, horizontal, and 2 meters, vertical, were obtained over 68 approaches using a laser tracker as a reference [66, 67, 68]. The MASPS deviates slightly from the RTCM messages for this application for three reasons:

(1) to ensure the higher update rates using existing data links, (2) to provide additional information required for the precision approach and landing application, and (3) to increase the integrity of the broadcast with a much stronger parity algorithm. The flight test results described above suggest that DGPS, using differential pseudorange corrections, can meet even Category III precision approach and landing requirements [68].

Wide Area DGPS   RTCA Special Committee SC159 is also preparing requirements for the use of Wide Area DGPS (WADGPS) as part of the FAA's future Wide Area Augmentation System (WAAS) to achieve a Category I precision approach and landing capability [69]. WAAS uses a braodcast through a geostationary satellite to provide corrections over a very wide area, such as the continental United States (CONUS). The accuracy of this approach suffers somewhat because of spatial decorrelation and limited broadcast bandwidth but is expected to provide the required Category I precision approach accuracy over a region such as CONUS. To achieve this, the broadcast messages differ significantly from the RTCM messages because information on ephemeris and ionospheric errors must be provided to correct for spatial decorrelation of these errors. More detail on the WAAS is provided in Section 5.7.3.

Differential Carrier Phase Techniques   There is some belief that differential carrier phase techniques are required to achieve Category III precision approach and landing accuracy performance using DGPS [2]. This application of DGPS has promoted the development of *differential GPS kinematic surveying* techniques toward achieving Category III performance. These techniques resolve the carrier phase ambiguities to provide dynamic accuracies on the order of a few centimeters. This is accomplished by augmenting the pseudoranges of Equation 5.1 with the term $c\lambda N_i$, where λN_i is the number (N_i is an integer) of ambiguous carrier wavelengths for satellite i, and to solve for these ambiguities [65]. Generally, these ambiguities are resolved using double-difference techniques, where the double differences are computed between the user and the reference station and between pairs of satellites, canceling all of the common errors [70]. Unfortunately, within the initialization accuracy, there are many possible solutions for the ambiguities, although only one solution is correct [71].

   The general technique for solving for the correct ambiguity is to search the uncertainty region for the correct solution using error minimization techniques. This requires a minimum amount of geometry change between the user and the satellites. The amount of change required depends upon the initial uncertainty, which is usually that of the double-differenced pseudorange measurement solutions. Unfortunately, because of multipath and receiver noise, the change required takes up to on the order of 30 seconds, even under the best conditions using enhanced L1-only C/A code performance [72]. Once the ambiguities are resolved, resolution after cycle slips or signal outages are instantaneous, provided that there are redundant satellites being tracked. There are a few techniques available to improve this initialization process. Two of the most promising are the use of dual-frequency receivers [72] and the use of local near-L1 transmitting pseudolites (pseudosatellites) [73]. The former approach uses the differential carrier between L1 and L2 to first resolve ambiguities using a larger beat frequency wavelength, and then transferring the solution to initialize the L1 ambiguity resolution. The larger wavelength ambiguity takes much less time to resolve. The second approach takes advantage of the rapidly changing geometry between the pseudolite and the user, providing more leverage to the resolution problem. The use of pseudolites for DGPS is discussed further in Section 5.7.4.

The use of carrier phase techniques for precision approach and landing is still in development. However, *carrier-smoothed-code* techniques are much more robust. These techniques solve for the ambiguity not as an integer but as a floating-point number. The ultimate accuracy of DGPS for precision approach still remains unknown.

5.5.10 GPS Accuracy

Error Budgets GPS accuracy depends upon user receiver implementation. There are PPS and SPS implementations, there are P code and C/A code users, there are L1/L2- and L1-only users, there are DGPS users, there are differential carrier phase users and there are combinations of all of the above. Table 5.7 presents error budgets for three of these implementations—one for the authorized P code L1/L2 PPS user from the MAGR (see section 5.5.8, Figure 5.24) specification [52], one for the unauthorized C/A code L1-only user for airborne supplemental navigation equipment [40], with and without SA, and one for the special Category I precision approach and landing DGPS user [41].

The position errors in Table 5.7 are obtained from the system pseudorange errors (UEREs) as follows:

1. The horizontal CEP (circular error probable) budgeted for the MAGR is given as [52]

$$CEP = 0.8326 \times rms_{hor} = 0.8326 \times HDOP \times UERE \qquad (5.83)$$

where HDOP was taken to be approximately 1.39 with an elevation mask angle of 5°.

2. The vertical LEP (linear error probable) budgeted for the MAGR is given as [52]

$$LEP = 0.6745 \times rms_{ver} = 0.6745 \times VDOP \times UERE \qquad (5.84)$$

where VDOP was taken to be approximately 1.97 with an elevation mask angle of 5°.

TABLE 5.7   GPS avionics user error budgets

	ERROR (meters)			
Error Source	Authorized L1/L2 User	Unauthorized C/A Code L1 User with SA	Unauthorized C/A Code L1 User Without SA	DGPS Special Category I User
Space/control segment/ reference station	6.0	30.8	6.0	1.21
User				
Ionosphere Compensation	2.2	10.0	10.0	0.0
Troposphere Compensation	2.0	2.0	2.0	0.02
Multipath	1.2	1.2	1.2	1.2
Receiver noise and resolution	1.47	7.5[1]	7.5[1]	0.5
Other	0.5	0.5	0.5	0.05
System UERE (RSS)	6.98	33.33	14.07	1.78
Horizontal position error	8.10 CEP	100 2drms	42.2 2drms	—
Vertical position error	9.28 LEP	—	—	5.52–95%
Time error (UTC)	100 ns, 1σ	—	—	—

[1]Lower receiver noise errors are obtainable using carrier aiding of the code loop, resulting in lower code loop bandwidth [121 (Chapter 8), 133 (Chapter 5), 135].

3. The 2drms budgeted for the commercial avionics user is given as [40]

$$2\text{drms} = 2 \times \text{rms}_{hor} = 2 \times \text{HDOP} \times \text{UERE} \tag{5.85}$$

where HDOP was taken to be approximately 1.5 with an elevation mask angle of 7.5°.

For the special Category I DGPS user [41], the space/control segment error budget is reduced to residual SA, residual clock, and spatial decorrelation errors amounting to 0.5, 0.01, and 0.01 meters, respectively. The reference station budget is set at 1.1 meters, resulting in a total space/control segment/reference station error budget of 1.21 meters, root-sum-squared. The users' error budget presented in Table 5.7 is an example. The special Category I user has the choice in allotting his budget between sensor (GPS receiver) error and flight technical error (FTE), which defines the pilot's or auto-pilot's ability to fly the prescribed flight path. The total vertical error budget is 9.76 meters, with a probability of 95%. Thus, the GPS receiver error budget depends upon the assigned FTE for a given aircraft. Using a 95% probability VDOP (ratio of 95% vertical navigation error to 1-meter rms pseudorange error) of 3.1, the 95% probability vertical error for the pseudorange error budget in Table 5.7 is 5.52 meters, leaving a 95% probability budget for FTE of 8.05 meters. An aircraft with a good autopilot could use a receiver with larger errors. The HDOP is not specified for special Category I DGPS users because the 95% probability horizontal navigation error budget is so large—33.54 meters, which is easy to achieve using DGPS techniques.

Time Accuracy The time error budget with respect to universal coordinated time (UTC) for the MAGR is a GPS system level specification for time transfer [52]. The control segment's budget for maintaining the difference between GPS time and UTC is 90 nsec, 1-sigma. The 1-sigma GPS time error due to pseudorange error is

$$\sigma_{\Delta t} = \text{TDOP} \times \frac{\text{UERE}}{c} = 1.12 \times \frac{6.98}{c}$$
$$= 26.1 \quad \text{ns} \tag{5.86}$$

assuming a TDOP of 1.12 with an elevation mask angle of 5°. The MAGR is required to output time via a one pulse per second (1-PPS) accurate to the specified 100 nsec, 1-sigma. The remaining error budget to achieve this is

$$\sqrt{100^2 - 90^2 - 26.1^2} = 34.9 \quad \text{ns} \tag{5.87}$$

The unauthorized user time-transfer accuracy is dominated by the SA errors. Using the same value of TDOP (1.12) yields the following:

$$\sigma_{TT} = \sqrt{\left(1.12 \times \frac{33.33}{c}\right)^2 + (90 \times 10^{-9})^2} = 153.6 \quad \text{ns} \tag{5.88}$$

Velocity Accuracy GPS velocity accuracy is not guaranteed, nor is it usually specified, except possibly in classified military specifications. However, measured results have been published and are very much a function of the dynamics of the host vehicle at the time of the measurement. Tests performed on the authorized MAGR have yielded velocity accuracies better than 0.1 meter/sec in constant dynamic or inertially aided maximum dynamic conditions, and better than 1 meter/sec if unaided [60, 74]. This velocity accuracy is generally accepted as the norm under these conditions. For the unauthorized user, on the other hand, the velocity accuracy is dominated by SA pseudorange rate errors, which have been measured to be between 0.3 to 0.9 meter/sec, rms, horizontal, under stationary conditions [75]. Carefully implemented DGPS yields the same velocity accuracy as the authorized user. Receiver implementation (P, or C/A, and/or L1/L2 or L1-only) has no effect on velocity accuracy. The use of differential carrier phase DGPS should yield much better velocity accuracy.

GPS Accuracy Summary Figure 5.30 provides a summary of GPS position accuracy for the various receiver implementations. The moving survey accura-

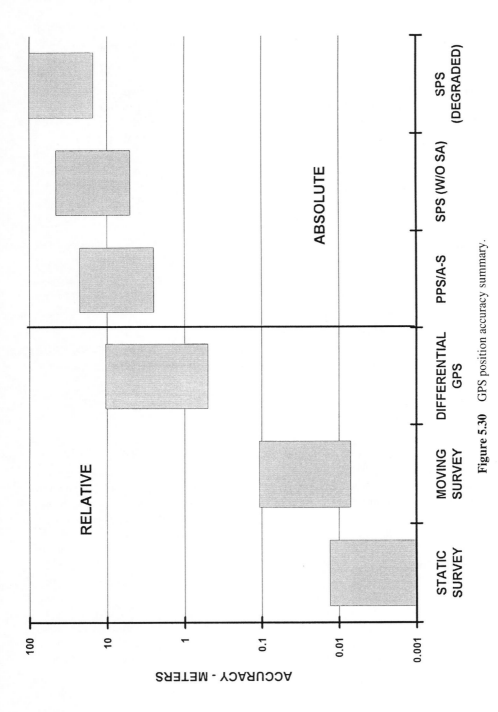

Figure 5.30 GPS position accuracy summary.

cies are indicative of what might be achieved if differential carrier phase techniques are developed for precision landing applications. Otherwise, accuracies indicated for differential GPS are more applicable.

5.6   GLOBAL ORBITING NAVIGATION SATELLITE SYSTEM (GLONASS)

The GLONASS satellite navigation system was developed by the Russian Federation; it was started by the former Soviet Union (USSR) [76]. It was declared operational in 1996 with a full constellation of satellites. GLONASS offers many features in common with the NAVSTAR GPS, but with significant implementation differences [77, 78]. In particular, the orbital plan also consists of 24 satellites. However, rather than 4 in each of 6 planes, GLONASS has a plan with 8 in each of 3 planes (designated planes 1–3) separated by 120° and with spacing of 45° within the plane. The GPS spacing is not uniform. The orbit altitudes also differ from that of GPS; the ground tracks repeat approximately every eight days rather than approximately one day for GPS. GLONASS satellites also transmit two spread-spectrum signals in the L-band (L1 and L2) at around the same power levels as GPS at frequency bands that are approximately 20 to 30 MHz higher than GPS [79]. However, satellites are distinguished by radio-frequency channel rather than spread-spectrum code (Frequency Division Multiple Access, FDMA, instead of CDMA). Common codes are used for all of the satellites. Both a C/A code and a P code are transmitted in quadrature on the L1 signal [80]. Otherwise, the basic principle of operation is identical to that of GPS.

5.6.1   GLONASS Orbits

Table 5.8 provides a summary of the GLONASS orbital parameters compared to those of GPS [77, 78]. Note that the orbit period and semimajor axis are less

TABLE 5.8   GPS and GLONASS nominal orbit parameters

Parameter	NAVSTAR GPS	GLONASS
Period (minutes)	717.94	675.73
Inclination	55°	64.8°
Semimajor axis (meters)	26560	25510
Orbit plane separation	60°	120°
Phase within planes	Irregular	±30°
Ground track repeat (orbits)	2	17
Longitude drift per orbit	180°	169.4°

than those of GPS that cause the GPS satellites to repeat their ground track each day. Because of this, the GLONASS ground tracks precess around the Earth and repeat every 17 orbits lasting 8 whole days plus 32.56 minutes. This is equivalent to 16 GPS orbits.

5.6.2 GLONASS Signal Structure

Broadcast Frequencies   The GLONASS satellites broadcast two signals: Link 1, L1 and Link 2, L2. According to figures made available to the International Frequency Registration Board (IFRB) in Geneva [81], and updated in November 1994 for GLONASS-M [82], GLONASS transmits a maximum power spectral density of -44 dBW/Hz in the frequency band of 1597–1617 MHz (L1) and -57 dBW/Hz in the frequency band of 1240–1260 MHz (L2). A shaped-beam antenna is used to produce uniform power spectral density on the ground [77]. The L1 C/A and P code signals are transmitted at the same frequency, in quadrature, just as the NAVSTAR GPS L1 signals are.

In the initial plan for GLONASS-M, each satellite was to be assigned a unique frequency according to the following equation [82]:

$$f_{1i} = f_1 + 0.5625i \quad \text{MHz}, \qquad i = 0, 1, \ldots, 24 \tag{5.89}$$

for satellite i of 24 satellites ($i = 0$ is for testing), where the base L1 frequency is

$$f_1 = 1602.0 \quad \text{MHz} \tag{5.90}$$

That is, when all 24 satellites were to be in the constellation, each would have a frequency assigned to it with 562.5 kHz separation between satellite signals.

Similarly the L2 P-Code signal is transmitted at a frequency assigned to the satellite. Each satellite is assigned a unique frequency according to the following equation [82]:

$$f_{2i} = f_2 + 0.4375i \quad \text{MHz}, \qquad i = 0, 1, \ldots, 24 \tag{5.91}$$

for satellite i of 24 satellites ($i = 0$ is for testing), where the base L2 frequency is

$$f_2 = 1246.0 \quad \text{MHz} \tag{5.92}$$

That is, when all 24 satellites are in the constellation, each will have a frequency assigned to it with 437.5 kHz separation between satellite signals. Note that the ratio of the L1 and L2 frequencies is 9/7, including the frequency separations. Also note that they are an integer multiple of a common frequency of 62.5 kHz.

However, because of interference issues, the Russian Federation revised the frequency plan as follows for GLONASS-M [79, 83]:

1. Until 1988, GLONASS-M will not use carrier frequencies for $i = 16$ through 20 for normal operations. Frequencies for $i = 0, 1, \ldots 12, 22,$ 23 and 24 will be used. Frequencies for $i = 13, 14,$ and 21 will only be used under exceptional circumstances. This revision is to prevent transmission into radio-astronomy antennas in that band, and will be realized by re-using frequencies on anti-podal satellites (satellites visible on the opposite side of the Earth).

2. From 1988 to 2005, GLONASS-M will use frequency channels $i = -7$ through +12 for normal operation and use $i = 13$ only under exceptional circumstances.

3. After 2005, GLONASS-M will use frequency channels $i = -7$ through +4 for normal operation and use $i = +5$ and +6 as technical channels only for limited periods of time during orbital insertion or other periods of exceptional circumstances.

The shift down in frequency is to avoid interference from future Mobile Satellite Services (MSS) terminals.

Signal Modulation   The L1 and L2 signals are both bi-phase modulated with the PRN codes and navigation data. The PRN code and navigation data characteristics are as follows:

1. *C/A code.* The GLONASS C/A code is comprised of a nine-state shift register with tap feedback that produces a 511-bit maximal-length sequence. It is clocked at a rate of 511 kHz so that it repeats every millisecond. A functional block diagram of the C/A code generator is shown in Figure 5.31.

Every satellite generates the same C/A code. The 1-msec C/A code epochs are coherently synchronized to the satellite's time, which is maintained to within

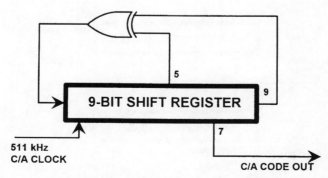

Figure 5.31   GLONASS C/A code generator.

1.953 msec of GLONASS system time. The resultant signal spectrum is a line spectrum centered at the assigned satellite frequency with an envelope equal to that given in Equation 5.54 with a T_c of $1/511,000$ sec, where the lines are spaced 1 kHz apart, and a spectral null occurs at multiples of 511 kHz. Since the assigned satellite frequencies are spaced only 562.5 kHz apart, there is an overlap of signal spectra.

Even though the spectra of the C/A codes of the different satellites overlap, it has very little effect on signal acquisition and tracking, because the user receivers, when correlating with the code, will track the correct carrier. Spectral interference will occur, but will be well below the thermal noise level. Adjacent frequency numbered satellite signals will have a cross-correlation level not to exceed 48 dB [79]. Because of the separation in frequency, even if full code correlation between signals occurred for an instant, postcorrelation integration reduces the effect to that level. The C/A code only appears on the L1 signals [77, 80].

2. *P code.* The CIS has never published the GLONASS P code. However, it has been determined independently [80]. The GLONASS P code is comprised of a 25-stage shift register with tap feedback that would produce a 33,554,431-bit maximal-length sequence, except for the fact that it is short-cycled to 5,110,000 bits and reset to all 1's. It is clocked at a rate of 5.11 MHz so that it repeats once per second. A functional block diagram of the P code generator is shown in Figure 5.32. Every satellite generates the same P code. The 1-sec code epochs are synchronized to the 1-msec C/A code epochs to ease the handover from one code to the other.

The resultant P code signal spectrum is a line spectrum centered at the assigned satellite frequency with an envelope equal to that given in Equation 5.53 with a T_c of $1/5,110,000$ sec, where the lines are spaced 1 Hz apart, which

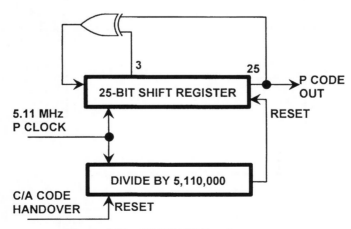

Figure 5.32 GLONASS P code generator.

essentially makes it a continuous spectrum. A spectral null occurs at odd multiples of 5.11 MHz. Since the assigned satellite frequencies are only 562.5 kHz and 437.5 kHz apart for the L1 and L2 frequencies, respectively, the P code spectra of the different satellites overlap a great deal. For the same reasons as with the C/A code, this does not pose a problem when acquiring and tracking the signals. The P code is modulated on both the L1 and the L2 signals [80].

3. *Navigation data.* The navigation data is modulo-2 added to both of the codes prior to the bi-phase modulation of the carriers. Because of that, these data do not alter the spectrum of the signals. The effective data rate is 50 bps. However, it is differential and return-to-zero encoded, so the modulation is actually at 100 symbols per second [84].

Signal Power The GLONASS ICD for the L1 C/A code signal indicates a minimum received power of −161 dBW, which is 1 dB less than specified for the GPS L1 C/A code [79], although this level has been updated to −160 dBW [82], which may be the total received C/A code plus P code power. The L1 P code signal level is not published. The L2 P code received signal power is −166 dBW [82].

5.6.3 The GLONASS Navigation Message

The GLONASS navigation message differs significantly from its GPS counterpart. It is made up of lines, frames, and super frames [79]. Each line is 2 seconds long containing 100 bits: 85 bits of digital data in 1.7 seconds containing 8 bits of a Hamming (85,77) parity code followed by a 30 symbol time mark at the 100 bps rate. Each frame contains 15 lines over 30 seconds. A super frame is 5 frames over 2.5 minutes.

The first four lines of each frame contain ephemeris and time information for the satellite broadcasting the message. The fifth line contains a day number and system time correction. Lines 6 through 15 contain almanacs for five satellites, two lines per almanac. The fifth frame contains only four almanacs.

Ephemeris Data The GLONASS ephemeris data are broadcast as ECEF cartesian coordinates in position and velocity with lunar/solar acceleration perturbation parameters that are valid over about 0.5 hour [84]. The assumption is that the user integrates via a fourth-order Runge-Kutta method the motion equations that include the second zonal geopotential harmonic coefficient. Details of these equations are given in the GLONASS ICD [79].

Almanacs Even though the ephemeris data differs completely from that of GPS, the almanac parameters are quite similar as modified Keplerian parameters.

Clock Corrections Clock correction parameters are also similar to that of GPS in terms of clock offsets and clock drift.

5.6.4 Time and Coordinate Systems

GLONASS Time The GPS control segment provides corrections so that GPS time can be related to UTC(USNO) modulo 1 sec to within 90 nsec, whereas the GLONASS control segment provides corrections so that GLONASS time can be related to UTC(Moscow) to within 1 μsec [79]. GPS time does not follow the leap second corrections that UTC occasionally makes. GLONASS time does [79].

GLONASS Coordinate System The GLONASS system transmits ephemeris and almanac data describing the satellite's antenna phase center in the Earth-fixed reference PZ-90, which differs from WGS 84 by under 15 meters. A preliminary estimate of the coordinate transformation between the two systems (ECEF) is a translation of 2.5 meters in the y direction and a rotation of 0.4 minutes about the z-axis [85].

5.6.5 GLONASS Constellation [86, 87]

At times, the GLONASS system has a full constellation of 24 satellites. In 1996 the system was usually operating with 21 or 22 satellites transmitting healthy signals [87]. However, none of these satellites were of the new GLONASS-M variety.

5.7 GNSS INTEGRITY AND AVAILABILITY

Acceptance of a global navigation satellite system (GNSS) as a sole means navigation aid in the U.S. National Airspace System (NAS) will necessitate meeting stringent availability and continuity of function requirements that are usually unachievable without some sort of augmentation. Key issues are safety and performance assurance that relate to satisfying accuracy, integrity, availability, and continuity of function requirements. The following definitions pertain to these issues:

1. *Accuracy* pertains to the capability of the system, with or without augmentation, to meet the navigation accuracy requirements specified by phase of flight in the U.S. Federal Radio Navigation Plan [24].
2. *Integrity* relates to the probability of detecting anomalous signals that could induce navigation errors beyond defined protection limits and to providing timely warnings to the users [40, 41].
3. *Availability* of a navigation system is the ability of the system to pro-

vide required guidance at the initiation of the intended operation. It is an indication of the ability of the system to provide usable service within the specified coverage area. Signal availability is the percentage of time that navigation signal broadcasts are available for use. Availability is a function of both the physical characteristics of the environment and the technical capabilities of the transmitter facilities [40, 41].

4. *Continuity* of a system is the ability of the total system (comprising of *all* elements necessary to maintain aircraft position within the defined airspace) to perform its function without interruption during the intended operation. More specifically, continuity is the probability that the system will be available for the duration of a phase of operation, presuming that the system was available at the beginning of that phase of operation [41].

However, current GNSS systems (GPS or GLONASS) do not meet these requirements for most phases of flight, especially for the more stringent precision approaches, without augmentation. In this regard, the FAA has defined the following GNSS user services [88]:

1. *Multisensor system* implies that the GNSS and any augmentations can be used for navigation, but only after it has been compared for integrity with another approved navigation system in the aircraft.

2. *Supplemental system* implies that the GNSS and any augmentations can be used alone without comparison to another approved navigation system. However, another approved navigation system must be on board the aircraft and usable when the GNSS is not available.

3. *Required navigation performance (RNP) system* is one that meets all the requirements without need for any other navigation equipment on board the aircraft. An RNP system may include one or more navigation sensors in its definition (e.g., GPS with an inertial reference system, IRS).

GNSS does not add much to the aircraft's navigation system if it is only certified as a multisensor system service. It can add accuracy as long as the system it is being compared with meets RNP requirements. This service is also useful for test purposes. GNSS can, by itself, be certified as a supplemental system through the use of receiver autonomous integrity monitoring (RAIM) [40] and possibly oceanic en route [69, 89]. In 1996, based on FAA requirements, as an RNP system, GNSS requires augmentation, such as combining GPS either with GLONASS, an independent WAAS-type system, and pseudolites or with another type of sensor, such as an IRS [69, 89].

5.7.1 Receiver Autonomous Integrity Monitoring (RAIM)

All GNSS RAIM schemes are based on making self-consistency checks of some sort. This idea is not new. Prudent navigators have used redundant observations to verify their position fixes since antiquity. The thing that is different now is that computer technology has made it possible to use relatively sophisticated mathematical methods in performing the consistency checks. Before getting into the details of one RAIM method, it should be noted that catastrophic failures are easy to detect with primitive methods, so they are not discussed further here. It is the more subtle or incipient failures that are treated here; those where a somewhat out-of-tolerance signal in space causes the user position error to wander outside some specified limit for the phase of flight in progress. If the GNSS is a *supplemental system*, it is sufficient for RAIM to simply detect the failure and sound an alarm accordingly. If the GNSS is an *RNP system*, it is necessary for RAIM to both detect and isolate and exclude the failed source. This added burden of isolation and exclusion complicates the RAIM (now called *fault detection and exclusion*, FDE) problem considerably [90, 91, 135].

RAIM Basics For tutorial purposes it is useful to begin with a simple two-dimensional example. Suppose that we have three range measurements, each defining a line of position (LOP) in a plane. Three possible situations are shown in Figure 5.33. In Figure 5.33*a* we have the usual situation with good geometry and consistent measurements (at least within the expected measurement noise). The result is three intersections (fixes) that are close together. The observer would then conclude "no failure" in this situation. In Figure 5.33*b* we see another possible situation in which we have favorable geometry but the fixes are relatively far apart. The observer must conclude that something is wrong here, and the decision is "failure." Note, though, that the information is insufficient to tell us which measurement is at fault; that is, we can do simple error detection here, but we cannot do fault isolation with just one redundant measurement. Finally, in Figure 5.33*c* we see an extreme case of poor geometry. Two of the LOPs are parallel. We can conclude here that measurements 1 and 2 are consistent (their LOPs are close together), but there is no valid check on measurement 3. An error in it would go unnoticed. Thus the decision as to a possible failure in this case is inconclusive. The observer must simply say, "No

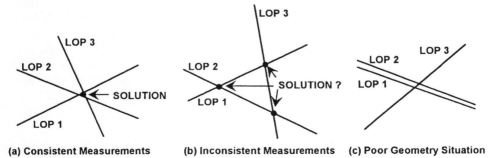

(a) Consistent Measurements   (b) Inconsistent Measurements   (c) Poor Geometry Situation

Figure 5.33 Examples of three LOP intersections.

valid integrity check is possible because of poor geometry." All three situations shown in Figure 5.33 have their counterparts in the more complex GNSS RAIM setting. Of course the meanings of "close together" and "far apart" need to be quantified. Also, statistical performance criteria relative to the reliability of the observer's decision need to be developed. More will be said of these items later.

Work on autonomous means of GNSS failure detection began in earnest during the latter part of the 1980s. It was also during this period that the acronym RAIM (for receiver autonomous integrity monitoring) was coined. We will not attempt to document all of the technical papers on RAIM that appeared during this period. One has only to browse through the proceedings of the meetings of the Institute of Navigation to assess the degree of activity that took place during this period and on into the 1990s. A summary of three different methods is given in [92].

One of many RAIM schemes will now be described; it is easily understood and can be thought of as a baseline or reference method. While it is a good scheme, there is no claim that it is the best.

RAIM Detection Algorithm   In 1987 a RAIM technique that is known as the *least-squares-residuals* method was presented [93]. It begins with the assumption that the receiver has simultaneous redundant pseudorange measurements (five or more), and that the position-fixing problem has been linearized in the usual manner (see Sections 5.5.2 and 5.5.7). First, the all-in-view least-squares solution is formed. It is well-known and is given by Equation 5.71. The sum of the squares (weighted, in general) of the components of the measurement residuals vector $\delta\overline{\mathbf{M}}$ (from Equations 5.69 and 5.71) is the scalar quantity

$$\text{SSE} = \delta\overline{\mathbf{M}}^T \mathbf{W} \delta\overline{\mathbf{M}} \tag{5.93}$$

SSE is the basic observable in the sum-of-squared-residual-errors RAIM method. The decision rule is as follows: If, for a predetermined threshold TH,

$$\text{SSE} \leq \text{TH} \tag{5.94}$$

decide "no failure," but if

$$\text{SSE} > \text{TH} \tag{5.95}$$

decide "failure."

The intuitive rationale for this rule is simply that if the measurements are consistent, we can expect the residuals to be small; on the other hand, if the measurements are inconsistent, we can expect SSE to be large because of a poor least-squares fit. Once the threshold value TH is set, the decision rule is quantified. With this RAIM algorithm, it is easy to set the threshold to yield an alarm rate that is independent of geometry in the absence of a satellite malfunction.

This is usually set at the maximum allowable rate. The RAIM algorithm then accepts whatever detection probability that results from this threshold setting.

RAIM Specifications There are four key parameters that must be included in the RAIM specifications:

1. Alarm limit (also called *alert limit*). Alarm limit refers to the maximum allowable radial error before the alarm is sounded.
2. Time response of the alarm (i.e., delay to alarm time). Too much delay can be disastrous in critical situations.
3. Maximum allowable alarm rate in the absence of a satellite malfunction. There must be a limit to nuisance alarms.
4. Detection probability. This must be close to unity if the RAIM algorithm is to be effective.

These specifications compete with each other to some extent. For example, tightening the false alarm rate specification makes it more difficult to meet the detection probability requirement. The "elastic" in the system that makes it possible to meet all of the stated four requirements is *availability*. The RAIM algorithm can (and indeed must) reject poor detection geometries. RAIM availability, of course, suffers from such rejections.

RAIM for a Supplemental System Extremely high integrity availability is not essential for use of GNSS as a supplemental system. However, nearly 100% availability and continuity of function are needed for an RNP system. Many studies have shown that RAIM alone will not provide this high degree of availability when operating with just the GPS (or GLONASS) 24-satellite configuration. This is especially true when one considers the extra burden on RAIM in having to isolate the faulty satellite as well as detect the failure. But GPS alone can be certified as a supplemental system, with an availability approaching 94% for the nonprecision approach phase of flight and up to 99% if aided with barometric altitude [94]. For the less stringent phases of flight, the availability is much higher. For a supplemental system, continuity of service is not required because, upon the sound of an alarm, the other approved navigation system can be used. However, because of the length of outage times, continuity of service requirements could never be met for an RNP system without some sort of augmentation.

RAIM for an RNP System In the RNP application, RAIM will have to be augmented with additional measurement information from outside GPS. Many such possibilities exist, such as using the combination of GPS and GLONASS, and the marketplace (and perhaps politics) will determine the mix of sensor information to be used in any particular application. Also, it is likely that the ultimate GPS integrity protection will be provided by a combined WAAS/RAIM system. The two systems are complementary, and there is much to be gained by having the two systems work together [95]. (For a discussion of the WAAS, see Section 5.7.3.)

5.7.2 Combined GPS/GLONASS

To improve the availability and continuity of service of RNP service using GNSS, augmenting GPS with GLONASS has been suggested. This would essentially guarantee the signal redundancy required for RAIM, even for fault isolation. However, there are problems with combining the two systems.

Technical Problems   There are numerous technical problems with combining the two systems that need to be resolved by the user. These include the fact that the two systems operate on different time scales and that they are referenced to different geodetic systems. The combined approach also increases the cost of the user avionics receivers, which now must receive signals from both systems and process two different sets of navigation data using different ephemeris algorithms. Even beyond the fact that the GLONASS system operates at a different frequency, receiving the FDMA signals results in a more complex receiver design than the receipt of the CDMA GPS signals.

Institutional Problems   Institutional problems also exist. First, the GLONASS system has not yet proved to be reliable. Over the years, there have been more GLONASS satellite failures than the number of GPS satellites that have been launched [96]. In addition, it is more susceptible to satellite communications transmissions that exist in frequency bands near and above the band allocated for GLONASS [97, 126].

GPS/GLONASS RAIM   The fact that the two systems operate on different time scales can be solved by the avionics user by simply adding the time difference to his solution state vector. This does, however, require an additional satellite signal source because it adds another unknown to the solution. Furthermore, at least two satellites are required from both systems in the solution. If there is only one satellite, any error in its pseudorange will simply be assigned to the solution for the time difference based upon the position solution determined from the other system. However, this requirement for an additional satellite would not exist continuously, since the time scales of both systems are quite stable and a reliable time difference solution would remain valid over a long time. Continuous monitoring may be required for solution integrity, however, as well as detecting interfrequency errors in the receiver.

The problem with operating with two coordinate systems can be solved over a period of time and updated as necessary with data-base parameters. For most phases of flight, the differences appear to be small enough, so they do not matter [85]. For precision approach applications, the use of differential corrections would cancel the differences, including the differences in the time scales.

The combined system can also be used in conjunction with the WAAS described in Section 5.7.3, in which case the differences can be broadcast via the WAAS. If there are a number of failures in either system, the availability of RAIM and continuity of service could also suffer because, as stated at the beginning of this section, the two systems' orbits are not synchronized. That is, if the GLONASS system were to augment the GPS system on one day, it may not on the next because the ground tracks of the satellites moved with respect to each other. This could be a problem if the GLONASS system continues to be unreliable and the number of satellites in orbit do not maintain an operational status.

5.7.3   Wide Area Augmentation System (WAAS)

The WAAS Concept   The WAAS is being developed by the FAA and is expected to provide a *test signal* by 1998 [125]. In parallel the Europeans are developing the European Geostationary Navigation Overlay Service (EGNOS) [127] and Japan is developing the MTSAT Satellite-Based Augmentation System (MSAS) [128]. Both of these systems will be very similar to WAAS. Japan will use their own satellites (MTSAT-1 and MTSAT-2). The Europeans will share the Inmarsat-3 satellites with the FAA. ICAO has named the generic WAAS-type system a Satellite-Based Augmentation System (SBAS). It is a safety-critical system consisting of a signal-in-space and a ground network to support en-route through precision approach air navigation. The WAAS augments GPS with the following three services: a ground integrity broadcast that will meet the RNP integrity requirements for all phases of flight down to Category I precision approach, wide area differential GPS (WADGPS) corrections that will provide accuracy for GPS users so as to meet RNP accuracy requirements for all phases of flight down to Category I precision approach, and a ranging function that will provide additional availability and reliability that will help satisfy the RNP availability requirements for all phases of flight down to Category I precision approaches [69].

Figure 5.34 illustrates the WAAS concept [69, 98]. The WAAS uses geostationary satellites (GEOs—Inmarsat-3's and successors) to broadcast the integrity and correction data to users for all of the GPS (and GEO) satellites visible to the WAAS network. This broadcast is at the GPS L1 frequency modulated with a C/A code in the same family as the GPS C/A codes. This family of codes contains 1023 codes, of which all but 256 are balanced codes and of which 36 are assigned or reserved for GPS [32]. Nineteen of the remaining codes have been reserved for the wide area augmentation broadcasts [98]. Thus, a slightly modified GPS avionics receiver can receive these broadcasts. Since these codes will be synchronized to the WAAS network time, which is the reference time of the WADGPS corrections, the signals can also be used for ranging. A sufficient number of GEOs provides enough augmentation to satisfy RNP availability and reliability requirements.

The first two launches of Inmarsat-3 satellites, the first such satellites avail-

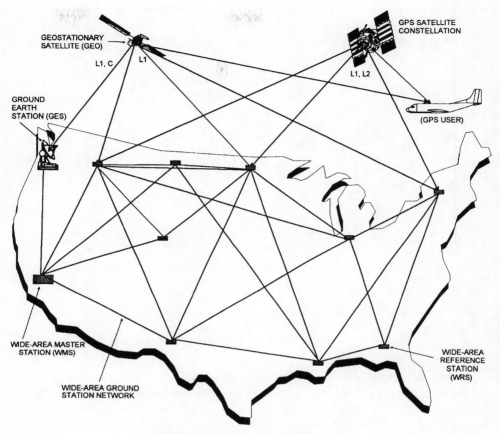

Figure 5.34 WAAS concept.

able for wide area augmentation at the L1 frequency were in 1996 [87]. Four satellites are planned with an edge-of-Earth coverage shown in Figure 5.35. Note that many areas have double coverage, while some areas (e.g., Europe) have triple coverage. However, at least double coverage is required everywhere in the service volume to provide the required RNP availability and reliability [29, 30]. Thus, for CONUS, at least one or two additional GEOs are required [30]. Unlike the Inmarsat-3 communications satellites, these additional GEOs may be small single-mission navigation satellites [100].

In the WAAS concept, a network of monitoring stations (wide area reference stations, WRSs) continuously track the GPS (and GEO) satellites and relay the tracking information to a central processing facility [69, 98]. The central processing facility (wide area master station, WMS), in turn, determines the health and WADGPS corrections for each signal in space and relays this information, via the broadcast messages, to the ground Earth stations (GESs) for uplink to

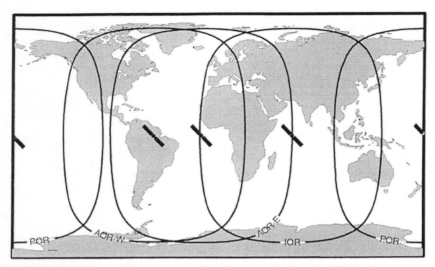

Figure 5.35   Inmarsat-3 four ocean-region deployment showing 5° elevation contours.

the GEOs. The WMS also determines and relays the GEO ephemeris and clock state messages to the GEOs. The signal is converted to the L1 frequency on the GEO satellite is then broadcast to the avionics user by the GEO satellite.

WAAS Navigation Payloads   The navigation payload of Inmarsat-3 is added to the normal communications payload to provide the wide area broadcast. This payload is simply a bent-pipe transponder that converts a C-band uplink to both a C-band downlink and the L1 broadcast using the normal communications C-band receivers and transmitters [32]. In the payload, input from the C-band receiver is converted to IF, filtered, and then converted to the frequency required for the C-band transmitter, plus converted to L1 for continuous transmission via the L1 HPA and antenna. Uplink power and the L1 gain are controlled so that the L1 HPA always operates in saturation. This minimizes transmitted power variations and, along with the requirement for a strong uplink signal, reduces the effects of uplink interference and maximizes the ability to sense it [101]. Power control is realized via an encrypted TT&C.

Future WAAS navigation payloads may take on a different form—more like that of the GPS satellites as described in Section 5.5.3 but without the military mission features of the GPS satellites [100]. Instead of being a simple signal transponder, the payload will act as a data transponder, incorporating its own stable clock. In this way, the uplinked data can be encrypted and, thus, provide a more secure data and signal channel.

Message Format and Content [102, 129]   The integrity message contains the status of each GPS satellite as *use/don't use* information as well as WADGPS error corrections and GEO ephemeris and clock data. The messages set into a format with the capability to include both GPS and GLONASS data, although GLONASS data are not planned in the FAA system. The magnitude of the WADGPS corrections can be also be used as error statistics for the users that are not applying the corrections in the appropriate phases of flight.

The message data rate differs from that of the standard GPS signal [102, 103, 104, 129]. It has a symbol rate of 500 symbols per second. A rate 1/2 forward error-correcting (FEC) convolutional code of length seven is used to reduce the effective data rate to 250 bps, but allowing a 5-dB gain in effective energy-to-bit ratio over an uncoded 250 bps transmission [105]. Each message block, shown in Figure 5.36, contains 250 bits, lasting one second. Each frame contains 24 bits of parity to provide a strong burst error detection capability as required for high integrity. The higher data rate provides two capabilities. The first is a required capability to provide an integrity alarm to within 5.2 sec of a signal-in-space fault during Category I precision approach [98]. The second is to broadcast WADGPS corrections at a rate commensurate with SA and ionospheric delay errors.

The various message types are listed in Table 5.9 [129]. There are two types of correction data—fast and slow. The types 2 through 5 fast corrections are intended to correct for rapidly changing errors such as GPS SA clock errors, while the slow corrections are for slower changing errors due to the atmospheric and long-term satellite clock and ephemeris errors. The fast GPS clock errors are common to all users and will be broadcast as such. Corrections designated with the maximum positive number indicate *not-monitored* satellites, while those designated with the maximum amplitude negative numbers indicate *don't-use* satellites, which is the integrity indication. Procedures for using these messages are given in the RTCA MOPS [104, 129].

For the slower corrections, the users are provided with ephemeris and clock error estimates for each satellite in view (message types 24 and 25). Users are separately provided with a wide area ionospheric delay model and sufficient real-time data to evaluate the ionospheric delays for each satellite using that model (message types 18 and 26). This model is comprised of vertical ionospheric delays at a set of grid of points that a user can interpolate to the ionospheric pierce points of his pseudorange observations.

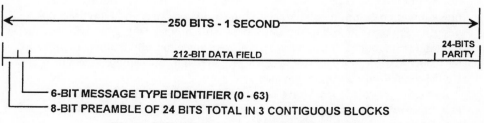

Figure 5.36   WAAS message data block format.

TABLE 5.9   WAAS message types

Type	Contents
0	Do not use this GEO for anything (for WAAS testing)
1	PRN Mask assignments, set up to 52 of 210 bits
2–5	Fast corrections
6	Integrity information
7	UDRE acceleration information
8	Estimated standard deviation message
9	GEO navigation message, $(X, Y, Z,$ time, etc.)
10–11	Reserved for future messages
12	WAAS network/UTC offset parameters
13–16	Reserved for future messages
17	GEO satellite almanacs
18	Ionospheric grid point masks
19–23	Reserved for future messages
24	Mixed fast corrections/long-term satellite error corrections
25	Long-term satellite error corrections
26	Ionospheric delay corrections
27	WAAS service message
28–62	Reserved for future messages
63	Null message

Since tropospheric refraction is a local phenomenon, all users must compute their own tropospheric delay corrections using a standardized model. The GEO broadcast messages will not include any explicit tropospheric corrections.

PRN masks are used to designate which PRN belongs to which correction slot. These masks improve the efficiency of the broadcast by preventing the continual inclusion of PRNs for the integrity data and corrections. The integrity data and corrections are provided sequentially based upon PRN numbers that are assigned to various types of satellies (GPS, GLONASS, GEO, and future GNSS satellites).

WAAS/Fault Detection/Exclusion Interaction   The users are only required to use fault detection (RAIM) or fault detection and exclusion (FDE) in conjunction with the WAAS in two cases: Fault detection is required during Category I precision approach, if available, and fault exclusion is required anytime WAAS integrity is not available in the other phases of flight [104]. Otherwise, neither fault detection nor fault exclusion is required, since these functions are provided by the WAAS broadcast. The WAAS provides the following enhancements to RAIM availability and performance [95, 106]:

First, in providing *use/don't use* information, the WAAS broadcast eliminates the requirement for the RAIM or FDE to perform fault detection and isolation (or satellite exclusion). The WAAS network is isolating the faults.

Availability is increased because it removes the requirement for a fifth or sixth satellite with good geometry.

Second, during precision approach, RAIM must be used if it is available—enough satellites are available with good geometry. Its purpose is not to detect satellite failures but to detect rare anomalous propagation events such as local ionospheric, tropospheric, and interference effects. If RAIM is not available, the performance of the signal in space indicated by the WAAS can be used for integrity. The probability of the nonavailability of RAIM coupled with the rare events is small enough to provide the necessary integrity.

Using GPS alone, these enhancements still would not increase the availability and continuity of function to that required for RNP service, primarily because of satellite coverage. However, the WAAS can provide one more enhancement that will do so—additional satellites with a ranging capability, which is the subject of the next section.

WAAS Ranging   Since the signals broadcast by the WAAS geostationary satellites are modulated with a C/A code, they can also be used for ranging if the timing of the signals are controlled with enough accuracy. Even if the signals are not controlled exactly, pseudorange corrections from their own data messages will provide the required accuracy. The effect of this ranging capability is a very good GPS constellation augmentation, although geostationary satellites, in addition to the Inmarsat-3 satellites, will be required for some areas of the earth, including the central part of CONUS. The effect of adding four ranging geostationary satellites on the coverage over CONUS is shown in Figure 5.37. These four satellites are the three (not including the one over the Indian Ocean) illustrated in Figure 5.35 plus one located at W120°. This availability can be compared to that presented earlier in Figure 5.9. The 99.999% availability of HDOP (2.37 average, and 3.03 worst-case) meets nonprecision approach availability requirements, although double coverage over all locations is required to obtain reliable coverage in the case of a long-term geostationary satellite failure. The availability presented in Figures 5.9 and 5.35 does not include the availability of continuity. That is, they only present instantaneous availability and do not take into account possible loss of availability over the entire flight phase, a subject discussed in reference [130].

The 99.9% availability of VDOP is 2.52 average and 3.13 worst case. These values may not appear to be acceptable for Category I precision approach. However, it is not the availability of VDOP that is important but the availability of vertical accuracy [106]. By applying a weighted-least-squares solution as described in Section 5.5.7, the concept of straight VDOP is not valid. However, the square root of the vertical component of the covariance matrix given in Equation 5.72 is valid. It has been shown that the 99.9% availability of vertical accuracy can be met using the four GEOs described above [30]. Furthermore, the weighted-least-squares approach can also be extended to RAIM [106].

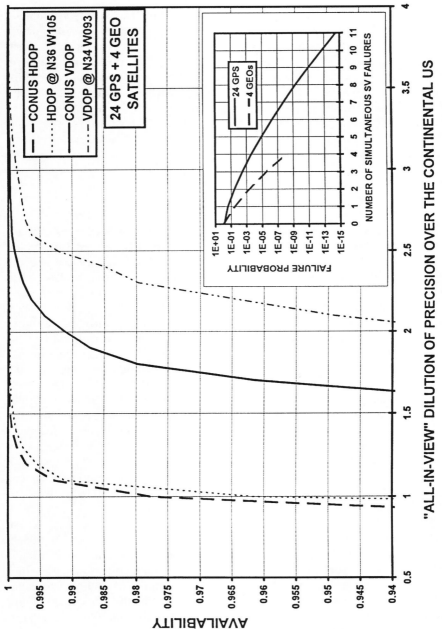

Figure 5.37 Availability of HDOP and VDOP when GPS is augmented with four GEOs.

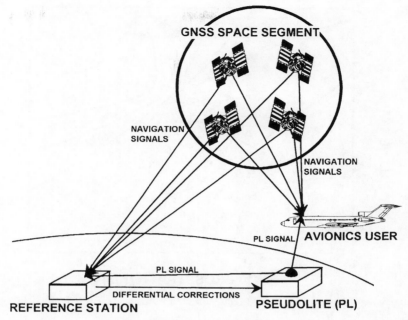

Figure 5.38 Pseudolite DGPS concept.

5.7.4 Pseudolite Augmentation

Pseudolite (PL) is an acronym for *pseudosatellite*. A PL is comprised of a GNSS-like signal generator at a fixed known location that broadcasts DGPS corrections as well as a ranging code. The concept is illustrated in Figure 5.38 and is analogous to the WAAS approach described in Section 5.7.3 with the exception that the signal is generated on the ground. The advantages of PLs for precision approach and landing applications are as follows:

1. They provide a data link for local DGPS corrections and integrity information that can be received with a slightly modified GPS avionics receiver [107, 108].

2. They provide additional ranging signals, just as the WAAS, resulting in a significant VDOP availability enhancement [109, 110, 111, 112, 113]. This VDOP enhancement is illustrated in Figure 5.39 for two PLs augmenting GPS at the FAA Technical Center in Atlantic City, New Jersey. VDOP is reduced from 2.3 for GPS only to less than 1 in the area of the runways for GPS augmented with two PLs [113]. Similar enhancements are shown in [112] for runways at O'Hare International Airport in Chicago.

3. They provide a rapid change in geometry that is extremely important for

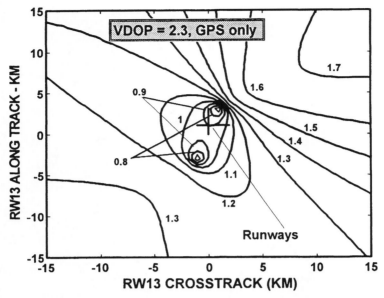

Figure 5.39   Illustration of VDOP reduction with 2 PLs at the FAA Technical Center.

kinematic carrier phase ambiguity (both integer and floating point) resolution techniques [73] and multipath mitigation. This approach is described in Section 5.5.9.

4. Their signals in space can be more accurate than satellite signals because of the nonpresence of ephemeris and ionospheric delay errors and reduced tropospheric delay errors [111].

Along with these advantages there are two significant disadvantages that require attention: 1. The proximity of the PL to the avionics user causes potential interference to the reception of satellite signals—the well-known *near/far* problem [107, 108, 110, 114]. The PL signals become quite strong when the avionics receiver is near the PL if the PL power is set for reception at a distance. For example, the received PL signal power increases inversely with the square of distance from the PL. For a precision approach and landing applications, the PL power would be set for reception at about 20 nmi. Thus, at 0.1 mile, the received signal is 40,000 times stronger (46 dB). 2. A PL's location on the ground could present a problem with the antenna location on the aircraft for simultaneous reception of the pseudolite and the satellites [113, 115].

Solutions to the Near/Far Interference Problem   The interference problem can be solved by altering the signal structure of the PL [113, 115]. Two signal diversity techniques have been tested: the use of a pulsed signal and an offset in frequency from that of the received satellite signals. Pulsing is required to prevent the capture of a receiver due to the excessive dynamic range required for a close-in signal [108]. The pulsing with a relatively low duty cycle reduces the interference to any signal in the reception band. The strong signal simply *punches* holes in the lower powered signals, reducing their received power by only the duty cycle percentage. For example, if the duty cycle is set at 0.1, the signal loss of the satellite signal is only $10 \log_{10} 0.9 = 0.458$ dB, although an additional loss in C/N_0 is realized because the pulse power also enters the correlator. This pulse power is reduced significantly by precorrelation pulse clipping or pulse suppression, techniques that are already implemented in GPS receivers. Pulsing is also required to prevent multiple PLs interfering with each other. This adds a requirement of pulse timing so that pulses from two different PLs are not received simultaneously, in addition to the requirement that received pulses must be asynchronous with the reception of GPS data bit edges. This type of timing is possible [113, 115].

Even though pulsing can reduce interference significantly, the strong signal within the clipped pulses can still cross-correlate with the received satellite signals, if indeed the PL signals carry GPS-like C/A codes [108, 110, 113, 114, 115, 116]. This causes another dynamic range problem because the cross-correlation margin between C/A codes is only on the order of 22 to 24 dB [18, 32]. The 10 dB reduction in average power of the pseudolite signal due to a 10% duty cycle pulsing does little to prevent cross-correlation [114]. This is where the frequency offset helps. It has been shown that the cross-correlation can also occur at frequency offsets but is proportional to the 1 kHz spectral line component levels of another C/A code in the same family [18, 32]. As it happens, however, the spectral line components near the *null* of the spectrum are down on the order of 70 to 80 dB, as can be observed in Figure 5.15. Thus, if the PL were to transmit in the *null* of the GPS satellite C/A code spectra, which are all at the same frequency to within 5 kHz due to Doppler differences, the cross-correlation would be insignificant [113, 115]. Cross-correlation peaks within 4 dB of the 0 offset case can still occur [116]. However, if the carrier frequency/code frequency ratio of 1540 is maintained, these peaks disappear very rapidly and simply create an interfering noise that the pulsing mitigates.

Test results back up these interference mitigation theories [113]. Figure 5.40

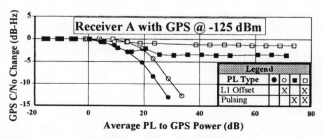

Figure 5.40 Effects of signal structure on PL interference to GPS satellite signal reception.

presents the degradation in GPS signal reception C/N_0 as a function of average received PL to GPS power ratio for four cases: no PL pulsing or frequency offset, either pulsing or frequency offset, and both. The improvements using the mitigation techniques are obvious. Using only a frequency offset buys very little mitigation. Using pulsing alone results in a degradation of about 3.5 dB for average PL to GPS power ratios of 16 to 70 dB (peak power is 10.5 dB higher). Cross-correlation could occur in this case. This could be part of the degradation, but mostly due to the fact that the receiver uses multi-bit sampling that allows some of the pulses to pass through the correlator. The best performance is achieved when both techniques are used, resulting in about 1.5-dB degradation. This is very acceptable considering the advantage gained in navigation performance using PLs. The effect on low-cost "hard-limiting" GPS receivers is even less. With this signal structure, the differential correction/integrity data rate can be as high as 1 kbit/sec with standard BPSK modulation, and up to 2 kbits/sec using quadrature phase shift keying (QPSK) data modulation [113, 115, 131].

Solutions to the Antenna Location Problem A combination of solutions needs evaluation to solve this problem if it is indeed a problem [113, 115]. First of all, if the signal diversity techniques are used to solve the interference problem, then the receivers become insensitive to the PL's transmitted power level as long as the average power is strong enough to receive at the maximum distance (e.g., 20 nmi). Thus, it might be possible to increase the power of the PL so that it can be received via the reduced antenna gain at negative elevation angles. Second, PL placement (high and off to the side) may be possible so that the elevation angle is near zero rather than large negative. Third, although not desirable, an additional antenna could be added to the bottom of the aircraft. This is not desirable for three reasons: an additional antenna also requires another hole in the aircraft's fuselage, it requires an additional LNA and associated cables, and it adds an additional *lever arm* and delay to the GNSS solution.

Initial test results have shown that the antenna location problem can be solved without an additional antenna [113, 132]. These results show PL signal data message dropouts, with no loss of lock, only occur at larger distances when the aircraft is maneuvering. No message dropouts occurred when the aircraft was on final approach. These preliminary test results are promising.

5.8 FUTURE TRENDS

In this chapter, satellite radio navigation is described as it existed in 1996, except for some developments to enhance GNSS integrity, availability, and accuracy for commercial aviation. However, there are other developments that will enhance GNSS even further. Two of these developments are discussed in this section: the relationship of future personal communications systems to GNSS and the evolution of a future civil GNSS.

With the proposed development of satellite-based personal communication systems such as Iridium, the possibility of *position reporting* via these systems is attractive. This brought about a concept in which the communication system signals themselves could be used as navigation signals. However, this concept failed to mature for two reasons. First, the number of satellites required for operation of the communication system was not enough to provide adequate continuous navigation, especially for dynamic users. Second, since GPS receivers have become so much smaller and less expensive, such a receiver could easily be embedded into the personal communication receiver/transmitter, without the development of communications signals with a navigation signal capability.

However, this does not mean that a future civil GNSS system could not take advantage of these future communications systems. With a suitable orbit configuration from a geometric point of view, the same satellites could be used for both purposes [99, 100, 117, 118, 119, 120]. The investigations of the feasibility of a future civil GNSS, at least as an augmentation to GPS, have been recommended by the RTCA Task Force 1 [2] and the FANS GNSS subgroup of ICAO [120]. Early investigations by Inmarsat recognized substantial cost savings for such a system if navigation payloads were to be hosted on future communication satellites, provided that they were placed in intermediate circular orbits (ICOs) [99, 100]. In fact, a navigation payload in each of these communication satellites (12–15) would provide an excellent augmentation to GPS, increasing accuracy, integrity, availability, and continuity of service significantly. Then, as a future option, additional low-cost navigation satellites could be launched to eventually provide a stand-alone civil GNSS as a future replacement for GPS, in the event that GPS is no longer available.

PROBLEMS

5.1. A GPS user receiver is tracking four GPS satellites—PRNs 1, 13, 19, and 22. Via the reception of navigation data, the receiver receives the following ephemeris parameters:

(a) Parameters common to all four satellites:

$$\sqrt{A} = 5153.619629 \text{ meters}$$
$$e = 0$$
$$i_0 = 0.3055555556 \text{ semicircles}$$
$$\omega_0 = 0 \text{ semicircles}$$
$$t_{0e} = 345,600 \text{ seconds in the GPS week}$$

(b) Individual parameters:

Satellite PRN	Ω_0 Semicircles	M_0 Semicircles
1	−0.3239929371	−0.1055577595
13	0.6760070629	0.5808311294
19	−0.9906596038	−0.3166133151
22	−0.6573262705	−0.2525022039

All other received parameters are zero. This user's measurements were taken such that a common time of applicability of the satellites' ephemerides is 531,000 sec in the GPS week. Evaluate the position of each satellite at that time of week.

Ans.:   The satellite positions are as follows:

$$X_1 = 13,672.46475 \text{ km}$$
$$Y_1 = -6,720.41440 \text{ km}$$
$$Z_1 = 21,755.97535 \text{ km}$$
$$X_{13} = -2,370.46666 \text{ km}$$
$$Y_{13} = -23,498.04734 \text{ km}$$
$$Z_{13} = -12,150.94171 \text{ km}$$
$$X_{19} = -18,962.99343 \text{ km}$$
$$Y_{19} = 6,971.55345 \text{ km}$$
$$Z_{19} = 17,240.21601 \text{ km}$$
$$X_{22} = -10,899.89991 \text{ km}$$
$$Y_{22} = -14,301.92165 \text{ km}$$
$$Z_{22} = 19,546.60953 \text{ km}$$

5.2. The user's estimated position (near Denver) in ECEF coordinates is as follows:

$$X = -1,268.4451896 \text{ km}$$
$$Y = -4,739.4160255 \text{ km}$$
$$Z = 4,078.0482708 \text{ km}$$

which corresponds to a position of N39° 44″, W104° 59″, at an altitude of 1,609.344 meters. (Assume a spherical earth with a radius of 6378.163 kilometers.) Compute the slant ranges and azimuth and elevation angles from his estimated position to the four satellites described in Problem 1. Compute the HDOP, VDOP, PDOP, TDOP, and GDOP at the user's estimated position.

Ans: The slant ranges are as follows:

$$R_1 = 23,230.69260 \text{ km}$$
$$R_{13} = 24,829.03623 \text{ km}$$
$$R_{19} = 24,969.68770 \text{ km}$$
$$R_{22} = 20,578.69009 \text{ km}$$

The azimuth and elevation angles are as follows:

$$\text{Az}_1 = 45.201° \qquad \text{El}_1 = 24.955°$$
$$\text{Az}_{13} = 171.127° \qquad \text{El}_{13} = 8.759°$$
$$\text{Az}_{19} = 305.645° \qquad \text{El}_{19} = 7.436°$$
$$\text{Az}_{22} = 302.780° \qquad \text{El}_{22} = 66.743°$$

The DOPs are as follows:

$$\text{HDOP} = 1.241$$
$$\text{VDOP} = 1.631$$
$$\text{PDOP} = 2.050$$
$$\text{TDOP} = 0.823$$
$$\text{GDOP} = 2.208$$

5.3. After traveling some distance the user then measures the following actual ranges to the four satellites:

$$R_1 = 22,280,304.178 \text{ meters}$$
$$R_{13} = 25,351,375.133 \text{ meters}$$
$$R_{19} = 25,373,230.135 \text{ meters}$$
$$R_{22} = 20,867,137.653 \text{ meters}$$

What is the user's new position in ECEF coordinates? What are the azimuth and elevation angles to the satellites and the HDOP, VDOP, PDOP, TDOP, and GDOP at this new position, which has an approximate position of N44° 58″, W93° 15″, at an altitude of 200 meters? Assume a spherical earth with a radius of 6378.163 kilometers.

Ans.:   The user's position solution (near Minneapolis) is as follows:

$$X = -255.843602 \text{ km}$$
$$Y = -4,505.54881 \text{ km}$$
$$Z = 4,507.55905 \text{ km}$$

The azimuth and elevation angles are as follows:

$$Az_1 = 51.406° \qquad El_1 \ = 36.316°$$
$$Az_{13} = 182.344° \qquad El_{13} = 3.909°$$
$$Az_{19} = 310.241° \qquad El_{19} = 3.709°$$
$$Az_{22} = 288.152° \qquad El_{22} = 59.474°$$

The DOPs are as follows:

$$HDOP = 1.304$$
$$VDOP = 1.603$$
$$PDOP = 2.066$$
$$TDOP = 0.8432$$
$$GDOP = 2.231$$

6 Terrestrial Integrated Radio Communication–Navigation Systems

6.1 INTRODUCTION

Since the 1970s, many radio communication and navigation systems have used the same portion of the frequency spectrum and common technology, such as time synchronous operation, digital modulation, spread spectrum wave forms, coding and user-borne clock oscillators. Synchronous operation, in conjunction with signal time-of-arrival measurement, has lead to a direct method for measuring the range between transmitter and receiver locations in systems using this technology. For these reasons, integrated relative and absolute communication–navigation systems, which provide both digital communication and navigation functions by means of the same wave form, have been widely developed. These systems typically use the content of digital data and the time of arrival of the messages measured by the receiver, to determine the receiver platform's position, through some form of multilateration. In general, the positions are determined in a *relative* sense within an arbitrary grid, although the unit positions can be referenced to an *absolute*, geodetic coordinate system, such as latitude, longitude, and altitude, through the use of reference stations whose positions are independently known in the absolute coordinate system. In addition, the position data may be combined in a Kalman filter with dead-reckoning sensor data, such as from an inertial platform, for the purpose of position extrapolation and calibration of the dead-reckoning sensor errors.

Several types of terrestrial integrated communication–navigation systems have been developed. One is a *decentralized* system, in which the operation is not dependent on any central site or node, and each user in a community of members determines its own position. Such a system is also called *nodeless*. A second type is a *centralized* system, wherein the operation is dependent on a central site (node) and may be controlled by it and wherein the determination of the positions of the users in the community is performed by that central site. Frequently, it is desired to have the positions of a large number of users known and tracked at the central site, such as in military or civil command and control systems. Typically, in such a system, users may obtain their positions by automatic, periodic, or occasional requests from the central node; hence such a system is considered *nodal*. Systems are being developed that exhibit both nodal and nodeless characteristics and thus become hybrid systems. However, the fundamental design of these systems is typically based on either the decentralized or centralized concepts.

Typical examples of these systems are represented by the Joint Tactical Information Distribution System Relative Navigation (JTIDS RelNav) function and the position location reporting system (PLRS) and its enhanced versions, whose principles of operation are described in this chapter. The former is representative of a decentralized system and the latter is representative of a centralized system. Applications of these types of systems cover a wide spectrum, including the handover of targets between units operating within a common grid, rendezvous of aircraft or other units, command and control from the viewpoint of a military commander having knowledge of the position of his forces, and such specialized purposes as search and rescue and medical evacuation.

The systems described in this chapter were mature and operational in 1996. For example, by 1996, over 3600 PLRS and enhanced PLRS user units had been produced and deployed on a variety of U.S. Army and Marine Corps vehicles, including tanks and helicopters and as manpack units, and about 1500 more were planned for the future. By 1996, about 500 airborne JTIDS terminals had been installed on such aircraft as the U.S. E3A, E2C, B-1, F-14D, F-15C, and JSTARS, as well as on several aircraft of other NATO countries. About 400 more such terminals had been planned for later installation in various military aircraft and ships. Also in 1996, a major development was under way by a consortium of several countries for a JTIDS-like smaller and modular MIDS terminal that includes the relative navigation function. This reduction in terminal size will make it possible to install it in a large variety of other aircraft.

6.2   JTIDS RELATIVE NAVIGATION

6.2.1   General Principles

The relative navigation (RelNav) function of JTIDS is a decentralized position location and navigation system wherein each user independently determines its position, velocity, and altitude from data received from other users. Member units in a JTIDS community make transmissions in time slots assigned on a precise common time base maintained by on-board synchronized clocks. Among the many transmitted message types are round-trip timing (RTT) messages and precise position location and identification (PPLI) messages. The RTT messages provide maintenance of the precise clock synchronism that supports one-way radio ranging, and PPLI messages provide the time-of-arrival (TOA) range measurements and the source position information that are the foundation of the RelNav function.

JTIDS RelNav is based fundamentally on trilateration, which may be visualized geometrically as scribing arcs of known radius (derived from the TOA of

PPLI messages) centered on the positions of the transmitters and intersecting at the position of the receiver. The algebraically equivalent process involves the solution of three simultaneous quadratic equations in two unknowns (the third equation serving to resolve the ambiguity in the solution of the first two). The solution may also be obtained by an iterative linear process in which an initial position estimate is adjusted in the direction of first one and then another of the sources until a position satisfying the ranges to all three is found. This essentially is the process mechanized in the JTIDS RelNav Kalman filter. Each PPLI message is processed independently and the observations need not be simultaneous. The importance of this is obvious if the user is moving and the observations are sequential as in JTIDS time slots.

JTIDS units typically transmit PPLI messages at intervals ranging from 3 to 12 seconds. A navigating user's processor employs an estimate of its velocity to extrapolate travel during the time between received PPLI observations. It uses the extrapolated position estimate at the instant of each observation to compute a range error vector, namely, the difference between the measured range (TOA) and the range computed from the estimated position and the received source position. The velocity estimate is obtained from the aircraft dead-reckoner system, such as inertial or air data, and the errors of the dead-reckoner system contribute to the observed range error vectors so that, over time, the dead-reckoner errors can be estimated. JTIDS RelNav is, therefore, typically operated as a hybrid multisensor navigation system in which the range measurements are used to derive corrections to the dead reckoner in a multi-state Kalman filter. Chapter 3 describes the basic concepts of hybrid multisensor navigation systems.

JTIDS RelNav may also operate without input from a dead reckoner in what is called a *TOA-only* mode, but in this mode the extrapolated position estimate is based simply on the immediately preceding two positions and is suitable only for very low-dynamic platforms, such as surface ships.

6.2.2   JTIDS System Characteristics

JTIDS is a synchronous, time-division multiple-access digital communication system operating in the 960- to 1215-MHz band. The time slot and message structure are shown in Figure 6.1. The first 32 pulses of each message are a synchronization preamble to establish precise receiver sampling times for the information chips modulated on the following data pulses. The unoccupied portion of the time slot allows the transmission of certain longer message types and guard time for RF propagation to all users before the beginning of the next time slot.

The preamble and its digital matched filter detector establish message start time to a precision of a few hundredths of a microsecond. Precise determination of message start is necessary to the sampling and decoding of the 200-ns data chips and synchronized clocks are necessary for slot number definition and crypto-decoding. With message TOA very precisely known on the receiver's

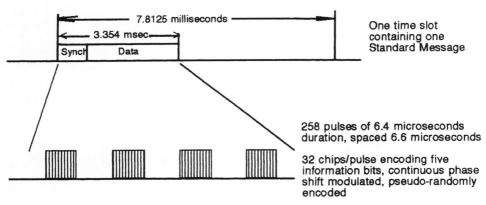

Figure 6.1   JTIDS signal structure.

clock, very precise synchronization of all the individual clocks, although not necessary for the communications function of JTIDS, allows the receiver to convert message TOA to an accurate one-way radio range to each of the transmitters and thus support a precise multilateration navigation function.

6.2.3   Clock Synchronization

Two means of maintaining clock synchronism are provided: round-trip timing (RTT), which operates independently of RelNav, and a passive technique intrinsic to the RelNav function. There is also provision for synchronization to an on-board external time reference (ETR) such as an atomic clock or a global positioning system (GPS) receiver.

The net time reference (NTR) transmits first in any new net and establishes the system time to which all other units synchronize by the exchange of round-trip timing interrogation (RTTI) and round-trip timing reply (RTTR) messages either directly with the NTR or with another unit already synchronized to the NTR. RTTIs are very short messages containing only the addresses of the interrogator and of the desired donor. The donor responds at a fixed time later in the same time slot with an RTTR message containing the address of the interrogator and the time of arrival (TOA) of the RTTI as measured on the donor's clock. The interrogator measures the TOA of the reply and computes the adjustment to its own clock necessary to make the donor's reported TOA equal the TOA of the reply at the interrogator. Figure 6.2 illustrates the message exchange and the associated calculation. A series of RTT transactions over a period of a few minutes provides an estimate of the interrogator's clock drift rate; i.e., the frequency error of its clock oscillator and the frequency error estimate is then used to retune the oscillator driving the clock. The estimation of clock bias and frequency errors is carried out in a small Kalman filter that provides error

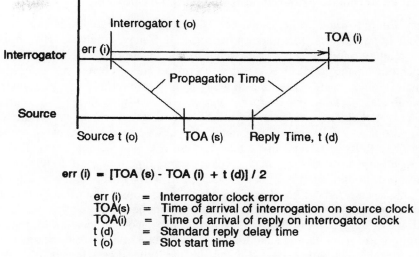

$$\text{err (i)} = [\text{TOA (s)} - \text{TOA (i)} + \text{t (d)}] / 2$$

err (i)   = Interrogator clock error
TOA(s)   = Time of arrival of interrogation on source clock
TOA(i)   = Time of arrival of reply on interrogator clock
t (d)   = Standard reply delay time
t (o)   = Slot start time

Figure 6.2   The round-trip timing (RTT) process.

uncertainties in its covariance matrix. The clock bias uncertainty (variance) is converted to a time quality number that is transmitted in the PPLI messages. A synchronizing user performs RTT transactions with the donor of the highest time quality exceeding its own.

A technique of passive synchronization is provided for users constrained to radio silence. A RelNav user's clock error adds linearly and equally to the observed one-way radio ranges to all the PPLI sources. The RelNav Kalman filter of passive users minimizes this common range bias by assigning it to the clock state carried in the RelNav Kalman filter. The clock errors of passive users are, therefore, correlated with source time and position errors and are magnified by geometric dilution of precision (GDOP). (See Chapters 4 and 5 for a discussion of GDOP.) Generally, the time and position errors of passive users are greater than those of active users because the simultaneous solution for both time and position is the equivalent of hyperbolation, rather than tri-lateration, and the associated GDOP is greater. Passive users with an external time reference, of course, do not suffer this effect.

Synchronization to an external time reference (ETR) follows essentially the same procedure as RTT. The ETR supplies a time pulse and a data message declaring the time of the pulse. The JTIDS terminal observes the difference between the TOA of the pulse and the ETR's declared time of the pulse and adjusts its clock and frequency models accordingly. Synchronization to ETRs allows widely separated JTIDS ground units to maintain synchronism and readiness to communicate and support RelNav over extended periods without radio contact with other RTT sources.

6.2.4   Coordinate Frames and Community Organization

Coordinate Frames   The primary coordinate frame for RelNav calculations is geodetic latitude and longitude; however, JTIDS may simultaneously support a secondary purely relative grid (RelGrid) coordinate frame unique to JTIDS. Operation in the RelGrid is at times useful for the exchange of target coordinates between units operating in areas with poor reference to local geodetic coordinates, such as over ocean or deep-penetration missions.

The RelGrid is a cartesian frame tangent to the Earth at its origin with U-coordinate east, V-coordinate north, and W-coordinate upward at the origin. The RelGrid is typically established by a single moving unit called the *navigation controller* (NC) that serves the same function in the RelGrid as do surveyed ground references or GPS-equipped users in the geodetic frame; it establishes a reference baseline from which other units make trilateration measurements. There must, therefore, be relative motion between the single NC and the dependent users. The NC uses a specified geodetic location for the grid origin to transform latitude and longitude from its on-board dead-reckoner system to RelGrid coordinates, and it appends the RelGrid coordinates to its PPLI message. Position and velocity errors of the NC dead reckoner lead to offset and drift of the RelGrid relative to the true Earth and azimuth errors introduce a rotation of the RelGrid about the W-axis. However, the objective of RelGrid operation is the calibration of the dead reckoners of the users relative to the dead reckoner of the NC rather than calibration with respect to the Earth.

RelGrid user units acquire an estimate of the RelGrid origin by computation from the geodetic and RelGrid coordinates in a PPLI message from the NC or from another unit established in the RelGrid. The new grid entrant derives its initial RelGrid coordinates from its own geodetic position estimate and the computed origin location. Thereafter, RelGrid and geodetic navigation computations proceed independently using the same source selection and Kalman filtering logics as described in Section 6.2.7. Though computations in the two coordinate frames are essentially independent, in a community with both geodetic and Rel-Grid sources there is some interaction through the common dead-reckoner error terms of the RelNav Kalman state vector (Table 6.1) as it attempts to reconcile own-unit dead-reckoner errors and the NC dead-reckoner errors reflected in the grid drift. If well disposed accurate geodetic references are available, there is no need to establish a relative grid.

Community Organization   A unique characteristic of JTIDS decentralized RelNav is that users navigating in the system become navigation reference sources for other users, and it is highly interactive in that the navigation errors of one user propagate to other users. RelNav employs a dynamic covariance-based user hierarchy to control these interactions and prevent reciprocal ranging and regenerative circulation of errors. Rank in the hierarchy is transmitted in the position quality and time quality fields of the PPLI message. Each user independently estimates position and velocity corrections on the basis of PPLI

TABLE 6.1   Kalman filter state vector for RelNav inertial dead-reckoner system

State-Vector Element	Description
1	Relative grid U-position error
2	Relative grid V-position error
3	Geodetic quaternion error—element 1
4	Geodetic quaternion error—element 2
5	Altitude error correction
6	RelGrid controller azimuth angle error
7	Clock bias
8	Platform Z-axis angle error
9	X-velocity error
10	Y-velocity error
11	X-axis tilt error
12	Y-axis tilt error
13	Altitude scale factor error correction
14	RelGrid controller U-axis velocity error
15	RelGrid controller V-axis velocity error
16	Clock frequency error

messages received from users of superior quality and establishes its own rank in the hierarchy on the basis of the qualities of its sources and the measurement geometry as reflected in the RelNav Kalman filter covariance matrix. Some units in the community must, of course, have independently known positions and qualities to get things started.

In the time hierarchy, the net time reference (NTR) transmits the highest time quality (15). Other units transmit time qualities derived from the time variance developed in their synchronization Kalman filters. Primary users employ only the RTT technique, while secondary users employ primarily passive synchronization, making recourse to RTT only under certain conditions of poor geometry. Explicit designation of secondary users is seldom made. Radio-silent users, unable to participate in RTT message exchanges, automatically assume secondary user status.

Within the RelNav hierarchy, the equivalent of the NTR is the RelGrid navigation controller (NC). RelGrid coordinates transmitted by the NC are by definition perfect as indicated by transmission of the highest relative position quality of 15. There is no equivalent of the NC in the geodetic frame; that is, there is no equivalent arbitrarily perfect reference designator. Designation as a position reference disables geodetic position update; however, this designation does not connote perfection. The accuracy of the geodetic position in PPLI messages is characterized by the geodetic position quality which, for all units, is determined initially by operator entry and subsequently by the RelNav Kalman filter covariance. The maintenance of the covariance based hierarchy in both time and position is the function of the source selection logic described in Section 6.2.6.

6.2.5 Operational Utility

JTIDS RelNav provides precise position registration in geodetic coordinates and (optionally) in its own unique U/V/W relative coordinate frame while requiring no additional hardware beyond the JTIDS data link terminal and its data bus interface with the host system. It provides calibration of on-board dead-reckoner errors (continuous in-air alignment) with consequently improved platform alignment for referencing on-board weapon launch and guidance systems and improved navigation accuracy during excursions out of JTIDS net coverage, such as low-level missions in forward areas.

Precise registration of user positions and calibration of dead-reckoner errors lead to improved registration between targets acquired by an on-board radar and target track reports received from surveillance systems, especially those from JTIDS-equipped airborne surveillance systems such as the E3 AWACS, E2C Hawkeye, and JSTARS. More importantly, accurate relative position and platform alignment provide improved registration of locally acquired targets exchanged between the mission elements themselves. Position accuracy of one-tenth to one-quarter mile is typically required to support reliable correlation of these target reports. This exceeds the relative accuracy of unaided inertial navigators after an hour of flight but is well within JTIDS RelNav accuracy. Accurate correlation of target reports between mission elements results in more efficient weapon/target allocation. Accurate knowledge of the locations of the cooperating mission elements via the exchange of PPLI messages also contributes to reduced risk of fratricide and allows greater freedom and precision of maneuver between supporting elements in low-visibility conditions.

6.2.6 Mechanization

The overall diagram of the RelNav function in Figure 6.3 shows its three major subfunctions: source selection, Kalman filter and navigation processing. Each received PPLI message is processed by the source selection function immediately following its reception. Host dead-reckoner data are processed in the navigation function to provide an estimated own-unit position at the time of receipt of each selected PPLI message. Source selection stores the selected PPLI observations and the associated navigation data to await processing by the RelNav Kalman filter. The computed range and direction to the source and the measured range from the TOA of the received PPLI message provide a range error vector which the filter uses to estimate position, velocity, and other dead-reckoner error states. These error estimates are applied to the internal dead-reckoner model in the navigation function and corrections are supplied to the host platform.

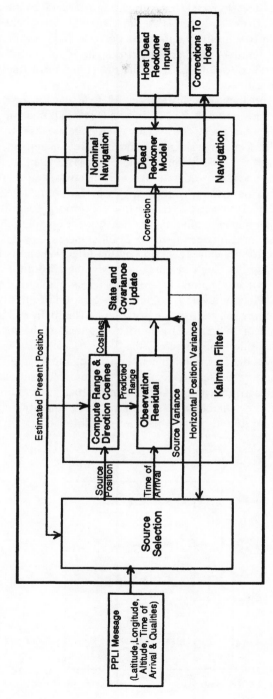

Figure 6.3 Overall relative navigation flow diagram.

Source Selection   The source selection function is crucial to the decentralized design of JTIDS RelNav; in fact, it is the fundamental instrument for the maintenance of community stability. It enforces a position and time quality hierarchy to prevent the regenerative circulation of errors and it also performs geometric tests to give preference to sources from the directions most likely to improve the user's position estimate.

Usually more PPLI messages are received than can be processed by the Kalman filter function and some must be rejected. The JTIDS computer performs many of its tasks on a slot-by-slot basis. However, there are tasks that do not need to be completed within one time slot, and, indeed, there is usually insufficient time for completion of all the tasks within a single slot. The Kalman filter involves many trigonometric and matrix operations requiring considerable computer time, but dead-reckoner error states are only slowly variable so the filter performance is not very time sensitive. It is assigned a low task priority, and the processing of a single observation may require many time slots for completion. Meanwhile, the source selection function operates on each PPLI message as it is received and buffers those selected to await processing by the filter. When the Kalman filter completes processing the batch of observations gathered during the preceding cycle, the current source selection buffer is transferred to the filter's input buffer and a new source selection cycle is begun. The duration of the filter cycle, and hence also of the source selection cycle, is variable (usually from 3 to 20 seconds) because it is dependent on both the total processing load on the computer and on the number of observations processed by the filter.

Figure 6.4 shows the functions and interfaces of the source selection process. The minimum range test rejects observations from sources so near as to threaten the linearity assumptions of the Kalman algorithm. The quality screening test is fundamental and maintains the quality-based hierarchy by a comparison of the time and position qualities of the received PPLI

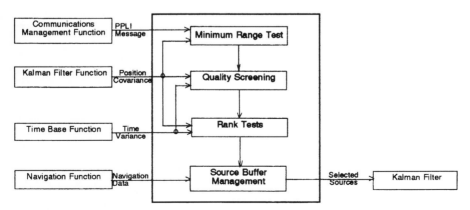

Figure 6.4   Relative navigation source selection flow diagram.

messages with those of the receiving user. In general, both the time and position qualities of the source must exceed the position quality of the receiving user. This test is, in a sense, antithetical to the concept of the Kalman filter which was designed to derive low-variancae estimates from higher-variance observations; however, it has been found to be essential to the operation of the network of interactive filters in a RelNav community. It recognizes that the simple qualities transmitted in the PPLI messages do not represent true variances (noises) in the estimation sense; rather, they represent uncertainties of estimation errors that are predominantly correlated bias errors. The limited PPLI message bits available to the RelNav function precluded the more sophisticated approach of transmitting the separate position terms of the RelNav filter covariance matrix.

Users enter the network with an operator-assigned initial position quality indicating the uncertainty of the initial position estimate. In airborne users, position quality will degrade with time in accordance with the error signature of the user's dead reckoner as modeled in the Kalman filter's time propagation of the covariance matrix. The Kalman filter processes PPLI observations to estimate the dead-reckoner errors (and reduce their covariance terms) and this will be reflected in a decreased rate of degradation of position quality. A dynamic balance supportable by the quality and geometry available from the PPLI message sources is soon established.

The geometric rank tests of Figure 6.4 recognize that the value of a PPLI observation is related not only to its position and time qualities but also to the direction from which it was received. Several sources of equally high quality all in approximately the same direction provide little more information than one such source, but each will consume source selection buffer locations and filter processing time and will crowd out observations of lesser quality but of greater value by virtue of their directions. The rank test uses the orientation, eccentricity, and semimjaor axis of the bivariate error ellipse defined by the horizontal position terms of the RelNav Kalman filter covariance matrix and the quality-based variance and direction of a received observation to compute a numerical rank that is stored with each observation. The rank is the approximate variance of a hypothetical observation lying directly on the extended major axis that would provide the same benefit as the received observation at its angle off-axis. In this context, benefit implies the reduction in the major axis of the error ellipse to be expected of Kalman filter processing of the received observation.

Kalman Filter   The JTIDS RelNav Kalman filter is an extended Kalman filter that estimates linear error states of the navigation process. Table 6.1 presents a typical state vector for use with an inertial dead reckoner. Two of the state-vector elements (7, 16) are time and frequency states used only by passive users. For active users, these two states are carried in a separate synchronization filter that uses RTT or ETR data. Five states (1, 2, 6, 14, 15) are relative grid states. Of the remaining nine states, the two horizontal geodetic position error terms are carried as quaternions, while the third dimension is carried as altitude error and an altimeter instrument scale factor term. The inertial filter models velocity errors in the north and west directions in the geodetic frame and the three platform misalignment or tilt states in the local-level frame. A filter for use with air-data computer/attitude and heading reference system (ADC/AHRS) inputs is also included and differs in that it models the dead-reckoner errors as two wind components, an airspeed instrument scale factor, and azimuth bias and azimuth gyro drift rate errors.

Figure 6.5 is a flow diagram of the Kalman filter function. Only the measurement geometry, measurement innovation, measurement validity, and filter characterization features that are peculiar to JTIDS RelNav will be discussed here. See Chapter 3 for a general discussion of Kalman filters used in multi-sensor navigation systems.

The source position and the own-unit position stored with it are subtracted vectorially to obtain a predicted range and three-dimensional direction cosines. The range error (measurement innovation) is obtained by subtracting the predicted range from the measured range. The measurement variance derived from the source qualities and the direction cosines are supplied to the Kalman gain function, and the range error is supplied to the state and covariance update function.

The validity tests are intended to detect divergence of the filter and to provide protection against inconsistent observations. This is particularly important to JTIDS RelNav, as compared to other radio-ranging systems because the ranging sources are typically other, sometimes erroneous, navigating users. The dilemma facing any validity test is whether the error lies with the local estimate or with the input data. The source selection function makes this decision more difficult by narrowing the group of sources to the few, typically three to five, with the best announced qualities (whether true or false). An observation is rejected if the measurement innovation (the computed range error) is large compared to the receiving user's filter variances (3-sigma reasonableness test). A series of observations exhibiting an average innovation exceeding 2-sigma for this test, triggers a proportional increase of the position terms of the filter covariance matrix. Eventually, if the recurrence rate of these covariance expansions exceeds a threshold, the process is abandoned and the filter is restarted.

The filter characterization function serves only to convert filter covariance data to forms more convenient to the source selection and PPLI generation functions. Position quality in the PPLI message is defined by the JTIDS message standard as representing the semimajor axis of the horizontal position error ellipse and the rank computation requires the semimajor and semiminor axes and the orientation angle of the error ellipse. Each Kalman cycle, the filter characterization module computes these terms and stores them for use during the next cycle.

Navigation Processing   Figure 6.6 is a diagram of the inertial navigation system implemented within the JTIDS terminal. The particular model (north slaved, wander azimuth, unipolar, or free azimuth) installed in a given terminal

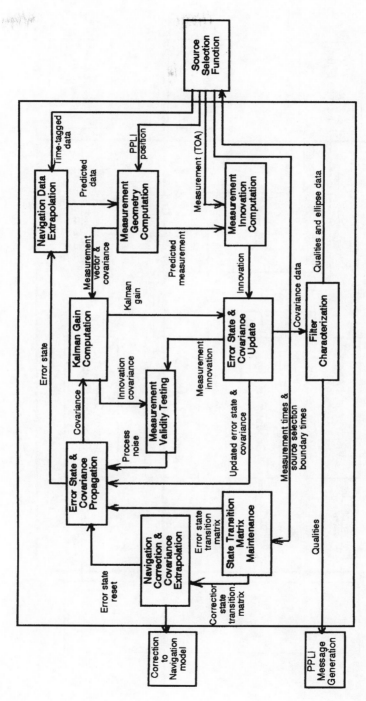

Figure 6.5 Relative navigation Kalman filter flow diagram.

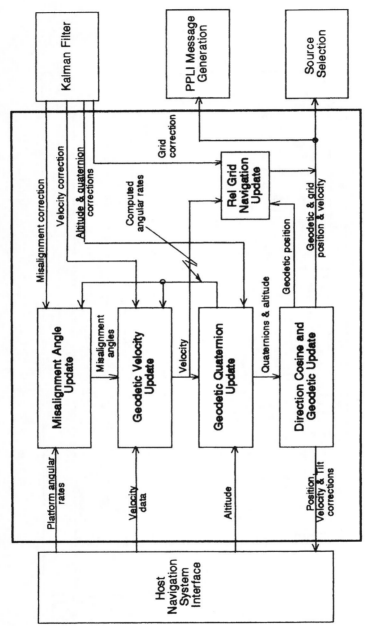

Figure 6.6 JTIDS relative navigation inertial navigation model.

is selected to correspond with that of the host system. JTIDS initial position is obtained from the host inertial system, but thereafter JTIDS uses only the velocity and baro-inertial altitude inputs to compute delta-velocities for use in the internal inertial mechanization. This model continuously applies the Rel-Nav filter corrections to velocity and acceleration (via platform misalignment calibration) to improve upon the solution provided by the host inertial system. Corrections to position, velocity, and tilts are also returned to the host data system.

As mentioned earlier, an air-data model is available for use in aircraft having only an ADC/AHRS, but JTIDS RelNav has seldom been installed in such aircraft. Its primary use is as a backup mode to continue RelNav operation and PPLI transmissions should the host's inertial system fail and force a switch to air data.

Integration with GPS   JTIDS RelNav processes position fixes from an interconnected GPS receiver as two uncorrelated, one-dimensional Kalman updates in the north and east directions. The variance data in the GPS input are used in RelNav source selection and Kalman gain computations in the same manner as the qualities in received PPLI messages. If GPS data are of high quality—that is, if GPS GDOP and signal availability (Chapter 5) are within typical GPS system performance criteria—the JTIDS terminal will tend to operate exclusively on GPS data, and the transmitted PPLI messages will reflect GPS position accuracy. A few aircraft with interconnected JTIDS and GPS equipments can, via JTIDS RelNav, extend the benefits of GPS-based position fixing to an entire community of JTIDS users.

6.2.7   Error Characteristics

JTIDS RelNav is subject to errors of the dead reckoner, errors in the positions of the reference sources, errors of range measurement and RF propagation, and the amplifying effect of GDOP.

Dead-Reckoner Errors   The inertial filter explicitly models velocity and platform misalignment errors, but higher-order terms are modeled only as dynamic process noise. The air-data filter models wind and azimuth errors as Markov processes. To the extent that actual dead-reckoner errors depart from these assumptions, JTIDS RelNav will experience errors. As one example, winds aloft typically vary with altitude so a RelNav user coupled with an ADC/AHRS can be expected to exhibit temporarily increased errors following a substantial altitude change.

Equipment Delay Errors   Installation-specific data are used to compensate transmission times and received message TOAs for equipment and cable delays between the antenna and the signal processor. These delays are determined for each installation and are subject to measurement error, but they are usually significant only in ground or shipboard installations where the antenna may be remote from the terminal.

Clock Synchronization Errors   A typical JTIDS terminal clock runs at 80 MHz (clock quantization of 12.5 ns) and the JTIDS RF wave form uses a phase-modulation chip of 200 ns (5-MHz chip rate). These parameters and simple double-oven crystal oscillators have proved able to provide consistent clock synchronism of ±25 ns using RTT exchanges approximately every two minutes. Clock synchronization errors have not been found to contribute appreciably to RelNav error; however, oscillator stability under the extremes of temperature, pressure, and acceleration of fighter aircraft is an important consideration as has been demonstrated in several flight tests.

The clock error of a passive user is typically greater than that of an active RTT user; however, it does not contribute independently to position error. It is simply a fully correlated manifestation of the same source errors and GDOP.

Source Position Errors   The best PPLI position quality (15) implies a site survey of better than 50 ft, 1-sigma. More important than accurate survey, however, are the position qualities assigned at the ground reference sites. These must reflect a conservative estimate of the position uncertainty. Optimistic position qualities can result in user validity failures and recurrent filter resets leading to instability rather than just increased position error.

RF Propagation Errors   JTIDS RelNav applies an approximate compensation for atmospheric index of refraction (speed of propagation) by calculating an average index over the signal path of each selected PPLI message using the altitudes of transmitter and receiver and a standard atmosphere lapse rate for index. Nonstandard atmospheres introduce significant error only over the very longest of paths between high-altitude aircraft.

The pseudorandom coding of the phase-modulated chips of the JTIDS signal wave form conveys strong resistance to multipath ranging errors. The digital matched filter of the preamble detector will ignore as noise a delayed replica signal arriving more than 200 or 300 ns after the direct signal. Delays of less than 200 ns (chip overlap) cause slight broadening and delay of the peak of the preamble correlator output.

Geometric Dilution of Precision (GDOP)   GDOP is a source of error in all multilateration and hyperbolation systems (see Chapters 4 and 5) but is particularly important to JTIDS RelNav. JTIDS RelNav GDOP is much more variable, both geographically and temporally than, for example, that of GPS or Loran-C. It may range from unity to over one hundred within the service area of a JTIDS community. JTIDS ground reference sites are typically major command and control centers or airfields whose locations result more from tactical, political, and geographic considerations than from an intent to support the JTIDS RelNav function. Also, local GDOP is a function of user altitude as the community of ground references within line-of-sight changes and as transient airborne sources temporarily contribute to filling holes in the basic GDOP contours provided by the reference sites. Furthermore, the GDOP at a given user position is determined by the geometry to only those few sources selected by the quality, rank, and validity tests of the source selection logic, and, despite their intended purpose, these tests may or may not always choose the geometrically optimum set of sources. Thus it is impractical to describe a generally meaningful GDOP contour map as can be done for other less dynamic systems. In many instances, however, GDOP is the dominant error contributor and must be explicitly included in JTIDS RelNav performance analysis.

6.2.8   System Accuracy

The JTIDS system specification defines RelNav performance tests that compare the measured position errors to a criterion called the available position quality (APQ). Short track segments are selected, and, from all the received PPLI messages recorded by the test unit during that segment, the best *a posteriori* solution and its error ellipse (the APQ) are computed for each point along the segment at which the test unit transmitted a PPLI message. The deviations of the test unit's computed positions from the true positions, as measured by the test-range tracking system, are compared to the available position quality as representing the 1-sigma bound of the error under the immediate local conditions.

   Results of flight tests by the military services have not yet been published in the open literature; however, the accuracy can be quite closely predicted from the system design parameters and conservative estimates of the error sources. Results of several computer simulations have been reported [1, 9, 13, 14], and indicate that JTIDS RelNav can, with high confidence, be expected to achieve airborne user position accuracies of 100 to 300 ft over a range of reasonably assumed error budgets, flight scenarios, and GDOP.

6.3   POSITION LOCATION REPORTING SYSTEM

6.3.1   General Principles

The position location reporting system (PLRS) and its derivative systems provide centralized position location and reporting and data communications for communities of hundreds of cooperating users in a tactical environment. Time-of-arrival (TOA) measurements between units in a community, aided by barometric pressure measurements, are processed at a central site to establish position tracking of a large number of users. The positions of a few participants are used as grid references. At the central site, both ranges and clock offsets are derived from mutual pairs of TOA measurements (Section 6.3.5). With the clock offsets established, additional ranges are derived from one-way TOA measurements. The positions of users are then tracked, using adaptive predictor-corrector filterng. All positions are available to the cooperating users and to command centers.

PLRS also provides short message data exchange for both manual and automated users. All control, measurement reporting, and data exchange are cryptographically secured in a synchronous, anti-jam communications network. Master stations (MSs) establish control circuits between radio sets (RSs) and the MSs via a control network. This control network supports the position location, navigation aid, and friendly unit identification. From a message flow standpoint, the control network provides an "order-wire" capability that can be used for data exchange between users. In addition to the order wire, the control network also supports user access to a wide range of position location, navigation, and identification information.

The system has a range of capabilities which support the conduct of coordinated military operations. For the individual tactical user on foot, in a surface vehicle, or in an aircraft, the system determines and displays to him his accurate position in real time. It alerts him if he enters a restricted area. It also provides the user with guidance to predesignated points, to other users, or along corridors in accordance with requests, as well as providing a free text data exchange capability.

For a tactical commander, the system provides the identification, location, and movement of all cooperating users within an assigned area of responsibility. In addition to allowing the commander to monitor the movement of forces, the system also has the ability to input and modify control measures such as coordination points, safe corridors, and restricted zones.

For all participants, the system (which operates beyond the line of sight via integral relays) incorporates electronic counter–counter measures (ECCM) and provides cryptographically secure digital data communications. Each user has the capability of sending preassigned short messages to provide data to or request information from the system and to exchange short free text messages.

A single synchronous community can support over 900 users with a varied distribution of manpack, surface vehicle, and airborne platforms. System performance is provided within the primary ground operating area, and airborne users can be located and tracked within a 300-km square extended operating area. It can interoperate with other communities in adjacent or overlapping geographical areas. It operates in the UHF band at frequencies from 420 to 450 MHz.

6.3.2 Major System Elements

PLRS employs two categories of hardware: master station (MS) and radio sets (RS). The MS provides centralized network management as well as automatic processing and reporting of position, navigation, and identification information for each participating RS. RSs, which are individually identifiable to the MS, perform reception, transmission (including relay), range measurement, and various signal-processing and message-processing functions necessary for position location and communication operations within the system.

PLRS is usually deployed with two or more identically equipped MSs. The MSs monitor each other's operation, cooperate in the network position location and communications functions, and assume adjacent community control either by planned action, directed by the MS operators, or automatically upon an MS failure.

The identification and position determination of RSs by PLRS is fully automatic. When the RS operator turns on the equipment, it automatically becomes and remains a member of the PLRS network. However, to permit the RS operator to provide data and requests to the MS and to receive and display information from the MS, a separate user input/output (I/O) device is employed with each RS. For manpack and surface vehicular RSs, this I/O device is a small hand-held device called the *user readout* (URO) module. For airborne RSs the I/O device is a pilot control display panel (PCDP). The PCDP provides a larger, brighter display and an interface to a bearing indicator so that the pilot can "fly the needle" in response to automatic bearing updates from the MS.

6.3.3   Control Network Structure

Tactical deployments require operations beyond line of sight (LOS) from the MS. The approach taken in PLRS to satisfy this non-LOS requirement is to use relays. In most deployments one and sometimes two or three relay levels are needed to establish a path between a remote RS and a MS. To satisfy this need, an integral relay capability is built into every RS. Any RS can be automatically utilized to maintain contact with any other RS. This reduces the need for dedicated relays and improves speed of adaptation to changing deployments. To maintain communications and provide organized reporting of data for position location calculation, a concept of control network organization, called a *PORT* structure, is used. This is a communications structure (Figure 6.7) consisting of

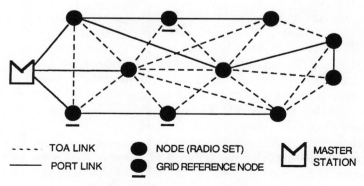

- - - - TOA LINK   ● NODE (RADIO SET)   MASTER
———— PORT LINK   ⬤ GRID REFERENCE NODE   STATION

Figure 6.7   PLRS network structure.

a set of PORT links that connects RSs (nodes) to the MS either directly or via one to three relay nodes. Network control and measurement reporting is transferred over the PORT path. In addition to the bilateral PORT links, one-way TOA links are utilized to provide the additional multilateration structure needed for position location and tracking. Since the timing of RS clocks is established using paired TOA data along the PORT paths (Section 6.3.5), one-way TOAs can be converted to true range estimates. In a typical PLRS deployment over half of the range measurements are based on one-way TOAs.

To initialize the position location function and to maintain a relationship between the internal coordinates and the external military grid reference system (MGRS) coordinates, the MGRS positions of three or more cooperating RSs are input to PLRS. These are normally input as three-dimensional fixed reference positions, and the RSs then become grid reference nodes. The MS may be, but is not necessarily, one of the grid reference nodes. In addition the system can operate without any fixed reference RSs as long as the positions of three or more RSs are regularly input to the position tracking function. In this latter case the positions may be input and updated by RSs which are moving. This is termed a *dynamic baseline operation*. External position sources such as the global positioning system (GPS) (Chapter 5) can be used to provide the position reference information to PLRS, but external data sources require the appropriate coordinate conversion from the respective geoids and datums to MGRS.

6.3.4   Waveform Architecture

PLRS performs its functions in the face of either deliberate or accidental interference. One of the design characteristics that makes this possible is the use of a spread spectrum type signaling wave form. Specifically, the information transmitted is spread to a bandwidth of approximately 3 MHz by modulation with a pseudonoise (PN) code sequence. The chip width of the PN code is 200 nsec. Each time a RS or a MS transmits a burst, that signal burst is spread by the code. The spread spectrum signaling format provides a low-density signal spectrum that reduces detection by would-be interceptors and offers minimum interference to other co-channel users. In addition, the effect of this modulation is to encode the signal and thus help protect it from those who might try to extract information if the signal is detected. The burst of signal that is sent by a PLRS RS consists of two portions: a preamble portion and a message portion (Figure 6.8). The receiver examines the preamble, using a digital matched filter that accepts or rejects the signal on the basis of the degree of correlation between the pseudonoise code received and that expected. If the preamble is accepted, then the message, which consists of addresses, commands, measurements, queries, and/or replies, can be decoded.

All of the burst transmissions appear to be identical. That is, whether the particular transmission being sent is a reply to a query, a request for information, or whatever, it has the same bandwidth and burst duration, and it cannot be distinguished from any other transmission without having the proper receiver

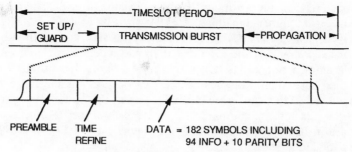

Figure 6.8   PLRS timeslot signal structure.

and crypto key. Also, ranging may be accomplished with any of the bursts sent, no matter what their purpose, as far as the message portion of the transmission is concerned.

PLRS is a synchronous time division multiple access (TDMA) system, which also employs frequency division multiple access and a spread spectrum wave form (Figure 6.9). PLRS employs a network that is fully synchronous in three respects: all RSs maintain timing such that cryptographic resynchronization is seldom required; all RSs perform actions in a programmed cyclic manner such that reprogramming RS assignments for relay, ranging, and reporting is seldom required; and each RS's time base is maintained with sufficient accuracy such that one way time of arrival measurements can be translated to ranges by the MS. Each of these aspects reduces the number of required control transmissions and makes time available for other system functions or for increased system capacity.

PLRS employs time division multiplexing to permit a large number of users to utilize the same frequency. Each of the RSs in a network takes turns transmitting its burst while other RSs listen. These timeslots are assigned by the MS, based on the particular requirements of each user. For example, for a given

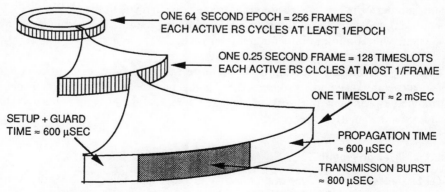

Figure 6.9   PLRS time division multiple access (TDMA) organization.

tracking accuracy, an aircraft-mounted RS needs more timeslots than a manpack RS because of its higher dynamics.

In the synchronous TDMA approach, every RS has its own time-base generator that keeps track of the time that it should receive, transmit, or perform other programmed operations. Once this time base is synchronized with the network time base, then messages can be sent or received and range measurements made. The MS's time base normally acts as the prime timing source for the RSs under its control, and the MS corrects each of the RS clocks whenever they require it. In this way, the timing oscillators included in the RSs need to have only moderate stability.

The network utilizes the resources of time, frequency, and code to multiplex the many operations necessary for system operation. The structured use of time allows a convenient and efficient method for gathering time-of-arrival data and for managing the multiple relay levels within the network. The use of the frequency resource provides additional anti-jam protection and allows for noninterfering, coordinated operation to increase system data capacity.

The Time Resource Each RS is commanded by the MS to perform transmissions and receptions at specific times. The time division structure simultaneously accommodates both minimum and maximum network access rate requirements. For position location and control, these vary from manpack RSs requiring update rates approximately once every minute to high-performance fixed-wing aircraft with desired updates 30 times per minute.

Timeslot The fundamental time division is the timeslot. The timeslot length is 1.95 msec. The burst transmission accounts for 800 μsec, and 600 μsec is allocated to RF propagation delay. The remaining time is required for processing overhead such as message encoding, validation, and guard time.

The Frequency Resource Each transmission occurs on a particular frequency channel. When operating in the hop mode, each channel is pseudorandomly hopped across the 420- to 450-MHz band to provide the network with additional anti-jam performance. There are eight frequency channels in the PLRS, thereby increasing the capacity of the network by allowing simultaneous transmissions to occur with a minimum of mutual interference.

The Code Resource In addition to the time and frequency separation, PLRS uses a pseudonoise code resource. These codes provide a different spread spectrum pattern for each transmission in the network, thereby eliminating cross talk and reducing interference.

6.3.5 Measurements

TOA Measurements One-way ranging is made possible by the fully synchronous nature of the PLRS network. Each RS employs a set of time markers

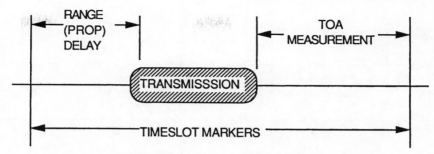

Figure 6.10   PLRS synchronous ranging.

to designate when a transmission is to start and when a reception must be completed. Thus a RS needs only to measure the time delay from the end of the reception to a time marker at the end of the timeslot. This yields a digital number (TOA measurement) precisely related to the range between the two RSs (Figure 6.10). These TOA measurements are compensated for local equipment delays and sent to the controlling MS.

Spread spectrum signals are especially amenable to range estimation because of the high chip rate codes employed in their modulation and because those code sequences have excellent correlation functions. Over the entire length of a code sequence, there is only one point at which a code will correlate with itself, and that point is only about one chip increment in length. Standard TOA measurement methods are used for spread spectrum systems that allow them to resolve time to a small fraction of a chip.

Time Offset and Range Measurements   To provide time synchronization, the system performs time-difference measurements. When one RS transmits, a second RS measures to TOA (Figure 6.11). The RSs report their TOA measurements to the MS. The MS uses these two TOA measurements to determine the timing offset between the two RSs, and the MS commands the RS to correct its clock timing, when required. All timing information is stored in the MS's computer.

The MS also uses these pairs of TOA measurements to estimate the range between the two RSs. These two-way range estimates are more accurate than one-way range estimates, since the clock offset errors are eliminated.

Altitude Measurements   Each PLRS RS contains a barometric transducer that measures air pressure (Chapter 8). These air pressure measurements are sent to the MS, along with the TOA measurements. The MS then converts these pressure measurements to altitude estimates. The use of relative barometric pressure to aid vertical location is especially useful when the line-of-sight range vectors have small vertical components.

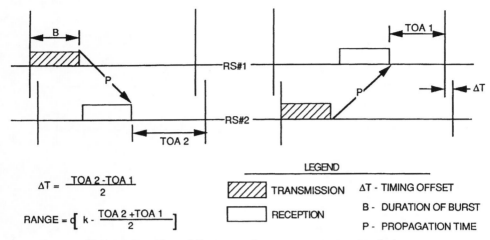

$$\Delta T = \frac{TOA\,2 - TOA\,1}{2}$$

$$RANGE = c \left[k - \frac{TOA\,2 + TOA\,1}{2} \right]$$

LEGEND

▨ TRANSMISSION	ΔT - TIMING OFFSET	
▢ RECEPTION	B - DURATION OF BURST	
	P - PROPAGATION TIME	

Figure 6.11   Time difference and range measurement technique.

6.3.6   Position Location and Tracking

The initial location of position within PLRS is based on the use of three ranges and an altitude to unambiguously locate a new RS in three dimensions (Figure 6.12). The altitude, based on a barometric measurement, establishes the RS on a horizontal surface; two ranges, based on TOA measurements, are then used to establish a pair of points on that surface; and the third range is used to resolve the ambiguity. Velocity is then established using a sliding three-point method for filter initialization. All of the position locations within PLRS are established and tracked at the MS based on measurements taken by the RSs. These data are provided to the MS via user measurement reports. Each user measurement

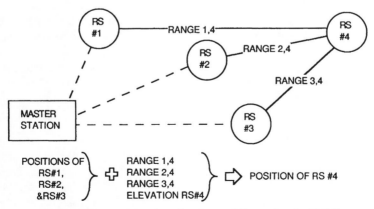

Figure 6.12   Position location by multilateration in PLRS.

report message may contain up to three TOA measurements and a barometric measurement.

Choice of an algorithm for position location update is strongly influenced by the RS dynamics and by the multi-level relay requirements. The majority of RSs are typically not in line of sight (LOS) to the MS and in a large network it is unlikely that most RSs have LOS to any known reference RSs. Thus a typical RS must be located by using measurements from other RSs which are themselves being tracked. Since all position measurements cannot be made simultaneously, it is necessary to extrapolate the position of one (or both) RSs cooperating in a range measurement. A portion of a position correction is ascribed to each RS involved. The amount of correction applied is a function of the track uncertainty of each RS along the line of path between the two RSs.

Because of the availability of multiple links and their usefulness in tracking other RSs, a single RS may have up to ten or more TOA measurements taken at various times, utilized in a single position report period. To maintain a simple algorithm that adapts to the variable data base and minimizes computer memory requirements, each TOA measurement is processed as it is received at the MS. This provides a sequence of partial updates and makes the tracking algorithm relatively independent of which RS is the primary beneficiary of a given TOA.

6.3.7   Tracking Filter

There are four adaptive predictor-corrector filters used in PLRS. All of these filters are simplified versions of the discrete Kalman filter (Chapter 3) and are implemented in the software of the MSs. Together the filters take the raw measurement data from each RS's measurement report message, consisting of TOA and barometric pressure transducer values, and convert them to updated estimates of the user's three-dimensional position and velocity. Figure 6.13 depicts the interconnectivity of the four filters. All filters take one piece of input data at a time. The mean sea level (MSL) filter's purpose is to furnish an offset calibration for all the nonreference RS barometric pressure transducers by using reference RS barometric pressure transducer data as input Reference RSs are at known altitudes. The output of the MSL filter is used as a constant by the altitude filter. The purpose of the altitude filter is to obtain vertical position and vertical velocity estimates based on barometric pressure transducer input when there is little or no vertical information in the TOA data (because of unfavorable geometry between users). The altitude filter output is also required to aid the position initialization algorithm in initialization of the track review and correction estimation (TRACE) filter. The position initialization algorithm provides checking of entries, initial track acquisition, ambiguity resolution, and initial position and initial velocity estimation for the TRACE filter. The central logic oscillator control (CLOC) filter is used to obtain estimates of each RS's clock offset and drift rate with respect to the MS's clock. CLOC provides, as required, commanded corrections to each RS's clock offset and/or

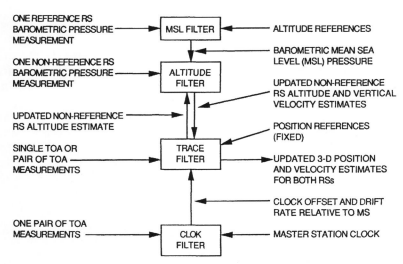

Figure 6.13   Tracking filters interconnectivity and data sources.

frequency to keep all RSs nominally synchronized with the MS, and thereby with one another.

Finally, the purpose of the TRACE filter is to take each TOA measurement and partially update the two cooperating RS's position and velocity estimates in three-dimensional space. Without the supporting processing from the other three filters, TRACE would not be able to operate successfully. Link value accounting is one of the unique byproducts of the TRACE processing. If a particular TOA link assignment is not aiding the position location accuracy (due to geometry or excess TOAs in a particular direction) of either RS then that TOA link assignment is replaced by one that may be more beneficial to overall system accuracy.

These filters provide the partially updated positions necessary to permit the MS to report a fully updated position about half a second after MS receipt of each user's measurement report.

6.3.8 Network and Traffic Management

The MS is a processing facility providing centralized technical control and monitoring of a community of PLRS users. The MS performs dynamic network management of all RSs under its control. The MS is the central control point of the PLRS network, allowing a technical control operator to maintain a real time overview of the network within his area of responsibility. The RSs accept and implement MS issued commands and report status and communicant data (i.e., information on mutual RS connectivity). These reports are used to support automatic adaptive routing. The MS operator, as the communications technical controller, is responsible for network initialization and monitoring community performance, using a graphic display. Normal operation requires only minimal operator intervention.

Control Network Traffic   Four types of messages are transmitted over the control network: user data input, user data output, network control, and measurement report messages. All messages are sent to their appropriate destination, through relays as necessary, without data content change. Error control is used in order to insure a low ($< 10^{-5}$) message error rate. User data input messages originate at any RS via an input/output (I/O) device. An I/O device may also be used to originate queries requiring user data output messages to be sent back to the RS. This establishes two-way communication between any RS and the MS. Two way communications between any pair of RSs can also be established using these message types.

Network control messages and measurement report messages are used by the MSs to maintain communication with, and exercise control over, each RS. The network control messages contain commands to the RS, (e.g., link assignments and timing correction commands). Measurement report messages contain TOA and altitude measurements and status information. The actual routing of messages to the proper destination is implicit in the network structure. The MS, with its knowledge of the connectivity, assigns transmit and receive times to provide proper relaying of messages to their destinations. RSs performing a relay function make no distinction between different message types.

User-to-User Traffic   The basic PLRS provides for user-to-user data exchange using two distinct approaches. The first is by way of the MS, using the data input and data output messages mentioned above. This approach can be used by any RS operator to send short alerting or coordination messages to any other user in the community. These messages are stored and forwarded by the MS(s). The other approach provides local groups of up to eight users the ability to send short messages to each other, without going through the MSs.

6.3.9   System Capacity and Accuracy

Each MS contains a data base which defines navigation aids and user parameters for the entire multiple MS community. This data base contains the identification, configuration, data access authorization, position, and current status for each participating RS. Up to 900 RSs can reside in the data base. A nominal community consists of 125 to 250 RSs controlled by each MS, but an MS is capable of controlling a maximum community of 460 RSs with some reduction in support rate and in positional accuracy. An MS can produce 50 position updates per second, spread across the RSs under its control.

In deployments with static RS positions, minimal clutter from buildings and trees, and surveyed grid reference RS locations the average RS radial position location error is about 5 meters. This ideal accuracy has been repeatedly demonstrated in system testing with communities of 150 RSs. For most deployments, the average radial error is 5 to 15 meters for static ground-based users, 10 to 30 meters for mobile ground-based users, and 15 to 50 meters for airborne users.

6.3.10   PLRS User Equipment Characteristics

The PLRS user equipment is the radio set (RS). Each RS consists of a receiver-transmitter, a user readout device, a power source, and an antenna. The user readout device serves as a control panel for the RS, and for limited data exchange. The RS generates and processes PLRS messages. Centralized control of the RS by a microprocessor within the message processor supports partitioning into the RS functions shown in Figure 6.14.

The RF/IF function performs frequency conversion, amplification, and filtering of the transmitted and received signals. During receive, this function performs an adaptive A/D conversin of the incoming signals. During transmit, the digital output of the signal processor is used to generate a continuous phase shift modulation (CPSM). The transmitted power of the RS is over 100 w. The signal processor function performs preamble detection/generation, interleaving/deinterleaving, error correction/error detection encoding/decoding, pseudonoise code generation, data correlation, and time-of-arrival measurement. The secure data unit performs encrypting/decrypting of transmitted and received data, message validation, and provides outputs for transmission security. The message processor contains a microprocessor that is the central controller of the RS. It controls all processes done within the RS. Additionally, the message processor generates and decodes link messages and provides the data interface format for the user readout.

6.3.11   System Enhancements

While several derivatives of PLRS have been developed, the most advanced functional extension is the enhanced position location reporting system (EPLRS). EPLRS maintains all of the basic PLRS capabilities while greatly increasing the user-to-user data communications capability. The EPLRS utilizes

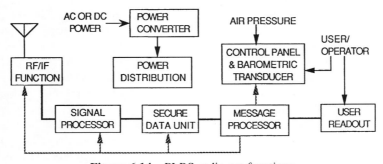

Figure 6.14   PLRS radio set functions.

the PLRS control network for monitoring and controlling large communities of user RSs. In addition to the positioning, position-reporting, navigation aid, cryptographic key distribution, and status-reporting functions, the control network is also utilized for distributing communications circuit assignments and monitoring user-to-user communications performance. The EPLRS RSs support communication network management by implementing the commands, monitoring and reporting circuit status, establishing new circuit paths, and controlling the flow of data packets into and out of the network.

Both duplex (point to point) and group-addressed (broadcast) types of service are available via the same user RS. Each RS can support up to 30 user circuits (needlines) simultaneously with a composite (receive plus transmit) information rate of 4 kbps. Each duplex circuit is capable of supporting acknowledged data rates of up to 640 bps in each direction. Each group addressed circuit is capable of supporting nonacknowledged data rates up to 1280 bps. The primary user interface is via the Army Data Distribution System Interface (ADDSI), which uses permanent virtual circuit protocols based on CCITT x.25. A 1553B data bus interface is also used for compatibility with existing aircraft and vehicular systems.

The EPLRS concepts have been proven through live testing with 160 RSs. In addition, extensive computer modeling and large-scale user community simulation have been used to confirm extension of performance to operations with over 500 user RSs. EPLRS is interoperable with the basic PLRS allowing for mutual support and coordinated operations between basic PLRS- and EPLRS-equipped users.

6.4 FUTURE TRENDS

In most JTIDS-equipped aircraft, there are similarities and redundancies in the computations performed separately by JTIDS RelNav, GPS and INS units. Coupled with the continuing miniaturization and cost reduction of digital processing, this suggests the future development of fully integrated navigation systems embodying all the functions of JTIDS, GPS, INS, and the air-data computer in one unit, occupying less space and consuming less power than the individual separate units, and also providing the optimum combination of the measurements from these multiple sensors (Chapter 3).

Miniaturization and lower equipment cost will also lead to small, expendable, *receive-only* JTIDS units for use in such vehicles as cruise missiles and unmanned reconnaissance aircraft. These units will perform JTIDS RelNav functions for midcourse self-positioning, in addition to any midcourse correction signals from control centers.

In 1996, an intensive international effort, called MIDS, was underway toward further miniaturation and modularization of JTIDS terminals which, in turn, will lead to a trend of much wider implementation of JTIDS and its relative navigation function on military aircraft worldwide in view of the low weight and size of these JTIDS terminals.

In the PLRS system, there is a strong trend toward distributing the position and range/bearing calculations from the central processors to the individual user units. This trend is based on the decreasing cost of digital processing and by the worldwide availability of GPS signals, along with low-cost GPS receiver modules. Using Kalman filtering techniques, the user units could then combine the GPS-based positions with the PLRS TOA range-based positions to improve both the accuracy and consistency of the positioning and navigation functions for the user.

PROBLEMS

6.1. A JTIDS RelNav RTT exchange indicates an accumulated clock error of +60 nsec (ahead) since the last clock update two minutes earlier. If the terminal clock oscillator operates at 80 MHz, what is its frequency error relative to the oscillator of the RTT source? Is the frequency high or low?

Ans.: 0.04 Hz high.

6.2. The PLRS time of arrival signal processing splits a PN chip into 16 equal parts. What is the precision of a one-way range measurement in meters:

Ans.: 3.7 meters.

6.3. If the clock offset error between two PLRS RSs involved in a range measurement is 15 nsec, what error in one way range measurement does this cause?

Ans.: 4.5

6.4. In Problem 6.2, what error would result if this were a two-way range measurement?

Ans.: 19 meters.

6.5. In Problem 6.3, what error would result if this were a two-way range measurement:

Ans.: None.

6.6. In the military grid reference system used for reporting positions to a user readout in PLRS, a location is reported as 4 decimal digits of easting and 4 decimal digits of northing within a designated 100-km grid square. What is the maximum error introduced due the precision of this report?

Ans.: $5\sqrt{2}$ meters

7 Inertial Navigation

7.1 INTRODUCTION

Inertial navigation is a technique for determining a vehicle's position and velocity by measuring its acceleration and processing the acceleration information in a computer. Compared with other methods of navigation, an inertial navigator has the following advantages:

1. Its indications of position and velocity are instantaneous and continuous. High data rates and bandwidths are easily achieved.
2. It is completely self-contained, since it is based on measurements of acceleration and angular rate made within the vehicle itself. It is nonradiating and nonjammable.
3. Navigation information (including azimuth) is obtainable at all latitudes (including the polar regions), in all weather, without the need for ground stations.
4. The inertial system provides outputs of position, ground speed, azimuth, and vertical. It is the most accurate means of measuring azimuth and vertical on a moving vehicle.

The disadvantages of inertial navigators are the following:

1. The position and velocity information degrades with time. This is true whether the vehicle is moving or stationary.
2. The equipment is expensive ($50,000 to $120,000 for the airborne systems in 1996).
3. Initial alignment is necessary. Alignment is simple on a stationary vehicle at moderate latitudes, but it degrades at latitudes greater than 75° and on moving vehicles.
4. The accuracy of navigation information is somewhat dependent on vehicle maneuvers.

The techniques of inertial navigation evolved from fire-control technology, the marine gyrocompass, and conventional aircraft instrumentation (Chapter 9) [9]. The earliest practical applications—and the heaviest expenditure of funds—were for ballistic-missile-guidance systems and for ship's inertial nav-

313

igation systems (SINS). In the late 1950s, increased procurement of military aircraft led to the development of aircraft inertial navigators. Many of the disadvantages of inertial systems can be overcome through aiding with other sensors such as GPS [54], radars, or star-trackers [29]. Chapter 3 discusses multisensor navigation systems.

In 1996, inertial navigation systems were widely used in military vehicles. Many ships, submarines, guided missiles, space vehicles, and virtually all modern military aircraft are equipped with inertial navigation systems due to the fact that they cannot be jammed or spoofed. Large commercial airliners routinely make use of inertial systems for navigation and steering [60].

7.2 THE SYSTEM

In the earliest inertial navigation systems, gimballed platforms isolated the instruments from the angular motions of the vehicle. The gyroscopes acted as null-sensors, driving gimbal servos that held the gyroscopes and accelerometers at a fixed orientation relative to the Earth. This permitted the accelerometer outputs to be integrated into velocity and position. In the late 1970s and early 1980s, the invention of large-dynamic-range gyroscopes and of more powerful airborne computers permitted the development of "strapdown" inertial systems in which the gyroscopes and accelerometers were mounted directly on the vehicle. The gyroscopes track the rotation of the vehicle, and algorithms in the computer (Section 7.4.1) transform accelerometer measurements from vehicle coordinates to the navigation coordinates where they can be integrated. In strapdown systems, the transformation generated by the computer performs the angular-stabilization function of the gimbal set in a platform system. In effect, the attitude integration algorithms permit the construction of an "analytic" platform.

Figure 7.1 shows a block diagram of a terrestrial inertial navigator. A platform (either gimballed or analytic) measures acceleration in a coordinate frame that has a prescribed orientation relative to the Earth. Usually, the stabilized coordinate frame is locally level (two horizontal axes, one vertical). The computer, which may be the aircraft's central computer or a navigation computer, calculates the aircraft's position and velocity from the outputs of the two horizontal accelerometers. The computer also calculates gyroscope torquing signals that maintain the platform in the desired orientation relative to the Earth. In a strapdown system, the analytic platform is "torqued" computationally. A vertical accelerometer is usually added in order to smooth the indication of altitude, as measured by a barometric altimeter or air-data computer (Chapter 8). The calculation of velocity from the accelerometer outputs is described in Section 7.5; the calculation of position from the velocities is described in Section 2.4.

In a platform system, the gimbal-isolated structure, on which the gyroscopes and accelerometers are mounted, is called the *stable element*. The gimbals (Figure 7.2) allow the aircraft to rotate without disturbing the attitude of the sta-

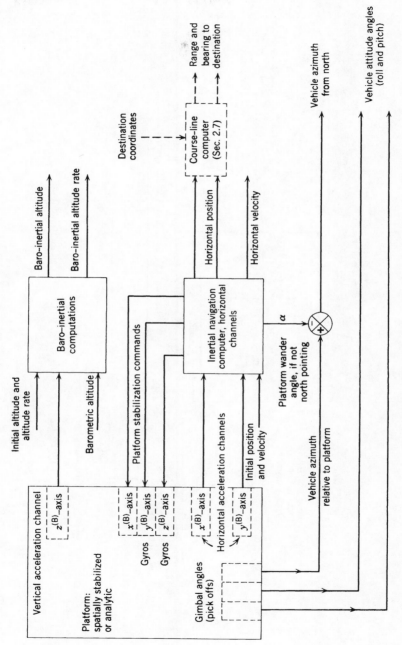

Figure 7.1 Block diagram of an inertial navigator.

315

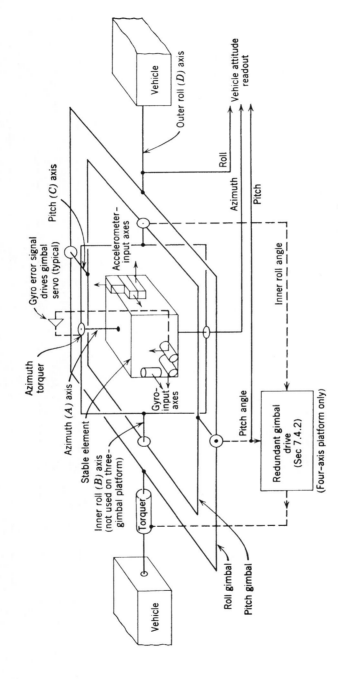

Figure 7.2 Four-axis stable platform of an inertial navigator.

ble element. The gimbal angles are measured by transducers, usually resolvers (Section 7.4.2), whose outputs indicate the aircraft's roll, pitch, and heading to the displays, auto-pilot, and sometimes to the computer. In strapdown systems, attitude angles are mathematically extracted from the analytic platform transformation matrix (Section 7.4.1).

When the inertial system is turned on, it must be aligned so that the computer knows the initial position and groundspeed of the vehicle and so that the platform (gimballed or analytic) has the correct initial orientation relative to the Earth. The platform is typically aligned in such a way that its accelerometer input axes are horizontal, often with one of them pointed north. As the vehicle accelerates, maneuvers, and cruises, the accelerometers measure changes in velocity, and the computer faithfully records the position and velocity.

The inertial navigator also contains power supplies for the instruments, a computer, often a battery to protect against power transients, and interfaces to a display-and-control unit. The system may be packaged in one or more modules. Typical gimballed systems in 1968 weighed 50 to 75 lb (excluding cables), of which 20 lb were for the platform. Steady-state power consumption was approximately 200 w. First-generation strapdown navigators (early 1980s) weighed 40 to 50 lb and consumed 100 to 150 w. In 1996, strapdown systems weighed 20 to 30 lb and consumed approximately 30 w.

7.3   INSTRUMENTS

This section discusses the sensing instruments (gyroscopes and accelerometers) as they relate to stable platforms and strapdown systems.

7.3.1   Accelerometers

Purpose   An accelerometer is a device that measures the force required to accelerate a proof mass; thus, it measures the acceleration of the vehicle containing the accelerometer. Figure 7.3 shows a black-box accelerometer whose input axis is indicated. The instrument will supply an electrical output proportional to (or some other determinate function of) the component along its input axis of the inertial acceleration minus gravitation. If the instrument is mounted in a vehicle whose inertial acceleration is a and if the vehicle travels in a Newtonian gravitational field G, (Section 2.2), then the force acting on the proof mass m_p is

$$F = m_p a = F_R + m_p G + F_D$$

$$\frac{F_R}{m_p} = a - G - \frac{F_D}{m_p} = f \qquad \text{(accelerometer output)} \qquad (7.1)$$

where F_R is the force exerted on the proof mass by the restoring spring or

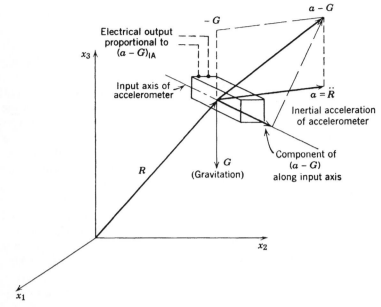

Figure 7.3 Black-box diagram of an accelerometer.

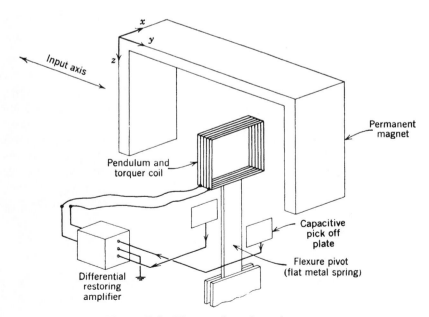

Figure 7.4 Flexure-pivoted accelerometer.

restoring amplifier, as shown in Figure 7.4, and F_D is the unwanted disturbing force caused by friction, hysteresis, mechanical damping, and the like. Thus, if the instrument is designed with negligible disturbing forces, the restoring force is a measure of $(a - G)$ along the instrument's input axis. As explained in Section 7.5, accelerometers are used to calculate the vehicle's acceleration a; their outputs must be corrected for gravitation G in the computer.

If the accelerometer rests on a table, then $a = 0$ (neglecting the rotation of the Earth) and the unit measures $-G$. If the accelerometer is falling in a vacuum, then $a = G$, and the output is zero. If the instrument is being accelerated upward with an acceleration of 7 g, then $a - G = 7$ g $- (-1$ g$)$, and the instrument reads 8 g (1 g is a unit of acceleration equal to approximately 32.2 ft/sec^2 = 981 cm/sec^2; if an acceleration must be specified more exactly than 0.5%, it should be stated in fundamental units of length and time).

On the rotating Earth, a stationary accelerometer at a position **R** is accelerating centripetally at $\mathbf{\Omega} \times (\mathbf{\Omega} \times \mathbf{R})$ in inertial space due to the Earth's rotation rate $\mathbf{\Omega}$. The accelerometer's output therefore measures $\mathbf{\Omega} \times (\mathbf{\Omega} \times \mathbf{R}) - \mathbf{G} = -\mathbf{g}$, which is the ordinary definition of gravity, as discussed in Section 2.2. A stationary plumb bob on the Earth's surface points in the direction of **g** not **G** [24].

Construction   Several accelerometer designs are used in aircraft inertial navigators:

1. Pendulum, supported on flexure pivots, electrically restrained to null.
2. Micro-machined (silicon) accelerometer with electrostatic nulling.
3. Vibrating beam accelerometer whose frequency of vibration is a measure of tensile force and hence acceleration.

The flexure-pivot accelerometer, shown schematically in Figure 7.4 is most commonly used in aircraft systems. The sensitive element consists of a pendulum with a torquer coil and pickoff supported by a torsional spring or flexure. The pickoff measures displacement of the pendulum from null and is often mechanized with an optical sensor and shadow mask or with capacitors. The torquer coils restore the pendulum to null, the torquer current being a measure of the restoring torque and, hence, of the acceleration. Mathematically, let $f = a - G$. The torque T on the pendulum is

$$T = T_R - k\theta + mbf_1 - mbf_2\theta = I(\ddot{\theta} + \ddot{\phi}) \tag{7.2}$$

where
  T_R is the residual torque applied to pendulum by friction in supports and connecting wires and by electrical forces, Newton-meters

θ is the pickoff angle, rad

k is the spring stiffness, mechanical and electrical, Newton-meters/rad

mb is the pendulosity, kg-meters

I is the moment of inertia of the pendulum about the pivot axis, kg-meters^2

$\ddot{\phi}$ is the angular acceleration of the case about pivot axis, rad/sec^2

f_1, f_2 two components of the linear acceleration of the case relative to inertial space, meters/sec^2. f_1 is along the input axis.

The damping is neglected for illustration. In the steady state

$$\theta = \frac{mb}{k} \left[\frac{f_y + T_R/mb - I\ddot{\phi}/mb}{1 + mbf_z/k} \right] \tag{7.3}$$

If the stiffness k is high enough, θ is small, and the instrument measures only f_1, independent of the presence of a cross-axis acceleration f_2. Sensitivity to f_2 is called *cross-coupling* and is most serious in a vibration environment when θ and f_2 oscillate in phase and rectify. This rectification is often referred to as *vibropendulous* error. The term T_R/k is the angle offset due to the presence of an unwanted torque on the pendulum; it causes an accelerometer bias. $I\ddot{\phi}/k$ is an angular offset of the pendulum due to angular acceleration of the case around the pivot axis; $\ddot{\phi}$ is negligible when the accelerometer is mounted on a mechanical platform, but it is an important source of error in strapdown systems where $\ddot{\phi}$ and θ oscillations can rectify. If position calculations are referred to the center of percussion of the pendulum, the sensitivity to angular acceleration is reduced (see size effect, Section 7.4.1). The distance from the center of mass to the center of percussion is I/mb.

Flexure-pivoted accelerometers are simpler to construct than the older floated instruments since they do not require adjustment for buoyancy [23, pp. 288–289]. Because they are undamped, they exhibit high-frequency mechanical resonances. Resonances must therefore be controlled relative to both vibration inputs and rebalance-loop characteristics. Undamped accelerometers offer the greatest bandwidth (important for strapdown systems) but must almost always be supported on a shock-mounted sensor block (Section 7.4.1) in order to suppress high-frequency vibration and to prevent shock damage. Accelerometers that include fluid damping exhibit reduced bandwidth and additional thermal sensitivity due to changes in the fluid characteristics.

In navigation-grade accelerometers, a restoring loop maintains the pendulum near null. The restoring servo must be linear and repeatable from 10 to 25 μg to 40 g, a range of six to seven orders of magnitude. A digital output can be obtained by either digitizing the analog output (the current in the torquer coil) or by pulse-rebalancing with a digital restoring servo. When rebalancing with pulses of uniform height, pulse width measures incremental velocity ΔV.

In either case, a properly initialized digital counter accumulates the pulses and stores the velocity change. The rebalance pulse train must not excite accelerometer resonances.

The pivot or flexure supporting the pendulum must provide minimal restraint for the pendulum in the direction of the input axis while exhibiting high stiffness in the other two directions. The spring constant of the pivot/flexure generates a restoring force that reduces the gain of the electronic restoring loop. The spring constant should be repeatable in order to ensure accuracy, but the high-stiffness restoring loop dominates. The pivot must not exhibit hysteresis, which may cause accelerometer biases. Generally, high-quality accelerometers can operate over wide temperature ranges (−55°C to 90°C) provided that temperature is measured and bias and scale factor are thermally compensated (Section 7.4.1) in the computer. Heating of the torquer coil due to rebalance current can lead to rectification of vibration inputs and must often be compensated. A pulse-rebalance torquer maintains constant heating.

A new generation of accelerometer employs silicon micro-machining [55]. A typical silicon accelerometer structure is shown in Figure 7.5. Single-crystal silicon forms the frame, hinges, and proof-mass. Anodic bonding joins this piece to metallized wafers which enclose the accelerometer and also serve as electrodes for sensing proof-mass motion and for rebalancing. Electrostatic centering of the proof-mass obviates the need for magnetic materials and coils. Due to the very small gaps achievable between the covers and the proof-mass, gas-film damping suppresses mechanical resonances. This permits the accelerometer to operate in high-frequency vibration environments.

SILICON ACCELEROMETER

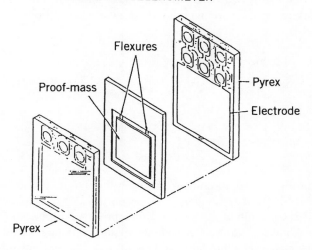

Figure 7.5   Silicon accelerometer (courtesy of Litton Guidance and Control Systems).

The silicon accelerometer can be rebalanced using either voltage or charge forcing. In voltage forcing, a potential is applied to the pendulum and to one or both electrodes. The voltages establish electric fields that induce charge on the nonconductive pendulum. This causes a net force to act on the proof mass. Thus, in the case of *voltage forcing*, the force generated is a function of the square of the applied voltage and of the gap between the pendulum and the electrode. Thus, non-linearities in force-versus-deflection may be incurred and may require compensation. The rebalance force can also be generated by applying charge to one or both electrodes of the device, by applying a precise current for a precise period of time. In the case of *charge forcing*, a fixed amount of charge generates a force that is independent of the pendulum's position, thereby permitting linear operation. However, proper charge metering requires complex electronics, particularly when small amounts of charge are to be transferred. Silicon accelerometers have less bandwidth than flexure-pivoted devices. An electrostatically induced spring rate results if the pendulum is not properly centered or if its position deviates from null. This causes scale factor or bias errors. A more detailed discussion of silicon accelerometers may be found in reference [55]. Silicon accelerometers are easy to manufacture using standard semiconductor technology, are rugged, and resist shock. In 1996, silicon accelerometers were used in some medium accuracy inertial measurement units (IMUs) and inertial-grade devices had been demonstrated.

Though the restrained-pendulum accelerometer is used in most operational aircraft inertial navigators, the micromachined vibrating beam or vibrating string accelerometer is sometimes used [38, 39]. One version consists of a proof-mass that exerts a tension T on one or more vibrating beams (fabricated of metal, quartz, or other dimensionally stable materials). The frequency of oscillation of each beam is proportional to the square root of T, which varies with acceleration. By using two beams in push-pull, under an initial tension T_0, a frequency-difference measurement can determine acceleration:

$$\text{Frequency of the first string} = \nu_1 = k_1\sqrt{T_0 + mga}$$
$$\text{Frequency of the second string} = \nu_2 = k_1\sqrt{T_0 - mga}$$

$$\therefore \nu_1 - \nu_2 = k_1\sqrt{T_0}\left[\frac{mga}{T_0} + \frac{1}{8}\left(\frac{mga}{T_0}\right)^3 + \cdots\right] \qquad (7.4)$$

If the tension T_0 is large in comparison with the maximum acceleration load *mga*, then the difference frequency is proportional to acceleration, with a decreasing series of higher-order corrections, in terms of odd powers of acceleration. The vibrating beam accelerometer requires a means of supporting the proof-mass in such a way that only the beam provides a support force along the input axis. Vibrating beam accelerometers are often sensitive to vibration and cross-axis inputs. One of their advantages is the ability of obtaining a digital output simply by counting the output frequency.

Multi-axis accelerometers that measure three components of acceleration with a single proof-mass have been developed. However, design difficulties exist in supporting the proof-mass and in constructing the geometry to keep inter-axis coupling sufficiently low; hence, these devices had not gained popularity in 1996. Unsuccessful attempts have been made to use the supporting force on a gyroscope to measure acceleration, thus converting the gyro into a combined gyro accelerometer. Other types of accelerometers, such as pendulous integrating gyro accelerometers (PIGAs) are used in space launch vehicles.

Error Model   A typical error model for an accelerometer (including the restoring-amplifier electronics) expresses the steady-state instrument output u as

$$u = D + H + k_0 + k_1 f_1 + k_2 f_1^2 + k_{12} f_2 + k_{13} f_3 + k_\theta \theta f_2 + k_{41} T + \cdots \qquad (7.5)$$

where

f_1	is a component of $\mathbf{a} - \mathbf{G}$ along the input axis
f_2, f_3	are cross-axis components of $\mathbf{a} - \mathbf{G}$
T	is the deviation from calibration temperature
D	is the dead zone, or threshold below which the instrument will not sense acceleration. This is typically caused by mechanical stiction and is much smaller than k_0. This term is negligible in most modern inertial quality accelerometers.
H	is hysteresis (generally thermal)
k_0	is the accelerometer bias (k_0 is slightly different each time the instrument is powered-up; the mean value is usually biased out in the computer or the instrument itself; the uncompensated residual causes navigation errors.)
k_1	is a linear scale factor, whose stability is essential in the design of the instrument
k_2	is a nonlinear calibration coefficient (it is often desirable that this be negligible, in order to simplify the navigation algorithms)
k_{12}, k_{13}	are coefficients of cross-axis sensitivity
k_θ	is the vibropendulous coefficient
θ	is the pendulum deflection angle
k_{41}	is a linear temperature coefficient, for small deviations around the operating temperature

Dynamic rectification effects can also exist as a result of vibration and of saturation of the restoring amplifier. Typical accelerometer specifications control the values of k_1 and k_2 and their permissible variation due to temperature, vibration, cross-axis acceleration, and magnitude of input acceleration. Maximum values of D, H, k_2, k_{12}, k_{13}, k_{41}, and k_θ are usually specified.

Accelerometer Testing   Accelerometers are statically tested and calibrated in the Earth's gravity field, using a dividing head. The dividing head causes the input axis to rotate in a vertical plane, around a horizontal axis, thus sensing a component of gravity that varies from 0 to ± 1 g. Scale factor and bias are determined from such a test [Chapter 4 of Ref. 41]. Cross-axis sensitivity and variation of scale factor with cross-axis acceleration can be determined by tilting the pivot axis of the dividing head and repeating the test. To calibrate scale factor on a dividing head, a gravimetric survey should be performed at the calibration station. Without a survey, sea-level gravity can be predicted from Equation 2.6 within 0.02 cm/sec^2. Centrifuge tests are used to calibrate instruments that operate at acceleration levels greater than 1 g [61a,e]. Ref. 61h discusses precision centrifuge testing of accelerometers.

Aircraft instruments should be capable of operation at acceleration levels as high as 12 g (during military maneuvers) with an accuracy of better than 100 parts per million (ppm), if 1 nmi/hr navigation accuracy is to be achieved. The presence of vibration or shock on the sensor assembly requires that the accelerometer be scaled to sense considerably more than the maximum expected linear acceleration, in order to prevent saturation in the presence of acceleration combined with vibration. The accelerometer's frequency response can be restricted as long as the proof-mass does not strike the stops. Wide bandwidth is essential in accelerometers that input to a flight-control system or that are used to strapdown navigators.

The performance parameters of a typical aircraft accelerometer (circa 1996) scaled for 30-g maximum acceleration were the following:

D	negligible
H	25 μg
k_0	25 μg (after thermal modeling)
k_1	Stable to 50 ppm (after thermal modeling)
k_2	< 5 μg/g^2
k_{12}, k_{13}	Stable to within 25 μrad
k_θ	< 2 μg/μrad-g
k_{41}	30–100 μg/$^\circ$C (stable to better than 0.5 μg/$^\circ$C)

7.3.2 Gyroscopes

Purpose   The purpose of the gyroscopes ("gyros") in an inertial navigation system is to space-stabilize the accelerometers. In gimballed platforms, the gyros measure rotation of the platform, which is angularly isolated from the vehicle's motions. The gyros rotate at inertial angular rates from 0.005 deg/hr to 50 deg/hr, the maximum torquing rate on fast aircraft; a range of 10,000.

The gyroscopes are used as error detectors to sense small rotations of the platform relative to the navigation coordinates. A gimbal servo-loop restores the error to near zero (see Figure 7.2 and Section 7.4.2).

In strapdown systems, the gyroscopes are fixed to the vehicle and follow its angular motion. A gyroscope on a military aircraft must sense angular rates as low as 0.005 deg/hr and as high as 400 deg/sec (1,440,000 deg/hr), a range of 8.5 orders of magnitude. Strapdown gyroscopes on civil aircraft need only sense an 8-order-of-magnitude range of angular rates.

Construction   Many types of gyroscopes have been invented. Since the 1930s, directional gyroscopes have been used in cockpits as heading references (Chapter 9). They are spinning-wheel, large-angle, unfloated instruments with ball bearings that have drift rates on the order of 50 deg/hr, and hence are useless for navigation.

From the 1940s to the 1960s, single-degree-of-freedom (SDF) floated gyros were perfected. A spinning wheel was mounted inside a single gimbal that was floated at neutral buoyancy. A magnetic pickoff sensed rotations (several minutes of arc) and a magnetic torquer precessed the rotor according to a rebalance algorithm. These gyros achieved drift rates less than 0.01 deg/hr but had to be used on a stable element since they were very sensitive to cross-coupling. They were used principally in space launch vehicles [61b].

From the 1950s to the 1970s, floated two-degree-of-freedom (TDF) gyroscopes were perfected for aircraft. A spinning wheel was mounted in two gimbals and floated at neutral buoyancy. Two orthogonal pickoffs (usually magnetic) sensed rotation of the float and two orthogonal torquers (also magnetic) precessed the float according to a rebalance algorithm. These gyros achieved drift rates of 0.01 deg/hr and were almost always used on a gimballed platform. The motors, fluids, and seals caused perennial maintenance problems. These instruments were described in the first edition of this book [23].

During the same period, electrostatically suspended TDF gyroscopes were developed for submarines [56]. Their spinning wheels are electrostatically centered, achieving 0.001 deg/hr accuracy but at costs orders of magnitude higher than floated TDF gyros. Due to their high accuracy, electrostatic gyroscopes are used for high-precision, deep-penetration, long-time-of-flight aircraft. Hydrostatically suspended gyros, using pressurized bearings to support the gimbals, were also developed but little used.

From the 1970s to the 1980s, "dry-tuned" or "tuned-rotor" or "dynamically tuned" gyroscopes were perfected, as described in Section 7.3.4. Navigation gyros of this type have insufficient range for strapdown use and were mostly used in gimballed platforms. Strapdown tuned-rotor gyros were used in 1996 for lower-cost, moderate accuracy attitude and heading reference systems (AHRS), which also serve as coarse navigators (Chapter 9 and Section 7.7.4).

In the 1980s, optical angular sensors were perfected after 30 years of development. They are the mainstay of aircraft inertial navigators in the 1990s and are described in Section 7.3.3. These instruments are called *gyroscopes* to emphasize their function. Research efforts on new gyros are described in Section 7.3.5.

7.3.3 Optical Gyroscopes

Optical gyroscopes were universally used in strapdown aircraft inertial navigators in 1996. These gyros offer extremely high dynamic range, linearity, bandwidth, ruggedness, and reliability. By the 1980s, most of the key problems inherent in the ring laser gyro (RLG) had been solved or circumvented. Thus, the RLG surpassed the mechanical gyro as the rotation sensor of choice for inertial navigation systems. Strapdown RLG systems have become the predominant inertial navigators for commercial and military aircraft. New generations of laser gyroscopes have also been developed. These include multioscillator laser gyros that employ optical biasing as a means of circumventing lock-in (a key limitation in laser gyros), and fiber-optic gyros. All optical gyros make use of the Sagnac effect, a relativistic phenomenon that permits the observation of rotation relative to inertial space. Optical gyros can be configured as resonators or as interferometers, as discussed below. Systems designed with optical gyros are much simpler than those using mechanical gyros.

The Sagnac Effect   The Sagnac effect [43] is a general relativistic phenomenon relating to the propagation of light in a rotating reference system. When laser beams circulate in a closed path that is rotating in inertial space, the optical length seen by the co-rotating beam appears longer than that seen by the counter-rotating beam. The Sagnac effect permits observation of rotation in one of two different ways.

- In a *resonator* (such as an RLG), the counterpropagating beams form resonant modes within the cavity. These create an electromagnetic standing wave that remains fixed in inertial space [1]. When the housing of the gyro rotates, a detector can count nodes of the standing wave, each of which represents a fixed increment of angle (see Figure 7.7).

- In an *interferometer*, counterpropagating beams are launched into an optical path and recombined as they exit. The interference generated by the recombination depends on the optical phase difference (proportional to the optical path difference) between the two beams and therefore provides a measure of rotation. In 1996 most fiber optic gyros were configured as interferometers. Fiber-optic gyros may also be constructed as resonators but the absence of a gain medium and the relatively high losses of the fiber rendered this type of device impractical in 1996.

Two-Mode Ring Laser Gyros   The RLG has undergone extensive development since the late 1970s. In 1996, the two-mode RLG [8, 59] was the most prevalent optical inertial sensor, although multioscillator gyros were penetrating the marketplace as fully strapdown sensors (i.e., no dither). The conventional two-mode and the multi-oscillator gyros subscribe to many of the same principles but also differ in fundamental respects.

Two-mode RLGs (Figure 7.6) are planar by design so that only linearly polarized modes can be resonant in the cavity. Suppression of one of the two polarizations ensures stable operation. The two-mode RLG therefore employs a single linearly polarized clockwise (cw) and a single linearly polarized counterclockwise (ccw) beam. Higher-order modes are suppressed through proper alignment and apertures. A block of glass is bored to form a three-or-more-sided polygonal path. High-quality mirrors at each vertex complete the resonant cavity. The bores are filled with a gas mixture (generally helium and neon) that

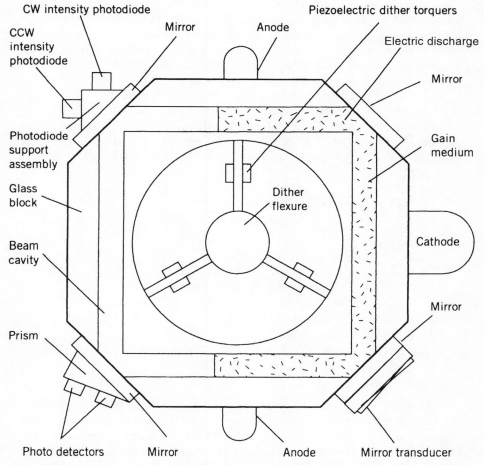

Figure 7.6   Two-mode ring laser gyro (courtesy of Litton Guidance and Control Systems).

serves as a laser gain medium. The laser is excited by an electrical discharge-generated by one or more cathodes and one or more anodes in contact with the gas. The laser beams that resonate within the cavity are electrically "pumped." A high gain-to-loss ratio permits the RLG to achieve good accuracy.

The RLG provides an angle readout via a partially transmissive mirror at a vertex. A set of combining optics (typically a prism) coherently recombines (heterodynes) the clockwise (cw) and counterclockwise (ccw) beams in order to permit the observation of the standing-wave pattern (also referred to as *interference fringes*) created by the counterpropagating modes. Photoelectric detectors measure the intensity of the interference fringes.

As discussed in the previous section, the standing-wave pattern does not rotate in inertial space. Thus, rotation of the RLG relative to the standing wave may be observed as a change in intensity sensed by the body-fixed detectors, as illustrated in Figure 7.7 for a fictitious "circular" RLG. In this figure, two counterpropagating waves create a standing-wave pattern. When the gyro is rotated, the detector moves with respect to the interference pattern and senses dark and light areas. Each dark/light cycle represents one-half wavelength of the laser beam along the circumference of the path. The number of dark/light transitions can therefore be geometrically related to the angle of rotation as indicated in Figure 7.7. The count of transitions yields the total rotation angle. At a typical laser wavelength of 630 nm, each dark/light cycle would represent one arcsecond of rotation for a 5-cm radius ring. The scale factor of the instrument depends on the ratio of enclosed area to path length, as shown in Figure 7.7 for the "circular" RLG. A similar analysis can be made for any closed polygonal laser path:

$$N_{\text{fringes}} = \frac{4A}{\lambda L}\, \Delta\theta \tag{7.6}$$

where

A	is the enclosed area of the laser path
L	is the path length of the laser beam
λ	is the wavelength of the laser
$\Delta\theta$	is the rotation angle increment
N_{fringes}	is the number of fringes traversed, measured in units of half a wavelength

RLG Quality   The laser is based on stimulated emission of photons. However, the gas medium that supplies the gain for the laser also occasionally emits photons which are unrelated to the laser signal. This is known as *spontaneous emission*, and leads to noise and random walk in the RLG angle output. Spontaneous emission is described statistically through quantum mechanics and cannot be eliminated. To reduce its impact on gyro performance, the active signal must

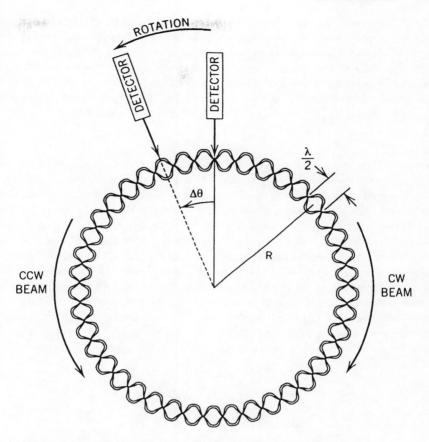

- Interference between CW and CCW beams creates standing
  wave pattern
- Standing wave pattern stays fixed in inertial space
- As gyro case rotates, detector moves around ring and
  counts minima
- Scale factor correspondence:

$\lambda/2$ circumferential displacement = 2π rad optical phase shift

$\Delta\theta$ mechanical rotation   = $(4\pi R/\lambda)$ $\Delta\theta$ optical phase shift

Figure 7.7   Circular ring laser gyro.

be as large as possible. A gyro with high gain and low loss is said to have a high "finesse" (analogous to Q in a resonator). To increase finesse, it is important to incorporate high-quality mirrors into the RLG. Low loss minimizes the impact of spontaneous emission and reduces the "quantum limit," which is a measure of the best noise performance (and hence angle random walk) achievable with the gyro.

For reasons discussed below, it is essential to minimize the backscatter generated by the mirrors. Greater angle of incidence leads to decreased backscatter. A trade-off must be made between the number of mirrors used and the resulting angle of incidence. For example, a three-sided gyro employs only three mirrors but exhibits a 30-deg angle of incidence while a four-sided gyro has a more favorable (45-deg) angle of incidence but requires a fourth mirror with its attendant losses. Gyros with more than 4 sides are not made.

Lock-in   The most severe problem encountered in the RLG is that of lock-in. In the 1960s, it was observed that the RLG was insensitive to low angular rates, as illustrated in Figure 7.8. The cause of the lock-in phenomenon is backscatter within the cavity, usually resulting from imperfections in or particulates on the mirror surfaces. At low rates, the two counterpropagating beams in the resonator are very close in frequency (less than a few hundred out of 5×10^{14} Hz) because their optical path lengths are nearly equal. Coupling of one beam into the other (which results from backscatter) causes the two modes to "lock," together thereby making the gyro insensitive to the actual rate. In Figure 7.7, backscatter amounts to friction between the standing-wave pattern and the cavity. When the gyro is rotated at low rates, the standing-wave pattern "sticks" to the cavity instead of remaining fixed in inertial space. The detector therefore does not shift with respect to the interference fringes, and the gyro does

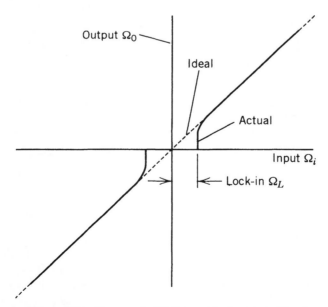

Figure 7.8   Two-mode RLG input/output (no dither).

not observe the rotation. At high rates, the "friction" is overcome because the frequencies separate and the gyro is capable of measuring rate.

In a two-mode RLG, mechanical biasing is employed to overcome lock-in. The usual means for accomplishing this is mechanical "dither," which is a large-amplitude sinusoidal motion applied to the gyro body. Typically, peak dither rates are 100 deg/sec. The output of the gyro must then be compensated for the dither motion so that the true rotation of the vehicle can be determined. There are many effective techniques for compensating. One of the drawbacks of dither is increased random walk. As the sinusoidal motion crosses through zero velocity, a small lock-in error occurs. Since the gyro reverses direction twice per dither cycle, these errors accumulate as a random walk process. Dither-induced random walk decreases with the square root of dither rate but is usually the dominant source of random walk.

An alternative method (which avoids the random walk problem) of biasing the RLG employs a turntable that applies a constant rotation to the gyro. An angular encoder measures the relative angle between the instrument and its base. This technique is referred to as *rate biasing*. Rate-biased systems with small-path-length RLGs have been delivered for missile applications. High-performance systems also use this method in order to avoid excess random walk, to provide partial error cancellation as the instruments rotate in space, and to improve calibration [35]. Because of the mechanical complexity involved in rate-biased systems, they are rarely used in aircraft.

Mechanical Design Most RLG systems in 1996 employed dither to circumvent lock-in. However, dither places serious constraints on the mechanical design of the system. High-frequency (typically several hundred Hertz), high Q mechanical flexures apply the dither. Coupling of dither to mounting structures has many undesirable effects such as acoustic noise, vibration, and energy loss. Thus, hard-mounted dithered systems are generally not practical and a low-frequency suspension (typically 30 to 50 Hz) isolates the sensor assembly from the aircraft. Dither torques in the three instruments excite coning rotations (discussed in Section 7.4.1), which cause errors in the navigation of the block solution [20]. Coning drift increases as the square of dither amplitude.

Cavity Length Control The RLG operates as a resonant cavity. The gas mixture, which sustains the laser, exhibits gain at certain optical frequencies that excite the stimulated emission, resulting in lasing action. Therefore, the length of the cavity must be tuned to be an integral number of wavelengths. For a helium-neon gas mixture, the wavelength is approximately 630 nmeters. Obviously, a cavity whose length is accurate and stable to 1% of a wavelength would be impractical to design. Thus, cavity length is controlled actively by continuously adjusting mirror positions in order to maximize total laser intensity. Piezoelectric transducers mounted on the back of one or more mirrors induce minute displacements of the mirror faces.

Since mirrors can only move a few wavelengths, the cavity must be made of a low-expansion glass so that the mirror travel is sufficient to compensate for expansion over the entire temperature range. Otherwise, "mode hops" must be performed, wherein the path length of the gyro is quickly changed by one wavelength to another control point. Unfortunately, data are lost or corrupted during a mode hop. Frequent mode hops or mode hops during high dynamics must be avoided. Unless low-expansion glass is used, mode hops could occur as frequently as once every $3°C$.

Gas Mixture   The RLG cavity should be designed to avoid gas flow within the cavity. A net gas flow causes gyro bias and can be a dominant error source in any RLG. To reduce flow, the temperature gradients across the glass block should be limited to $1-2°C$.

Because of their small size, helium atoms diffuse easily into many materials. In very small RLGs, the volume-to-surface ratio is low and, helium diffusion limits gyro life. The glass that forms the laser cavity must have low thermal expansion and low helium permeability.

RLG Scaling Laws   The performance of a ring laser gyro depends on its size. The parameters that describe gyro performance include the random walk coefficient, bias stability, resolution (also known as quantization), and scale factor stability. Because the ratio of area enclosed by the beam to path length determines the sensitivity of the gyro, most of the performance parameters improve with path length. The following scaling laws are provided as guidelines:

For quantization, $\dfrac{1}{L}$;

for bias stability, $\dfrac{1}{L^3}$;

for scale factor nonlinearity, $\dfrac{1}{L^{5/2}}$;

for random walk, $\dfrac{1}{L^2}$ caused by spontaneous emission; and

$\dfrac{1}{L^3}$ caused by dither.

Reference [59] provides a more detailed discussion of error sources and mechanisms. Due to the strong path-length dependencies, quality (particularly of mirrors) becomes more critical as path length is reduced. Thus small RLGs, while offering packaging advantages, generally do not provide cost advantages.

Vibration Sensitivity   Although the laser is insensitive to vibration, the RLG may have dynamic errors. For example, if the RLG flexure (which permits dither) exhibits cross-axis compliance, then the three gyros in an inertial system no longer form a rigid body, and large navigation errors may result during vibration. The flexure design must be compliant (with a very high Q) about the input axis while being extremely rigid about the other two axes. An angular pick-off senses the dither motion thereby correcting for dither. Still, vibration-induced errors will result if the pickoff mechanism is sensitive to translational acceleration.

For example, if the strain sensors that measure dither are slightly asymmetrical, they will erroneously indicate an angular motion of the gyro. If the false angular signal is synchronous with a true angular motion about a perpendicular axis, the strapdown equations will generate a coning-like error which is called *pseudo-coning*.

Multioscillator Gyro   The two-mode RLG requires dither or turntable rotation, imposes constraints on the mechanical designs, and causes increased noise (both vibratory and acoustic), random walk, and coning. Therefore, methods of optically biasing ring laser gyros have been attempted since the 1960s. This has led to the class of RLGs known as *multioscillators* [49]. These are fully strapdown, wide bandwidth, high-resolution, angle-sensing devices. They have no moving parts and generate no acoustic noise. A description of their operation is given below.

Construction   In one form of multioscillator RLG, a left-hand circularly polarized (LCP) mode and a right-hand circularly polarized (RCP) mode are each split apart in frequency creating two gyros acting within the same resonator. The LCP and RCP modes are separated with an optically active crystal that rotates polarization states and consequently introduces a differential phase between the LCP and RCP waves (reciprocal splitting) [49]. A more attractive alternative makes use of an out-of-plane geometry that causes polarization rotation. This is likened to the rotation of an image as it is subject to a series of reflections. The geometric technique of polarization separation is preferred, since it does not require the addition of a crystal within the beam path.

Once the LCP and RCP modes are split, they may be treated as two separate gyros each possessing clockwise and counterclockwise beams. As such, lock-in may occur in each of the gyros thereby precluding low-rate measurements. To avoid this, the clockwise beam is biased away from the counterclockwise beam. This can be accomplished with a doped glass element in the beam path which, when subjected to a magnetic field, causes a differential phase shift between the clockwise and counterclockwise beams. The shift is in opposite directions for the LCP and RCP modes. The phenomenon responsible for the phase shift is known as the *Faraday effect*, and the glass element that produces it is known as a *Faraday rotator*. The frequency splitting in a multioscillator gyro is illustrated in Figure 7.9a. It is noted that in this multioscillator, four laser modes simultaneously resonate within the cavity.

As illustrated in Figure 7.9b, when a mechanical rotation is applied to the multioscillator, the rate sensed by one of the two "gyros" (LCP in Figure

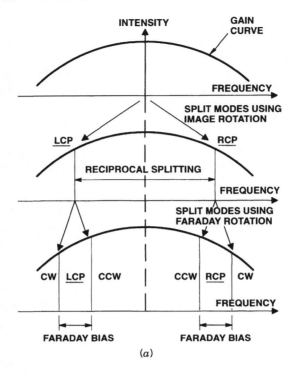

(a)

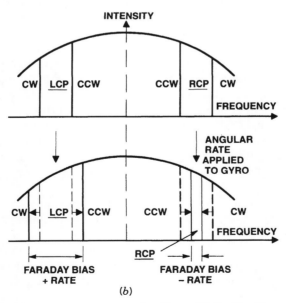

(b)

Figure 7.9 (a) Mode splitting in a multioscillator RLG; (b) effect of rate on multioscillator RLG.

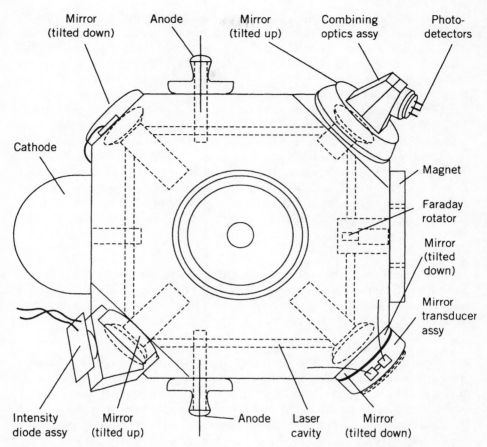

Figure 7.10   Multioscillator RLG (courtesy of Litton Guidance and Control Systems).

7.9) increases, while the rate sensed by the other "gyro" (RCP in Figure 7.9) decreases. The subtraction of the two gyro outputs cancels the Faraday bias while doubling the true angular rate measurement. The multioscillator readout is much the same as that of the RLG except that two sets of fringes (one from each polarization) are counted. The difference in the number of fringes is proportional to the rotation angle. The resonant multioscillator cavity resembles a conventional RLG cavity but must have at least four sides arranged so that the beam does not circulate in a plane. A Faraday rotator lies within the beam path and a magnet applies the field required to generate the Faraday rotation. Figure 7.10 depicts such a multioscillator gyro. As with conventional RLGs, cathodes and anodes support the electric discharge, which pumps the laser, and combining optics detects the interference fringes.

While lock-in is avoided in multioscillators, other difficulties arise. Interaction between scatter sources on the surfaces of the Faraday rotator and of the mirrors causes mode coupling, which can lead to increased gyro bias. High-quality mirror and rotator coatings minimize this problem. To ensure cancellation of common mode errors, it is important to balance the LCP and RCP intensities. This may be accomplished by dynamically adjusting the cavity length either to maximize total gyro intensity or to control the difference between the LCP and RCP intensities.

The elimination of mechanical dither makes the multioscillator gyro exceptionally well suited for low noise, flight control, and pointing applications. The elimination of dither leads to a low random walk coefficient. The scale factor stability is exceptionally good due to the absence of scatter-induced lock-in effects present in dithered gyros. The doubling of the scale factor allows smaller instruments to be used, and the lack of dither-induced mechanical noise permits superior angle measurement and enhanced flight control potential. The mechanical designs are simplified due to the absence of high-frequency, high-Q dither flexures.

Fiber-Optic Gyro   Fiber-optic gyros (FOGs) may be constructed as resonators (much as RLGs) or interferometers. Resonant FOGs have been attempted but suffered from a high loss-to-gain ratio and excessive scatter. In 1996, most operational FOGs were interferometers [11].

Principle of Operation   The interferometric fiber-optic gyro (IFOG) consists of a light source, a coupler, a fiber coil, and a detector as shown in Figure 7.11. Light is launched from a broadband laser source and coupled through a fiber-optic coil in both the clockwise and counterclockwise directions. Because of the Sagnac effect, the optical paths seen by the two beams differ in proportion to the angular rate applied to the gyro. Upon recombination, the two beams interfere and the intensity measures the phase difference between the beams. Reference [32] shows that the phase difference is proportional to

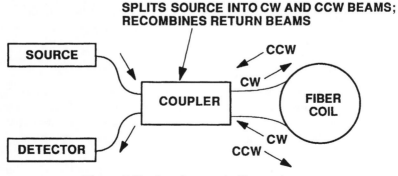

Figure 7.11   Interferometric fiber-optic gyro.

$$\phi_s(t) = \int_{t-\tau}^{t} \omega(t')dt' \tag{7.7}$$

where ω is the instantaneous inertial angular velocity along the axis of the coil and τ is the time for the light beam to traverse the coil. Thus, the fiber-optic gyro's output characteristic is that of a rate-integrating gyro with a short memory as opposed to a ring laser gyro, which is a rate-integrating gyro over a longer period of time (i.e., as long as interference fringes are being counted).

Fiber-Optic Gyro Modulation   In an interferometer, small phase shifts (corresponding to low angular rates) cause minute intensity changes (see Figure 7.12). To increase the rate sensitivity, it is necessary to modulate the fiber-optic gyro so that the phase shift between beams is an odd multiple of $\pi/2$. In early FOGs, the beams were phase-modulated mechanically by the piezoelectric mandrel that served as the spool for the fiber-optic coil. An electrical excitation applied to the piezoelectric material stressed and stretched the optic fiber, thus causing a change in its index of refraction. The result was a modulation of the beam phase in the fiber. The development of integrated optics permits the replacement of the piezoelectric mandrel with an electro-optic modulator within the beam path as shown in Figure 7.13. Light passing through the modulator is phase-shifted in proportion to the voltage applied. Modulation must be applied with a period approximately equal to the transit time. A detailed discussion of FOG modulation may be found in reference [52]. It may take the form of a sinusoidal wave form, but, in 1996, state-of-the-art devices often employed complex digital modulation to achieve maximum sensitivity and to avoid problems with distortion.

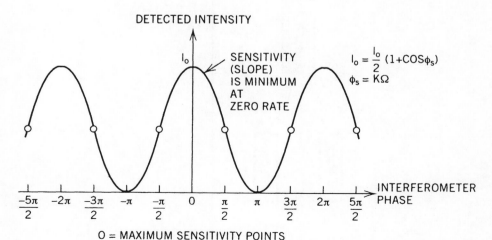

Figure 7.12   Detected intensity versus interferometer phase.

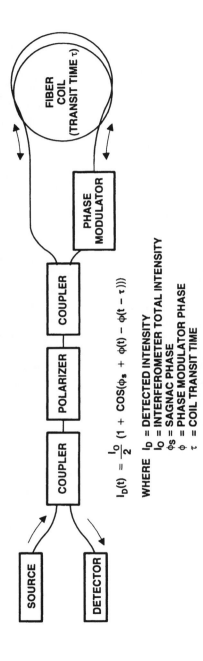

$$I_D(t) = \frac{I_o}{2}(1 + \cos(\phi_s + \phi(t) - \phi(t - \tau)))$$

WHERE I_D = DETECTED INTENSITY
       I_o = INTERFEROMETER TOTAL INTENSITY
       ϕ_s = SAGNAC PHASE
       ϕ = PHASE MODULATOR PHASE
       τ = COIL TRANSIT TIME

Figure 7.13   IFOG with phase modulator.

Closed-Loop Fiber-Optic Gyro Operation Gyro modulation improves sensitivity at low rates. However, electronics nonlinearity, intensity variation, photodetector sensitivity, preamp gain, and background intensity all contaminate the open-loop output of the FOG. For this reason, closed-loop operation of the FOG is advantageous for higher accuracy and greater dynamic range [52]. Angular-rate-induced phase shift may be nulled by applying a phase rebalance with the modulator. However, the modulator is not selective as to direction of beam travel. A step voltage applied to the device will sustain a differential phase shift between the clockwise and counterclockwise beams for one transit time of light through the coil. A steady-state voltage will result in no net steady-state differential phase. A persistent differential phase can only be generated by a repeated increase in the voltage applied to the modulator. Thus, to null the rate-induced shift with a phase modulator, it is necessary to increase the voltage applied at least every transit time. Since available voltages are bounded, the phase cannot increase indefinitely. Thus, periodic voltage "resets" (of sub-microsecond duration) with corresponding phase magnitude of 2π are applied to maintain the voltage supplied to the phase modulator within prescribed limits. The magnitude of each reset must be exactly 2π of phase to ensure that the gyro is not perturbed. A block diagram of a typical closed-loop FOG mechanization is given in Figure 7.14. As in the case of modulation, a digital implementation of the rate rebalance loop is attractive because it permits more precise control, tracking, and integration of the rate rebalance signal. Control of the reset amplitude is usually accomplished through the use of a secondary servo, which compares the effect of a nominal $\pi/2$ step to that of triple the nominal step. If the step

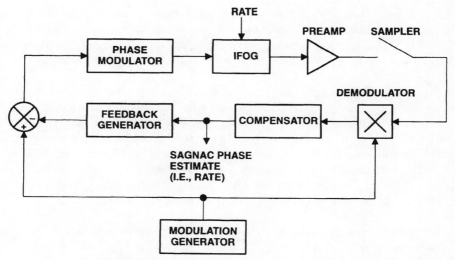

Figure 7.14 Closed-loop FOG operation.

were exactly $\pi/2$, the triple step would be $-3\pi/2$, which should have the same effect. However, if the step was not exactly $\pi/2$, the difference between the step and triple step would adjust the voltage on the phase modulator to achieve $\pi/2$ [52].

Polarization Nonreciprocity   The construction of the fiber-optic gyro usually leads to a high degree of reciprocity. That is, in the absence of external influences (angular rate, modulation), the clockwise and counterclockwise beams each experience equal phase shifts, leading to zero differential phase. However, one common error source, which may be nonreciprocal, is the coupling of different polarizations within the fiber/coupler circuit. Such coupling is usually highly temperature-dependent and cannot be modeled. Polarization nonreciprocity must be minimized through the use of high-quality polarizers, short-coherence-length sources, and polarization-maintaining fiber and/or depolarizers. In 1996, most fiber-optic gyros employed a broadband source such as a superluminescent diode (SLD) or an active gain fiber source. Narrow-band laser diodes are generally unsuitable for use in FOGs.

Vibration/Thermal Sensitivity   The index of refraction and the physical length of the fiber-optic gyro coil are affected by ambient temperature and pressure. These cause a rate error known as the *Shupe effect*. Thermal Shupe effect leads to a gyro bias that is a function of temperature and temperature gradient changes, while mechanical Shupe effect converts periodic translational vibration into periodic angular rate. Both of these effects may be reduced through clever coil-winding methods. Thermal compensation may further improve performance.

Electronics   Short-fiber-length (50 to 1000 meters) gyros require fast electronic components that generate modulation, process data, and rebalance the gyro phase. Digital modulation, demodulation, and loop-processing are the most effective.

Advantages of Fiber-Optic Gyros   The FOG requires no mechanical biasing and is rugged enough to be operated in a hard-mounted configuration. Short-fiber-length FOGs offer small size, weight, and cost. The fiber-optic gyro provides extremely fine quantization (<0.01 arcsec) thereby permitting its use as a rate-integrating device and as a low-noise rate sensor. In 1996, one-deg/hr FOG systems were in production for attitude and heading reference systems (AHRS) [36]. They are adequate for many GPS-inertial systems. Navigation-accuracy FOGs have also been produced and demonstrated.

Size, Weight, and Performance   In 1996, optical gyros suitable for inertial navigation weighed from 500 to 2000 g per axis. Laser gyros employed path lengths of between 15 and 35 cm, while navigation-grade fiber-optic gyros utilized approximately 1 km of fiber. In most cases, optical gyros are sold with their supporting electronics. The drift rate of an optical gyroscope is given by

$$\text{Drift rate} = C_0 + C_1 T + D + H + M_{11}\Omega_1 + M_{12}\Omega_2 + M_{13}\Omega_3 + k_{11}B_1 + k_{12}B_2$$
$$+ k_{13}B_3 + W \tag{7.8}$$

The sources of error are classified below. The Ω_i are the components of angular rate about orthogonal system axes and T is the difference between the operating and calibration temperature of the gyro.

Bias and Random Drifts   Bias terms C_0 and C_1 are driven by gas flow effects in RLGs, scatter effects in multioscillators, and polarization and electronics effects in FOGs. They change with age but bias is extremely stable from turn-on to turn-on. Long-term bias aging is usually compensated in a system using Kalman filter observations of bias error (Section 7.7.3). Thermal hysteresis H may be incurred due to the buildup of gradients or stresses in the gyros during thermal cycling.

Deadband   The deadband or threshold term D specifies the rate below which the gyro is insensitive. It is due to lock-in in RLGs and electronics errors in FOGs.

Scale Factor and Nonorthogonality   In laser gyros, scale factor error M_{11} is due to mode coupling effects and is usually negligible, limited only by the accuracy of calibration. In FOGs, scale factor error is driven by the wavelength of the light source and the index of refraction of the fiber. Nonorthogonality errors M_{ij} are due to mechanical misalignment between the gyros and the sensor assembly. Compensation of nonorthogonality is performed in the system computer.

Magnetic Sensitivity   Magnetic sensitivity k_{ij} is due to the interaction of magnetic fields with polarization states and with the propagation medium. Laser gyros and FOGs are usually enclosed in a high-permeability shield that attenuates external magnetic fields.

White Noise   The white noise W of an optical sensor is usually a significant error source. Noise due to spontaneous emission of photons in light sources and due to backscatter in dithered RLGs sets the ability to measure the gyro output within a set period of time. For example, a gyro whose power spectral density of rate noise is 0.12 deg/hr-$\sqrt{\text{Hz}}$ will measure angular rate with a standard deviation of 0.0055 deg/hr when using an integration time of eight minutes. Rate noise can be converted to angle noise by dividing by 60. Thus, a spectral density of (0.12 deg/hr-$\sqrt{\text{Hz}}$) is equivalent to an angle spectral density of (0.002 deg/$\sqrt{\text{hr}}$)^2.

After compensation, the residual errors are given by

C_0	0.005 deg/hr uncompensated gyro bias
C_1	5×10^{-5} deg/hr-°C uncompensated bias thermal sensitivity
D	0.003 deg/hr residual deadband
H	0.003 deg/hr bias thermal hysteresis which depends on previous temperature history
M_{11}	2×10^{-6} uncompensated scale factor error
M_{12}, M_{13}	5×10^{-6} rad uncompensated input axis orthogonality error
k_{11}, k_{12}, k_{13}	0.002 deg/hr-gauss bias magnetic sensitivity
W	0.12 deg/hr-$\sqrt{\text{Hz}}$ the gyro white noise density

Testing Optical gyroscopes are tested using rate tables with thermal chambers to measure scale factor and bias at various temperatures. Testing is simplified due to the excellent scale factor stability of these gyros. Vibration testing is sometimes performed to verify construction quality and durability and to measure vibration rectification errors. Ref. [61i] describes test procedures for single-axis laser gyroscopes.

7.3.4 Mechanical Gyroscopes

Prior to the advent of the optical gyroscope, mechanical devices formed the basis of inertial navigation systems. References [47, 23] describe various types of mechanical gyroscopes.

Spinning Wheel Gyros The principle of operation is that in the absence of applied torque, a rapidly rotating mass will tend to maintain its orientation in inertial space. If a torque acts on the mass, then it will precess at a constant rate. If a rigid body of angular momentum $\mathbf{H}$ ($\mathbf{H} = \mathbf{I}\omega_s$, where $\mathbf{I}$ = moment of inertia of the mass about the axis of rotation, and ω_s = spin rate) were acted upon by a torque $\mathbf{T}$, then the body would precess at an inertial angular velocity ω:

$$T_x = A(\dot{\omega}_x + \ddot{\theta}_x) + H(\omega_y + \dot{\theta}_y) + (C - A)\omega_y\omega_z$$
$$T_y = A(\dot{\omega}_y + \ddot{\theta}_y) - H(\omega_x + \dot{\theta}_x) - (C - A)\omega_x\omega_z \qquad (7.9)$$

where

T_i	are components of the applied torque
A	is the rotor transverse moment of inertia
C	is the rotor polar moment of inertia
H	is the angular momentum of the rotor
ω_i	are the case angular velocities in inertial space
θ_i	are the pickoff angles

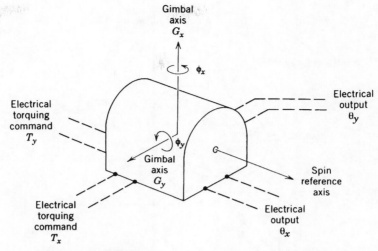

Figure 7.15 Black-box model of mechanical gyroscope.

If the angular momentum is high enough, Equation 7.9 can be simplified as represented in Figure 7.15:

$$T_x = H\omega_y$$
$$T_y = -H\omega_x \tag{7.10}$$

The simplification neglects anisoinertia (C-A), mass unbalance, and gimbal moments of inertia. Many of these effects must be considered particularly in strapdown systems that experience high dynamics and change their orientation with respect to gravity. Other errors are present at higher frequencies where the rebalance loops cease to faithfully maintain the rotor at its null position. Detailed discussions of errors may be found in references [9, 10].

Equation 7.10 shows that a torque applied around an axis perpendicular to the spin axis generates a precession rate T/H around an axis perpendicular to the other two. By definition, the rate ω resulting from a deliberately applied torque is called a *precession*, whereas that due to an accidentally (and unwanted) applied torque is called *drift*.

If a constant torque were applied to a nonrotating mass, the result would be a constant angular acceleration T/I. After a time t, the nonrotating mass would turn through an angle $Tt^2/2I$, whereas the gyro would turn through $Tt/I\omega_s$. By increasing the spin rate ω_s, a gyro can be made much stiffer than an inert mass of the same moment of inertia. References [48, 51] discuss the dynamics of mechanical gyros in great detail.

Tuned-Rotor Gyros Figure 7.16 schematically illustrates a two-degree-of-freedom (TDF) gyro. A balanced rotor supported in flexure-gimbals is free to

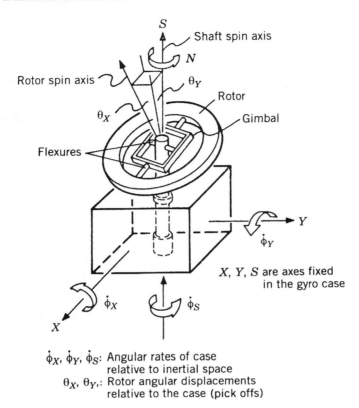

$\dot\phi_X$, $\dot\phi_Y$, $\dot\phi_S$: Angular rates of case
                relative to inertial space
   θ_X, $\theta_{Y,}$: Rotor angular displacements
                relative to the case (pick offs)

Figure 7.16   Schematic representation of TDF tuned-rotor gyro.

rotate about two axes relative to the shaft. Preloaded bearings support the shaft within the case and a motor drives the rotor at a precise spin speed of approximately 200 revolutions/sec. Pickoffs (usually magnetic) measure the angular displacements (θ_x and θ_y) of the rotor relative to the case. Mechanical stops prevent damage to the gyro due to excessive motion of the rotor. The pickoff outputs drive servo loops, which control torquers that restore the rotor to its null position. The gyroscopic equations relate the torque applied (measured by the current supplied to the torquer coils) to the angular rate sensed by the gyro. Angular rate measurements about two perpendicular axes are obtained. Additional descriptions of tuned-rotor gyro design are given in reference [31].

Suspension Tuning   The suspension includes a gimbal and two sets of flexures. Its function is to provide translational support for the rotor while decoupling the case and the rotor for rotations about any axis perpendicular to the spin direction. When the gimbal-flexure-rotor assembly spins, a dynamically induced spring rate is generated [10]. The tuned condition is achieved when the dynamic spring rate exactly cancels the mechanical spring rate attributed to the flexure. The gimbal inertias are adjusted such that their tuned frequency exactly matches the motor frequency. When ideally tuned, the rotor will appear to be completely free to rotate about axes perpendicular to the spin axis.

Rebalance Servo   To keep the gimbal angles within seconds of arc, a rebalance servo drives the pickoff signals to zero. Magnetic torquers act on the rotor to provide the restoring force. As in the case of accelerometers, the rebalance loops can be analog or digital (pulse rebalance). Gyro resonances and rebalance loops must be designed to achieve sufficient bandwidth while ensuring stability. Torquer calibration includes orthogonalization relative to the gyro spin axis and relative to the other torquers. In older designs, such calibration was generally performed electrically with a resistor matrix. Newer instruments rely on mathematical compensation in the navigation computer.

Torquing of strapdown gyros is difficult for several reasons. To achieve high rate capabilities and high bandwidth, either large torquers must be used or a rotor with low inertia must be used; both degrade performance. In the first instance, excessive power dissipation, thermal sensitivity, and thermal gradient sensitivity cause drift. In the second case, accuracy is sacrificed because of the reduced gyroscopic effect. The angular momentum of inertial-quality gyros is 200,000 to 2,000,000 gm-cm^2/sec. For strapdown navigation, optical gyros (Section 7.3.3) have nearly displaced mechanical gyros. Lower-accuracy strapdown inertial measurement units still employ miniature two-degree-of-freedom-tuned gyros.

Size, Weight, and Performance   Inertial-quality TDF gyros range from micromachined 30-g instruments to 300-g tuned instruments excluding power supplies and control electronics. They consume milliwatts to 5 w. The drift rate of one axis of a mechanical gyro can be represented as

$$\text{Drift rate} = C_0 + C_1 a_1 + C_2 a_2 + C_3 a_3 + C_{12} a_1 a_2 + C_{13} a_1 a_3 + C_{23} a_2 a_3 + C_{41} T$$
$$+ C_{51} B_1 + C_{52} B_2 + C_{53} B_3 \tag{7.11}$$

The sources of error are classified below. The a_i are the components of case acceleration along the spin axis and gimbal axes of the gyro, T is the difference between the operating and calibration temperature, and the B_i are the components of the ambient magnetic field.

Bias and Random Drifts   Bias drift C_0 is caused largely by suspension torques and by the back reactions of pickoffs. The bias drifts differ slightly each time the instrument is turned on (day-to-day and long-term repeatability) and will fluctuate randomly with time because of pivot friction, pigtail hysteresis, brinelled bearings, and power supply variations. A turn-on to turn-on bias shift can result from the way in which the shaft bearings align themselves at each spin-up. Bias and random drifts are specified in deg/hr. Mechanical gyros in aircraft are usually rebiased on a regular schedule, based on the number of flights, flying hours, or elapsed time. In most systems, biases are estimated in flight by the navigation Kalman filter (Chapter 3).

Mass-Unbalance Drift   The C_1, C_2, and C_3 are the mass-unbalance drift coefficients. Mass-unbalance drift is proportional to vehicle acceleration and is caused by inadequate mass balance of the assembly or by a defective spin motor. If $H = 2 \times 10^6$ g-cm^2/sec and the rotor weighs 250 g, a mass shift of 1 μin. causes a drift coefficient of 0.06 deg/hr-g. The absolute values of C_i and their stability are usually specified. Compensation is sometimes performed in the system computer using accelerometer measurements.

Anisoelastic Drift   The C_{12}, C_{13}, and C_{23} are the anisoelastic drift coefficients, usually specified in deg/hr-g^2. If the wheel suspension is not isoelastic, the mass center of the rotor does not deflect along the direction of acceleration and a torque results. A difference in stiffness of 1 lb/μin. will cause a drift coefficient of 0.04 deg/hr-g^2 if $H = 2 \times 10^6$ g-cm^2/sec and the rotor weighs 250 g. Furthermore, a vibration that has in-phase components along and normal to the spin axis will cause rectified drift.

Higher-Order g-Sensitivity   If the deflections along the principal axes are nonlinear functions of load, the anisoelastic drift coefficient will vary with g^3 and higher-order terms. These terms are not ordinarily discernible in aircraft systems.

Temperature Coefficient of Drift   The temperature-dependent drift in a gyro results from dimensional changes in the mechanical assembly or temperature dependent terms in the magnetics. These coefficients are quoted in deg/hr-°C of temperature off calibration and of the temperature gradient. For maximum accuracy, mechanical gyros are often heated and maintained at a precise temperature. A temperature model can also be derived during calibration and subsequently applied in the system computer for drift compensation.

Magnetic Field Coefficient of Drift   External magnetic fields can act on the motor or suspension causing torques that depend on the field strength and on the orientation of the gyro in the field. The source of the field can be the Earth, nearby equipment (e.g., radars), platform torque motors, or sources within the gyro. The magnetic field coefficient is quoted in deg/hr-gauss.

In a typical navigation-grade mechanical gyro (circa 1996), the coefficients in field usage are

C_0	0.1 deg/hr, stable within 0.02 deg/hr for periods of weeks
C_1, C_2, C_3	0.5 deg/hr-g, stable within 0.1 deg/hr-g for periods of weeks
C_{12}, C_{13}, C_{23}	0.1 deg/hr-g^2
C_{41}	0.02 deg/hr-°C
C_{51}, C_{52}, C_{53}	0.005 deg/hr-gauss

Gyro scale factor linearity 0.05%

Testing   There are many methods for conducting static drift tests on gyros. In the simplest, the gyro is mounted on a rigid table and connected as a single-axis or two-axis rate gyro, with its pickoff(s) caged to its torquer(s). The indicated gyro output expressed as a rate, minus the calculated Earth rate, gives the drift rate. This method depends on knowledge of the torquer scale factor and requires the subtraction of two large numbers to calculate the small drift rate. This test is usually performed on mechanical aircraft gyros because they have calibrated torquers and because the test maintains the gyros in a fixed orientation relative to gravity. Measurements of the gyro drift rates in various orientations relative to the gravity field can be used to solve a set of simultaneous equations of the form of Equation 7.11 to yield the drift coefficients in that equation [61g].

Vibration tests of a gyro are often desirable, particularly to determine the anisoelastic coefficients, which are a function of the frequency of vibration. Centrifuge tests also characterize gyros but are difficult to perform accurately. Sled tests and Scorsby tests (Section 7.4.3) are used to test strapdown blocks but not individual gyros.

7.3.5   Future Inertial Instruments

For precise inertial navigators (better than 2 nmi/hr RMS error), optical gyroscopes are likely to remain in use for decades. Efforts will continue to avoid mechanical dithering and to improve the reliability of the laser cavities. Approaches that combine three RLGs into one block of glass may also be pursued for some applications [57].

As worldwide, continuous, precise satellite fixes become available at low cost, they will be coupled to moderate-accuracy (5 nmi/hr error) inertial navigators of the kind that were called "attitude and heading reference systems" (AHRS) from the 1970s to the 1990s. Micro-machined gyroscopes, combined gyroscope-accelerometers [19], and FOGs are likely to dominate in this arena, which may become the largest quantity market for inertial navigators during the period when GPS is in service.

Micro-machined gyroscopes are likely to be vibrating beams of various designs that detect the Coriolis force on the oscillating tines when the gyroscope rotates in inertial space. They are likely to be packaged as a microchip with integral signal conditioning and rebalance electronics.

Hemispherical Resonator Gyro   This gyro has been in development since the 1960s [50]. It employs a quartz resonator in the shape of a wineglass to support acoustic modes that are inertially stabilized. By measuring the motion of the acoustic nodes relative to the glass, it is possible to infer rotations. Manufacture of the hemispherical resonator gyro (HRG) is complicated by the requirement for very high mechanical Q's (in the millions), high-resonator uniformity, high resolution, high-impedance readout electronics, and high-quality vacuum. The HRG generally exhibits significant vibration sensitivity. These factors have limited its use in the navigation market. HRGs have been used in space applications.

7.4 PLATFORMS

7.4.1 Analytic Platform (Strapdown)

Mechanization   In a strapdown navigator, gyroscopes and accelerometers are rigidly mounted to a sensor assembly that is usually mounted to the vehicle on a set of shock mounts. The gyroscopes track the rotation of the body and drive an algorithm that calculates the orientation of the vehicle. The accelerometer outputs are transformed to the navigation axes by the computed rotation matrix. This leads to the *analytic platform*, a computed set of stabilized axes, which are analogous to the stable element axes in a gimballed system. The transformed accelerometer outputs are integrated to velocity in the analytic platform coordinate system. In a strapdown system the gyroscopes do not act as null-sensors (as in gimballed units) but sense the inertial angular rate of the vehicle. An extremely high dynamic range (0.005 deg/hr to 400 deg/sec or more) is required in many applications. Further, the calculation of system orientation and the transformation of accelerations require complex computations. Strapdown navigation systems have been made possible by optical gyroscopes and high-throughput computers [44, 45]. Since the mid-1980s, navigation performance has been similar to the best gimballed systems. Strapdown units offer additional advantages such as extended bandwidth, reduced mechanical complexity, wide temperature operation, and improved reliability. Table 7.1 shows typical characteristics of a strapdown inertial navigation unit. The error propagation of a strapdown navigator follows the same laws as gimballed navigators, but errors depend more heavily on trajectory, since the instrument orientation varies as the aircraft maneuvers.

Strapdown Computations   The purposes of these computations are (1) to calculate the vehicle's attitude relative to the navigation coordinates using the gyro measurements, (2) to transform the accelerometer measurements from vehicle axes into navigation coordinates, and (3) to perform the dead-reckoning computations of Equation 2.5.

TABLE 7.1   Typical inertial navigator specification (1996)

Parameter	Value
Navigation accuracy	0.8 nmi/hr
Velocity accuracy	2.5 ft/sec RMS
Pitch-and-roll accuracy	0.05 deg rms
Azimuth accuracy	0.05 deg rms
Alignment time gyrocompass	3–8 min
stored heading	30–90 sec
Size	500–1000 in.^{3}
Weight	20–30 lb
Power	30–150 w
Acceleration capability	30 g
Angular rate capability	400 deg/sec
Mean time between failures in a fighter environment	3500 hr

Attitude Integration   In three dimensions, it is not possible to add rotation angles. The readers may convince themselves of this point by manipulating a three-dimensional object with labeled *x*-, *y*-, and *z*-axes. For example, as illustrated in Figure 7.17, a 90-deg rotation about the *x*-axis followed by a 90-deg rotation about the *z*-axis yields a different final orientation as compared to a 90-deg rotation about the *z*-axis followed by a 90-deg rotation about the *x*-axis. This example involving large angles illustrates the principle of *noncommutativity* discussed in many references [4, 16]. Noncommutativity also applies in the case of small angles. Attitude computations must therefore take into account the properties of rotations.

In three dimensions rotations may be described by three or more parameters. Three-parameter definitions include the Euler angles, which specify three rotation angles taken in a specific order (thereby emulating a gimbal set). Unfortunately, the Euler angles suffer from singularities (as do all three parameter systems) and extreme nonlinearity and are ill-suited for attitude integration. They are, however, commonly used as attitude readout parameters. The rotation vector is another three-parameter description of rotation. Such a vector specifies an instantaneous axis of rotation and the angle of rotation about this axis (any orientation can be transformed to any other by a single-axis rotation). The rotation vector is a useful concept for small angles but is difficult to manipulate for large rotations.

The most common means of describing rotation in strapdown systems employ more than the minimum three parameters [37]. A calculation using direction cosines [21] was briefly in use for slowly rotating vehicles in the 1960s. In the late 1960s, quaternions supplanted the direction cosines, and, in the early 1970s, preprocessing of the gyro outputs was introduced to speed up the computations.

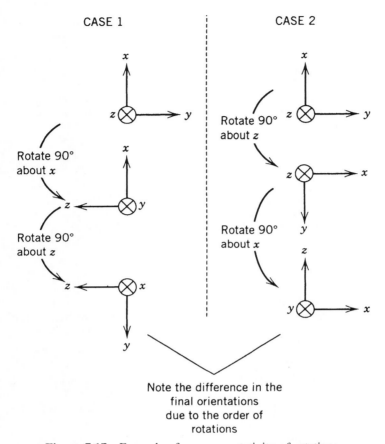

Note the difference in the
final orientations
due to the order of
rotations

Figure 7.17   Example of noncommutativity of rotations.

Direction Cosine Formulation   A direction-cosine matrix is mathematically well behaved and suited for integration. Vectors are easily transformed using these matrices.

Let the vehicle's coordinate frame (the body frame) be denoted by B, and the navigation coordinate frame by N. The direction cosine matrix transforming from the body coordinates to the navigation coordinates is C_B^N. The exact relationship between C_B^N and the instantaneous angular rate is

$$\frac{dC_B^N}{dt} = \dot{C}_B^N = [C_B^N][\omega_{BN}^{(B)}] = [\omega_{BN}^{(N)}][C_B^N] \tag{7.12}$$

where $\omega_{BN}^{(B)}$ is the instantaneous angular rate vector of the body frame with respect to the navigation frame as measured in body coordinates, ω_{BN}^N is the same angular-rate vector, resolved into navigation coordinates, and the $[\omega]$

matrix is:

$$[\boldsymbol{\omega}] = \begin{bmatrix} 0 & -\boldsymbol{\omega}_3 & \boldsymbol{\omega}_2 \\ \boldsymbol{\omega}_3 & 0 & -\boldsymbol{\omega}_1 \\ -\boldsymbol{\omega}_2 & \boldsymbol{\omega}_1 & 0 \end{bmatrix} \tag{7.13}$$

The solution to the differential equation (7.12) is updated at specific intervals determined by the computer's workload. Thus, strapdown gyros must be "rate-integrating gyros" that measure the integral of the component of angular velocity along their input axes (sometimes called "incremental angles" $\Delta\theta_i$) during the computing interval. From the $\Delta\theta_i$, the computer calculates the aircraft's attitude change during the interval. The $\Delta\theta$ outputs of the integrating gyros are numerically scaled as angles though they do not represent geometric angles because they do not form a true vector; consecutive $\Delta\theta$ are neither additive nor commutative. The smaller the angle, the more closely $\Delta\theta$ approximates a vector and represents the change in attitude. Note that a rate gyro, that samples the instantaneous rate sometime during the computing interval, would introduce large attitude errors because the rates change during the iteration interval.

To a first approximation,

$$\dot{C} \approx \frac{C_n - C_{n-1}}{\Delta t} \approx C_{n-1} \frac{[\Delta\boldsymbol{\theta}]}{\Delta t}$$

$$C_n \approx C_{n-1}([1] + [\Delta\boldsymbol{\theta}]) \tag{7.14}$$

This is a method of calculating the nine elements of the direction cosine matrix C at each read-time from the set of three gyro measurements. It is practical only in the slowest-rotating applications such as some land vehicles and spacecraft because of the following problems:

1. The C matrix will gradually become nonorthonormal and nonorthogonal unless explicitly corrected. Orthonormality requires that the sum of the squares of any row or column equal unity, and orthogonality requires that the dot products of any two rows or columns equal zero. Mathematically,

$$\sum_k C_{ik} C_{jk} = \delta_{ij}$$

$$\sum_k C_{ki} C_{kj} = \delta_{ij} \tag{7.15}$$

where δ_{ij} is the Kronecker delta ($\delta_{ij} = 1$ for $i = j$, $\delta_{ij} = 0$ for $i \neq j$).

2. There are nine simultaneous equations to be propagated at each gyro-read interval.

3. Rate-integrating gyros prevent the loss of angular information in the presence of angular acceleration during the gyro-read interval. However, a change in the direction of the angular rate vector during the interval leads to noncommutativity or coning errors, which render Equation 7.14 inaccurate.

Quaternion Formulation   The first two problems described above are avoided through the use of quaternions. A quaternion is a four-element entity consisting of a scalar part λ and a vector part ρ with the following representation:

$$\tilde{q} = [\lambda, \rho]$$

It defines the instantaneous axis of rotation. The product of two quaternions is defined as

$$\begin{aligned}\tilde{q}_1 * \tilde{q}_2 &= [\lambda_1, \rho_1] * [\lambda_2, \rho_2] \\ &= [\lambda_1\lambda_2 - \rho_1 \cdot \rho_2, \lambda_1\rho_2 + \lambda_2\rho_1 + \rho_1 \times \rho_2]\end{aligned} \tag{7.16}$$

For a rotation quaternion, the norm is unity (given by $\lambda^2 + \rho \cdot \rho = 1$). Thus, there is a single normalization constraint (the sum of the squares of the four elements must be equal to unity). A rotation quaternion may be expressed as

$$\tilde{q} = \left[\cos\left(\frac{\phi}{2}\right), \mathbf{1}_\phi \sin\left(\frac{\phi}{2}\right) \right] \tag{7.17}$$

where $\mathbf{1}_\phi \sin(\phi/2)$ is the vector part ρ and $\cos \phi/2$ is the scalar part λ. Thus, any rotation can be expressed as a single rotation about an inclined axis. In this case, $\mathbf{1}_\phi$ is the unit vector along the inclined axis of rotation and ϕ is the angle of rotation about that axis.

The quaternion inverse is given by

$$\tilde{q}^{-1} = [\lambda, \rho]^{-1} = [\lambda, -\rho] \tag{7.18}$$

The differential equation for a quaternion is

$$\frac{d\tilde{q}}{dt} = \frac{1}{2}\tilde{q}[0, \omega] \tag{7.19}$$

where $\boldsymbol{\omega}$ is the instantaneous angular rate vector. The exact solution to equation (7.19) is given by

$$\tilde{q}((n + 1)\Delta t) = \tilde{q}(n\Delta t) * \tilde{q}_u$$

$$\tilde{q}_u = \left[\cos\left(\frac{\Delta\phi}{2}\right), \sin\left(\frac{\Delta\phi}{2}\right) \frac{\Delta\boldsymbol{\phi}}{\Delta\phi} \right] \qquad (7.20)$$

This equation must be solved using the incremental angles measured by the gyroscopes. To a first approximation, the vector change in angle is related to the angular rate vector sensed in the form shown below:

$$\Delta\boldsymbol{\phi} \approx \Delta\boldsymbol{\theta} = \int_{t-\Delta t}^{t} \boldsymbol{\omega} \, dt \qquad (7.21)$$

A better approximation is needed in aircraft systems as discussed in the next paragraph. The quaternion update algorithm is executed at a rate (typically 50 to 500 times per second) such that the magnitude of the $\Delta\boldsymbol{\phi}$ vector will remain small (0.01 sec at 50 deg/sec = 0.5 deg). In this case, a second-order expansion is used for $\cos \Delta\phi/2 \approx 1 - ((\Delta\phi)^2/8)$ and for $\sin (\Delta\phi/2)/(\Delta\phi/2) \approx 1 - (\Delta\phi)^2/24$. Quaternion integration algorithms usually make use of the normalization constraint to control error growth in the computations. The sum of the squares of the quaternion elements are subtracted from unity to yield the normalization error.

Coning Errors   Coning refers to a motion in which the axis of rotation is itself moving in space. In this type of motion, the axes of the body trace a cone in space. Reference [16] demonstrates that a gyroscope whose input axis describes a cone will sense an average angular rate equal to the solid angle swept per unit time. However, there is, in fact, no net rotation taking place about that axis. In the other two axes perpendicular to the coning axis, the actual motion may be described as "oscillatory signals in phase-quadrature." Because the attitude integration in a strapdown system takes place at a finite iteration frequency, the oscillatory components will not be faithfully reproduced (particularly if the coning frequency approaches or exceeds the iteration frequency), and a net error will be generated. This error is attributed to the approximation $\Delta\boldsymbol{\phi} \approx \Delta\boldsymbol{\theta}$ in Equation 7.21. The net coning error ϵ_c depends on the coning frequency f_c and the iteration interval Δt and is proportional to the coning rate Ω_c:

$$\epsilon_c = \left(1 - \frac{\sin(2\pi f_c \Delta t)}{2\pi f_c \Delta t}\right) \Omega_c \qquad (7.22)$$

For a circular cone of angle α and frequency f_c,

$$\Omega_c = 4\pi f_c \sin^2\left(\frac{\alpha}{2}\right) \tag{7.23}$$

The apparent drift rate ϵ_c increases with the frequency and amplitude of the coning motion. Coning errors may be reduced by raising the quaternion iteration frequency but this is costly in terms of computer throughput [37]. Instead, an algorithm preprocesses the gyro data at a higher rate than the quaternion integration to improve the approximation in Equation 7.21 in order to follow the actual motion of the rotation axis closely. The algorithm computes an average rotation over the slower quaternion update interval. The preprocessing algorithms are sometimes called *coning algorithms* [28, 37, 46]. An example of a coning algorithm is given below.

Every "fast" preprocessing cycle

$$\mathbf{C} = \mathbf{C} + \delta\boldsymbol{\phi} \times \Delta\boldsymbol{\theta}_n + \frac{1}{12}\,\Delta\boldsymbol{\theta}_{n-1} \times \Delta\boldsymbol{\theta}_n$$

$$\delta\boldsymbol{\phi} = \delta\boldsymbol{\phi} + \Delta\boldsymbol{\theta}_n$$

Every "slow" quaternion update cycle

$$\Delta\boldsymbol{\phi} = \delta\boldsymbol{\phi} + \mathbf{C}$$
$$\delta\boldsymbol{\phi} = \mathbf{0}$$
$$\mathbf{C} = \mathbf{0} \tag{7.24}$$

where

n is the fast iteration counter
$\mathbf{C}$ is the vector coning correction
$\Delta\boldsymbol{\theta}_n$ is the three axis integrated rate over the nth fast cycle
$\delta\boldsymbol{\phi}$ is the resettable integration of $\Delta\boldsymbol{\theta}$
$\Delta\boldsymbol{\phi}$ is the rotation vector used to propagate the quaternion over one "slow" cycle

At the end of the "slow" cycle $\Delta\boldsymbol{\phi}$ updates the quaternion as in Equation 7.20. Typically, the preprocessing algorithm executes up to 2000 times per second, while the quaternion algorithm executes 50 to 250 times per second.

Direction Cosine Formation   A direction cosine matrix may be calculated exactly from a rotation quaternion. If the quaternion is properly normalized, the resulting direction cosine matrix will always be orthogonal and orthonormal:

$$
C_B^N = \begin{bmatrix} \lambda^2 + \rho_x^2 - \rho_y^2 - \rho_z^2 & 2(\rho_x\rho_y - \lambda\rho_z) & 2(\rho_x\rho_z + \lambda\rho_y) \\ 2(\rho_x\rho_y + \lambda\rho_z) & \lambda^2 - \rho_x^2 + \rho_y^2 - \rho_z^2 & 2(\rho_y\rho_z - \lambda\rho_x) \\ 2(\rho_x\rho_z - \lambda\rho_y) & 2(\rho_y\rho_z + \lambda\rho_x) & \lambda^2 - \rho_x^2 - \rho_y^2 + \rho_z^2 \end{bmatrix} \qquad (7.25)
$$

Euler angles may be extracted from this matrix as discussed in Section 7.5.1. In a quaternion mechanization, the Euler parameters serve strictly as system outputs for attitude (e.g., to show roll and pitch).

Incremental Velocity Transformation   Accelerations sensed in the body frame are transformed through the body-to-navigation direction-cosine matrix in order to compute acceleration in a stabilized coordinate system. However, because this transformation cannot be performed on a continuous basis, the accelerometer outputs are integrated in the body frame to form incremental velocities (ΔV) which are then transformed to the navigation frame:

(Between transformations)

$$
\Delta \mathbf{V}^{(B)} = \int_{t-\Delta t}^{t} \mathbf{a}^{(B)}(t)\, dt \qquad (7.26)
$$

(Transformation)

$$
\Delta \mathbf{V}^{(N)} \approx C_B^N \Delta \mathbf{V}^{(B)} \qquad (7.27)
$$

The use of incremental velocities instead of instantaneous acceleration is important to preserve the correct velocity in the presence of changing acceleration during a sampling interval. The approximation in Equation 7.26 can lead to sculling errors, as discussed on page 359.

Quaternion Navigation Updates   The gyroscope outputs update the quaternion as given in Equation 7.20. However, gyroscopes measure rotation relative to inertial space, while navigation is generally performed in Earth-fixed, local-level coordinates, as discussed in Section 7.5. Thus, the attitude quaternion must be modified to account for the rotation of the Earth and the travel of the vehicle around the Earth. The combination of these two terms describes the rotation of the locally level plane with respect to inertial space. A quaternion torquing algorithm (analogous to gimbal torquing) is used in addition to the body update algorithm to permit tracking of attitude with respect to the locally level plane. The algorithm that employs torquing rate commands (also known as *tilt* corrections) is given below:

$$\tilde{q}_{BN}((n+1)\Delta t) = \tilde{q}_L * \tilde{q}_{BN}(n\Delta t) \tag{7.28}$$

where

$$\tilde{q}_L = \left[\cos\left(\frac{\Delta\phi_L}{2}\right), \frac{\Delta\mathbf{\phi_L}}{\Delta\phi_L} \sin\left(\frac{\Delta\phi_L}{2}\right) \right]$$

with

$$\Delta\mathbf{\phi}_L \approx \mathbf{\omega}_L * \Delta t$$

where $\mathbf{\omega}_L$ is the torquing rate vector. The rotation rate of the navigation frame is usually less than 50 deg/hr. Thus, the iteration frequency of the navigation updates can be substantially lower than that of the body updates. Further, due to the very small angles involved (Earth rate over a 10-Hz interval represents only 7.5 μrad of angle change), a first-order expansion of the trigonometric functions

$$\cos\frac{\Delta\phi_L}{2} \approx 1$$

$$\frac{\sin\Delta\phi_L/2}{\Delta\phi_L/2} \approx 1$$

is usually all that is needed in Equation 7.28.

Packaging   The gyros and accelerometers are rigidly mounted on a block with temperature sensors. The block is usually shock-mounted to the vehicle to control the bandwidth of the motion sensed at the instruments [33]. Shock-mounting limits vibration rectification errors within the instruments themselves (accelerometers, in particular). In choosing the natural frequency of the isolation mount, trade-offs must be made between bandwidth, navigation accuracy, and computational throughput. In all cases, the isolation system must be designed to avoid vibration-induced coning and sculling errors. If the strapdown instruments are used for flight control, tight coupling to the vehicle (e.g., high bandwidth) is desirable, though it increases navigational errors.

In the 1970s, some early strapdown blocks were mounted on a gimbal-like turntable for preflight calibration and employed heaters for thermal stabilization. Optical gyroscopes have eliminated the need for turntables and heaters in the majority of applications. A typical strapdown sensor assembly is shown in Figure 7.18.

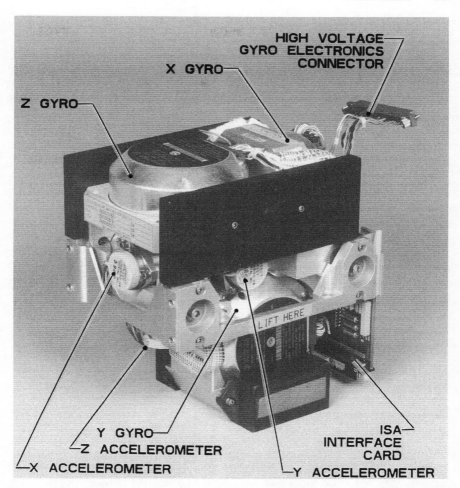

Figure 7.18 Typical strapdown inertial navigator or RLG sensor assembly (courtesy, Honeywell Inc.).

For fault tolerance, blocks may contain more than the minimum of three single-axis gyroscopes and accelerometers. With four of each, fail-safe operation is obtained allowing the detection but not necessarily the isolation of a single failure. *Isolation* is the process of determining which instrument failed and changing the algorithm to use the remaining instruments for navigation. With six gyroscopes and six accelerometers, detection and isolation of up to two failures of each type of sensor is possible. Some systems have been constructed with six RLGs and accelerometers on a single isolated sensor assembly. They have been applied to space-launch vehicles. However, the added complexity of maintaining six redundant channels has made them costly and bulky, so they are not widely used. Aircraft operators prefer redundant IMUs, each having three single-axis gyroscopes and accelerometers.

Calibration Instruments must be calibrated at the factory and compensated during flight for best performance. In the factory, systems are mounted on rate tables, and specific motions are executed to excite various error sources. For example, rotation about a given axis will excite gyro scale factor and misalignment errors. Tipping an accelerometer into gravity will excite accelerometer scale factor and misalignment errors. In certain calibration procedures, a large number of measurements are taken, and the data are reduced in a least-squares program to fit the anticipated error model [5]. Reference [34] presents an attractive method for isolating individual error sources in minimum time. Often, the calibration steps are repeated over a wide temperature range in order to generate thermal models for the instrument parameters.

Most strapdown systems do not require re-calibration since optical gyroscopes exhibit excellent long-term stability and since in-flight Kalman filters recalibrate the instruments during operation. Curve fits of the following parameters are made in the factory as a function of temperature and the model coefficients are stored in nonvolatile calibration memory:

- Gyro bias
- Gyro scale factor
- Gyro misalignment
- Accelerometer bias
- Accelerometer scale factor
- Accelerometer misalignment

For mechanical gyros, mass unbalance is also calibrated.

In flight, the strapdown sensors (generally not temperature controlled) are compensated based on measured temperature, using the models stored at the factory. The navigation filter is able to calculate residual instrument errors if an independent source of velocity or position information (e.g., GPS, Doppler radar, or star-tracker) is available. A Kalman filter (Chapter 3) often fine-tunes gyro and accelerometer biases in flight. More complex Kalman filters may also calibrate scale factor and misalignment states.

Size, Weight, and Performance Instrument error models are given in Section 7.3 (for accelerometers in 7.3.1, for optical gyroscopes in 7.3.3, for mechanical gyroscopes in 7.3.4). Strapdown systems have added errors due to trajectory, angular acceleration, and computing cycle time.

Coning Errors In military aircraft, large coning rates are generated in maneuvers known as S turns. Large roll and azimuth rotations occur 90 deg out of phase, leading to a large coning rate about the pitch axis. If ±45-deg turns occur four times per minute, a coning rate of 25,000 deg/hr will be generated. The observed system drift will depend on the effectiveness of the coning and quaternion algorithms [28, 37, 46] as well as on the scale factor and misalignment accuracies. Errors as small as a few parts per million or a few microradians are significant.

Sculling Errors   If the acceleration vector were measured, transformed to stabilized navigation coordinates continuously, and integrated to velocity, no error would be incurred. The computed $\Delta \mathbf{V}$ would be

$$\Delta \mathbf{V}^{(N)} = \int_{t-\Delta t}^{t} C_B^N(t) \mathbf{a}^{(B)}(t) \, dt \tag{7.29}$$

However, the accelerometers integrate acceleration in the body frame at discrete intervals. This yields the approximation of Equations 7.26 and 7.27.

In the presence of combined rotation and acceleration known as *sculling*, this approximation can lead to errors referred to as *sculling errors*. Figure 7.19 illustrates sculling and indicates how errors are generated. In the presence of a rotation synchronous with an oscillating acceleration, an average acceleration is erroneously computed. For most mechanizations, the residual acceleration error for small angular motions may be expressed as

$$\text{Sculling error} = \frac{a_0 \theta_0}{2} \left(1 - \frac{\sin \omega \Delta t/2}{\omega \Delta t/2} \right) \tag{7.30}$$

where a_0 is the amplitude of the oscillatory acceleration and θ_0 is the amplitude of the oscillatory angular motion in radians. Sculling error is usually expressed in μg.

To avoid large sculling errors, it is necessary to execute the velocity transformation at least four times faster than the anticipated frequencies of vibration. It is also possible to use a high-iteration-rate sculling algorithm. It computes the cross-product between the gyro data and the accelerometer data to improve the approximation in Equation 7.27. It is important to use wide-bandwidth accelerometers in order to reconstruct the oscillatory information.

Size-Effect Errors   It is impossible to co-locate the three accelerometers. As a result, each accelerometer senses acceleration at a slightly different point in space (typical separations are on the order of a few centimeters). In the presence of angular motion, each of the accelerometers will sense centripetal and/or tangential acceleration. The set of three accelerometer outputs will not be consistent due to their different physical locations. The result is a size-effect (*lever-arm*) error. The size-effect error rectifies with a magnitude proportional to the

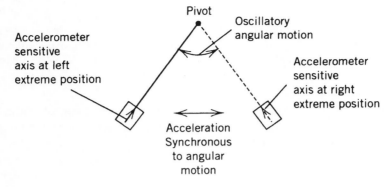

(a) Pendulum motion

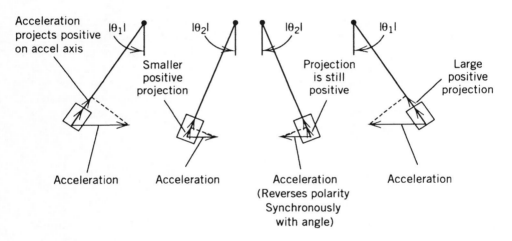

Sculling computes an erroneous
steady acceleration although
inputs are purely oscillatory

(b) Following the pendulum through its cycle

Figure 7.19   Illustration of sculling motion.

distance between accelerometer centers and the square of the angular rate. Size-effect errors do not occur in gimballed systems, since the accelerometers are isolated from angular motion.

Size-effect errors can be excited with low-frequency rocking motion. For example, if a system is rotated ±45 deg with a period of 4 sec and an accelerometer lever arm of 2 cm, the size-effect is:

$$\text{Size effect} = \frac{1}{2} \frac{2\,\text{cm}}{981\,(\text{cm/sec}^2)/g} \left(\frac{\pi}{4}\,\text{rad}\,\frac{2\pi}{4\,\text{sec}} \right)^2 \approx 1.6\,\text{mg}$$

This error is easily corrected in the computer.

7.4.2   Gimballed Platform

The purposes of a stable platform are the following:

1. To orient the accelerometers in a definite coordinate frame relative to the Earth or inertial space, despite angular motions of the vehicle.
2. To provide a convenient readout of vehicle attitude by reading each of the gimbal angles separately, thereby yielding convenient Euler angle outputs.
3. To protect the accelerometers and stabilization gyros from large angular motions of the vehicle, which would cause incorrect operation.
4. To protect the instruments from vibration, temperature, and magnetic environments.

Gimbal Order   A schematic diagram of a four-gimbal stable platform is shown in Figure 7.2. Though four-axis platforms were universally used on aircraft, the three-axis platform will be discussed first for clarity. A three-axis platform is formed by holding the inner-roll angle (*B* angle) of the four-gimbal platform rigidly at zero. The vehicle is free to rotate about the roll, pitch, or yaw axes without disturbing the stable element.

Any gimbal order can be chosen, subject to limitations imposed by the mission. The innermost and outermost gimbal axes have potentially unlimited freedom, whereas the middle axis (pitch in Figure 7.2) is limited to approximately ±70 deg. The condition in which the middle-gimbal angle approaches 90 deg is called *gimbal lock*. In this condition, the innermost and outermost axes coincide, thus depriving the platform of one of its degrees of freedom. All-attitude operation can be achieved with a four-axis platform as described on page 363.

Stabilization   In some simple systems, the orientation of the stable element is maintained by brute-force gyro stabilization, relying on the moments of inertia of the stable element and the angular momentum of the gyros to reduce the precession caused by small friction torques. Brute-force stabilization is used only on platforms designed for a few seconds of flight, where the precession rate of the platform (friction torque ÷ **H**) is acceptably low.

In most cases, servo stabilization is needed in order to maintain the orientation of the stable element (Figure 7.2). The error detectors, which provide inputs to the stabilization servos, are typically precision TDF gyroscopes. Their outputs are resolved and used to drive the gimbal servos.

The ability of the stable element to remain nonrotating relative to a space or Earth coordinate frame depends on the following factors:

1. Drift rate of the reference gyros
2. Accuracy of the gyro torquers
3. Angular motion of the vehicle (vehicle motion isolation)
4. Orthogonality of components on the platform

Gyro errors are discussed in Section 7.3.4. Base-motion isolation and cross-coupling are measured by nine ratios $(\omega_{pi}/\omega_{bj})$. Each of these ratios measures the angular velocity of the stable element about the x_ith-axis (ω_{pi}) in response to the angular velocity of the vehicle about the y_jth-axis (ω_{bj}). These ratios serve as a measure of platform performance, since they indicate how well the platform isolates the instruments from vehicle motion. The nine ratios provide measures of in-axis as well as cross-axis isolation as a function of frequency.

The degree of base-motion isolation is largely determined by the servo characteristics. For a single-axis platform using TDF gyros for stabilization, the platform response has the form

$$\frac{\omega_{\text{platform}}}{\omega_{\text{vehicle}}} = \frac{K_s(1 + T_1 s)}{1 + T_2 s + (T_3 s)^2} \tag{7.31}$$

where T_i are functions of the gimbal drives, moments of inertia, and of the damping. A more exact analysis would include the characteristic time of the gyro and the intergimbal coupling.

Another parameter of importance is the platform's response to gyro-torquing signals (e.g., those commanding the platform to remain locally level and north oriented). The torque commands change slowly, even when the vehicle executes a high-speed turn. As a result, there is seldom a problem of attenuation of command signals in the servos. The design of gimbal servos is discussed in detail in [40, 41] and Chapter 3 of [7].

Mechanical Design   From the mechanical designer's viewpoint, the platform serves the following three functions:

1. *Vibration control.* The gimbal structure should attenuate vibration (externally-induced and self-excited due to spinning gyros and other gimbal-mounted components), thus allowing the gyros and accelerometers to operate in a benign environment. The gimbals should exhibit no undue resonances and should not have vibration characteristics that are strong functions of gimbal angle. Attempts have been made to build gimbals of a laminated material that has good dimensional stability and can detune resonance peaks. Shock mounts are usually needed to supplement the gimbal's attenuation characteristics.

2. *Temperature control.* Most mechanical gyros and accelerometers must be held within a degree or less of their desired operating temperature (and temperature gradient) in order to maintain full accuracy. As a result, the gimbal system and housing must provide for suitable conduction and convection (usually using cooling air) of heat from gyros, accelerometers, motors, and gimbal-mounted electronics. An air-to-fluid heat exchanger or Peltier-effect refrigerator can prevent contaminated cabin-cooling air from entering the platform cavity. Temperature control of the stable element must be independent of changes in the gimbal angles. This requires ingenuity in the method of circulating coolant within the platform cavity. Rapid warmup may impose stringent requirements for preventing large temperature gradients in the stable element. The measurement of gyro temperature and the computation of compensations eases the problem.

3. *Magnetic field control.* Gyros and accelerometers are often magnetically sensitive. Each instrument so affected is usually separately shielded. However, care must be taken to prevent large externally (cables, radars) or internally (direct-drive gimbal torque motors) generated fields from affecting the instruments.

Misalignments between instruments cause navigation errors, as discussed in [25]. For example, a gyro-to-gyro misalignment β causes an apparent drift rate $\beta\omega_m$, where ω_m is the platform precession rate. Orthogonality errors can be corrected mechanically during assembly, or they can be computer-compensated in the outputs of the instruments. Aircraft platforms are typically orthogonalized within minutes of arc. It is important that the platform be designed so that the orthogonality angles do not change in the presence of vibration and repeated thermal cycling.

Four-Axis Platform   A four-axis, all-attitude platform is shown schematically in Figure 7.2. Its inner three gimbals are identical to those of a three-axis platform. In normal flight, B is servoed to zero; A, C, D are the degrees of freedom. At near-vertical aircraft attitudes ($C \approx 90$ deg), $A = 0$, and B, C, D are the degrees of freedom. Much ingenuity has been exercised in designing control circuits to execute the transition from A, C, D freedom ($B = 0$) to B, C, D freedom ($A = 0$).

The four-gimbal platform trades the gimbal-lock problem for the "flip" problem exhibited when the C-axis approaches 90 deg and D must flip through an angle of 180 deg. Whereas a three-gimbal platform would tumble in the gimbal-lock position, the four-gimbal platform does not, as long as the gimbal servos are fast enough to prevent the gyros from hitting their stops during the flip maneuver. Electric power and servo signals must be carried from gimbal to gimbal. Axes of limited freedom can be supplied with coiled wires. Other axes require slip rings. Figure 7.20 shows a four-axis aircraft navigation platform.

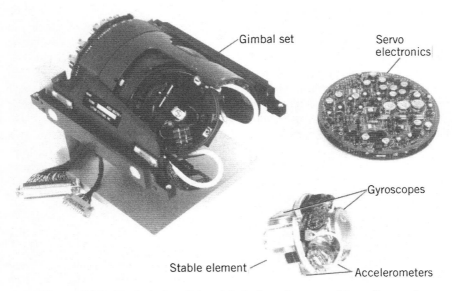

Figure 7.20   Typical aircraft inertial platform (courtesy, Litton Systems).

7.4.3   Inertial Specifications

A typical performance specification for an inertial system contains the following:

1. Size and weight.
2. Cooling and/or heating requirements.
3. Power consumption during warm-up and cruise. Increased servo power required during maneuvers (platforms only). Power regulation (voltage and frequency) and susceptibility to transients and momentary dropouts.
4. Maximum rates and accelerations along and around each axis.
5. Vibration specifications for survivability and for meeting specified navigation performance.
6. Shock-mount returnability and maximum angular deflection.
7. Reliability including MTBF (mean time between failure) and recalibration interval (if applicable).
8. Self-test capability and equipment integrity.

Inertial systems for civil use are packaged in accordance with ARINC specifications [60]. Military systems must conform to the applicable Department of Defense standards and specifications such as [62]. The equipment used to test platforms or strapdown systems include the rate table (two- or three-axis) which applies angular rates about different axes and the vibration table which supplies oscillatory motion to the system. Rate tables are sometimes used in *Scorsby mode* whereby a sine motion on one axis is simultaneously applied with a cosine on the other. This test induces a coning rate in the instrument package. Rate table and vibration tests are also performed over temperature using thermal chambers. For some applications, centrifuges and rocket-propelled sleds may be used to test system parameters and sensitivities. Tests of individual instruments are discussed in Sections 7.3.4 and 7.4.1.

7.5 MECHANIZATION EQUATIONS

The mechanization equations calculate velocity and position from the outputs of the horizontal accelerometers in a platform or from the transformed accelerations in a strapdown system. This section discusses those portions of the mechanization equations that calculate velocity of the aircraft relative to the ground. Section 2.4 discusses the navigation computation equations that convert ground velocity into position.

7.5.1 Coordinate Frames

Several coordinate frames must be defined for the purposes of mechanizing an inertial navigator, Section 2.3. Figure 7.21*a* shows the *I* coordinate frame (ECI) with *x*-, *y*-, and *z*-axes centered at the mass center of the Earth and nonrotating in inertial space (nonrotating relative to the stars). Such a coordinate frame can be regarded as inertial if the measurement accuracy is not more precise than 5×10^{-5} deg/hr and 2×10^{-7} g [26]. The *E* coordinate frame has its origin at the mass center of the Earth and is fixed to the Earth with its *z*-axis along the spin axis of the Earth. This frame is also known as the Earth-centered, Earth-fixed (ECEF) frame. It rotates at $\Omega = 15.04107$ deg/hr relative to the ECI frame, to an accuracy of 5×10^{-5} deg/hr [26].

The geographic coordinates in which the vehicle position is calculated are labelled *G*. Figure 7.21*a* shows the latitude-longitude coordinate frame. $x^{(G)}$ points level and east; $y^{(G)}$ points level and north; $z^{(G)}$ points vertically up (along the **g** vector; see Section 2.2).

The navigation (or platform) coordinates *N* lie along the orthogonal accelerometer input axes in the case of a gimballed platform. In a strapdown system the *N* coordinates are the axes of the analytic platform defined by the coordinate transformation matrix or quaternion. As discussed in Section 7.4.1, these axes represent a set of orthogonal accelerometers whose *x* and *y* axes are level and whose *y*-axis makes an angle α west of true north (see Figure 7.23). The use of an analytic platform renders the transformed strapdown acceleration outputs equivalent to the outputs of mechanically stabilized accelerometers. All

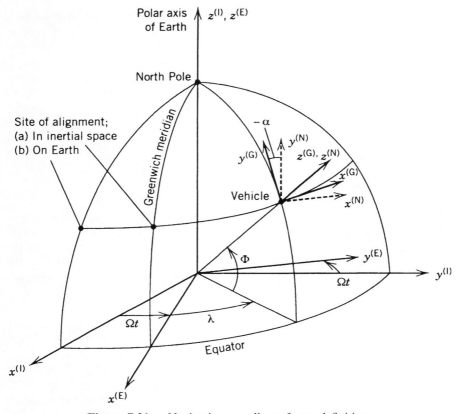

Figure 7.21a   Navigation coordinate frame definition.

navigation mechanization equations then become common to both forms of systems. The α angle is known as the *wander azimuth* and its use facilitates navigation in the polar regions. The navigation frame is related to the Earth frame by the following matrix:

$$C_E^N = \begin{bmatrix} -S_\lambda C_\alpha - C_\lambda S_\Phi S_\alpha & C_\lambda C_\alpha - S_\lambda S_\Phi S_\alpha & C_\Phi S_\alpha \\ S_\lambda S_\alpha - C_\lambda S_\Phi C_\alpha & -C_\lambda S_\alpha - S_\lambda S_\Phi C_\alpha & C_\Phi C_\alpha \\ C_\lambda C_\Phi & S_\lambda C_\Phi & S_\Phi \end{bmatrix} \quad (7.32)$$

where
  Φ is the latitude
  λ is the longitude
  α is the wander azimuth

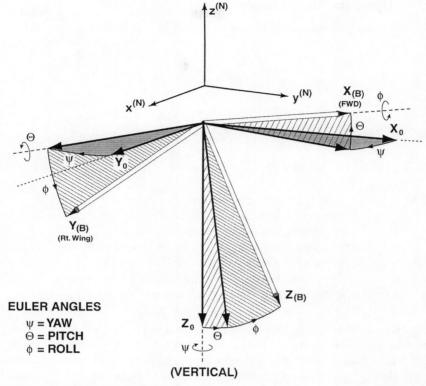

EULER ANGLES

ψ = YAW
Θ = PITCH
ϕ = ROLL

(VERTICAL)

Figure 7.21b   Euler angles: X_0, Y_0, Z_0 reference axes at $\psi = \theta = \phi = 0$. X_B, Y_B, Z_B vehicle body axes.

C_i are cos(i)
S_i are sin(i)

In a strapdown system an additional coordinate transformation relates the body frame (illustrated in Figure 7.21b) to the navigation frame:

$$C_B^N = \begin{bmatrix} S_\psi C_\theta & S_\psi S_\theta S_\phi + C_\psi C_\phi & S_\psi S_\theta C_\phi - C_\psi S_\phi \\ C_\psi C_\theta & C_\psi S_\theta S_\phi - S_\psi C_\phi & C_\psi S_\theta C_\phi + S_\psi S_\phi \\ S_\theta & -C_\theta S_\phi & -C_\theta C_\phi \end{bmatrix} \qquad (7.33)$$

where

ψ is yaw
θ is pitch
ϕ is roll

7.5.2   Horizontal Mechanization

The differential equations that must be solved for navigation are given by

$$\left.\frac{d\mathbf{V}}{dt}\right|_N = \mathbf{u} - (\boldsymbol{\omega} + \boldsymbol{\Omega}) \times \mathbf{V} + \mathbf{g} \tag{7.34}$$

where

$\left.\dfrac{d\mathbf{V}}{dT}\right|_N$ is the vector derivative of $\mathbf{V}$ in navigation (platform) coordinates

$\mathbf{V}$ is velocity of the vehicle relative to the Earth

$\mathbf{u}$ is the vector whose three components in the navigation axes are proportional to the accelerometer outputs

$\boldsymbol{\omega}$ is the inertial angular velocity of the navigation frame

$\boldsymbol{\Omega}$ is the angular velocity of the Earth in inertial space

$\mathbf{g}$ is gravity; the horizontal component of $\mathbf{g}$ is nominally zero at the surface of the Earth, as discussed in Section 2.2

The solution of these equations lead to the computation of ground speed using the mechanization illustrated in Figure 7.22 where the navigation components of $\mathbf{V}$ are calculated from the accelerometer outputs $\mathbf{u}$. The Coriolis correction (generated by velocity in the presence of platform angular velocity) is given by

$$\text{Coriolis acceleration} = (\boldsymbol{\omega} + \boldsymbol{\Omega}) \times \mathbf{V} \tag{7.35}$$

The navigation frame must rotate at $\boldsymbol{\omega}_i$ in inertial space; hence, $\boldsymbol{\omega}_i$ "torques" the analytic platform as discussed in Section 7.4.1 for strapdown systems or torques the mechanical platform as discussed in Section 7.4.2 for gimballed systems. The forms of $\boldsymbol{\omega}_i$ and $\boldsymbol{\Omega}_i$ depend on the choice of navigation coordinates, some of which are noted below:

1. Locally level, north pointing, $x^{(N)}$ along $x^{(G)}$, $y^{(N)}$ along $y^{(G)}$, $z^{(N)}$ along $z^{(G)}$:

$$\Omega_x = 0, \qquad \omega_x = \Omega_x - \frac{V_y}{R_M + h}$$

$$\Omega_y = \Omega \cos \Phi, \quad \omega_y = \Omega_y + \frac{V_x}{R_P + h}$$

$$\Omega_z = \Omega \sin \Phi, \quad \omega_z = \Omega_z + \dot{\lambda} \sin \Phi \tag{7.36}$$

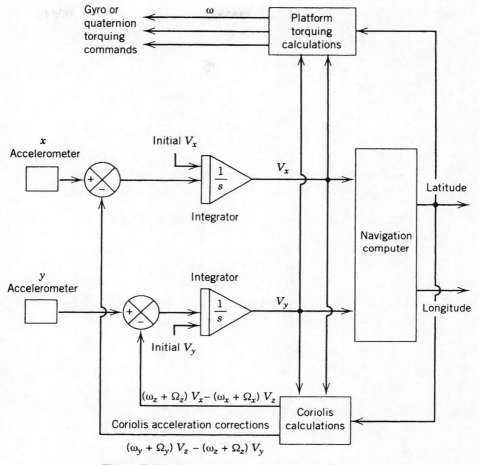

Figure 7.22 Mechanization of an inertial navigator.

where

Φ is the latitude of vehicle

$\dot{\lambda}$ is the rate of change of vehicle longitude

$\mathbf{V}_x$ is east velocity

$\mathbf{V}_y$ is north velocity

R_M is the radius of curvature of the Earth in the meridional plane, Equation 2.2

R_P is the radius of curvature of the Earth in the vertical east–west plane, Equation 2.3

2. Locally level, wander azimuth [27][42]; $x^{(N)}$ and $y^{(N)}$ are in a level plane, with $y^{(N)}$ at an angle α from $y^{(G)}$; $z^{(N)}$ lies along $z^{(G)}$ (Figure 7.23).

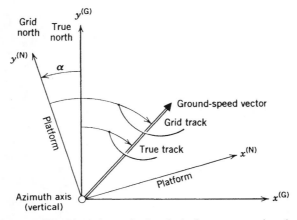

Figure 7.23 Plan view of a level platform at an azimuth α.

$$\Omega_x = \Omega \cos \Phi \sin \alpha$$
$$\Omega_y = \Omega \cos \Phi \cos \alpha$$
$$\Omega_z = \Omega \sin \Phi \qquad (7.37)$$

$$\omega_x = -V_y \left(\frac{\cos^2 \alpha}{R_M + h} + \frac{\sin^2 \alpha}{R_P + h} \right) + \frac{V_x \sin 2\alpha}{2} \left(\frac{1}{R_P + h} - \frac{1}{R_M + h} \right) + \Omega_x$$

$$\omega_y = V_x \left(\frac{\cos^2 \alpha}{R_P + h} + \frac{\sin^2 \alpha}{R_M + h} \right) - \frac{V_y \sin 2\alpha}{2} \left(\frac{1}{R_P + h} - \frac{1}{R_M + h} \right) + \Omega_y$$

$$\omega_z = \dot{\alpha} + \Omega_z \qquad (7.38)$$

where V_x and V_y are the level components of the velocity vector in the navigation coordinates and α is the azimuth of the $y^{(N)}$-axis relative to north (Figure 7.23).

3. Tangent plane; $y^{(N)}$ north at point of tangency (Figure 2.3):

$$\omega_x = \Omega_x = 0$$
$$\omega_y = \Omega_y = \Omega \cos \Phi_0$$
$$\omega_z = \Omega_z = \Omega \sin \Phi_0 \qquad (7.39)$$

where Φ_0 is the latitude at the point of tangency, usually the takeoff point.

The first coordinate frame offers the advantages of simplicity at low latitudes, the readout of familiar latitude-longitude coordinates from the computer, and the measurement of vehicle heading directly from the direction cosine matrix in a strapdown system or from the platform azimuth angle in a gimballed system. Unfortunately, this mechanization exhibits polar singularities at high latitudes.

The second coordinate frame introduces the complexity of an extra coordinate, the wander angle α, in exchange for the following advantages:

1. Operation at all latitudes, with the correct choice of mechanizing $\dot{\alpha}$.
2. Omission of the azimuth torquing by setting $\omega_z = 0$ (*free azimuth*).
3. In gimballed platforms, some improvement in navigation performance can be achieved by continuous azimuth rotation at a constant rate $\dot{\alpha}$. The instrument axes point in various directions thus tending to cancel the effect of their drift. In one platform [6] $\dot{\alpha}$ was 50 times Earth rate, so that instrument drifts whose correlation time was greater than one-half hour tended to be smoothed. The azimuth rotation can be reversed periodically to reduce the effects of an azimuth-gyro torquer error. Such performance improvements cannot be achieved in strapdown systems because the instruments remain fixed to the body.

Wander-azimuth systems do not read vehicle heading directly from the direction cosine matrix or the platform azimuth gimbal. The vehicle's heading relative to north is platform-indicated azimuth minus α. Heading becomes indeterminate at the poles. The majority of inertial navigation systems implement a wander azimuth, since the benefits outweigh the additional computational complexity involved. As an illustration of a typical mechanization, consider the wander-azimuth system of Equations 7.37 and 7.38:

1. Integration of Equations 7.34 yields **V** from the accelerometer outputs.
2. The torquing signals are calculated from **V** according to Equations 7.37 and 7.38 and the azimuth condition $\omega_z = \dot{\alpha} + \Omega \sin \Phi$. Notice that vehicle latitude and the wander angle are required for this calculation. Longitude and azimuth exhibit polar singularities and therefore should not be used as fundamental variables for worldwide operation. Instead, the direction cosines C_{ij} of the N axes relative to the E axes are introduced (Equation 7.32). These parameters are mathematically well behaved at all latitudes even when longitude and azimuth are not. The direction cosines obey the differential equations

$$\dot{C}_{i1} = (\omega_y - \Omega_y)C_{i3} - (\omega_z - \Omega_z)C_{i2}$$
$$\dot{C}_{i2} = (\omega_z - \Omega_z)C_{i1} - (\omega_x - \Omega_x)C_{i3}$$
$$\dot{C}_{i3} = (\omega_x - \Omega_x)C_{i2} - (\omega_y - \Omega_y)C_{i1} \qquad (7.40)$$

$$i = 1, 2, 3$$

and latitude-longitude coordinates are calculated from

$$\sin \Phi = C_{33}$$

$$\tan \lambda = \frac{C_{32}}{C_{31}} \qquad (7.41)$$

α is calculated only when it is convergent:

$$\tan \alpha = \frac{C_{13}}{C_{23}} \quad \text{or} \quad \sin \alpha = \frac{C_{13}}{\cos \Phi} \quad \text{or} \quad \cos \alpha = \frac{C_{23}}{\cos \Phi} \qquad (7.42)$$

3. The wander angle α can be constructed in many ways. Clearly, if α itself had to be calculated as part of the position-velocity mechanization, the system would not be convergent over the poles. However, $\dot{\alpha}$, defined as $\omega_z - \Omega \sin \Phi$, can always be calculated in a convergent manner. The wander angle α is not used explicitly in the navigation solution. The wander azimuth may be extracted for readout purposes using Equation 7.42. Wander-azimuth mechanizations include, as special cases, all locally level systems. For example, in a north-pointing system, α is constrained to be zero.

4. In early navigation systems, where computational throughput was at a premium, updating of the direction cosines was performed at a low iteration rate (a few times per second). In 1996, inertial systems included sufficient computer power to perform the updates 50 or more times per second. Direction-cosine algorithms generally have employed 32 bit-integer word length, but microprocessors in the 1990s used floating-point instructions to enhance accuracy.

An insight into the nature of the inertial mechanization can be obtained by considering the steady-state situation in level flight.

1. The outputs of the level accelerometers are virtually zero, except for the Coriolis acceleration, and accelerations due to air turbulence and equipment vibration. Barring inadvertent rectification within the accelerometers, vibration-like outputs do not indicate any net position change, no matter where in the aircraft the instruments are located.

2. The dV_i/dt in the computer are zero, since the computer-calculated gravity vector $\mathbf{g}$ and Coriolis correction $(\boldsymbol{\omega} + \boldsymbol{\Omega}) \times \mathbf{V}$ just cancel the gravity and Coriolis accelerations measured by the accelerometers.

3. The velocity registers do not change.

4. The computed position changes slowly as the vehicle moves over the Earth.

5. The torquing signals are nearly constant, changing slowly as the latitude and heading of the vehicle change.

Section 2.7 shows how inertially derived position is used for steering. Locally level mechanizations are desirable in terrestrial navigation systems for the following reasons:

1. In a gimballed, north-pointing system, roll, pitch, and azimuth are directly available as gimbal angles. In a strapdown system, they are mathematically extracted from the quaternion/direction cosine representation. For wander-azimuth mechanizations, the platform (mechanical or analytic) yaw angle minus the computer-calculated α equals true azimuth.

2. No gravity corrections are needed in the horizontal accelerometer channels. A small component of gravity is measured by the horizontal accelerometer channels when they are tilted from the locally level plane or when gravity anomalies exist (see Section 2.2).

3. For a gimballed platform, the gyros require accurate mass balance in only one orientation, since horizontal acceleration is intermittent.

In some instances (particularly space applications), inertial systems may use nonlocally level navigation frames. Reference [58] discusses an Earth-fixed cartesian mechanization.

In pure inertial systems without external updates, constant instrument errors tend to cause oscillatory position, velocity, tilt, and azimuth errors as discussed in Section 7.6. Stationary noise in the instruments tends to cause errors that grow with the square root of time. For aircraft systems that operate for only a few Schuler periods (multiples of 84.4 minutes), a pure inertial system is often adequate. Growth of errors can be reduced by periodic position, velocity, or azimuth updates (e.g., radio or celestial). Reference [13] discusses the effects of the frequency of fixing on error propagation. Chapter 3 discusses multisensor navigation systems in which frequent external fixes are analytically combined with inertial measurements.

7.5.3   Vertical Mechanization

It might appear that altitude could be calculated by double integration of the measured vertical acceleration. Unfortunately, such a mechanization would be unstable, as will now be shown. Assume a constant-velocity flight at a speed that is negligible compared with orbital speed. Let:

h_a Actual altitude

h_c Computed altitude

A_0 Constant uncompensated accelerometer error (bias)

u Accelerometer output $= d^2 h_a / dt^2 + g_a + A_0$

a Radius of the Earth

g_c Computed gravity $\simeq g_0(1 - 2h_c/a)$; a linearization of the inverse-square field at the Earth's surface

g_a Actual gravity $\simeq g_0(1 - 2h_a/a)$

ν^2 $2g_0/a$

Then

$$h_c = \iint (u - g_c)\, dt\ dt \tag{7.43}$$

Expanding and letting $\Delta h = h_c - h_a$,

$$\Delta \ddot{h} - \frac{2g_0}{a} \Delta h = A_0 \tag{7.44}$$

whose solution is

$$\Delta h(t) = \Delta h_0 \cosh \nu t + \frac{\Delta \dot{h}_0}{\nu} \sinh \nu t + \frac{A_0 a}{2g} (\cosh \nu t - 1) \tag{7.45}$$

For the first few seconds of flight:

$$\Delta h(t) \approx \Delta h_0 \left(1 + \frac{\nu^2 t^2}{2} \right) + \Delta \dot{h}_0 t \left(1 + \frac{\nu^2 t^2}{6} \right) + \frac{A_0 t^2}{2} \left(1 + \frac{\nu^2 t^2}{12} \right) \tag{7.46}$$

An initial altitude error (Δh_0) or altitude-rate error ($\Delta \dot{h}_0$) or an accelerometer error (A_0) will lead to an exponential growth in computed altitude (doubling in the first 78 minutes), thus making the indicated altitude and altitude-rate useless after the first few minutes. The instability of the vertical channel will result, no matter how carefully the vertical component of gravity is mechanized as a function of computed altitude.

On the other hand, a barometric altimeter is very stable for long periods of time but suffers from a noisy output and a long time lag (due to aerodynamic noise and the dynamics of air flowing through tubes, ports, and orifices; see Section 8.2). As a result, instantaneous readings of altitude and altitude-rate are impossible. The best features of inertial and barometric altimeters can be combined in a baro-inertial altimeter shown in Figure 7.24. By comparing baro-

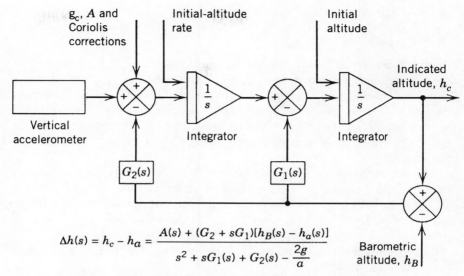

$$\Delta h(s) = h_c - h_a = \frac{A(s) + (G_2 + sG_1)[h_B(s) - h_a(s)]}{s^2 + sG_1(s) + G_2(s) - \dfrac{2g}{a}}$$

Figure 7.24   Mechanization of a stable baro-inertial altimeter.

metric altitude with inertial altitude and feeding back through suitable filters, the indicated altitude can have (1) the long-term stability of the barometric altimeter, (2) a frequency response that is much faster than that of a pure barometric altimeter but that attenuates barometric noise, and (3) relative insensitivity to low-frequency accelerometer errors.

The transfer function shown in Figure 7.24 (where G_1 and G_2 are constant gains and the system is of second order) demonstrates that in a baro-inertial altimeter, a constant accelerometer error A_0 causes an altitude error or approximately $\Delta h = A_0/\omega_N^2$ where ω_N is the natural frequency of the altimeter loop. A bias change of 1 mg causes only a few inches of altitude error for a one-minute natural period. As a result, the Coriolis correction can usually be omitted and vertical gravity can be computed as a constant, independently of altitude and latitude. Nonetheless, most modern inertial navigation systems implement a more complete gravity model. This permits more accurate vertical channel operation particularly at high altitudes. Baro-inertial altitude and altitude-rate are used in weapon delivery computers and in the calculation of flight path angle, γ

$$\gamma = \dot{h}/\sqrt{V_x^2 + V_y^2}.$$

Further information regarding standard baro-inertial loops may be found in reference [3]. If the statistical properties of the accelerometer and altimeter noise are known, an optimum filter (Chapter 3) can be designed to give the best estimate of altitude and altitude rate. More sophisticated mechanizations may include estimates of the accelerometer bias and barometer scale factor [2].

7.6   ERROR ANALYSIS

7.6.1   Purpose

After mechanizing an inertial navigation system, the designer must investigate the propagation of errors in the proposed system, based on mathematical analysis and on the results of past tests. The error analysis establishes the maximum permissible component tolerances; to simplify the mechanization equations (and thus the computer complexity) when advantageous; and to predict compliance with the position, velocity, attitude, and azimuth specifications established by the user.

The sources of error that should be considered in performing the analysis depend on the system design, the components used, and the required accuracy of the navigation system. Some typical sources of error are enumerated below (instrument errors are discussed in more detail in Section 7.3).

1. Gyro drift errors caused by temperature variation, acceleration, magnetic fields, and vibration.
2. Gyro scale factor errors including nonlinearity and asymmetry.
3. Accelerometer bias errors caused by variations in temperature or by rectification of vibration inputs. Accelerometer resonances and dynamic response are important considerations.
4. Accelerometer scale factor errors including nonlinearity and asymmetry.
5. Sensor assembly errors such as nonorthogonality of the gyro and accelerometer input axes, transmission of vibration inputs through the shock mounts, and thermal gradients that affect instrument performance and alignment.
6. Computational errors due to roundoff, truncation, readout accuracy, as well as approximations inherent in the algorithms. Detailed evaluation of computational error propagation often requires a bit-by-bit simulation.
7. Initial condition errors (position, velocity, tilt, and azimuth) must be considered as they propagate with time.

7.6.2   Simulation

Error analysis is performed by assuming a flight path and processing the resulting acceleration and angular-rate profile in a mathematical model of the inertial system and computer. The indicated positions and velocities are compared to the assumed flight path, and the error histories are recorded. Azimuth and tilt errors can similarly be derived. Analyses are done either with fixed instrument errors or with statistically distributed errors in a large number of runs (*Monte Carlo* simulations).

7.6.3   Error Propagation

A complete simulation assesses overall system performance for particular trajectories. It yields a time-varying covariance matrix (Chapter 3) that quantifies the growth of errors and the correlations among errors. However, to understand the physical causes of errors and to troubleshoot systems, a simplified closed-form analysis is invaluable. For simple flight profiles (e.g., straight legs, 90-deg-turns), deterministic initial-condition and instrument errors are readily propagated with time, using analytic approximations. The solution is complicated by the correlation that exists between in-flight instrument errors and the initial conditions. In gimballed systems, some instrument errors during alignment tend to cancel the same instrument errors incurred in flight. However, in strapdown systems, the instrument axes rotate with respect to the analytic platform axes and error cancellation may not occur. Therefore, the propagation of errors is far more dependent upon the trajectories in strapdown systems.

For closed-form analysis, the mechanization equations can be linearized for a vehicle moving at low speed (relative to the surface speed of the rotation of the Earth). The result is that undamped oscillations are present in the position, velocity, tilt, and azimuth errors. The frequencies of these oscillations are (1) Ω, Earth rate; (2) $\omega_s = \sqrt{(g/a)}$, the Schuler frequency (84.4-minute period); and (3) beats between $\omega_s + \Omega \sin \Phi$ and $\omega_s - \Omega \sin \Phi$, which occur at $2\Omega \sin \Phi$. These oscillations result naturally from a correct mechanization of the dynamic relations between position and tilt. The expression *Schuler tuning* dates back to the early twentieth century, when attempts were made to eliminate maneuver-induced errors in ship gyrocompasses. The term was retained during the late 1940s, when the gyrocompass evolved into an inertial navigator.

For short times of operation (less than four hours) and at speeds at least as high as Mach 4, the Schuler frequency predominates. It alone was observed in the early days of experimentation with inertial systems because long-time operation was impossible with the crude instruments then available. Schuler oscillations are often compared to those of an Earth's radius pendulum whose bob remains fixed at the center of the Earth.

If long-period oscillations are not of interest, the behavior of a navigation system can be predicted from the single-axis uncoupled model and transfer functions shown in Figure 7.25. Here the Earth-rate torquing signals applied to the platform (which cause 24-hour oscillations), and the errors in the Coriolis corrections are neglected. The resulting system errors are shown in Table 7.2. Notice that a constant gyro drift during flight causes a linearly increasing position error (with superimposed Schuler oscillation) as well as oscillatory velocity and tilt errors at the Schuler frequency. A 100-μg accelerometer bias causes a 0.3-nmi peak error, and a 0.017-deg/hr gyro drift causes a 1-nmi/hr position error growth rate. Initial position, velocity, and tilts give oscillatory bounded errors. A 4.5-knot initial velocity error (as might occur when aligning on an aircraft carrier) causes a 1-nmi peak position error. Initial

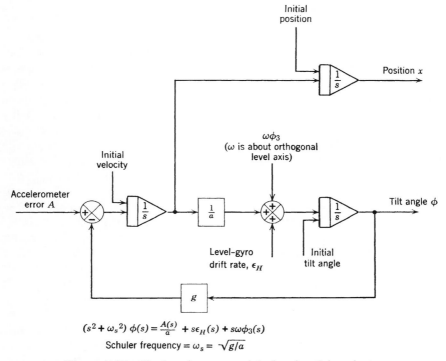

$$(s^2 + \omega_s^2)\,\phi(s) = \frac{A(s)}{a} + s\epsilon_H(s) + s\omega\phi_3(s)$$

$$\text{Schuler frequency} = \omega_s = \sqrt{g/a}$$

Figure 7.25   Single-axis error model of an inertial navigator.

azimuth misalignment rotates the coordinate frame about the vertical and gives an error proportional to distance traveled from the point of alignment. Instrument nonorthogonality, accelerometer nonlinearity, and gyro mass unbalance cause oscillatory impulse responses in position and velocity after each maneuver of the vehicle. Single-axis error analyses are widely available [41, pp. 172–173; 12, pp. 335–346]. Because of the presence of Schuler oscillations, the rms velocity error is 25% greater than the average increase in position error and the peak velocity error is twice as large. As vehicle speed approaches orbital speed, many of these simple relationships break down, and the analysis requires a complete simulation. Reference [12] discusses this case for constant speed, using undamped inertial altitude measurements in the loop. Complete analyses are readily run on digital computers.

The analysis of the vertical channel of an inertial navigator shows that altitude calculations, using the vertical accelerometer channel at slow speeds without external damping, are unstable (Section 7.5.3). If barometric, radar, or GPS damping is not used, the altitude error, shown in Equation 7.45, increases rapidly with time. At orbital speeds, altitude errors are marginally stable [12]. It is clear that error propagation must also depend on the choice of navigation axes. In a locally level platform mechanization, the horizontal channels can

TABLE 7.2   Error propagation in terrestrial inertial navigators

Deterministic Error	Time less than 4 hr Speed < Mach 4	Time = Several Days Speed < 200 knots
Initial position Δx_0	$\Delta x = \Delta x_0$	$\Delta x = \Delta x_0 \cos \Omega t$ (latitude) $\Delta x = \Delta x_0$ (longitude)
Initial velocity ΔV_0	$\Delta x = \dfrac{\Delta V_0}{\omega_s} \sin \omega_s t$	$\Delta x = \dfrac{\Delta V_0}{\omega_s} \sin \omega_s t$
Initial tilt ϕ_{y0}	$\Delta x = a\phi_{y0}(1 - \cos \omega_s t)$	$\Delta x = a\phi_{y0} \cos \Omega t$
Initial azimuth ϕ_{z0}	$\Delta x = y\phi_{z0} + a\phi_{z0}\left(t - \dfrac{\sin \omega_s t}{\omega_s}\right)$	$a\phi_{z0} \sin \Omega t$
Accelerometer bias A	$\Delta x = \dfrac{A}{\omega_s^2}(1 - \cos \omega_s t)$	$\Delta x = \dfrac{A}{\omega_s^2}(1 - \cos \omega_s t)$
Gyro drift ϵ	$\Delta x = a\epsilon\left(t - \dfrac{\sin \omega_s t}{\omega_s}\right)$	$\Delta x = \dfrac{a\epsilon}{\Omega} \sin \Omega t$ (latitude) $= at \cos \Phi(\epsilon_z \sin \Phi + \epsilon_y \cos \Phi)$ (longitude)

Note: Symbols: x, y = east and north distances traveled; $\Delta x, \Delta y$ = errors in x and y; Φ = latitude; t = time, measured from the instant at which the system is switched into the "navigate" mode; ω_s = Schuler radian frequency = $\sqrt{g/a} \approx 2\pi$ rad/84.4 min ≈ 0.00124 rad/sec; g = magnitude of gravity on the surface of the Earth ≈ 32.1 ft/sec$^2 \approx 9.81$ meter/sec^2; a = radius of Earth ≈ 3440 nmi $\approx 20.9 \times 10^6$ ft; Ω = Earth's inertial angular velocity ≈ 15 deg/hr.

remain stable, since they are decoupled from altitude. However, in a system that implements a platform at an arbitrary orientation with respect to gravity, all channels may be affected by the vertical instability so that altitude damping (barometric or other) must be used to stabilize the system.

7.6.4   Total System Error

Section 2.8 discusses the definition of mean and circular-error-probability (CEP) horizontal errors, and their measurement from flight-test data. References [15, 17] discuss the errors present in strapdown systems in the late 1980s and early 1990s, including test results over a large number of flights. Strapdown RLG-based inertial navigators conforming to the military specification of Ref. [62] demonstrated a CEP better than 0.3 nmi/hr accuracy during 4097 flight tests performed by the Springfield, IL, Air National Guard during 1991 and 1992. These systems have also demonstrated an MTBF approaching 4000 hours in military environments [15].

7.7   ALIGNMENT

An inertial system solves the five simultaneous second-order differential equations for the attitude and horizontal position of its platform. Hence, ten initial conditions are required in order to initialize the computer, as discussed below:

1. *Two initial position coordinates.* When aligning an inertial system in a stationary vehicle on the ground, the initial conditions are obtained from a survey of the launch site. Airborne position fixes can be obtained by visual or radar checkpoints or by the use of radio-navigation aids.

2. *Two initial velocity coordinates.* When aligning on a stationary vehicle, the initial velocity relative to the Earth is nominally zero. When aligning on a moving aircraft, velocity is measured with a Doppler radar (Chapter 10), a succession of radio-aid fixes (Chapter 4) or a GPS velocity measurement (Chapter 5). When aligning on an aircraft carrier, a ship's inertial navigation system is usually the reference and a lever-arm correction must be made for the aircraft's position on the deck.

3. *Three platform orientation coordinates.* Terrestrial inertial navigation systems are initialized by leveling (two coordinates) and azimuth aligning (one coordinate) to a convenient reference. Leveling is accomplished by rotating the inertial platform (analytic or mechanical) until the average acceleration along each of the two horizontal platform axes read zero, at which time the navigation frame is deemed level. In a mechanical platform, this corresponds to a null output of the two horizontal accelerometers. In a strapdown system, none of the accelerometer outputs are necessarily nulled. Rather, their outputs, resolved onto the horizontal axes of the analytic platform, are driven to zero. Azimuth alignment is accomplished by any of several procedures discussed below.

4. *Three orientation rates.* These initial conditions are implicitly determined by the values of the gyro drifts at the instant of switching from the "align" mode to the "navigate" mode. These rates are not always measurable during alignment and can cause navigation errors.

The presence of initial-condition errors causes position errors, which propagate according to Table 7.2. To achieve 1 nmi/hr navigation accuracy, initial errors should be limited to 5 arcmin in azimuth, 10 arcsec in tilt, and 1 knot in velocity.

Several methods of alignment are used for aircraft:

1. *Leveling and gyrocompassing using gyros and accelerometers.* This procedure is analyzed in Sections 7.7.2 and 7.7.3. It can be performed on a stationary or moving vehicle, except near the Earth's poles.

2. *Transfer alignment, relative to a master platform.* A secondary inertial navigation system may be aligned to a master unit (e.g., an aircraft unit to a shipboard unit or a low grade unit to a higher accuracy system). The transfer alignment can be accomplished via acceleration/velocity matching and/or via attitude matching, Section 7.7.4. Large lateral accelerations, such as exist on the deck of an aircraft carrier, are usually needed to increase the signal-to-noise ratio during acceleration or velocity matching.

3. *Memory alignment, in which the attitude angles recorded during a previous alignment are stored.* This procedure saves time, because it can be done in seconds, but it is limited by the angular repeatability of the aircraft landing gear and inertial-system shock-mounts. The aircraft must not be moved between the time of precise alignment and the time of memory alignment. This method is not suitable for use on ships or in flight.

4. *Runway alignment when the aircraft taxis on a runway or taxiway of known heading prior to takeoff.* Alignment procedures can take advantage of turns and stops to estimate errors. Stops at different headings provide opportunities for refining the platform heading estimates and fine-tuning the gyro-bias estimates.

5. *Optical alignment.* An inertial system can be aligned relative to an external optical line of sight within a few seconds of arc. It is impractical, however, on ships and aircraft because of the necessity for optical access and the inconvenience of establishing theodolite sites. Optical alignment of terrestrial aircraft guidance systems is sometimes used in conjunction with platforms that have star-trackers mounted on the stable element. Navigation systems for space applications often make use of optical alignment.

6. *Portable platform alignment for rapid alignment.* This procedure was developed in the 1960s. It used a portable platform that was aligned in a ready-room and carried aboard the aircraft at flight time. Portable platforms are rarely if ever implemented because they must be precisely installed in an accessible location and because they accumulate drift while being carried to the aircraft.

Fast warm-up is an important military requirement for interceptors, strategic bombers, and carrier-based aircraft. In older gimballed systems, warm-up time was limited by the interval taken by the inertial instruments to reach thermal equilibrium, typically 30 minutes. The warm-up rate could not be too high because of the detrimental temperature gradients induced by rapid application of heat. Use of the system before it had warmed up resulted in large navigation errors. In some systems, heaters were placed on the instruments in order to achieve rapid thermal stabilization. Those systems were capable of full performance within a few minutes. In the early 1980s, the shift to strapdown systems based on optical gyroscopes, and high-speed microprocessors eliminated the need for thermal stabilization. Since optical gyros exhibit much lower thermal sensitivity than their mechanical counterparts, it is possible to model errors such as bias, scale factor, and misalignments using polynomial functions of temperature in a flight computer. Typically, second- through fourth-order functions are implemented to model instrument errors during warm-up.

$$y = a_0 + a_1 T + a_2 T^2 + \cdots \qquad (7.47)$$

Strapdown accelerometers are also designed with low thermal sensitivity and repeatability that permit thermal modeling. A factory calibration procedure (Section 7.4.1) derives the temperature coefficients that are loaded into a non-volatile memory. Sensors measure temperatures at key locations within the instruments. During system operation, the calibration coefficients are applied to polynomials of temperature, Equation 7.47. Thermal compensation in 1996 permitted better than 1 nmi/hr performance with extremely rapid reaction (four minutes). Only the most accurate systems included heaters for thermal stabilization.

7.7.1   Leveling

The purpose of leveling is to orient the platform (analytic or mechanical) with respect to gravity.

Coarse Leveling
Gimballed Platform   In leveling a gimballed platform, the gimbal servos rotate the stable element in order to null the outputs of the horizontal accelerometers. To reduce alignment time, the gimbals are first slewed to null the pickoffs (e.g., synchros) or the accelerometers. This "coarse leveling" can be performed at slew rates of several hundred degrees per second, limited only by the construction of the mechanical gyros. Synchro nulling is used when the attitude of the platform within the vehicle at rest is nominally level. The accuracy of the coarse-leveling process is limited by (1) errors in mounting the platform to the vehicle, (2) shock-mount angular deflection and nonreturnability, (3) synchro errors, and (4) rocking of the aircraft on its landing gear and tires, induced by wind gusts or by personnel climbing on the aircraft.

Analytical Platform   In a strapdown system, the orientation of the accelerometers is determined by the attitude of the vehicle and cannot be adjusted. Instead, leveling consists of rotating the transformation quaternion or direction cosine matrix that relates the instrument axes to the locally level navigation axes. Leveling is accomplished by adjusting the analytic platform in such a way that the accelerometer outputs transformed to horizontal platform axes are nulled. Estimates of the vehicle's pitch θ and roll ϕ are made using the accelerometer outputs (body frame readouts):

$$\theta = \tan^{-1}\left(\frac{-A_y^{(B)}}{-A_z^{(B)}} \right)$$

$$\phi = \tan^{-1}\left(\frac{A_x^{(B)}}{\sqrt{A_y^{2(B)} + A_z^{2(B)}}} \right) \tag{7.48}$$

where x is forward, y is transverse (positive out the right wing), and z is down. Based on the pitch and roll so-derived, the system's body-to-navigation coordinate transformation (quaternion or direction cosine matrix) is initialized. The process of coarse-leveling a strapdown system requires only a few seconds, the time it takes to obtain a good measure of acceleration. The accuracy of coarse leveling is limited by accelerometer errors and noise as well as by motion of the vehicle.

Fine Leveling In a gimballed system, the outputs of the accelerometers generate gyro torque commands. The gyros, in turn, drive the gimbal servos until two accelerometers are level and, on a stationary vehicle, measure no average acceleration. The maximum precession rate of the gyro torquers limits the alignment speed.

In a strapdown system, the accelerometer outputs are transformed through the attitude matrix to form the analogue of gimbal-mounted accelerometer outputs. The computed horizontal accelerations are used to "torque" the analytic platform (i.e., rotate the platform axes by recomputing the attitude quaternion or direction cosine matrix) to maintain zero level-axis accelerations.

Figure 7.26 shows one axis of a fine leveling loop for a stationary vehicle. Tilt errors cause computed velocity to build up. Reducing tilt (via gyro torquing in a gimballed system or via quaternion torquing in a strapdown system) rotates the platform axes into alignment with the navigation axes. The addition of feedback to the velocity integrator provides the damping that forces the oscillations to decay.

The choice of gains K_1 and K_2 represents a tradeoff between rapidity of leveling and noise immunity. A Kalman mechanization computes time-varying optimal gains based on the system noise and error model. Nonetheless, fixed precomputed gains usually yield acceptable leveling performance with-

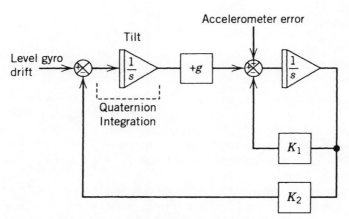

Figure 7.26 Second-order fine leveling loop.

out a Kalman filter. Leveling on a stationary vehicle takes approximately 30 to 60 seconds. Vehicle motion lengthens the required leveling time. Leveling-loop errors result from level-axis accelerometer or gyro errors, or from other horizontal disturbances. During ground alignment, rocking of the aircraft on its landing gear may be induced by wind gusts, machinery, and personnel boarding the aircraft.

The response of the second-order leveling loop shown in Figure 7.26 is given by

$$\delta\phi(s) = \frac{s(s + K_1)\phi_0 + K_2 A + K_2 s V_0 + (s + K_1)\epsilon}{s^2 + K_1 s + K_2 g} \qquad (7.49)$$

where
$\delta\phi$ is the tilt error
ϕ_0 is the initial tilt
A is an uncompensated accelerometer bias
V_0 is the initial velocity error
ϵ is an uncompensated gyro bias error
s is the Laplace transform operator

7.7.2 Gyrocompass Alignment

In addition to orienting the platform (mechanical or analytic) with respect to gravity, it is also necessary to initialize its orientation relative to north. Azimuth alignment consists of either rotating the platform around the vertical to a desired direction (e.g., true north or a great-circle flight track) or to determine the orientation relative to the desired direction. The latter is applicable to wander azimuth systems in which the initial wander angle α_0 is calculated during alignment. Gyrocompass alignment usually consists of the following steps:

1. Coarse leveling
2. Fine leveling
3. Gyrocompassing during which fine leveling continues

In strapdown systems, steps 2 and 3 are executed simultaneously, using a Kalman filter to provide the proper weighting. Gyrocompass times are typically 3 to 8 minutes for stationary alignment of military vehicles (some of which have quick-alignment requirements), 1 to 2 minutes for in-flight alignment with high-quality velocity references, and 10 minutes for stationary alignment on commercial airliners.

Gyrocompass alignment on a stationary vehicle is based on the fact that if a platform is to remain level with respect to a rotating Earth, then it must rotate at Earth rate. For a level platform at latitude Φ and azimuth α, the steady-state platform level rates are (see Figure 7.21a)

$$B_x = \Omega \cos \Phi \sin \alpha \qquad \text{about } x^{(N)}$$
$$B_y = \Omega \cos \Phi \cos \alpha \qquad \text{about } y^{(N)}$$
$$B_z = \Omega \sin \Phi \qquad \text{about } z^{(N)} \qquad (7.50)$$

A stationary gyrocompass loop can be mechanized as an extension of the leveling loops as shown in Figure 7.27. Tilt rate states B_x and B_y estimate the platform rates necessary to maintain a level platform while B_z maintains azimuth. Azimuth α can be computed from

$$\alpha = \tan^{-1}\left(\frac{B_x}{B_y}\right) \qquad (7.51)$$

If latitude is known, then an estimate of the north gyro bias error may also be obtained:

$$\text{North gyro bias error} = \sqrt{B_x^2 + B_y^2} - \Omega \cos \Phi \qquad (7.52)$$

Computed gyro bias error in the north direction is often resolved into instrument axes and applied as compensation (known as *mini-bias*) upon entering the navigation mode. In a single-position gyrocompass alignment, gyro bias error in the east direction cannot be distinguished from azimuth bias. Any uncompensated component of gyro bias in the east axis will result in a heading error.

An alternative mechanization may be used for gyrocompassing in which the platform is rotated so as to null α (i.e., to make $B_x = 0$). This can be accomplished via a mechanical rotation in a gimballed platform or by quaternion rotation in a strapdown system. It cannot be done accurately above about 70-deg. latitude.

The choice of fixed gains in the gyrocompass filter depends on the noise characteristics of the instruments and on the operating environment. Reference [23] presents a detailed discussion of conventional gyrocompass loops and their response characteristics. Since the 1970s, systems have employed variable-gain Kalman filters for the fine leveling and alignment process. Kalman filters can be mechanized to align at zero velocity (on stationary vehicles) or can be mechanized to accept external velocity or position fixes. A simple zero-velocity ground-align filter implements the model shown in Figure 7.27 but includes noise sources to calculate the optimum gains. Initial level errors and accelerometer noise limit the first stages of alignment (i.e., the first 30 seconds). The tilt gains peak in the first few seconds, then die out as the system levels. In the steady state, heading estimation is limited by the gyro noise. The tilt-rate gains start out low then grow within the first minute. Heading is captured during this time. The tilt rate gains then tail off slowly to allow fine tuning of the tilt rate estimates (and hence heading). A system with a gyro white noise of 0.12 deg/hr-$\sqrt{\text{Hz}}$ (typical of a navigation grade optical gyroscope) and gyro-bias accuracy of better than 0.01 deg/hr can gyrocompass to a heading accuracy of better than 1 mrad in less than 8 minutes at low and mid-latitudes.

The state equations for the diagram in Figure 7.27 are given by

$$
\begin{bmatrix} \dot{x} \\ \dot{y} \\ \dot{V}_x \\ \dot{V}_y \\ \dot{\phi}_x \\ \dot{\phi}_y \\ \dot{B}_x \\ \dot{B}_y \end{bmatrix} = \begin{bmatrix} 0 & 0 & 1 & 0 & 0 & 0 & 0 & 0 \\ 0 & 0 & 0 & 1 & 0 & 0 & 0 & 0 \\ 0 & 0 & 0 & 0 & 0 & g & 0 & 0 \\ 0 & 0 & 0 & 0 & -g & 0 & 0 & 0 \\ 0 & 0 & 0 & 0 & 0 & 0 & 1 & 0 \\ 0 & 0 & 0 & 0 & 0 & 0 & 0 & 1 \\ 0 & 0 & 0 & 0 & 0 & 0 & 0 & 0 \\ 0 & 0 & 0 & 0 & 0 & 0 & 0 & 0 \end{bmatrix} \begin{bmatrix} x \\ y \\ y \\ V_y \\ \phi_x \\ \phi_y \\ B_x \\ B_y \end{bmatrix} \qquad (7.53)
$$

This matrix equation is used in the Kalman filter equations [14] (see Chapter 3), along with estimates of the initial state uncertainties and noise terms. Noise sources that must be considered include velocity noise (due to aircraft motion and accelerometer quantization) which drives the position states, accelerometer white noise and gyro quantization which drive the velocity states, and level-axis gyro white noise which drives the tilt states.

In summary, a typical alignment consists of leveling (sometimes broken down into coarse and fine leveling), and gyrocompassing. In the alignment process, the gravity vector and the Earth-rate vector serve to define the coordinate reference frame. Modern strapdown navigators of the 1-nmi/hr variety align in 3 to 10 minutes on a stationary vehicle while achieving level accuracies of approximately 5 arcsec and heading accuracies of 1 mrad at mid-latitudes. Heading accuracy degrades in the polar regions due to the weak north component of Earth rate.

Chapter 3 describes multisensor navigation systems that mix inertial measurements with GPS or Doppler radars. These systems level and gyrocompass the platform coordinates continuously. In the event of loss of power or equipment failure, the multisensor system raises its gains for several minutes to realign, after which navigation proceeds with optimal gains. In-flight align times are typically on the order of one to two minutes but depend on the maneuvers executed and on the quality of the reference sensor.

7.7.3 Transfer Alignment

Some vehicles contain a high-quality inertial navigation system and one or more lower-cost units, for example, mounted on weapons that are to be released in-flight. Transfer alignment is the process of matching the slave platforms to the master, using natural or deliberately-induced maneuvers of the vehicle. Transfer

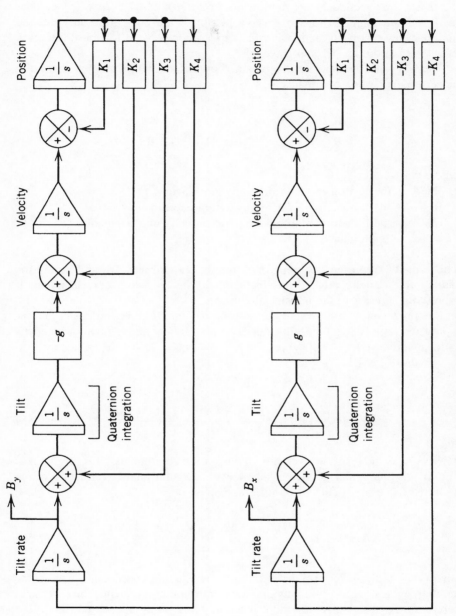

Figure 7.27 Two-axis, fourth-order gyrocompass loops.

alignment consists of simultaneous leveling and gyrocompassing. Accuracy is limited by the flexural vibration of the vehicle, which causes the outputs of the master and secondary accelerometers to differ slightly even when the two platforms are aligned. Since the slave platforms are not co-located with the master, a lever-arm correction is necessary to compensate for the differences in velocity at the master and slave.

The lever-arm velocity can be obtained by differentiating the lever-arm distance in navigation coordinates:

$$\mathbf{R}^{(N)} = C_B^N \mathbf{R}^{(B)}$$
$$\dot{\mathbf{R}}^{(N)} = \mathbf{V}^{(N)} = C_B^N (\boldsymbol{\omega} \times \mathbf{R}^{(B)}) + C_B^N \dot{\mathbf{R}}^{(B)} \tag{7.54}$$

where
$\mathbf{R}^{(B)}$ is the lever-arm vector in body coordinates
$\mathbf{R}^{(N)}$ is the lever-arm vector in navigation coordinates
C_B^N is the transformation matrix between body and navigation coordinates

The term $\dot{\mathbf{R}}^{(B)}$ represents bending or flexing of the airframe. Difficulties in estimating $\dot{\mathbf{R}}^{(B)}$ usually preclude its inclusion in the lever-arm compensation. It represents an error in the transfer alignment.

In strapdown systems where the body-to-navigation direction-cosine matrix is updated rapidly (e.g., 200 times per second), an alternative form of compensation is preferred:

$$\mathbf{R}^{(N)}(n\Delta t) = C_B^N \mathbf{R}^{(B)}$$
$$\mathbf{V}_{LN}((n + \tfrac{1}{2})\Delta t) = \frac{\mathbf{R}^{(N)}((n + 1)\Delta t) - \mathbf{R}^{(N)}(n\Delta t)}{\Delta t} \tag{7.55}$$

where $\mathbf{V}_{LN}$ is the lever-arm velocity in navigation coordinates. This mechanization avoids potentially noisy measurements of angular rate. Lever-arm acceleration may also be calculated by differencing $\mathbf{V}_{LN}$.

$$\mathbf{A}_{LN}(n\Delta t) = \frac{\mathbf{V}_{LN}((n + \tfrac{1}{2})\Delta t) - \mathbf{V}_{LN}((n - \tfrac{1}{2})\Delta t)}{\Delta t} \tag{7.56}$$

where $\mathbf{A}_{LN}$ is the lever-arm acceleration in navigation coordinates. This computation avoids the use of an angular acceleration measurement. The lever-arm correction is particularly important when aligning on an aircraft carrier where an aircraft INS is being aligned to the ship's inertial navigation system. The limitations of the lever-arm compensation include the lever-arm flexure as well as uncertainties in the distances involved.

7.7.4   Attitude and Heading Reference Systems (AHRS)

Some inertial systems are configured only to provide attitude. Such systems use lower accuracy gyroscopes. Inertial navigation systems may also revert to AHRS operation in case a fault is detected in the system or degraded navigation performance becomes apparent. In AHRS mode, it is assumed that horizontal accelerations are transient conditions (Chapter 9). The system is maintained in a "loose" leveling mode using gyros to track rapid attitude changes. In the steady state, the accelerometers maintain the platform level. A magnetic compass or other heading reference maintains long-term azimuth (Chapter 9). A second-order AHRS leveling loop is mechanized in the same manner as the fine-leveling loop discussed in Section 7.7.1, illustrated in Figure 7.26, and discussed in Section 9.3. Time constants on the order of 30 seconds are used, and cutouts are usually implemented to open the leveling loops when large accelerations or angular rates are detected. Reference [30] discusses an AHRS implementation and reference [53] discusses a low-cost GPS/inertial system designed to provide attitude.

7.8   FUNDAMENTAL LIMITS

The accuracy of inertial systems cannot be improved indefinitely, even with the best instruments. Reference [26] shows that the most severe limit on position measurement is the uncertainty in the Earth's gravity field in the region of operation, since accelerometers cannot distinguish between kinetic acceleration and gravity. Angular errors in the measurement of inertial space are primarily limited by the precession of the equinoxes and the migration of the Earth's pole. These errors are 5×10^{-5} deg/hr, equivalent to 100 ft position error [26].

Measurements of azimuth and tilt of the vehicle are typically limited by the angular returnability of the shock-mounts, the installation accuracy, and the flexure of the vehicle. These errors typically range from 1 minute of arc to 0.5 deg.

7.9   FUTURE TRENDS

Trends in inertial instruments are discussed in Section 7.3.5. Several trends are emerging for inertial navigation systems:

1. *Miniaturization of strapdown systems.* Technological advances permit smaller gyros (e.g., fiber optic) to be used. Advances in computer and electronics packaging permit the miniaturization of system electronics, resulting in lower volume and power consumption.
2. *Integrated systems*, which include inertial sensors and GPS receivers, provide high-bandwidth, low-noise navigation and attitude data with long-term accuracy ensured by GPS [54].

- Low-cost civil navigators (using 2-10 nmi/hr inertial sensors) will rely heavily on GPS, fiber-optic gyros, and micromachined accelerometers. They will be used in many military vehicles as well.
- High-performance military navigators with precise inertial instruments will be used in hostile conditions where reliable GPS signals may not be available.
- Inertial systems are being incorporated into landing aids (see Chapter 13).

3. *Multi-function inertial systems.* In addition to providing data for inertial navigation, the sensors are capable of supplying high-quality information for flight control, weapon delivery, and compensation of synthetic aperture radar. Excellent angle resolution is also possible for pointing and tracking applications. Rigid mounting of the angular sensors is required.

4. *Throwaway* inertial systems with GPS aiding will be used for a wide variety of applications including munitions and intelligent weapons. These low-cost systems will be based on fiber-optics and micro-machined sensors. Multi-axis gyro/accelerometer combinations employing silicon micro-machining may be acceptable.

When Com-Nav satellites, that supply intermittent radio fixes, supplant GPS, precise inertial navigation will be needed between fixes on commercial aircraft. These inertial navigators may take the form of micromachined panel-mounted instruments.

PROBLEMS

7.1. The following questions relate to accelerometers:

(a) An accelerometer is mounted on the table of a centrifuge with its input axis horizontal and radially outward. What is its steady-state output in terms of the radial dimension and the angular velocity of the centrifuge about the vertical?

Ans.: $\omega^2 R$.

(b) The input axis is rotated 45 deg downward so that it lies in the vertical plane containing the centrifuge axis. What is the steady-state output?

Ans.: $(\omega^2 R + g)/\sqrt{2}$.

(c) The pendulosity of an accelerometer is $mb = 10$ g-cm, and the moment of inertia of the pendulum about its pivot axis is 20 g-cm^2. Calculate the stiffness of the restoring electronics that would give a deflection of 10 arcsec at 5 g. Calculate the undamped natural frequency.

Ans.: 4.9×10^3 dyne-cm/arcsec, 1.1 kHz.

7.2. The following questions relate to gyroscopes:

(a) What is the scale factor of a square ring laser gyroscope operating at

$\lambda = 630$ nm and with a pathlength of 30 cm in terms of seconds of arc per count (half-wavelength)?

Ans.: 1.73 arcsec/count.

(b) What path length is needed to obtain the same scale factor with an equilateral triangle RLG operating at the same wavelength?

Ans.: 39.0 cm.

(c) What is the maximum gyro random walk coefficient that will permit measurement of angular rate in a three-minute averaging period with a measurement standard deviation of less than 0.005 deg/hr?

Ans.: 0.0011 deg/$\sqrt{\text{hr}}$ (= 0.066 deg/hr-$\sqrt{\text{Hz}}$).

(d) A mechanical gyroscope has $H = 2 \times 10^6$ g-cm^2. What residual torque is allowable to achieve a drift rate of 0.005 deg/hr?

Ans.: 0.05 dyne-cm.

7.3. The following questions relate to analytic platforms:

(a) Given the navigation coordinates shown in Figure 7.21 with $\alpha = 0$, an aircraft is flying level in the north direction. What is the body-to-navigation quaternion (assume x-body is foward, y-body is out the right wing, and z-body is down)?

Ans.:

$$\left(\frac{1}{\sqrt{2}}, \frac{1}{\sqrt{2}} \, \mathbf{1}_x + \frac{1}{\sqrt{2}} \, \mathbf{1}_y \right).$$

(b) What is the direction cosine matrix corresponding to the condition in (a)?

Ans.:

$$\begin{bmatrix} 0 & 1 & 0 \\ 1 & 0 & 0 \\ 0 & 0 & -1 \end{bmatrix}.$$

(c) Given a 50-Hz sinusoidal acceleration with amplitude $a_0 = 2.5$ g along the z-axis and a sinusoidal angular displacement in the same phase of amplitude $\theta_0 = 300$ μrad about the y-axis, what is the sculling rectification of the x accelerometer?

Ans.: 375 μg.

(d) For the conditions in (c), calculate the net error resulting from a velocity transformation algorithm executing at 250 Hz. What is the error if the transformation executes at 100 Hz?

Ans.: 24 μg, 136 μg.

(e) An aircraft performing S-turn maneuvers generates a coning rate of 25,000 deg/hr about the wing axis. Assuming the maneuvers last 10 minutes, what tilt error is generated by a 5 μrad gyro nonorthogonality error? Assuming no other error sources, what is the position error exhibited at the end of the maneuver? What is the position error 42.2 minutes after the end of the maneuver?

Ans.: 331 μrad, 0.11 nmi, 2.4 nmi.

7.4. The following questions relate to gyrocompass alignment:

(a) At a latitude of 30 degrees, what is the approximate azimuth error resulting from a 0.01 deg/hr east gyro bias error in a stationary ground alignment?

Ans.: 0.044 deg ≈ 2.6 arcmin ≈ 0.77 mrad.

(b) What would the heading error be at 80 degrees latitude?

Ans.: 0.22 deg ≈ 13 arcmin ≈ 3.8 mrad.

(c) Given a gyro random walk coefficient of 0.002 deg/$\sqrt{hr}$, what is the minimum alignment time required to attain an azimuth standard deviation of better than 0.7 mrad at 45° latitude?

Ans.: 261 sec.

(d) A strapdown system at 45-degrees latitude is aligned at a heading of 30 degrees. The x body-axis gyro has a bias error of 0.02 deg/hr. What is the resulting heading error during gyrocompass alignment? What bias corrections are computed for the gyros using mini-biasing?

Ans.: −0.054 deg, x gyro 0.015 deg/hr, y gyro −0.0086 deg/hr.

(e) A strapdown system at 34-degree latitude is aligned at a north heading. The x gyro has a bias error of 0.01 deg/hr and the y gyro has a bias error of 0.007 deg/hr. What is the heading error of the system? After alignment the system is switched to navigate mode with no mini-biases used and immediately rotated 180 degrees in azimuth. What are the north and east position error growth rates neglecting Schuler oscillations? What are these rates if minibiases are used?

Ans.: −0.032 deg, north 0.84 nmi/hr, east −0.6 nmi/hr; north 0.84 nmi/hr, east 0.00 nmi/hr.

8 Air-Data Systems

8.1 INTRODUCTION

An air-data system consists of aerodynamic and thermodynamic sensors and associated electronics. The sensors measure characteristics of the air surrounding the vehicle and convert this information into electrical signals (via transducers) that are subsequently processed to derive flight parameters. Typical flight parameters calculated by air vehicles include calibrated airspeed, true airspeed, Mach number, free-stream static pressure, pressure altitude, baro-corrected altitude, free-stream outside air temperature, air density, angle of attack, and angle of sideslip. This information is used for flight displays, for autopilots (flight-trajectory control and control-loop gain adjustment), for weapon-system fire-control computations, and for the control of cabin-air pressurization systems.

Air-data systems are an outgrowth of the airspeed indicator and altimeter used in early aviation. In those primitive, pneumatically driven instruments, the computation was performed by nonlinear spring mechanisms incorporated into specially designed bellows, which expanded or contracted in response to changes in sensed pressures, thereby moving the dials of the flight instruments. In the late 1950s, analog computers were interposed between the pressure sensors and flight instruments. The transducers and computation elements found in typical analog air-data computers of that era were documented in the first edition of this book [27]. In those designs, servo-driven cams or nonlinear potentiometers computed parameters such as altitude, airspeed, and Mach number. In the 1990s, all computation and data management are digital and based on microprocessor technology. Indeed, the miniaturization made possible by this technology is tending to obsolete the air-data computer as a separate entity. New avionics architectures are incorporating air-data functions into other subsystems such as inertial/GPS navigation units or are packaging the air-data transducers into the flight-control computers.

Regardless of how the air-data functions are packaged, they provide flight-critical information and therefore are implemented with appropriate redundancy and automated fault detection and isolation. Air-data systems can be found in every class of air vehicle, from combat fighters and helicopters to manned spacecraft such as Space Shuttle *Orbiter*. Their implementation in commercial jet transports uses ARINC standard specifications that are also used in business aviation. Each type of aircraft has unique challenges, primarily in regard to the accuracy of measuring the basic aerodynamic phenomena.

8.2   AIR-DATA MEASUREMENTS

8.2.1   Conventional "Intrusive" Probes

All of the air-data parameters that are relevant to flight performance are derived by sensing the pressures, temperatures, and flow direction surrounding the vehicle. Free-stream pressures and temperatures are required for the computation of static air temperature, altitude, airspeed, and Mach number. Because air is moving past the aircraft, the pressure at various places on the aircraft's skin may be slightly higher or lower than free stream. Figure 8.1 illustrates the probes typically deployed around the skin of an aircraft. They sample static pressure (via static ports), total pressure (via the pitot tube), total temperature (via the temperature probe), and local flow direction (via the angle-of-attack and sideslip vanes). All of these sensing elements, except for the flush-mounted static port, are intrusive because they disturb the local airflow. In flight testing of new aircraft, integrated air-data booms are often used to mount combinations of these probes forward of the flow which normally contacts the aircraft skin. Figure 8.2 illustrates such an instrumentation boom that would normally be located at the aircraft nose. The angle-of-attack and angle-of-sideslip vanes are self-aligning, measuring the direction of local flow. Total pressure must be measured at the front opening on the pitot tube which extends directly into the air flow, but at an angle with respect to the relative wind. That angle, defined by sideslip β and angle-of-attack α does not usually produce any significant inaccuracies as long as α and β are within ± 10 degrees. In applications where α and β excursions are large, then special booms containing a gimbaled pitot tube can be used. Such tubes contain wind vanes or may be servo driven to align with the relative wind. Typical applications of self-aligning pitot tubes are in developmental testing for experimental aircraft, or for high-angle-of-attack fighter aircraft. Non-intrusive probes are discussed in Sections 8.5.2 and 8.5.3.

8.2.2   Static Pressure

Static pressure is the absolute pressure of the still air surrounding the aircraft. To obtain a sample of static air in a moving aircraft, a hole (static port) or series of holes are drilled in a plate on the side of the fuselage or on the side of the pitot tube probe which extends into the free airstream. These sensed static pressures will differ from the free-stream values for reasons noted above. That difference is referred to as the *static defect*. The location of static ports is selected by wind-tunnel tests and by tests at numerous locations on actual aircraft. The location of a static port on helicopters or on fixed-wing aircraft that operate at very high angles of attack is especially difficult because of unusual local flow phenomena. Even with an optimum static source location, a large static defect usually remains, which is a function of Mach number, angle of attack, and aircraft configuration (flap deployment, wing stores, etc.). Because

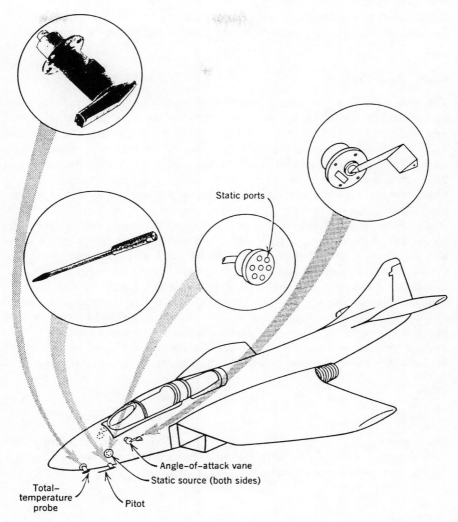

Figure 8.1   Air-data system, input probes and vanes (probes courtesy, Rosemount, Inc.).

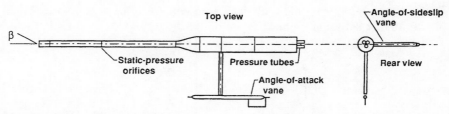

Figure 8.2   Typical nose-mounted air-data boom with pressure probes and flow-direction vanes.

static defect is predictable, it can be corrected in the air-data computations. Techniques for correcting such errors are covered in Sections 8.3 and 8.4.

8.2.3 Total Pressure

The terms *total pressure, stagnation pressure,* or *pitot pressure* refer to the pressure sensed in a tube that is open at the front and closed at the rear. The pitot tube is illustrated in Figures 8.1 and 8.2. When the static pressure ports and the pitot tube are combined into a single probe, the instrument is called a *pitot-static tube*. Such tubes are electrically heated to prevent ice formation. Pipes in the aircraft, referred to as the *pneumatic plumbing*, carry the sensed pressure to transducers associated with the air-data computations and also to direct-reading airspeed indicators. Since the 1970s, large aircraft carry direct-reading, pneumatic instruments at the crew stations to backup the computer-driven instruments. In subsonic flight, a pitot tube's recovery of total pressure is reasonably accurate for typical variations in angle of attack and Mach number; hence, compensations to correct static defect are generally not required. At supersonic speeds, the pressure sensed within the tube is ideally the pressure that develops behind a normal shock wave. Design and calibration of the pitot tube orifice to achieve the desired shock wave is difficult, so measurement errors in total pressure are higher at supersonic speeds and must be compensated.

Total pressure is used to compute calibrated airspeed V_c and Mach number M. Impact pressure q_c is measured with a differential pressure transducer:

$$q_c = p_t - p \tag{8.1}$$

where p_t is the total pressure and p is the static pressure. Impact pressure differs from dynamic pressure q by compressibility factors. Dynamic pressure is often used in aerodynamic calculations because of its simplicity:

$$q = \tfrac{1}{2}\rho V^2 \tag{8.2}$$

where ρ is the air density and V is the true airspeed, often referred to as V_t. There are many forms for the analytical representation of q_c. It is usually shown [4, 24, 30] for subsonic flight as Equation 8.3

$$q_c = p\left[1 + \left(\frac{\gamma - 1}{\gamma}\right)\frac{\rho}{2p}V^2\right]^{\gamma/(\gamma-1)} - p \tag{8.3}$$

and for supersonic flight as Equation 8.4:

TABLE 8.1   Standard atmosphere properties

Atmospheric Constant	Metric Units	English Units
Standard pressure at sea level, p_0	760 mm or 1013 millibars	29.921 in.
Standard temperature at sea level, T_0	15°C, Kelvins	59°F, 518.4 Rankines
Standard mass density at sea level, ρ_0	0.12497 Kg–sec^2/meter^4	0.002378 slugs/ft^3
Standard speed of sound at sea level	661.4748 knots	661.4748 knots
Standard temperature gradient below 10,769 meters (35,332 ft)	0.0065°C/meter	0.003566°F/ft

$$q_c = \left(\frac{1+\gamma}{\gamma}\right)\frac{\rho V^2}{2}\left\{\frac{(\gamma+1)^2/\gamma[(\rho/p)V^2]}{4(\rho/p)V^2 - 2(\gamma-1)}\right\}^{1/(\gamma-1)} - p \qquad (8.4)$$

where γ is the ratio of specific heat of air at constant pressure to specific heat at constant volume; for air, $\gamma = 1.4$.

From Equations 8.2 through 8.4, we can obtain the definitions of the various types of airspeed encountered in aviation practice and theory: *Calibrated airspeed V_c* is defined from Equations 8.3 and 8.4 as the airspeed V that would result from the measured value of q_c if the aircraft were at standard sea level conditions (see Table 8.1). *Indicated airspeed V_i* is identical to V_c, except that it represents the readings of an instrument that has not been correct for pitot-static and other errors. Another term often used by aerodynamicists is *equivalent airspeed V_e*. This is a theoretical parameter that can be computed if desired:

$$V_e = V_t \sqrt{\frac{\rho}{\rho_0}} \qquad (8.5)$$

Equivalent airspeed approaches calibrated airspeed at low Mach numbers. With some algebraic manipulation, the parts of Equations 8.3 and 8.4 that multiply $\rho V^2/2$ can be shown to be functions of Mach number M [2, 30]. The departure of q_c from q can be observed from Figure 8.3, which plots the ratio q_c/q versus Mach number. There are many ways to manipulate the variables in the above equations to derive airspeeds but a set of standard equations is usually used (Section 8.3).

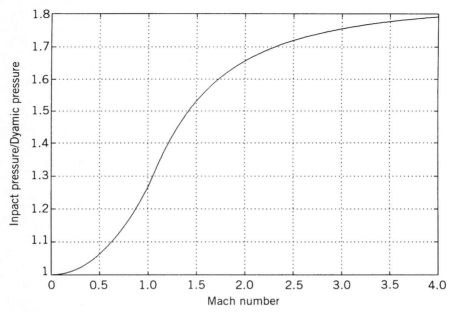

Figure 8.3   Ratio of impact pressure to dynamic pressure versus Mach free stream.

8.2.4   Air Temperature

Outside air temperature, referred to as *static air temperature* T_s and sometimes as *OAT*, is required for the computation of true airspeed. It is also used in the computation of air density, which is required for some types of fire-control aiming solutions. The temperature measured by a thermometer on the exterior of a moving aircraft is higher than the free-stream air temperature because of frictional heating and compression of the air impinging on the thermometer (altered by radiation from the thermometer to the sky and airframe). A typical temperature probe, illustrated in Figure 8.1, is installed to point along the local streamline and compresses the impacting air to zero speed, thus causing a "total," or "stagnation," temperature to exist at the thermometer. To avoid time lags in the temperature measurement, a leakage hole at the rear of the probe allows for a rapid air change. Probes may be mounted on the wing tips, vertical tail, forward fuselage, or other locations where the local Mach number is the same as the free-stream Mach for all expected flight attitudes and speeds. The temperature actually measured by the thermometer is T_m [2, 30].

$$T_m = T_s \left[1 + \frac{\gamma - 1}{2} \, \eta M^2 \right]$$

$$= T_s [1 + 0.2 \eta M^2] \tag{8.6}$$

where absolute temperature is in Kelvins or Rankines, M is the local Mach number, and η is the recovery factor of the probe. The recovery factor accounts for frictional heating, re-radiation, and nonisentropic compression of the air. η is measured empirically, and when independent of Mach number and angle of attack, it does not contribute to any errors in the computation of static air temperature. Temperature probes are available with η values ranging from about 0.7 to greater than 0.99. For high values of η, special shields reduce radiation heat exchanges. Electrical heaters used for deicing must be isolated from the thermometer element. The thermometer is usually a small coil of wire whose resistance varies with temperature. The resistance variation is detected in a bridge circuit, whose excitations and signal processing are located in a signal-conditioning box or the computer. Moisture, water ingestion, and icing are significant error sources that are reduced by a variety of design techniques, including heaters. Substituting numbers into Equation 8.6 shows that the 0.2 M^2 term accounts for 12.8% of measured temperature at $M = 0.8$. A 1.0°C error in temperature will result in a true airspeed error greater than 1.0 knot at typical transport–aircraft flight conditions.

8.2.5 Angle of Attack and Angle of Sideslip

Angle of attack is the angle, in the normally vertical plane of symmetry of the aircraft, at which the relative wind meets an arbitrary longitudinal datum line on the fuselage. Usually that datum line is the aircraft's X-axis, the orthogonal line out the right wing from the center of gravity is the Y-axis, and the Z-axis is downward to complete a right-hand coordinate frame. The angles of attack α and sideslip β can be defined in terms of air velocity components along these axes, V_x, V_y, and V_z.

$$\alpha = \arctan\left(\frac{V_z}{V_x}\right)$$

$$\beta = \arctan\left(\frac{V_y}{V_x}\right) \tag{8.7}$$

The pivoted vane, illustrated in Figure 8.1, measures local flow angle and is the most commonly used method of measuring α or β. Maximum angle-of-attack boundaries define the aircraft's low-speed flight envelope. When α sensors are installed on aircraft, they are usually part of an independent stall warning or stall control system. Since such systems are *flight critical* (also called *safety critical*), redundant sensors are usually installed. Other methods of measuring α and β are discussed in Section 8.5. Engine inlet controls also use α measurements that tend to be safety-critical in supersonic flight.

Sideslip sensors are usually used only in developmental flight test instrumentation. In normal operation, sideslip is approximated by a body-mounted lateral accelerometer (y-axis) and displayed on the pilot's *ball bank* indicator. Automatic flight control systems compute sideslip from inertial measurements and include sideslip control as part of their lateral-directional control loops. Many yaw dampers use β mechanizations based upon the estimated rate of change of β from measurements of lateral acceleration, bank angle, and inertial yaw rate rather than obtaining the desired information from a sideslip sensor. With fast computers, α and β can be estimated from the aircraft force and moment equations and the more commonly available inertial and airspeed measurements. Analytically derived α may supplement a vane sensor for redundancy.

8.2.6   Air-Data Transducers

Measurements of pressures and temperature must be accomplished by transducers that convert the sensed parameters into mechanical motions or electrical signals that are compatible with the various user subsystems. Figure 8.4 summarizes the evolution of transducers, starting with the early pneumatic indicators of altitude and airspeed. In these early instruments, the sensor mechanism and the computation function were combined. The dial whose displacement was proportional to altitude or airspeed (Figure 8.4a) included the computation of altitude or airspeed from the sensing of static pressure and impact pressure. That computation was accomplished by the use of nonlinear springs inherent in the pressure capsules or bellows and associated linkage mechanisms. By the 1950s, avionic systems began to require electrical signals proportional to altitude, airspeed and other air-data parameters. At first, attempts were made to attach electrical transducers such as synchros and potentiometers to these pneumatic instruments but technical problems associated with achieving required sensitivities soon led to servo-driven shafts whose rotation was proportional to the desired air-data parameters. These shaft-driven devices, illustrated by an altitude computing instrument in Figure 8.4b and a Mach computing instrument in Figure 8.4c, combined the sensing and the computation. These illustrations represent the force-balance sensors which dominated central air-data computers in the 1960s and early 1970s. More details of the physical mechanization of such sensors are given in the first edition of this book [27].

Figure 8.4d shows a typical digital computation of altitude using a class of pressure sensor that is usually referred to as a digital sensor. In reality, it is an analog sensor whose analog of pressure is frequency rather than displacement or voltage. Frequency can be encoded into a numerical value with greater precision than voltage. The introduction of digital air-data computers in the early 1970s produced a major improvement in accuracy, reliability, size, and weight. Accuracy improved because sensors could store precise calibration data, which remained with the sensing element in the form of a read-only memory (ROM) chip. Digital computation of the required functions is considerably more accurate and repeatable than electromechanical analog computation. Reliability increased because of the elimination of the motor-driven parts and their dif-

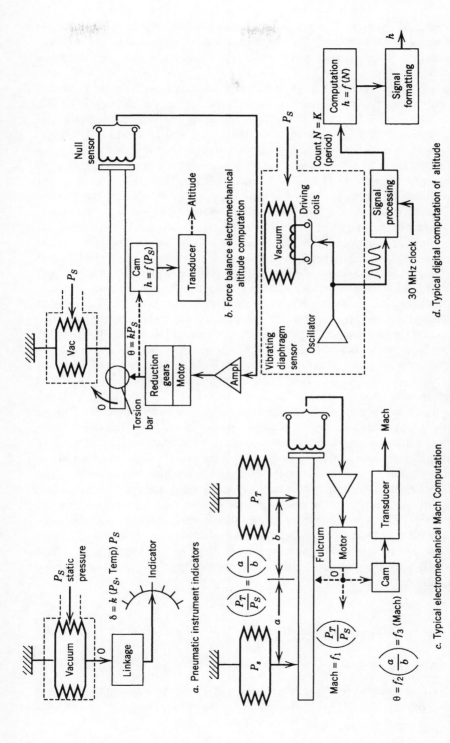

Figure 8.4 Evolution of air-data instrumentation: (*a*) Pneumatic instruments; (*b*) force-balance electromechanical altitude computation; (*c*) electromechanical Mach computation; (*d*) digital computation of altitude.

401

ficult test and calibration procedures. Size/weight advantages were a natural consequence of the elimination of electromechanical hardware, while improvement continues to be realized as computing and input-output chips become more powerful.

Figure 8.4*d* is a conceptual view of a vibrating-diaphragm sensor produced by Sperry Corp. (now Honeywell) for such aircraft as the Boeing 757, 767, 737-300/400, the F-15 and F-16 fighters, and other military, commercial, and general aviation aircraft. Another class of pressure sensor used in digital air-data computers is based on silicon piezo-resistive technology. One of the more difficult problems associated with achieving accuracy has been the temperature sensitivity of silicon. Designs that will work over very wide temperature ranges require balancing techniques in the bridge circuitry that detects the voltage unbalance generated by the silicon transducer. Some solutions use two bridges, one to measure the output of the pressure sensor plus its temperature drift, and a second bridge that does not receive a pressure input but only responds to the temperature changes. Another approach uses the piezo-resistive element as one arm of the bridge circuit and a second silicon element with "identical" temperature drift characteristics in the balance arm of that bridge. Accuracy of any precision pressure sensor is determined largely by the quality of the temperature compensation. Most silicon sensors produce electrical outputs that must be passed through analog-to-digital (A/D) converters. The conversion must be accurate to at least 16 bits (1 part in 65536), which is beyond the capability of 1996 successive-approximation converters. Because high bandwidth is not a requirement for air-data parameters, the slower, more precise dual-slope A/D converters were used.

High performance of all types of sensors over a wide temperature range requires compensation. If the calibration data are stored, then the temperature behavior of the sensing device must be stable and repeatable. This ultimately becomes a materials technology issue. Beryllium-copper capsules, used in the vibrating diaphragm sensor illustrated in Figure 8.5*a*, must be heat treated and "aged" before their properties become sufficiently stable for use in a production sensor. Silicon fabrication processes are the key to meeting cost and performance goals. The trend for future applications seems to favor solid-state sensors because they lend themselves to automated fabrication processes. The size advantage of the solid-state/silicon sensor is illustrated in Figure 8.5*b*, which shows two sensors produced by the same manufacturer to the same accuracy specifications.

8.3 AIR DATA EQUATIONS

8.3.1 Altitude

To determine altitude from static pressure, international standards have been established [5, 11, 32, 47, 48]. The standard atmosphere model gives the rela-

Figure 8.5 (*a*) Vibrating-diaphragm pressure sensor (courtesy, Honeywell, Inc.); (*b*) silicon piezo-resistive pressure sensor (courtesy, Honeywell, Inc.).

tionship between a height Z and properties of the atmosphere as the solution to a differential equation relating the difference in pressure dp between two altitudes, Z and $Z + dZ$, to the weight of that column of air, where ρ is the local air density and g is the acceleration of gravity:

$$dp = -g\rho dZ \qquad (8.8)$$

Solutions of this equation are given in [5, 11, 32]:

$$Z = \frac{p_0}{\rho_0 g \, \log_{10}(e)} \left(\frac{T}{T_0} \right) \log_{10} \left(\frac{p_o}{p} \right) \tag{8.9}$$

where the temperature T varies as a function of altitude, defined by the standard altitude model [5, 11, 30]. The standard atmospheric constants are given in Table 8.1, with pressures defined in terms of the height of a manometer column or in millibars (1.0 mbar = 0.1 kpascal = 0.0145 psi). Substituting the appropriate constants into equation 8.9, gives

$$\log_{10} p = 2.880814 - \frac{Z}{67.4072T}$$

for p in mm, Z in meters, and T in °C. The standard temperature gradient or lapse rate given in Table 8.1 defines a value of T for each altitude below 10,769 meters (35,332 ft), where the standard temperature becomes −55°C = 218 K = −67°F = 392.4 Rankines. Above this altitude, temperature remains constant until about 20 km and then begins to increase before decreasing again as the Earth's atmosphere effectively disappears. These higher-altitude temperature models play a role in hypersonic flight.

When standard temperatures are used for T, the value of Z for each p is called the *pressure altitude*. Air-data computers usually compute pressure altitude from look-up tables derived from Equation 8.9 with appropriate constants incorporated, including the standard function for T versus altitude, or pressure. A combination of the stored table of pressure and density for values of altitude spaced at 10- to 100-ft intervals will give more than adequate results by interpolation. Pressure altitude is the standard parameter used for vertical navigation in controlled airspace. However, density altitude is needed to assess an aircraft's performance margin. By reading or computing static air temperature and pressure, a corrected air density can be found:

$$\rho_{corrected} = \rho_0 \left(\frac{p}{p_o} \right) \left(\frac{T_0}{T} \right) \tag{8.10}$$

The altitude corresponding to that density can be extracted from the look-up table; it is density altitude. An alternative way of computing density altitude is to enter the actual value of static air temperature for T in Equation 8.9 and compute Z using the existing value of static pressure p.

Another type of altitude produced by an air-data system is baro-corrected altitude, computed from pressure altitude by adding an offset. The offset is the deviation from standard sea-level pressure as computed from the difference between the barometric reading at a ground station and the standard barometer at that altitude. Baro correction is entered either in millibars or inches of

mercury. Throughout the world, aircraft flying below 18,000 ft use a local barometric correction, while those flying above 18,000 ft set their altimeters to 29.92
in. of mercury.

8.3.2 Mach Number

The Mach number M is computed from the corrected pitot and static pressure
measurements, P_t and P_s, respectively, where $P_s \equiv p$ and $p_t = p_s + q_c$. The equations that define M can be derived from Equations 8.3 and 8.4 by recognizing
that

$$V_t = C_s M \tag{8.11}$$

where C_s is the speed of sound [2, 30]:

$$C_s = \sqrt{\gamma \left(\frac{p_0}{\rho_0} \right) \left(\frac{T}{T_0} \right)} \tag{8.12}$$

Substituting for V_t into Equations (8.3) and (8.4); with $\gamma = 1.4$, gives the expressions for subsonic and supersonic M that are computed by air-data systems. The
subsonic Mach equation is

$$\frac{P_t}{P_s} = \left(\frac{q_c}{P_s} + 1 \right) = (1 + 0.2M^2)^{3.5} \tag{8.13}$$

which can be rearranged to solve for M explicitly:

$$M = \left[5 \left(\frac{q_c}{P_s} + 1 \right)^{2/7} - 1 \right]^{1/2} \tag{8.14}$$

The supersonic Mach equation is

$$\left(\frac{P_t}{P_s} \right) = \left(\frac{q_c}{P_s} + 1 \right)$$
$$= \frac{(1.2M^2)^{3.5}}{[(7M^2 - 1)/6]^{2.5}} \tag{8.15}$$

This expression cannot easily be solved explicitly for M, although it can be
solved as a polynomial expansion of q_c/P_s. Many air-data computers store

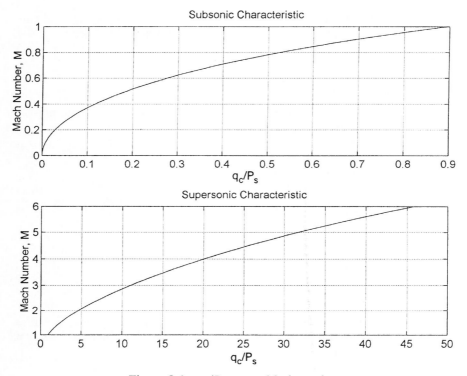

Figure 8.6 q_c/P_s versus Mach number.

tables of q_c/P_s versus Mach obtained from Equations 8.13 and 8.15, using inter-polation routines to extract M from the ratio of corrected pitot and static mea-surements. Figure 8.6 illustrates the Mach versus pressure-ratio relationships for both the subsonic and supersonic regimes, noting that at $M = 1.0$, both equations give the same result.

8.3.3 Calibrated Airspeed

Calibrated airspeed V_c is defined from Equations 8.3 and 8.4 by letting $V = V_c$, $p = p_0$, and $\rho = \rho_0$. This will allow an explicit solution for V_c. Thus, for the subsonic case,

$$V_c = C_{s0} \sqrt{5\left[\left(\frac{q_c}{P_0} + 1\right)^{2/7} - 1\right]} \qquad (8.16)$$

where C_{s0} is the standard day, sea-level speed of sound (Table 8.1).

8.3.4 True Airspeed

True airspeed is usually computed from the relationship $M = V/C_s$. The value of M is computed from Equations 8.13, 8.14, and 8.15. The value of C_s from the computation of static air temperature was given as Equation 8.12, and T_s is

$$T_s = \frac{T_{measured}}{1 + 0.2\eta M^2} \tag{8.17}$$

If T is in Kelvin, then $C_s = 38.96695 \sqrt{T_s}$ knots.

In low-speed flight of aircraft and helicopters, the Mach number is not an appropriate flight parameter; hence alternative methods of computing true airspeed V_t are needed. The usual method is

$$V_t = V_c \frac{f}{f_0} \sqrt{\frac{\rho}{\rho_0}} \tag{8.18}$$

where f and f_0 are compressibility factors containing the same parameters as in Equation 8.3 [2]. The ratio f/f_0 will approximate 1.0 at low speeds. Tables of ρ versus altitude are stored in the air-data computer.

8.3.5 Altitude Rate

Air-data computers usually provide an output identified as *altitude rate*. It is an outgrowth of early pneumatic rate-of-climb indicators, which differentiated a pressure by feeding it to a constriction in the tubing with an air chamber. This analog differentiation suffered from undesirable lags, so these instruments were replaced with "instantaneous vertical speed indicators" that mixed a signal derived from a Z-axis accelerometer with the constricted pneumatic measurement. With the advent of central air-data computers, altitude-rate signals were computed from altitude. Since the 1980s, altitude rate has been computed digitally from vertical acceleration and rate of change of pressure altitude (see Section 7.5.3).

8.4 AIR-DATA SYSTEMS

8.4.1 Accuracy Requirements

The most stringent demands on the accuracy of civil air-data systems come from the need to estimate pressure altitude with sufficient precision to support vertical separation in crowded airspace (Table 8.2). The key to accurate pressure altitude is the calibration of static defect errors. By contrast, calibrated

TABLE 8.2 Typical air-data computer accuracy requirements

Parameter	Accuracy	Resolution
Altitude, h, Z	±10 to ±15 ft. sea level ±20 ft. at 10,000 ft. ±40 ft. at 30,000 ft. ±80 ft. at 50,000 ft. $> \pm100$ ft. at h > 60,000 ft.	0.0365 millibars = 1.0 ft sea level
Baro-corrected altitude, h_o	Same as Altitude +0.25 mb (0.25 mb = 6.8 ft at sea level)	Same as altitude
Total pressure, P_i	±0.68 mb = ±0.02 in. Hg	0.01 mb
Impact pressure, q_c	Same as total pressure	Same as total pressure
Calibrated airspeed, V_c	±5 knots at 60 knots ±2 knots at 100 knots ±1 knot, $V_c > 300$ knots	0.0625 knot, $V_c > 100$
Mach number, M	$\pm0.015M$ at $M = 0.2$ $\pm0.005M$ at $M = 0.6$ $\pm0.003M$ to $0.005M$ at $M = 0.6$ to 0.95, depending upon altitude $\pm0.01M$, or 1.0% above $M = 1.0$	$< 0.0001M$ based upon computer quantization, not transducer sensitivity
True airspeed, V	±4 knots, $V > 100$ knots	< 0.1 knot
Total air temperature, T_t	$\pm0.5\circ$C	$0.25\circ$C
Static air temperature, T_s	$\pm1.0\circ$C	$0.25\circ$C
Angle-of-attack and sideslip	±0.25 deg	< 0.05 deg

airspeed, true airspeed, and Mach number are not used for navigation. Their accuracy is driven by flight performance/stall envelopes in takeoff and landing conditions. The components of true airspeed are often subtracted from those of inertially measured ground velocity to calculate instantaneous wind velocity in cruise conditions, which is useful for the on-board computation of fuel-optimal trajectories in flight-management computers. Mach number is used for programming of stabilizer position for speed stability, defining high-speed performance boundaries such as flutter onset, and for defining optimal cruise paths. Many commercial air-data specifications require outputs for maximum operational velocity (VMO) and maximum operational Mach (MMO) [3]. Table 8.2 summarizes typical accuracy requirements imposed on civil air-data computers since the mid-1970s that were unchanged into the 1990s. Accuracies specified for military air-data systems have, in general, followed the civil specifications, except for supersonic Mach number and airspeed. Those specifications are tailored to the aircraft mission and to probe calibration technologies. Mach number accuracy drops off at very high altitudes. In the supersonic flight regimes of fighters, Mach number accuracies of about 1.0% are typical. The level of performance in Table 8.2 implies calibration of the pitot and static sources. For civil transports, such calibration data are usually available. For military supersonic aircraft, complete probe calibration data are often lacking, especially for high acceleration maneuvers, and when aircraft stores (external pods, missiles and tanks) alter flow characteristics. Extensive flight testing is required to obtain this type of calibration data, and the high cost of such testing often compromises accuracy.

8.4.2   Air-Data Computers

A typical central air-data computer (CADC) is a box containing: (1) the pressure transducers, associated excitation circuitry, and signal-conditioning circuitry, (2) the computer; and (3) the output drivers that are compatible with interfacing subsystems (Chapter 15). The box containing these elements includes fittings that allow the pitot and static pressure lines to connect to the computer's internal pressure transducers. Figure 8.7 illustrates a generic digital CADC. Its inputs are the pitot and static pressure tubes, a temperature probe signal (for the bridge circuit), a barometric setting (either from an analog baro-set potentiometer or from the aircraft's flight management system), and various aircraft configuration discretes. These discretes include flap deployment and external stores status for use in compensating raw pressure and vane readings. They also include "program pin" status, designated connector pins that allow a standard CADC to serve more than one aircraft type. Thus, pin i could recognize that the installation is in aircraft i, thereby activating its own pitot-static correction algorithm, or activating a unique signal interface. Input processing usually involves A/D conversion and packing of discrete signals into logic words, plus implementation of special serial data interfaces such as an RS 232 interface to an external software load and test environment. Input processing must also provide for the circuitry associated with the pressure transducers and vane signal conditioners. They include oscillators, clocks, counters, and precision excitation voltages, plus the interface to the internal computer bus that provides the pathway to the computer's memory. Output processing includes the port to the system data bus. In military applications, this has been a *Mil-Std.* 1553 bus, while commercial applications have used the ARINC 429 broadcast bus standard (although civil multiplex buses such as ARINC 629 have also appeared, Chapter 15). Special outputs such as synchro drivers for electromechanical instruments must also be accommodated by a CADC that is compatible with many aircraft types. One of these outputs, shown on Figure 8.7, is for the air-traffic-control transponder's Mode-C altitude-reply code. This format is needed by early vintage transponders, some of which remain in service despite the arrival of newer generation transponders that can receive the required altitude information over standard avionics busses. The CADC is also commonly used as the source of probe heater controls.

The air-data equations are solved in the processor subassembly, which contains the CPU, memory for the operational flight program (usually stored in electrically erasable programmable read-only memory, EEPROM), data storage memory (usually static random access memory, RAM), and nonvolatile RAM

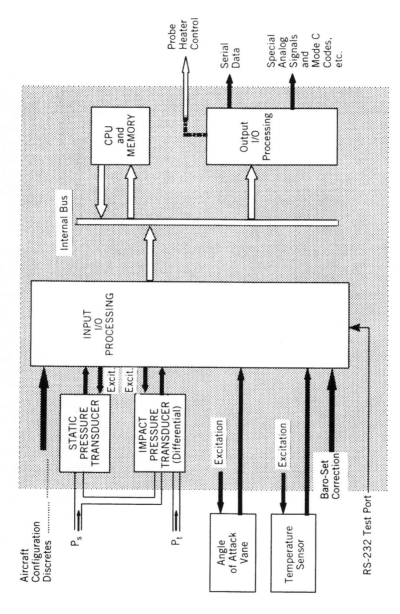

Figure 8.7 Functional block diagram of digital air-data computer.

410

for storage of in-flight maintenance information. Since the computational elements of a CADC represent a small part of the hardware, new system architectures have eliminated the CADC as a separate subsystem (Section 8.4.3).

The air-data software includes a considerable number of built-in test and monitoring algorithms for establishing the validity of all sensor inputs and processing. Other than this "overhead" software, typical computations performed by a generic CADC can be summarized with the following simplified sequence:

- Read probe data from transducers $(P_t, P_s, \alpha, \beta, T_t)$. Read internal transducer temperatures T_{int} and aircraft discretes D.
- Preprocess transducers (e.g., variable-frequency vibrating diaphragm)

   Circuitry measures count N in time t

   Frequency is $N/t = F$; $F_p = P_s$ frequency, $F_q = q_c$ frequency

   P'_s is uncompensated static pressure, $f_1(F_p, T_{\text{int}}, R)$, where R is the stored calibration curve for the specific transducer

   P'_q is uncompensated impact pressure, $f_2(F_q, T_{\text{int}}, R)$

- Correct P'_s, q'_c, α_1 where α_1 = measured α

   $\alpha = \alpha_1 + \Delta\alpha(\alpha_1, D)$

   $P_s = p = P'_s + f_3(\alpha, M, D)$

   $q_c = P'_q + f_4(\alpha, M, D)$

- Compute air-data parameters; initialize without Mach compensations, and allow solution to converge after 1 or 2 iterations. If the M compensations are large, and solution does not converge, compute a raw Mach solution from uncompensated p and q_c and use it in the pressure compensations.

   $M = f_5(q_c/p)$ from Equations 8.13–8.15

   $T_s = f_6(T_t, M)$ from Equation 8.17

   $C_s = f_7(T_s)$ from Equation 8.12

   $V_c = f_8(q_c)$ from Equations 8.3, 8.4, 8.16

   $V = f_9(C_s, M)$ from Equation 8.11

   Pressure altitude is $Z = h = f_{10}(p)$ from Equation 8.9

   Baro-corrected altitude is $h_B = h + f_{11}$ (baro correction)

   Density is $\rho = f_{12}(p, T_s)$ from Equation 8.10

   Density altitude is $f_{13}(\rho, h)$ from Equation 8.9

Computer throughput requirements vary with the choice of table look-up or polynomial equation solutions, iteration rate, and whether multi-rate computation executives are used. In the 1970s, a 100-KOPS, 16-bit microprocessor could perform the CADC functions using 10% to 50% of available time. In the 1990s, all CADC computations could execute in less than 1% of available time. The program and the aircraft-related compensation constants can be stored in less than 20,000 bytes of memory.

8.4.3 Architecture Trends

As microprocessors became smaller and cheaper, it became possible to package them with probes and transducers. The result is a distributed air-data system that replaces the CADC. A key feature is the packaging of signal-processing functions with or adjacent to the probes. Mechanization of such systems may be with "smart probes" whose integral electronics provide the probe and transducer calibrations, plus the digital interface. Figure 8.8 illustrates this concept with dual-redundant probes and vanes. Such an architecture provides corrected pressure, temperature, and angle-of-attack data to a flight control computer that computes altitude, Mach, calibrated, and true airspeed. It can also compute other standard air-data parameters and transmit them to flight management computers, mission computers, or other subsystems.

A major advantage of distributed architecture is the elimination of pneumatic plumbing to the CADC boxes. Reliability and maintenance problems, including water drains, are eliminated. Electrical wiring weighs less than tubing and the electrical transmission of pressure information eliminates lags associated with long lengths of tubing. These lags are negligible with minimum-volume pressure transducers. For example [39], a tube length of 12.70 meters, tube inner diameter of 9.5 mm, and a transducer volume of 100 cm^3 (large for contemporary designs) will produce a lag time-constant of 0.008 sec at an altitude of 10,000 ft (3048 meters), and 0.187 sec at an altitude of 80,000 ft. (24384 meters). Even if the lag were significant, it could be corrected with blended inertial measurements.

The Boeing 777 has a partly distributed air-data architecture. Miniature air-data processing modules are located in the vicinity of the probe on the air-

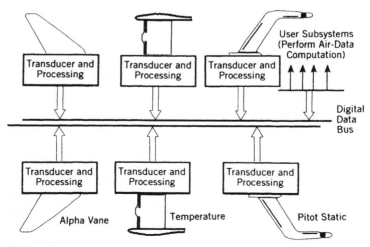

Figure 8.8 Smart-probe architecture for distributed air-data system.

craft structure. The modules contain transducers and signal-processing circuitry. They compensate the transducers, control probe heaters, and interface with the aircraft's data bus. A module's output transmits to the aircraft's integrated inertial/air-data unit [36] via an ARINC 629 data bus (Chapter 15).

8.5 SPECIALTY DESIGNS

8.5.1 Helicopter Air-Data Systems

Helicopter air-data systems differ from their fixed-wing counterparts primarily in the implementation of airspeed measurements at low speeds, including the inference of winds while the aircraft is hovering. Unlike fixed-wing aircraft, where a knowledge of airspeed is essential for safe flight, a helicopter's airspeed is not an essential pilotage quantity, except for certain engine failure conditions where hover capability is lost. Ground velocity from Doppler, inertial, and/or GPS is often used as an approximation. Military helicopters require low-airspeed measurements for fire control with ballistic (unguided) weapons.

The conventional pitot tube and pressure transducer become ineffective as airspeed drops below about 40 knots. At the lower speeds, impact pressure is equal to dynamic pressure q, and the sensitivity of this pressure to a change in airspeed V is obtained by differentiating Equation 8.2.

$$\frac{\partial q}{\partial V} = \rho V \tag{8.20}$$

When V is zero, as in ideal hover with zero wind, the sensitivity is zero. For example, if $V = 1.0$ knot, the change in force on a 1.0×1.0 cm pressure transducer's surface area with a 1.0 knot speed change is

$$\Delta F = \Delta q(\text{Area}) = \rho V(\Delta V)(1.0 \text{ cm}^2) = 3.31 \text{ mg} \qquad (0.007 \text{ lb})$$

Since this transducer must also measure pressures in excess of 100 lb/ft^2 during cruise flight, its dynamic range must be $1.475 * 10^5$, which is beyond the capability of conventional airspeed-sensing instruments. Hence, different technologies are used for low airspeed measurements as explained below.

Static-source errors in helicopters tend to be difficult to compensate because of rotor downwash that differs significantly in and out of ground effect. Fixed-wing aircraft do not compensate their static source errors in ground effect (during landing and takeoff), and neither do helicopters.

In the mid-1970s, the U.S. Army flight tested many devices that were designed to measure low airspeed omnidirectionally [1, 14, 15, 16, 17, 45]. Of them, the rotating anemometer, the vortex counter, and the downwash flow detector are described below. Also, an analytical method that infers the airspeed vector from the position of the helicopter controls and other on-board measurements of aircraft states is discussed.

Rotating Anemometer   This device increases the magnitude of the pressure change caused by a change in airspeed when the aircraft airspeed is near zero. Such systems are called *low omni range airspeed systems*. A variation of this concept embeds airflow sensors and associated pressure transducers within the rotor blade. Blade-mounted sensors have been tested experimentally, and they have been considered for the United States Army's *Comanche* helicopter and Russian attack helicopters. Figure 8.9 is a schematic representation of a rotating anemometer. The dynamic pressure seen at port *A* of Figure 8.9 is

$$q_A = \tfrac{1}{2}\rho[\Omega R + V_x \cos\psi - V_y \sin\psi]^2 \tag{8.20}$$

and at port *B*, the pressure will be

$$q_B = \tfrac{1}{2}\rho[\Omega R + V_x \cos(\psi+\pi) - V_y \sin(\psi+\pi)]^2$$
$$= \tfrac{1}{2}\rho[\Omega R - V_x \cos\psi + V_y \sin\psi]^2 \tag{8.21}$$

where
  Ω is the rotation rate of sensor in rad/sec
  R is the radius arm
  Ψ is the rotation angle of sensor with respect to reference frame
        $= \Omega t$

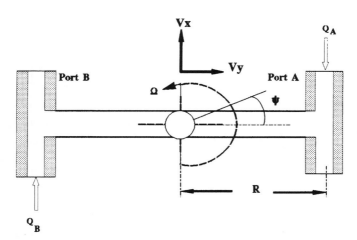

Figure 8.9   Geometric relationships for rotating anemometer sensor.

V_x, V_y are forward and lateral components of airstream velocity with respect to aircraft

Solving for $q_A - q_B$ yields

$$q_A - q_B = 2\rho\Omega R[V_x \cos \Omega t - V_y \sin \Omega t] \tag{8.22}$$

Thus, the pressure difference, $q_A - q_B$, is proportional to the port speed ΩR of the rotating probe. If $V_x = 1.0$ ft/sec and $V_y = 0$, the pressure seen by the transducer at sea level ($\rho = \rho_0$) will be magnified significantly over what would be seen by a conventional pitot tube. For example, if $R = 0.5$ ft, $\Omega = 12$ rev/sec $= 24\pi$ rad/sec, and $V = 1.0$ ft/sec, then for the rotating sensor,

$$2\rho\Omega R * (1.0 \text{ ft/sec}) = 0.179 \text{ lb/ft}^2$$

and for a pitot tube,

$$\tfrac{1}{2} \rho * (1.0 \text{ ft/sec})^2 = 1.188 * 10^{-3} \text{ lb/ft}^2$$

The amplification obtained by rotation is therefore $0.179/0.00119 = 150$.

In addition to obtaining improved sensitivity at low speeds, the rotating probe measures omnidirectional airspeed, including backward velocities. V_x and V_y are extracted from Equation 8.22, which then permits true airspeed V_t and sideslip angle β, to be obtained using the relationships

$$V_t = \sqrt{V_x^2 + V_y^2}$$

$$\beta = \arctan \frac{V_y}{V_x} \tag{8.23}$$

The rotation axis is assumed to be near vertical so, at large bank angles, the V_x and V_y measurements are no longer accurate, and the solution is ignored. In level flight, V_x and V_y are used to estimate the wind vector which is important in fire control equations. Figure 8.10 illustrates an internal view of a rotating anemometer sensor.

Vortex Sensing   The sensor measures vortices shed by fluid flow over a deliberately-inserted obstruction. The frequency of vortices is proportional to the air speed. This method has been used to measure low airspeed in helicopters and in ground-vehicle fire-control systems [1, 17, 46]. A version was used on early models of the AH-64D aircraft.

The theory of the vortex sensor dates back to Von Karman in about 1912

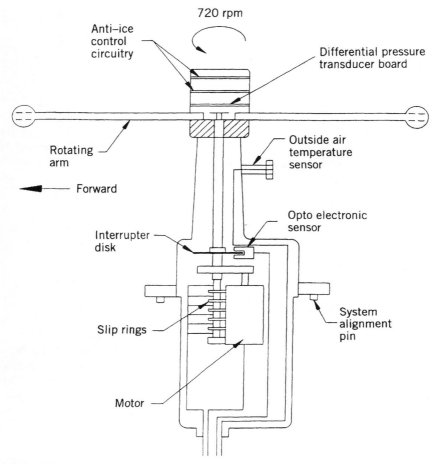

Figure 8.10 Omnidirectional airspeed sensor, rotating anemometer type (courtesy, Pacer Systems, Inc.).

[41, 19]. The frequency F of vortex formation from each side of the obstruction is given by

$$F = S \left(\frac{V}{d} \right) \tag{8.24}$$

where S is the Strouhal number, V is the air velocity, and d is the width of the obstruction. The Strouhal number has been experimentally determined for a variety of obstruction widths and fluid properties [35]. Theory and experimental work have shown that the sensitivity threshold for this type of sensor is about

1.0 knot. One method of measuring vortex frequency directs an ultrasonic beam through the vortex trail. The rotational velocity of the vortices combine vectorially with the sonic ray velocity, causing the sonic rays to be deflected. This causes an amplitude modulation of the received energy at the vortex frequency. To measure the horizontal velocity vector, an orthogonal sensor is required. As in all helicopters, the airspeed sensors should be mounted above the rotor for minimum downwash effects.

Swiveling Pitot Tube Below Rotor   The swiveling pitot tube was developed in the United Kingdom; it is currently in use on the AH-1S, AH-64D, and other attack helicopters. It was tested extensively by the United States Army in the 1970s [15, 16, 45]. A gimballed pitot tube contains a vane arrangement that causes the tube to align with the airflow within the downwash field emanating from the rotor blades. Changes in the airflow field vector are correlated with changes in true airspeed. With appropriate angular pick-offs to measure vane orientation, the true airspeed is estimated using a calibration associated with each aircraft and its rotor system.

The theory of its operation can be seen in Figure 8.11 for low-speed, forward flight. The induced flow velocity, V_i is normal to the rotor tip path plane. V_i sin i is proportional to the thrust component that overcomes aircraft drag and causes a forward velocity. The vector diagram is expressed by

$$V_i \sin(i) + V_H = V \cos \alpha \qquad (8.25)$$

A swiveling probe aligns with the resultant flow velocity **V**, sensing both its magnitude and angle, α and β. The principle of the probe is that V_i sin i is a

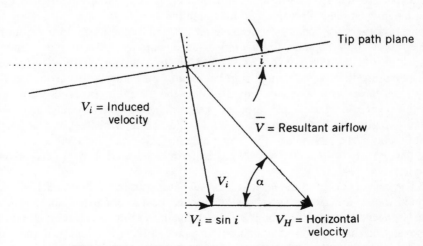

Figure 8.11   Flow field vectors for swiveling probe.

repeatable function of horizonal airspeed, irrespective of thrust, weight, vertical speed, sideslip angle, center of gravity, but varies only with ground proximity. Hence, a radar altimeter measurement is required to accommodate the ground effect. The basic sensing equations are

$$V \cos \alpha = f(V_H)$$
$$V_x = V_H \cos \beta$$
$$V_y = V_H \sin \beta \qquad (8.26)$$

where β is the yaw angle also measured by the swiveling probe. Placing a pitot tube in the downwash flow field avoids the need to measure the low pressures existing near hover since the minimum downwash airflow V_i will always be greater than about 15 knots. Also, aligning the pitot tube with the airflow eliminates alignment errors in both the pitot and static pressures. Figure 8.12 is a cutaway view of the swiveling probe.

Analytical Estimation of True Airspeed   A predictable airspeed vector results from each combination of collective and cyclic control trim position, and pitch and roll attitude. Methods have been developed [6, 31] that estimate a helicopter's airspeed vector from measurements of these quantities. Augmenting these estimates with inertial velocity accommodates dynamic, nontrim conditions. Flight tests have demonstrated accuracies of 4 knots, 2-sigma using this approach [6].

8.5.2   Optical Air-Data Systems

Laser Velocimetry   Nonintrusive optical methods of flow visualization have been part of wind tunnel test instrumentation [7, 25]. Since the 1970s, optical techniques have also emerged as viable air-data systems, motivated by the radar-observability penalty of intrusive probes and by the unsuitability of intrusive probes for hypersonic flight. Optical sensors are located within the vehicles and look out through the local flow into the free stream. Laser velocimeters that measure the Doppler shift from backscatter of naturally occurring aerosol particles have been tested on aircraft since the 1970s in experiments related to the detection of clear air turbulence. Laser radars (Lidars) have also been used experimentally to detect microbursts and severe wind shears. During the 1980s and early 1990s, several laser velocimeters were marketed to calibrate intrusive air-data systems; NASA has experimented with them for hypersonic applications [7].

The basic concept of the laser velocimeter is illustrated in Figure 8.13, a one-dimensional view of the system geometry. In most applications, three orthogonal sensors are used in which the laser beam is split into three component beams. Each is focused at a standoff distance sufficiently removed from the aircraft to be in undisturbed flow (typically several meters away). The lens con-

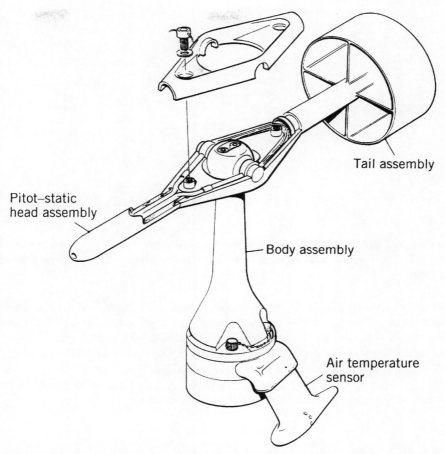

Figure 8.12   Airspeed and direction sensor, swiveling probe (courtesy, GEC Avionics).

figurations that converge the beams, with optimum polarization and geometric characteristics for maximizing backscatter response, are generally proprietary with suppliers of such systems.

The reflected or "backscattered" signal is Doppler shifted from the transmitted frequency by an amount proportional to the relative velocity between the aircraft and the undisturbed atmosphere. Backscattered signals are mixed with the transmitted signals using interferometers. Test results show accuracies of one knot or better at altitudes where particle (aerosol) density is adequate. Aerosol densities and particle sizes vary with altitude, time, and volcanic eruptions. Testing has shown that there is adequate aerosol density up to about 10,000 meters. Blending the inertial velocity vector with an optically derived true airspeed vector allows operation at higher altitudes.

In 1996, laser velocimeters met the civil and military eye-safety standards,

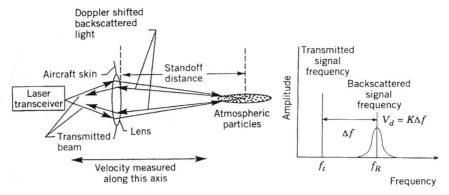

Figure 8.13   Laser Doppler velocimeter.

although there may be some question regarding the intensity at the focused region. The trend is toward improved signal processing and lower-power laser beams.

Particle Time-of-Flight Method   Another laser-based airspeed-measurement technique estimates the time of flight required for aerosol particles to traverse the distance between two laser beams, Figure 8.14. Two sheets of laser light, separated by a distance *d* of a few centimeters, are transmitted through a window in the aircraft's skin. The light sheets are in the YZ-plane, so they measure V_x. Airborne particles generate signals in each of the two detectors that can be timed or correlated; hence true airspeed can be determined.

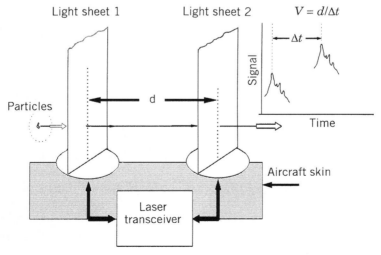

Figure 8.14   Particle time-of-flight method of measuring airspeed.

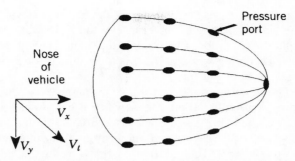

Figure 8.15   Flush-mounted air-data system with distributed pressure ports.

Bogue [7] has reported good performance on an F-104 aircraft. In addition to particle reflections, the detector sees ambient radiation from sky or ground. Thus, selection of laser wave length, optical filters, and signal-processing are difficult design issues. A successful product would have to measure particles penetrating the light sheets at a distance of 5 to 10 meters from the aircraft skin. Figure 8.15 shows one axis.

Altitude Measurement   In conventional air-data systems, altitude and air density are derived from the static pressure measurement. Optical air-data techniques have not demonstrated any methods that will measure pressure, but there are approaches to measuring density ρ and temperature T from which pressure p can be inferred (Equation 8.10). Bogue [7] describes test results with a fluorescence and Rayleigh scatter sensor. In 1996, performance was not adequate for measuring pressure altitude.

8.5.3   Hypersonic Air Data

Air-data measurements in hypersonic flight do not provide primary flight control parameters. They usually support aerodynamic research to confirm structural loading and aero-thermal models. In controlling air-breathing engines (scramjets) in hypersonic flight, pressure and flow-direction measurements are critical [20]. From the earliest experiences with hypersonic vehicles, it was recognized that conventional air-data measurements would be inadequate, primarily because of the thermal loads and shock wave effects on pitot tubes and vanes. Nonintrusive probes were studied in the early days of space flight [40]. The X-15 research aircraft, which flew to 350K ft and Mach 6.7, used a servoed nose ball and conventional pressure transducers [42, 44, 20].

In the X-15, a spherical ball (Q ball) was located in the forward part of the aircraft's conic nose. The ball was servo controlled with hydraulic actuators to align with the relative wind by nulling differential pressures derived from pairs of lateral and vertical pressure ports on the ball. The ball's rotation measured angles of attack and sideslip. Flush-mounted static pressure ports were located at the side of the conic nose. Performance of the ball was adequate to monitor hypersonic and supersonic flight. Nevertheless, the X-15 was augmented with conventional pitot-static probes for subsonic flight and landing. Mach accuracy was 5% to 10%, becoming worse at the low total pressures existing at its peak altitude.

The Space Shuttle *Orbiter* does not use an air-data system during launch or reentry. It deploys conventional pitot tubes when it has slowed to about Mach 4.0. A variation of the ball nose sensor is used in the Shuttle *Orbiter*'s shuttle entry air-data system (SEADS), where an array of flush pressure orifices are distributed around the nose section [20, 38, 43]. This array includes 14 orifices on the nose cap and 6 aft of the nose cap. Figure 8.15 illustrates the concept of locating multiple pressure ports around an aircraft forebody to extract total pressure and flow direction. This type of configuration is also referred to as a *flush air-data system* (FADS) and has been used at NASA Dryden in high angle-of-attack research flights [7]. In general, the differences in pressure between upper and lower orifices measure angle of attack and the differences between left and right orifices measure sideslip angle. Algorithms that combine all sensors to determine total pressure are refined during wind tunnel testing. In-flight determination of α and β with SEADS has been accurate to about 0.25 degree; P_t has been accurate to 0.5%. However, static pressure remains the largest source of error which results in a Mach number accuracy of about ±5% [20, 38].

8.6   CALIBRATION AND SYSTEM TEST

8.6.1   Ground Calibration

Typical equipment used to calibrate both the sensors and the central air-data computers is shown in Figure 8.16. The elements in this figure are usually part of automatic test equipment (ATE), with the pressure sources packaged in a separate pneumatic test module. Manometers are the most precise way to measure the pressures that are applied to sensors, but most test sets use secondary standards, often a specially calibrated version of the sensor used in the flight equipment ("unit under test"). Pressure measurements with an accuracy of 0.005 in. of H_g (0.127 mm of H_g = 0.17 mbars) are required.

When calibrating a sensor, the valves shown in Figure 8.16 would respond to test software that commands the pneumatic test module to produce a sequence of input pressures. The sensor outputs are fed to a test computer ("monitor of displays and outputs" on the figure), compared against stored values, and recorded. At the factory, the sensor is mounted inside a temperature chamber, and the pressure test sequences are applied for various temperatures. The result is a table of errors versus pressure for each test temperature. These tables are stored in a calibration ROM that is mounted on the transducer or in the computer.

When the unit under test is a complete air-data computer, the ATE applies

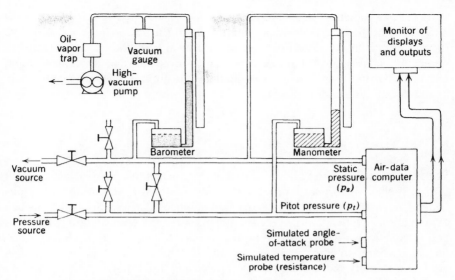

Figure 8.16   Typical test and calibration system.

a sequence of pressure inputs corresponding to discrete values of altitude, airspeed, and Mach. It also sets simulated temperature-probe resistances and angle of attack to specified values. The output of the computer, usually in the form of a serial digital word stream, is read by the ATE. A dynamic test is also run in which a continuously changing pressure sequence is applied corresponding to a flight profile. The performance capability of the pneumatic test module often limits the severity of the test dynamics.

8.6.2   Flight Calibration

Flight calibration is often a limit to the attainable system accuracy. To calibrate for static defect and for local flow corrections to α vanes, test aircraft are usually equipped with booms that extend forward of the aircraft's nose (Section 8.2.1). These booms contain specially calibrated pitot static tubes and flow direction vanes. On-board laser velocimeters have been used to augment this process (Section 8.5.2). With contemporary inertial/differential-GPS systems, precise aircraft ground velocity is also available.

8.6.3   Built-in Test (BIT)

Digital air-data systems can detect nearly 100% of their own failures. At startup, processor tests and memory tests verify computer operation. Continuous BIT checks all interfaces, "wraps around" outputs into inputs to check A/D and D/A conversions, and verifies that the processors are operating properly. BIT monitors a sensor's function such as excitation voltages and oscillator or voltage reasonableness. However, it cannot detect a slow degradation of accuracy resulting from a mechanical deterioration of a sensor (e.g., a leak in the tubing). Since air-data functions are often critical to flight safety, systems are redundant. Thus, transport aircraft use dual or triplex air-data systems. Combat aircraft often use single air-data computers but provide pneumatic altimeters and airspeed indicators as backup.

8.7   FUTURE TRENDS

With the growing popularity of distributed architectures in which redundant "smart probes" incorporate processors, the CADC is disappearing and flight-control, navigation and flight-management computers are executing the air-data computations. Optical air data measurements offer attractive solutions for difficult applications where intrusive probes are precluded. Micro-machined transducers are likely to come into widespread use. The size and cost of electronics are being steadily reduced by new computer and input-output chips.

PROBLEMS

8.1. **(a)** At an altitude of 20,000 meters (65,617 ft), the static pressure measurement is 41.41 mm of Hg, What is the static temperature?

(b) If Mach number is 4.0 at this altitude, what is the impact pressure q_c in mm of Hg?

(c) Under the above conditions, what is the true airspeed in knots?

Ans.: (a) 235 K; (b) 831 mm Hg; (c) 2388 knots.

8.2. **(a)** If an aircraft is flying at Mach = 0.8, what is the impact pressure to static pressure ratio q_c/p_s?

(b) If the aircraft has an uncorrected static source error of ±5% and a pitot tube error of ±5%, what is the range of computed Mach numbers?

Ans.: (a) q_c/p_s = 0.524; (b) M = 0.766 to 0.835.

8.3. **(a)** A digital autopilot is to maintain altitude to an accuracy of 0.5 meters. Its spec calls for a resolution of 5.0 cm. The aircraft has an altitude ceiling of 20,000 meters. To meet the specification, how many bits are required to encode the altitude measurement?

(b) To meet the 5-cm resolution, what pressure measurement sensitivity will be required for the static pressure transducer?

(c) What is wrong with this specification?

Ans.: (*a*) 19 *bits*; (*b*) 301.24 ∗ 10⁻⁶ *mm* Hg = 5.822 ∗ 10⁻⁶ *psi*; (*c*) *the dynamic range is marginal for static pressure sensors and the ability to sense 5.0-cm altitude changes at 20,000 meters is beyond the capability of the best devices. The altitude-control algorithm should use blended inertial information to compensate for the limited pressure resolution.*

8.4. At what airspeed will a conventional pitot tube become more sensitive than a rotating anemometer type of airspeed sensor and at what speed will the pressure at the pitot tube transducer exceed the pressure measured by a rotating anemometer?

Ans.: When $V > 2\Omega R$, the pitot tube sensitivity will be greater, while when $V > 4\Omega R$, the pitot tube pressure will be greater. (For Ω = 12 rev/sec and R = 6 in., V = 150.8 ft/sec = 89.3 knots for equal pressures.)

9 Attitude and Heading References

9.1 INTRODUCTION

A heading reference is required for steering and navigation. It may be as simple as a gravity-leveled magnetic compass or as elaborate as an inertial navigator (Chapter 7). An aircraft also requires some form of attitude reference. In the simplest case, it may be the visible horizon, but, if the aircraft is to be flown in poor weather, then an instrument must substitute for the visible horizon.

An automatic pilot (flight-control system) requires measurements of body rates and attitude. Attitude and rate instruments stabilize other avionic sensors such as Doppler radars (Chapter 10), navigational radars (Chapter 11), and weapon delivery systems.

In the late 1990s the trend was toward obtaining attitude and heading information from inertial navigators including low-cost inertial systems called *Attitude and Heading Reference Systems* (AHRS). Rate measurements (for crew-display or flight-control computers) were from separate instruments whose bandwidth was wide enough for flight control (high bandwidth increases the apparent bias and raises the sensitivity to vibration, so the unit might not also be suitable for navigation).

Cockpit displays in inexpensive aircraft incorporate self-contained vertical and directional gyroscopes that are viewed directly by the crew. In complex aircraft, the cockpit displays of attitude are driven from remotely located sensors and are displayed on "glass" instruments, Figure 9.1. The vertical situation display, Figure 9.1*a*, can emulate a mechanically servoed "8 ball." It is driven by the level-axis outputs of an inertial navigator or AHRS. The horizontal situation display can emulate a radio-magnetic compass or, as shown in Figure 9.1*b*, can combine heading and weather-radar data (military aircraft can superimpose target data). It is driven by the azimuth output of an inertial navigator or AHRS and by the magnetic compass. Complex aircraft usually carry at least one set of self-contained vertical and directional gyroscopes for emergencies [15].

426

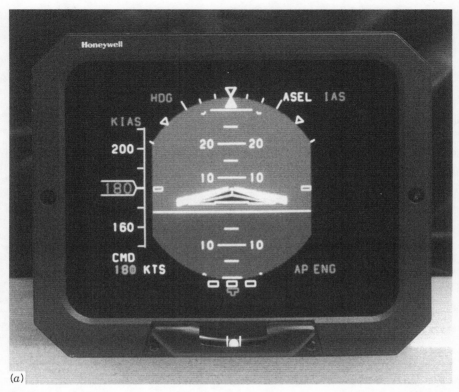

(a)

Figure 9.1   Electronic displays (*a*) Vertical Situation Display (VSI), courtesy of Honeywell; (*b*) Horizontal Situation Display (HSI), courtesy of Smiths Industries Aerospace.

9.2   BASIC INSTRUMENTS

9.2.1   Gyroscopes

All practical attitude references require gyroscopes. As discussed in Chapter 7.3.4, the mechanical gyroscope is a spinning wheel (a source of angular momentum) that will retain its direction in inertial space in the absence of applied torques. When a torque is applied perpendicular to the spin axis, deliberately or accidentally, the gyroscope will precess relative to inertial space. A gyroscope therefore remembers the direction of a line in space if the disturbing torques are kept negligibly small. At the same time the reference direction of the line can be changed through the application of accurately controlled torque. The product of precession rate in radians per second and angular momentum in gram-centimeters squared per second is the applied torque in dyne-centimeters; the precession is about an axis at right angles to that of the applied torque [9].

    The gyroscopes used in heading- and attitude-reference systems are similar

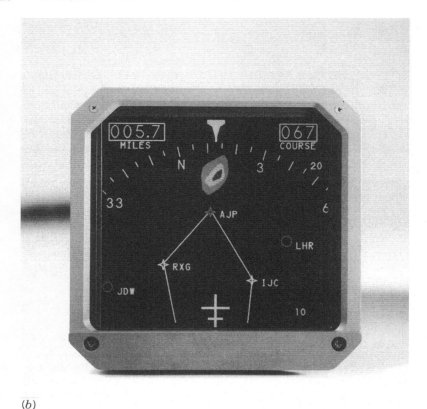

(b)

Figure 9.1   (*Continued*)

to the mechanical gyros once used in inertial systems except that their cost is far lower and their drift rates are much larger. Two-degree-of-freedom gyroscopes (sometimes called *free gyros*) and rate gyros are widely used, as discussed in Section 7.3.4.

9.2.2   Gravity Sensors

Many devices have been used over the years for the erection of gyroscopic vertical references [9]. Experience with reliability, accuracy, and cost has eliminated most of them. As discussed in Section 7.3.1, gravity sensors and accelerometers are synonymous; the former term refers to a low-accuracy instrument for vertical references.

Simple pendulums with electromagnetic pickoffs have been satisfactorily used. Damping can be provided by immersing the pendulum in a viscous liquid, and mechanical stops can be used to limit the deflection. Because the viscosity of liquids decreases with temperature, viscous-liquid damping requires temperature control. Alternatively, the damping can be set based on good leveling-loop response at minimum temperature, resulting in increased error due to damping at maximum temperature.

The simplest gravity sensors in vertical references are electrolytic levels. The electrolytic level contains a conductive liquid in which three electrodes are immersed. When level, the bubble covers two of the electrodes equally to result in equal electrical resistances to the third. A tilt moves the bubble so that it covers one electrode more and the other less, thus unbalancing the resistances. The use of conventional ac bridge circuits provides an electrical output that is nearly proportional to tilt, for small angles.

The sensitivity of the electrolytic level varies with the curvature of the tube; the linear range, with the length; and the damping, with the viscosity of the liquid. Despite some problems related to consistent manufacture, the electrolytic level has been a popular gravity sensor for simple heading- and attitude-reference systems. The electrolytes are chosen with regard to wetting, resistivity, corrosion, outgassing, freezing, and precipitation of solids. Electrolytic levels are commonly linear within 1% to 10% of full scale over their operating ranges and have null stability on the order of 1% of the maximum output. They can have ranges from a few minutes of arc to about 15 degrees. The higher-range devices are used in acceleration-corrected vertical references. In many instances, the electrolytic level feeds power directly to a two-phase gyroscope torquer, although in others an amplifier is interposed. Their small size and weight make them suitable for attachment directly to gyroscopes.

The late 1990s may see the introduction of self-contained micro-machined vertical gyros for direct cockpit diplay. When combined low-cost gyros and accelerometers become available in the twenty-first century, the cockpit display may serve as the attitude reference for the entire aircraft.

9.3 VERTICAL REFERENCES

The basic vertical reference in all forms of heading and attitude devices is the Earth's gravitational field, Section 2.2. The direction of the Earth's gravity vector can be sensed with great accuracy on a stationary platform by a simple pendulum (plumb line), spirit level, or accelerometer. However, when the platform moves, all of these devices indicate the vector sum of vehicle acceleration and local gravity. The angle δ between the true and apparent vertical is

$$\tan \delta = \frac{a_H}{g + a_V} \tag{9.1}$$

where a_H and a_V are the horizontal and vertical accelerations of the aircraft and g is the acceleration due to gravity (1 g = 32.2 ft/sec^2), Figure 9.2.

In high-performance aircraft, sustained a_H can be 5 g; therefore, the direc-

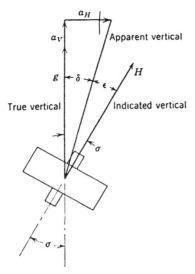

Figure 9.2   Geometry of vertical determination.

tion of the apparent vertical can be arctan $5g/1g = 79$ deg away from the true vertical. Hence, the major problem to be solved in all vertical-reference devices is the determination, to adequate accuracy, of the direction of the gravity vector in the presence of horizontal acceleration.

There are three basic approaches, depending on the desired accuracy and cost:

1. The simplest approach depends on the assumption that, for a minute or more, the average acceleration will be small; therefore, any technique that permits the filtering out of high-frequency components in the angle δ will allow a reasonable determination of the direction of g.

2. The vertical reference can be partially corrected for known maneuvers. The simplest corrections characterize a vertical gyro. More complete corrections characterize a 1970s-era AHRS, as discussed in Section 7.7.4.

3. The vertical indicating system can be a complete inertial navigator (See Chapter 7) that indicates the vertical and heading anywhere on Earth, in the presence of any maneuvers. 1990s-era AHRS can be mechanized as complete inertial navigators.

The first approach is usable where an accuracy of about one degree of arc is satisfactory: the second, for applications where an accuracy of a few tenths of a degree is satisfactory; and the third, where the greatest possible accuracy is needed. The first two approaches are discussed in the following paragraphs; the third is discussed in Section 7.7.4.

9.3.1 The Averaging Vertical Reference

As stated in the preceding section, the simplest approach to the determination of the true vertical is to time-average the direction of the apparent vertical (vector sum of gravity and vehicle accleration). Conceptually, this could be done with a gyroscope having a high angular momentum and a small pendulum, heavily damped to the gyroscope, with the indicated vertical being the direction of the spin axis.

A gravity sensor (Section 9.2.2) measures the angular deviation between the spin axis of the gyroscope and the apparent vertical (Figures 9.2 and 9.3). A torque proportional to the deviation is applied to the gyroscope, precessing it toward the apparent vertical.

The response of this simple vertical gyroscope in one axis can be written as

$$\tau_1 \dot{\epsilon} + \epsilon = \tau_1 \dot{\theta} + \tau_1 \omega_D \qquad (9.2)$$

where

τ_1 is the time constant of the leveling loop ($\tau_1 = H/k$, where H = angular momentum of the gyroscope, in gram-cm^2/sec, and k = sensitivity of the applied torque to tilt, in dyne-cm/rad).

ϵ is the deviation angle between the gyroscope spin axis (indicated vertical) and the apparent vertical.

$\dot{\theta}$ is the rate of rotation in inertial coordinates of the apparent vertical

ω_D is the drift rate of the gyroscope

Let σ be the angle from the gyroscope spin axis to the true vertical. Then, for small angles,

$$\sigma = \epsilon + \delta = \epsilon + \frac{a_H}{g} \qquad (9.3)$$

Thus

$$\tau_1 \dot{\sigma} + \sigma = \frac{a_H}{g} - \tau_1 \dot{\theta} + \tau_1 \omega_D \qquad (9.4)$$

This equation shows that error in the indication of local vertical exhibits first-order response [12] to the forcing functions of horizontal acceleration a_H, gyro drift ω_D, and rotation of the local vertical in space, $\dot{\theta}$. The errors caused by the forcing functions can be improved by introducing either of two nonlinearities:

1. *Erection cutoff.* When a large deviation ϵ_D is sensed, the precession of the gyroscope is interrupted.

2. *Precession rate limiting.* The precession rate is proportional to ϵ until a large value ϵ_E is reached, after which the precession rate is limited.

This simple approach to determining the direction of the vertical is adequate for many purposes; however, it has a number of faults that limit the achievable accuracy:

1. A standoff constant error, $\sigma_{ss} = \tau_1(\omega_D - \dot{\theta})$, exists. Since the direction of the local vertical changes as a result of both the rotation of the Earth and the aircraft's own motion around the Earth, there is a constant error due to each of these. To keep the error small, the value of τ_1 must be small but this leads to poor smoothing of the acceleration errors. Usually the value of τ_1 will be on the order of one minute, and the error will be on the order of 0.5 deg.

2. Asymmetry in the gravity-sensor response (for positive versus negative tilt) results in rectification errors in the presence of oscillatory inputs. This can be especially serious if the positive and negative precession-rate limits are different.

3. If precession-rate limiting is used, the gyroscope will be precessed away from the true vertical at the limit rate during the entire time of a steady acceleration. If erection cutoff is used, an error will accumulate—at a rate equal to gyroscope drift plus the rate of rotation of the local vertical—as long as the acceleration is large enough to cause erection cutoff.

4. When erection cutoff is used, there is need for a special mode of operation that provides for initial leveling, because no precession rate would exist in the presence of a large initial tilt.

In some vertical-reference devices the gravity sensor is simply a switch (e.g., electrolytic level; Section 9.2.2) that applies a modest positive-, negative-, or zero-precession torque depending upon the deviation. Such a reference has a standoff error dependent upon the switch's dead zone and gives poor acceleration-error averaging. Significant improvement can be obtained in the averaging vertical reference by using a higher-order leveling loop. One form of second-order leveling system is obtained by first-order filtering of the gravity-sensor signal. This is most commonly done by providing mechanical damping between the gravity sensor and the vertical gyroscope. The response equation can be written

$$\tau_1\tau_2\ddot{\sigma} + \tau_1\dot{\sigma} + \sigma = -\tau_1\tau_2\ddot{\theta} - \tau_1\dot{\theta} + \frac{a_H}{g} + \tau_1\tau_2\dot{\omega}_D + \tau_1\omega_D \qquad (9.5)$$

where τ_1 is again the time constant of the leveling loop (without gravity-sensor damping) and τ_2 is the response time constant of the damped gravity sensor. There is no improvement in the standoff error [$\sigma_{ss} = \tau_1(\omega_D - \dot{\theta})$], but the filtering of high-frequency ($\omega > 1/\sqrt{\tau_1\tau_2}$) acceleration errors is improved. Practically, however, this added filtering would permit the use of a smaller value of τ_1, with a consequent improvement in standoff error.

A second approach to second-order leveling is to include the integral of the gravity-sensor error signal. The sensor signal and its time integral are summed to determine the gyroscope precession torque. The response of this loop can be written

$$\tau_1\dot{\epsilon} = -\left(\epsilon + \frac{1}{\tau_3}\int \epsilon\, dt\right) - \tau_1\dot{\theta} + \tau_1\omega_D \tag{9.6}$$

or

$$\tau_1\tau_3\ddot{\sigma} + \tau_3\dot{\sigma} + \sigma = -\tau_1\tau_3\ddot{\theta} + \tau_3\frac{\dot{a}_H}{g} + \frac{a_H}{g} + \tau_1\tau_3\dot{\omega}_D \tag{9.7}$$

In this case τ_3 should be large in comparison with τ_1 to avoid amplification near the resonant frequency ($\omega_N = 1/\sqrt{\tau_1\tau_3}$). Error integration eliminates the standoff errors due to the constant rotational rate of the gravity vertical $\dot{\theta}$ and constant gyro drift ω_D. There is no improvement in acceleration-error filtering except that standoff-error elimination allows the use of a larger value of τ_1 for improved filtering. The time constant cannot be made too long because maneuvering and turbulence may not permit steady-state error conditions to be reached.

A third-order leveling system results if both damping of the gravity sensor and error integration are used. Such a system provides good high-frequency filtering and eliminates standoff error; it can be particularly useful when the random drift of the gyroscope is high. It is necessary to provide relatively high leveling gain in the mid-frequency region.

9.3.2 Rate Compensations

As was shown above, there is a standoff error in the averaging vertical reference because of rotation of the local vertical, $\dot{\theta}$. An alternative to the shaping of the leveling-loop response (to attenuate or eliminate this error) is to apply a computed torque proportional to $\dot{\theta}$ (Figure 9.3). This requires a computed precession-rate correction V_E/R_E around the gyroscope axis normal to the ground track and $\Omega_E\cos\Phi$ around the North–South gyroscope axis.

Horizontal Earth-rate, $\Omega_E\cos\Phi$, can be set manually, based on estimated latitude. Vehicle speed is usually available from air data (Chapter 8) or Doppler radar (Chapter 10). Thus, a small standoff error will result from inexact knowledge of Earth rate and vehicle speed. The more complex the compensations, the more closely the vertical references approach an AHRS or inertial navigator.

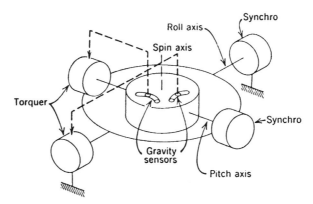

Figure 9.3   Elementary vertical gyroscope.

9.3.3   Acceleration Corrections

At best the vertical gyroscope, as discussed in Section 9.3.2, exhibits an error equal to the average horizontal acceleration (over some smoothing time) divided by the acceleration due to gravity. Even when an aircraft is being flown straight and level, the small rate of turn will usually give substantial acceleration error. As an illustration, consider a craft traveling at a velocity V at a constant latitude Φ (due east or west). The resultant error in the vertical as indicated by a simple averaging reference will be

$$\sigma = \frac{V^2}{gR_E} \tan \Phi \quad \text{rad} \tag{9.8}$$

If $V = 1000$ ft/sec (593 knots) and the latitude is 70 deg, the steady-state error in the indicated vertical is 0.20 deg.

Much greater errors occur when substantial maneuvers are made, unless they are always made with sufficient acceleration to cause the leveling to be cut off. In the latter case, however, the reference system may be almost continuously in the cutoff condition, and large drift errors may accumulate.

Both of these problems are greatly reduced in a system that provides corrections for known accelerations (Figure 9.4) and that, if relatively long periods of maneuvering flight are expected, uses wide-range gravity sensors to assure that leveling will be accomplished during a substantial fraction of the time. Such a system is a 1970s-era AHRS or, if all corrections are made, an inertial navigator.

Equation 7.34 shows that when a vehicle moves at a velocity **V**, relative to the rotating Earth, the inertial acceleration can be divided into three parts:

1. Acceleration relative to the surface of the Earth, $d\mathbf{V}/dt$.

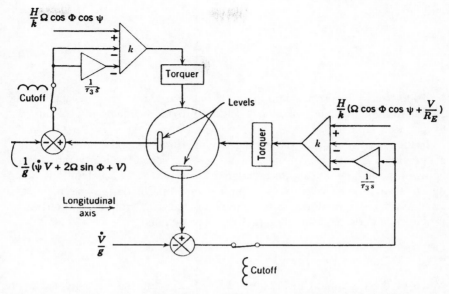

Figure 9.4 Gyro vertical with speed and Earth-rate corrections. R_E is the radius of the Earth, τ_3 is the integration time-constant, V is the horizontal speed.

2. Coriolis acceleration caused by rotation of the Earth, $2\mathbf{\Omega} \times \mathbf{V}$. Its horizontal component is $2\Omega V \sin \Phi$. Its magnitude is 0.14 ft/sec^2 at 1000 ft/sec and 70 deg latitude, resulting in a 0.24 deg standoff error in the indicated vertical.

3. A correction, that depends on the coordinate frame in which the gyroscope is constrained. In a stabilized local-level frame, this correction is $(\mathbf{\omega} - \mathbf{\Omega}) \times \mathbf{V}$, the remainder of the second term in (7.34) when the Coriolis correction is subtracted.

The objective of correcting a vertical reference for acceleration is to build a system whose accuracy and cost are intermediate between those of an inertial navigator and a simple averaging vertical. A cost advantage over the inertial system comes about primarily from the reduced quality of the sensors required.

Longitudinal acceleration relative to the Earth can be obtained by differentiating either the GPS carrier frequency (Chapter 5), the Doppler-radar-determined velocity (Chapter 10) or, with less accuracy, the airspeed data (Chapter 8). Lateral acceleration relative to the Earth can be computed as the product of forward speed and rate of turn, $\dot{\psi}V$. GPS, Doppler radar and airspeed are sufficiently accurate to calculate these corrections. The real problems in computing lateral acceleration lie in determining the rate of turn to adequate accuracy.

An error $\Delta\dot{\psi}$ in knowledge of the rate of turn will result in an error in computed acceleration ($\Delta a = V\Delta\dot{\psi}$); the resultant error in the vertical indication will be

$$\sigma = \frac{\Delta a}{g} = \frac{V \Delta \dot{\psi}}{g} \tag{9.9}$$

At 1000 ft/sec, the error would be 0.008 deg per deg/hr error in heading rate. Noninertial-quality rate gyroscopes seldom provide adequate rate of turn data to lead to a satisfactory correction.

In the absence of accurate heading-rate measurements, the heading angle of a directional gyroscope can be used to resolve vehicle velocity into components in stable coordinates; differentiating them will yield acceleration components. Resolution of velocity into components is particularly convenient in an attitude-reference platform, where the computed acceleration components are in the same coordinates as the gravity-sensor outputs.

9.3.4   Maneuver Errors

Much of the discussion in Sections 9.3.1, 9.3.2, and 9.3.3 is directed at the problem of reducing errors in the vertical gyroscope caused by vehicle accelerations acting on the gravity sensors. There is a similar source of error in the leveling of the directional gyroscope (Section 9.4.6) and the remote magnetic sensor (Section 9.4.4) used in slaving many directional gyroscopes to north.

In addition to the maneuver errors discussed above, there are drift errors resulting from vehicle accelerations acting on unbalanced gyroscopes and from the anisoelastic characteristics of the gyroscopes. There is also a gimbal error (Section 9.4.6) in the directional gyroscope that results from vehicle attitude. Finally, there are drift errors that come about in both the directional and vertical gyroscopes when an aircraft-attitude change produces a change in the outer-axis orientation with respect to the gyroscope. Such a change results in a change in the torque (both friction and torquer output) on the outer gimbal axis.

9.4   HEADING REFERENCES

The best airborne heading references are inertial navigators. Less expensive, smaller and less accurate heading references are (1) those that depend on the Earth's magnetic field ("magnetic compass"), (2) those that depend on the use of a gyroscope to retain a preset azimuth ("directional gyroscope"), and those that use subinertial gyroscopes to maintain a three-axis reference (AHRS).

By far the greatest number of stand-alone aircraft heading references depend on the Earth's magnetic field. Many of these magnetic compasses are coupled to directional gyroscopes or AHRS to improve performance. Aircraft directional gyroscopes are sometimes erroneously called "gyrocompasses." True gyrocompasses are pendulous two-degree-of-freedom gyros with a horizontal axis, tuned for 84- to 100-minute periods. They are used only for marine applications where vehicle-speeds are small.

9.4.1   Earth's Magnetic Field

Magnetic fields $\mathbf{B}$ are measured in gauss (volt-sec/cm^2 = weber/cm^2) and tesla (10^4 gauss). In geomagnetism, units of "gamma" (10^{-5} gauss) are used [5]. To first order, the Earth's magnetic field is produced by a bar magnet (*dipole*) oriented 11 degrees away from the spin axis and passing within a few hundred kilometers of its center. In the Northern Hemisphere, the north magnetic pole was at 79° N latitude and 105° W longitude and the south magnetic pole was at 65° S and 138° E at the beginning of 1996 [7]. The field strength varies from 0.3 gauss near the equator to 0.6 gauss near the poles. At any point, the *dip angle* is the depression of $\mathbf{B}$ from the horizontal (zero dip at the magnetic equator and 90 deg at the poles). The *magnetic declination*, also called *magnetic variation* or *magvar*, is the angle between the horizontal component of $\mathbf{B}$ and true north. Navigators use the term "magvar" in order not to confuse *astronomical declination* with *magnetic declination*. In the conterminous United States, magnetic north points 10 deg west of true north on the East Coast to 20 deg east of true north on the West Coast.

More exactly, the magnetic potential function of the Earth is described by a spherical harmonic series [5]. The more fine-grain the detail, the higher the order of the model. National models are produced by the United States, Canada, the United Kingdom, France, and Russia. The U.S. Geological Service uses aircraft and satellite measurements of magnetic field to produce the publicly available twelfth-order World Magnetic Model (WMM with 168 coefficients) of the "main field" (caused by the Earth's core) using an algorithm called GEOMAG [7], all available from the U.S. Geological Survey's National Geomagnetism Information Center in Boulder, Colorado. The model defines the geomagnetic potential function in terms of latitude, longitude, and altitude, from which the three components of $\mathbf{B}$ can be calculated. The real field has local anomalies (e.g., caused by ore bodies, dc power lines, and solar flares) that would require thousands of coefficients and are therefore described in a separate model or in tables. The World Magnetic Model also contains an eighth-order spherical harmonic model of the linear rate of change of the field, based on worldwide (mainly Northern Hemisphere) surface observations and, increasingly, on historical satellite data. The International Association of Geomagnetism and Aeronomy combines the national models and produces two composite global models. One of its worldwide models is the Definitive Geomagnetic Reference Field (DGRF), issued every five years as an archival description of the Earth's field during the past five years [6]. The other worldwide model is the International Geomagnetic Reference Field (IGRF) which predicts the field for the next five years [6].

Navigators are primarily concerned with magvar, which an airborne computer can add to magnetic heading to calculate the direction of true north. Magvar cams were incorporated into the airborne analog computers of the 1950s and 1960s. When airborne digital computers were less powerful, navigation manufacturers processed the spherical harmonic equations of WMM or IGRF on the ground to calculate the magnetic variation in one-degree or two-degree squares that were stored in tables. In the late 1990s they solved polynomial expansions in flight in terms of latitude, longitude, and altitude. An airborne model would be fourth to ninth order (15 to 80 coefficients) and, for 0.5-deg accuracy, would include rates of change of the larger coefficients. The discrepancy between the twelfth-order model and the raw data has a standard deviation of less than one degree. Large local anomalies that exist at an airport may be added from tables to permit precise initial alignment. Navigators are interested in the *dip angle* insofar as it sets a polar-latitude limit to the operation of magnetometers due to a weak horizontal component (typically unusable for measuring magvar above 75 to 80 deg magnetic latitude).

Magvar has a diurnal change (on the order of ±0.1 deg), a long-term change due to migration of the magnetic poles (on the order of 0.1 deg/year), and random changes because of magnetic storms (on the order of ±0.1 deg). The magnitude of the magnetic field changes about 0.5% per year. The DGRF model shows the linear rates of change of the coefficients for the past few years, while the WMM and IGRF models show them for the years of prediction.

9.4.2 Aircraft Magnetic Effects

Aircraft have magnetic fields that add to the Earth's field and are sensed by the airborne magnetometers. The induced error, called *deviation*, can be 10 deg (in a ship or automobile, it can reach 40 deg). The aircraft's magnetic field is caused by its own structure and machinery.

1. *Hard-iron* portions of the aircraft's structure behave like permanent magnets and retain their polarity as the aircraft rotates. Hard-iron effects can be represented by three orthogonal permanent magnets (Chapter 2, Ref. 2, p. 85). They can either be measured and compensated in the computer or three permanent magnets can be placed near the magnetometers to cancel the vehicle's structural magnetism (as was done before the 1980s). A strapdown magnetometer is compensated by constant bias corrections in body axes. A leveled north-pointing compass needle is compensated in the form

$$\Delta\psi = A\sin\psi + B\cos\psi \tag{9.10}$$

where ψ is heading and $\Delta\psi$ is heading error. In most cases, the compensation can be restricted to level flight, so no pitch and roll corrections need be made. The coefficients A and B are measured during "swinging" or by multisensor navigation systems during flight.

2. *Soft-iron* portions of the aircraft's structure are made of high-permeability magnetic material. Soft iron is magnetized by the Earth by induction. As the aircraft rotates, the soft iron remains polarized so that its north-facing poles

always face the north pole. In a leveled compass, soft iron causes a heading error (Chapter 2, Ref. 2, p. 87).

$$\Delta\psi = C \sin 2\psi + D \cos 2\psi \qquad (9.11)$$

The error can be compensated by measuring the coefficients C and D during swinging and correcting the compass analytically. Analytic compensation of a strapdown compass is difficult, because the induced soft-iron field is hard to predict at arbitrary angles. Ships, which are made of soft iron, use 9 soft iron rods to compensate for structural magnetism. Precomputer aircraft contained soft-iron compensators. Fortunately, aircraft are made of aluminum or nonmagnetic composites.

3. Electrical equipment—motors and wiring—produce fixed magnetic fields that rotate with the aircraft and have the same effect as hard iron. The magnetic field depends on the power-on status of the airborne electrical equipment. Hence, full compensation requires a one-time measurement of the field of each electrical device and the storage, in tables, of the heading errors caused by it and its wiring. For example, a 20-amp current in a 1-meter-diameter loop causes a 30-gauss field at 10-cm distance. This is more than 100 times the Earth's field and must not be allowed at the magnetometer's location. Close spacing and twisting of wires reduces their magnetic fields.

Computer programs have been written to model the distortion of the Earth's field caused by the aircraft's own structure and machinery [3].

Accurate determination of the errors induced by the aircraft is difficult and involves rotating the aircraft through a series of headings while error measurements are made (this is known as *swinging* the aircraft; Section 9.4.5). An area that is sufficiently free of magnetic disturbances is required for this purpose. Aircraft swinging normally involves parking at a series of headings, measured by a theodolite, at 45-deg intervals to obtain enough data points to identify the errors with sufficient accuracy. As a substitute for actual swinging of the aircraft, *electrical swinging* techniques have been developed for use with remote magnetic sensors (Section 9.4.5). Aircraft are rarely swung in roll and pitch to measure the induced soft-iron errors.

Soft-iron errors are usually negligible (<0.1 deg) in large aircraft using remote sensors; however, they can be significant in small aircraft. The error can be considered as characteristic of the design, and a standard correction can be made for each aircraft model.

9.4.3 The Magnetic Compass Needle

Magnetic compasses (a form of magnetometer) are frequently the only heading references in small aircraft and are carried as standby references in larger aircraft [15]. The simplest compasses measure the orientation of a nearly free permanent magnet ("needle") attached to a buoyant body floating in a liquid (because the liquid is usually alcohol, it was called a *whiskey compass*). The float is pendulous to keep the magnet horizontal during unaccelerated flight. The floating body carries heading marks that are visible through a window.

A nearly level needle has an azimuth error

$$\Delta\psi = (\text{N–S tilt}) \tan \gamma \qquad (9.12)$$

In an area where the dip angle γ, is 60 deg, the azimuth error is 1.7 times the compass' tilt around the magnetic north–south axis, which can easily be 15 deg. Tilt of the compass around the magnetic east–west axis does not effect the azimuth indication. The needle is usually tuned to oscillate around magnetic north with a period of 5 to 30 seconds in order to average out aircraft motions.

The major deficiencies of the needle compass for aircraft use are (1) the damping action of the liquid causes a significant lag during turns, (2) the pendulous magnet is not level during turns and hence tracks the dipped magnetic field, (3) there is no means of obtaining an electrical output, and (4) the compass must be located where it can be directly read by the crew.

For many years, the first two deficiencies were overcome by using a combination of two separate instruments: the magnetic compass and the directional gyroscope (Section 9.4.6). With this combination, the compass was read for straight flight; however, before entering a turn, the pilot would manually set the directional gyroscope to match the magnetic compass ("caging" the gyroscope to place its spin axis in the aircraft floor plane). During the turn heading was read from the directional gyroscope with neither a lag error nor an error due to turn acceleration. These free directional gyroscopes are satisfactory heading references for approximately 10 minutes of time after caging.

The most common heading reference is a directional gyroscope that is automatically aligned to the direction of the horizontal magnetic field by a remote magnetic sensor, Section 9.4.4.

9.4.4   Magnetometers

More complex magnetometers are remotely located where the vehicle's field is negligible and drive a readout in the cockpit (e.g., a horizontal situation indicator, Figure 9.1*b*). The earliest remote magnetometer was the *flux gate* [4], Figure 9.5. Figure 9.6 shows its principle of operation [4]. A wound toroid is excited at approximately 1 kHz. Two orthogonal sensor coils are read out at the second harmonic (2 kHz). They measure the components B_x and B_y of the Earth's magnetic field in the plane of the toroid. Flux gates were the most common airborne magnetometers, even in the 1990s. Two or more orthogonal flux-gates comprise a 3-axis magnetometer. An array of three orthogonal sensors gives redundant coverage of each axis and can function after any single failure.

Remote flux-gate compasses, whose output is analog, have three-phase coils, the flux magnitude in each coil being proportional to the component of the Earth's magnetic field along the coil axis [4]. The flux in each coil is modu-

Figure 9.5 Magnetic Flux Detector, *Compass Engine*, courtesy of KVH Industries, Inc.

lated as a result of an applied ac field from an excitation coil; this applied field causes saturation of the magnetic core of the coils at twice the frequency of the excitation. The three-phase winding is connected to the stator of a synchro receiver. A typical combination of remote magnetic sensor and receiver synchro will have errors of both one and two cycles per revolution, each of about 0.2 deg magnitude.

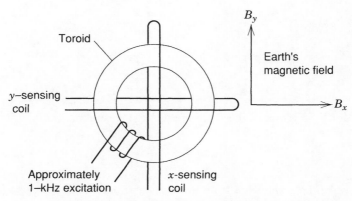

Figure 9.6 Principle of flux gate magnetometer.

The two-cycle error in the combined sensor and receiver—together with any soft-iron error due to the aircraft structure—can be corrected by adding proper impedances in series with the transmission wires. Such a correction is independent of the strength of the Earth's field.

The single-cycle instrument error is usually corrected together with the aircraft hard-iron error. However, this is not exact because the instrument error (in angular terms) is independent of the Earth's horizontal field—whereas the aircraft error is inversely proportional to the Earth's horizontal field (since the permanent magnet part of the aircraft's field is constant). In the 1990s, the trend was to do all corrections in the airborne computer.

Strapdown magnetometers came into use during the 1980s. The body-axis components of the Earth's field are

$$
\begin{bmatrix} H_{forward} \\ H_{starboard} \\ H_{floor} \end{bmatrix} = H \begin{bmatrix} 1 & 0 & 0 \\ 0 & \cos\phi & \sin\phi \\ 0 & -\sin\phi & \cos\phi \end{bmatrix} \begin{bmatrix} \cos\theta & 0 & -\sin\theta \\ 0 & 1 & 0 \\ \sin\theta & 0 & \cos\theta \end{bmatrix}
$$

$$
\cdot \begin{bmatrix} \cos\psi & \sin\psi & 0 \\ -\sin\psi & \cos\psi & 0 \\ 0 & 0 & 1 \end{bmatrix} \begin{bmatrix} \cos\gamma \\ 0 \\ \sin\gamma \end{bmatrix} \tag{9.13}
$$

where ϕ, θ, ψ are the roll, pitch, and azimuth angles of the aircraft, respectively. These three body-axis components can be measured continuously. To solve for heading, Equation 9.13 is inversed:

$$
\begin{bmatrix} \cos\psi & \sin\psi & 0 \\ -\sin\psi & \cos\psi & 0 \\ 0 & 0 & 1 \end{bmatrix} \begin{bmatrix} \cos\gamma \\ 0 \\ \sin\gamma \end{bmatrix} \tag{9.14a}
$$

$$
= \frac{1}{H} \begin{bmatrix} \cos\theta & 0 & -\sin\theta \\ 0 & 1 & 0 \\ \sin\theta & 0 & \cos\theta \end{bmatrix}^{-1} \begin{bmatrix} 1 & 0 & 0 \\ 0 & \cos\phi & \sin\phi \\ 0 & -\sin\phi & \cos\phi \end{bmatrix}^{-1} \begin{bmatrix} H_{forward} \\ H_{starboard} \\ H_{floor} \end{bmatrix}
$$

Therefore:

$$
\begin{bmatrix} \cos\psi\cos\gamma \\ -\sin\psi\cos\gamma \\ \sin\gamma \end{bmatrix} = \begin{bmatrix} A \\ B \\ C \end{bmatrix} \tag{9.14b}
$$

To solve for ψ in terms of the three body-axis measurements,

$$
\tan\psi = -\frac{B}{A} \tag{9.15}
$$

where A and B are calculated from the measured body-axis components of the Earth's field. Hard iron is compensated in body-axes by subtracting 3 constants from the measured field. Soft iron is difficult to compensate, as explained in Section 9.4.3, so the magnetometer is best located far from structural iron.

For small roll and pitch angles, Equation 9.13 simplifies to

$$\frac{H_{forward}}{H} = \cos \gamma \cos \psi - \theta \sin \gamma$$

$$\frac{H_{starboard}}{H} = -\cos \gamma \sin \psi + \phi \sin \gamma$$

$$\frac{H_{floor}}{H} = (\theta \cos \psi + \phi \sin \psi) \cos \gamma + \sin \gamma$$

$$= \text{tilt around E-W axis)} \cos \gamma + \sin \gamma \qquad (9.16)$$

In that case, azimuth is computed easily subject to θ and ϕ errors:

$$\tan (\text{measured azimuth}) = -\frac{H_{starboard}}{H_{forward}} \qquad (9.17)$$

Problem 9.6 shows that the heading error due to the magnetometer being off vertical is (north-south-axis tilt)∗ tan (magnetic dip).

In the 1980s, many small solid-state magnetometers became available with remote readouts for vehicle use [1]. They were based on the Hall, magnetostrictive, and Faraday effects. Superconducting laboratory magnetometers have been too costly and complex for routine aircraft use. The most common heading reference in the 1990s was a directional gyroscope (or AHRS) that is slaved to magnetic north by a remote magnetometer.

9.4.5   Electrical Swinging

Swinging is the process of compensating a compass at various headings. Small aircraft are swung by parking them at a succession of headings in an area free of construction iron. For large aircraft, electrical-swinging techniques have been developed that are as good as mechanical swinging.

The electrical-swinging technique depends on the fact that it is possible to create, at the remote magnetic sensor, a magnetic field of arbitrary magnitude and direction by passing direct currents through the sensor-output coils. It is thus possible to create a magnetic field at the sensor with an effect equivalent to that of rotating the sensor in the Earth's field. This requires good knowledge of the characteristics of the sensor and the strength of the local Earth's field. The required information is obtained by a preliminary test of the sensor out of the aircraft, followed by an electrical swing in the aircraft. Swinging must be done with electrical equipment "on" and "off" to correct for the magnetic fields of motors and wiring.

Multisensor navigation systems can perform continuous swinging by recalibrating the magnetometer in level flight at many headings, as measured by the inertial sensor or a sequence of GPS fixes, for example. Reference [16] describes a Kalman filter for in-flight calibration of a compass. Continuous recalibration should account for the power-on status of electrical machinery and for newly installed magnetic materials.

9.4.6   The Directional Gyroscope

The directional gyroscope is a two degree-of-freedom device whose spin axis is nominally horizontal (Figure 9.7). The gimbal arrangement places the outer axis normal to the floor plane of the aircraft and the inner axis in the floor plane. Aircraft heading is read about the outer axis, and there is no data pickoff on the inner axis. A single electrolytic level is attached to the gyroscope housing to sense departure of the spin axis from a level condition; its output is applied to a torque motor on the outer axis. A torque motor is provided on the inner axis to precess the gyroscope about the outer axis for correction purposes and to slave the gyroscope to alignment with a remote magnetic sensor (if used). Figure 9.8 shows the application of magnetic slaving to a directional gyroscope. Some specific aspects of this design are as follows:

1. Since the inner axis is not used for indication, only drift about the outer axis is of importance. For this reason, the inner-axis-bearing design and adjustment are made to favor inner-axis torque (outer-axis drift).

2. Because mass unbalance along the spin axis results in drift about the outer axis, symmetry of the wheel and motor assembly is important. This is particularly true of instruments that may have to operate for long times without magnetic reference slaving.

3. Inner-axis gimbal bearings are sometimes designed so that the outer race

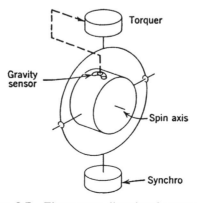

Figure 9.7   Elementary directional gyroscope.

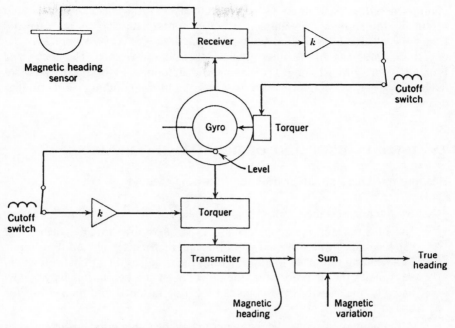

Figure 9.8 Directional gyroscope slaved to a magnetic sensor.

is kept rotating (either continuously or in an oscillatory manner) to reduce friction.

4. Because of the limited angle required and the great importance of inner-axis friction, center-of-rotation contacts are common for electrical connections across the inner axis, although slip rings must be used on the outer axis.

5. Multiple pancake synchros, one of which may be the receiver for a remote magnetic sensor, can be used on the outer axis. If several transmitters are needed, a heading-repeater servo or a single-heading transmitter and isolation amplifier is used.

Directional gyroscopes that are designed for continuous slaving to a remote magnetic sensor can have large free drift. For applications requiring extensive periods of unslaved operation, directional gyroscopes have been designed with a free drift of 0.25 deg/hr to 0.5 deg/hr over a wide range of temperatures.

When sustained operation without magnetic slaving is required, proper account must be taken of the vertical component of the Earth's rate and the aircraft's motion over the Earth. These effects are fully compensated in an inertial navigator and partially compensated in an AHRS.

In nonlevel flight, the outer-gimbal-axis torque may have a component that is in the horizontal plane and normal to the gyroscope spin axis. This torque component will cause drift about the vertical axis. For this reason, it is desirable to cut off the spin-axis leveling loop during turns. The error is at twice the frequency of angular vibration with peak magnitude of $\lambda^2/4$, where λ is the departure, in radians, of the outer-gimbal axis from the vertical. The peak error is about 1° for a tilt of 15 deg. The slaving of a directional gyroscope to a remote magnetic sensor is accomplished by using acceleration smoothing similar to that discussed in Section 9.3.1.

9.5   INITIAL ALIGNMENT OF HEADING REFERENCES

Heading references are aligned by the following methods:

1. By reference to the Earth's magnetic field (Sections 9.4.4 and 9.4.6).
2. Use of a transfer gyroscope. This is a portable, low-drift gyroscope that can be aligned to a surveyed north line and then carried aboard the aircraft as a heading reference to allow setting the installed directional reference. Accuracies on the order of 0.1 deg have been attained in this way. Alternatively, the transfer gyroscope can itself be fastened into the aircraft for use in flight.
3. Alignment relative to the known runway heading prior to and during the takeoff run. Accuracy better than 0.5 deg has been attained by this method.
4. An astrocompass (Chapter 12) can reset the directional gyro. This technique was widely used from 1930 to 1970.
5. Heading can be transferred from an inertial navigator to a directional gyro.

9.6   FUTURE TRENDS

Self-contained vertical gyroscopes, directional gyroscopes, and magnetometers will continue to be used on small aircraft, including target drones and unmanned aerial vehicles. They will continue to be used on larger aircraft as backup instruments for emergency use. In large commercial and military aircraft, the inertial navigation system (Chapter 7) will continue to be the source of attitude and heading information to the electronic displays and flight-management system.

Micro-machined gyros and accelerometers will probably be packaged as self-contained vertical and heading references, resulting in ultra-compact AHRS designed to operate with external sensors (Doppler, GPS, magnetometer). With the growing use of GPS, low-performance inertial navigators will overlap attitude and heading references in civil aircraft and in military aircraft that do not deliver weapons. In many military helicopters, AHRS will continue to be the heading and attitude reference for Doppler navigation radars (Chapter 10). Three-axis strapdown magnetometers will increasingly serve as the absolute heading source for simple AHRS, perhaps backing up GPS' indication of ground track angle. The search for lower-cost, more reliable 3-axis gyros, accelerometers, and magnetometers will continue. The growth of data bases will parallel the expansion of aviation into the Southern Hemisphere.

Higher-order magnetic models, supplemented by tables of local anomalies, will come into use as more magnetic data are collected worldwide. When intermittent-fix com-nav satellites come into use in the twenty-first century, they will motivate the re-introduction of precise inertial navigators.

PROBLEMS

9.1 An aircraft directional gyroscope has an angular momentum of 10 million gram-cm^2/sec, and the wheel weighs 500 grams. What stability of axial positioning of the wheel is required to obtain an acceleration-sensitive drift of 0.25 deg/hr-g.

Ans.: 2.5×10^{-5} cm.

9.2 An aircraft is maneuvering in such a way that its horizontal acceleration is 20 ft/sec^2 and its vertical acceleration is -10 ft/sec^2 (downward). What is the departure of the apparent vertical from the true vertical?

Ans.: 42 deg.

9.3 A vertical gyro is to be constructed using a wheel whose angular momentum is 10^7 gram-cm^2/sec, a gravity sensor with a sensitivity of 0.5 volt/deg, and a coil that provides 5×10^3 dyne-cm of torque per volt applied. What voltage gain is required in an amplifier between the gravity sensor and torquer to yield a 1-min leveling-time constant?

Ans.: 1.12.

9.4 It is desired that the response of a second-order vertical reference to a peak sinusoidal horizontal acceleration of 2 ft/sec^2 at a period of 15 sec not exceed 0.05 deg. It is also desired that the leveling be critically damped. What should the response time constant for the gravity sensor be? What should the leveling-loop time constant be? (HINT: first determine the ratio of the time constants for critical damping.)

Ans.: 0.165 min; 0.66 min.

9.5 Under the flight condition of Problem 9.4, it is desired that the directional gyro slaving error due to the acceleration not exceed 0.1 deg for the worst heading and at a latitude giving a dip angle of 60 deg. What time constant is required in a first-order slaving loop?

Ans.: 2.45 min.

9.6 For small roll and pitch, differentiate Equation 9.17 with respect to roll and pitch to show that:

$$\Delta \psi = (\phi \cos \psi - \theta \sin \psi) \tan \gamma$$

Show that the expression in parentheses is the component of the magnetometer's tilt-angle around the north-south axis.

10 Doppler and Altimeter Radars

10.1 DOPPLER RADARS

10.1.1 Functions and Applications

The primary function of a Doppler radar is to continuously determine the velocity vector of an aircraft with respect to the ground. If the measurement is made in, or has been converted to, an Earth-referenced coordinate frame and resolved about north and east, the velocity components can be integrated into distance traveled from a known point of departure and the aircraft's geodetic present position and course and distance to destination can be calculated. Thus, a Doppler radar can be the primary sensor of a dead reckoning navigation system or one of the sensors in a multisensor system. The velocity is determined by measuring the Doppler shift of microwave signals transmitted from the aircraft in several narrow beams pointed toward the surface at relatively steep angles, backscattered by the surface and received by the Doppler radar receiver.

A Doppler radar has the following advantages over other methods of velocity measurement or dead reckoning navigation:

1. Velocity is measured with respect to the Earth's surface. This is in contrast to air data systems which measure velocity with respect to the air mass and to most terrestrial radio navigation systems in which velocity measurement is based on differencing of successive position measurements.

2. It is self-contained, that is, it requires no ground-based stations or satellite transmitters.

3. The airborne transmitter power requirements are extremely small, which leads to low weight, size and cost of equipment.

4. Its radar beams are narrow and pointed toward the ground at steep angles, which leads to extremely low detectability.

5. It is an all-weather systems, except in extreme conditions of rain.

6. It operates over both land terrain and water (except for completely smooth water surfaces).

7. Its average velocity information is extremely accurate.

8. It is particularly suitable for the measurement of three-dimensional velocity and at low velocities, as required for helicopter navigation and hovering.

9. International agreements are not required, since ground equipment is not needed.

10. Pre-flight alignment and warm-up are not required.

The disadvantages of a Doppler radar are the following:

1. For autonomous dead reckoning navigation, it requires an external airborne source of heading information, such as a gyro-magnetic compass, an attitude-heading reference system (AHRS), or an inertial platform.

2. It requires either internal or external vertical reference information for conversion of its velocity information into an earth referenced coordinate frame; however, this vertical information need not be of high quality.

3. Position information derived from Doppler radar dead reckoning degrades as the distance traveled increases.

4. The instantaneous or short-term velocity information is not as accurate as the average or smoothed velocity. This difference is not significant for general navigation but may be significant for other applications.

5. For over-water operation, accuracy is somewhat degraded due to backscattering characteristics and water motion.

The techniques of Doppler radar velocity measurement and navigation evolved from airborne radar development, notably airborne moving target indication, and from automatic dead reckoning navigation systems using airspeed meters. The earliest applications of Doppler radar were in military aircraft for dead reckoning navigation and weapon delivery. These systems have been installed in thousands of military aircraft of all major nations. Typical examples in the United States are the B-52, F-111, B-1A, E3A, P2V, P3V, S3A, E2A, and many thousands of helicopters. In the 1960s, the world's international commercial airlines began using Doppler radar systems for dead-reckoning navigation, primarily for over-ocean operations. These were interfaced with gyro-magnetic heading references and course line computers. In 1996, inertial and inertial-radio systems had replaced Doppler radars for that application.

Doppler radars have been used for the velocity measurement required for the soft landing of planetary and lunar space vehicles, such as the *Surveyor* and the *Apollo* Lunar Excursion Module (LEM).

In 1996, the most important wide use of Doppler radars was in various types of military helicopters, for such applications as navigation, hovering, sonar dropping, target handover for weapon delivery, and search and rescue. They have also been used on unmanned aerial vehicles (UAVs) and drones, as well as on military fixed-wing aircraft. By 1996, about 40,000 Doppler radars had been produced and deployed on aircraft worldwide.

In many aircraft, Doppler radars are employed in conjunction with inertial platforms, wherein the velocity data from the Doppler radar are used for damping of the inertial navigation system's Schuler oscillations (Sections 7.6.3 and 3.4). The difference in the characteristics of the velocity data from these two sensors, such as the small long-term velocity error of the Doppler radar and the small short-term velocity error of the inertial system, has lead to the desirability of combining the data from these two sensors in some form of optimum estimation filter (Chapter 3). Similarly, in some configurations Doppler radar velocity data are mixed with data from a position sensor, such as the Global Positioning System (GPS) receiver (Sections 10.1.5 and 3.8).

The Doppler radar velocity measurement function has been incorporated into coherent forward looking search and tracking radars for precision velocity update of the aircraft's inertial system; in 1996 this approach was used widely on high performance military aircraft (Section 11.5).

To accomplish dead-reckoning navigation by Doppler radar, a complete Doppler system needs to contain the functions shown in Figure 10.1. The Doppler radar measures the aircraft velocity with respect to its antenna frame coordinates. The heading reference and vertical reference, or a combined attitude and heading reference system (AHRS), or an inertial reference unit (Chapter 7) determines the direction of the aircraft antenna with respect to the horizontal plane and north. The navigation computer then resolves the aircraft velocity components obtained from the Doppler radar about the vertical and true north and continuously integrates the horizontal velocity components into distance traveled from the point of departure. The resulting present position can then be compared with destination coordinates to provide other desired navigational quantities, such as bearing and distance to destination, Figure 10.1.

10.1.2   Doppler Radar Principles and Design Approaches

The Doppler Effect   Operation of a Doppler radar is based on the Doppler effect which was predicted in 1842 by the Austrian scientist Christian Doppler in connection with sound waves and was later found also to be exhibited by electromagnetic waves. The Doppler effect can be described as the change in observed frequency when there is relative motion between a transmitter and a receiver. Furthermore, this change in frequency, called the *Doppler shift*, is directly proportional to the relative speed between transmitter and receiver. In the case of electromagnetic waves (unlike the case of sound waves), it makes no difference in the proportionality relationship whether the transmitter, the receiver, or both, are moving. If the relative velocity of the transmitter and receiver is much smaller than the speed of light (as in the case of aircraft), the Doppler shift is expressed by

$$\nu = \frac{V_R f}{c} = \frac{V_R}{\lambda} \tag{10.1}$$

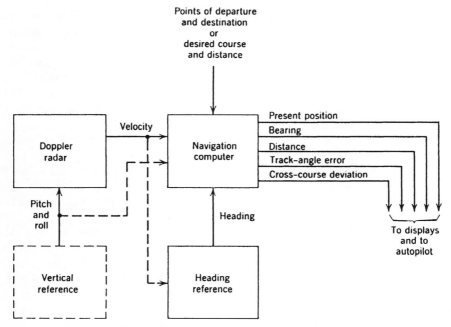

Figure 10.1 Doppler navigation system.

where

ν	is the Doppler shift
f	is the frequency of the transmission
c	is the speed of light
V_R	is the relative velocity between transmitter and receiver
$\lambda = c/f$	is the wavelength of transmission

From Equation 10.1 it is seen that, if the value of λ is known and ν is measured, the relative velocity can be determined.

To measure the aircraft's velocity, a radar transmitter-receiver is mounted on the aircraft and radiates electromagnetic energy toward the Earth's surface by means of several beams, one of which is shown in Figure 10.2. Some of the energy is backscattered by the Earth and is received by the radar receiver on the aircraft. If the aircraft is moving with a total velocity V, the beam measures V_R:

$$V_R = 2V \cos \gamma = 2V\mathbf{b} \qquad (10.2)$$

where γ is the angle between the direction of the velocity vector V and the

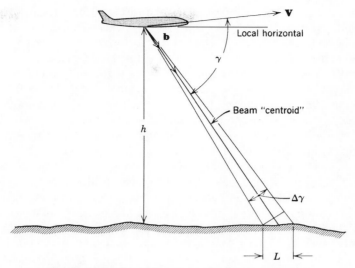

Figure 10.2 Basic Doppler radar beam geometry.

direction of the beam centroid, and **b** is the unit vector along the beam centroid. V_R is the component of relative aircraft velocity along the beam centroid. The factor 2 appears in Equation 10.2, since both the transmitter and the receiver are moving with respect to the Earth, from which the energy is backscattered. When Equation 10.2 is substituted into Equation 10.1, the following expression results:

$$\nu = \frac{2Vf}{c} \cos \gamma = \frac{2V}{\lambda} \cos \gamma \qquad (10.3)$$

Equation 10.3 is the fundamental expression for the measurement of velocity by means of a Doppler radar. It is also the basis for the operation of the synthetic aperture, Doppler beam sharpening, and precision velocity update (PVU) modes of airborne radars (Chapter 11). It states that each Doppler radar beam measures the component along the beam of the aircraft's velocity with respect to the Earth. Two faulty arguments have been advanced in the past which question whether proper operation of Doppler radar is possible—*the smooth Earth paradox* and *the mountain paradox*. The first paradox argues (falsely) that as an aircraft moves parallel to flat terrain at constant altitude, the range to the ground does not change and since there is no rate of closure (range rate) there can be no Doppler shift and Doppler radar operation is not possible at all. The second paradox argues (falsely) that if an aircraft is flying horizontally above upsloping terrain, its range to the ground along the beam continuously decreases and this range rate, with respect to the surface, gives rise to a significant error in velocity measurement. Both arguments are incorrect for essentially the same reason:

the radar backscattering is produced by the discrete and irregular objects on the ground (pebbles, leaves, etc.) and there is indeed relative motion between the aircraft and each of these scatterers. If the surface were perfectly smooth, reflection at the surface would be specular and no reflected energy would reach the aircraft. Hence, if the scattering medium is sufficiently rough to give rise to a signal at the receiver, the signal will exhibit a Doppler shift according to Equation 10.3. Also, there is absolutely no error due to upsloping terrain (*mountain paradox*), since the backscattered signal comes from the individual, discrete, *stationary* objects on the ground and it therefore experiences the correct Doppler shift ([4] [40] [41]).

Since the three orthogonal components of velocity are of interest, a minimum of three noncoplanar beams are required to measure the three components. A beam configuration designed to accomplish this is shown in Figure 10.3. Since such a beam configuration has both forward- and rearward-looking beams, it is called a *Janus* configuration after the Roman god who had the ability to look backward as well as forward. In a Janus system, the Doppler shift obtained with, say, the right-forward beam can be subtracted from that obtained with the right-rearward beam in order to determine the heading velocity component V_H. Since

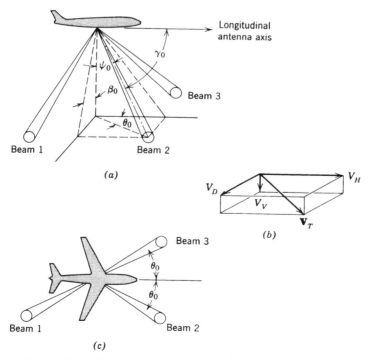

Figure 10.3   Three-beam lambda Doppler radar configuration.

the forward-looking beam (beam 2 in Figure 10.3) gives rise to an increase in frequency (positive Doppler shift), and the rearward-looking beam (beam 1) gives rise to a decrease in frequency (negative Doppler shift), the subtraction process of the two Doppler shifts actually results in an addition process. For a condition of no drift, roll, or pitch, wherein the forward and rearward Doppler shifts are equal, the equation for the total Doppler shift from such a forward-rearward (Janus) pair of beams takes the form $\nu = (4V/\lambda)\cos\gamma$. In typical microwave Doppler radars, the value of this Doppler shift is on the order of 30 Hz per knot of speed.

The configuration shown in Figure 10.3 has been called a *lambda-configuration*, since the plan view of the beams has the form of the Greek letter λ, as seen from Figure 10.3c. The Doppler frequency in each of the beams is proportional to the algebraic sum of the projections of the three orthogonal velocity components along the beam. The mathematical expressions for the computation of the three orthogonal velocity components in aircraft coordinates are obtained from the beam direction cosines between the velocity components and the beam centroids and the addition or subtraction of the beam Doppler frequencies. For this particular (lambda-) configuration the beam Doppler frequencies are given by

$$\nu_1 = \frac{2}{\lambda}\,(-V'_H\cos\gamma_H + V'_D\cos\gamma_D + V'_V\cos\gamma_V)$$

$$\nu_2 = \frac{2}{\lambda}\,(V'_H\cos\gamma_H + V'_D\cos\gamma_D + V'_V\cos\gamma_V)$$

$$\nu_3 = \frac{2}{\lambda}\,(V'_H\cos\gamma_H - V'_D\cos\gamma_D + V'_V\cos\gamma_V)$$

where

$$\cos\gamma_H = \cos\alpha_0\cos\theta_0$$
$$\cos\gamma_D = \cos\alpha_0\sin\theta_0$$
$$\cos\gamma_V = \sin\alpha_0$$

From the above expressions the three orthogonal velocity components are

$$V'_H = \frac{(\nu_2 - \nu_1)\lambda}{4\cos\alpha_0\cos\theta_0} \tag{10.4}$$

$$V'_D = \frac{(\nu_2 - \nu_3)\lambda}{4\cos\alpha_0\sin\theta_0} \tag{10.5}$$

$$V'_V = \frac{(\nu_1 + \nu_3)\lambda}{4 \sin \alpha_0} \tag{10.6}$$

where

V'_H is the along-heading velocity component, in aircraft coordinates

V'_D is the cross-heading (drift) velocity component, in aircraft coordinates

V'_V is the vertical velocity component, in aircraft coordinates

ν_n is the Doppler shift in beam n

α_0 is the depression angle of the beam centroids from the plane of the antenna, assumed equal for the three beams, $\psi_0 = 90° - \alpha_0$ (Figure 10.3)

θ_0 is the azimuth angle of the antenna beams, that is, the acute angle between the projections of the longitudinal axis of the antenna and the antenna beam centroids on a ground plane parallel to the plane of the antenna, assumed equal for the three beams (Figure 10.3); the relationship among $\gamma_0, \phi_0,$ and θ_0 is $\cos \gamma_0 = \sin \phi_0 \cos \theta_0$

$\gamma_H, \gamma_D, \gamma_V$ are the angles between V'_H, V'_D, V'_V and the beam centroids; that is, $\cos \gamma_H, \cos \gamma_D, \cos \gamma,$ are the direction cosines

Although only three beams are required to provide the three components of velocity, most modern Doppler radars employ four beams, because planar array antennas naturally generate four such beams. The Doppler frequency of the fourth beam (ν_4) can be combined with ν_3 to obtain another estimate of V'_H (replace ν_2 and ν_1 in (10.4) and ν_3 and ν_4, respectively). The two estimates of V'_H can then be averaged to obtain a more accurate value of this component. The difference of the two estimates of V'_H should be very small; a large difference indicates that there is an error in the measurement of the Doppler frequency and hence that the Doppler velocity data are suspect and should not be used until the cause of this difference is corrected. This technique is used as part of the BITE (built-in-test equipment) of many Doppler radars. Similarly ν_4 can be combined with ν_1 to form another estimate of V'_D by replacing ν_2 and ν_3 in Equation 10.5 with ν_1 and ν_4, respectively. Finally, ν_4 can be combined with ν_2 in Equation 10.6 (replace ν_1 and ν_3 with ν_2 and ν_4) to form a second estimate of V'_V. In each case the two estimates are averaged to obtain more accurate values of the respective velocity component. The fourth beam is thus redundant, since only three beams are needed to form a complete solution, as shown by Equations 10.4 to 10.6.)

In a fixed-antenna system, after V'_H, V'_D, and V'_V have been obtained, they must then be combined with pitch and roll information in order to generate the aircraft velocity components in Earth coordinates, V_H, V_D, and V_V (as described later in this section). The total velocity vector magnitude V is the resultant of the three orthogonal components (Figure 10.3b). Hence, a Doppler radar with either a three- or four-beam configuration is capable of measuring the three velocity components and their sense of direciton.

A variety of beam configurations have been used for Doppler radars, including Janus (two-way looking) and non-Janus (one-way looking) configurations. The Janus configuration has a very important advantage over a non-Janus configuration, namely a much lower sensitivity of velocity error to knowledge of the vertical attitude of the aircraft. Specifically, the expressions for the velocity error as a function of error in pitch angle for Janus and non-Janus systems, for a condition of no pitch, no drift, and no vertical velocity are as follows:

$$\epsilon_V = \frac{\delta V}{V} = (\tan \gamma_0)\delta P \qquad \text{(non-Janus)} \qquad (10.7)$$

$$\epsilon_V = \frac{\delta V}{V} = 1 - \cos \delta P \qquad \text{(Janus)} \qquad (10.8)$$

where
  ϵ_V is the fractional horizontal velocity error
  δV is the absolute horizontal velocity error
  δP is the error in pitch angle

Based on the conditions cited above, Equations 10.7 and 10.8 are equally applicable to fixed and physically stabilized antenna systems. From Equations 10.7 and 10.8, for a non-Janus system having a γ-angle of 70°, which is a reasonable value, the horizontal velocity error is 4.7% per degree of error in pitch angle, whereas for a Janus system the horizontal velocity error is only 0.014% per degree of error in pitch angle. Because of these considerations, all modern dedicated Doppler radar designs use some form of Janus configuration. However, when Doppler velocity is extracted from forward-looking search and mapping radars, a non-Janus configuration results (Chapter 11.5).

The choice of γ_0 (nominal angle between antenna longitudinal axis and central beam direction) for a typical Doppler system represents a compromise between (1) high sensitivity to velocity (Hertz per knot) and overwater accuracy, which increases with smaller γ_0-angles, and (2) high signal return over water, which increases for larger γ_0-angles. Most equipments use a γ_0 of somewhere between 65° and 80°. The choice of β_0-angle depends on the desired sensitivity to drift (Hertz per degree), which tends to increase with increasing β_0.

There are two different basic types of Doppler radar mechanizations that can be used for drift angle measurement. These are the *fixed-antenna* system, which is used in most modern systems, and the *track-stabilized* (drift-angle-stabilized) antenna system (Figure 10.4). Zero pitch and roll angles are assumed for the example in Figure 10.4. Figure 10.4a depicts a condition of no-drift

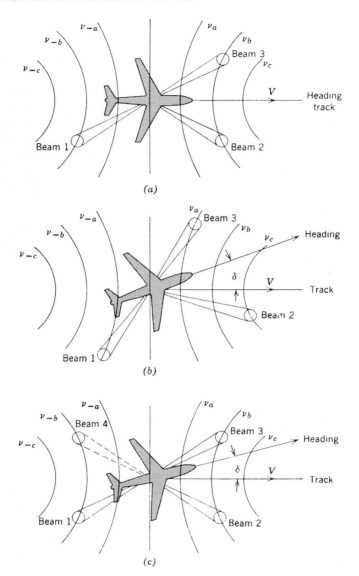

Figure 10.4   Comparison of fixed (heading-stabilized) and track-stabilized Doppler systems.

angle ($\delta = 0$) and no-climb angle so that the velocity vector V is located along the intersection of the local horizontal plane and the local vertical plane through the longitudinal axis of the aircraft. In other words, the aircraft flight is horizontal and the aircraft's ground track (or track) is the same as the aircraft heading. The hyperbolas in Figure 10.4, marked ν_a, ν_b, ν_c, ν_{-a}, etc., are lines of constant

Doppler shift, the positive subscripts representing positive Doppler shifts and the negative subscripts representing negative Doppler shifts. These hyperbolas, called *isodops*, are generated by the intersections of constant Doppler cones with an assumed flat Earth. In Figure 10.4a, it is clear that subtraction of the Doppler shifts from beams 1 and 2 will provide a measure of the along-heading velocity V_H as given by Equation 10.4. Subtraction of the Doppler shifts from beams 2 and 3, as in Equation 10.5, will indicate a zero cross-heading-velocity component V_D or a zero drift angle (since $v_b - v_b = 0$). Similarly, addition of the Doppler shifts of beams 2 and 3 will indicate a vertical velocity of zero. All of the ground intersections of the beams are located on the same equivalent isodops. Under drift conditions, as indicated in Figure 10.4b and 10.4c, the aircraft track direction is no longer coincident with the aircraft heading direction. If the antenna is fixed to the aircraft, a case depicted in Figure 10.4b, the beams will move with the aircraft, and the ground intersections of beams 1, 2, and 3 will be located on different isodops. Thus, subtraction of the Doppler shifts from beams 3 and 2 will indicate a nonzero cross-heading velocity as given by Equation 10.5; and subtraction of the Doppler shifts of beams 1 and 2 will determine the along-heading-velocity component, as given by Equation 10.4.

Operation of the track-stabilized antenna system concept, as used in earlier designs, is depicted in Figure 10.4c. In such a system, the difference between the Doppler shifts from beam pairs 1–3 and 2–4 is used to drive a servo, which turns the antenna in azimuth until this difference is nulled. This occurs when the ground intersections of these beam pairs lie on the same isodop, thereby placing the antenna axis along the aircraft track. In these systems, Doppler signals from beam pairs 1–3 and 2–4 are typically obtained on a sequential basis and compared, thereby allowing the time sharing of receiving equipment. The drift angle can be read out directly as the angle between the antenna axis and the aircraft center line (e.g., from a synchro mounted on the antenna), and the ground speed can be obtained directly by averaging the Janus Doppler shifts from beam pairs 1–3 and 2–4. Since track angle is defined as heading angle plus drift angle, this system is also called a drift-angle-stabilized antenna system.

With regard to horizontal stabilization, two generic design approaches are possible (Figure 10.5). One uses an antenna fixed to the vehicle frame (Figure 10.5a) and the other uses a gimballed antenna that is physically stabilized to the local horizontal (Figure 10.5b). In 1996, the majority of dedicated Doppler radars configurations used the fixed-antenna approach. In this configuration, if the aircraft pitches or rolls, the antenna and hence the Doppler beam cluster center line will move with the vehicle, as shown by the dashed line in Figure 10.5a. To convert the velocity components from airframe coordinates V_H', V_D', and V_V' to the Earth-referenced coordinates needed for navigation, the former must be resolved about aircraft pitch angle P and roll angle R obtained from a vertical reference sensor (vertical gyro, AHRS, or INS). The conversion is given by

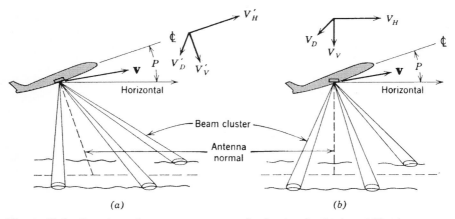

Figure 10.5   Doppler radar antenna geometry for fixed and attitude stabilized antennas.

$$V_H = V'_H \cos P + V'_D \sin R \sin P + V'_v \cos R \sin P \qquad (10.9)$$

$$V_D = V'_D \cos R - V'_V \sin R \qquad (10.10)$$

$$V_V = -V'_H \sin P + V'_D \sin R \cos P + V'_v \cos R \cos P \qquad (10.11)$$

where the velocity components are those defined in Figure 10.3.

The second type of Doppler radar is one whose antenna is continuously slaved to the local horizontal by means of the information from a vertical sensor. Although only a few Doppler radar designs employed this technique in 1996, it was used widely in previous years. Having computed V_H, V_D, and V_V, from Equations 10.9–10.11, the drift angle can be obtained from the arctan (V_D/V_H) and the ground speed V_g from the magnitude of V_g:

$$V_g = (V_H^2 + V_D^2)^{1/2} \qquad (10.12)$$

The Doppler Spectrum   Since each Doppler-radar antenna beam (Figures 10.2 and 10.3) has a finite beam width in the γ-direction, the return signal associated with the beam comes from a spread of γ-angles. Furthermore, the backscattering medium (the Earth) is composed of a multitude of randomly situated scattering centers, and the return signals from each have, in general, different amplitudes and different phases. In view of the frequency spread and randomness of the amplitude and phase of the scattering centers, the Doppler signal associated with each beam is in the form of a noiselike frequency spectrum (Figure 10.6). The spectrum is equivalent to band limited noise, the primary Doppler spectrum being superimposed on a substantially flat (uniform power spectral density) background noise. The shape of this spectrum is related to the antenna beam shape and is roughly Gaussian. The amplitude is a function of the radar

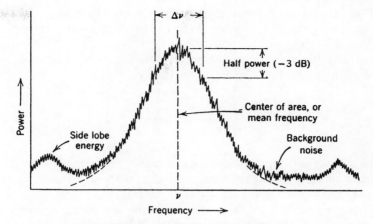

Figure 10.6 Typical Doppler spectrum.

parameters and the terrain backscattering coefficient in the radar-range equation discussed in Section 10.1.3. The amplitude modulation and the frequency modulation due to the scattering centers in the beam affect the shape and width of the spectrum. It is shown in the next section that amplitude effects are small, except at very low altitudes. The frequency characteristics will be considered further in the present discussion. The small spectra on each side of the main spectrum in Figure 10.6 represent energy returned by way of antenna side lobes. In typical systems, their level is so low that they are not sensed by the Doppler acquisition circuits; they are outside the frequency passband during Doppler tracking and therefore do not affect system performance.

The Doppler frequency of interest, which is proportional to the velocity component along the particular beam centroid, is the mean or center of area of the Doppler frequency spectrum. It is this center of spectral power that defines the beam centroid for the velocity measurement.

The Doppler spectrum width is obtained, to a first approximation, by differentiation of Equation 10.3 with respect to γ and is given by the expression

$$\Delta \nu = \frac{2V}{\lambda} \, \Delta\gamma \sin \gamma \qquad (10.13)$$

where $\Delta\nu$ is the half-power Doppler spectrum width and $\Delta\gamma$ is the half-power two-way beam width of the antenna in the γ-direction. The approximations involved in the above expression for the spectrum width involve ignoring the differential effect of the inverse square law of the radar-range equation between the near and far regions of the illuminated area and assuming a constant backscattering coefficient of the target area. The former is a valid approximation for all practical Doppler radars, those using narrow two-way γ-beam widths, typically near 4 degrees. The latter is a good approximation for operation over typical land terrain. However, the scattering properties of water cause a change in the shape and average frequency of the spectrum, giving rise to an overwater calibration-shift error (see Section 10.1.4).

The relative (fractional) spectrum width is frequently of interest and is found by dividing Equation 10.13 by Equation 10.3; namely,

$$\frac{\Delta \nu}{\nu} = \Delta \gamma \tan \gamma \qquad (10.14)$$

For typical practical Doppler radars, $\Delta \nu / \nu$ ranges between 15% and 25%.

Because of the appreciable spectrum width, the instantaneous central frequency of the Doppler signal is subject to random fluctuations about its mean value, giving rise to a noise (or *fluctuation*) error in tracking the centroid and, hence, in velocity and distance measurement. A certain amount of smoothing time is therefore required to determine the velocity to a desired accuracy; that is, the accuracy of measurement increases with smoothing time. In general, it is necessary to select a velocity smoothing time whose value represents a compromise between velocity accuracy and the data rate required on the basis of system dynamics (e.g., the maximum acceleration of the vehicle). If too large a velocity smoothing time constant is selected, the Doppler radar will follow the vehicle accelerations with too great a lag. For the navigation problem the effective smoothing time for the average velocity or distance measurement is the total time flown. Hence, for the typical Doppler navigation problem, the fluctuation error due to this noiselike nature of the information is completely overshadowed by other errors after only a few miles of flight. (A quantitative discussion of this error is given in Section 10.1.4.) In multisensor navigation systems, such as Doppler-inertial systems (Chapter 3), the Doppler-radar information may be intentionally smoothed further, since the high-frequency information is supplied by the inertial sensor. It is the function of the frequency tracker to determine the mean or center of power of the Doppler spectrum, that is, to determine the single-frequency ν that is proportional to the desired velocity component.

The Doppler correlation time τ_ν is proportioanl to the reciprocal of the spectrum width $\Delta \nu (\tau_\nu \approx 2/\Delta \nu)$. This is based on the fact that the power spectrum and the autocorrelation function are Fourier transforms of each other. The Doppler correlation time is the period during which the frequency and phase of the signal are invariant or predictable. At the end of this period the signal is nearly uncorrelated. This means that an independent Doppler measurement is made during each correlation time.

Scanning Noise   At very low altitudes, there is a small amount of spectrum broadening due to amplitude modulation effects, over and above the basic spectrum width discussed above. This spectrum broadening has been called *scanning noise*, and the additional spectrum width is therefore called *scanning noise spectrum.* It will be shown that the effect is quite small in conventional Doppler radars, even at relatively low altitudes.

The time required for one set of scatterers that is illuminated by the entire beam intersection to be replaced by a new set is the scanning-noise correlation time of the signal τ_s. Twice the reciprocal of this time is approximately its frequency spectrum width $\Delta\nu_s$. From the geometry of Figure 10.2, this correlation time τ_s is given by

$$\tau_s = \frac{L}{V} = \frac{h\Delta\gamma}{V\sin^2\gamma} \tag{10.15}$$

where L is the diameter of the beam intersection and h is the altitude. Hence, the scanning noise spectrum width $\Delta\nu_s$, is given by

$$\Delta\nu_s = \frac{2V\sin^2\gamma}{h\Delta\gamma} \tag{10.16}$$

The ratio of the scanning noise spectrum width and the basic spectrum width R is found to be

$$R = \frac{\Delta\nu_s}{\Delta\nu} = \frac{\lambda\sin\gamma}{h(\Delta\gamma)^2} \tag{10.17}$$

Substitution of practical values for the parameters shows that $\Delta\nu_s$ is insignificant except at very low altitudes. Because in an actual antenna the beam diameter in the near field (range is less than D^2/λ, where D is antenna diameter) does not continue to decrease to a "point," the scanning noise spectrum width does not continue to increase for altitudes below the extent of the near field. On the basis of conventional near-field antenna considerations, the scanning noise spectrum width will be just equal to twice the basic spectrum width at the altitude at which the antenna near field begins. This means that, in typical Doppler radars, R is never larger than 2 and that the total spectrum width is never larger than $\sqrt{5}\Delta\nu$, even at extremely low altitudes.

Operating Frequency   In 1996, Doppler radars transmitted at a center frequency of 13.325 GHz in the internationally authorized band of 13.25 to 13.4 GHz. This frequency represents a good compromise between too low a frequency, resulting in low-velocity sensitivity (Hertz per knot) and large aircraft antenna sizes and beam widths, and too high a frequency, resulting in excessive absorption and backscattering effects of the atmosphere and precipitation. (Earlier Doppler radars operated in two somewhat lower frequency bands, i.e., centered at 8.8 and 9.8 GHz, respectively, but, in 1996, these bands were no longer used for stand-alone Doppler radars.)

Polarization The two types of polarization that have been used for Doppler radars are linear and circular odd (opposite rotation received). The latter has the advantage of efficient duplexing techniques. While circular even (same rotation received) has well-known rain discrimination characteristics, it suffers from an appreciable backscattering loss over water. In 1996, linear polarization was used.

Doppler Radar Functions A typical dedicated Doppler radar contains four major functions: the antenna, transmitter, receiver, and frequency tracker (Figure 10.7). The transmitter generates the signal to be radiated via the antenna system toward the ground; the signal is backscattered by the ground, intercepted by the antenna system (either by the same antenna as that which transmitted the signal or by a separate receiving antenna), and fed to the radar receiver. The received signal is mixed (heterodyned) with the transmitted signal or a local-oscillator signal and the resulting Doppler shift difference signal is amplified in the radar receiver. The latter produces Doppler spectra from the various beams, of the form discussed previously. These are fed to the frequency tracker, which determines the mean frequencies of the Doppler spectra and hence the velocity components represented by them. The data converter converts the frequencies into the proper form of outputs, such as orthogonal velocity components or ground speed and drift angle.

Types of Transmission Perhaps the most important design characteristic of a Doppler radar is the type of transmission or modulation used. The two types of transmission generally used for modern dedicated Doppler radars are continuous wave (CW) and frequency modulated–continuous wave (FM-CW). Non-coherent (self-coherent) pulse and coherent pulse modulations were widely used in earlier designs. The former is no longer used in modern systems because of its signal inefficiency. The latter is only rarely used. The so-called self-

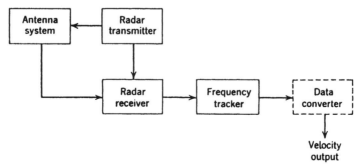

Figure 10.7 Functional diagram of Doppler radar.

coherent systems represented an innovative solution to the problem of how to achieve a Doppler frequency measurement with a noncoherent pulse radar whose transmitter was not phase coherent from pulse to pulse, such as a magnetron. This was achieved by directly heterodyning the signals backscattered in the forward-looking beams with those of the rearward-looking beams, which originated from the same pulses and were therefore phase coherent with each other. Hence, a stable Doppler frequency was obtainable from such a noncoherent radar. Many thousands of these systems were developed and installed in aircraft. When coherent pulse transmitters became available, dedicated coherent pulse Doppler radar systems were implemented and are still used in some Doppler radar designs. Since the modulation used in these systems has very high duty cycle, they are also called interrupted continuous wave (ICW) systems. The complexity of these systems is somewhat greater than that of pure continuous wave (CW) and frequency modulated–continuous wave (FM–CW) systems described in the remainder of this section.

Continuous Wave Transmission   Pure continuous wave transmission is inherently the simplest and most efficient type of transmission. No modulators of any kind are required, the spectrum-utilization efficiency is essently 100%, and no altitude holes exist. However, pure continuous wave transmission systems are faced with the difficulty of transmitter-receiver isolation, as well as an inherent lack of discrimination against echoes from nearby objects and from the aircraft structure itself. Lack of isolation can result in large undesirable carrier and noise leakage signals, which can lower the gain of the receiver and increase the total effective noise level, thus reducing the signal-to-noise ratio, particularly at the lower Doppler frequencies, near zero beam velocity. Pure continuous wave systems may be limited by the signal-to-leakage ratio, rather than by the signal-to-receiver noise ratio. This is of importance for operation at higher altitudes, where the backscattered signal is small in comparison with the leakage signal. Reflection and backscattering from nearby objects (stationary or vibrating structural members, e.g., the radome), nearby turbulent air (supersonic shock waves), and precipitation will cause undesirable noise power, which may be in the frequency band of interest and whose level is proportional to that of the transmitted power. To improve the basic transmitter-receiver isolation, separate antennas for transmission and reception (space duplexing) are used in pure continuous wave systems. A pure continuous wave system inherently provides operation down to zero altitude. The block diagram of a transmitter-receiver of a basic continuous wave (CW) Doppler radar is shown in Figure 10.8. If it is desired to improve the receiver-noise figure over and above that of the homodyne (*zero-frequency intermediate frequency*) configuration shown in Figure 10.8, a genuine intermediate-frequency (IF) receiver configuration can also be used. Also, if it is required to maintain sense of velocity direction, as in helicopter operation, some form of offset reference frequency or quadrature detection technique must be incorporated into the system.

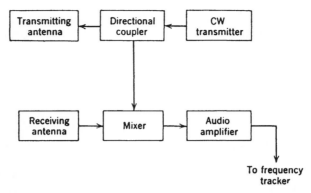

Figure 10.8 Block diagram of transmitter-receiver of CW Doppler radar.

Frequency Modulation–Continuous Wave Systems The frequency modulation–continuous wave (FM-CW) type of transmission combines some of the advantages of pulse and pure continuous wave systems. Therefore, this technique is used in the majority of dedicated modern Doppler radars. The problems of transmitter-receiver isolation and discrimination against nearby echoes are reduced or eliminated in such a system on a frequency basis, much as they are eliminated in pulse systems on a time basis. In an FM-CW system, the transmitter is sinusoidally frequency modulated, and the receiver is designed to use only the Doppler shift of a particular sideband (other than the zero-order side band) of the beat between the received and transmitted signals. Since the modulation index of the beat spectrum and hence the amplitude of all but the zero-order side bands decreases very rapidly with decreasing range and becomes zero at the receiver mixer terminals, high transmitter-receiver isolation and suppression of returns from nearby objects are achieved. Since the amplitudes of these side bands vary as *Bessel* functions, they are called *Bessel* side bands. FM-CW systems require a simple low-power sine-wave modulator/transmitter and hence their transmitter-receiver approaches pure continuous wave systems in simplicity. Since the power in only one of the side bands is used, the efficiency is not as great as that of continuous wave systems. It is maximized by the use of the optimum transmitted modulation index.

A problem with *single-antenna* FM-CW systems is that the combination of internal line length and mismatch of practical microwave components (antenna, switches, radome, etc.) will generate a leakage signal, which limits the achievable signal-to-leakage ratio. To alleviate this, a special leakage elimination filter is usually used. Another problem of a single-antenna FM-CW system is the fact that full transmitter-power feedthrough, which is determined by practical duplexer-isolation characteristics, is continuously applied to the receiver, regardless of the isolation obtained by the sideband processing. However, in 1996, both of these problems were overcome in FM-CW Doppler radars and single-antenna configurations were used widely.

FM-CW Doppler radars, as well as pulse modulated radars, are subject to altitude hole problems because of the relationship between modulation frequency and echo delay. These altitude holes can be eliminated by continuously changing the modulation frequency or by using different modulation frequencies over different preset altitude ranges.

The important parameters that affect the performance of FM-CW systems are (1) the order of the side band, (2) the modulation frequency, and (3) the modulation index. The order of the sideband (i.e., whether the first, J_1, or a higher-order Bessel side band is used) determines low-altitude performance, effective transmitter-receiver isolation (signal-to-leakage ratio), and, to a less extent, the signal efficiency (signal-to-noise ratio) of the system. The first-order J_1 system inherently permits operation down to zero-feet altitude and exhibits a flat (constant) signal-to-noise ratio versus altitude characteristic for the lower altitude region of operation [16, 37]. However, it has the lowest effective transmitter-receiver isolation and, hence, the lowest achievable signal-to-leakage ratio. The second- and third-order J_2 and J_3 systems have a greatly improved transmitter-receiver isolation performance (in view of the slope of the Bessel functions near zero) and still permit reasonably good low-altitude performance. In general, then, the higher the order of sideband used, the better the isolation but the worse the low-altitude performance. In 1996, the first-order (J_1) Bessel side band was used in most Doppler radars designed for helicopters to take advantage of its relatively flat S/N characteristic at low altitudes. This approach minimizes the occurrences of unwanted returns from vibrating structures near the Doppler radar antenna while retaining adequate S/N for returns from terrain below the aircraft.

The choice of modulation frequency affects the location of the first altitude hole, transmitter-receiver isolation, and low-altitude signal performance. The use of a very low modulation frequency can cause the first altitude hole to appear above the maximum altitude of operation, thus avoiding the existence of any altitude holes over the range of interest. However, unless a low-order sideband is used at the same time, low-altitude performance is limited. A high-modulation frequency (the modulation wavelength being much smaller than the maximum altitude of operation) results in many altitude holes over the altitude range but in a much higher signal-to-noise ratio at the lower altitudes (though a somewhat lower signal-to-leakage ratio). In 1996, a relatively low modulation frequency, 25 to 30 kHz, was used in most Doppler radars designed for helicopters.

A high-modulation frequency, when used in conjunction with a homodyne (zero-frequency IF) receiver, inherently results in an intermediate frequency of sufficiently high value from a viewpoint of detector noise temperature. However, since the received Doppler spectrum is "folded" about zero frequency, the homodyne approach does not provide information on sense of velocity as required in helicopters or vertical takeoff and landing aircraft, unless additional (e.g., quadrature) circuitry is added (see Figure 10.9).

The choice of transmitted modulation index affects low-altitude performance,

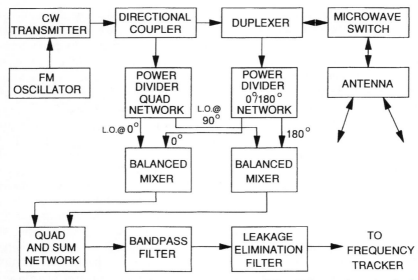

Figure 10.9 Block diagram of transmitter-receiver of FM-CW Doppler radar for helicopters.

transmitter-receiver isolation, and the signal-to-noise ratio. Therefore, the modulation index is generally selected so as to be compatible with the order of the sideband and the modulation frequency that have been selected, primarily from the viewpoint of maximizing received power. A mathematical indication of this behavior can be obtained from the expressions for the received modulation index M and the radar-range equation modulation–efficiency factor E of an FM-CW Doppler radar. These are

$$M = 2m \sin\left(2\pi f_m \frac{h \sec \psi}{c}\right) \qquad (10.18)$$

and

$$E = J_n^2(M) \qquad (10.19)$$

where
   M is the received modulation index (after heterodyning)
   m is the transmitted modulation index
   f_m is the modulation frequency
   h is altitude
   ψ is the central beam looking angle with respect to vertical

c is the speed of light
$J_n(M)$ is a Bessel function of order n and argument M
n is the order of sideband

For high-modulation frequency FM-CW systems, it is necessary to average Equation 10.19 over one-half cycle of the argument in Equation 10.18. The maximum efficiency is obtained for the value of m, which makes this average a maximum. An approximate expression for this optimum value of m for the nth sideband is [20]:

$$m = \frac{n+2}{2} \tag{10.20}$$

The block diagram of a FM-CW Doppler radar for use in helicopters is shown in Figure 10.9. A directional coupler diverts a small amount of transmitter power to a power divider whose outputs are two signals in phase quadrature. These outputs serve as the reference or local oscillator (LO) for the two balanced mixers. The RF energy backscattered from the ground and received by the single antenna is directed via the duplexer to a power divider that generates two outputs of opposite phase which are then mixed with the two LO outputs in the two balanced mixers. The outputs of the two balanced mixers have now been translated to zero frequency and the Doppler frequency–shifted spectra of the sidebands below the carrier now appear as images to the true signals on the sidebands above the carrier. The two mixer outputs, which are in phase quadrature, are phase shifted an additional 90 degrees relative to each other and then summed, resulting in phase cancellation of the unwanted image signals while retaining the wanted signal. This approach preserves the sense of the Doppler frequency shift, since the latter can change sign during hover and backward flight of helicopters. The resultant signal contains the desired frequency spectrum above (positive shift) or below (negative shift) of each one of the modulation (Bessel) sidebands. The desired sideband is selected by filtering out all other sidebands. Single transmit/receive antennas tend to have high transmitter-to-receiver leakage resulting in a large unshifted signal at base band and at each of the sidebands. This leakage is removed in a filter that is centered at the desired sideband frequency. During low speed and hover operation, however, the true signal spectrum will have a very small frequency shift and will therefore occur close to the unwanted leakage signal. The filter that removes the leakage must therefore be very narrow (1 to 2 Hz) to avoid affecting the true signal spectrum. The output to the frequency tracker is thus the Doppler frequency-shifted spectrum offset from zero frequency by a multiple of the modulation frequency, which is typically the first (J_1).

Frequency Trackers   The function of the frequency tracker is to determine the centroid of power (mean frequency) of the noise like Doppler spectrum obtained from the ground echo (Figure 10.6). Practically all modern Doppler radars use some form of closed loop frequency discriminator as the frequency-tracking device. The Doppler signal is fed to one or more mixers, which mix it with a signal from a variable-frequency oscillator (tracking oscillator) and feed the output to the device that performs the discriminator function. The output from the latter is fed to an integrator and used to control the frequency of the tracking oscillator, which, in turn, can then provide the frequency-tracker output signal (Figure 10.10). During acquisition, frequency trackers usually use a sweeping operation by changing the tracking oscillator frequency linearly over the entire Doppler band of interest.

Two Doppler frequency-tracker configurations that have been used are shown in Figure 10.11. The one shown in Figure 10.11a is called the *two-filter* tracker. Actually, a single filter is used, but the Doppler spectrum is mixed with a tracking oscillator signal, which is square-wave frequency modulated over the extent of the spectrum width. The mixer output is fed to a low-pass filter, which therefore looks successively at the upper and lower halves of the spectrum. The filter output is phase detected against the frequency-modulated oscillator signal. The phase-detector output is then fed to the integrator and, having sense of direction, will drive the tracking oscillator until its output frequencies just straddle the center of power of the Doppler spectrum. The average frequency of the tracking oscillator output is thus the frequency that corresponds to the center frequency of the spectrum, which, in turn, is proportional to the desired veloc-

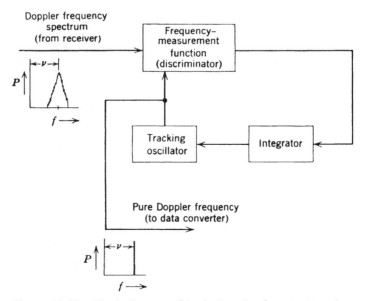

Figure 10.10   Block diagram of basic Doppler frequency tracker.

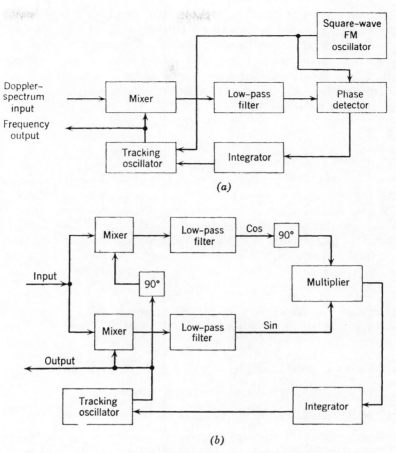

Figure 10.11 Typical Doppler frequency tracker configurations.

ity. This type of frequency tracker implementation normally does not provide sense of velocity direction through zero velocity, but it is simpler to mechanize and hence was used in Doppler radars for fixed-wing applications where negative velocities do not occur.

The frequency tracker shown in Figure 10.11b is called the sine-cosine tracker. Here the input spectrum is mixed simultaneously with the tracking-oscillator signal and its quadrature signal. The mixer outputs, which are actually the folded Doppler spectra at zero frequency, are fed through low-pass filters in two separate channels (sine and cosine) and then again phase shifted by 90°. The latter 90° phase shift may be obtained by placing a low-pass and a high-pass filter in the sine and cosine channels, respectively. The signals from the two channels are then multiplied; the multiplier output is integrated and drives the tracking oscillator, thus closing the loop. By virtue of the two quadrature channels, sense of direction for the tracking-oscillator drive is maintained. When the loop is nulled, the tracking oscillator frequency represents the average frequency of the input Doppler spectrum. Typically, the input spectrum is offset from an intermediate frequency so that sense of velocity direction is obtained from the output. The sine-cosine frequency tracker is in widespread use in modern Doppler radars and particularly in those designed for helicopters where negative velocities occur.

10.1.3   Signal Characteristics

General Criteria   The performance of a Doppler radar is generally expressed by the Doppler signal-to-noise ratio (S/N) that is available at the input to the frequency tracker. The lowest S/N generally occurs at maximum altitude, speed and pitch and roll, and over terrain with the lowest radar backscattering characteristics. The performance of a Doppler radar is also affected by the sensitivity of the frequency tracker, that is, the minimum S/N at which the tracker will operate properly.

Doppler Signal-to-Noise Ratio (S/N) The Doppler S/N is a function of the following variables:

1. The range to the terrain with respect to the beam of interest

2. The velocity with respect to the beam of interest

3. The RF backscattering properties of the terrain

4. The attenuation (absorption) and backscattering properties of the atmosphere

5. Radar parameters such as wavelength, receiver noise figure, transmitted power, antenna gain, beam-looking angle, transmitter and receiver path losses, and transmitter-receiver leakage noise

6. The efficiency of the type of transmission (modulation) used

By modifying the basic radar-range equation [1], the Doppler signal-to-noise ratio per beam (using the same type of antenna for transmission and reception) for a coherent system in which each beam is demodulated separately is given by

$$\left(\frac{S}{N}\right)_d = \frac{P_t G_0 \lambda^2 \sigma_0 L_a L_r w E}{16\pi^2 R^2 B_d [KTNF + P_t K_i (N/S)_t] \cos \psi} \qquad (10.21)$$

where

$(S/N)_d$ is the Doppler signal-to-noise ratio (the ratio of the total Doppler signal power to the noise power, in the bandwidth of interest B_d)

P_t is the average transmitted power

G_0 is the one-way maximum antenna gain relative to an isotropic radiator

λ is the wavelength of transmission

E is an efficiency factor (including the spectral modulation efficiency in pulse and FM-CW systems, gating improvements, gating losses, and noise foldover losses)

L_r are losses in the radar transmitter and receiver paths, such as the wave-guide plumbing, duplexer, radome and other radio-frequency components

L_a is the attenuation in the atmosphere

w is an antenna pattern factor (normally between 0.5 and 0.67)

σ_0 is a scattering coefficient or backscattering cross section per unit area of the scattering surface

ψ is the incidence angle of the center of the beam with respect to a normal to the surface

R is the range to the scattering surface

NF is the noise figure of the receiver

K is the Boltzmann's constant, 1.38×10^{-23} joule per Kelvin

B_d is the bandwidth of interest, usually the -3-dB bandwidth of the Doppler spectrum and is proportional to velocity along the beam

K_i is the effective transmitter-receiver isolation coefficient

$(N/S)_t$ is the ratio of transmitter generated noise to the transmitter power

T is the absolute temperature, normally taken as $290°$ Kelvin

In most well-designed systems, K_i is so small that the term $P_t K_i (N/S)_t$ becomes negligible. Isolation is obtained by different means in different systems, such as by separate antennas in CW systems, by time separation in pulse systems, and by frequency separation (and possibly also antenna separation) in FM-CW systems. If this term is indeed negligible, Equation 10.21 becomes

$$\left(\frac{S}{N}\right)_d = \frac{P_t G_0 \lambda^2 \sigma_0 L_a L_r w E}{16\pi^2 R^2 B_d K T N F \cos \psi} \qquad (10.22)$$

Because the scattering surface fills the entire radar beam in a Doppler navigation radar, the signal-versus-range dependence in the radar range equation is the basic inverse-square law ($1/R^2$) as given by Equation 10.22. However, for certain FM-CW and incoherent-pulse systems, the signal-versus-range dependence can vary considerably from the inverse-square law.

In FM-CW systems, E is a function of $J_n^2(M)$ [see Equations 10.18–10.20 for the definition of $J_n^2(M)$]. In low-rate frequency modulation systems (modulation wavelength of the same order as the altitude of operation, $E = J_n^2(M)$. In these systems, since M varies with propagation delay (and hence range) and with modulation frequency and since the Bessel functions of different orders have greatly different shapes, the signal-versus-altitude dependence in specific altitude regions can vary markedly from the basic inverse-square law of Equation 10.22. In contrast, in high-rate frequency modulation systems (modulation wavelength much smaller than maximum altitude of operation), $J_n^2(M)$ must be averaged over one-half cycle of the argument of Equation 10.18 [20]. Typical values of E for these systems range from -6.4 dB for J_1 systems to -11.1 dB for J_4 systems. The *average* signal-versus-altitude dependence of these systems follows the inverse-square law of Equation 10.22.

Terrain-Scattering Coefficient   It is seen from Equation 10.22 that an important factor in determining the Doppler signal-to-noise ratio is σ_0, which is the parameter of nature that determines the amount of power backscattered by the surface to the Doppler radar receiver. The σ_0 factor is defined as the backscattering cross section per unit area (at the target surface), normal to the direction of propagation, intercepting that amount of power which, when scattered isotropically, would produce an echo equal in power to that actually observed per unit area of the target surface. Figure 10.12 shows curves of σ_0 versus ψ angle for a radar system operating in the frequency band currently assigned to Doppler radar, namely 13.250 to 13.400 GHz. Included in Figure 10.12 are curves for various types of terrain including land and water. It is seen from the curves in Figure 10.12 that for normal wooded-land terrain σ_0 is nearly constant with beam incidence angle ψ and that it has a value of -7.5 dB for the incidence angles ψ of interest ($10°$ to $30°$). However, for water surfaces, σ_0 decreases radically as ψ increases and assumes different values for different conditions of sea state or water roughness. The sea-state scale shown in Figure 10.12 was developed by the United States Army to specify σ_0 for their helicopter Doppler radar development programs and is typical of curves used by other development agencies. For the typical Doppler-radar incidence angles ψ of $10°$ to $30°$, σ_0 is considerably smaller for most sea states than for land and decreases markedly for the smoother sea states. Therefore, a conservative Doppler-radar design must be based on a σ_0 for the smoothest sea state over which the aircraft is expected to navigate by means of the Doppler radar. (It is known, however, that very smooth sea states are relatively rare.) For physically roll-and-pitch-stabilized antenna systems, the value of ψ remains essentially constant and equal to the chosen design value. For fixed-antenna systems, a conservative design must be based on σ_0 and range R for the largest ψ-angle that would be expected for the largest combination of pitch and roll angles of the aircraft.

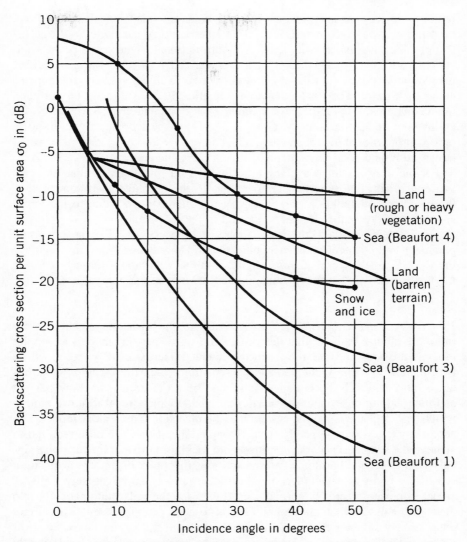

Figure 10.12   Radar backscattering coefficient versus incidence angle for different terrains at Ke-band.

Bandwidth, Antenna Gain, and Losses   The bandwidth B_d in Equation 10.22 is the effective bandwidth of the Doppler frequency tracker. In most systems, this is selected to be the -3-dB Doppler spectrum width. From Equation 10.13 we have, for a coherent or post-tracker Janus system, $B_d = \Delta\nu = (2V/\lambda)\Delta\gamma \sin\gamma$, where γ is the angle between the velocity vector and the angle of radiation, and $\Delta\gamma$ is the -3-dB two-way beam width in the γ-direction. (When the antenna beam pattern has a predominantly Gaussian shape in the region of interest, as is typical for Doppler radar antenna beams in the γ-direction, the two-way beamwidth is 0.707 of the one-way beamwidth.) Thus, the Doppler signal-to-noise ratio is inversely proportional to the speed of the vehicle. In view of this, the design of a Doppler radar as regards signal performance must be based on the maximum expected vehicle velocity components along each beam. The product $G_0\lambda^2$ in Equation 10.22 is directly proportional to antenna area A from the basic expression for antenna gain. Specifically, for an antenna of 55% efficiency, $G_0\lambda^2 = 7A$. Thus, the Doppler signal-to-noise ratio is proportional to antenna area. (This is strictly true only for a radar using transmitting and receiving antennas having the same area.)

For the transmission wavelength normally used for Doppler radar (2.2 cm) and considering the relatively short ranges to the ground (when compared to those of forward looking search radars), the attenuation L_a due to the atmosphere and to typical rain rates are found to be very small.

From Equation 10.22, the noise figure of the receiver NF is an important parameter; it has therefore been an objective in Doppler-radar design to achieve the lowest possible receiver-noise figure. Similarly, the radio frequency losses L_r of the microwave circuitry must be kept as low as possible.

Frequency-Tracker Sensitivity   In addition to the available signal-to-noise ratio, the other parameter that determines the signal performance of a Doppler radar is the sensitivity of the frequency tracker. This is usually expressed by two quantities, namely, the acquisition sensitivity (i.e., the Doppler signal-to-noise ratio at which the Doppler signal can be acquired and tracking begins) and the tracking or dropout sensitivity (i.e., the Doppler signal-to-noise ratio at which tracking stops, and the system may be placed into a memory mode). The acquisition sensitivity depends on the Doppler signal-to-noise ratio required to achieve the specified accuracy and to avoid locking on to extraneous noise signals such as second harmonic spectra. In typical Doppler radars, the acquisition sensitivity is set at a Doppler signal-to-noise ratio of approximately 5 dB. A signal-to-noise detector is normally used to place the radar automatically into the tracking mode when this Doppler signal-to-noise ratio is present. This circuit continuously samples the received signal level and the system noise level (or its equivalent) and measures their ratio so as to determine whether a sufficiently high signal-to-noise ratio is present for acquisition and tracking. The tracking, or dropout, sensitivity is the Doppler signal-to-noise ratio level at which the signal-to-noise detector is set to cause the frequency tracker to stop tracking. In typical Doppler radars, the frequency-tracker dropout sensitivity is approximately 3 dB.

Full Doppler radar accuracy (particularly the Doppler fluctuation error) is frequently not obtained unless the Doppler signal-to-noise ratio is between 7 and 10 dB. Therefore, systems utilizing Doppler velocity data only for critical or sensitive functions (versus integrated velocity for navigation) should require minimum Doppler signal-to-noise ratios of near 10 dB.

10.1.4 Doppler Radar Errors

Classification of Errors Doppler radar velocity errors can be classified as either random (varying with time) or systematic (independent of time). Random errors are those errors that vary during a flight or flight leg. Systematic errors are those that are constant, although perhaps unknown, for the duration of the flight. Known systematic errors can be calibrated out, either before the start of a mission, after equipment installation in the aircraft, or even at the factory. All uncompensated systematic errors must be included in an error analysis. There are two types of random errors: those with relatively long correlation times and the Doppler-fluctuation noise that has a correlation time τ_c (at the output of the frequency tracker) on the order of 0.1 sec. Both of these are typically assumed to be exponentially autocorrelated. Since 15 minutes would be the least desirable correlation time for a velocity error in a Doppler radar used in a Doppler-inertial system—because of the effects of the Schuler period (Section 7.6.3)—it has been of special interest to keep Doppler velocity errors having correlation times near 15 minutes as low as possible. Another classification of errors is that of percentage-of-speed or scale-factor errors, and errors independent of speed or speed-offset errors. Most errors are scale-factor types. In this section, each of these various errors is treated separately, followed by a discussion of the total Doppler radar velocity error and the Doppler-navigation system errors. The coordinate system that is most suitable for describing the errors of a Doppler radar with a fixed antenna is (H', D', V'), as described previously (Equations 10.4 to 10.6).

Doppler-Fluctuation Error The Doppler fluctuation error e_f is due to the noiselike nature of the Doppler signal spectrum, which, in turn, is caused by the backscattering properties of terrain. The standard deviation of the basic Doppler velocity fluctuation error per beam of a coherent system, assuming a perfect frequency tracker, can be expressed by [3]:

$$\epsilon_f = \frac{\sigma_\nu}{\nu} = \frac{\sigma_{\nu,T}}{\nu} = \frac{1}{\nu}\left(\frac{\Delta\nu''}{2T}\right)^{1/2}$$

$$= \left(\frac{\Delta\gamma''\lambda\tan\gamma}{4VT\cos\gamma}\right)^{1/2} = \frac{K_1}{(VT)^{1/2}} \tag{10.23}$$

where
  ϵ_f is the standard deviation of the fractional velocity fluctuation error
  σ_ν is the standard deviation of the absolute velocity fluctuation error
  T is smoothing time
  $\sigma_{\nu,T}$ is the average fluctuation averaged over time T for n independent measurements, $\sigma_\nu/(n)^{1/2}$

σ_ν is the instantaneous statistical fluctuation of $\nu = \Delta\nu''/2, n = T/\tau_\nu$

$\Delta\nu''$ is the Doppler spectrum width, which is twice the 1-sigma half-width

$\Delta\gamma''$ is the equivalent 2-sigma, two-way antenna beamwidth.

If the beam has a Gaussian shape, which is typical for Doppler radar beams in the region of interest, the 2-sigma two-way beamwidth is equal to $1/1.18$ times the -3-dB beamwidth). In Equation 10.23, τ_ν represents the Doppler correlation time $2/\Delta\nu''$ (at the input to the frequency tracker; in contrast to the correlation time t_c at the output of the frequency tracker, discussed previously). K_1 is a constant combining the various radar parameters and constants in the expression.

From the standpoint of navigation between two points (separated by many antenna lengths and many frequency tracker time constants), T is the total time flown. It is seen from this equation that V and T occur only as a product and hence can be replaced by the distance flown D:

$$\epsilon_f = \left(\frac{\Delta\gamma''\lambda \tan \gamma}{4D \cos \gamma} \right)^{1/2} = \frac{K_1}{(D)^{1/2}} \tag{10.24}$$

Equation 10.24 indicates that the basic Doppler velocity fluctuation error is inversely proportional to the square root of the distance traveled. A physical explanation of this can be obtained by realizing that it is the total number of independent scatterers seen by the Doppler radar beam that determines the amount of smoothing afforded and hence the final velocity fluctuation error.

Equations 10.23 and 10.24 express the Doppler velocity fluctuation error under ideal or error-free frequency measurement conditions. The performance of a practical system will differ from an ideal one by some factor N, which has been called the performance factor:

$$\epsilon_f = \frac{NK_1}{(VT)^{1/2}}$$

$$= \frac{NK_1}{D^{1/2}}$$

$$= \frac{K}{D^{1/2}} \tag{10.25}$$

where N is a factor that relates measured values to theory. For practical equipment, N has a value somewhere between 1 and 2. The all-digital frequency tracking circuits used in 1996 resulted in N being nearly equal to 1. A typical Doppler radar operating at 13.3 GHz and with a beamwidth of $6°$ has a fluctuation error of 0.051% after 10 mi and 0.016% after 100 mi of travel. Thus, the fluctuation error is negligible after only a few miles of travel when compared to other instrumentation errors of the system. Doppler radar velocity accuracy specifications are usually cited for a condition of "after 10 mi of flight."

Equations 10.23 and 10.24 give the fluctuation error per beam, assuming single beam tracking. Modern systems measure each of the four beams sequentially and then combine the four beam velocities to arrive at V_H', V_D' and V_V'. Each beam is tracked for 25% of the time, which increases ϵ_f by $(4)^{1/2}$, but four beams are combined and thus ϵ_f decreases by $(4)^{1/2}$. Equation 10.24 thus applies to $\epsilon V_H'$, $\epsilon V_D'$, and $\epsilon V_V'$ as well, except that $\Delta \gamma''$ is replaced by $\Delta \gamma_H$, $\Delta \gamma_D$, and $\Delta \gamma_V$, and γ by γ_H, γ_D, and γ_V:

$$\epsilon V_H' = \left(\frac{\Delta \gamma_H \lambda \tan \gamma_H}{4D \cos \gamma_H} \right)^{1/2} = \frac{K_2}{(D)^{1/2}} \tag{10.26}$$

$$\epsilon V_D' = \left(\frac{\Delta \gamma_D \lambda \tan \gamma_D}{4D \cos \gamma_D} \right)^{1/2} = \frac{K_3}{(D)^{1/2}} \tag{10.27}$$

$$\epsilon V_V' = \left(\frac{\Delta \gamma_V \lambda \tan \gamma_V}{4D \cos \gamma_V} \right)^{1/2} = \frac{K_4}{(D)^{1/2}} \tag{10.28}$$

When quasi-instantaneous short-term velocity information is considered, the smoothing time T in Equation 10.23 represents the integration time constant of the Doppler radar frequency tracker. Clearly, the longer the time constant, the lower will be the fluctuation error. However, the shorter the time constant, the better will be the system response to velocity changes (accelerations). In typical aircraft systems, this time constant is approximately 0.1 sec.

For certain applications, notably in Doppler-inertial systems, it is frequently of interest to find the power spectral density per unit of speed of the Doppler-fluctuation error, called P_0. The fluctuation component of the velocity error at the output of the frequency tracker is assumed to be white noise over the frequency range of interest in Doppler-inertial system analysis. This is based on the fact that the Doppler signal received in the frequency tracker has the properties of band-limited noise and, to a first approximation, this spectrum has a Gaussian shape with a half-width proportional to speed. This spectrum width determines the standard deviation of the fluctuation. The result of this is a velocity error spectrum with power spectral density at zero frequency which is proportional to speed; that is, the power density is equal to $P_0 V$. This noise is then filtered by the frequency tracker and the radar velocity readout circuitry, which act as low-pass filters. The relationship of P_0 to the standard deviation of the relative Doppler fluctuation error ϵ_f is expressed by

$$\epsilon_f = \frac{\sigma_\nu}{V} = \left(\frac{P_0\pi}{VT}\right)^{1/2} = \left(\frac{P_0\pi}{D}\right)^{1/2} \quad \text{or} \quad P_0 = \frac{K^2}{\pi} \tag{10.29}$$

where P_0 is the velocity error angular spectral density per knot of speed (the statistical "power" value of the error source has the dimension of knots^2 and P_0 is in units of knots^2 per (radian/second) per knot, which is in units of distance, namely, knot-seconds. Typical values of P_0 for operational systems in 1996 were between 0.003 and 0.005 knots^2 per (radian/second) per knot for V'_H and V'_D and approximately 2.5 times smaller for V'_V.

Errors in Beam Direction   As seen from Equations 10.4 to 10.6, the nominal beam angle γ_0 must be accurately known and maintained in order to permit accurate measurement of velocity. The basic fractional velocity error ϵ_b for an error in beam direction $\delta\gamma$ is

$$\epsilon_b = (\tan \gamma_0)\delta\gamma \tag{10.30}$$

An error in beam direction of one minue of arc and a nominal γ_0-angle of 70° yields an error of 0.08% of ground speed. (When four beams are used for velocity determination, the total effect of the random error in direction in each of the four beams is reduced by the square root of four.) Beam direction errors resulting from radome refraction effects, and temperature effects in certain antennas (notably linear and planar waveguide arrays) will contribute to ϵ_b. This scale-factor error is primarily systematic and, in most cases, can be largely removed by some form of ground or flight calibration procedure. Because of the smaller effective γ-angle, the equivalent error in vertical velocity is an order of magnitude smaller than that in ground speed.

The temperature error in slotted array antennas is proportional to the deviation from the calibration temperature and the linear coefficient of expansion of the antenna material. In typical systems, this error is less than 0.05% for the horizontal velocity component and approximately one-tenth of this for the vertical velocity component.

Error in Transmission Frequency   As seen from Equation 10.4 to 10.6, the knowledge and maintenance of the transmission frequency f (or wavelength λ) directly affects the value of the measured Doppler frequency and hence the measured velocity accuracy. The long-term frequency stability of modern solid-state microwave sources is typically in the range of 10^{-4} to 10^{-6}, resulting in a negligible error. Moreover, linear and planar slotted array antennas for Janus systems can be designed so as to make the Doppler calibration constant completely independent of transmission frequency, that is, dependent only upon slot spacing.

Error in Frequency Measurement (Frequency Tracker Bias)   This error is a function of frequency tracker design and is usually caused by unbalance in the frequency tracker discriminator. Also, a nonuniform noise power density in the frequency tracker bandwidth will cause a bias error in the frequency measurement whose value is generally a function of Doppler signal-to-noise ratio. In 1996, frequency trackers use digital signal processing techniques for which this error is typically less than 0.05 knots at 6 dB or higher S/N.

Altitude-Hole Error   Certain pulse and FM-CW systems exhibit a residual error due to spectrum-weighting effects in the altitude-hole regions due to the effect of the modulation periods, even if some form of modulation wobbling is used. Typical values for the residual altitude hole error using modulation wobbling are less than 0.02%.

Land-Terrain Error   Over land terrain a small error results from (1) range-difference effects over the beam width, (2) the nonlinear function of converting ray angles within the beam width to Doppler frequencies (see Equation 10.3), and (3) the small change in scattering coefficient with looking angle over the beam width (Figure 10.12). The first two effects are very small and may be eliminated by flight calibration. The third effect is exactly the same type as the overwater calibration shift error described in the next paragraph. Because of the very small change in scattering coefficient over the beam width for typical land terrain (Figure 10.12), this error is normally quite small, unless a very small antenna with a large beamwidth is used. A 6×12 in. antenna would have an error of about 0.1%.

Overwater Errors   The three different types of overwater errors of Doppler radars are (1) the calibration-shift error, (2) the sea-current error, and (3) the surface wind induced water-motion error.

The overwater calibration-shift error (or *sea bias*) results from the change in the scattering coefficient σ_0 versus incidence angle ψ over the antenna beam width, as it relates to the direction of changing Doppler frequencies, that is, normal to the isodops. The phenomenon is evident from Figure 10.12 which shows a plot of σ_0 versus ψ for various sea states. A Doppler radar with a nominal ψ-angle of 20° and a beam width of 5°, covering a ψ-angle range of 17.5° to 22.5°, has a significant change in scattering coefficient σ_0 over water. The slope m in the σ_0 curve will cause the Doppler spectrum to be weighted in the direction of lower Doppler frequencies and will therefore cause the frequency tracker to read out too low a velocity. This is illustrated in Figure 10.13, which shows plots of typical (artificially smoothed) Doppler power spectra over land and water. The mean of the Doppler spectrum obtained over land (i.e., frequency ν_l), represents the correct frequency for the speed of the vehicle. The lower power spectrum is the Doppler spectrum over water for the same vehicle speed, having a mean frequency ν_w. The difference between ν_l and ν_w is the *overwater calibration shift error* or *sea bias*. For an antenna pattern having a

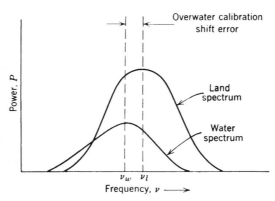

Figure 10.13 Doppler spectra (smoothed) over land and water.

Gaussian shape and for a linear function of the logarithm of σ_0 (in decibels) versus ψ within the beamwidth, which is a good approximation, as seen from Figure 10.12, the resultant overwater spectrum has a Gaussian shape like the overland spectrum but with its centroid of power shifted by $\nu_l - \nu_w$ (Figure 10.13). Figure 10.13 shows that at any nominal ψ_0 the slope of σ_0 versus ψ changes for different sea states, and, since the overwater calibration shift error is a function of this slope, the error has different values for different sea states.

An exact determination of the calibration shift error is obtained by integration of the elemental powers returned by the antenna beams as a function of Doppler frequency (along the isodops) and as a function of the scattering coefficient and incidence angles for different terrains and sea states (Figure 10.12). For typical Doppler systems (for level flight), an excellent approximation of the uncompensated calibration shift error in percent is given by

$$\epsilon_w = \frac{-0.0728 \sin^2 \gamma_0}{\sin 2\psi_0 (\Delta\gamma')^2 m} \qquad (10.31)$$

where
  ψ_0 is the nominal (central) beam incidence angle (Figure 10.4)
  $\Delta\gamma'$ is the 3-dB one-way γ-beam width, in degrees
  m is the slope of the γ_0 versus ψ curve at the ψ_0 angle, in decibels per degree
  γ_0 is the angle between the longitudinal axis of the aircraft and the beam centroid (Figure 10.3)

For typical Doppler radar parameters, the uncompensated error given by Equation 10.31 is approximately

$$\epsilon_w = 0.1(\Delta\gamma')^2 m \qquad (10.32)$$

Equations 10.31 and 10.32 are also valid over land terrain but are typically negligible because m is generally small. Over water, however, if no compensation techniques were used, ϵ_w could take on peak values anywhere between 1% and 5% over an extreme spread of sea states, depending on the radar parameters used. Note also the strong dependence upon $\Delta\gamma'$ and thus on antenna size.

Several techniques have been developed to compensate for the overwater calibration shift or sea bias error. Early Doppler radars often employed a manual land-sea switch operated by the flight crew. When the switch is in the sea position, an overwater calibration shift correction is added to the Doppler radar velocity output corresponding to the most frequently occurring sea state expected on the missions flown. The residual error is the difference between the actual sea state and the one used for calibration. Based on a Gaussian distribution of the probability of sea-state occurrence and a properly chosen land-sea switch setting, the residual overwater calibration-shift error for this land-sea switch technique is near 0.3% to 0.6% (1-sigma).

A fully automatic technique for sea bias compensation used in early Doppler radars was *lobe switching* [19]. In this technique, each antenna beam is oscillated periodically by a small amount in the γ-direction at a low rate (e.g., 20 Hz). If the oscillations are square wave, the return signal consists of two Doppler spectra existing alternately in time at the switching rate. The frequency tracker (which bears some similarity to the two-filter tracker discussed in Section 10.1.2) effectively places a narrow filter at the point where the two spectra have equal power or crossover and reads out the corresponding frequency as the aircraft's velocity. The crossover point at a particular aircraft speed is the same for both land and water, since the returned energy for the two spectra were derived from the same group of scatterers and at the same incidence angle. A similar technique using *simultaneous lobing* by means of a "monopulse" type antenna achieves essentially the same effect [34].

The lobe-switching and simultaneous lobing techniques achieve a large reduction in the overwater (and overland) calibration-shift errors but can cause a significant increase in cost and complexity. For these reasons they are being replaced by newer techniques, such as beam shaping, wherein the calibration shift is reduced by the use of a special beam geometry. In this approach the beam geometry is shaped to cause the centroid of the beam to remain at the same γ-angle even when the slope m of σ_0 has changed. In 1996, antenna design and fabrication techniques provided considerable flexibility in shaping the beam to the desired geometry. In one technique, the beam is generated as the product of a function of γ-angle, $f(\gamma)$, and a function of ψ-angle, $f(\psi)$. The received signal is the product of $f(\gamma)$, $f(\psi)$, and $\sigma_0(\psi)$. For small variations in ψ, σ_0 can be replaced by $(m \times \psi)$. A change in m causes $[f(\psi) \times m \times \psi]$ to change but not $f(\gamma)$. Thus, the resultant spectrum shape is that of $f(\gamma)$, since $[f(\psi) \times m \times \psi]$ multiplies all elements of $f(\gamma)$ equally. The overwater calibration shift in the forward or H' direction is reduced to the extent that the beam shape approximates $[f(\gamma) \times f(\psi)]$. The calibration shifts in the D' and V' directions are not compensated by this technique but are generally small, since V'_D and V'_V are small compared to V'_H. In 1996, a residual overwater shift bias of 0.1% to 0.2% was achieved with this technique. As an aircraft flies over areas of changing sea state, the residual error will appear as a slowly varying random error of 0.05% to 0.1%.

Sea-current and tidal effects result in errors, because the Doppler radar measures velocity relative to the moving sea surface, giving rise to a navigation error relative to the Earth. Fortunately, the speed of sea currents is generally very low. According to available information, general random sea currents rarely exceed a speed of 0.4 knots. Also, since these random currents have random directions, their effects tend to average out for flights of any appreciable distance. Major currents such as the Gulf Stream have a maximum surface speed of less than 3 knots. Where necessary, manual compensation for flight over major sea currents can be made in the navigation computer in flight or before the start of a flight. The sea-current error is a bias, speed-offset error.

In some special applications, such as the dropping of sonobuoys in antisubmarine warfare, the velocity of the aircraft with respect to the sea is actually the desired quantity, and no sea-current correction need be made.

The *surface-wind water-motion error* is caused by wind-blown water particles at and above the surface of the sea. Since these surface water droplets are a portion of the scattering surface seen by the Doppler radar beams, their motion results in an error in the velocity measurement. Theoretical analysis and actual measurement of this error have revealed that the error is [24]

$$\epsilon_{SW} = 1.28 W^{1/3} \qquad (W > 2\,\text{knots}) \qquad (10.33)$$

where ϵ_{SW} is the surface-wind, water-motion error in knots and W is the surface-wind speed.

It has been found experimentally [23] that this effect is not observable at wind speeds of less than 2 knots. Experiments have also shown that the angular difference between the mean spray direction and the wind direction is generally below 30° [23]. For automatic correction of this error by means of surface-wind data, it is reasonable to assume that the direction of the error is in the direction of the wind. The surface-wind water-motion error is a bias speed-offset error. Since the direction of surface winds varies over any appreciable area, the surface-wind, water-motion error will be reduced by averaging. It is of interest to note that the water's *wave motion* as such does not produce any Doppler radar error, since the water particles are not actually moved forward in the wave action but undergo a periodic up and down movement.

Maneuver-Induced Errors Aircraft maneuvers, such as acceleration and turns, cause the components of velocity along the beams to change. The frequency trackers must follow these changing frequencies. To reduce short-term fluctuation noise, a smoothing time of about 0.1 sec is usually incorporated into the frequency tracker. As a result, a velocity lag exists in the presence of acceleration. If the tracker dynamics are first order, this velocity error is approximately equal to half the acceleration times the smoothing time. Fortunately, such errors are transient and can be further reduced by appropriate frequency-tracker instrumentation, such as the use of a double integrator in the tracker loop. In fixed-antenna systems, aircraft pitch and roll rates can give rise to changes of Doppler frequency like those produced by aircraft acceleration and will cause equivalent lag errors if the tracker dynamics are not capable of following them. Therefore, fixed-antenna systems are more demanding on tracker dynamics than physically pitch-and-roll-stabilized antenna systems (Section 10.1.2).

Error of Attitude Stabilization or Conversion from Vehicle to Ground Coordinates The basic Doppler velocity information is obtained in antenna coordinates, which, in the case of fixed antennas, are the same as aircraft coordinates. Velocity is required in ground coordinates for purposes of navigation and is therefore transformed through the pitch and roll of the vehicle in a stabilization computer, which receives the pitch-and-roll information from the vertical reference (Chapter 9). The resulting error then depends on the error in the vertical reference and the stabilization computer, as well as on the values of pitch, roll, and drift and on beam geometry. The error due to roll is relatively small as long as the drift angle is small. Furthermore, roll angles tend to average out over reasonable flight lengths; pitch angles generally do not, because of angle-of-attack changes due to aerodynamic loading and prolonged periods of climb and descent.

Using partial differentiation techniques in connection with Equations 10.9–10.11, which inherently assumes that the errors in pitch and roll are small, the following expressions can be derived for the errors in the three orthogonal velocity components in ground coordinates resulting from errors in pitch and roll in either a pitch-and-roll-stabilized system or a fixed-antenna system that is compensated for pitch and roll:

$$\delta V_H = (-V'_H \sin P + V'_D \cos P \sin R + V'_V \cos P \cos R)\delta P$$
$$+ (V'_D \sin P \cos R - V'_V \sin P \sin R)\delta R \qquad (10.34)$$

$$\delta V_D = (-V'_D \sin R - V'_V \cos R)\delta R \qquad (10.35)$$

$$\delta V_V = (-V'_H \cos P - V'_D \sin R \sin P - V'_V \sin P \cos R)\delta P$$
$$+ (V'_D \cos R \cos P - V'_V \cos P \sin R)\delta R \qquad (10.36)$$

where

$\delta V_H, \delta V_D,$ and δV_V	are absolute errors in the along-heading, cross-heading, and vertical-velocity components in Earth coordinates
$V'_H, V'_D,$ and V'_V	are along heading, cross-heading, and vertical-velocity components in aircraft coordinates
δP	is the pitch-angle error
δR	is the roll-angle error

Equations 10.34 through 10.36 can be further simplified by expressing the errors as a function of ground velocity components (V_H, V_D, V_V), namely,

$$\delta V_H = V_D(\sin P)\delta R + V_V \delta P \qquad (10.37)$$

$$\delta V_D = \delta R(-V_H \sin P - V_V \cos P) \qquad (10.38)$$

$$\delta V_V = -V_H dP + V_D(\cos P)\delta R \qquad (10.39)$$

For a pitch-and-roll-compensated system, the approximate errors ΔV_g in ground speed and $\Delta\delta$ in drift angle δ are

$$\Delta V_g = V_V(\delta P \cos d - \delta R \sin \delta \cos P) \qquad (10.40)$$

$$\Delta\delta = -\delta R\left(\sin P + \frac{V_V \cos P \cos \delta}{V_g}\right) - \frac{\delta P V_V \sin \delta}{V_g} \qquad (10.41)$$

Calibration Error   In 1996, the primary errors in the calibration constants of Doppler radars result from errors in the central antenna beam looking angles. The effect of errors in these angles on the velocity output can be obtained by differentiating Equations 10.4, 10.5, and 10.6 with respect to α_0 and θ_0. The departure of each beam angle from its nominal value can be measured by flight test or on an antenna range. Values below 0.05% are typical for this error.

Installation Error   When installed in the aircraft, the antenna must be accurately aligned with either the longitudinal axis of the aircraft or the reference axis of the heading reference. In fixed-antenna systems, an azimuthal error in antenna installation will contribute to both of the horizontal-velocity component scale-factor errors of the system. In track-stabilized systems, an azimuthal antenna misalignment results directly in a bias error in drift angle. Antenna-installation errors about the pitch-and-roll axis have the same effect as the attitude errors discussed previously. Generally, pitch-and-roll installation errors have a significant impact only on vertical velocity accuracy.

Error in Data Conversion and Readout   In 1996, electronic frequency-to-digital conversion devices exhibit errors between 0.01% and 0.05%. For use with digital navigation computers, Doppler frequency velocity data are converted into binary digital form with very small errors. In such a digital system, the error is a function of the time-base inaccuracy and of the number of binary bits used. Since these digital quantization errors are uniformly distributed, the standard deviation of the error is obtained by dividing the quantization error by $\sqrt{12}$.

Doppler Radar Velocity Errors   There are three types of errors:

1. Bias, speed-offset errors ϵ_b^o (expressed in knots), which cause position errors $= \epsilon_b^o T$, where T is the time since the last fix. Unknown offset bias errors can be calibrated for the mean bias expected on an ensemble of flights; ϵ_b^o is the deviation of the offset bias from the calibrated value.

2. Bias scale-factor errors ϵ_b^s (expressed in percent), which cause velocity errors $= \epsilon_{b_i}^s$ times V_i, where V_i is the velocity component V_H', V_D', or V_V', and $\epsilon_{b_i}^s$ is the corresponding scale-factor error. The scale factors are calibrated for the mean value expected in flight; $\epsilon_{b_i}^s$ is the deviation from the calibrated value.

3. Random velocity errors. These are generally errors assumed to have zero mean; that is, the mean is included in errors 1 or 2 above. If the random velocity errors have a mean square σ_r^2 and are assumed to be exponentially autocorrelated with a correlation time τ_r, which is much less than the flight time T, the distance error $\epsilon_D = (2\tau_r T \sigma_r^2)^{1/2}$. These random errors generally consist of two types; those with relatively long correlation times, and the Doppler-fluctuation error, having a correlation time of 0.1 sec. These errors are frequently expressed as a fraction of distance traveled (or a fractional speed error) after a total smoothing distance D, such that $\epsilon_D/D = \epsilon_V/V = (2\tau_r \sigma_r^2/DV)^{1/2}$. Hence, they cause a position error that is proportional to the inverse square root of D.

Since the individual error sources are statistically independent, the total velocity component errors are obtained by taking the square root of the sum of the squares (rss) of the individual errors. This rss combination of errors should then represent the average performance of many systems on many flights. The Doppler radar velocity error is normally expressed in terms of the three orthogonal velocity components σ_{V_H}, σ_{V_D}, and σ_{V_V}. Since each of these component errors is assumed to have a normal probability distribution, the standard deviation error σ is the 68% probability error (Section 2.7).

In 1996, typical lightweight small-size Doppler radars, operating over land and after 10 nmi of travel, have standard deviation velocity errors of less than 0.25% plus 0.1 kn.

The above errors describe the Doppler velocity error for overland operation. For overwater operation, errors due to water motion and the calibration shift error (due to the larger slope of the radar backscattering coefficient versus incidence angle) must be included. In 1996, Doppler radars typically employ

TABLE 10.1 **High-performance Doppler radar velocity errors (over land)**

Error	Value (1-sigma)
Doppler fluctuation (after 10 nmi of flight	0.073%
Altitude hole (residual after modulation wobbling)	0.02%
Readout (data conversion)	0.02%
Beam direction (antenna boresight and radome)	0.065%
Installation (affects primarily V'_D)	0.03%
Frequency tracker bias	0.05 knot
Typical total overland orthogonal velocity component error	0.10% + 0.05 kn (1-sigma)

beam shaping to automatically compensate for the overwater calibration shift, resulting in a residual error of 0.1% to 0.2%. The residual water-motion error after compensation for the known water motion effects in the operating area is typically 0.5 knots. The resultant velocity error due to overwater operation is (0.15% + 0.5) knots 1-sigma, which must be added to the overland errors to obtain the total overwater error.

High-performance Doppler radars have been designed that have total standard deviation errors less than half of those cited above. Most of the Doppler radars of that type included either the lobe switching [19] or simultaneous lobing [34] overwater calibration-shift compensation technique. These high-performance Doppler radars exhibited overland V'_H and V'_D velocity component errors as shown in Table 10.1.

For overwater operation, the residual water motion error of 0.5 kn plus the residual overwater calibration shift error of 0.035% must be added to the values in Table 10.1, resulting in a total overwater error of (0.11% + 0.5 kn) (1-sigma). The V'_V component velocity error is generally two to three times smaller than the V'_H and V'_D components.

Doppler-Navigation System Errors Doppler velocity information is measured in aircraft coordinates and must be transformed through aircraft heading, pitch and roll in order to generate the desired navigation information. The effects of pitch-and-roll errors on horizontal navigation are small and can generally be ignored. Heading errors affect horizontal velocity, and hence navigation directly. Quantitatively, a 1° error in heading represents a 1.75% cross-track position error, as given by the expression $\sigma_{Th}(\%) = 1.75\sigma_h(\text{deg})$ where σ_{Th} is the percent standard deviation of the cross-track distance error due to heading error, and σ_h is the standard deviation of the heading error in degrees. Heading error can thus have a major effect on Doppler-navigation system accuracy. In 1996, navigation computers are digital and contribute a negligible error.

The total position error of a Doppler-navigation system is thus determined by the errors of the two major components of the system, namely, the Doppler radar and the heading reference. The position error may be determined on the basis of these errors and statistical considerations of the problem as a two-dimensional error problem (Section 2.7). The two dimensions usually chosen are the along-track and cross-track directions, frequently called range and transverse directions, respectively. This assumes, as is generally justified for Doppler-navigation systems, that the covariance matrix of the errors is diagonalized along these directions, so these errors may be considered independent. The range error σ_R consists of the error in *Doppler along-heading velocity* σ_H, and a very small second-order contribution of heading error. The latter is so small that it can safely be neglected. Thus, σ_R can be taken as equal to σ_H. The *cross-track (transverse)* error σ_T consists of the error in *Doppler cross-heading velocity component* σ_D and the error due to the heading reference σ_{Th}. In this dimension, the heading error is first order and generally is the dominant error. Based on measured results, the range and transverse errors exhibit roughly normal probability distributions. The resulting position error therefore exhibits a two-dimensional normal or elliptical probability distribution (Section 2.7), the two dimensions being σ_R and σ_T. The standard deviation of σ_T is given by the root sum square of the two individual cross-track errors: $\sigma_T = (\sigma_D^2 + \sigma_{Th}^2)^{1/2}$. Measured results [8, 26] have shown that σ_R and σ_T, and hence the percent position error, of Doppler-navigation systems are inversely proportional to the square root of the distance traveled. This behavior may be attributed to the effects of the smoothing of certain slowly varying errors.

Flight test results of complete Doppler-navigation systems have shown 68% probability position errors of near 1% of distance traveled. The results of an extensive flight test program on a military system [24] showed 1-sigma range (along-track) errors near 0.25% and transverse (cross-track) errors near 0.5% for 1000-mi flight lengths, which amounts to a 68% probability position error of 0.63%. These systems used gyromagnetic heading references and analog computers of that period and had no automatic overwater calibration shift compensation techniques incorporated into them. In 1996, the performance improvement of heading references (Section 9.4) and digital navigation computers made it possible to capitalize on the inherent accuracy capability of Doppler radars. Also, the various automatic overwater calibration shift compensation techniques have greatly reduced the previously largest error of the Doppler radar itself, namely, the overwater bias error. The effect of these performance improvements was verified during a flight test in a fixed-wing aircraft of a fixed-antenna FM-CW Doppler radar with beam shaping (for overwater bias correction) and an inertial quality heading reference. This test exhibited an along-track error of 0.14% (1-sigma) and a cross-track error of 0.15% (1-sigma) [7]. The flight test consisted of 207 legs over land for a total distance of over 6000 nm.

In 1996, large quantities of military helicopters were outfitted with Doppler-navigation systems. In most cases, these systems used a magnetic heading reference, and their flight profiles consisted of short legs (10 to 20 nmi) between waypoints. Typical total position errors exhibited in flight tests were along-track error of 0.25% (1-sigma) and cross-track error of 1% (1-sigma) of distance traveled. The cross-track error is strongly dependent upon the accuracy with which the magnetic compass is calibrated or "swung." In some applications the compass "swing" is performed in flight by flying over known checkpoints and measuring the position error on that leg and on that course. The cross-track component of the position error, converted into degrees, is then inserted into the navigation computer and used to correct subsequent flights at that course angle. The process can also be performed using *electrical swinging* (see Section 9.4.6).

10.1.5 Equipment Configurations

One of the first Doppler radar equipments developed, the AN/APN-81, which became operational in approximately 1956, weighed approximately 290 lb, radiated 50 w of average power, and consumed 1700 w. When the navigation computer was added, the total weight of the Doppler navigation system was 700 lb. Doppler systems performing the equivalent function in 1996, weighed 12 lb. including the antenna, all electronics, and a MIL-STD-1553 data bus interface (Section 15.2); they radiate 20 mw of average power and consume 20 w.

Figure 10.14 shows the AN-ASN-157 Doppler-navigation system for helicopters. The antenna is at the bottom of the unit. It uses a single four-beam fixed antenna, a Gunn diode transmitter, and an integrated receiver/antenna beam-switching module. The modulation is FM-CW, with a modulation frequency of 30 kHz. Its weight is 12 lb. Variations of that type of system use multiple modulation frequencies to overcome altitude hole effects above 15,000 ft. Another single–unit FM-CW Doppler radar is the CMC-2012, designed for helicopter operation. It has a MIL-STD-1553 data bus interface. Its horizontal velocity range is −50 to +350 knots, and this particular radar has demonstrated a three-axis hover accuracy of 0.72m/min. In 1996, Doppler-navigation systems weighing 9 lb were operational in drones providing navigation and altitude above terrain. The latter systems usually employ the FM-CW technique described in Section 10.2.4.

Considerable effort was underway in 1996 on combining Doppler radars with GPS (Section 5.5) by embedding a GPS receiver module into the Doppler radar unit and using the data from both sensors [9]. This approach combines the high accuracy of GPS (Section 5.5.10) and the continuous dead-reckoning operation of Doppler navigation. Use of the GPS data solves the initialization problem of dead-reckoning navigation systems. For low altitude operation, such as in helicopters, the Doppler radar provides continuity of navigation when GPS signals are not available or loss of tracking occurs, due to terrain, foliage masking, severe maneuvers, or jamming. Also, the Doppler radar can provide accurate velocity for aiding the GPS-tracking loops during acquisition in GPS State-3 operation. Use of the GPS data overcomes the Doppler-navigation system limitation of increasing position error with distance traveled due to heading reference and Doppler velocity errors and water motion errors. In 1995, such a Doppler/GPS navigation set was successfully flight tested by the U.S. Army [30]: and by 1996 production of it was in progress.

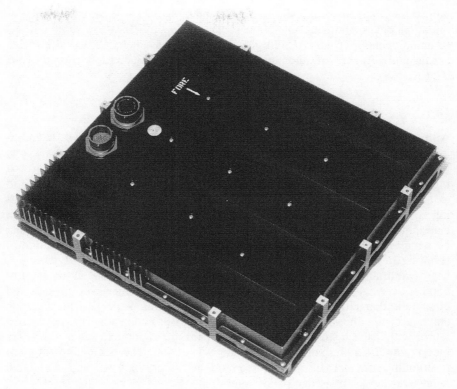

Figure 10.14   AN/ASN-157 Doppler radar (courtesy, GEC-Marconi Company).

10.2   RADAR ALTIMETERS

10.2.1   Functions and Applications

The basic function of the radar altimeter is to provide terrain clearance or altitude with respect to the ground level directly beneath the aircarft. The altimeter may also provide vertical rate of climb or descent and selectable low altitude warnings. In 1996, radar altimeters typically weighed 4 to 10 lb, exhibited a 1.5 ft or 2% altitude accuracy and transmitted 5-w peak pulse or 500-mw average CW power for a 5000-ft altitude capability. Altimeters built during the early 1960s weighed 15 lb or more and transmitted 100 w of peak pulse power. Performance characteristics are designed to match particular applications. High-performance low-flying military aircraft and cruise missile systems require accurate altitude tracking at vertical rates of over 2000 ft/sec while maintaining a high degree of covertness and jam immunity. Altimeters designed for terrain correlation for navigational purposes (Chapter 2) must process an extremely small ground illumination spot size at high altitudes to provide the required altitude resolution. Altitude marking radars are generally low altitude altimeters designed specifically to provide mark signals at specific altitudes for initiation of an automatic operation such as fuze triggering on submunitions or chute opening on lunar-landing systems. Radar altimeters for civil aviation are designed to support automatic landing, flare and touchdown computations (Chapter 13).

10.2.2   General Principles

Altimeters perform the basic function of any range measuring radar. A modulated signal is transmitted toward the ground. The modulation provides a time reference to which the reflected return signal can be referenced, thereby providing radar-range or time delay and therefore altitude. The ground represents an extended target, as opposed to a point target, resulting in the delay path extending from a point directly beneath the aircraft out to the edge of the antenna beam. Furthermore, the beam width of a dedicated radar altimeter antenna must be wide enough to accommodate normal roll-and-pitch angles of the aircraft, resulting in a significant variation in return delay. For example, a typical 50° beam-width antenna, at 5000-ft altitude, provides an altitude delay of 5000 to 5500 ft over flat terrain. The resulting altitude error becomes even greater, and potentially dangerous, when flying low over mountainous peaks with the wide antenna beam illuminating adjacent valleys. To provide range to the nearest return within the bounds of the antenna beam, many modern radar altimeters incorporate a leading edge range tracker servo loop. The tracker functions to position the gate in a pulse modulated radar or a filter in a frequency-modulated radar over the leading edge of the return.

Frequency Band   The frequency band of 4.2 to 4.4 GHz is assigned to radar altimeters. This frequency band is high enough to result in reasonably small sized antennas to produce a 40° to 50° beam but is sufficiently low so that rain attenuation and backscatter from rain have no significant range limiting effects.

Antenna Requirements   Typical installations include a pair of small microstrip antennas for the transmit and receive functions. The antennas typically provide 10-dBi gain with a 50 × 60 degree beam. A 30-in. spacing between transmit and receive antenna provides about 85-dB isolation. The antennas are spaced to provide isolation loss greater than the maximum expected ground return loss at low altitudes. At low altitude, the radar loop sensitivity is limited, as a function of altitude through a sensitivity range control (SRC) mechanization, to allow the altimeter to detect the ground return without detecting antenna leakage.

10.2.3   Pulsed Radar Altimeters

The basic pulsed radar altimeters of the early 1960s typically operated with 100-w cavity-tuned tube oscillators to provide performance to 5000 ft. These early radars evolved to 5-w solid state transmitters incorporating receiver pre-amplifiers in the 1970s and provided a high degree of reliability, low probability of intercept (LPI), small size, and high accuracy. During the 1980s, the discrete transmit/receive RF circuits were replaced with GaAs MMIC's (Microwave Monolithic Integrated Circuits). In 1996, receive/transmit functions were being accomplished with RF hybrid packages incorporating the GaAs MMIC functions. Advances in digital integration technology allowed replacement of the analog leading edge altitude tracker loops with digital altitude tracker loops. Although the basic radar altimeter functions have not changed significantly, incorporation of these technology advances resulted in vastly improved performance and reliability at a fraction of the cost and size of earlier systems.

The functional diagram of Figure 10.15 illustrates the basic operation of pulsed altimeters. The PRF generator provides the modulation wave form for the transmitter and a t_0 time mark for the tracker loop. The transmit signal is derived from a stabilized oscillator, modulated, and amplified to the desired power level. The receiver low-noise amplifier (LNA) typically has a 2- to 3-dB noise figure, resulting in a sensitivity level that allows relatively low transmit power (1 to 5 w). A separate stabilized local oscillator (LO) provides down conversion to the IF for band limiting, amplification, and envelope detection. The track loop positions the range gate at the leading edge of the signal return to provide "gate overlap" energy corresponding to the leading edge threshold reference. The Class-2 servo loop, consisting of range rate-and-range integration, is designed to provide the track rate required to follow terrain elevation change rates at the velocity limits of the particular aircraft. This typically results in a closed loop bandwidth of 100 Hz to provide a 2000-ft/sec range rate capability. The integrated gate position error is added to the altitude register that repositions the gate through the altitude to time delay converter. The time delay converter is typically a high-speed counter or analog ramp generator referenced to the transmit t_0 time mark. Additional functions not shown in Figure 10.15 include noise AGC to control IF gain, closed loop transmit power management control (PMC) to maintain minimum transmit power for (LPI), sensitivity range control on the receiver to provide necessary low-altitude desensitization for assurance against false lock onto antenna leakage or aircraft appendages, and altitude output conditioning, altitude rate derivation, and aircraft data bus interfaces. Figure 10.16 shows the APN-194 pulsed radar altimeter with antennas and indicator. This system, a standard for all U.S. Navy fixed-wing aircraft, provides 5000-ft altitude capability in a 6-lb package.

10.2.4   FM-CW Radar Altimeters

Figure 10.17 is the block diagram of a typical FM-CW radar altimeter. A linear frequency modulation is applied to the transmitter. Range delay is determined by measuring the frequency difference of the return signal with respect to the transmitter through the receive mixing function. To provide a nonzero IF (intermediate frequency), the transmit frequency is mixed with an IF offset local oscillator (LO) and filtered to provide the lower sideband ($f_t - f_{LO}$) for

494

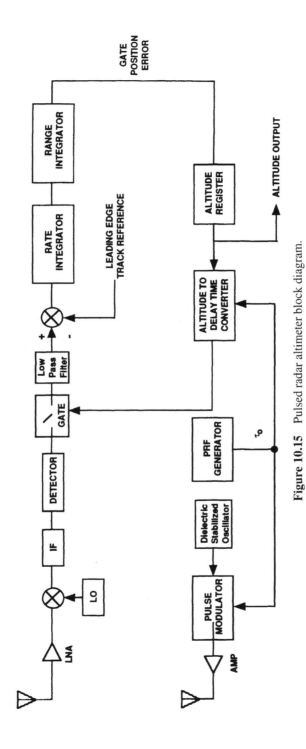

Figure 10.15 Pulsed radar altimeter block diagram.

Figure 10.16   APN-194 pulsed radar altimeter (courtesy, Honeywell, Inc.).

converting the radar return to an IF of f_{LO}. The balanced detector converts the return to base band. Altitude can then be determined with a frequency counter. To reduce errors due to Doppler shift of the return, a triangular wave form is commonly used to modulate the transmitter. Thus, a positive Doppler shift will produce a negative frequency error on the rising modulation slope and a positive error on the falling slope. By averaging the frequency count, the error due to Doppler shift is minimized.

The IF and low-frequency amplifiers must maintain a bandwidth sufficiently high to accommodate the frequency shift at maximum altitude. The track loop is commonly closed on the LO to provide a fixed return frequency shift resulting in narrow processing band width and high sensitivity. Altitude is then determined by counting the LO frequency required to hold the return within the narrow band width. At low altitudes (i.e., below a few hundred feet) the LO frequency is fixed, and altitude is determined by counting the relatively low return shift directly. Accuracy is a direct function of transmit modulation linearity. As linearity worsens, the return spectrum widens resulting in altitude error. Closed control loops are normally incorporated as part of the modulation scheme to provide linearity. Transmit frequency is sampled and compared against a linear slope reference to provide the feedback error signal necessary to control the modulator. Typically 500 mw to 1 w transmit power and accuracies of 1.5 ft below 100 feet and 2% above 100 feet are realized. Disadvantages of the FM-CW altimeter include the step error ($\Delta h = c/4\Delta F$) where c is the speed of light and ΔF is the total frequency deviation. The step-error effects can be reduced by wobbling the modulation frequency or the phase of the transmitter output. Standard FM-CW altimeters also have the tendency to lock-on to airframe appendages and strong targets on the surface. Recent tech-

496

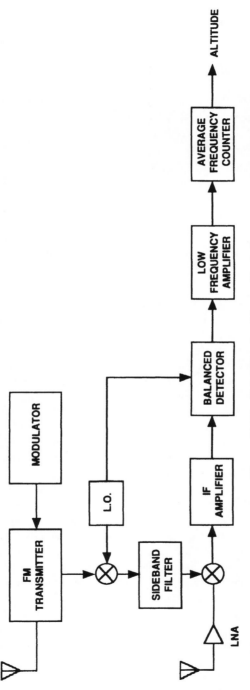

Figure 10.17 FM-CW radar altimeter block diagram.

Figure 10.18   ALA-52A FM-CW radar altimeter (courtesy, Allied Signal).

nology advances have resulted in incorporation of GaAs MMIC RF functions, FFT processors to provide frequency tracking and digital signal processing and control. Figure 10.18 shows the ALA-52A radio altimeter, which is designed for civil aviation, primarily for autoland applications.

Another FM-CW altimeter technique, which has been implemented for specialized applications uses a sinusoidally modulated FM-CW wave form. A particular sideband (Bessel sideband) is selected, and the relative phase difference between the transmitted and the returned signal modulation frequency is a measure of altitude [16]. In 1996, this approach was used on unmanned airborne vehicles. Previously, it was implemented in a combined Doppler velocity and altitude measuring radar.

10.2.5 Phase-Coded Pulsed Radar Altimeters

Pulse compression is typically achieved in pulsed radars through biphase modulation of the transmit pulse. Incorporation of coherent phase modulation within the transmit pulse allows the use of a wide pulse while maintaining a small ground spot size, high-range resolution, and the wide transmitted spectral width of a narrow pulse altimeter. The widened pulse provides higher radiated energy with a low peak power and high duty cycle. Additionally, coherency allows incorporation of a narrow-band predetection filter providing a substantial improvement in receiver sensitivity. While a noncoherent pulsed altimeter with a 1% duty cycle requires about 5 w of peak transmit power at 5000 ft-altitude, a coherent pulsed altimeter with a 25% duty cycle requires only 100 mw of peak transmit power at 5000 ft. The functional block diagram of a typical phase coded pulse radar altimeter is shown in Figure 10.19. Modern radar altimeters of this type typically incorporate GaAs MMIC technology in the transmitter/receiver RF circuits and control loops. Radar return processing is implemented digitally in software by a microprocessor, and high-speed code generation is provided by gate array technology. This technology leads to a tenfold increase in altitude capability at one-fifth the peak transmit power, within the same package size of pulsed altimeters developed in the 1970s. The microprocessor operation provides additional functional capability and versatility not possible with earlier systems. Control loop band widths, transmit pulse widths, and transmit power can be adaptively controlled to optimize performance at all altitudes. Built-in-test, jam detect, closed-loop power management control, frequency hopping, PRF jitter, automatic self-calibration are additional functions possible in radar altimeters of this type. An example of this type of altimeter is the HG9550 shown in Figure 10.20. This phase coded pulsed radar altimeter is designed for 50,000 ft altitude capability with a 1-w transmitter, and it incorporates "down look" altimetry at 4.3 GHz and "look-ahead" capability at 35 GHz.

10.3 FUTURE TRENDS

Based on the early work in 1995, extensive future development is expected on integrating dedicated Doppler radars with embedded GPS receivers (Section 5.5). This integration promises to provide a high-accuracy, cost-effective design that combines the high-positioning accuracy of GPS with the dead-reckoning operation of a Doppler radar. The continuing development of small, low-cost and lightweight GPS receivers and RF components for the Doppler radar will result in total Doppler/GPS navigation systems weighing well below 10 lb. These systems, using differential GPS technology (Section 5.5.9), will include a tactical (military) nonprecision landing capability that will gradually be upgraded to include Category I capability as well (Chapter 13).

In 1996, most operational Doppler navigation systems in helicopters accepted heading information from conventional magnetic compasses (Section 9.4). In the future, units will be developed that use a triad of strapdown magnetic sensors whose outputs can be combined with pitch-and-roll data to compute heading information that is largely free from vehicle maneuver-induced, transient

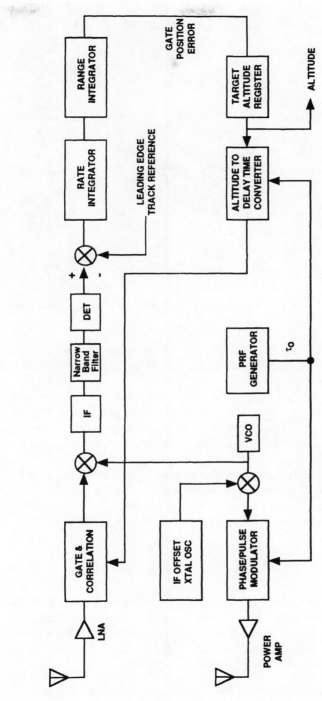

Figure 10.19 Phase-coded pulsed radar altimeter block diagram.

499

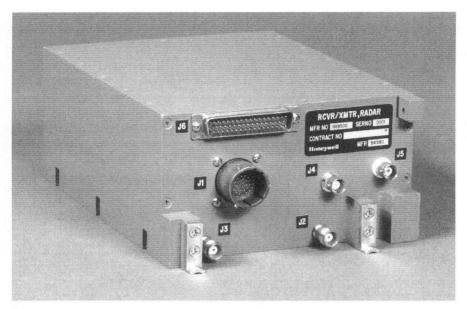

Figure 10.20   HG9550 phase-coded pulsed radar altimeter (courtesy, Honeywell, Inc.).

errors. Such sensors will increase Doppler-navigation accuracy when GPS data are not available. Also, low-cost inertial sensors will be used as heading references for Doppler/GPS systems.

The use of the phase shift of the frequency modulation side bands of the Doppler radar signal to measure altitude will continue and will probably be combined with some form of carrier phase modulation to obtain greater accuracy. Combined Doppler velocity and radar altitude sensors with embedded GPS receivers are likely to be developed in the future.

In the area of radar altimetry, millimeter wave, narrow beam, forward-looking, steered-antenna sensors, combined with downward-pointing wide-beam altimeter antennas, will make it possible for radar altimeters to provide a "look-ahead" capability for determination of altitude ahead of the aircraft.

PROBLEMS

10.1. Derive the fundamental Janus pitch-error expression (10.8), assuming a Janus system with its two beams located in a vertical plane containing the aircraft center line and making equal angles with the aircraft center line. Assume the ground-velocity vector to be along the aircraft center line and zero pitch, climb, and roll angles.

10.2. Calculate the Doppler signal-to-noise ratio per beam for a continuous-wave Doppler radar, using single side-band detection, transmitting 1w at 13.325 GHz from an antenna having a gain of 30 dB, a two-way beam width of 3.68°, and an antenna-pattern factor of 0.5. The beam is in a vertical plane containing the aircraft center line and makes an angle of 75° with it. The terrain has a cross section per unit area σ_0 of -14 dB. The receiver-noise figure is 9 dB. The aircraft is operating at an altitude of 50,000 ft and at a ground speed of 500 knots. Assume the pitch, roll, drift, and climb angles to be zero. Assume the sum of RF losses and atmospheric losses to be -3 dB. Assume infinite isolation between the transmitting and receiving antennas, and assume the transmitting and receiving antennas to have equal gain and beam width.

Ans.: 35 dB.

10.3. What is the efficiency factor, to be used in the radar-range equation, for an FM-CW Doppler radar using the J_2 side band, operating at an altitude of 10,000 ft, with a beam incidence angle of 15°, a modulation frequency of 20 kHz, and a transmitted modulation index of 1.5?

Ans.: 0.144.

10.4. (a) What is the single-beam percent 1-sigma Doppler velocity-fluctuation error after 10 nmi of flight for a Doppler radar having the following characteristics: a wavelength of 0.074 ft, a two-way 3-dB beam width in the γ-direction of 3.5°, and an angle between the aircraft center line and the beam centroid of 75°. Assume that the antenna-beam has a Gaussian shape in the region of interest. Assume a constant speed; straight and level flight; and pitch, roll, drift, and climb angles of zero. Assume a frequency-tracker performance factor of 1.25.

Ans.: 0.059%.

(b) What is the single-beam percent Doppler velocity-fluctuation error for this radar after one second of smoothing at a ground speed of 500 knots?

Ans.: 0.51%

10.5. What are the absolute errors in the along-heading, cross-heading, and vertical-velocity components of a pitch-and-roll-compensated Doppler radar, when the pitch-and-roll-angle errors are each 0.1°, the pitch angle is 10°, the ground speed is 500 knots, the drift angle is 5°, and the vertical velocity is 500 ft/min?

Ans.: 0.022 knot, 0.16 knot, 80 ft/min.

10.6. (a) Derive the expression for the side bands of a continuous-wave Doppler radar that is frequency modulated with a low-frequency sine wave (10.18).

    (b) Show that the resultant Bessel side bands of the first order each contain a Doppler frequency shift proportional to speed.

    (c) Show that the amplitude of the first or J_1 side band is essentially independent of distance to the backscattering surface for short distances.

10.7. Referring to the processing in the block diagram of Figure 10.9, show, by deriving the appropriate equations, that the sense of direction of the velocity component's Doppler shifts are preserved in the transmitter/receiver's output, as required for helicopter operation during hover and backward flight.

11 Mapping and Multimode Radars

11.1 INTRODUCTION

Airborne ground-mapping radars were originally developed in World War II as a means of bombing through clouds and weather—and at night, when the bombing-aircraft operator could not see his target visually. These radars performed two navigation functions. First, they permitted the aircraft to find its way over enemy terrain, without ground navigation aids or sight of the ground, to the threshold of the bombing run. Second, the radar then provided precise navigation during the bombing run by use of cursors set on the target point in a display. Later, a "beacon" mode was added to enable the radar to make fixes on beacon transponders placed at known ground positions in friendly territory and coded to establish their identity. This permitted much easier and more precise navigation by radar during the early phases of the inbound leg to the target and, more important, on return. It was also found that the radar could be used to see intense storms and navigate around them during times of darkness or flight through clouds.

These early radars illustrate the major themes that have concerned navigation radar designers ever since, namely, general navigation; precision navigation for landing, weapon delivery, air drop of personnel and material; and beacon navigation.

Radar-navigation economics also remain much the same. First, a ground-mapping radar is heavier, more expensive, and complex compared to navigation equipment aided by either ground-based systems, such as VOR, Loran, and Omega, or spaced-based equipment, such as GPS. Second, it is uneconomical in comparison with specialized self-contained navigation equipment, such as dedicated Doppler-navigation radars and inertial sensors. However, in regions like the Arctic and Antarctic, where there are few ground aids, wilderness areas, and hostile terrain in combat zones, radars may be the only source of accurate navigation data. Radar navigation became economically attractive with the advent of high-speed, programmable signal and data processors. Then a single radar could perform a multitude of functions solely through additions to software. For example, weather avoidance or weather penetration modes, which are direct descendants of the storm-detection functions of the World War II radars, can be performed by the multimode radar. Other navigation modes implemented through software in multimode radars are terrain avoidance (TA) and terrain following (TF) for low-level flight by military aircraft operation in hostile territory, determination of altitude (AGR), and determination of speed and drift (PVU) for improved dead reckoning.

A less obvious economic factor that is at work in favor of multimode radars is the steady increase in the size and expense of transport aircraft. The great cost of these aircraft and the large number of people they carry makes it imperative that very great safety of operation be obtained. This may make it attractive to add navigation landing-aid modes to the pure weather radars used by commercial airlines.

Another factor at work is the rapid improvement in the reliability and maintainability of current radars. The techniques for producing high reliability by proper design and manufacturing methods have become highly developed in recent years. Radar mean-time-between-failure (MTBF) has risen from tens to hundreds of hours over a very short time span. In addition, analog and digital circuit miniaturization has allowed fault detection and isolation technology to be incorporated down to the most critical complex component. Circuits cards perform self-tests as a background task and report the results to the system's built-in-test (BIT) function when requested. Thus, when the radar does fail, isolation to and identification of the failed complex component is instantaneous. The repair philosophy is to replace the failed module with a spare module. A remote repair facility utilizes the stored BIT codes to select a failed complex component for replacement. Thus, on-site repair work is reduced to a task requiring little skill, and the necessity of keeping highly skilled radar technicians at every airport along each route—an economic impossibility—is eliminated.

11.2   RADAR PILOTAGE

The most common method of position fixing by radar navigation is completely analogous to aircraft pilotage by visual reference. The pilot corrects his heading by observing the terrain around him as a maplike image on the radar display. However, the scale of usable reference points in radar is different. Visually, roads, lakes, parks, railroads, and rivers are most useful for reference when they are within 10 mi of the aircraft. Since radars have narrow antenna beamwidths, typically, 3 to 6 deg, they do not have the eye's field of view and cannot find as many reference points instantaneously. Thus, a sequence of radar modes is employed. Figure 11.1 illustrates the progressive magnification of terrain. First, the radar utilizes the real beam ground map (RBGM) modes to scan a large angular and range sector from a great distance with coarse resolution to find suitable large reference points for initial acquisition. Second, the radar employs the Doppler beam sharpening (DBS) mode to decrease its field of view about the selected ground point, magnify the radar image with moderate resolution, and refine the acquisition. Third, a fine resolution image with narrow field of

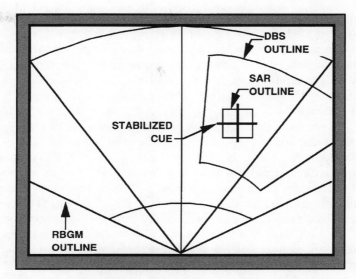

Figure 11.1   RBGM, DBS, SAR coverage shown on RBGM display.

view is generated using the synthetic array radar (SAR) mode and utilized for final designation and position update. The initial acquisition scan has resolution such that towns, cities, lakes, and rivers can be recognized at distances as great as 160 nmi from the aircraft. Subsequent, finer resolution radar images have resolution such that individual streets within a city or specific terrain features, such as road intersections, can be recognized at ranges as large as 50 nmi. Radar resolution required to identify terrain and cultural features is summarized in Table 11.1.

In the use of references points, radar has an advantage because in visual pilotage the distance to an object and its angular position with respect to the aircraft must be estimated. This can be done with adequate accuracy at low

TABLE 11.1   Radar image resolution requirements

Feature	Square Resolution or Cell Size
Coastline, outline of cities, detection of large discrete scatterers, outline of mountains	500 ft
Detection of major highways (some fading), variations in fields, large airfields	80 to 100 ft
City street structure, reliable detection of highways, large building shapes, small airfields. "Roadmap" type imagery	30 to 50 ft
Reliable detection of vehicles, shapes of houses, buildings	5 to 20 ft

altitudes; however, at high altitudes, distances are very difficult to judge with visual pilotage, even in clear weather. With radar pilotage, accurate range and angle measurements can be made by superimposing range and azimuth markers on the radar image feature selected for the update point and instructing the radar to compute the aircraft position vector relative to the update point.

The radar display format used for initial acquisition is the plan-position-indicator (PPI) format. A large azimuth angular region, typically 120 deg, about the aircraft velocity vector is scanned, and the radar collects a few pulses of noncoherent radar return over a large range interval, 20 to 160 nmi dependent upon the range scale chosen, at each antenna beam position. RBGM images are formed in the radar signal processor and stored digitally in its memory as an amplitude, range bin, azimuth angle matrix. To display these data to the pilot in a geometrically correct format, the range bin amplitudes for each antenna beam position, or *range trace*, are read sequentially from the processor memory and "painted" on the display along a radial sweep with azimuth angle corresponding to the antenna's position relative to a display reference angle, typically North up or aircraft velocity vector up, at the time of data collection. The range trace starts at a range slightly larger than the range corresponding to the width of the transmitted pulse, or "main bang," and proceeds to a range corresponding to the maximum number of range bins that are processed.

Typically, radars collect 256 to 512 range samples of data from each radar pulse and vary the range bin width to provide range scales from 20 to 160 nmi. Once the first range trace of data is displayed, the radar retrieves the second range trace of data and "paints" it on the display along a radial corresponding to the azimuth angle of the antenna when that data were collected. This retrieval and display "painting" process continues until all data collected during the scan are displayed. Then, the next scan of data is retrieved and displayed. To the pilot, this process has the appearance of a display "painting" the raw radar returns on the display in synchronism with the scan of the antenna as it sweeps back and forth across the aircraft's velocity vector.

Radars form images in angle and range coordinates, rather than azimuth and elevation angle coordinates, as in a visual image. Despite this, the radar image bears a direct resemblance to the topographical features of the terrain it represents—perhaps because it resembles a visual image of the terrain as seen from a great height above the terrain. The correlation between the radar image and the actual terrain varies with the characteristics of the radar, type of terrain, aircraft altitude, and viewing direction. The interpretation of the presentation is easy in most cases; in other cases, it requires experience and knowledge of how radar reflections build an image. Just as in visual pilotage, operators must be trained to read and interpret the radar presentation by study and practice, including prebriefing the pilot about the appearance of the radar images along his route.

The most striking and most easily identified RBGM terrain feature is a land-water boundary. The smooth surface of the water reflects most of the incident energy away from the radar, whereas the rougher land causes energy to be scattered in all directions, including the direction from which it originated (backscatter). As a consequence, a larger portion of the energy incident on land gets back to the radar antenna than that which returns from water, and land areas appear much brighter in the radar image than water areas.

Since radar returns from very short range and very long range must be processed each antenna beam position, a large amplitude dynamic range must be accommodated even for terrain with uniform reflectivity. When urban areas and cities are prominent features of RBGM images, amplitude dynamic range management can be difficult and result in problems of image interpretation. Signal processing was effectively performed "manually" on the display of World War II era radars. The pilot managed signal levels by adjusting the display gain. As a result, amplitude adjustments for land-water contrast saturated all land-area signals, and individual land features would appear to be of uniform brightness even though urban areas have much more radar reflectivity than the surrounding countryside. A skilled pilot surmounted this difficulty by adjusting display gain up and down to bring out differentiation of terrain features at widely different intensity levels. Modern radars are designed to accommodate the large amplitude dynamic range of the radar ground return and form a quality RBGM image. First, the antenna elevation pattern is designed proportional to $\csc^2 \epsilon$, where ϵ is the elevation angle. This allows the radar to illuminate the large RBGM range interval, or *swath*, with an energy profile that partially compensates for radar return variation with range. Second, sensitivity time control (STC), which is a preset variation in radar receiver gain with time delay, is used to reduce further the more intense short-range returns' amplitude. After STC has reduced the amplitude dynamic range of all range bins in the swath to a nearly constant level, assuming uniform terrain reflectivity, automatic gain control (AGC), measures the average intensity across all range bins and azimuth angles to compute a gain setting for the radar's receiver. Typically, receiver gain is adjusted such that the average intensity is 12 dB below a full-scale analog-to-digital converter (ADC). This gain setting will minimize radar-processing distortions. Prior to display, radar images are compensated for the display's amplitude nonlinearities.

Moderate resolution radar images are formed using a Doppler beam-sharpening (DBS) mode and fine resolution radar images are formed using a synthetic aperture radar (SAR) mode. Since radar display resolution and processor memory are limited, radar coverage must be sacrificed for improved resolution. Airborne radar displays have approximately 5×5 in of viewing area. Approximately 256 azimuth by 256 range cells of radar data can be displayed in a geometrically correct format in the data area of this display. Modern airborne radar processors contain approximately one million words of data memory. When the radar's memory is utilized to store the total radar image prior to display, only one complete radar image and one partial radar image (the one currently being generated) can be accommodated. Thus, each time the pilot selects a finer resolution radar imaging mode, the displayed data will be an image with greater magnification, but smaller coverage.

A DBS image is formed by generating resolution cells with constant azimuth angular resolution and constant range linear resolution. DBS offers from 4:1 to about 70:1 azimuth angular resolution improvement when compared to RBGM. Azimuth coverage is typically between 12 and 40 deg with range coverage matched to the azimuth linear extent at the DBS map center. For example, if a DBS map was formed with 12-deg azimuth angular coverage and 50-nmi map center range, the range coverage would be chosen equal to 10 nmi. Coherent processing is utilized to form several DBS range traces simultaneously. The DBS image is displayed in correct geometric format, a truncated PPI scan, centered about the RGBM designation point and with proper azimuth orientation relative to the display reference.

Synthetic aperture radar images are formed about the DBS designation point to magnify further the acquisition area for final designation and position update. A SAR image is formed by generating resolution cells with equal range and cross-range (azimuth) resolution. For example, a SAR resolution cell might have 20 × 20-ft dimensions. SAR radars are designed to provide resolution that is independent of the map center range. Thus, SAR offers an azimuth angular resolution improvement ratio relative to RBGM that is range variable. Since SAR images provide the finest resolution possible, designation error will be minimum. SAR images appear geometrically correct as squares on the pilot's display but rotated in azimuth to the proper aspect between display reference and mapping angle.

The physics of radar reflectivity causes radar images to have anomalies when compared to visual images. Urban areas are very effective in redirecting radar energy back toward its source and are brighter than rural areas. Towns become steadily more reflective than open country as range increases, because radar energy strikes the ground close to the grazing angle at long ranges and little energy from open country is scattered back. Conversely, the vertical sides of buildings become more effective backscatterers, because radar energy hits them at normal incidence at great ranges. As a result, cities are brighter than surrounding terrain at short ranges and stand out alone at long ranges. The difference between the brightness of urban and rural areas is most pronounced when the city is approached on the compass points, because most buildings are oriented to face the compass points and make better reflectors when their walls are viewed square on.

Mountainous terrain imagery is heavily impacted by radar shadowing. In SAR images mountains appear to be illuminated by a brilliant light originating from the aircraft position. Dark shadow areas appear behind hill crests, so far slopes, not illuminated by the radar, are dark. The near slopes are more brilliant than level ground due to more normal incidence of radar energy. Shadow orientation and length change with aircraft altitude and position relative to the mountains being illuminated, making position fixing from the image sometimes difficult, although any isolated peaks may be quite usable references.

The strong directional reflectivity of man-made ground features, motion of terrain shadows, and target scintillation due to the vectorial addition of the returns from the many scatterers within each resolution cell, make the technique of identifying SAR checkpoints somewhat different from that used in visual identification. In visual identification, a single, well-defined terrain feature is typically utilized for an update point. When the point is not visible, there is a good possibility that one is lost and is forcing terrain features to coincide with the chart terrain features when, in fact, they do not. Conversely, only a small number of terrain features need be checked against the chart in order to ensure reliable identification. With radar, the failure to find quite a few terrain features at any moment is almost certain, due to radar reflectivity variability. Several features within the scene must be checked against the chart to ascertain position. Fortunately, this is more easily done with radar than in the visual case, because, due to the maplike nature of the image, the navigation chart can be electrically superimposed directly on the radar image by any one of various alignment correlation algorithms within the processor and the fit between digital navigation data base and radar image at many points can be easily measured.

11.3 SEMIAUTOMATIC POSITION FIXING

Semiautomatic navigational position updating is performed on most aircraft today. Here, the pilot identifies the reference point in the radar image. Predicted aircraft range and azimuth angle to a predetermined ground identification point are generated and the radar is commanded to form a SAR image centered on the identification point (IP). An in-video *cursor* is displayed at the map center along with the radar image. If accumulated navigation errors result in a difference between predicted and actual identification point range and azimuth angle so that the identification point is not at the center of the radar image, the pilot can slew the cursor over the actual identification point, and command a position update. Then the radar processor computes the range and cross-range difference between the predicted and actual identification points locations. This position error vector is added to the original aircraft-to-IP position vector to determine the aircraft's actual position relative to the IP. A block diagram of this navigational update mechanization is given in Figure 11.2.

In general, a complete SAR image cannot be formed during a single coherent processing interval due to illumination limitations of the real airborne antenna. Typically the 3-dB beam width of airborne antennas ranges from 3 to 6 deg. Furthermore, a SAR radar processes less than the 3-dB beam width of the antenna each coherent processing interval to minimize amplitude modulation of the radar return by the real antenna pattern, the so-called *SAR scalloping* phenomenon. Thus, a radar might be commanded to form only 2-deg-wide SAR segments, called *SAR patches*, each coherent processing interval. If a 40-deg-wide map is commanded, the SAR radar would form 20 sequential SAR patches to generate the entire image.

The antenna scan sequence for a SAR image is ground stabilized so that aircraft maneuvers will not cause map distortion. Scan sequence construction is

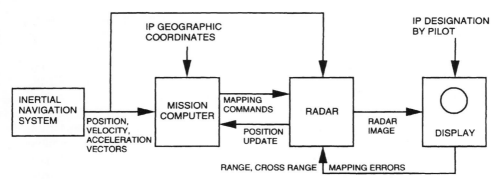

Figure 11.2   Semiautomatic position fixing by radar.

illustrated in Figure 11.3. The initial vector from the aircraft to the map center $P_c(t_0)$ is chosen as the difference between current aircraft position and commanded position (IP latitude and longitude) for the SAR map scan center. Then the SAR map apex is constructed to be on the current aircraft flight path and at a distance D from the current aircraft position equal to one-half the product of estimated scan time and aircraft velocity. The vector from SAR map apex to SAR map center R_c is generated, and an arc normal to R_c is constructed. SAR patch centers separated by the desired fractional antenna beam width are laid out along the arc to determine ground coordinates for each SAR coherent array. For each coherent array a vector $D(j)$ from map enter to the corresponding jth patch center is computed and stored in processor memory.

During SAR image generation, the vector $P_c(t_j)$ from present aircraft position to the fixed map center is computed, for each processor computation cycle

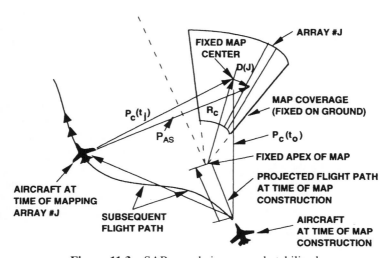

Figure 11.3   SAR map being ground stabilized.

using inertial navigation system (INS) velocity data that have been compensated for the relative motion between antenna and INS locations. (Ideally, the INS would be located coincident with the antenna phase center.) Motion compensation during a coherent array time is based on the antenna-to-patch center vector, P_{AS}, which equals the summation to $P_c(t_j)$ and $D(j)$.

The radar processor stores all P_{AS}, $D(j)$, and velocity vectors used to generate the SAR image. Therefore, when a particular range/Doppler cell is designated to perform a navigation update, the radar processor can construct the exact position vector from the aircraft's present position to the designated ground point. Some radars also provide monopulse measurement data to ground points each patch. In that case, the real antenna azimuth and elevation monopulse gradient vectors are stored for each patch for later retrieval to compute the monopulse angles to any designated SAR image cell. Thus, the radar can construct very accurate position, Doppler, and angle measurements to designated ground points. Even areas of low reflectivity, such as road intersections, can be utilized as IP's for navigation Kalman filter update.

The operating RF wavelength that best balances the conflicting requirements of resolution and weather-penetration ability is probability shorter than that used in radars searching for airborne targets. For one thing, resolution is more important in the ground-mapping case than in air-to-air detection. No matter how great the radar resolution becomes, there are always more terrain features of interest that could be seen if more resolution were available. It is rare that point targets, such as aircraft or missiles, are so closely spaced that distinction of the targets as separate objects becomes the overriding consideration in design.

Also, the radar-range equation is more favorable for mapping than for point target acquisition and tracking. In the basic point target radar-range equation, target return decreases inversely as the fourth power of the range to the target. In ground mapping, the resolvable element of terrain increases in width directly with range, because the element width is $R\theta_{HP}$, where R is the range and Θ_{HP} is the azimuthal beam width. This effect causes the terrain-return signal to decrease as $1/R^3$, rather than $1/R^4$, as in the point-target case. This, coupled with the fact that the terrain element usually has a greater radar cross section than interesting point targets have, results in less emphasis on power in the design of ground-mapping radar than in the design of search/track radars. Details of the calculation of radar range in both the search and mapping cases, as well as the effects of weather, are treated in many texts on radar [5, 6].

11.4 SEMIAUTOMATIC POSITION FIXING WITH SYNTHETIC APERTURE RADARS

Since general descriptions of the operation of synthetic aperture radars are given in various references [1, 2, 3, 4, 5, 6], a detailed description of SAR theory will not be given here. However, the fundamental principles of synthetic aperture radars will be discussed in order to treat their unique properties as navigational radars. Synthetic aperture radar image formation can be explained utilizing either antenna theory or Doppler-filtering theory. Antenna theory will be employed to disclose the basic principle of synthetic aperture radar image formation.

In forming a SAR image, the real antenna on the aircraft serves only as a single radiation element in a long linear synthetic array, as shown in Figure 11.4. The synthetic array itself is generated by motion of the aircraft as it flies by the ground area being imaged. As the aircraft moves forward, the radar antenna is pointed at the ground area, radar transmission pulses are scheduled for the time interval required to achieve the desired azimuth resolution, and radar return range samples for the total range swath are collected and stored, still correctly resolved in range. Image formation is accomplished only after all the data are collected.

At the end of each data collection time, called the *SAR array time*, the radar return samples from each resolvable element in range are retrieved from processor memory and analyzed together. In order that these returns may contain a record of the pulse-to-pulse change in phase of the echoes from that range element, the radar must be coherent. A coherent radar "remembers" the phase and frequency of each transmission by means of an internal oscillator phased to the carrier of the pulse. The radar return signals are heterodyned with this reference signal in a synchronous demodulator. The detector output, called *coherent video*, has an amplitude proportional to the product of the echo amplitude and the cosine of the echo-carrier phase angle with respect to the reference signal. The coherent addition of the pulse returns from a given range, accumulated during the SAR array time, forms the synthetic array beam for each range bin. This process is completely analogous to the addition of radar returns collected by the individual radiation elements of a real phased array occupying the space defined by the aircraft flight path.

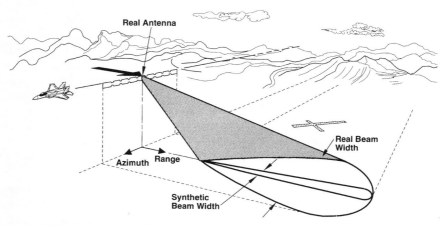

Figure 11.4   SAR image formation explained using antenna theory.

Synthetic array azimuth resolution is twice that of a real array with the same aperture length. A real array has both a one-way and a two-way radiation pattern. The one-way pattern is formed upon transmission as a result of the progressive difference in the distances from successive array elements to any point off the boresight line. This pattern has a $\sin x/x$ shape when amplitude weighting is not utilized. The two-way pattern is formed upon reception, through the same mechanism. Since the phase shifts are the same for both transmission and reception, the two-way pattern is essentially a compounding of the one-way pattern and therefore has a $(\sin x/x)^2$ shape. A synthetic array, on the other hand, has only a two-way pattern because the array is synthesized out of the returns received by the real antenna, which sequentially assumes the role of successive array elements. Since each element receives only the returns from its own transmissions, the element-to-element phase shifts in the returns received from a given point off the boresight line correspond to the differences in the round-trip distances from the individual elements to the point and back. This is equivalent to saying that the two-way pattern of the synthetic array has the same shape as the one-way pattern of the real array of twice the length, $\sin 2x/2x$. Since antenna beam width is inversely proportional to array length, the synthetic array azimuth resolution is twice that of the real array.

In practice, SAR azimuth linear resolution d_A, is very often limited by fluctuations in aircraft velocity. These fluctuations cause errors in the placement of the radiator at its successive positions where radar transmission occurs as the array is generated synthetically. These errors constitute a deterioration of the synthetic antenna pattern; that is, they introduce random phase errors in the array gradient. This is exactly analogous to phase errors in real arrays due to errors in location of the array elements or the phases of the signals fed to them. Just as with a real array, the larger the synthetic aperture, the larger the phase-gradient errors—and the more difficult they are to control. Another practical limitation is the stability of the atmosphere between the array and the targets. Turbulence in the atmosphere introduces optical path difference fluctuations between the target and the aircraft antenna at its successive array positions; this produces errors in the phase gradient of the array. Again, these errors increase in a random-walk manner with increasing aperture length, and impose a limit on the maximum useful aperture length.

Aside from these practical limits to synthetic aperture length, there are theoretical limits to the azimuth resolution that can be obtained. Doppler frequency-filtering theory is useful in explaining unfocused SAR and focused SAR resolutions. First, observe that if the SAR radar is carried on an aircraft that has constant velocity, radar return from each angular direction, on each side of the velocity vector, will have a unique Doppler frequency shift. If a narrow bandpass filter is used to process the ground return, only reflections from scatterers within a narrow angular increment in azimuth will contribute to the output. Thus, the SAR radar acts as if it had a very narrow beam-width antenna pointing in a direction corresponding to targets with Doppler shift equal to the frequency of the band center of the bandpass filter. Second, by moving the filter band center frequency from zero Doppler frequency shift to maximum positive and negative Doppler shifts, the synthetic antenna beam can be scanned over a 180° arc on one side of the ground track. (See also Doppler theory in Chapter 10.) By repeating the process on the other side of the ground track, it is possible to look in all azimuth directions. Third, by using a bank of narrow-bandpass filters, one can get the equivalent of a simultaneous array of contiguous azimuth antenna beams that create a complete high-resolution image of a ground range bin without scanning. Forming simultaneous Doppler filter banks for several contiguous range bins creates the complete SAR image.

Synthetic aperture radars are characterized as either focused or unfocused systems. If the SAR Doppler filter banks are not tuned to the Doppler filter frequency of the desired imaging ground area continuously throughout the SAR array time, the system is said to be *unfocused*. When the SAR Doppler filter banks are continuously tuned, the system is *focused*. The process of tuning is called SAR *motion compensation*.

11.4.1 Unfocused Systems

The azimuth and range resolutions of an unfocused system are determined by the length of time a ground point's radar return remains in a range/Doppler resolution cell. Consider azimuth resolution first. If a strong ground target's radar return drifts from one Doppler filter to another due to its changing Doppler frequency during the SAR array time, part of its energy will be processed in one Doppler filter and part in another. Thus, the SAR image will show a bright spot in two contiguous Doppler filters and SAR azimuth resolution will be degraded. Thus, unfocused SAR azimuth resolution is defined by the change in Doppler frequency to the desired ground point during the SAR array time.

As an aircraft moves, the Doppler frequency shift of a given ground point P_T is not constant since its azimuth angle changes. During a time period ΔT_A, the azimuth angle to ground point P_T will change by $\Delta\theta$ due to the aircraft motion. The time required for the azimuth angle to a ground point at range R to change by $\Delta\theta$ degrees is

$$\Delta T_A = \frac{R\,\Delta\theta}{V\,\sin\theta} \tag{11.1}$$

where V is the aircraft velocity and θ is the azimuth angle between the aircraft velocity vector and the line-of-sight vector to the ground point P_T.

The corresponding Doppler-frequency change is

$$\Delta F_d = \frac{2V}{\lambda}\,\sin\theta\,\Delta\theta \tag{11.2}$$

where λ is the radar transmission wavelength.

The SAR Doppler filter bank passbands must have Δf_d Hz width if the ground point's return is to stay in their passband during coherent integration. The SAR array time T_A is the reciprocal of the Doppler filter passband Δf_d. Therefore, the array time is

$$T_A = \frac{\lambda}{2 V \sin \theta \Delta \theta} \tag{11.3}$$

To a first order, a good radar image will be formed as long as the ground return stays in a single Doppler filter during the coherent integration time. Thus, ΔT_A must equal T_A. Equating 11.1 and 11.3 and solving for $\Delta \theta$ obtains

$$\Delta \theta_{\text{unfocused}} = \sqrt{\frac{\lambda}{2R}} \tag{11.4}$$

for the azimuth angular resolution. The linear azimuth resolution is obtained by multiplying $\Delta \theta$ by R. This yields

$$|d_A|_{\text{unfocused}} = \sqrt{\frac{R\lambda}{2}} \tag{11.5}$$

for the linear azimuth resolution.

So far, our definition of azimuth resolution has been coarse. A refined definition of azimuth resolution is the linear extent of the SAR Doppler filter's 3-dB frequency width when the coherent integration time is chosen to maximize the radar return energy in a resolution cell. This optimal SAR array time is

$$T_A = \frac{1.2 \sqrt{R\lambda}}{V}$$

and the resolution corresponding to this array length is approximately

$$d_A = 0.6 \sqrt{R\lambda} \tag{11.6}$$

Now, consider range resolution of an unfocused SAR radar. Unless compensated, aircraft motion toward the imaged ground point P_T will cause the radar return to move from one range cell to another. The time ΔT_r required for a ground element to move one SAR range bin of width d_r is

$$\Delta T_r = \frac{d_r}{V \cos \theta} \qquad (11.7)$$

This restricts the maximum length of the synthetic array to

$$(D_{SAR})_{\max} = V \Delta T_r = \frac{d_r}{\cos \theta} \qquad (11.8)$$

and the minimum synthetic array beam width to

$$(\Delta \theta)_{\min} = \frac{\lambda}{2(D_{SAR})_{\max} \sin \theta} = \frac{\lambda \cot \theta}{2 d_r} \qquad (11.9)$$

Modern SAR radars compensate for this *range-walk* phenomenon. One implementation progressively delays the start of analog-to-digital converter (ADC) range sampling from pulse to pulse by a time delay amount equal to the product of line-of-sight velocity divided by PRF.

Early SAR systems were employed to generate *synthetic aperture strip maps* for the reconnaissance mission. The radar antenna beam was pointed orthogonal to the flight path as shown in Figure 11.4, and the aircraft was required to fly straight and level during data collection. Radar returns were stored in a tape recorder and processed on the ground by an optical correlator to create an unfocused SAR strip map image [1]. Range walk was not compensated.

11.4.2 Focused Systems

If the center frequency of a Doppler filter is shifted to track the changing Doppler frequency of a ground point P_T, a focused SAR image is created. SAR array time is limited only by the time the target is in the antenna beam of the physical antenna carried by the aircraft. The maximum length of the synthetic aperture D_{SAR} is then

$$(D_{SAR})_{\max} = R \theta_{HP} \csc \theta = \frac{R \lambda \csc \theta}{D} \qquad (11.10)$$

where D is the physical (real) length of the aircraft antenna and θ_{HP} is its physical (real) beam width. Then the azimuth angular resolution of the focused radar is

$$|\Delta \theta|_{\text{focused}} = \frac{\lambda}{2d} \csc \theta = \frac{D}{2R} \qquad (11.11)$$

and the linear azimuthal resolution at range R is

$$|d_A|_{\text{focused}} = \frac{D}{2} \tag{11.12}$$

This result asserts that the linear azimuthal resolution of a focused synthetic aperture radar is not a function of operating wavelength, range to the identification point, or the angle to the identification point with respect to the velocity vector. Furthermore, in order to increase resolution, the aircraft antenna must be made smaller. Of course, in practice, there are limitations imposed by the problems of keeping the synthetic antenna "straight" and by atmospheric "seeing" conditions.

Another limitation is the motion of the target element out of its original range bin, just as in an unfocused system. This occurs when

$$(d_A)_{\text{max}} = \frac{R\lambda \csc \theta}{D} > \frac{d_r}{\cos \theta} \tag{11.13}$$

or

$$d_r < \frac{R\lambda \cot \theta}{D} \tag{11.14}$$

When this occurs, one reverts to Equation 11.8 to find $\Delta\theta$:

$$(\Delta\theta)_{\text{min}} = \frac{\lambda \cot \theta}{2d_r} \tag{11.15}$$

Again, modern SAR radars compensate for this *range-walk* phenomenon.

The limiting resolutions set by Equations 11.12 and 11.15 are much more likely to be reached in a focused system than in an unfocused one, because d_r must be made quite small to equal the linear azimuthal resolution developed in the first case. Relative to antenna theory, the operation of focusing is equivalent to physically bending the synthetic array aperture so that its focal length is equal to the range to the identification point—hence the name. The theory of focused antennas is well developed [8].

The azimuth resolutions realized by a real beam (RBGM) radar, an unfocused SAR radar, and a focused SAR radar are compared in Figure 11.5. Each radar is assumed to transmit a 0.1-ft wavelength (X-band) pulse train through a 10-ft long real antenna aperture. Note that the azimuth resolutions of the RBGM and unfocused SAR radars are range variable, while the focused SAR resolution is constant. This very desirable resolution property is achieved through motion compensation.

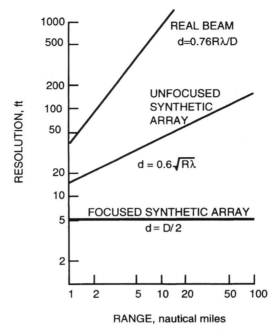

Figure 11.5   Real beam, unfocused SAR, and focused SAR azimuth resolution comparison.

11.4.3   Motion Compensation

Modern synthetic array radars employ motion compensation techniques to create focused radar images during times of aircraft maneuver. SAR beams are focused relative to fixed ground points illuminated by the real antenna beam by subtracting the phase shift due to antenna phase center spatial motion with respect to the fixed ground point from each complex-valued sample of radar return. Each range/Doppler cell has a different Doppler frequency history since the line-of-sight angles to each cell are unique. Thus, the optimum motion compensation approach is to compute and apply a unique phase correction history for each resolution cell. However, this mechanization would be prohibitive to implement from a required processor throughput point of view. Fortunately, several cells, termed the *depth of focus*, can use a common phase correction history.

Two signal-processing formats have been created to generate SAR images: rectangular and polar. The rectangular format is based upon the assumption that the loci of points of constant range on the ground are orthogonal to the loci of points of constant Doppler frequency on the ground. In this case, geometric distortion caused by SAR image creation in the range/Doppler domain is minimal.

The polar format is based upon the assumption that radar returns are actually collected at equally spaced angle increments from a constant range. First, stored radar returns must be resampled to convert from polar format to rectangular format. Then two-dimensional fast Fourier transforms (FFTs) are formed to generate the SAR image. Due to its added complexity, polar format is utilized only for very fine resolution SAR image applications where geometric distortion in rectangular format is severe. For the remainder of this chapter, the rectangular format will be assumed.

Two methods of motion compensation have also been developed. Originally, motion compensation was performed entirely based on on-board inertial sensors that measured the antenna's spatial motion during the SAR coherent processing interval. The radar data processor utilized INS acceleration and attitude vector measurements to compute the line-of-sight phase history for SAR motion compensation. Then, the stored raw radar return was retrieved and the phase history subtracted from the radar return on a pulse-to-pulse basis. Finally, range bin and Doppler filter FFT's were formed sequentially to generate the SAR image.

In the early 1980s, INS technology could not provide the velocity and angular accuracies needed for motion compensation of very fine resolution SAR images. Therefore, the idea of utilizing the actual radar return to compute motion compensation phase histories was explored. This adaptive process has been termed *autofocus*. Algorithms to focus fine resolution SAR images both with prominent discrete radar returns and with only diffuse terrain have now been successfully developed. Autofocus is based on a three step logic: First, the radar forms one or more coarse resolution SAR images utilizing motion compensation data generated from the aircraft's master INS, which may be very remote from the antenna phase center. During this step, the original unprocessed radar returns are not destroyed but are stored in radar processor memory for the final processing step. Extra processor memory is required to store the results of the first step. Second, the coarse image(s) are analyzed to either select a prominent point target, whose stored phase history will be utilized to compute a motion compensation vernier, or to compute the Doppler frequency drift of a visible geometric formation from image to image. Third, the raw radar return data are retrieved, a refined motion compensation correction is applied, and the final SAR image is formed. A generic autofocus functional diagram is given in Figure 11.6.

Since most multimode radars utilized rectangular format to create SAR images, SAR signal-processing will be illustrated for this format. Synthesis of a signal-processing architecture for generating minimum-time SAR images during aircraft maneuvers involves a division of the radar ground return into range/Doppler cells that can be focused by a single-point motion compensation correction. For the range dimension, it is convenient to divide the range swath into zones, whose widths are fixed at a value less than the smallest expected range depth of focus. Signal-processing time will not be adversely affected by a conservative choice for range zone width, say, 16 range bins. The cross-range swath illuminated by the antenna is divided into zones whose widths are less than the computed azimuth depth of focus. Since signal-processing time will

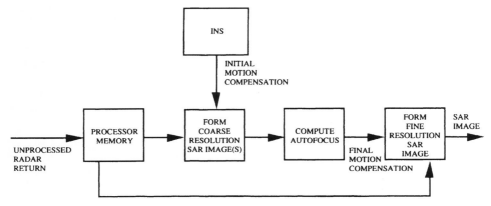

Figure 11.6 Autofocus motion compensation utilizing radar return.

increase in direct proportion to the number of azimuth zones, azimuth depth of focus will be computed each array to select the minimum number of required azimuth zones.

The signal-processing steps required to form a SAR image are shown in Figure 11.7. The signal processor compensates the phase of the nth radar return pulse, kth range zone after the Nth INS measurement by a correction $\phi(n, k)$ that is the sum of the patch center correction, and a range-dependent vernier correction.

The patch center correction is computed using a second-order Taylor series approximation of the patch center phase history. Line-of-sight velocity and acceleration values derived from INS data form the Taylor series coefficients. The range-dependent vernier correction is computed using a first-order gradient vector of Doppler frequency with respect to slant range vector.

After range-dependent phase correction, the radar return is processed through an azimuth prefilter and range compressed. Processing is performed in real time up to this stage. Range-compressed data are stored in processor memory until all the pulses required to form the SAR azimuth resolution cell have been collected. During the data collection period, an azimuth-dependent vernier correction is computed using a second-order Taylor series approximation of patch center phase history. Line-of-sight velocity and acceleration values derived from INS data are used as the Taylor series coefficients. Motion compensation data are stored in processor memory.

When all the radar pulses have been collected, the azimuth depth-of-focus is computed, and the number of azimuth zones with a constant motion compensation correction N_{AZ} are determined. Then the azimuth-dependent phase correction, azimuth compression, and subpatch selection functions are exercised N_{AZ} times. During the jth iteration, an azimuth-dependent phase correction $q_A(j)$ will be applied to the range compressed data. The azimuth cells that are focused by $q_A(j)$ will be selected after azimuth compression and magnitude detection

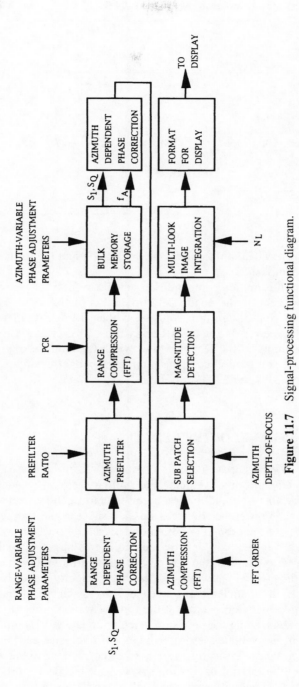

Figure 11.7 Signal-processing functional diagram.

521

TABLE 11.2   Motion compensation error budget

Parameter	Value	Sensitivity	Phase Error, Wavelengths
1. Velocity error, ΔV	0.3 M/S	$\dfrac{2V\Delta V}{\lambda R}\,(T_A^2/4)$	0.13
2. Accelerometer bias, a_b	200 μg	$(a_b/\lambda)(T_A^2/4)$	0.06
3. Accelerometer Scale factor error, S_F	300 PPM	$0.3\,\dfrac{S_F g}{\lambda}\,(T_A^2/4)$	0.08
4. Attitude error, ϵ	0.3 MRAD	$\dfrac{\epsilon g}{\lambda}\,(T_A^2/4)$	0.1
RSS total			0.21

will be performed. Then multilook image integration and the formatting for the display are accomplished.

The ability of this motion compensation scheme to form focused SAR images is directly dependent on the quality and timeliness of INS data. Typically, the motion compensation design criterion is less than 90 deg (0.25 wavelengths) of quadratic phase error during a SAR array time. This criterion results in less than 10% broadening of the Doppler filter width (azimuth resolution degradation). Table 11.2 is a motion compensation error budget example. Phase error is computed for a 1.8 sec SAR array time (T_A).

11.5   PRECISION VELOCITY UPDATE

Determination of ownship velocity is obviously an important concern in the navigation of any airborne vehicle. For military aircraft, especially those employing radar air-to-ground modes, precise determination of ownship velocity often makes the difference between mission success and failure. For example, SAR images of the ground are often de-focused by ownship velocity error, and shifted in a way that can destroy their utility as targeting tools; Doppler-based techniques to estimate the velocity of ground-moving vehicles are rendered futile if the error in knowledge of the radar velocity vector is comparable to the speeds of the vehicles being tracked; miss-distances of ballistic munitions are enlarged by velocity error at the time of weapon release. Fortunately, air-to-ground radars, using the precision velocity update (PVU) mode, can measure the error in the vehicle's estimate of its own velocity and, thus, enhance their own performance. Typically, radar-derived corrections to the velocity vector, supplied by the inertial navigation system, are fed back to the integrators within that device. If velocity error feedback to the INS is not an option, a feed-forward mechanization, in which the radar accumulates velocity vector corrections over time, may be used.

Radar measurements of aircraft velocity exploit the Doppler principle to obtain projections of the velocity vector on chosen lines of sight (See Chapter 10, section 10.1.2). Using boresight directions that span 3-dimensional space and intersect the ground, one can solve for the components of the velocity vector in some reference axis set, e.g., a local-level North-East-Down frame. Since the Doppler frequency shift on an electromagnetic pulse that is sent and received by an emitter, and reflected from a stationary reflector, is $(2/\lambda)$ times the radial velocity of the emitter in the direction of the reflection point, the relationship of aircraft velocity to Doppler measurements can be written:

$$f_D = \left(\frac{2}{\lambda} \right) \mathbf{V} \cdot \mathbf{L} \tag{11.16}$$

The vector $\mathbf{V}$ is the unknown aircraft velocity vector, the vector $\mathbf{L}$ is an arbitrarily chosen unit line of sight (LOS) intersecting the ground, and the scalar f_D is the ground clutter Doppler measurement. If three independent lines of sight are chosen and the measurements are made so rapidly that the velocity vector is effectively constant during that time, then it is possible to solve for the three components of the velocity vector by inverting the matrix equation

$$\left(\frac{\lambda}{2} \right) \mathbf{f} = A\mathbf{V} \tag{11.17}$$

in which f is a triplet (f_1, f_2, f_3) of ground clutter Doppler measurements and the 3×3 matrix A has as its jth row ($j = 1, 2, 3$) the unit boresight line of sight (referred to the same axis set as $\mathbf{V}$) associated with the jth Doppler measurement. Given the independence of the rows of A, the components of $\mathbf{V}$ are obtained by inverting A:

$$\mathbf{V} = \left(\frac{\lambda}{2} \right) A^{-1}\mathbf{f} \tag{11.18}$$

11.5.1   PVU Mechanization

Unlike the dedicated Doppler radars described in Chapter 10, where a bottom-mounted antenna is used to provide the measurement of the Doppler shifts, along three or four forward- and rearward-looking (Janus) and typically fixed beams, a multimode radar, either nose- or side-mounted, can be commanded to generate beams along almost any desired directions. In nose-mounted radars, these beams will, of course, be pointed only in the forward direction (non-

Janus). The Janus and non-Janus effects on Doppler velocity accuracy are discussed in Section 10.1.2.

In the PVU mode of a multimode radar, the quantity measured is the difference between the Doppler shift predicted by use of the velocity vector supplied by the on-board INS and that associated with the true aircraft velocity. This difference, together with the line-of-sight information, is then sent to a navigation Kalman filter, usually residing in the INS, for estimation of the error in the velocity vector supplied by the INS. Instead of processing a triplet of measurements at a time, updates are performed, in turn, for each commanded line of sight. This approach obviates the need for rapidity in making the sequence of measurements along three lines of sight; $\mathbf{V}$ can change appreciably between measurements. The technique is based on the usually well-founded assumption that the velocity *error* vector is slowly varying. Figure 11.8 shows the functional mechanization of a typical PVU mode of a multimode radar.

The mechanization commits to hardware the task of referencing the radar returns to a frequency equal to the sum of the carrier frequency and a Doppler shift that is consistent with the current aircraft velocity and a commanded antenna LOS. The radar's variable frequency oscillator (VFO) is set to the frequency value given by $2/\lambda$ times the projection of the INS velocity estimate on the commanded antenna LOS. Radar processing is such that, if the INS velocity is exact, and the commanded LOS is achieved exactly by the antenna boresight, then the radar return from the groundpoint where the boresight intersects the ground will have been shifted to zero Hz. Any departure from this value is evidence of error in the INS velocity, the achieved antenna beam boresight, or both.

An essential feature of PVU radar data processing is that it determines the range/Doppler cell corresponding to the antenna beam boresight direction that was actually achieved (as opposed to the one commanded). Typically, this is

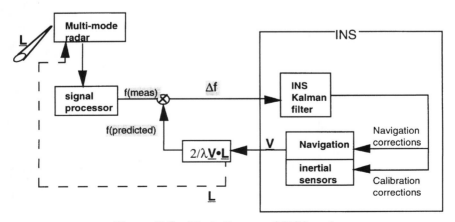

Figure 11.8   Block diagram of PVU mode.

done in a sequential manner. First, using main/guard sum channel data, a coarse estimate of beam boresight range is obtained by a sliding-window technique that finds the contiguous region containing the strongest returns; this estimate is then refined by averaging the values of elevation null returned by monopulse discriminants in a set of range-bins surrounding the coarse range estimate. Then, in a narrow swath surrounding the range associated with the newly found elevation boresight, the achieved azimuth is determined by averaging the values of azimuth null delivered by azimuth monopulse discriminants; the PVU measurement is taken to be the Doppler value associated with that cell. Discriminant techniques based on difference-to-sum ratios have the virtue that they are relatively immune to reflectivity differences in the area illuminated by the beam.

Since the Doppler observations are inevitably noisy, a method of filtering is needed. Typically, at each beam position, the Doppler measurement is taken to be the average of N, approximately 6, independent measurements taken in quick succession. In the absence of information regarding the slope and nonuniformity of the terrain, the clutter return is regarded as coming from the intersection of the radar beam and a horizontal surface. The shape of the clutter spectrum (intensity versus Doppler) is a function of the illumination pattern delivered by the radar beam.

11.5.2 PVU Measurement Errors

The quantity measured by the radar in the PVU mode is the difference Δf between anticipated and measured Doppler frequency along a commanded antenna line of sight:

$$\Delta f \equiv f_{\text{true}} - f_{VFO} \qquad (11.19)$$

where f_{true} is the actual Doppler frequency of the radar return along the antenna beam boresight, and f_{VFO} is the Doppler value set in the VFO. Those quantities are related to velocity and line of sight through

$$f_{\text{true}} = \left(\frac{2}{\lambda}\right) \mathbf{V}_{\text{true}} \cdot \mathbf{L}_{\text{achieved}} \qquad (11.20)$$

$$f_{VFO} = \left(\frac{2}{\lambda}\right) \mathbf{V}_{INS} \cdot \mathbf{L}_{\text{commanded}} \qquad (11.21)$$

The difference between the anticipated Doppler value (f_{VFO}) and the Doppler actually measured (f_{true}) is related to the following system error vectors: (1) the current error (ΔV) in the INS velocity vector and (2) the error (ΔL) in the achieved line of sight (i.e., the difference between the commanded LOS, whose calculation was based on estimates of heading, pitch, and roll supplied by the INS, and the LOS actually achieved by the antenna boresight):

$$\Delta f = \left(\frac{-2}{\lambda} \right) (\Delta \mathbf{V} \cdot \mathbf{L}_{\text{commanded}} - \mathbf{V}_{INS} \cdot \Delta \mathbf{L}) + \text{noise} \qquad (11.22)$$

where

$$\Delta \mathbf{V} \equiv \mathbf{V}_{INS} - \mathbf{V}_{\text{true}} \qquad (11.23)$$

$$\Delta \mathbf{L} \equiv \mathbf{L}_{\text{commanded}} - \mathbf{L}_{\text{achieved}} \qquad (11.24)$$

Equation 11.22 shows that the measured Doppler difference contains not only the desired projection of the current velocity error on a chosen line of sight but also contains a term given by the projection of the full INS velocity vector on the boresight pointing error.

Failure to take explicit account of the last term ($V_{INS} \cdot \Delta L$) in equation 11.22 leads to a biased estimate of the actual velocity error; as an example, for an aircraft traveling at 300 meters/sec, every milliradian of boresight pointing error contributes an unmodeled velocity error of 0.3 meter/sec. Military aircraft often rely on velocity estimation accuracy of 0.2 meter/sec or better. Boresight pointing errors arise from several sources, including (1) INS attitude, (2) knowledge of the relative orientation of the INS case and the radar antenna base, (3) antenna gimbal readout errors affecting the antenna control algorithm, (4) electrical boresight, and (5) compensation for radome refraction. The dominant errors can usually be modeled as biases in some coordinate frame, and are thus well-suited to Kalman filter estimation. PVU-derived velocity estimation accuracy better than 0.25 meter/sec (1-sigma) is typical for some tactical aircraft equipped with medium-accuracy navigators whose unaided performance is 0.8 meter/sec (1-sigma), approximately. Achieving this at speeds approaching Mach 1 calls for antenna pointing error calibration accuracy considerably better than 1 mrad.

Equation 11.22 can be rewritten to show explicitly the dependence of the PVU measurement on the contributors to antenna beam-pointing errors listed above. The error ΔL can be written in the form of a vector cross product:

$$\Delta \mathbf{L} = \Delta \theta \times \mathbf{L_c} \qquad (11.25)$$

where $\Delta \theta$ is some triplet of small angle errors, and for brevity L_c is used to denote $L_{\text{commanded}}$. Equation 11.22 can then be made to read

$$\Delta f = \left(\frac{-2}{\lambda} \right) [\Delta \mathbf{V} \cdot \mathbf{L}_c - \mathbf{V}_{INS} \cdot (\Delta \theta \times \mathbf{L}_c)] \qquad (11.26)$$

or

$$\Delta f = \left(\frac{-2}{\lambda} \right) (\Delta \mathbf{V} \cdot \mathbf{L}_c + \mathbf{V}_{INS} \times \mathbf{L}_c \cdot \Delta \theta) \qquad (11.27)$$

The vector quantity $\Delta \theta$ can be written as a sum of angle errors resident in different frames; for example,

$$\Delta \theta = \Delta \boldsymbol{\phi}_{INS} + C_{NB}(\Delta \boldsymbol{\phi}_B + C_{BA}\Delta \boldsymbol{\phi}_A) \qquad (11.28)$$

where $\Delta \boldsymbol{\phi}_{INS}$ is the vector of small errors (referred to the navigation frame) in the INS estimate of its own attitude; $\Delta \boldsymbol{\phi}_B$ is the triplet of small-angle errors (referred to the aircraft body frame) describing the uncompensated misalignment of the antenna base with respect to the INS case; $\Delta \boldsymbol{\phi}_A$ denotes the three small errors in the antenna gimbal readouts (roll, elevation, azimuth, in some sequence).

C_{NB} and C_{BA} are 3×3 orthogonal matrices; C_{NB} transforms vectors defined in the aircraft body frame into their representations in the navigation (INS) frame, while C_{BA} transforms $\Delta \boldsymbol{\phi}_A$ elements to the aircraft body frame. C_{NB} appears in the output message of the INS or can be constructed from the heading, pitch and roll angles in that message. The matrix C_{BA} is constructed from the antenna roll, elevation, and azimuth angles that are commanded in order to achieve the two conditions: (1) alignment of the antenna face normal with the commanded line of sight and (2) having the elevation monopulse axis horizontal (the azimuth axis then lies in the vertical plane containing the line of sight).

11.5.3 PVU Kalman Filter

The inner products appearing in Equations 11.26 and 11.27 allow immediate writing of the Kalman filter measurement equation; it is

$$\Delta f = H\mathbf{x} + \text{Noise} \qquad (11.29)$$

where the scalar Δf is the measured Doppler difference described above, $\mathbf{x}$ is the state vector in the INS Kalman filter, and H is the single-row measurement matrix. It is standard practice in INS software to use a formulation of the Kalman filter that has the vector of *estimation errors* (estimated whole value minus true) as the state vector. If the elements of $\mathbf{x}$ are ordered so that INS velocity errors $\Delta \mathbf{V}$, INS attitude errors $\Delta \boldsymbol{\phi}_{INS}$, antenna-to-body misalignments $\Delta \boldsymbol{\phi}_B$, and antenna gimbal angle errors $\Delta \boldsymbol{\phi}_A$ occupy the first 12 locations as follows:

$$\mathbf{x}^T = [\Delta \mathbf{V}^T, (\Delta \boldsymbol{\phi}_{INS})^T, (\Delta \boldsymbol{\phi}_B)^T, (\Delta \boldsymbol{\phi}_A)^T, \ldots] \qquad (11.30)$$

where superscript T denotes transpose, then measurement sensitivities to the errors in **x** are given by the row-matrix H, where

$$H = -[(\mathbf{L}_c)^T, (\mathbf{V}_{INS} \times \mathbf{L}_c)^T, (C_{NB}(\mathbf{V}_{INS} \times \mathbf{L}_c))^T, (C_{NB}C_{BA}(\mathbf{V}_{INS} \times \mathbf{L}_c))^T, 0, \ldots]$$

(11.31)

with zeros in the locations beyond 12. Each of the items separated by commas in the preceding two equations is a three-element row vector.

The evolution of the estimation errors appearing in Equation 11.30 is defined through the transition matrix definition. $\Delta\mathbf{V}$ and $\Delta\boldsymbol{\phi}_{INS}$ are propagated via standard navigation error differential equations. $\Delta\boldsymbol{\phi}_B$ is treated as a random bias, namely, as an unknown vector that is constant in the body frame. In the formulation described above, $\Delta\boldsymbol{\phi}_A$ is also treated as a random bias, a constant vector in the antenna face frame. Thus, only constant errors (readout errors that are independent of the actual value of the commanded angles) can be treated with this approach. Other antenna-resident errors can be pursued (e.g., angle-proportional or scale-factor errors) but only at the cost of introducing further complexity.

The noise term in Equation 11.29 can be regarded as being equivalent to a random angle jitter. Thus,

$$\text{Noise} = \mathbf{V} \times \mathbf{L}_c \cdot \Delta\theta_R$$

(11.32)

where $\Delta\boldsymbol{\Theta}_R$ is a random vector that includes the contributions of any and all noise sources in the data collection and processing; it would include I/Q calibration errors, numerical round-off errors, residual postcompensation radome correction errors, and many others. It is worth noting that the magnitude of the measurement noise varies with aircraft speed $|\mathbf{V}|$. In practice, it would be modeled as a zero-mean process with standard deviation given by

$$\sigma_{\text{meas}} = |\mathbf{V}| \, \sigma_{\text{jitter}}$$

(11.33)

where σ_{jitter} is a value (measured or assumed) for the standard deviation of the effective jitter.

When the antenna is electronically scanned, the mounting error $\Delta\boldsymbol{\phi}_B$ remains in the measurement equation; the gimbal readout errors disappear, of course. Electrical pointing errors specific to an ESA are so small that they can be ignored. The PVU measurement model is simpler. However, beam broadening, especially for large angles off the antenna face normal, often leads to reduced measurement accuracy, which can affect the rate of estimation convergence. Otherwise, the operation of the mode is the same as for the mechanically scanned antenna.

11.5.4 PVU Mode Observability Concerns

The PVU measurement model (Equations 11.26 and 11.27) assumes the Doppler difference measurement is a weighted sum of velocity and angle errors. The weights are functions of antenna line of sight, aircraft velocity vector, aircraft body attitude, and antenna face attitude with respect to aircraft body. To estimate and calibrate the individual errors, PVU measurements should be collected in such a way that those weights do not appear in fixed ratios; this avoids the condition in which certain errors appear in fixed combinations with others, with the result that isolation of individual errors becomes impossible. For example, if straight and level flying conditions are maintained during a sequence of PVU measurements, then the INS attitude error $\Delta\phi_{INS}$ and the antenna mounting error $\Delta\phi_B$ appear through their sum; that is the ratio of their weightings is 1:1, regardless of the chosen line of sight. Thus, some departures from level flight must be made in the course of forming measurements along various lines of sight. These maneuvers need not be severe; changes of $10°$ to $20°$ in pitch and roll are adequate. Similarly, in order to distinguish velocity error from angle errors, the relative values of the weights arising from inner products with $\mathbf{L}_c$ (velocity error) and $\mathbf{V} \times \mathbf{L}_c$ (angle errors) must be made to change during the PVU measurement sequence. This can be done, for instance, by changing $\mathbf{V}$, either through acceleration along track or by maneuvering. Similar concerns apply to ensuring observability of the antenna-resident errors $\Delta\phi_A$.

From a practical standpoint the decision to augment the state-vector of an INS Kalman filter with radar-specific states (e.g., $\Delta\phi_B$ and $\Delta\mathbf{\Phi}_A$ above) is often problematic. The INS software design may not allow for easy insertion of new states or for the extra processing demands entailed. There is operational consolation in the fact that, if estimation of antenna-pointing errors is to work at all, it will converge in the first few minutes of the mission (over land) if the radar is dedicated to PVU operation during that time. The newly estimated antenna calibration corrections can then be stored for use in the antenna control function during the rest of the mission, and the angle-error states in the INS Kalman filter can be ignored.

11.6 TERRAIN FOLLOWING AND AVOIDANCE

The term *terrain following* refers to an aircraft maintaining a fixed height relative to the terrain immediately beneath it. *Terrain avoidance* and *terrain clearance* are related techniques, wherein terrain avoidance is performed for turn rates exceeding those normally performed during terrain following.

Figure 11.9 illustrates the geometry of the terrain-following process. The terrain-following system, comprised of airframe, flight control system, radar forward-looking sensor, radar altimeter, and pilot fail-safe monitoring, maintains the aircraft at the desired terrain following height, h_0. The forward-looking sensor measures the range R and angle β to the terrain. Angle β is measured

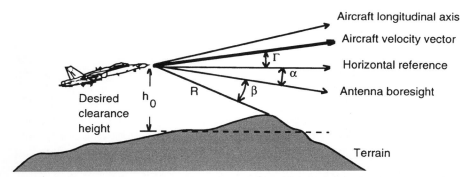

Figure 11.9   Terrain-following geometry.

relative to the radar antenna boresight, which in turn points at an angle α relative to horizontal during a terrain measurement interval. The terrain range and angle are then used to compute pitch (Γ, as defined in the figure) inputs to the flight control system that modify the aircraft trajectory to clear terrain at the desired clearance height h_0.

The motivation for terrain following (TF) is based on terrain masking, reducing aircraft exposure to ground-based and airborne air defense systems. The degree of exposure reduction is a function of achieved terrain following height, terrain roughness, and system lookdown capability. Early applications of terrain following started in the 1960s with the U.S. Air Force F-111 and Navy A-6. Both the U.S. B-1A and B-1B aircraft support TF, with the latter being accomplished with a multimode radar interleaving other radar modes in between TF profile measurement intervals.

Figure 11.10 is a functional block diagram of a modern terrain following system. Starting from the left, terrain is sensed by the forward-looking radar, with enough range and angle coverage to clear the highest expected terrain at the desired clearance height. The radar processor generates a terrain profile consisting of elevation versus range, which forms the input to the pitch-command processor. The flight control system (FCS) accepts pitch commands and modifies the control surface positions. The system loop is closed when the radar collects a new terrain profile, which then generates a new set of pitch commands, different from the previous commands, since the aircraft is at a different position relative to the terrain. The altimeter acts as a check by comparing the distance of the terrain beneath the aircraft with the corresponding radar measurements. The forward looking infrared (FLIR) sensor provides backup data on the presence of wires, with the radar providing height information designed to fly over the tops of towers, thus avoiding the wires. The pilot's interface consists of the display, flight controls, and TF vertical acceleration switch for choosing smooth, medium, or hard ride quality. Pilot work load is intensive, with the requirement to perform a positive-pitch "pull-up" maneuver the instant an unsafe flight condition occurs.

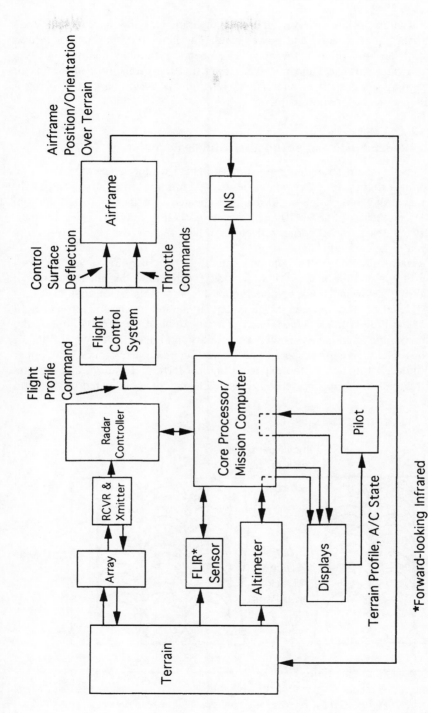

Airframe Position/Orientation Over Terrain

Control Surface Deflection

Throttle Commands

Flight Profile Command

Terrain Profile, A/C State

*Forward-looking Infrared

Figure 11.10   Terrain-following functional block diagram.

531

Flight safety is the primary system design parameter being maximized, while minimizing radar on-time and power level. The latter is important for a low-observable aircraft. Flight safety requires embedding fail-safe features into system hardware, software, and interfaces such that the probability of an unsafe failure going undetected in an hour of TF flight is very small, 10^{-8} being a typical specification requirement.

11.6.1 Radar Mode and Scan Pattern Implementation

A TF sensor encounters a wide range of terrain backscatter values during routine operation. A flight profile over snow, desert, and military installations can easily encounter a dynamic range of 30 dB, as indicated in Chapter 10. In addition, discretes targets as large as 10^4 meters^2 cross section may exist due to man-made structures acting as corner reflectors, or high radar-cross-section reflectors deployed as a countermeasure.

A terrain-following radar supports safe flight by scanning a volume containing the projection of the aircraft velocity vector onto the horizontal plane, defined as the ground track vector. Figure 11.11 depicts a nominal scan volume for level flight, with scan coverage equally split between the left and right sides of the ground track vector. The required extent of the scan volume in the horizontal dimension increases with uncertainty in either the velocity of the air mass or the aircraft with respect to an inertial navigation reference. The upper scan limit is a function of the expected highest terrain elevation, and the distance that elevation must be sensed for safe clearance, as indicated in Equation 11.34.

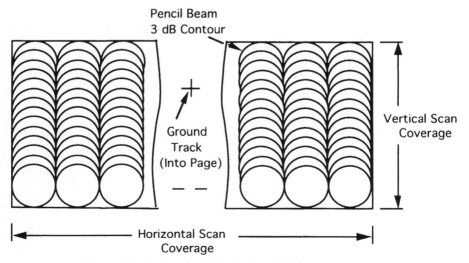

Figure 11.11 Terrain-following level flight scan coverage.

$$\theta_u = \tan^{-1}\left(\frac{\Delta Z}{V_H T_w}\right) \tag{11.34}$$

where

θ_u is the elevation scan uplook
ΔZ is the terrain clearance height above current altitude
V_H is the horizontal component of aircraft velocity
T_w is the minimum time required to safely clear maximum terrain height

The aircraft flight time from the instant of sensing to the time the aircraft passes over the highest terrain point is defined as the *warning time*, T_W. As described later, during terrain-following operation, the aircraft flight control law is based on maintaining a constant horizontal velocity, and either a positive or negative vertical acceleration causing consecutive *pull-up* or *push-over* parabolic trajectories. Safe flight requires that pull-up and push-over trajectories intersect at no more than one point in space, where the slopes of the two parabolas are equal. Combining the equality of slopes with the definition of warning time results in a relationship between maximum terrain elevation ΔZ, warning time T_W, and the two vertical accelerations, as shown in Equation 11.35.

$$\Delta Z = \frac{-A_P A_N T_W^2}{2(A_P - A_N)} \tag{11.35}$$

where

A_P is the positive vertical acceleration
A_N is the negative vertical acceleration

Substituting this result for ΔZ in Equation 11.34 leads to a requirement for the upper elevation scan limit, θ_U.

A terrain-following radar performs terrain data collection over the scan volume appropriate for either level or turning flight. Scan implementation is a function of antenna beam width. For the case of a pencil beam, scanning is performed in vertical bars such that consecutive beam pointing angles within a bar have overlapping 3-dB contours, as indicated in Figure 11.11. This overlapping provides smoothed estimates of terrain SNR and height as a function of beam position. The beam selection logic then uses these data to select which beam positions to use within a range cell for final height estimates. This beam selection logic ensures safely measuring the tops of towers, and reduces upward biasing of rain, as indicated in Figures 11.12 and 11.13.

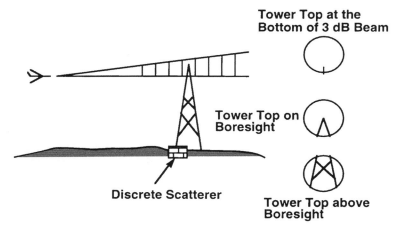

Figure 11.12   Beam selection logic for towers.

11.6.2   Terrain Measurement

Figure 11.14 depicts radar height estimate formation for a single-beam-pointing angle. The radar's video sampling interval and system bandwidth define the range cell.

The term *centroid* defines the angle, as observed from the radar and measured in a vertical plane, from the antenna boresight to the center of the range cell. The extent defines the angular width of the terrain within the range cell, measured in a vertical plane from the radar. If the extent were a fraction of a milliradian, then the terrain return due to a coherent radar pulse can be treated as being from a

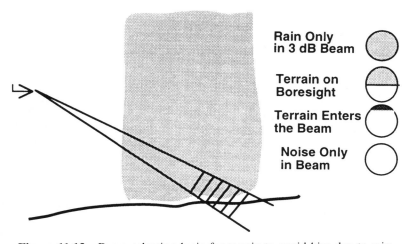

Figure 11.13   Beam selection logic for terrain to avoid bias due to rain.

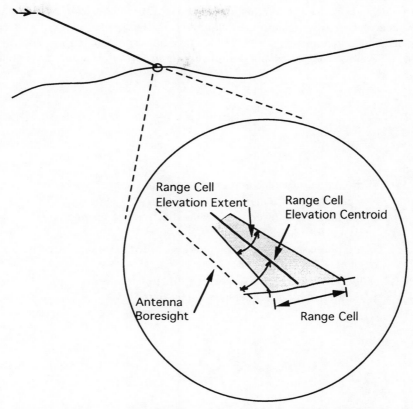

Figure 11.14   Terrain height measurement based on centroid and extent.

point target, and mutual interference from multiple scattering sources within the range cell, described by the term scintillation, would have negligible effect on centroid measurement. This point target assumption is invalid for most terrain conditions and practical range sampling intervals. Assuming an aircraft height of 200 ft, and 150-ft range cells, results in flat-terrain extents of 7.0 mr at 2000-ft slant range. In the worst case, scintillation could eliminate the effective radar cross section of all but the near-range scatterers within the range cell, causing an unsafe underestimation of terrain elevation by several milliradians. The effect is more pronounced (and unsafe) for terrain features with great vertical height, such as cliffs and towers. To obtain maximum radar illumination across the full elevation extent, the radar transmits over a wide bandwidth, either with sequential pulses at different frequencies or a single wideband pulse, whose returns are filtered independently using a digital or analog filter bank. Centroid and extent estimates are then made from these wideband radar returns.

Elevation centroid and extent processing has been successfully applied in several terrain-following systems. The radar return from each range cell is processed using in-phase (I) and quadrature-phase (Q) samples from both sum and difference monopulse channels. Equation 11.36 shows the processing of sum and difference video signals to form a centroid discriminant, C_{DISC}, and Equation 11.37 shows the corresponding processing to form the extent discriminant, E_{DISC}.

$$C_{DISC} = \frac{\sum_i |S| |D| \cos \alpha_{SD}}{\sum_i |S|^2 - N_S} \qquad (11.36)$$

where
$|S|$	is the magnitude of complex-valued sum channel video signal for ith frequency
$|D|$	is the magnitude of complex-valued difference channel video signal for ith frequency
α_{SD}	is the phase angle between complex-valued sum and difference signals for ith frequency
N_s	is the sum channel noise power estimate
i	is the transmitted frequency channel number

$$E_{DISC}^2 = \frac{\sum_i |D|^2 - N_D}{\sum_i |S|^2 - N_S} - C_{DISC}^2 \qquad (11.37)$$

where N_D is the difference channel noise power estimate.

These discriminants (C_{DISC} and E_{DISC}) are based on maximum-likelihood estimation, assuming a Swirling II fluctuation [18] model for terrain returns, and a Gaussian radar receiver noise model. As shown in Figure 11.15, the centroid and extent discriminants are coupled, deterministic functions of the ideal centroid and extent, due to the effective integration of sum and difference patterns across the terrain extent and the location of the terrain centroid relative to the difference channel null. A look-up table indexed on centroid and extent discriminant inputs performs the required inverse mapping. The final terrain elevation is formed by summing the centroid and half of the extent.

11.6.3 Aircraft Control

The terrain-following height control algorithm meets three basic requirements. First, the aircraft should clear terrain with negligible vertical velocity. Second, the autopilot should issue either a constant positive or constant negative vertical acceleration, relative to 1g, depending on terrain measurements. Third, the horizontal velocity should remain constant. As indicated in Figure 11.16,

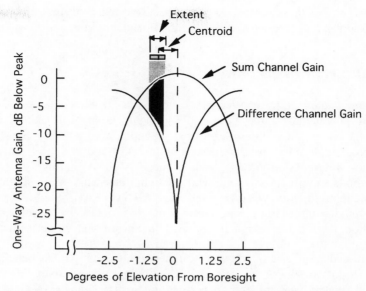

Figure 11.15   Terrain projected onto sum and difference gain patterns.

the combination of these requirements leads to the aircraft executing either a pull-up parabolic trajectory (corresponding to a positive vertical acceleration implemented by pitch-plane control surfaces) or a push-over parabolic trajectory (corresponding to a negative vertical acceleration).

For a final step in terrain height measurement, the processing places push-over parabolas over each of the terrain profile range and elevation pairs. Whenever the aircraft is higher than the desired terrain following height, the control law continues to command a push-over maneuver, until the pull-up parabola is

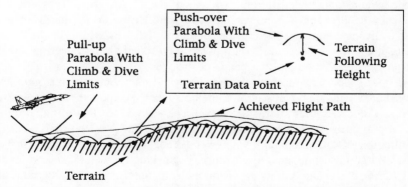

Figure 11.16   TF aircraft height control based on terrain heights.

tangent to one of these push-over parabolas. Note that both pull-up and push-over parabolas are augmented by airframe-dependent climb and dive limits.

11.7   MULTIMODE RADARS

Multimode radars perform a variety of functions using common hardware. The most prevalent examples are for military applications. Life-cycle cost economics has forced the military services to abandon their traditional approach of developing single mission aircraft, such as the F-14 and A-6, and concentrate instead on multimission aircraft, such as the F/A-18. A multimission aircraft radar's mode suite includes range-while-search, track-while-scan, single-target track, weapon support, weather avoidance, RBGM, SAR, fixed-target track, ground-moving target track, and terrain-following modes. In the multimission application, the radar is designed to provide the best balanced performance for all missions.

Traditionally, air-to-ground radar designers prefer high RF frequencies (Ku-band) that provide finer antenna azimuth resolution, given a fixed length aperture, and shorter SAR integration periods, given fixed range, speed, azimuth angle, and azimuth resolution. Low PRF wave forms are employed to yield range-unambiguous measurement of ground target location. In contrast, air-to-air radar designers select lower RF frequencies (X-band) to minimize atmospheric losses. High PRF wave forms are utilized to maximize average radiated energy and to provide velocity-unambiguous coverage. Ground return and approaching airborne targets, which are the most threatening, are separated in Doppler frequency when this wave form is employed. Medium PRF is also utilized to acquire and track airborne targets in a look-down engagement. Multimode radar designers must balance these two design preferences.

Multimode radars comprise an antenna, transmitter, receiver/exciter, and processor units. Power supply modules may be distributed among the functional units or packaged in their own unit. Antenna, transmitter, and exciter RF frequency choice is a compromise between the air-to-air and air-to-ground performance benefits gained at different extremes of the RF spectrum. Typically, a RF frequency band in the X-band regime is chosen. However, RF hardware technology advancements are focused on increasing RF bandwidth. In the twenty-first century multimode radars may possess RF bandwidths covering suitable portions of the X-band and Ku-band spectrums.

Air-to-air radar antennas are designed to yield maximum antenna gain, minimum beam width, and negligible sidelobe level. A "pencil" beam width is desired. Colocated guard horns are sometimes utilized for RFI elimination and ECCM purposes. Air-to-ground radar antennas are designed to provide uniform illumination of large ground swath over a range of depression angles. Typically, a large cosecant-squared elevation beam is employed. Sidelobe level is not as critical a design parameter. Monopulse or interferometer techniques are used to obtain azimuth target location information. A multimode radar antenna must possess all the properties of both designs. Basically, the multimode radar antenna is designed for the more stringent air-to-air application and has provisions to spoil its elevation beam when required for air-to-ground applications. Antenna elevation beam width has two values selectable by the radar processor.

Multimode radar transmitters and exciters must provide a large range of RF frequencies, pulse widths, and PRFs. Transition from one set of operating parameters to another must be rapid. The radiated RF wave form must be completely "programmable" by the radar processor. Power management of the transmitted energy is also desirable.

The receiver function must be "programmable" as well. A narrow bandwidth, large dynamic range receiver is required to support air-to-air modes, while a wide bandwidth, low dynamic range receiver is needed to support air-to-ground modes. The receiver must provide matched filtering for a variety of signal pulsewidths and PRF's. The analog-to-digital covnerter's precision and number of range samples taken each pulse repetition interval (PRI) must be selectable.

A multimode radar processor performs four functions, specifically, signal processing, data processing, radar units control, and system input/output management. All functions are completely defined and managed by software. A typical multimode radar software program comprises 100,000 to 200,000 instructions. Higher-order languages, such as JOVIAL and Ada, are employed for data processing, radar units control, and system input/output. Signal processing is accomplished through a structured software language. The hardware architecture comprises a cluster of signal-processing elements acting on radar return data in parallel, a cluster of data-processing elements partitioned to accomplish data processing, radar unit control, or system input/output, a large radar return data memory, and a large program memory. The number of processor modules depends on application.

11.8 SIGNAL PROCESSING

High-speed, software programmable signal processors are key to the implementation of multimode radars. Radar weight would be prohibitive, and aircraft power and cooling services could not support radar functioning if most radar signal conditioning and processing were not performed in a compact, lightweight special-purpose computer. SAR signal processing, which is illustrated in Figure 11.7, extracts radar imagery from the radar ground return. PVU signal processing, which is described in Section 11.5, extracts velocity vector error estimates from ground return. Terrain-following signal processing, which is defined in Equations 11.28 and 11.29, produces a terrain elevation angle profile versus range and isolates elevated point targets. Moving-target signal processing, which is discussed here, extracts target location information from the radar target return. These four signal-processing algorithms examine different portions of the radar Doppler frequency spectrum to perform their tasks.

Figure 11.17 is a generic functional block diagram for airborne or ground-moving target detection and location signal processing. This entire sequence of functions is implemented in software that is executed by parallel, "pipeline" processors. The execution of software modules is divided into three time-critical levels, namely, those computations that must be performed during a pulse repetition interval (PRI), those computations that must be performed during a coherent processing interval (CPI), and those computations that must be performed during an antenna dwell.

During a PRI, signal processing must compensate the range samples of azimuth difference channel and sum channel radar return for RF front-end imbalances, such as sum/difference phase/gain imbalance, and perform pulse compression. Each PRI pulse compressed range sample is stored in memory for later processing.

During a CPI, a fast Fourier transform (FFT) is utilized to form a Doppler filter bank spanning the PRF interval for each range bin and each radar channel of data. The resulting sum channel data grid is scanned to detect moving targets. A constant false alarm rate detector (CFAR) adapted to a noise probability density function (PDF) estimate is utilized to detect moving targets. Preliminary estimates of target range, Doppler frequency, and azimuth angle are formed for each potential detection and stored in processor memory.

If the radar wave forms are range and Doppler ambiguous, each potential target's range/Doppler measurement from each CPI must be expanded into the set of all possible unambiguous range/Doppler coordinates for the target on each CPI. The sets from all CPI's are compared during an antenna dwell to determine an unambiguous range/Doppler measurement.

11.9   AIRBORNE WEATHER RADAR

Any enterprise involving travel or navigation has to take account of the potentially catastrophic impact of adverse weather. Airborne radar is in some senses unique in its ability not only to make measurements (of ownship position, velocity and heading, etc.) that aid the navigation function but also to detect the presence and the features of weather formations. The radar observations involved in a weather mode are (1) magnitude detection of reflections from clouds and precipitation and (2) Doppler measurements of the motion of particles within a weather formation. Magnitude detection, accompanied by use of sophisticated reflectivity models, allows determination of particle type (rain, snow, hail, etc.) and precipitation rate. Doppler measurements, usually achieved by pulse-pair processing, can be made to yield estimates of turbulence intensity and wind speed. Reliable determination by airborne radar, of the presence and severity of the phenomenon known as wind shear, has been an important area of study in recent years.

Dedicated weather radars and multimode radars with a weather mode are usually nose mounted, and they include a weather display. Avoidance of haz-

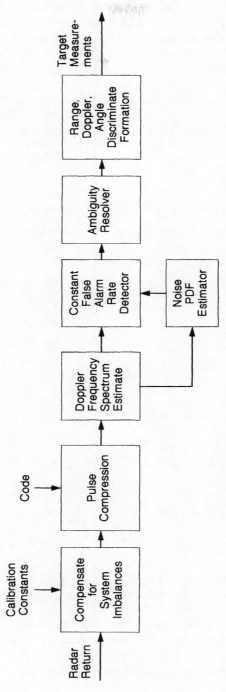

Figure 11.17 Generic moving target detection signal processing.

ardous weather conditions is the intent; however, in tactical aircraft the output of the weather mode is often used to adjust radiated power levels and processing thresholds for the presence of precipitation in the region between the radar and the target or area of interest. Typically, a real-beam map mode is employed, using a single value for elevation angle and scanning in azimuth. Radar returns are range gated and indexed by beam boresight angles. Processing is similar to that in which ground maps are formed but no attenuation is applied to the radar return, because of the relatively low reflectivity of clouds. The resulting display often shows weather formations in range (out to 100 km, approximately) and angle with respect to the aircraft nose, with indications of precipitation and hazard levels.

11.9.1 Radar Reflectivity of Weather Formations

The radar reflectivity of weather formations depends strongly on the precipitation type (rain, snow, hail, etc.), on the precipitation rate, and on the carrier wavelength. The radar cross section (RCS) of an individual droplet is well approximated by [19]

$$\sigma_i = \frac{\pi^5}{\lambda^4} \, |K|^2 \, d_i^6$$
$$K = \frac{(m^2 - 1)}{(m^2 + 2)} \tag{11.38}$$

where d_i is the diameter of the droplet and m is the complex index of refraction. For microwave radar wavelengths, $|K|^2$ has the value 0.93 for raindrops and 0.2 for ice crystals and snow [19]. Summation of the σ_i over a unit volume of the cloud gives the RCS (per unit volume) for the precipitation:

$$\eta = \frac{\pi^5}{\lambda^4} \, |K|^2 \sum_{\substack{unit \\ volume}} d_i^6 \equiv \frac{\pi^5}{\lambda^4} \, |K|^2 \, Z \tag{11.39}$$

Knowing the distribution of particle sizes for a given precipitation type and rate allows calculation of reflectivity after determining the value Z of the summation in Equation 11.39. Z has been determined experimentally for many conditions; Figure 11.18 shows a log–log plot of RCS (m^2 per cubic meter) versus precipitation rate (mm per hour), for rain and snow, with carrier wavelength as parameter. Most airborne weather radars operate in either C- or X-band. The λ^{-4} dependence of reflectivity on carrier wavelength favors X-band radars for this task.

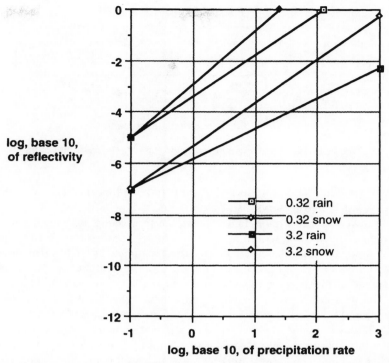

Figure 11.18   Log–log plot of reflectivity (meter^2/meter^3) versus precipitation rate (mm/hr).

11.9.2   Weather Radar Processing

Figure 11.19 depicts the processing stages involved in using radar data to detect weather formations and to measure precipitation rates and air-movement conditions. Frequency agility is used to minimize range ambiguities. The sequence (upper path in the figure) by which the presence of a given weather state is inferred from the magnitude of the radar return is (1) use the radar range equation to solve for the RCS η, after accounting for range and for parameters specific to the radar set (transmit power, antenna gain, etc.), (2) choose a value for $|K|^2$ that is consistent with some precipitation type (e.g., snow) and solve for Z, (3) knowing Z, use a look-up table or formula to infer the precipitation rate, (4) repeat (2) and (3) for any other precipitation types that may be present.

Determination of wind speeds within a weather formation has considerable safety value. Doppler-based processing can detect the presence of turbulence and provide a warning to the pilot if a pre-set safety level for wind speeds is exceeded. Even clear-air turbulence can be detected, despite the absence of a large mass of water molecules, if the sensitivity and transmit power are suffi-

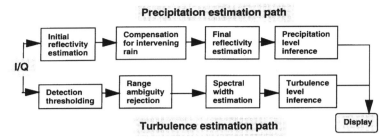

Figure 11.19   Weather mode radar processing for each range bin.

cient to deal with the low reflectivity levels provided by dust at low levels and sparsely distributed ice crystals at high altitude.

The lower path in Figure 11.19 shows the processing associated with measurement of air motion within a given volume. The pulse repetition frequency (PRF) is higher than that used in the upper path, in order to span the spectral width needed to represent faithfully the Doppler shifts of the radar returns. The spread of radial velocities within the air mass of interest is then calculated from the measured spectral width. Comparison with pre-stored values allows calculation of a safety index; for example, 5 meters/sec is often regarded as a threshold beyond which a turbulent air mass is regarded as posing a hazard for navigation. The mean velocity of mass within a given range-angle cell cannot easily be calculated without a moving target indicator (MTI) mode and accurate ownship velocity.

11.9.3   Radar Detection of Microburst and Wind Shear

Detection of microburst-induced wind shear has been a very active area of study in recent years. This phenomenon, involving strong downdrafts and high-velocity, horizontal airflow at low altitudes, is extremely hazardous to aircraft during landing and takeoff. Microbursts usually contain rain. The rapid transitions in lift conditions (see Figure 11.20) can confront the pilot with a life-or-death situation without enough reaction time to prevent disaster. As the aircraft approaches the center of the microburst, the outflow acts as a headwind. The pilot, after responding to an apparent condition of excessive lift, may suddenly find that there is not enough time or altitude in which to avert disaster in the reduced-lift conditions that rapidly occur, for the microburst outflow becomes a tailwind once the aircraft has passed through the center. There are plans to install radar wind-shear detection capability on all commercial airliners in the next few years.

Methods for radar detection of microbursts are based on recognition of air movement through Doppler processing. Some approaches attempt to detect the low-altitude, radial, horizontal airflow that accompanies them. Others key on detection of the fast downward movement of the vertical column of air and

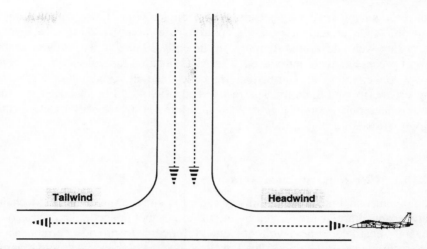

Figure 11.20 Microburst-induced wind shear which poses a hazard to aircraft at landing and takeoff by creating adjacent regions of opposite wind directions.

precipitation that characterizes a microburst. All have the objective of detecting those dangerous conditions at a range that allows enough time for evasive action.

Devising a radar mode for wind-shear detection presents the designer with numerous trade-offs. For example, a high-carrier frequency is desirable in order to maximize radar reflectivity, which is proportional to λ^{-4}; however, this choice (e.g., K_u-band) may reduce the performance of the long-range weather mode through increased atmopsheric attenuation. Another example relates to waveform selection. While medium PRF is best suited to detection of motion with velocities found in microbursts, range ambiguities with this wave form allow reflections from ground-moving vehicles (GMVs) to appear in the range bins of interest; this complicates the task of estimating air-mass velocities. Selection of a low PRF, such as 2 kHz, gives rejection of GMV returns through removal of range ambiguity but requires careful beam management in order to ensure that ground returns do not appear in the range interval of interest. Pulse compression selection poses another design decision—improved signal-to-noise ratio versus unwanted sidelobes.

11.10   FUTURE TRENDS

The first fifty years of radar history have witnessed dramatic technology advancements. Yet, greater advances are forecast for the future. The trend is toward a truly digital radar. Figure 11.21 illustrates the radar's progression from analog to digital technology. In the 1960s, radars employed analog technology almost exclusively. Analog-to-digital conversion (ADC) took place at the output of an analog crystal filter bank (after Doppler signal processing and target detection). Only elementary mode control logic, target tracking, and display formating were performed digitally. In the early 1980s, digital, programmable signal processors were introduced. The ADC function took place after receiver down conversion to an intermediate frequency, and all signal and data processing was performed digitally. The focus of the 1990s and in later years will be large-scale application of electronic scanned arrays, which will enable further progress toward a digital radar.

11.10.1 Electronic Scanned Arrays

A fundamental limitation in military radars is situation awareness. Mechanically scanned arrays are limited by inertia and cannot collect radar returns over several separated, spatial volumes quickly enough to provide current, accurate information to the pilot. This limitation is overcome by electronic scanned arrays (ESA's). In these, the antenna's beam is moved electronically by setting the phase angles of phase shifters located at each radiating element to provide a linear phase taper across the array surface. The slope of the phase taper determines the direction in which the antenna's beam will be pointed. Since the antenna's phase shifter settings can be changed in a few microseconds, an ESA's beam can be repositioned almost instantaneously. In addition, antenna reliability is dramatically improved, since ESA's have no mechanically moving parts, such as antenna gimbals, potentiometers, rotary joints, or hydraulic/electronic motors.

Two ESA technologies are being pursued. *Active ESA's* contain active components, such as transmission amplifiers, while *passive ESA's* do not. Active ESA's will be populated with radiation elements that contain individual transmitters and first stage receivers in addition to phase shifters. Gallium arsenide circuit technology will be utilized. The traditional single transmitter and a single mechanically scanned antenna unit will be replaced by a single active array antenna populated with thousands of small transmit/receive (T/R) modules. A typical airborne active array antenna could contain 1000 to 2000 T/R modules, each capable of transmitting 5 to 20 watts of power. Obviously, an active array radar will solve many of the reliability and maintainability problems experienced in conventional single transmitter radars due to TWT and power supply failures. 4 to 6% of the T/R modules could fail without noticeable degradation in antenna performance. Also, a new radar system control variable is introduced. Since the gain of each T/R module is controllable, active array radars will adaptively alter their transmit and/or receive antenna patterns to maximize system performance. No longer will a mechanical switch be required to provide a cosecant "fan" beam for air-to-ground operation. In the future, any number of antenna beam patterns will be generated and altered instantaneously. Active ESA's will inherently provide the radar with a digital front-end.

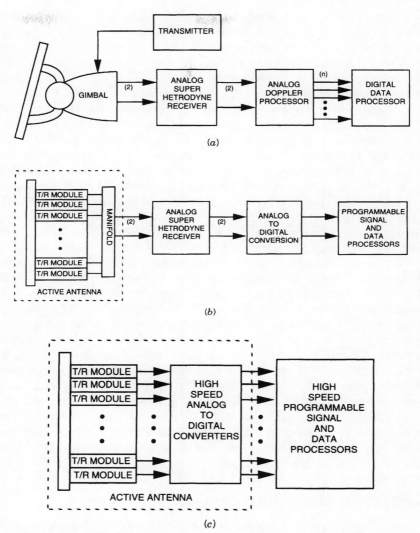

Figure 11.21 (*a*) 1960s radar functional diagram; (*b*) 1990s radar functional diagram; (*c*) 2000s digital radar functional diagram.

11.10.2 Radar Processing

In the early 1980s, high-speed, programmable processors with architectures optimized for the complex-valued signal processing required by radars were introduced. Airborne radars possessed 128 to 256 thousand words of memory, performed signal processing with 2 to 6 million complex operations per second throughput, and accomplished data processing with 300 to 600 thousand instructions per second throughput. As a result, a full spectrum of air-to-air and air-to-ground radar modes employed in multimode operation could be implemented cost-effectively in a single radar through software. Dramatic improvements in the packaging density, throughput, and memory of radar processors was achieved between 1985 and 1995. In 1996, it was not uncommon for production radars to possess 4 to 16 million words of memory, perform signal processing with 40 to 60 million complex operations per second (MCOPS) throughput, and accomplish data processing with 2 to 16 million instructions per second (MIPS) throughput. Sixteen-bit processor precision was standard. While computer capacity has been increasing, circuit packaging advances, such as multichip modules (MCM's), have allowed the processor's physical size to decrease. By the year 2000, processor memory capacity, signal processing throughput, and data processing throughput will increase by a factor of ten. Gallium arsenide circuit technology may be utilized to achieve throughput goals. Thirty two-bit or higher processor precision will become the industry standard. Fiber-optic interconnects among circuit packages on a module will offer a promising solution to the pin fanout problem attendant with advances in circuit dense packaging.

With the great computation power and memory capacity of future radars, adaptive signal- and data-processing algorithms will be implemented. For example, current airborne radars cancel ground return by computing its expected Doppler spread and range extent using INS and antenna pointing information, and blanking the appropriate range/Doppler cells. Each filter in a Doppler filter bank has the same frequency domain transfer function, which is selected a priori as the best balanced design for all missions. In the future, radars will have the capability to select adaptively, based on the radar return, the frequency domain transfer function of each Doppler filter to optimize performance relative to its position in the range/Doppler grid at the current instant in time. Thus, Doppler filters close to main-lobe clutter will have asymmetric transfer functions adaptively chosen to minimize clutter feedthrough, while Doppler filters in the "clear" region will have symmetric transfer functions adaptively chosen to minimize thermal noise feedthrough.

11.10.3   Radar Receiver/Exciter Function

In the next century, analog receiver and exciter functions will be replaced with digital equivalents to achieve a totally digital radar. With the rapid evolution toward very fast ADCs, it is possible that by that time, ADCs will be fast enough to sample RF signals above the Nyquist rate, though with low precision. Then it will be possible to incorporate ADCs in each active array T/R module, or subgroupings of T/R modules. Thus, the radar system will reduce to two units: a "digital" active array antenna and a very high speed digital computer. The digital active array antenna will transmit the selected radar wave form in the desired spatial direction. Each T/R module will amplify the radar return it receives and convert the result into a digital format. The very high speed digital computer will receive the digital outputs from the T/R modules and synthesize digital antenna patterns, such as the classical sum channel, azimuth difference channel, elevation difference channel, and guard channel, each stabilized for aircraft motion. RF and IF signal shaping and filtering will be accomplished digitally. Finally, the very high speed digital computer will generate the transmission wave form and antenna pointing commands, detect targets and determine their locations, track targets, generate synthetic aperture radar (SAR) images, format them for display, perform navigation functions, and execute periodic built-in test (BIT).

One advantage of the digital implementation is that segments of the active array aperture can be controlled independently. Thus, a digital radar will be able to search several separate volumes of space with different RF wave forms simultaneously, though with less overall transmitted power and larger antenna beam width, since the subaperture sizes are smaller. Air-to-ground and air-to-air radar modes will run simultaneously.

11.10.4 Interfaces and Packaging

In the future, fiber optics will be a basic ingredient of the digital radar. High-speed, large bandwidth busses will be required to connect the very high speed computer with the digital radar's active array antenna (see also Chapter 15). Noise immunity (as well as weight) will be critical. Fiber optics appears to offer the best solution to this problem. An active array with a fiber-optics feed network will provide true time delay signal collection so that high-performance SAR resolution will not be degraded at large azimuth angles.

Smart skins is a future radar packaging concept that envisions radar components being a component of the aircraft structure. In the future, the conventional physical design using SEM-E modules housed in individual units will give way to conformal sheets of electronics that are "wall papered" to the back side of the aircraft's skin or onto a bulkhead.

11.10.5 Displays

Radar information will be displayed on color flat-panel displays in the future. Liquid-crystal display (LCD) technology is expected to dominate this application. Nevertheless, electroluminescent, plasma, and field-emitter display technologies are being pursued for possible future use.

PROBLEMS

11.1. Calculate the width of the Doppler spectrum of the coherent terrain return detected by a radar with a beamwidth of 0.05 rad, a beam direction of 30 deg with respect to the velocity vector, having a carrier frequency of 10 GHz, and moving 90 meter/sec parallel to the terrain.

Ans.: 150 Hz.

11.2. An airplane traveling at 100 meters/sec velocity parallel to the Earth commands its radar to form a SAR image of terrain at 100-km slant range and 60-deg azimuth angle relative to the velocity vector. The radar has 10-GHz carrier frequency and a 1-meter diameter antenna. What is the smallest unfocused azimuth resolution that can be achieved? If motion compensation is applied, what azimuth resolution will be realized?

Ans.: 60 meters, unfocused
0.5 meter, focused.

11.3. How long will a X-band radar traveling at 100 meter/sec velocity parallel to the Earth collect radar ground returns to form a 10-meter resolution SAR image of the terrain at 100-km slant range and 45-deg azimuth angle relative to the velocity vector?

Ans.: 7 sec.

11.4. Find the angular protusion of a hill through the clearance plane of a terrain avoidance radar when the linear protrusion of the hill through the clearance plane is 600 ft, the slant range to the hill is 3000 ft, and the radar measures an angle of 30 deg between the antenna boresight axis and the hill.

Ans.: 13 deg.

11.5. What is the aircraft clearance in Problem 11.4? Assume that the antenna boresight axis is horizontal.

Ans.: 2100 ft.

11.6. How much error will result if the slant range to a navigation fix point of 10 nmi is used as the ground range when the aircraft altitude is 30,400 ft?

Ans.: −1.8 nmi.

12 Celestial Navigation

12.1 INTRODUCTION

This chapter discusses the principles of celestial navigation, and in particular, high-precision stellar-inertial navigation. Since the time of the ancient mariners, stars have been used as a means of navigating on the Earth's surface using night-time manual celestial fixes. Until the 1970s, manual celestial fixes were used on transoceanic commercial aircraft. Since the 1960s, high-precision stellar-inertial navigation systems have been developed with automatic daylight and night-time star-tracking capability. These stellar-inertial navigators are highly useful for military aircraft in that they provide accurate position and attitude information, are autonomous, nonradiating, and invulnerable to jamming. They have been used on such aircraft as the SR-71 and U-2 reconnaissance aircraft and the B-2 and B-58 bombers. The development of these high-precision daylight stellar-inertial navigators has been brought about by (1) improved star-light detection devices, (2) techniques for improving star signal to sky background noise ratios, (3) advances in the throughput and memory of airborne digital processors, and (4) the development of algorithms for optimally combining the star observation with the inertial navigator output.

12.1.1 Evolution of Celestial Navigation

The nearly fixed positions of the stars in space, and the predictable motion of the Earth, make stars an obvious measurement source for navigation. The North Star (Polaris) can be used to estimate latitude in the Northern Hemisphere using the horizon as a local-level reference at the observer's position; this is a technique that was used by the ancient mariners. Estimation of both latitude and longitude requires, in addition to a local-level reference, an accurate measure of time (year, month, day of month, and time of day) along with a star catalog that defines star locations. Navigators on early transoceanic aircraft flights used bubble-level sextants to manually measure star angles relative to the local vertical. Using these lines of sight from two or more stars, along with a star catalog and an accurate time reference, the navigator deduced his position in Earth latitude and longitude; this was known as a *celestial fix* or a *star fix*.

The advent of inertial navigation resulted in the stabilized local-level platform on which gyros and accelerometers were mounted; these instruments were used to stabilize the platform to the local horizontal and to provide linear acceleration for computing vehicle position and velocity (Section 7.4). Since the position error growth of these free inertial navigation systems was excessive on long flights, it was natural to periodically update their position-outputs with position-updates derived from manual star fixes. It then became possible to implement completely automatic star fixing by physically integrating the sextant with the stabilized local-level platform which inherently provided the necessary local-level reference. The telescope, with a star-light detector, was mounted on the stabilized platform through a gimbal structure. Real-time, automatic star-pointing and star-tracking commands were generated by the on-board navigation system computer from which position errors were deduced. These automatic star fixes were used to update (correct) the inertial navigation system position-outputs and to correct the stabilized local-level platform in azimuth; in some cases instrument errors were also corrected.

The augmentation of inertial navigation with periodic stellar updates provided an ideal application for optimal filtering techniques (e.g., Kalman filtering) which combines multiple sensor inputs into "best-estimate" navigation outputs (Chapter 3). The implementation of an optimal filter (estimator) requires the use of accurate mathematical models for the propagation of the system error dynamics and for the measurement process.

The star measurement (observation) is the angular difference between the measured and computed lines of sight to a star. While this angular difference, measured in a local-level frame of reference, was used to deduce position error measurements, more precisely it is only a measure of the inertial attitude error of the reference system involved (whether or not the latter is local level). This error angle is known as the ψ (psi) angle. Consequently, star observations can no longer be treated as position error measurements when used in conjunction with accurate stellar-inertial navigation systems; rather they must be treated as ψ measurements. The implications of this have a significant impact on the design and performance of the stellar-inertial navigation system. This impact is discussed in Section 12.3.

12.1.2 General System Description

Operational stellar-inertial navigation systems have the star sensor physically mounted on the stabilized local-level platform through a gimbal structure providing two degrees of freedom in azimuth and elevation. The star sensors have a narrow field of view and maintain a centered star image through gimbal drive commands and are called *star-trackers*. They have a day and night tracking capability, given reasonable cloud cover conditions. A window is provided on the top side of the aircraft for star viewing by the star sensor. The integrated star sensor/inertial platform package is mounted directly below the window. Window location and size are important system considerations.

A star catalog defining star locations and magnitudes is stored in the navigation system computer along with star sensor calibration coefficients. Using an accurate time source, star-pointing commands (including line-of-sight corrections) are computed and the differences between computed and measured lines of sight are determined. These angular differences are processed as inertial attitude reference errors by the navigation filter. The filter outputs are used to update the navigation system outputs as well as update instrument calibration coefficients. The navigation outputs of position, velocity, and attitude (in the required coordinates) are distributed to other systems, subsystems and displays on-board the aircraft (Chapter 15).

To reduce cost and size and to improve reliability, an advanced configuration, not yet operational in 1996, utilizes a stellar sensor rigidly mounted on a strapdown inertial measurement unit (IMU). The IMU is gimballed with at least two degrees of freedom for coarse telescope pointing and for instrument calibrations. The star sensor typically utilizes a focal plane array (Sections 12.4.4 and 12.4.5). The inertial attitude error is calculated by computing the angles to the off-centered star image. This configuration has become viable through technology advances such as the ring laser gyro, the charge coupled device (CCD) array, and powerful airborne computers. A variation of this configuration, which had undergone limited testing by 1996, is to rigidly mount the strapdown IMU and the telescope to the airframe (Section 12.5.4).

12.2   STAR OBSERVATION GEOMETRY

The geometry and mathematics of star observations and celestial fixes are described in this section. The discussion applies to both the navigational approach used by the ancient mariners and the mathematics required to automatically point the telescope and to integrate the star measurement into a stellar-inertial navigator. For the latter, vector-matrix techniques are used. (For a development utilizing spherical trigonometry, see [1].)

Suppose that we know a star's coordinates in an inertial coordinate frame. It may be assumed that the star's position is fixed in this inertial frame (see Section 12.6 for a more detailed discussion of a star's position). Let the Z^I-axis of the inertial frame lie along the Earth's rotation vector, the X^I, Y^I axes lie in the Earth's equatorial plane with the X^I-axis pointing in the direction of the First Point of Aries, Υ (the intersection of the equatorial and ecliptic planes where the sun crosses the equator from south to north). Star catalogs and astronomical almanacs tabulate a star's position in this inertial coordinate frame in terms of its right ascension α and its declination δ as shown in Figure 12.1. In terms of α and δ, a unit vector in the inertial frame pointing to the star is given by

$$\mathbf{u}^I = \begin{bmatrix} \cos\alpha & \cos\delta \\ \sin\alpha & \cos\delta \\ & \sin\delta \end{bmatrix}$$

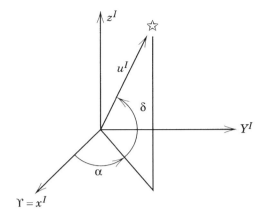

Figure 12.1   A star's right ascension and declination measured from an inertial coordinate frame.

The Earth is rotating with respect to the inertial frame with a rate Ω_E, about the inertial Z^I-axis (Ω_E is the Earth's sidereal rotation rate of 15.04°/hr). An Earth-fixed coordinate frame is defined such that $Z^E = Z^I$, and X^E, Y^E lie in the equatorial plane with X^E passing through the Greenwich meridian (i.e., longitude = 0). At any instant of time, t, the E frame is displaced from the I frame, by a rotation about the Z^I-axis given by GHAϓ (the Greenwich Hour Angle of Aries). The latter is available from astronomical almanacs as a function of t or can be computed (see Section 12.6). The transformation from the I frame to the E frame is

$$\mathbf{C}_I^E = \begin{bmatrix} \cos\,(\text{GHA}ϓ) & \sin\,(\text{GHA}ϓ) & 0 \\ -\sin\,(\text{GHA}ϓ) & \cos\,(\text{GHA}ϓ) & 0 \\ 0 & 0 & 1 \end{bmatrix}$$

and a unit star position vector in the E frame is

$$\mathbf{u}^E = \mathbf{C}_I^E \mathbf{u}^I \tag{12.1}$$

Consider an abserver at a fixed location on the Earth given by a latitude Φ and a longitude λ (Figure 12.2). Define a local-level coordinate frame L located at the observer, where $\mathbf{X}^L$ is along the local East, $\mathbf{Y}^L$ is local North, and Z^L is the local vertical. The transformation from the E frame to the L frame is

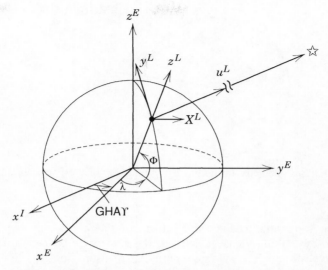

Figure 12.2   Local-level geometry for a star observation.

$$\mathbf{C}_E^L = \begin{bmatrix} -\sin \lambda & \cos \lambda & 0 \\ -\sin \Phi \cos \lambda & -\sin \Phi \sin \lambda & \cos \Phi \\ \cos \Phi \cos \lambda & \cos \Phi \sin \lambda & \sin \Phi \end{bmatrix}$$

and a star unit position vector in the L frame is

$$\mathbf{u}^L = \mathbf{C}_E^L \mathbf{C}_I^E \mathbf{u}^I \qquad (12.2)$$

It is seen from Figure 12.2 that not only has a rotation occurred in going from the E frame to the L frame, but a translation of the origin of the frames has also occurred. The local-level geometry is redrawn in Figure 12.3 in the plane containing the center of the Earth, the observer, and the star. From Figure 12.3 it is observed that the light from the star to the center of the Earth is nearly parallel to the light from the star to the observer on the surface of the Earth. For the star Sirius, which at 8.7 light years is one of the closest stars to Earth, the angular separation (parallax) due to the Earth's radius is less than 20 micro-arcsec.

Because of this, any star measurement made relative to inertial space will yield no information with regard to position of an observer on Earth. Hence, the stars have the same orientation angles, α and δ (Figure 12.1), in inertial space regardless of the observer's position on the Earth's surface.

Star measurements made with respect to an Earth-fixed frame provide no information about the observer's position along the local vertical axis, Z^L.

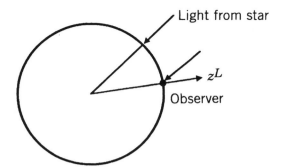

Figure 12.3   Star geometry in the plane of the center of the Earth, the observer, and the star.

However, it is possible to determine information about the observer's horizontal position by making measurements of the star's angular position relative to an Earth-fixed frame. The star's azimuth, Az, measured positive East of North about Z^L, and elevation, El, measured positive above the local horizontal and about the rotated X^L-axis provide information about the observer's latitude and longitude. (Astronomers refer to the angle above the horizontal as *altitude*; however, the term "elevation" will be used here to avoid confusion with the linear vertical distance above a reference.) In terms of Az and El, the star unit position vector in L coordinates is

$$\mathbf{u}^L = \begin{bmatrix} \sin(Az)\cos(El) \\ \cos(Az)\cos(El) \\ \sin(El) \end{bmatrix} \qquad (12.3)$$

Equations 12.2 and 12.3 provide the relationships for Az and El as a function of α, δ, GHAΥ, Φ, and λ. Given $\mathbf{u}^L$, elevation and azimuth are obtained from

$$El = \sin^{-1}[\mathbf{u}^L(3)]$$

$$Az = \tan^{-1}\left[\frac{\mathbf{u}^L(1)}{\mathbf{u}^L(2)}\right] \qquad (12.4)$$

Therefore, given an observer's location, a star catalog (almanac), and the time, the azimuth and elevation of a selected star can be computed from Equation 12.4, and a telescope can be pointed toward that star. Suppose that one only knows the observer's location approximately but sufficiently close to unambiguously identify a selected star through computation of its expected azimuth and elevation. The actual azimuth and elevation of the star can be measured. From these Earth-fixed measurements, one can determine the correct observer latitude and longitude.

From Equations 12.2 and 12.3 one obtains the following relationships:

$$\cos(Az)\cos(El)\cos \Phi + \sin(El)\sin \Phi = \sin \delta \qquad (12.5)$$

$$\cos \Phi \cos(\alpha - \text{GHA}\Upsilon - \lambda)\cos \delta + \sin \Phi \sin \delta = \sin(El) \qquad (12.6)$$

These equations can be solved for Φ and λ, and ambiguities in the solution can be resolved by approximate knowledge of the observer's position.

While the previous equations allow for the computation of latitude and longitude given measurements of azimuth, elevation, and time, they were difficult to implement manually due to the lack of precise instruments to measure azimuth. Celestial navigation was done with sextants and bubble-level devices that measure the angle between a star and the horizontal.

A single measurement of elevation of a star is not sufficient to locate an observer's position; it places the observer on a cone with the star at the vertex. The intersection of this cone with the Earth yields a circle (approximately) as a line of position (LOP) for the observer. This LOP can be determined from Equation 12.6. If only a single elevation measurement is available, then the intersection of the LOP with the line connecting the observer's estimated position and the projection of the star onto the Earth serves as an updated estimate of position.

The measurement of the elevation of a second star yields a second circular LOP. If the observer is stationary, then the intersection of the two LOP's yield two possible positions for the observer. The ambiguity in position can be resolved either with a priori approximate knowledge of position or by a third star elevation measurement.

If the observer is moving with respect to the Earth, the following is required to determine position: (1) a sequence of star elevation measurements and (2) the time and approximate location of the observer for each measurement. If the integral of the observer velocity error over the star measurement interval is small relative to the error in the approximate position, then the position error over the measurement interval can be treated as a constant and the sequence of star elevation measurements can be used to estimate the position error.

With multiple measurements (at least two), a nonlinear least squares solution for the assumed constant position error can be obtained. The position accuracy is a function of the separation of the star measurements and is highest when at least two of the measurements are from stars whose separation is near $90°$.

12.3 THEORY OF STELLAR-INERTIAL NAVIGATION

Since the 1970s, navigation systems have used Kalman filtering techniques (Chapter 3) to integrate all sensed navigation data to obtain the best-estimate navigation solution. This use of the Kalman filter to integrate stellar measurements requires that the star observations and errors be correctly modeled mathematically. As a result, high-accuracy stellar-inertial navigation systems do not explicitly implement "position fixes" using star sightings; rather they use star angle measurements as observations. This section describes what the star observation actually measures for a Kalman filter implementation and gives a linearized error formulation that is particularly well suited for the application of Kalman filtering to stellar-inertial navigation. It then provides a basis for analyzing and determining the effectiveness of stellar observations through the use of "observability" and "information" concepts.

12.3.1 Modeling and Kalman Filtering

The application of Kalman filter theory to the stellar-inertial navigation problem begins with a precise mathematical definition of the star observation (measurement). The inertial line of sight (LOS) to any given star is precisely known from the star catalog. Therefore the angular error which is measured by a star sensor is the angular difference between the computed (error corrected) LOS to the star and the physical star sensor LOS.

The *error angle vector* which rotates the computational (or computed) coordinate frame into the physical (or platform) coordinate frame is called the ψ- *angle vector* [2, 6]. The following error angle symbols and definitions are used:

$\delta\theta \triangleq$ vector angle defined from the true coordinate frame to the computed frame

$\phi \triangleq$ vector angle defined from the true coordinate frame to the platform frame

$\psi \triangleq$ vector angle defined from the computed coordinate frame to the platform frame

For terrestrial navigation, the $\delta\theta$ error is a direct consequence of latitude and longitude position errors, and the ϕ error is a direct consequence of platform tilt and azimuth (or heading) errors. These definitions apply equally to stabilized or strapdown navigators and are illustrated in Figure 12.4. In the figure, R is the actual position of the vehicle, l is the true local level at the actual position, l_p is the actual physical or platform reference at the actual position, $\hat{R}$ is the computed position of the vehicle, and l_c is the computed level reference at the computed position (see also Figure 3.4 of Chapter 3).

Since the star sensor is physically mounted on or referenced to the platform and since the star LOS is computed based upon a level reference at the computed position, the angular error measured by the star sensor is clearly a function of both ϕ and $\delta\theta$. From these definitions, ψ is given by

$$\psi = \phi - \delta\theta \tag{12.7}$$

Since these angles are small, they are treated as vectors and can be resolved

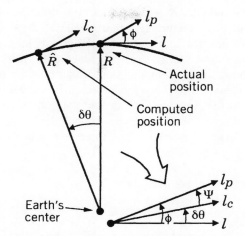

Figure 12.4 ψ, ϕ, and $\delta\theta$ definitions.

into any coordinate frame. From Equation 12.7 it is clear that star sensor measurements are not equivalent to position measurements. In fact, a large position error $\delta\theta$ could be canceled out by an equal platform tilt error ϕ caused by an accelerometer bias (see Problem 12.1).

Most inertial navigation error equations are formulated using the ϕ and $\delta\theta$ error variables. This results in coupling between position, velocity, and attitude errors. If instead the variables position, velocity, and ψ are used, then only position and velocity are coupled. An independent differential equation can be written for the ψ equation. This becomes important for stellar augmented systems, since it is the angle ψ that the sensor measures. In the ϕ, $\delta\theta$ mechanization, by Equation 12.7, the star observation is a measure of the angle $\psi = \phi - \delta\theta$. If approximations are made in the defining linearized differential equations for ϕ and $\delta\theta$, which are valid individually but not completely consistent with each other, the difference between the propagated ϕ and $\delta\theta$ angle vectors may not be precisely equal to the angle vector ψ, resulting in the stellar measurements being used incorrectly to estimate error parameters.

To implement the Kalman filter, we define a dynamic error model that is consistent with the position, velocity, and ψ formulation [2].

Dynamic Error Model

$$\delta\dot{R} = F(\Omega_{EC})\delta R + \delta V \tag{12.8a}$$

$$\delta\dot{V} = G(\hat{R})\delta R + F(\Omega_{EC} + 2\Omega_{IC})\delta V + F(\hat{a})\psi + \delta a \tag{12.8b}$$

$$\dot{\psi} = -F(\Omega_{IC})\psi + \delta\omega \tag{12.8c}$$

In Equation 12.8, all errors are expressed in computational coordinates; δ indicates an error of a variable, R is the vehicle position vector, a is the specific force vector, $F(\cdot)$ is a skew symmetric matrix equal to the negative of the cross-product matrix, $G(\hat{R})$results from perturbing the gravity term about the best estimate of R. Errors in the Ω terms are zero, since the Earth rate Ω_{IE} is known precisely, as is the rate of the computational frame relative to the Earth's frame Ω_{EC}; the ψ equation is forced by the gyro drift rate $\delta\omega$. The super dot on the left-hand side of the equation indicates a time derivative relative to inertial space.

These results yield the linearized system error model used to formulate the *extended Kalman filter* (Chapter 3) [4, 5, 7, 8]:

System Dynamics or Process Model

$$
\begin{bmatrix} \delta\dot{R} \\ \delta\dot{V} \\ \dot{\psi} \\ \delta\dot{a} \\ \delta\dot{\omega} \\ \delta\dot{b} \end{bmatrix} = \begin{bmatrix} F(\Omega_{EC}) & I & 0 & 0 & 0 & 0 \\ G(\hat{R}) & F(\Omega_T) & F(\hat{a}) & \hat{C}_A^C & 0 & 0 \\ 0 & 0 & F(\Omega_{IC}) & 0 & \hat{C}_G^C & 0 \\ 0 & 0 & 0 & \hat{K}_a & 0 & 0 \\ 0 & 0 & 0 & 0 & \hat{K}_G & 0 \\ 0 & 0 & 0 & 0 & 0 & \hat{K}_B \end{bmatrix} \begin{bmatrix} \delta R \\ \delta V \\ \psi \\ \delta a \\ \delta\omega \\ \delta b \end{bmatrix} + \begin{bmatrix} U_R \\ U_V \\ U_\psi \\ U_a \\ U_\omega \\ U_b \end{bmatrix}
$$

$$(12.9a)$$

Stellar Measurement or Observation Model

$$\delta Y = [0 \vdots 0 \vdots \hat{C}_S^C \vdots 0 \vdots 0 \vdots I][\delta R \vdots \delta V \vdots \psi \vdots \delta a \vdots \delta\omega \vdots \delta b]^T + V \qquad (12.9b)$$

In Equation 12.9, δa, $\delta\omega$, and δb represent the dynamics of the random accelerometer errors, gyro errors, and stellar subsystem boresight errors, respectively; the U and V vectors represent "white" process and stellar observation noise, respectively; $\Omega_T \triangleq \Omega_{EC} + 2\Omega_{IE}$; the matrices $\hat{C}_A^C$, $\hat{C}_G^C$, and $\hat{C}_S^C$ are used to denote best estimates of the transformations (including nonorthogonalities) from the accelerometer triad, gyro triad and stellar sensor to the computational frame, respectively. The sensor input axis misalignments (including nonorthogonalities) are included in the dynamic sensor models.

From Equation 12.9a it is clear that accelerometer errors δa integrate directly into vehicle position and velocity errors, as expected. It is also clear that vehicle position and velocity errors do not couple back into the ψ error. Consequently, star sensor measurements cannot be used to estimate or correct for those position errors that occur due to dynamic accelerometer errors or errors caused by gravity uncertainties (not included in Equation 12.9 for simplicity). Further, since the gyro induced drift errors $\delta\omega$ are the only error variables that drive the ψ equation, it is these gyro errors that the stellar updates are effective in estimating and compensating for, as well as the position and velocity errors that occur due to the $F(a)\psi$ acceleration error term caused by ψ (see Equations 12.8b and 12.9a). Note that the ψ error does not oscillate at the Schuler frequency as do the position and velocity errors.

In general, the star measurements are effective in estimating parameters that contribute to the initial $\psi(0)$ error, the position and velocity errors that propagate as a result of this initial $\psi(0)$ error, and in-flight gyro induced position and velocity errors. As time progresses, the navigation solution propagated by a Kalman filter tends to become independent of the initial state. Consequently, star measurements must be used early in the mission if they are to be effective in estimating initial position errors, accelerometer bias errors, and the position and velocity errors generated by these accelerometer biases. It should be pointed out that care must be taken in establishing initial error variances and their respective cross-correlations, which are used to estimate $\psi(0)$ effects, and which are functions of the method and/or technique used in establishing the initial alignment.

An example of position error propagation is shown in Figure 12.5, using an error analysis program, for a 14-hour flight. After takeoff, the aircraft climbs to a cruise altitude where it executes numerous turns. The first star measurement is made at the end of the pre-flight alignment for one case and ten minutes after the alignment for the second case. Three star observations per minute are made thereafter each with an observation error of 1.5 arcsec standard deviation. The circular error probable (CEP) time histories display the familiar 84-minute Schuler oscillation in position error and the 24-hour Earth rate loop oscillations (Chapter 7).

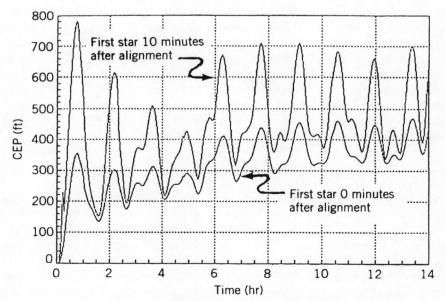

Figure 12.5 Stellar-inertial navigation performance.

Since the star observation measures the star inertial line of sight and is not a direct measure of position, the position error is not bounded. This is seen by the small position error growth on which the Schuler oscillation is superimposed.

The effect of the time of the first star observation can also be seen in Figure 12.5. The earlier start of star observations has reduced the position CEP by over 100 ft over the duration of the flight. The explanation for this is as follows: At the end of alignment, the two horizontal components of the ψ error are the result of accelerometer bias errors. The correlations that exist in the filter covariance matrix between ψ and the accelerometer errors make these accelerometer errors observable. When a star observation is made, reductions in the accelerometer errors are achieved producing slower position error growth and correcting for accumulated position and velocity error growth due to these accelerometer errors. When the star observation is delayed by ten minutes (as may be the case where there is cloud cover and it takes the vehicle ten minutes to clear this cover), then the various process noise terms feed into the ψ and accelerometer errors and the correlations between ψ and the accelerometer errors are reduced. When the first star observation is obtained, it has less effect on the estimated position, velocity, and accelerometer errors. As time passes, correlations with accelerometer errors disappear, and the star observations are no longer useful in estimating accelerometer bias errors, or the position errors that stem from them.

If a filter is to be implemented using the ϕ, $\delta\theta$ formulation rather than the ψ-angle vector formulation described above, a number of special simulation tests should be conducted to verify proper implementation and operation; for example, the propagated ϕ and $\delta\theta$ solutions must be such that the Schuler magnitude and phase of these angle vectors exactly cancel when differenced, since $\psi = \phi - \delta\theta$ does not have a Schuler oscillation. Simulation tests should start with all zero initial filter covariance conditions and dynamic accelerometer error process noise. The effect of the process noise is to generate "in-flight" accelerometer errors which integrate into position and velocity errors. As previously discussed, star observations should have no effect on reducing these in-flight induced position and velocity errors. Two runs, one with star observations and one without, should yield the same results. Such tests can verify correct filter design and performance for the ϕ, $\delta\theta$ formulation.

12.3.2   Information and Observability

In defining an optimally integrated system, there are two problems that must be addressed. The first problem is how to make optimal use of each measurement in a given sequence of measurements; this problem is solved by Kalman filtering each measurement. The second problem deals with the determination of the optimal sequence of measurements to be taken, restricted to the sensors available; this is the problem of maximizing observability. There is no known general solution to the second problem. The use of information and observability concepts is quite useful. These concepts can be used to understand the basic observability problems encountered in Kalman filtering applications and to show analytically how to obtain a sequence of measurements that will provide performance superior to that of another measurement sequence. For example, observability problems associated with the Kalman filter are greatly intensified in a stellar-inertial navigation system if the telescope is rigidly mounted to the IMU even though the IMU is gimballed to provide star pointing. Simulations show very significant performance degradations going from the gimballed to the nongimballed telescope configuration. Though not intuitively obvious, an observability analysis shows that components of the $\boldsymbol{\psi}$ error may become near unity correlated with the telescope boresight errors, resulting in an inability to observe or estimate either error. This analysis provides the insight necessary to understand and solve the observability problem associated with the strapdown telescope. (Problem 12.2 addresses the essence of this problem using information and observability concepts and is discussed later in this section.)

The terms observability and information are usually used qualitatively when describing system characteristics. There are, however, precise mathematical definitions for these functions that become very useful in the understanding and design of stellar-inertial navigators. One form of the observability matrix [4] is given by

$$J(N) = \sum_{i=1}^{N} A^{-T}(i)H^T(i)H(i)A^{-1}(i), \qquad (12.10)$$

where

A is the system transition matrix
H is the system measurement matrix
N is the number of measurements to process

In Equation 12.10, if the matrix $J(N)$ is positive definite then the system is observable, and the filter will be able to estimate all errors in the state vector.

The information matrix, in discrete form, is given by [4, 9]

$$L(N) = \sum_{i=1}^{N} A^{-T}(i)H^T(i)W^{-1}(i)H(i)A^{-1}(i)$$

$$+ A^{-T}(N,0)P^{-1}(0)A^{-1}(N,0) \qquad (12.11)$$

where

$P(0)$ is the initial (a priori) estimation error covariance matrix
W is the measurment error covariance matrix

If a linear system has zero process noise, and if a Kalman filter is used to optimally estimate the state of the system, then the state error covariance matrix P is equal to the inverse of the information matrix L:

$$P(N) = L^{-1}(N). \tag{12.12}$$

In comparing the information matrix $L(N)$ with the observability matrix $J(N)$, it is clear that a measure of the new information brought into the system via observations or measurements is directly proportional to the observability measure of the system and inversely proportional to the measurement uncertainty. This also provides analytical closed form performance predictions for postulated observation sequences processed by the Kalman filter.

For an application of this theory to the design of stellar-inertial navigators, see Problem 12.2 which is a simplified version of the strapdown telescope problem. From part e of this problem, it is clear that the ψ errors and the stellar boresight errors ϵ become unity correlated and inseparable without a capability to rotate the ϵ errors relative to the ψ errors; after a few measurements, the variances no longer decrease. From parts c and e it is clear that system observability is established by this rotation and that, using Equation 12.12, the estimation error variances become proportional to the measurement noise variance and inversely proportional to the number of stellar measurements taken, thereby approaching zero. While the same star and star sensor were used to obtain the N measurements described in parts a and c of Problem 12.2, the sequence described in part a provided little information following the first measurement, while the sequence described in part c drives the errors to zero.

Hence, in the application of estimation theory, there is the problem of optimally processing the measurements for a given measurement sequence (which is the Kalman filter) and the selection of the optimum measurement sequence, for which the general solution is not known but for which the notions of information and observability provide good insights [9].

12.4   STELLAR SENSOR DESIGN CHARACTERISTICS

12.4.1   Telescope Parameters

The telescope gathers energy for a light detector and defines the line of sight (LOS) to a star by a physical arrangement of optical elements. Telescopes can be categorized into three major classifications: (1) refractive, which uses lenses as the primary focusing element [11], (2) reflective, which uses a curved mirror for focusing, and (3) catadioptric, which mix lens and mirror systems. With the advent of diffractive optical elements, such as binary optics and holographic optical elements, there are many types of optical systems available for star sensor use; some of these mix several different types of optics in the same telescope. Regardless of the type of optics in the telescope, each design is characterized by an optical aperture, an effective focal length, and a field of view (Figure 12.6a) [12, 13]. The optical aperture b is the physical diameter of the first optical element in the telescope. The effective focal length F of a system is given by $F = (b/2)/\tan(\beta/2)$, where β is the angular size of the image seen from the focal plane. For small angles β, F is approximated by b/β as shown in Figure 12.6b.

The image of a star at the focal plane of a telescope with perfect, unobstructed optics is a bright spot surrounded by a series of concentric dark and light rings. This image is called a *diffraction-limited image*. The angular size of the image is determined by the diameter of the optical system, the wavelength of the incoming light, and the number of rings included in the image:

$$\sin\theta = \frac{n\lambda}{b} \tag{12.13}$$

where

θ is the angle from the center of the image to a dark or light ring defined by n

n is a constant for each ring in the image

λ is the wavelength of the incoming light

b is the diameter of the aperture

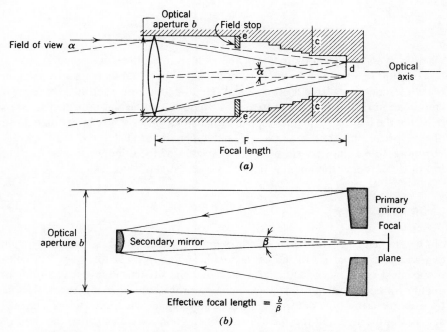

Figure 12.6   (a) A simple optical system; (b) a Cassegrain telescope.

TABLE 12.1   Values of n and the percent of energy in each ring of circular aperture telescopes

Ring	n	Percent of Energy in Each Ring
Central	0	86.4
First dark	1.22	
Second bright	1.635	7.3
Second dark	2.233	
Third bright	2.679	2.8
Third dark	3.238	
Fourth bright	3.699	1.5

The values of n for a circular aperture telescope are given in Table 12.1. The second column in Table 12.1 is the percentage of the energy in each ring relative to the total star energy [14]. For example, the second bright ring of a star image for a telescope with a 3-in. aperture and 0.5-micron light (0.5×10^{-4} cm) has an angular diameter of 3 seconds-of-arc and contains 93.7% of the total star energy.

The physical size of the star image depends on the angular size of the image and the effective focal length of the telescope:

$$d = 2F\theta = \frac{2n\lambda F}{b} \tag{12.14}$$

where d is the physical diameter of the rings in the diffraction image.

The ratio of focal length to aperture is called the f/number; for example, an $f/10$ telescope may have a 30-in. focal length and a 3-in. aperture. By using a combination of mirrors and lenses, it is possible to fold the optical path, thus reducing the length of the telescope to dimensions only slightly larger than those of the aperture and still retain a reasonable f/number [11].

Figure 12.6a shows a simple optical system of focal length F and aperture b. For a telescope, the angular field of view is given by α, where

$$\alpha = d/F \text{ rad} \tag{12.15}$$

Due to aberrations [12, 13], the symmetry and sharpness of the image degrade when it is not in the center of the telescope field of view. The telescope must be designed in such a way that the off-axis image has sufficient quality for detection and star image location processing; this is particularly true for area- or array-type detectors and for fully strapdown applications. The size of the photosensor and the focal length of the telescope determine the field of view over which a quality image must be maintained.

12.4.2   Star-Signal Power

The signal power of a star P_S, as seen by a light detector at the focal plane of the telescope, is the effective area of the telescope entrance times the irradiance of the star:

$$P_S = HA \qquad (12.16)$$

where

H is the usable irradiance from a star found from the spectral distribution of the star radiation H_λ, the photosensor spectral response $(\sigma(\lambda))$, and the star magnitude (units are watts per square centimeter [15])

A is the effective light-gathering area of the entrance aperture of the telescope, including obstructions and allowances for optical transmission losses (units are square centimeters)

Optical transmission losses also reduce the star-signal power at the focal plane of the telescope. Mirrors and lenses do not transmit or reflect light perfectly. For antireflection-coated optics, the light loss may be 1% or 2% at each air-to-glass or glass-to-air optical surface within the telescope, including windows and sensor covers or surfaces. For uncoated optics, the loss per surface is about 4% or 5%.

Star-signal power depends on the spectral irradiance of the stellar light, the spectral response of the light detector, and the brightness of the star. Figure 12.7 shows the spectral response of a B8-type blue star and two types of light detectors; for example, the B8 type, or blue-colored, star (which emits most of its energy at blue wavelengths) appears much brighter to a light detector that is sensitive to blue light than to a light detector that is sensitive to red light. This is shown in Figure 12.7 by the area common to both star and light-detector spectral-response curves. For this reason, the irradiance term H must be given in terms of the light-detector spectral response. A consistent method of calculating star irradiance based on blackbody standards is given below [16, 17]. The irradiance equation is

$$H = 5.18 \times 10^{-16} t_e 10^{-0.4m} \int_0^\infty \left(\frac{H_\lambda}{H_{\lambda\,\text{max}}} \right) \sigma(\lambda) d\lambda \qquad (12.17)$$

where

m is the bolometric magnitude of the star

t_e is the effective star temperature, in degrees Kelvin

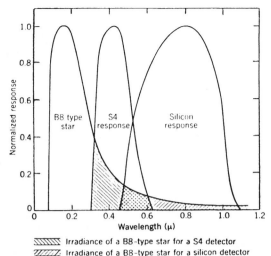

Figure 12.7   Spectral sensitivity to detector material types.

| H | is irradiance, in watts/square centimeter, for a point source |
| $\int_0^\infty (H_\lambda/H_{\lambda\,max})\sigma(\lambda)\,d\lambda$ | is the integral product of the normalized spectrum $(H_\lambda/H_{\lambda\,max})$ and the detector-normalized spectral response $\sigma(\lambda)$ |

Equation 12.17 is used in conjunction with Table 12.2 to solve for star irradiance. Table 12.2 gives the value of the integral term for 5 different types of light detectors and 17 effective star temperatures. For example, the integral value of a +1.22 bolometric-magnitude star of type B8 ($t_e = 11,700$ K) is 0.0663 when a silicon light detector is used and 0.1170 when a photomultiplier with an S-4 photocathode [17] is used. Substituting into Equation 12.17, the irradiance is found to be 1.31×10^{-13} w/cm^2 for the silicon light detector and 2.30×10^{-13} w/cm^2 for the S-4 detector. The irradiance from the star found by Equation 12.17 applies at the top of the atmosphere. For values at sea level, it is necessary to correct for the transmission loss caused by the atmosphere which is 10% to 20%, depending on the wavelength of light [17].

The irradiance of ten typical target stars for various light detectors is given in Table 12.3.

12.4.3   Sky Background Power

The power from the sky background at the focal plane of a telescope depends on the background radiance, the effective entrance area, and the field of view of the telescope:

TABLE 12.2   Normalized stellar-light detector integrals

t_e, Kelvin	Star Type	Type of Detector[a]				
		Silicon	Visual	S-4	S-11	S-20
20,000	B1	0.0126	0.0094	0.0337	0.0303	0.0344
14,000	B5	0.0382	0.0264	0.0819	0.0755	0.0858
11,700	B8	0.0663	0.0421	0.1170	0.1097	0.1254
10,700	A0	0.0843	0.0519	0.1362	0.1291	0.1477
10,400	A1	0.0908	0.0549	0.1422	0.1351	0.1547
9,900	A2	0.1024	0.0608	0.1513	0.1448	0.1663
9,400	A3	0.1163	0.0669	0.1611	0.1551	0.1785
8,100	A7	0.1645	0.0865	0.1838	0.1810	0.2108
7,400	F0	0.1980	0.0984	0.1932	0.1932	0.2270
6,500	F5	0.2488	0.1117	0.1935	0.1982	0.2372
5,800	G1	0.2960	0.1202	0.1836	0.1928	0.2360
5,700	G2	0.3030	0.1211	0.1838	0.1930	0.2368
5,000	G9	0.3485	0.1201	0.1587	0.1713	0.2181
4,900	K0	0.2551	0.1194	0.1506	0.1644	0.2112
4,200	H5	0.3908	0.1096	0.1131	0.1292	0.1754
3,500	M1	0.3816	0.0758	0.0630	0.0748	0.1150
3,400	M2	0.3747	0.0702	0.0560	0.0671	0.1056
Sky light		0.1737	0.0915	0.1198	0.1352	0.1579

[a]Normalized stellar-light detector integrals

$$P_b = NA\alpha^2 \tag{12.18}$$

where

N is the effective background radiance, found from background brightness and the spectral characteristics of the source and light sensor, in watts/cm^2-deg^2. Unlike H (for a point source), N describes the power radiated by an extended source

A is the effective area of the entrance aperture of the telescope, in square centimeters

α^2 is the angular area of the telescope field of view, in degrees squared.

As compared to (12.16), the background radiance in (12.18) has an added dimension, namely, that of degrees squared because the light is gathered from an extended source rather than a point source.

Normally, the night-sky background radiance is so low that its effect may be ignored in comparison with star power. The daylight sky, however, is extremely bright in comparison with a star [17, 18, 19, 20]. Typically, the sea-level sky radiance, 45° from the sun on a clear day is about 2000 ft-lamberts. The background radiance of the daylight sky is given by

TABLE 12.3 Irradiance table for 10 bright stars

Star					Bolometric	Visual	Type of Detector[a]				
Common Name	Bayer Name	Year 1900 Hour Angle	Declination	Type	Magnitude	Magnitude	Silicon	S-20	S-4	S-11	
Alpheratz	α And	0^{h}3.2^m	28°32′	B8	1.22	2.12	1.31	2.46	2.30	2.15	
Polaris	α UMi	1^{h}22.6^m	88°46′	F8	2.06	2.10	1.295	1.14	1.02	0.932	
Hamal	α Ari	2^{h}1.5^m	22°59′	K2	1.64	2.04	1.96	1.04	0.719	0.806	
Betelgeuse	α Ori	5^{h}49.8^m	7°23′	M2	−0.70	+0.7	12.50	3.53	1.87	2.24	
Canopus	α Car	6^{h}40.8^m	−52°38′	F0	−0.77	−0.77	15.42	17.65	15.05	15.05	
Sirius	α CMa	7^{h}34.1^m	−16°35′	A1	−1.94	−1.43	29.20	49.80	45.70	43.50	
Procyon	α CMi	10^{h}3.0^m	5°29′	F5	0.35	0.35	6.08	5.80	4.74	4.84	
Regulus	α Leo	22^{h}52.1^m	12°27′	B8	0.41	1.31	2.76	5.21	4.86	4.56	
Fomalhaut	α PsA	16^{h}23.3^m	−30°9′	A3	0.87	1.19	2.55	3.92	3.54	3.41	
Antares	α Sco		−25°13′	M1	0.13	0.98	6.18	1.85	1.02	1.21	

[a]Effective irradiance, 10^{-13} w/cm^2.

$$N = (0.511 \times 10^{-8})B_S \int_0^\infty \left(\frac{N_\lambda}{N_{\lambda\,\text{max}}} \right) \sigma(\lambda)\, d\lambda \quad \text{w/deg}^2\text{-cm}^2 \qquad (12.19)$$

where

$\int_0^\infty (N_\lambda/N_{\lambda\,\text{max}})\sigma(\lambda)\, d\lambda$ is the integral product of the normalized sky spectrum and the normalized spectral response of the detector $\sigma(\lambda)$

B_S is the daylight sky brightness for a standard observer (human eye), in lumens per steradian per square foot (π ft-lamberts = 1 candle per square foot = 1 lumen per steradian per square foot)

The value of the normalized integral for various light detectors and a daylight sky of 1000 ft-lamberts is given in the last row of Table 12.2 under the heading "sky light." Based on this table and Equation 12.19, a 1000 ft-lambert sky gives a background radiance of 2.8×10^{-7} w/cm^2-deg^2 for a silicon detector. By comparison, outside the Earth's atmosphere, a silicon detector sees about 2.2×10^{-15} w/cm^2-deg^2 of sky background.

Another important property of the daylight sky is its radiance gradient, which can be shown to range from 1.7×10^{-9} to 3.4×10^{-9} w/cm^2-deg^2/deg [17, 18, 19, 20].

Example 12.4.1

To appreciate the significance of the daylight star sensor problem, calculate (a) the sky background power and (b) the star-light power for the following conditions: a 7-cm effective telescope aperture with a 1.5 arcmin field of view, a 1000 ft-lambert sea-level sky radiance, a 100% efficient silicon detector located at the focal plane, and the viewing star, Alpheratz.

Part a   From Equations 12.18 and 12.19 and from Table 12.2,

$$P_b = \left[(0.511 \times 10^{-8}) \left(\frac{1000}{\pi} \right) (0.1737 \text{ w/cm}^2\text{-deg}^2) \left(\frac{49\pi}{4} \text{ cm}^2 \right) \right.$$

$$\left. \cdot \left(\frac{1.5}{60} \text{ deg} \right)^2 \right. = 6.8 \times 10^{-9} \text{ w}$$

Part b   From Equation 12.16 and Table 12.3,

$$P_s = (1.31 \times 10^{-13} \text{ w/cm}^2) \left[\frac{\pi}{4} (49) \text{ cm}^2 \right] = 5 \times 10^{-12} \text{ w}$$

Note that the sky background noise power is approximately three orders of magnitude larger than the star-light power. An essential feature in any daylight star sensor is the method used to increase the star signal-to-sky background noise ratio, see Problem 12.3.

12.4.4   Star-light Detection

The radiation sensed by the detector in a star sensor arrives as photons of light, each of which has a quantity of energy determined by Planck's constant and the wavelength of light. For a radiation power of P (in watts), and a wavelength, λ (in micrometers), the number of photons of light, ρ (in photons per second), arriving at the sensor is given by [14]

$$\rho = 5.04 \times 10^{18} P\lambda \tag{12.20}$$

Only a certain number of these photons arriving at the detector are converted to signal-generating electrons. The quantum efficiency of the detector is defined as the ratio of the photons converted to signal-generating electrons to the total number of photons arriving.

Radiation detectors used in star sensors may be classified by the way in which photons of light are converted to signal-generating electrons, for example, photovoltaic, photoconductive, and photoemissive. Of particular interest are the photovoltaic or photoconductive detectors that are packaged as charge-coupled devices (CCD) [21]. These can produce an array of small cells or pixels. In its simplest form, the CCD is a closely spaced monolithic array of metal-insulator-semiconductor (MIS) capacitors. By the application of a proper sequence of vertical and horizontal clock voltage pulses, the charge packets accumulated in each capacitor (or pixel) can be sequentially shifted through the array of capacitors. Large arrays of CCD detectors can be implemented without the maze of wires and bonding required if individual elements were used to construct the array.

The most important class of MIS capacitors is the metal-oxide semiconductor (MOS) made from silicon and silicon dioxide. By 1995, CCD arrays were available for the visible to mid-infrared wavelength (0.4 to 20 μm), with array sizes ranging from 128×128 to 1024×1024, with clock voltage pulse frequencies greater than 15 MHz, and with full well (pixel) capacities in the neighborhood of 1×10^6 electrons. The use of CCD arrays allows a small instantaneous field of view for sky background, for improved signal-to-noise ratio, while maintaining a large telescope field of view.

Reference [22] provides an excellent treatise on radiation detectors, including a chapter devoted to the CCD array. It contains a comprehensive discussion of, and the noise sources for, photovoltaic detectors, photoconductive detectors, and photoemissive detectors. For a general discussion on the advantages and disadvantages associated with imaging tubes, such as the vidicon, and the photomultiplier tube, see [1].

12.4.5 Focal Plane Array Processing

In some star sensors, the star signal is modulated to distinguish it from the constant sky background level. One technique for doing this is by rotating an optical wedge, which is placed a short distance from the focal plane, and synchronously demodulating the signal. Another is to scan the star area with a very narrow field-of-view telescope.

Another technique to remove the sky background noise when using a CCD array is by direct subtraction. The telescope is pointed, say one degree, away from the star (to a region where there are no bright stars), and a frame of data is gathered and stored. The telescope is then pointed at the star, and another frame is gathered and stored. These two data frames are then differenced, pixel by pixel [23]. The difference frame contains the star without the sky background noise. Further, the effects of differences between individual pixel responsivities are minimized. Another implementation of the noise subtraction technique is to calibrate the individual pixel responsivities and then subtract a computed sky background mean from each pixel; the sky background mean is computed by averaging over a large number of pixel elements. This technique has the advantage of increasing the signal-to-noise ratio (S/N), by $\sqrt{2}$.

The frame time, associated with the CCD, is the time over which the signal and noise can be integrated before the pixels reach their full well capacities.

Example 12.4.2
Suppose that the star Alpheratz is being viewed in a 2000 ft-lambert sky background with the telescope of Example 12.4.1 and the focal plane array of Problem 12.3. What is the maximum frame averaging time given that 90% of full well capacity is not to be exceeded? Assume a full well pixel capacity of 1×10^6 electrons with a quantum efficiency of 50%, that 93.7% of the star energy is contained within a one pixel diameter, and that there is a 20% loss of star power due to the atmosphere.

From Example 12.4.1, the star power arriving at a single pixel is

$$P_S = (5 \times 10^{-12}\,\text{w})(0.937)(0.80) = 3.75 \times 10^{-12}\,\text{w}$$

and from Problem 12.3, for 2000 ft-lambert sky background, the mean noise power per pixel is

$$P_b = 2(7.5 \times 10^{-12})\,\text{w} = 15 \times 10^{-12}\,\text{w}$$

From Equation 12.20 the total photons arriving at the detectors, assuming an average wavelength of 0.8 μm, is given by

$$\rho = (5.04 \times 10^{18})[(15 + 3.75) \times 10^{-12}](0.8) = 76 \times 10^{6} \text{ photons/sec}$$

If the quantum efficiency in converting photons to electrons is 50%, then the time required to reach 90% of full well capacity, or frame time t_F, is given by

$$t_F = \frac{0.9 \times 10^{6} \text{ electrons}}{(0.5 \text{ electrons/photon})(76 \times 10^{6} \text{ photons/sec})} = 24 \quad \text{msec}$$

To further improve the star S/N, a number of difference frames are averaged. Given the average difference frame, star detection is evaluated using a likelihood ratio (threshold) test [10]. The threshold level is a function of the star brightness and noise variance and is performed on subsums of pixels of a size to contain the star image. For example, if the star image, or spot size, is 2 pixels in diameter, then all 2×2 pixel sums are tested against the threshold. Other functional tests can be implemented in conjunction with the star detection threshold test. Once the star is detected its position is known to within a small group of pixels. The precise location of the star (subpixel accuracy) is determined using a centroiding technique [24] which is analogous to a "center-of-gravity" computation. This involves constructing weighted averages in both directions using a group of pixels that includes the group, found by detection, to contain the star. Other techniques include the "polynomial least-squares fit" described in [24], a "matched-filter" approach described in [25], and a template-matching approach based upon knowledge of the star signal point spread function.

Once the precise location of the star has been determined within the array, it can be passed on to the Kalman filter as a measurement (observation) of the star LOS. Alternatively, in a "star tracker" implementation, the results of the centroiding can be used as inputs to the azimuth and elevation gimbal servo-loops to center the image on the array [23]. The gimbal angle resolver readouts then provide the line-of-sight observation to the Kalman filter.

Factors that affect centroiding accuracy include S/N, image spot size, and fill factor (dead space between pixels). The optimum image spot size for precise centroiding has a diameter between one and two pixels [24, 26] and the centroiding performance degrades rapidly for fill factors less than 60% [24]. Nonuniform responsivity of pixels and nonsymmetrical star imaging can be calibrated and compensated for, if required. The centroiding accuracy as a function of S/N is described in Figure 12.8 for pixel fill factors near 100%, for image spot diameters in the neighborhood of 2 pixels, and for on-axis imaging. The band between the upper and lower curves in the figure accounts for the effects of the random variation of image center to pixel center.

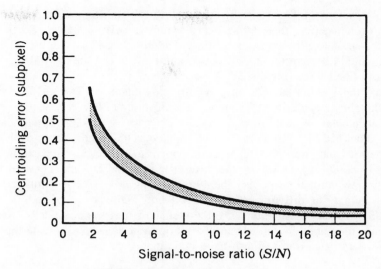

Figure 12.8   Centroiding accuracy.

12.5   CELESTIAL NAVIGATION SYSTEM DESIGN

Star observations provide no information with respect to altitude, and since the vertical channel of an inertial navigator is unstable, vertical damping must be provided by some other information source. Typically, a barometric altimeter or radar altimeter is used to provide altitude measurements. These altitude measurements can be processed in the Kalman filter in a fashion similar to the star measurements to provide altitude corrections to be fed back to the navigator.

12.5.1   Time Reference

All stellar-inertial systems require an accurate measure of time in order to initialize and maintain the transformation from inertial coordinates to Earth-fixed coordinates. Also, the location of the stars in inertial coordinates at the instant of a star measurement requires time in order to apply corrections to the star's location (Section 12.6.4).

Time is initialized with a chronometer. The chronometer may be a ground-based portable device that is interfaced to the stellar-inertial system before take off or it may be airborne equipment. Chronometers are accurate time-keeping devices that have previously received a time initialization. One method of time initialization is the receipt of a radio signal from WWV (Fort Collins, Colorado) or WWV-H (Hawaii), which are maintained by the U.S. National Bureau of Standards [27]. These radio signals provide universal time (UT), or Greenwich mean time which is based on the rotation of the Earth. For high-accuracy systems, UT needs to be corrected to UT1 to account for irregularities in earth rotation (less than 0.9 sec). In the future, a GPS receiver (Section 5.5) could be used to provide accurate time. Given UT, GHAϒ (Figure 12.2) can be computed [28] and the initial transformation between the inertial frame and the Earth-fixed frame obtained.

Since the Earth's rotation is generally slowing down (which occurs with unpredictable irregularities), UT is not a uniform time. For very accurate astronomical calculations, a uniform time scale is necessary. In 1984, dynamical time (TD), as defined by atomic clocks, was adopted as the standard for astronomical ephemerides and published in the *Astronomical Almanac*. Star correction calculations use TD [28]. However, the difference between UT and TD is insignificant for navigational purposes. Therefore, as long as GHAϒ is computed using UT, the remaining differences between UT and TD may be ignored.

After time has been initialized, the stellar-inertial system clock is used to maintain time, and the known Earth's rotation rate is used to update the transformation from inertial to Earth-fixed coordinates.

12.5.2 Star Observation and Pointing Errors

The azimuth and elevation angles to a selected star are computed using Equation 12.4. These angles, compensated by estimates of mechanical and optical deviations and deflections of the line of sight (LOS), are functions of the right ascension and declination of the star, the Greenwich Hour Angle of Aires for the given time, an estimate of the vehicle position in latitude and longitude, and an estimate of the platform attitude at this location. The measurements of the star LOS are functions of the actual location of the vehicle; the actual attitude of the physical platform upon which the star sensor is mounted; and all actual mechanical and optical deviations and deflections in the line of sight. The angular difference between the computed LOS and the measured LOS, δY, provides the Kalman filter star observation, as described in Section 12.3.1. From Equation 12.9b, this filter observation δY is given by

$$\delta Y = \boldsymbol{\psi} + \delta \boldsymbol{b} + \boldsymbol{v}$$

where $\boldsymbol{\psi}$ is the inertial angular error, $\delta \boldsymbol{b}$ denotes the total vector of mechanical and optical LOS error compensation residuals, and $\boldsymbol{v}$ is the independent random observation error.

The errors $\boldsymbol{\psi}$ and $\delta \boldsymbol{b}$ are included in the state vector of the dynamic error model used to propagate the navigation system errors (see Equation 12.9a). They are described in more detail in what follows for the stabilized platforms and strapdown IMU configurations. The optical line-of-sight deflection errors are independent of the configuration and are described later.

Stabilized, Local-Level Implementation Errors   For the stabilized, local-level platform with gimballed telescope mounted on the stabilized element, the pointing error includes platform tilt and heading errors ϕ, computed position errors $\delta\theta$, telescope gimbal angle resolver errors, telescope/stellar sensor boresight errors, detector image location errors, and optical LOS deflection errors. The effects of the ϕ and $\delta\theta$ error uncertainties are modeled mathematically by $\psi = \phi - \delta\theta$.

With a detached star sensor, the errors of the gimbal angle resolver of the stable platform and the mechanical and flexure alignment uncertainties between the two housings must be modeled. Because of the mechanical and flexure alignment uncertainties the detached implementation is not used in high-accuracy applications.

Strapdown Implementation Errors   The pointing error uncertainties for the strapdown configuration include ψ error ($\phi - \delta\theta$), stellar sensor boresight error, star image location error and optical LOS deflection error. No gimbal resolver errors are involved since the telescope/star sensor is hard-mounted to the inertial instrument cluster. If gimbals are used to point the IMU-telescope structure, the entire cluster is rotated; the relationship between the telescope LOS and the inertial instrument axes remain fixed, and the orientation of the instrument cluster is determined using the inertial instruments, not the gimbals.

For the fully strapdown (ungimballed) configuration, a wide field of view is required by the telescope. Consequently, there is a requirement to locate star images for large off-axis star measurements. This introduces a source of error in image location due to coma which must be accounted for in the error modeling. Star sensor boresight error observability is a significant problem for this configuration, as well as star location accuracies, given the large star catalog required (Section 12.6.2).

Implementation-Independent Errors   Deflections of the LOS arise from nonparallelism of top and bottom window surfaces [12, 13, 29, 30]. Nonparallel viewing-window surfaces act as prisms and deflect the incoming light ray from the original path. If Snell's law of refraction is applied to a light ray impinging on a window at an incident angle of 50 deg, a 5 arcsecond window wedge will cause a 10 arcsecond change in the angle at which the ray emerges and hence in the LOS to a star.

A 1-atm pressure difference between the window surfaces will cause sag in the area cut by the aperture of the telescope [29, 30]. A 12-in. diameter fused silica-window with a thickness of 0.8 to 1 in. has negligible sag-and-distortion effect, but the 1-atm of pressure difference creates a refraction effect at the window interface [12, 13]. The magnitude of the effect is sensitive to both the pressure and the temperature of the inside and outside air. For a 15-psi pressure differential, inside air at 130°F, and a viewing angle of 50 deg with respect to the window, the deflection of the line of sight is about 60 arcseconds.

Factors that cause light gradients or uncontrolled reflections include bubbles within the glass as well as dust and oil on the window. The presence of bubbles and dust (if not thick) is not a serious problem, because the telescope, being focused at infinity, looks through them. An oil film, however, can act as an interference coat of the wrong wavelength and cause light reflections and/or severe transmission losses [29, 30]. Surface flatness over the viewing aperture of the telescope [29, 30] also affects image quality, with values ranging from 0.1 to 2 wavelengths, depending on the desired quality of the image.

Studies have shown that the magnitude of atmospheric refraction requires compensation [17, 31]. At sea level, refraction varies from 0 arcseconds at 0-deg co-elevation angles to 70 arcseconds at 50-deg co-elevation angles. The deflection due to atmospheric refraction increases with increasing air mass between the star and the observer and varies slightly with air temperature and water-vapor content.

Depending upon its nature, turbulence has both good and bad effects on star viewing. Slow-moving and stagnant air full of temperature discontinuities degrades viewing. Local turbulence causes shimmer—a motion of the star image within the telescope [32]. This motion may approach 30 arcseconds at frequencies as high as 100 Hz for telescope apertures under 4 in. [17]. There may be rapid or highly turbulent air along the outer window or skin of the aircraft. Studies [30, 33] of boundary layers and turbulence in aircraft up to Mach 2.5 show shimmer disks of up to 12 to 15 sec of arc with frequencies of up to 1000 Hz. The effect of this shimmer is to spread the star irradiance over a greater area at the focal plane of the telescope. To compensate, the detector must be large enough to encompass the image, or sensitive enough to detect the portion of the light energy intercepted by the small detector.

Atmospheric turbulence far from the telescope does not cause shimmer; instead, it produces scintillation—a varying change in the brightness of the star image ("twinkle" to the human eye) [32, 34]. Scintillation, which is due to atmospheric diffraction, is inversely proportional to telescope aperture and can change the brightness of a star as much as 35% for a telescope aperture of less than 5 in. Its amplitude is proportional to the secant of the co-altitude pointing angle, that is, to the mass of air the light must penetrate.

During supersonic flight, the line of sight to a star is refracted through a shock wave, which is a boundary layer of air separating less dense air from the more dense air behind the wave. Although precise data are lacking, the amplitude of shock-wave refraction is under 2 arcseconds for angles of up to ±45 deg to the wave created by a 2.85-Mach aircraft at 45,000 ft [30, 35]. The line-of-sight deviation increases at more oblique angles to the shock wave.

12.5.3 Stabilized Platform Configuration

In the stabilized design, the inertial measurement unit (IMU) provides a stabilized platform upon which a two-gimbal star sensor is mounted. The IMU contains accelerometer and gyro triads (Chapter 7). The stabilized design is depicted in Figure 12.9.

The accelerometers determine the local-level reference with respect to the

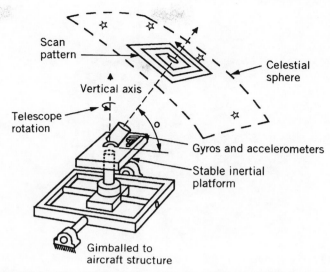

Figure 12.9   Three-gimballed platform with two-gimballed star sensor.

Earth and the gyro torquing current is used to track the orientation of the stable platform relative to an inertially fixed frame in which the stars are located. Along with time, this provides the information required to generate the azimuth and elevation angles to point the telescope at a selected star. Therefore, the star observations are effective in calibrating gyro bias drifts. The telescope gimbal angle resolver errors cause line-of-sight measurement errors, but the stable platform gimbal angle resolver errors do not.

A typical platform and star tracker gimbal structure is shown in Figure 12.10, and a simplified block diagram of the stabilized stellar-inertial navigator is shown in Figure 12.11. The latter shows the functions and the flow of data.

To integrate a stellar observation into the navigation process, the vehicle position (latitude and longitude) and attitude of the platform with respect to the Earth are computed from the inertial navigation equations (Chapter 7). This information, along with time and the on-board stored star catalog, is used to select the star to be observed. The telescope pointing commands (AZ, EL) are computed and the telescope slewed to the proper location. A preprogrammed search pattern is conducted around this location until the star is detected. The location of the detected star can be determined by centering the image and measuring the azimuth and elevation angles of the telescope gimbal angle resolvers. The difference between the computed azimuth and elevation angles and the measured azimuth and elevation angles is the Kalman filter observable δY. The components of δY, in telescope coordinates, are orthogonal to the telescope line of sight and defined in telescope coordinates:

Figure 12.10   Typical platform/tracker gimbal structure (courtesy, Northrop Grumman Electronics System and Integration Division).

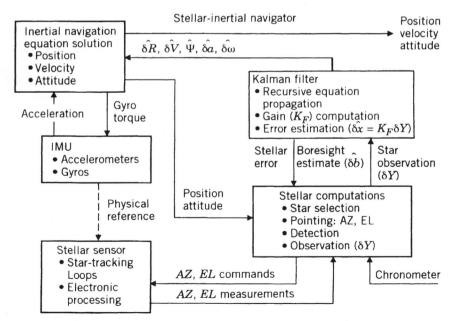

Figure 12.11   Simplified block diagram of stabilized stellar-inertial navigator.

$$\delta Y^s = \begin{bmatrix} \delta Y_1 \\ \delta Y_2 \end{bmatrix} = \begin{bmatrix} EL_{\text{MEAS}} - EL_{\text{COMP}} \\ \cos(EL)[AZ_{\text{MEAS}} - AZ_{\text{COMP}}] \end{bmatrix} \qquad (12.21)$$

and in computational coordinates,

$$\delta Y = \hat{C}_S^C \delta Y^S$$

where δY is as defined in Equation 12.9b.

A high performance, terrestrial stellar-inertial navigator is shown in Figure 12.12. This navigator utilizes a three-degree-of-freedom locally level stabilized platform with a two-gimbal star tracker mounted on it.

Typically, two to three stars per minute are tracked for high-performance applications with rms star location accuracies of several seconds of arc or better. To benefit from this level of stellar performance, gravity deflections of the vertical must be compensated to the level of several arcseconds or less.

The viewing window for the stellar-inertial navigator should be as small as possible both from airframe considerations and for stellar performance (large windows bend and so must be thick). While the size of the window is a function of the field of regard and the size of the telescope aperture, it is also a function of the telescope gimbal arrangement. Figure 12.13 shows an offset-gimbal configuration that reduces the requirements for viewing-window size. This offset gimbal arrangement is implemented in the daylight stellar-inertial navigator pictured in Figure 12.12.

12.5.4 Strapdown IMU Configurations

Since the 1980s IMUs have been strapdown to reduce cost and size, and increase reliability (Section 7.4). Since the 1990s, the development of stellar-inertial navigators were therefore directed toward the use of the combination of strapdown IMUs and focal plane array star sensors hard-mounted to the IMU (Sections 12.4.4. and 12.4.5). A strapdown IMU configuration is depicted in Figure 12.14. In this configuration, the inertial cluster is mounted in a two-axis gimbal structure with the telescope hard-mounted. The gimbals are used for approximate pointing of the telescope, and for pre-flight and in-flight instrument calibration and system alignment (Section 12.7). In one mechanization of the configuration of Figure 12.14, the inertial cluster is periodically rotated 180 deg in azimuth (a 180-deg-rotation about both pitch and roll axes effectively rotates 180 deg in azimuth). This has the effect of making the star sensor boresight error observable so that system performance is relatively independent of this error (Problem 12.2, Parts c and e). Typically, azimuth rotations every 5 minutes maintain good observability and the rotations take less than 30 seconds to perform.

The gimbal structure can be removed producing a fully strapdown stellar-

Figure 12.12   A high-performance stellar-inertial navigator (navigation system for the B-2 Bomber; Courtesy, Northrop Grumman Corp.).

inertial system. A major design difference between stabilized and strapdown IMU configurations is that for the latter, the image is not stabilized on the focal plane; rather, it is computationally stabilized. The fully strapdown configuration requires the use of a very wide field-of-view telescope with a significant and accurate off-axis imaging capability. The size of the focal plane array must be consistent with the FOV requirements. Star sensor boresight stability becomes a critical issue, and the contents of the on-board catalog are increased from 50 to 80 stars to the order of 10,000 (Section 12.6.2), the location of many of which are not known with a high degree of accuracy.

The block diagram of Figure 12.11 also provides a functional description of the computations involved for the strapdown stellar-inertial navigator configurations, except for the "stellar sensor" block. For the strapdown configurations, star tracking is not implemented. The Kalman filter observation δY is determined by computing the angular difference between where the center of the

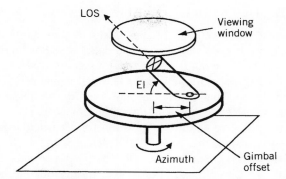

Figure 12.13   Offset tracker gimbal arrangement.

star image is expected to fall on the focal plane array and where the computed star image centroid actually falls on the focal plane array. This is the case for both the gimballed and nongimballed strapdown IMU configurations; gimbal angle measurements are not involved in either case. The angular orientation of the telescope is determined by the IMU through a direction cosine or quaternion computation using the inertial cluster gyro inputs.

12.6   STAR CATALOG CHARACTERISTICS

A necessary part of any stellar-inertial navigation system is a star catalog that provides data on star locations and star magnitudes. The star location data are required to point the telescope and determine the expected angular position of the star in the star sensor field of view. The star magnitude is required to determine the star integration time and detection threshold in order to achieve the desired probabilities of detection and false alarm.

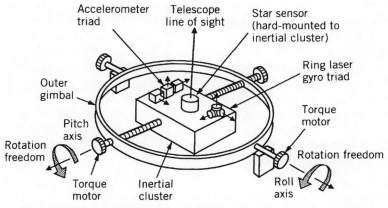

Figure 12.14   Strapdown stellar configuration with gimbal pointing.

12.6.1 Star Catalog Contents

A *star catalog* is a stored data file of stars designated by name or ID containing data fields describing the star, the critical data for stellar-inertial navigation being star position and magnitude. Star position, as discussed in Section 12.2, is referenced to the vernal equinox and the celestial equator. However, the vernal equinox and the celestial equator are not fixed in inertial space, primarily due to gravitational effects of the sun, moon, and planets. Therefore, it is necessary for a star catalog to define star position for a particular instant of time or epoch. The current standard epoch is 2000 January 1 at 12 hours dynamical time (TD). This epoch is designated J2000. The unit of time is the Julian year or Julian century. To locate a star at any other instant of time, it is necessary to account for both the star's motion and the Earth's motion from J2000 to the current instant of time. Reference [28] describes how to determine the time period from the time of interest to J2000. Corrections for motion are described later in this section.

In addition to star position and proper motion data referenced to a particular epoch, a star catalog contains star magnitude data. The magnitude data are usually specified for more than one frequency band including the visual band. For stellar-inertial applications, the magnitude of interest is the magnitude in the band of the star sensor. Frequently, this band does not match identically with the magnitude data provided by a star catalog. Therefore, some additional computations may be required to convert the magnitude data given in a star catalog to the magnitude data applicable to the system star sensor detector.

12.6.2 Star Catalog Size

A subset of all the stars in a large star catalog needs to be selected to form the on-board star catalog for a stellar-inertial navigator. The number of stars required for the on-board star catalog is a function of the system implementation. For the stabilized design (Section 12.5.3) or the strapdown design with two gimbals (Section 12.5.4), both of which have the ability to point the sensor telescope, a very small subset of a star catalog is required. Typically, 50 to 80 of the brightest stars in the sky can be selected to provide worldwide, year-around coverage that always provides at least three stars available for the system to select. The fully strapdown design (no gimbal structure) (Section 12.5.4) requires a much larger on-board star catalog because the system must select from the stars that happen to be in the field of view. The number of stars depends on the size of the star sensor field of view and the frequency at which star measurement updates are required. A fully strapdown celestial navigator may have to acquire stars down to magnitude 6.0 and require an on-board star catalog of about 10,000 stars.

Variable Stars   Variable stars are those whose magnitudes are not constant. A large number of stars vary in their brilliance, some in a regular manner, others less regularly, and some are completely unpredictable and irregular in their light variation [37]. Some representative variable stars are given in Table 12.4.

Magnitude variations cause a problem for determining the integration time and detection threshold. If the magnitude used in the star catalog is brighter than the actual star magnitude at the time of star acquisition, the integration time will be too short, the detection threshold too high, and the star will likely not be detected. If the magnitude used in the star catalog is dimmer than the actual star magnitude at the time of star acquisition, then the star may not be selected as the best available star; if the star is selected, the signal-to-noise ratio will be higher than expected and the uncertainty assigned to the star location measurement will not be as small as it should be.

For some variable stars it is possible to model the magnitude variation and compute the star magnitude at the time of the star acquisition. For a star like Betelgeuse, whose magnitude varies slowly (Table 12.4), the magnitude can be updated at time intervals that are long relative to the flight time. Except for bright, slowly varying stars it is probably best to exclude variable stars from the on-board star catalog.

Double Stars   Double stars are two stars that appear very close to one another. Some representative double stars are given in Table 12.5 [38].

Double stars present two problems for a stellar-inertial navigation system. A double star like Mizar with a separation of only 14.4 arcsec will likely be beyond the resolving capability of the star sensor. Consequently, a centroid computed on Mizar would yield a location that is a weighted composite of the two stars. Unless a weighted composite location is stored in the star catalog, the star location error will not be modeled properly.

A double star like α Capricornus presents a different problem. The two stars are sufficiently far apart, so they do not present a star centroiding problem. However, α Capricornus presents an initial acquisition problem because at the end of alignment there exists a relatively large heading error. Depending upon the time of the first star measurement (which might be delayed due to cloud cover), the attitude uncertainty may grow larger. The attitude uncertainty may cause difficulty in distinguishing between the two stars of α Capricornus. One

TABLE 12.4   Variable stars

Name	Magnitude Variation	Period
Betelgeuse	0.4–1.3	5.7 years with variations
Mira	1.0–10.2	320–370 days
δ Cephei	3.28–4.63	5.37 days
Algol	2.20–3.47	2 days, 20 hours, 49 minutes

TABLE 12.5   Double stars

Name	Magnitudes	Separation
α Capricornus	3.5, 4.0	376 arcsec
Mizar	2.4, 4.0	14.4 arcsec
Mizar—ζ Ursa Major	2.4, 4.0	11.8 arcmin
Alcor—80 Ursa Major		

way to deal with double stars in the star catalog is to use a flag that excludes the double star for acquisition if the system attitude uncertainty is above a threshold comparable to the double star separation.

12.6.3   Planet and Moon Avoidance

Planets do not represent good candidates for angular measurements for a stellar-inertial navigation system because of the following: (1) dimness of the planet or its proximity to the sun (only Venus, Mars, Jupiter, and Saturn are viable candidates for measurement), (2) planets are not point sources of light like stars and hence, they spread over a larger portion of the detector than a star necessitating different centroiding techniques, and (3) because of the size of the planets, the location measurement is less accurate than that of a star.

Planets also represent a problem in that they wander across the stars and, at any particular instant in time, can be confused with a star. One way to handle the planet problem is to precompute the orbits of the planets and determine the time periods when a planet is in close proximity to a star. The star catalog can include data fields that indicate when the star is unavailable for acquisition due to possible confounding with a planet.

A similar situation exists with the moon. It does not represent a candidate for a measurement but does obscure stars that are candidates. Approaches to solving the moon interference problem include (1) precomputing the time periods when the moon interferes with a star and storing the data in the star catalog and (2) implementing an on-board computation of the orbit of the moon and excluding those stars that are obscured.

12.6.4   Star Position Corrections

Star position corrections are those that must be applied to the star in order to account for changes in position occurring between the epoch of the star catalog and the time of the desired star acquisition. Causes for star corrections include

1. *Proper motion.* The motion of the star in inertial space.
2. *Precession.* Long-term, regular motion of the celestial pole about the pole of the ecliptic.

3. *Nutation.* Short-term, periodic motion of the celestial pole about the mean position defined by the precessional motion.

4. *Annual aberration.* Apparent change in star position due to the Earth's orbital velocity.

5. *Diurnal aberration.* Apparent change in star position due to the Earth's rotational velocity.

6. *Parallax.* Change in position of the star due to displacement of the Earth in its orbit and the finite distance to the star.

The first four items require corrections for high-accuracy stellar-inertial navigation systems and are discussed below. The last two, diurnal aberration and parallax, are sufficiently small so that they can be ignored.

Proper Motion   Proper motion is the simplest of the star corrections to apply in that the star catalog will contain each star's rate of change of right ascension and declination. If T is the time since the epoch of the star catalog (if the epoch is J2000, then T will be negative for all times prior to J2000), then the star position corrected for proper motion is

$$\alpha = \alpha_0 + \dot{\alpha}T$$
$$\delta = \delta_0 + \dot{\delta}T$$

where α_0, δ_0 and their rates of change are the star catalog entries for the standard epoch.

Precession   The direction of the rotational axis of the Earth and the plane of the Earth's orbit are not fixed in space. Therefore, a star's coordinates change, not because the star moves on the celestial sphere, but because the inertial coordinate frame defined by the equator and the ecliptic moves on the celestial sphere. These changes are called *precession* and *nutation.* Long-term changes are called precession (providing the mean position), and the short-term periodic displacement of the celestial pole about its mean position is called nutation [39].

Due to precession, the celestial pole slowly turns around the pole of the ecliptic, with a period of about 26,000 years; as a consequence, the vernal equinox regresses by about 50 arcsec per year along the ecliptic. Moreover, the plane of the ecliptic itself is not fixed in space. Due to the gravitational attraction of the planets on the Earth, it slowly rotates, the speed of this rotation presently being 47 arcsec per century.

Three precessional angles, ζ, z, and θ, are computed as power series expansions in the time interval T [28]:

$$\zeta = T * (2306''.2181 + T * (0''.30188 + T * 0''.017988))$$
$$z = T * (2306''.2181 + T * (1''.09468 + T * 0''.018203))$$
$$\theta = T * (2004''.3109 - T * (0''.42665 + T * 0''.041833))$$

where T is the time in Julian centuries before or after J2000 and the angles are in arcseconds ($''$).

These precessional angles are small angles defining three rotations of the inertial axes from J2000 to the time defined by T. The transformation matrix for precessional motion is given by

$$\mathbf{P} = \begin{bmatrix} \cos(z)\cos(\theta)\cos(\zeta) & -\cos(z)\cos(\theta)\sin(\zeta) & -\cos(z)\sin(\theta) \\ -\sin(z)\sin(\zeta) & -\sin(z)\cos(\zeta) & \\ \sin(z)\cos(\theta)\cos(\zeta) & -\sin(z)\cos(\theta)\sin(\zeta) & -\sin(z)\sin(\theta) \\ +\cos(z)\sin(\zeta) & +\cos(z)\cos(\zeta) & \\ \sin(\theta)\cos(\theta) & -\sin(\theta)\sin(\zeta) & \cos(\theta) \end{bmatrix}$$

If $\mathbf{u}$ is a star unit position vector corrected for proper motion and given by

$$\mathbf{u} = \begin{bmatrix} \cos\alpha\cos\delta \\ \sin\alpha\cos\delta \\ \sin\delta \end{bmatrix}$$

then the star position vector $\mathbf{u}_p$ corrected for precession is

$$\mathbf{u}_p = \mathbf{P}\mathbf{u}$$

For a 10-hour flight, the precession angles ζ, z, and θ change by less than 0.03 arcsec; therefore the precession rotation matrix $\mathbf{P}$ needs to be computed only once at initialization for the duration of the flight, and star position vectors can be corrected for precession by multiplication by the computed rotation matrix $\mathbf{P}$.

Nutation   Nutation is a periodic oscillation of the rotational axis of the Earth around its mean position, which advances by precession around the pole of the ecliptic. The effect of nutation is small, never exceeding 15 arcsec. Nutation is due principally to the action of the moon and can be described by a sum of periodic terms [28]. Nutation is partitioned into two components, one parallel to the ecliptic denoted by $\Delta\psi$ (called the *nutation in longitude*, Λ) and a component perpendicular to the ecliptic denoted by $\Delta\epsilon$ (called the *nutation in obliquity*). The symbol $\Delta\psi$ is used for the nutation in longitude to remain consistent with astronomer's notation and should not be confused with the $\boldsymbol{\psi}$ error angle (Section 12.3.1).

The theory of nutation provides series expansions for $\Delta\psi$ and $\Delta\epsilon$ in terms of (1) the time interval from the current time to the star catalog standard epoch, (2) the mean elongation of the moon from the sun, (3) the mean anomaly of the Earth, (4) the mean anomaly of the moon, (5) the moon's argument of latitude, and (6) the longitude of the ascending node of the moon's mean orbit on the ecliptic. Algorithms are given in [28].

In addition to the nutation terms, the obliquity of the ecliptic is required. The obliquity of the ecliptic is the angle between the equatorial pole and the ecliptic pole. An expression for the mean obliquity of the ecliptic ϵ_0 is given in [28]. The true (instantaneous) obliquity of the ecliptic ϵ is given by

$$\epsilon = \epsilon_0 + \Delta\epsilon$$

The position of the star on the celestial sphere does not change; rather, the true equator changes from the mean equator moving the true equinox with it. The celestial longitude has increased from Λ to $\Lambda + \Delta\psi$, which means that the equinox has moved through a rotation of $-\Delta\psi$ about the ecliptic pole. Furthermore, the obliquity has increased by an amount $\Delta\epsilon$, which means that the equator has moved through a rotation of $-\Delta\epsilon$ about the equinox.

Given ϵ_0, $\Delta\psi$, and $\Delta\epsilon$, a transformation matrix $\mathbf{N}$ can be computed that rotates a star position vector from the mean equator and equinox of date to the true equator and equinox of date. The expression for $\mathbf{N}$ is

$$\mathbf{N} = \begin{bmatrix} 1 & -\Delta\psi\cos\epsilon_0 & -\Delta\psi\sin\epsilon_0 \\ \Delta\psi\cos\epsilon_0 & 1 & -\Delta\epsilon \\ \Delta\psi\sin\epsilon_0 & \Delta\epsilon & 1 \end{bmatrix}$$

A star unit position vector $\mathbf{u}_{np}$ that has been corrected for nutation can be determined from a star unit position vector $\mathbf{u}_p$ that has been corrected for precession and proper motion by

$$\mathbf{u}_{np} = \mathbf{N}\mathbf{u}_p = \mathbf{N}\mathbf{P}\mathbf{u}$$

Nutation in longitude changes by less than 0.08 arcsec for a 10-hour period and the nutation in obliquity by less than 0.015 arcsec. Therefore, for flights of less than ten hours, the error will be less than 0.1 arcsec if the nutation matrix $\mathbf{N}$ is computed at initialization and treated as a constant thereafter.

Annual Aberration   Annual aberration is the apparent change in the star position due to the velocity of the Earth in its orbit about the sun. The maximum correction due to annual aberration occurs when the star direction is along the ecliptic pole. The maximum correction is the ratio of the Earth's orbital speed to the speed of light and is approximately 20.5 arcsec.

The velocity vector for light coming from a star is $\mathbf{V}_{\text{light}} = -c\mathbf{u}_{np}$ where $c =$

the speed of light. Let $\mathbf{V}_{earth}$ be the velocity vector for the Earth in its orbit. The Earth moves in an elliptical orbit with eccentricity e with the sun at one focus. The velocity $\mathbf{V}_{earth}$ lies in the plane of the ecliptic and has two components, a transverse component $\mathbf{V}_0$ perpendicular to the radius vector from the sun to the Earth and a component $e\mathbf{V}_0$ along the semiminor axis of the orbit [39]. To compute the two components of $\mathbf{V}_{earth}$ in inertial coordinates, it is necessary to compute for the current time the following: (1) the eccentricity of the Earth's orbit, (2) the longitude of perihelion of the Earth's orbit, and (3) the longitude of the sun. Algorithms are found in [28].

The apparent direction of the star light on the moving Earth is $\mathbf{V}_{app} = \mathbf{V}_{light} - \mathbf{V}_{earth}$. The star position vector must be corrected to the apparent direction. The orbital velocity of the Earth is changing slowly enough so that for flights of ten hours or less, it can be precomputed during system initialization and treated as a constant during the flight.

12.7   SYSTEM CALIBRATION AND ALIGNMENT

12.7.1   Factory Calibration

Three aspects of factory calibration of stellar-inertial navigation systems are discussed in this section: (1) calibration of accelerometer errors, gyro errors, and star-tracker boresight errors, (2) star-tracker optical calibration, and (3) focal plane array calibration.

Calibration of Accelerometer, Gyro, and Star-Tracker Boresight Errors   The desirable way to calibrate a stellar-inertial navigation system for accelerometer errors, gyro errors, and star-tracker boresighting errors is to utilize a Kalman filter similar to the real-time software. In the calibration filter, the error state vector is usually larger than that of the real-time navigation filter to include error sources whose effects would be excessive if left uncalibrated. However, those variations about the calibrated values are small enough so that they are dropped from the real-time software for throughput and observability consider-ations. A system can be calibrated by having real-time data acquisition software record the outputs of the instruments which can be read by the calibration filter. This aids the calibration of early systems when not all of the error mechanisms are understood by allowing repeated playback of the same recorded data with modifications to the error modeling until a satisfactory model is obtained.

The stellar-inertial navigation system is calibrated with a fixture that provides at least one simulated star as well as the rotational capability to (1) point the star sensor at the simulated star(s) and (2) orient the accelerometers and gyros to sense varying amounts of the Earth's gravity vector and angular rotation vector. Figure 12.15 shows such a fixture and a stellar-inertial navigation system under test. It contains two precision gimbals to rotate the navigation system and a sin-gle simulated star mounted above the unit on a yoke whose gimbal position can

Figure 12.15   Stellar-inertial factory calibration fixture (courtesy of Northrop Grumman Electronics Systems and Integration Division).

also be precisely controlled. The three gimbals can be moved in 0.5-deg increments and are accurate to 0.5 arcsec. The gimbal positions are under computer control so that a predetermined gimbal trajectory can be executed while the unit under test records the instrument outputs. The gimbal trajectory is designed to provide system error observability as defined in Section 12.3.2.

The calibration values are downloaded for use by the real-time software to correct the instrument outputs during flight. Error states are included in the real-time filter to estimate time variations in some of the factory calibrated values.

Star-Tracker Optical Calibration   The above described procedure provides the calibration of the star tracker boresight errors. It assumes that the position error of the star-tracker is independent of the location of the star within the field of view; however, with wide-angle telescopes, this is not the case. The position error of the star location varies with the star location within the field of view. For example, the star signal distribution across the detector element varies with star position, and this signal distribution variation influences the star peak location estimate, producing an error that is a function of star location. A calibration fixture of the type shown in Figure 12.15 has the ability to accurately place a star across the field of view, and it allows a grid of star locations to be generated. The grid of star placements allows for the location dependent errors to be calibrated relative to an offset or boresight error. This calibration can be performed on the star tracker prior to the combined calibration of the inertial sensors and the boresight errors.

Focal Plane Array Calibration   Focal plane arrays contain individual defective pixels and entire lines of defective pixels. These defective pixels can be determined by exposing the star sensor to a light source that is nearly uniform across the focal plane array and examining the focal plane array response. A defective pixel table can then be generated. It becomes another calibration table for use by the real-time software, which disregards those exposures where portions of the star signal are lost. The nondefective pixels possess input amplitude-to-output amplitude transfer functions that vary from pixel to pixel. This pixel variation looks like noise on both the sky background and the star signal and can have considerable effect on lowering the signal-to-noise ratio.

If the sky background is removed by pixel averaging (Section 12.4.5), pixel-to-pixel variations may be calibrated. The focal plane array is exposed to a uniform light source for varying exposure times so that its output varies from low level to near saturation. Enough data are gathered at each exposure time to average out the random noise effects. A curve is fit to the measured transfer function of each pixel. The coefficients from this curve form a set of amplitude calibration coefficients to be applied by the real-time software. The pixel-to-pixel variations can then be removed from the sky background and from the star signal. By computing the mean sky background from a neighborhood of the focal plane array not containing the star, the need for frame differencing disappears. Disadvantages to this approach are (1) the need for a uniform light source, (2) a lengthy calibration procedure, (3) the volume of data stored for the calibration coefficients, (4) the throughput required to apply the corrections, and (5) the need to recalibrate if the pixel-to-pixel variations change with time. However, for a fully strapdown stellar-inertial navigation system, use of this technique may be required.

12.7.2   Pre-flight and In-flight Calibration and Alignment

Navigation systems employ pre-flight and/or in-flight calibration and alignment (C&A) processes. These processes use Kalman filter techniques and, in general, pre-flight C&A, in-flight C&A, and in-flight navigation use essentially the same Kalman filter software algorithms.

Pre-flight C&A   To perform a pre-flight C&A, the stationary system is placed in the navigate mode. Utilizing the fact that the vehicle is stationary relative to the Earth, constant position and zero velocity observations are integrated into the navigation process through the filter. The result is that estimates of heading and level alignment are provided along with estimates of instrument calibration errors. For the fully strapdown cluster, and for small initial position errors, the heading error and the level errors will rapidly become nearly unity correlated with the component of gyro drift in the east direction, and the level component of accelerometer bias error, respectively. The repeatabilities of the long-term instrument errors determine the pre-flight alignment accuracy for the nongimballed strapdown system.

Example 12.7.1   What is the initial 3-sigma search area for the nongimballed strapdown stellar-inertial navigator given that the system is pre-flight aligned at a 45-deg latitude location and that the selected star has zero-degree elevation. Assume that the factory calibration for the gyros and accelerometers are repeatable to 0.01 deg/hr (1-sigma) and 75 μg (1-sigma), respectively.

The azimuth search dimension is given by

$$AZ = \frac{(2)(3)(0.01°/\text{hr})}{(\cos \Phi)\Omega_E} \simeq 20 \, \text{arcmin} \qquad (3\text{-}sigma)$$

The elevation search dimension is given by:

$$EL = (2)(3)(75\mu g)/g \cong 1.5 \, \text{arcmin} \qquad (3\text{-}sigma)$$

Therefore, for the zero-elevation star, the initial 3-sigma search area is 20×1.5 (arcmin)^2.

For the case of the stabilized stellar-inertial navigator, a 180-deg rotation about azimuth provides observability of the east gyro drift and ψ_{AZ} alignment error. The azimuth (or heading) alignment error following rotation is reduced by a factor of 5 or 10. The level tilt errors remain essentially the same.

For the two-gimbal strapdown stellar-inertial system, ±180-deg rotations about both the pitch and roll axes provides a relatively complete inertial cluster calibration and a relatively precise alignment including both heading and level. The alignments are in the neighborhood of arc seconds, for the conditions specified in Example 12.7.1.

In-flight C&A   For many military applications, the use of a pre-flight C&A is not considered feasible because of rapid takeoff requirements. This is especially true for the stabilized local level configurations which may require warm-up in excess of thirty minutes. Therefore, the procedure is to do system warm-up during the taxi and climb out period. The in-flight C&A begins at altitude, again with the stellar-inertial navigator in the navigate mode. The Kalman observations for velocity must be generated by other avionics equipment such as a Doppler radar (Chapter 10) or GPS (Chapter 5) and heading and attitude from an AHRS (Section 7.7.5). The alignment accuracy determines the size of the initial search pattern for the selected star.

12.6   FUTURE TRENDS

In the near term, the trend in celestial navigation is to eliminate all rotating devices. For example, the solid state focal plane array has eliminated the need to mechanically scan. Some form of spatial and temporal sharing of the focal plane array, such as electronically scanned discrete optics, will probably be required to replace mechanical gimbals for high-performance daylight applications; this would be accompanied by a design technique that maintains an extremely stable star sensor boresight over long time periods.

In the long term, celestial navigation will largely be replaced by satellite navigation systems such as GPS. Specialized military aircraft will continue to use stellar-inertial systems where jamming of satellite navigation receivers is a credible threat and/or where second-of-arc pointing accuracy is required, for example, for on-board laser systems. These stellar-inertial systems will tend toward fully strapdown configurations.

PROBLEMS

12.1. Suppose that the computed vehicle position is in error by 1000 ft in latitude, $\delta\Phi$, and a zenith star measurement is made. What is the accelerometer bias error in the North direction $\delta\beta_N$ that yields a zero star sensor observation? Assume a perfect star sensor and an Earth radius $|R|$ of 4000 statue miles.

For $\psi = 0$, from Equation 12.7, $\delta\theta = \phi$. From the geometry in Figure 12.4:

$$\delta\theta_E = \frac{\delta\Phi}{|R|} = \frac{1000}{(4000)(5280)} = 47.3 \quad \mu\text{rad}$$

where $\delta\theta_E$ is the east component of $\delta\theta$ (small rotation about East axis), and from inertial reference considerations,

$$\phi_E = \frac{\delta\beta_N}{g}$$

where g is Earth's gravity, and ϕ_E is the East component of ϕ.

Ans.: The accelerometer bias error is

$$\delta\beta_N = g\phi_E = g\delta\theta_E = g(47.3 \ \mu\text{rad}) = 47.3 \ \mu g$$

12.2. A star sensor is used to estimate the physical reference errors in level, ψ_1 and ψ_2, using a star located at the zenith of the observer's position. Suppose that the star sensor has boresight errors ϵ_1 and ϵ_2 with equal variances, σ_ϵ^2, which are collinear with ψ_1, and ψ_2 with equal variances, σ_ψ^2. Further, suppose that the errors ψ_i and ϵ_i are random variables, $E\{\psi_i\} = E\{\epsilon_i\} = 0$, and are uncorrelated at time zero. The system is described by

$$\mathbf{P}(0) = \begin{bmatrix} \sigma_\psi^2 \mathbf{I}_2 & 0 \\ 0 & \sigma_\epsilon^2 \mathbf{I}_2 \end{bmatrix} \quad \mathbf{A}(i) = \begin{bmatrix} \mathbf{I}_2 & 0 \\ 0 & \mathbf{I}_2 \end{bmatrix}$$

$$\mathbf{H}(i) = [\mathbf{I}_2 \vdots \mathbf{I}_2] \quad \mathbf{W} = \sigma_\omega^2 I_2, \quad \mathbf{Q} \equiv 0$$

where $\mathbf{I}_2$ is a 2×2 identity matrix.

(a) Define the information matrix following N measurements.

Ans.: From the system description, and Equation 12.11, the information matrix is given by

$$\mathbf{L}(N) = \frac{N}{\sigma_\omega^2} \begin{bmatrix} \mathbf{I}_2 & \mathbf{I}_2 \\ \mathbf{I}_2 & \mathbf{I}_2 \end{bmatrix} + \begin{bmatrix} \dfrac{\mathbf{I}_2}{\sigma_\psi^2} & 0 \\ 0 & \dfrac{\mathbf{I}_2}{\sigma_\epsilon^2} \end{bmatrix}$$

(b) Is this system configuration observable? Hint: Show that $\mathbf{J}(N)$ is singular.

(c) Suppose that $N/2$ measurements are made. Then the star sensor is rotated 180 deg about the azimuth, and another $N/2$ measurements are made. Define the information matrix following the total N measurements. Hint: For the first $N/2$ measurements, $\mathbf{H}_1 = [\mathbf{I}_2 \vdots \mathbf{I}_2]$, and for the second $N/2$ measurements, $H_2 = [I_2 \vdots -I_2]$.

(d) Is this system configuration observable?

(e) Show that the Kalman filter error variances following the processing of N star observations for part (a) is given by

$$\sigma_{\psi_{1,2}}^2(N) = \frac{(N\sigma_\epsilon^2(0) + \sigma_\omega^2)\sigma_\psi^2(0)}{N(\sigma_\epsilon^2(0) + \sigma_\psi^2(0)) + \sigma_\omega^2}$$

$$\sigma_{\epsilon_{1,2}}^2(N) = \frac{(N\sigma_\psi^2(0) + \sigma_\omega^2)\sigma_\epsilon^2(0)}{N(\sigma_\epsilon^2(0) + \sigma_\psi^2(0)) + \sigma_\omega^2}$$

and for part (c) is given by

$$\sigma^2_{\psi_{1,2}}(N) = \frac{\sigma^2_\psi(0)\sigma^2_\omega}{N\sigma^2_\psi(0) + \sigma^2_\omega}$$

$$\sigma^2_{\epsilon_{1,2}}(N) = \frac{\sigma^2_\epsilon(0)\sigma^2_\omega}{N\sigma^2_\epsilon(0) + \sigma^2_\omega}$$

Hint: $P(N) = L(N)^{-1}$

Note that for part (a), as N becomes large, the ψ and ϵ variances both approach the same lower limit of $\sigma^2_\psi(0)\sigma^2_\epsilon(0)/(\sigma^2_\psi(0)+\sigma^2_\epsilon(0))$. For part (c), as N becomes large, both error variances approach σ^2_ω/N; this lower limit approaches zero for increasing values of N. Further, for the case where the initialization errors $\sigma_\psi(0)$ and $\sigma_\epsilon(0)$ are much larger than the star sensor noise, σ_ω, these lower limits are approached after one measurement.

12.3. Suppose that in Example 12.4.1 a focal plane array (FPA) is used; let a 30 × 30 CCD silicon array be placed at the focal plane of the telescope. What is the ratio of the sky background power to the star signal power, given that the star power is focused on a single pixel, and each pixel covers an "area" of approximately (3 arc sec)^2 ? Hint: $P_b = 7.5 \times 10^{-12}$ w.

Ans.: 1.5

12.4. What is the physical size of a pixel for the telescope of Problem 12.3 given that the star light has a wavelength, $\lambda = 1\times 10^{-6}$ meters, that 93.7% of the star light energy is to be contained within a one-pixel diameter, and that the telescope f/number = f/4.

Ans.: 18 μm × 18 μm.

13 Landing Systems

13.1 INTRODUCTION

Every successful flight culminates in a landing. Although the majority of landings are conducted solely with visual cues, aircraft must frequently land in weather that requires electronic assistance to the pilot or to the autopilot. This chapter describes the landing maneuver and the electronic systems that provide lateral and vertical guidance to the aircraft relative to the runway chosen for landing. The emphasis is on current systems proposed or in use; historical notes are added when needed for proper perspective.

On a normal flight, an aircraft takes off, climbs to cruising altitude, and flies to the vicinity of its destination. There it begins its descent and intercepts the projected runway center line, then makes a final approach and landing with position errors of a few feet in each axis at touchdown (Section 1.5). The approach and landing are the riskiest phases of flight; approximately one-half of the catastrophic accidents occur during these flight phases of which two-thirds are attributed to errors made by the flight crew (Table 13.1). Considering the "exposure times" in the approach and landing phases, the catastrophic-accident probability for any one operation is on the order of 10^{-8} [1].

13.2 LOW-VISIBILITY OPERATIONS

Considerable interference to civil and military operations results due to reduced visibility in terminal areas. For example, Figure 13.1 shows that the visibility at London's Gatwick Airport requires Category II operational capabilities for 115 hours per year and Category III capabilities for 73 hours per year during primary operating hours [2]. Although these hours represent only a few percent on an annual basis, the cost of diverting to an alternate landing site is large. In their 1992–93 fiscal year, KLM estimated savings of $10.3 million (U.S. $) by being able to land 500 flights at the Amsterdam (Schiphol) airport in Category III conditions. During the period 1996 to 2009, KLM estimates costs of $53 million (U.S. $) just for passage if Category III capabilities were not available at Schiphol [3]. While the successful landing of aircraft depends on many factors other than ceiling and visibility, such as crosswinds and storm activity, the term *all-weather operations* often refers only to operations in condition of reduced visibility [4c].

Table 13.1. Hull-loss accidents by flight phase for the world-wide commercial jet fleet, 1959 to 1990

Primary Causal Factor (% in phase)	Total Hulls Lost[a]	Takeoff and Climb (15)[b]	Cruise (60)	Descent and Initial Approch (21)	Final Approch and Landing (4)	Load and Taxi (NA)
All causes	440	145 (33)[c]	20 (5)	88 (20)	178 (40)	9 (2)
Flightcrew	276	68 (25)	5 (2)	68 (25)	133 (47)	2 (1)
Unknown	72	29 (40)	5 (7)	10 (14)	26 (36)	2 (3)
Airplane	40	26 (64)	3 (8)	3 (8)	6 (15)	2 (5)

[a]These totals exclude sabotage and military action.
[b]Flight-phase duration as percentage of average 1.5-hr flight.
[c]Numbers in parentheses are percentages of the row total.

Instrument meteorological conditions (IMC) are those in which visibility is restricted to various degrees defined by law in certain countries. Aircraft operating in IMC are supposed to fly under Instrument Flight Rules also defined by law. During a landing, the decision height (DH) is the height above the runway at which the landing must be aborted if the runway is not in sight. Usually, the better the electronic aids, the lower is the DH. The International Civil Aviation Organization (ICAO) defines three categories of visibility for landing civil aircraft with the aid of an instrument-landing system [4c]:

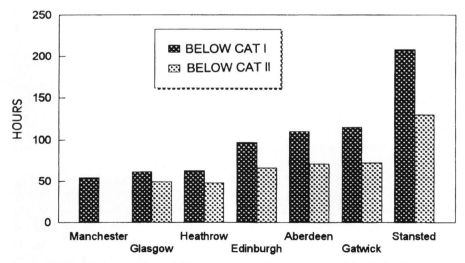

HOURS PER YEAR BELOW CAT I AND CAT II
(0700 - 2300 HRS)

Figure 13.1 Periods of low visibility at major airports in the United Kingdom, 1972 to 1982.

Category I. Decision height not lower than 200 ft; visibility not less than 2600 ft, or Runway Visual Range (RVR) not less than 1800 ft with appropriate runway lighting. The pilot must have visual reference to the runway at the 200-ft DH above the runway or abort the landing. Aircraft require ILS and marker-beacon receivers beyond other requirements for flights under IFR. Category I approaches are performed routinely by pilots with instrument ratings.

Category II. DH not lower than 100 ft and RVR not less than 1200 ft (350 m). The pilot must see the runway above the DH or abort the landing. Additional equipment that aircraft must carry include dual ILS receivers, either a radar altimeter (Section 10.2) or an inner-marker receiver (Section 13.5.5) to measure the decision height, an autopilot coupler or dual flight directors, two pilots, rain-removal equipment (wipers or chemicals), and missed-approach attitude guidance. An auto-throttle system also may be required.

Category III. This category is subdivided into [5b]

* *IIIA.* DH lower than 100 ft and RVR not less than 700 ft (200 m)—sometimes called *see to land;* it requires a fail-passive autopilot or a head-up display.

* *IIIB.* DH lower than 50 ft and RVR not less than 150 ft (50 m)— sometimes called *see to taxi;* it requires a fail-operational autopilot and an automatic rollout to taxing speed.

* *IIIC.* Zero visibility. No DH or RVR limits. Category IIIC had not been approved anywhere in the world in 1996.

Aircraft are certified for decision heights, as are crews. When a crew lands an aircraft at an airport, the highest of the three decision heights applies. An abort at the *decision height* is based on visibility. Another abort criterion is equipment failure; *alert height* is the altitude below which landing may continue in case of equipment failure. To meet alert height restrictions, either the avionics must be fault-tolerant or the crew must be able to take over manually. A typical alert height is 100 ft.

Additional aircraft equipment may include automatic systems for landing, rollout, and braking. Automatic landing systems must demonstrate satisfactory touchdown dispersion limits in stringent environmental conditions including headwinds to 25 knots, tailwinds to 10 knots, crosswinds to 15 knots, moderate turbulence, and wind shears of 8 knots per 100 ft of height from 200 ft to touchdown. Accidents have resulted when much higher shears (e.g., microbursts) have been encountered.

Ceilometers and transmissometers are used at airports to measure terminal-area visual conditions. One version of the latter instrument consists of a light source and a paired photocell, placed 250 or 500 ft apart near the runway. Its indication of visibility is the RVR. As landing minimums are reduced, the measurement of ceiling is less important than slant-range visibility and RVR. Airports at which Category III landings are permitted must be equipped with the standard lighting pattern in Figure 13.2; three transmissometers; outer, middle, and inner marker beacons; and a suitably calibrated instrument-landing system with redundant transmitters. ICAO describes the conditions for permitting Category I, II, and III landings worldwide [4c]. The aircraft and airport equipment required, crew training and means of compliance are published also in FAA advisory circulars [5] and by national aviation authorities in major countries.

13.3   THE MECHANICS OF THE LANDING

13.3.1   The Approach

Day and night landings are permitted under visual flight rules (VFR) when the ceiling exceeds 1000 ft and the horizontal visibility exceeds 3 mi, as judged by the airport control tower. In deteriorated weather, operations must be conducted under Instrument Flight Rules (IFR) [5e]. An IFR approach procedure is either nonprecision (lateral guidance only) or precision (both lateral and vertical guidance signals). Category I, II, and III operations are precision-approach procedures.

An aircraft landing under IFR must transition from cruising flight to the final approach along the extended runway center line by using the standard approach procedures published for each airport [6]. Approach altitudes are measured barometrically, and the transition flight path is defined by initial and final approach fixes (IAF and FAF) using VOR, VOR/DME, Tacan (Chapter 4), and marker beacons (Section 13.5.5). In addition radar vectors may be given to the crew by approach control (Chapter 14).

From approximately 1500 ft above the runway, a precision approach is guided by radio beams generated by the ILS. Large aircraft maintain a speed of 100 to 150 knots during descent along the glide path beginning at the FAF (outer marker). The glide-path angle is set by obstacle-clearance and noise-abatement considerations with 3 deg as the international civil standard. The sink rate is 6 to 16 ft/sec, depending on the aircraft's speed and on headwinds. The ICAO standards [4a] specify that the glide path will cross the runway threshold at a height between 50 and 60 ft. Thus, the projected glide path intercepts the runway surface about 1000 ft from the threshold, as shown in Figure 13.3 [7].

When the aircraft reaches the authorized decision height, the law requires that the crew abort the landing unless it sees the runway or its lights. Note that a DH is not defined for Category IIIC. Figure 13.2 shows the ICAO standard Category II and III lighting pattern for civil airports [4b]. The light bars provide azimuth, roll-and-pitch cues to the pilot. Center line and edge lights provide rollout cues. The typical runway commissioned for precision approaches is 150 ft wide by 8,000 to 12,000 ft long.

For a nonprecision approach, a *minimum descent altitude* (MDA, 250 to 1000 ft above the runway) is defined below which the aircraft may not descend

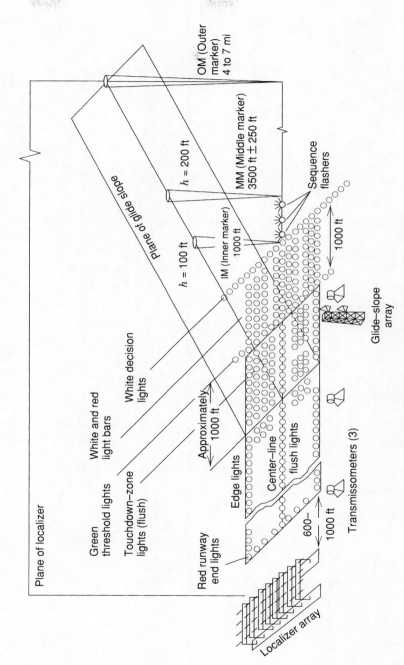

Plane of localizer

Green
threshold lights

White and red
light bars

White decision
lights

Touchdown-zone
lights (flush)

Edge lights

Center-line
flush lights

Red runway
end lights

Plane of glide slope

$h = 100$ ft

$h = 200$ ft

$h = 1000$ ft

IM (Inner marker)
1000 ft

MM (Middle marker)
3500 ft ± 250 ft

OM (Outer
marker)
4 to 7 mi

Sequence
flashers

1000 ft

Glide-slope
array

Transmissometers (3)

600–
1000 ft

Approximately
1000 ft

Localizer array

Figure 13.2 Category III runway configuration.

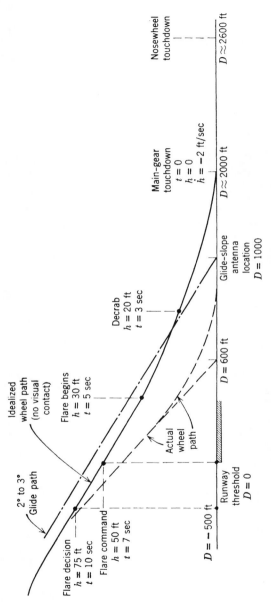

Figure 13.3 Wheel path for instrument landing of a jet aircraft.

Idealized
wheel path
(no visual
contact)

Flare begins
h = 30 ft
t = 5 sec

2° to 3°
Glide path

Flare decision
h = 75 ft
t = 10 sec

Flare command
h = 50 ft
t = 7 sec

Decrab
h = 20 ft
t = 3 sec

Actual
wheel
path

Main-gear
touchdown
t = 0
h = 0
ḣ = −2 ft/sec

Nosewheel
touchdown

Glide-slope
antenna
location
D = 1000

D = −500 ft

Runway
threshold
D = 0

D = 600 ft

D ≈ 2000 ft

D ≈ 2600 ft

without visual contact with the runway. The choice of an MDA depends on local obstructions, aircraft type, available navigational aids, and runway lighting [5d].

13.3.2   The Flare Maneuver

Land-based aircraft are not designed to touch down routinely at the 6 to 16 ft/sec sink rate that exists along the glide path. Thus, a flare maneuver must be executed to reduce the descent rate to less than 3 ft/sec at touchdown.

During approach, the angle of attack is maintained at a value that causes a lift force equal to the aircraft's weight, and the speed is adjusted for a specified stall margin, typically 1.3 times the stall speed plus a margin based on reported wind speed and shear. The flare requires an elevator deflection to increase the angle of attack $\Delta\alpha$ in order to produce an upward acceleration

$$\ddot{h} = \frac{\Delta\alpha C_{L\alpha}}{m} \frac{\rho V^2}{2} A$$

where

$C_{L\alpha}$	is the derivative of the lift coefficient with respect to angle of attack
$\rho V^2/2$	is dynamic pressure (Chapter 8)
A	is the wing area
m	is the aircraft's mass

The upward acceleration causes the descent velocity to decrease.

The speed reduction of a typical jet aircraft during flare is 5% to 10% [7]. Reference [7] shows that the flare begins at a wheel height of 30 ft for jet aircraft, requiring that the pilot or autopilot apply the control command 2 sec earlier (because of the lag due to pitch-axis inertia) and that the flare decision point be 1 or 2 sec still earlier. Thus, from the time of visual cue, the flare is prolonged over a 2500-ft horizontal and 75-ft vertical distance, as shown in Figure 13.3.

13.3.3   The Decrab Maneuver and Touchdown

In a crosswind V_{cw}, an aircraft will approach with a crab angle b such that its ground-speed vector lies along the runway's center line. At an approach airspeed V_a and a headwind V_{hw}, $\sin b = V_{cw}/(V_a - V_{hw})$; b is usually less than 5 deg and is always less than 15 deg. Near touchdown, the aircraft can decrab, slip, or execute a combined decrab-slip maneuver, as explained in Section 13.4.3. After the decrab, the wind causes the aircraft to begin drifting across the runway. For large aircraft, decrab typically occurs 2 to 3 sec prior to touch-

Table 13.2 Acceptable range of variables at touchdown for jet transport aircraft

Variable	Limits (95%)	Reason for Limits
Longitudinal position from threshold (ft)	800–2300	Touchdown on runway with adequate braking distance
Airspeed (kts)	110–145	Lower limit to maintain control; upper to limit braking effort
Lateral position from runway's center line (ft)	±27	Touchdown with main gear more than 5 ft from runway edge
Lateral velocity (ft/sec)	±8	Limit risk of leaving runway after touchdown
Sink rate (ft/sec)	0–5	Limit landing gear/tire damage
Pitch attitude (degrees)	0–5	Limit risk of a noise-wheel landing or tail drag
Roll attitude (degrees)	±5	Limit risk of damage to wing tips or engine nacelles

down. It must occur late enough for the yaw rates and adverse roll to be nulled, but not so late as to result in appreciable lateral-drift speed or displacement. The *slip* consists of (1) lowering the upwind wing to compensate for drift and (2) holding enough opposite rudder to prevent the aircraft from turning and to keep it flying at the same heading as the runway; slip is initiated 200 to 400 ft above the runway. Reference [8] reports simulations showing that pilots will tolerate roll angles as large as 3 degrees below 100-ft altitude if the visibility is less than 700 ft, and up to 5 degrees if the visibility is greater than 1200 ft. Many large transports cannot roll more than 10 deg without striking a wingtip or engine on the runway. Reference [8a] discusses the aerodynamics of decrab in detail. Sections 13.4.2 and 13.4.3 discuss autopilot-coupled flare and decrab just prior to touchdown. The constraints to be satisfied for an acceptable landing are shown in Table 13.2 [9].

13.3.4 Rollout and Taxi

Approximately 600 ft after main-gear touchdown [7], a large jet aircraft lowers its nose wheel and subsequently behaves like a ground vehicle. It must complete its rollout within the boundaries of the runway, locate an exit, and follow the taxiway to the apron. As operational visibility decreases, the ground movement requirements become more severe. Some methods for guiding aircraft on taxiways include the following:

1. Runway stopping-distance can be measured with distance-measuring equipment (DME, Chapter 4) located with the localizer.
2. Systems that modulate taxiway lights to guide an aircraft along a specific taxi route. Lighted signs are located at intersections; reprogrammable signs could be installed.
3. Surface movement radars (Airport Surface Detection Radar in the United States) that aid in avoiding taxiway and runway-incursion accidents. Iden-

tifying aircraft and vehicles is a problem, because these radars have no transponder mode.

4. Transponder-based systems that derive positions on the airport from the aircraft's transponder replies to interrogators located with taxiway lights. The surveillance data are relayed to controllers.

5. Radio broadcast of on-board derived position and velocity (e.g., from GPS). If all vehicles were so equipped, the ground controller could more effectively prevent runway incursions and collisions on the ground.

6. Milliwatt marker-beacon transmitters placed at all runway thresholds would give a visual and audible alarm on the flight deck of any aircraft that taxied onto an active runway. This would reduce the likelihood of a lost pilot causing a catastrophic collision.

It is mandatory that fire and rescue vehicles be able to locate disabled aircraft on the field in low visibility. Aircraft must be moved to ramps, hangars, maintenance areas, and fueling areas during periods of low-visibility landing. Except in military operations, a visibility limit exists below which aircraft operations are no longer practical because of the inability to move passengers and cargo from the airfield. The need for Category III (especially IIIC) landing by civil air carriers is tied to these supporting capabilities.

13.4 AUTOMATIC LANDING SYSTEMS

Air carrier aircraft that are authorized for precision-approach below Category II must have an automatic landing (*auto-land*) system. Extended-vision devices such as head-up displays improve the pilot's ability to monitor the approach and can substitute for autoland, at least to Category IIIA. In the 1990s many air-carrier aircraft such as MD-80, MD-11, A-300 Series, Boeing 757, 767, 777 were equipped with auto-land systems that were certified to Category IIIA or IIIB [5b]. Below the DH, neither a manual nor an automatic go-around is permitted because of the risk of ground contact; the landing must continue.

In 1967, the first fully automatic landings were made in airline passenger service [10]. The Blind Landing Experimental Unit (BLEU) of the Ministry of Defence in the United Kingdom had been very active during and after World War II. The economic incentives of expanding commercial air operations after the war provided the motivation for auto-land system development. Even with the certification of air carrier aircraft for automatic landings, DHs were not lowered below 100 ft until the reliability and failure-detection capability (*integrity*) of the aircraft (e.g., propulsion, flight control, auto-land) and ground guidance had been proven and standards agreed.

In the early 1970s the British *Trident* and the U.S. *L-1011* aircraft (including their auto-land systems) were certified for Category IIIa operations in revenue service using guidance information from the ILS and from other on-board sensors [11]. Many airliners were certified thereafter. In 1996, Category III operations were "automatic-to-touchdown," based on the use of auto-land systems that do not require pilot intervention [5b]. These Categories have been developed, defined, and agreed internationally in a generic manner to allow innovation in meeting the requirements.

13.4.1   Guidance and Control Requirements

The FAA regulations for Category II approval [5a] require that the coupled autopilot or crew hold the aircraft within the larger of ±12 ft or ±35 μamps (measured by the ILS glide slope) of the glide path from 700 ft down to the 100-ft DH. This corresponds to a vertical error of ±38 ft (2-sigma) at the 700-ft height and ±12 ft (2-sigma) at the 100-ft height on a 3-deg glide path. For Category III certification, the allowable touchdown dispersions also are specified. The demonstrated touchdown dispersions should be limited to 1500 ft longitudinally and ±27 ft laterally on a 2-sigma basis [5c].

Paralleling the improvement in accuracy was the development of systematic methodologies for conducting hazard analyses. Thus, in addition to the operational concepts, an acceptable level of risk needs to be specified as the starting point for a hazard analysis of the aircraft and ground navigation systems. At its simplest, the probability of a catastrophic event during a Category III landing should not be greater than it is for a visual approach and landing [12]. Historical accident rates and the characteristics of modern turbojet aircraft suggest that less than one fatal accident should occur in 10^7 flight-hours, due to all systems, a number generally agreed upon by the designers of civil transports. Thus the probability of a catastrophic event due to an individual subsystem (e.g., auto-land) is closer to one event in 10^9 flight-hours, a time interval that is likely to exceed the cumulative fleet life of any transport model.

Airport equipment in Category II, IIIA, and IIIB, differ only in the extent of failure detection and automatic reconfiguration following failure.

13.4.2   Flare Guidance

During the final approach the glide-slope gain in the auto-land system is reduced in a programmed fashion (called *gain scheduling*). Supplementary sensors must supply the vertical guidance below 100 ft.

In the 1990s, automatic landing systems used height and height-rate data from redundant radar altimeters (Section 10.2), frequently complemented with inertial sensor data, to accomplish this flare maneuver. Low-range radar altimeters have a nominal accuracy of 1.5 ft (2 sigma) below 100 ft altitude and 2% of full scale at higher altitude (Section 10.2.4). They have a low-noise output that allows the derivation of vertical rate information. The ILS cannot develop similar information, since the glide-slope receiver provides deviations from the zero-DDM glide path which is not at a constant height across the width of the localizer course/runway (Section 13.5). Therefore, accurate calculation of aircraft height using the glide-path deviations and distance information from a stop-end DME (Section 4.4.6) is not practical.

The flare command is initiated at a wheel height of about 75 ft, well before the runway threshold. However, the radar altimeter may not be usable prior to threshold due to uneven terrain or tidal variations. Typically, auto-land systems cover the gap between flare initiation and the beginning of altimeter coverage over the paved runway (50-ft height) with data from on-board inertial-navigation systems (Chapter 7).

The flare is an exponential path tangent to a horizontal plane several feet below the runway, which ensures positive touchdown. The desired height-rate at touchdown is included in the calculation of steering signals, so that the touchdown point and vertical velocity at touchdown are repeatable. A typical vertical rate command is $(h/\tau + \dot{h} + 2 \text{ ft/sec})$, where h and $\dot{h}$ are the measured instantaneous altitude and altitude rate [8b].

13.4.3   Lateral Guidance

In 1996, ILS localizers met Category-III performance requirements for landing and rollout guidance at many locations. Tracking of the localizer is aided by heading (or integral-of-roll), roll, or roll-rate signals supplied to the autopilot (Section 2.7.3) and by rate and acceleration data from on-board inertial systems. During most of the approach in a crosswind, the wings are level and the aircraft points into the wind (*crabbed*) so that its velocity vector tracks the localizer. The localizer gain must be scheduled to decrease as the runway is approached because the surfaces of constant DDM (Section 13.5) converge to make the error signal more and more sensitive to distance displacement.

The aircraft should touch down with its axis along the runway's center line to avoid side forces on the landing gear. To do this in Category I and II conditions, the crew can slip or decrab manually before touchdown. In lower visibility, when landing operations may be automatic to touchdown, the slip and crab are mechanized by feeding the appropriate elevator and rudder commands to the autopilot approximately 200 ft above the runway. This establishes as much slip as needed, but not more than about 5 deg roll for a typical commercial airliner, thus reducing the crab angle that was established on approach. If the crosswind is so high (above about 10 knots) that the roll angle exceeds about 5 deg, a decrab maneuver is inserted about 15 ft above the runway to point the aircraft axis along the runway. The velocity vector must be within 3 deg of the runway axis at touchdown to keep the aircraft on the runway. For example, a 40-ft lateral touchdown error (FAA 3-sigma requirement) and a 3 degree lateral flight-path error could cause an aircraft with a 20-ft landing-gear spread to leave a 150-ft wide runway in about 2.5 sec.

13.5 THE INSTRUMENT LANDING SYSTEM

The instrument landing system (ILS) is a collection of radio transmitting stations used to guide aircraft to a specific airport runway, especially during times of limited visibility. High-density airports may be equipped on more than one runway—Chicago's O'Hare airport had an ILS installed on 12 runways in 1996.

First commercially used in 1939, ILS has been in civil use for the equivalent of Category I landings since 1947 [13] and had been certified for Category II (1960s) and Category III (1970s) operations at many major airports. About 1500 ILSs are in use at airports throughout the world with U.S. airports, accounting for more than 1000 of these. In 1996, nearly 100 airports worldwide had at least one runway certified to Category III. There were approximately 117,000 aircraft worldwide with one or more ILS receivers.

The ILS development worldwide was aided by the publication of signal standards by the International Civil Aviation Organization (ICAO, created in 1944). The ICAO Standards and Recommended Practices (SARPs) [4a] for the ILS, subscribed to by the 180 member states, standardize the signal in space of the ILS ground transmitting equipment, no matter where they are manufactured, to ensure compatibility on international flights. The FAA and RTCA publish manuals for ILS equipment in the U.S. national airspace system that ensure conformance to ICAO standards [18]. In Europe, the Joint Aviation Authority (JAA) and EUROCAE do the same [31]. U.S. receivers are standardized by Aeronautical Radio, Inc. [32].

Typically, an ILS includes

- the localizer antenna is centered on the runway beyond the stop end to provide lateral guidance
- the glide slope, located beside the runway near the threshold to provide vertical guidance
- marker beacons located at discrete positions along the approach path; to alert pilots of their progress along the glide-path
- radiation monitors that, in case of ILS failure alarm the control tower, may shut-down a Category I or II ILS, or switch a Category III ILS to backup transmitters (Figure 13.2).

Increasingly, distance measuring equipment (DME) (Chapter 4) is located with the ILS, and distance readouts in the cockpit are used instead of marker beacons.

13.5.1 ILS Guidance Signals

The localizer, glide slope, and marker beacons radiate continuous wave, horizontally polarized, radio frequency energy. The frequency bands of operation are: localizer, 40 channels from 108–112 MHz; glide slope, 40 channels from 329–335 MHz; and marker beacons, all on a single frequency of 75 MHz. An audible Morse-code identification signal is transmitted on the localizer frequency; a voice channel from the control tower also may be provided. The localizer, glide slope, and DME frequencies are paired such that, for example, an ILS with a localizer frequency of 109.5 MHz has a glide slope frequency of 332.6 MHz and has channel 32× for the DME, if installed. This *hard pairing* reduces the pilot's work load, since only the localizer frequency need be entered to tune the ILS and associated DME.

The localizer establishes a radiation pattern in space that provides a deviation signal in the aircraft when the aircraft is displaced laterally from the vertical plane containing the runway center line [18d]. This deviation signal drives the left–right needle of the pilot's cross-pointer display (or flight director) and may be wired to the autopilot/flight-control system for "coupled" approaches.

To form the localizer course, an RF carrier is generated in the transmitter and amplitude modulated with discrete 90-Hz and 150-Hz tones to create equal-amplitude sidebands placed 90 Hz and 150 Hz above and below the carrier frequency. This "carrier-with-sidebands" (CSB) signal is radiated from the localizer array in a broadside "sum" pattern that provides general coverage of the approach course area. Simultaneously, some of the sideband energy is separated from the carrier and shifted in phase to provide a "sidebands-only" (SBO) signal. The SBO signal is radiated from the same array in a "difference" pattern, and the sharp null, characteristic of the difference pattern, is aligned with the runway center line. The lobes of the difference pattern (and thus the 90-Hz and 150-Hz sidebands) are opposite in phase on either side of the center line null (Figure 13.4).

The net result is that, on centerline where the SBO signal is nulled, only the equal-amplitude sidebands of the CSB signal are detected and the receiver provides an "on-course" output. Either side of center line, the SBO signal amplitude increases rapidly and unbalances the CSB sidebands such that a 90-Hz sideband will dominate to the left of course and a 150-Hz sideband to the right. The total signal modulation M is defined as $M = (A+B)/C$, where A and B are the amplitudes of the 150- and 90-Hz signals respectively, and C is the amplitude of the carrier [14]. The difference in depth of modulation (DDM) is $(A - B)/C$. The ILS receiver measures the magnitude of this DDM and outputs a "deviation-from-course" signal. Thus, the cross-pointer course-deviation (CDI) needle or flight director shows "fly right" when the 90-Hz tone dominates and "fly left" when the 150-Hz tone dominates.

The deviation signal is proportional to azimuth angle usually out to 5 deg or more either side of the center line. The ICAO standards [4a] require that a DDM of ±0.155 cause a full-scale needle deflection and that this DDM value occur at 350 ft on either side of the center line at the approach threshold. Thus, with different runway lengths, the angular course width from the localizer will be 3 deg to 6 deg, as set by adjusting the radiated sideband power.

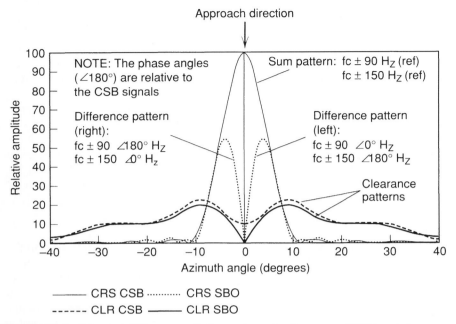

Figure 13.4 Sum and difference radiation patterns for the course (CRS) and clearance (CLR) signals of a directional localizer array.

The ICAO standard is that full-scale deviation signals be provided to 35 deg either side of center line to aid acquisition of the zero-DDM course. If the localizer CSB pattern is narrow (e.g., 5 deg to 10 deg), to reduce reflections from buildings and other aircraft, the system requires the addition of "clearance" signals to provide the ±35 deg coverage. These clearance signals are added by providing a second set of CSB and SBO signals radiated from a shorter array with a broad-coverage pattern. A two-frequency "capture-principle" method may be used to reduce the interference effects on the course signal due to clearance signal reflections from, for example, large hangars at the wide angles.

The frequencies of the two RF carriers are spaced a nominal 8 kHz and are offset symmetrically about the channel center frequency. The two sets of signals (course CSB and SBO and clearance CSB and SBO) are within the 25-kHz passband of the receiver and are detected. A basic characteristic of the linear AM detector is that, with two modulated carriers present, the demodulated course information greatly favors (*captures*) the even slightly stronger carrier signal. The airborne receiver, when near the center line, receives two carrier frequencies each with 90- and 150-Hz modulations. The weaker clearance signals reflected into the course region are received along with the stronger course signal and thus have less influence on the detector output. Because this "capture principle" can reduce the course roughness due to clearance-signal reflections by a factor of 3 to 10, two-frequency systems are needed at most airports where higher-category landing minimums are desired. For Category IIIB and IIIC operations, the localizer signal is used along the runway for rollout guidance. ILS ground equipment being procured by the FAA in the mid 1990s are two-frequency designs that can support Category II and III operations. The localizer accuracy required to support the three ILS performance categories is shown in Table 13.3 [4a].

With regard to the vertical guidance signals, the glide slope in most respects can be considered as a localizer on its side. The minimum coverage extends approximately 10 nmi from threshold, to 8 deg on either side of center line, and up to 5000 ft for the nominal 3 deg path angle. Given a nominal

Table 13.3.   ILS guidance errors allowed by ICAO standards (ft)

Point Along Approach Center line	ILS Element	Category I		Category II		Category III	
		Bias (Max)[a]	Bends (95%)	Bias (Max)	Bends (95%)	Bias (Max)	Bends (95%)
Outer Marker (5 nmi)	GS	122	77	121	77	65	77
	Loc.	136	249	93	249	41	249
Middle Marker (3000 ft)	GS	15	10	15	6	8	6
	Loc.	48	43	33	14	15	14
Inner Marker (1000 ft)	GS	8	5	8	3	4	3
	Loc.	42	37	29	12	13	12
Threshold[b] (0)	GS	NS[c]	NS	4	2	2	2
	Loc.	NS	NS	26	11	12	11
Main-Gear Touchdown[d]	GS	NS	NS	NS	NS	NS	NS
	Loc.	NS	NS	NS	NS	10	10

Note: The listed values include all significant tolerances for the signal-in-space and for the airborne receiver, which apply on the center line course and on the 3-deg glide path. The localizer-to-threshold distance is taken as 12,000 ft. All values are rounded to the nearest integer.

[a] In typical ILS operations, the bias is maintained at a small value.

[b] Accuracies are specified at threshold for measurement convenience but have no operational meaning.

[c] NS = no standard.

[d] Assumes touchdown 2000 ft beyond threshold.

angle and height at runway threshold, the linear glide path extension intercepts the runway surface approximately 1000 ft inside the threshold (Figure
13.2) [18e].

The sideband arrangement is for the 90-Hz signal to predominate above
the glide path and the 150-Hz signal to dominate below (Figure 13.5) [16].
The relative signal amplitudes are such that full-scale needle deflection and a
DDM = ±0.175 exist at ±Θ/4 (±0.75 deg for Θ = 3 deg glide path) [4a]. The

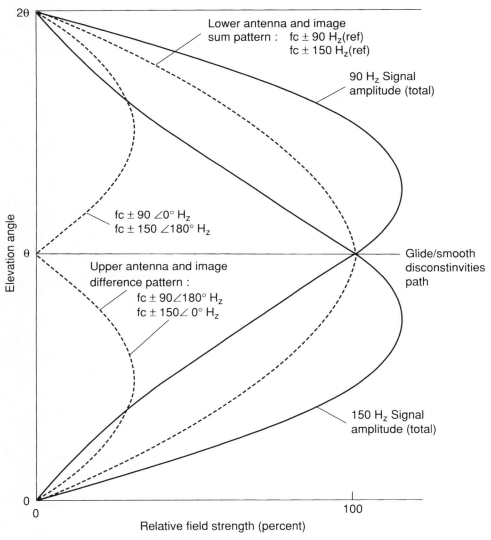

Figure 13.5 Vertical beam patterns of the null-reference glide-slope array.

vertical deviation indicator (VDI) needle reads full-scale "fly up" for all down-ward departures exceeding $\Theta/4$ and full-scale "fly down" for upward departures exceeding $\Theta/4$ to as high as 3Θ. The glide-slope accuracies required to support the three ILS performance categories are shown in Table 13.3 [4a].

13.5.2   The Localizer

The typical localizer is an array of antennas usually located 600 to 1000 ft beyond the stop end of the runway. The array axis is perpendicular to the run-way center line, and the localizer zero-DDM course is aligned with the runway center line extended (Figure 13.2). Localizer arrays range from 40 to 130 ft in length on which are mounted from 6 to more than 20 antennas. Some local-izers are bi-directional and provide lateral steering signals for both the normal *front-course* approach and for a *back-course* approach in the opposite direction (at which the steering signals are reversed).

Although there are three different types of antenna elements in use, all mod-ern arrays and new procurements by the U.S. FAA use the *log-periodic-dipole (LPD)* antenna which provides uni-directional radiation (Figure 13.6). It is 9 ft long, can be installed without guy cables, and has good broad-band characteris-tics. Each dipole is driven by a balanced transmission line within the horizontal supporting structure which, in turn, is fed from a matching network at the for-ward end [5g]. Also within this structure, a parallel transmission line samples signals from the seven dipole arms for the integral monitor. Plans exist to mod-ify this antenna in order to make it bi-directional for use where a back course is desired.

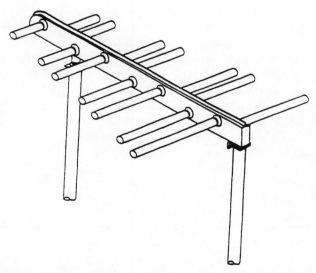

Figure 13.6   Log-periodic dipole antenna used in many localizer arrays.

Figure 13.7 Category IIIB localizer (courtesy, Wilcox Corporation).

Clearance signals, where required (Section 13.5.2), may be provided by a separate array or, in some designs, both course and clearance signals are radiated from several elements of a single array. A Category III localizer with LPD elements is shown in Figure 13.7.

13.5.3 The Glide Slope

There are five different types of glide-slope arrays in common use; three are image systems and two are not. Image arrays depend on reflections from level ground in the direction of approaching aircraft to form the radiation pattern. The three image systems are the *null-reference* system, with two antennas supported on a vertical mast 14 and 28 ft above the ground plane; the *sideband-reference* system, with two antennas 7 and 22 ft above the ground plane; and the *capture-effect* system, with 3 antennas 14, 28, and 42 ft above the ground plane.

The operating principles of each image array are similar. A two-element array is formed by the CSB antenna and its image below the ground plane. The height of the CSB antenna is chosen so that the direct signal and the image signal will add and form a "sum" radiation pattern. Similarly, the SBO antenna is placed on the mast so that the array formed with its image provides a "difference" radiation pattern with the null at the desired glide-path angle.

Because each antenna and its image form a two-element interferometer, multiple lobes are produced in the broad-side (vertical) radiation patterns. Thus, false nulls exist at 2Θ, 3Θ, 4Θ, etc., but the steering errors show "fly down" on both sides of 2Θ, "fly away" at 3Θ, and "fly up" on both sides of 4Θ. The first stable null is at 5Θ, where the steering signals are again correct. The steepness of that stable false glide path (15 deg for a 3 deg glide slope) makes it easy to distinguish from the true glide path.

The most capable of the image glide slopes, the *capture-effect array* (Figure 13.8), is designed to cancel the CSB signal below about 1.5 deg in order to reduce the energy reflected from rising terrain under the approach course. To provide a strong fly-up signal, a 150-Hz clearance signal is radiated below the glide path on a separate carrier, offset in frequency, and use is made of the "capture principle" (Section 13.5.2) to reduce the interference due to reflections of the clearance signal from the terrain.

As shown in Figure 13.9, the zero-DDM surface for image arrays is the hyperboloid of revolution whose axis of symmetry is approximately the glide slope mast [17]. Note that, due to the use of the antenna images, this zero-DDM surface does not extend to ground level, although the origin of its asymptotic cone is near the bottom of the mast. The half-angle of the asymptotic cone is 87

Figure 13.8 Category IIIB capture-effect glideslope and Tasker transmissometer (courtesy, Wilcox Corporation).

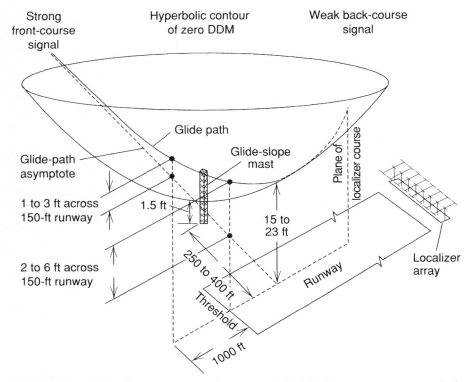

Figure 13.9   Glide-slope pattern near the runway [17]. DDM contours are symmetrical around the vertical, but signal strength drops rapidly off course.

deg for a 3 deg glide path. A vertical plane containing the runway center line intersects the zero-DDM hyperboloid in a hyperbola whose asymptote is the ideal (linear) glide path. Because the glide-slope array is offset from center line, the zero-DDM contour is 1 to 3 ft above the (straight-line) asymptotic glide path at the inner marker, 2 to 6 ft above at the threshold of a 150-ft-wide runway, and 15 to 23 ft above the runway opposite the glide-slope mast (for an array mounted 400-ft from center line).

The performance of the image glide slopes depends on the characteristics of the terrain in front of the array. A minimum ground plane extending 1000 ft in front of the array and varying not more than 12 in. from level ($\frac{1}{4}$ wavelength) is required. To accommodate sites where this ground plane cannot be provided, two non-image systems have been developed:

1. The *wave-guide glide slope* is a broadside array of slotted-wave-guide radiators with a vertical aperture of some 70 ft. There were a small number in service in the United States in 1996; one was installed on Runway 23 at Buffalo International Airport, New York. By tilting the array out of

the vertical by the amount of the glide-path angle, the zero-DDM surface may be planar rather than hyperbolic.

2. The *end-fire glide slope* (EFGS) became available for installation at difficult sites late in the 1970s. This system uses two horizontal coaxial cables, each with 96 radiating slots, to form an array of 96 antennas. The cables are laid almost perpendicular to the runway's center line and separated such that the radiation from one slot in the front cable and the corresponding slot in the rear cable are in phase and form a conical radiation pattern which contains the glide path (Figure 13.10). The 450 ft spacing of the front and rear cables creates a zero-DDM surface, which is a cone with a 6-deg vertex angle in the far-field. The pattern created by any single end-fire pair would not be satisfactory for glide-path guidance, because the cone would intersect the ground at 3 deg on either side of the localizer course. However, 96 of them merge to form a broad glide-path. Two additional slotted cables and the two-frequency "capture principle" provide strong fly-up commands on either side of the localizer course area.

The cable radiators of the end-fire array are installed on stands 40 in. high and are sited alongside the runway near the desired touchdown point (Figure 13.11). In 1996, 15 such systems were in operation worldwide with approximately 3 systems added each year.

The closest portion of the end-fire antennas is only 25 ft from the edge of the runway. The sideband reference mast is usually about 185 ft from the edge, and the null-reference and capture-effect antenna masts are 275 to 325 ft from the edge. The tall wave-guide array is laterally offset from the runway by about 1,000 ft.

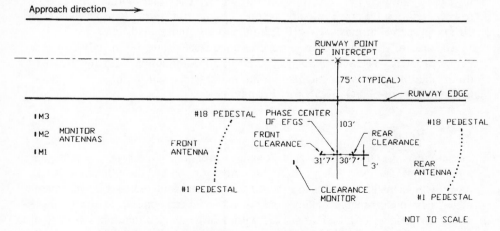

Figure 13.10   Standard end-fire glide-slope system layout.

←——— Approach direction

Figure 13.11   Front slotted-cable radiator of an end-fire glide slope (courtesy, Watts Antenna Company).

13.5.4   ILS Marker Beacons

Marker beacons provide pilot alerts along the approach path. Each beacon radiates a fan-shaped vertical beam that is approximately ±40° wide along the glide path by ±85° wide perpendicular to the path (half-power points). The outer marker (OM) is placed under the approach course near the point of glide-path intercept (distance varies from 4 to 7 nmi from the threshold). It is modulated with two 400-Hz Morse-code dashes per second. The middle marker (MM) is placed near the point where a missed-approach decision would need to be made for a Category I approach procedure (nominally 3000 ft from the threshold). The middle marker is modulated with one 1300-Hz dash-dot pair each second. The inner marker (IM) may be required at runways certified for Category II and III operations and is placed near the point where the glide path is 100 ft above the runway (nominally 1000 ft before threshold). The inner marker has six dots per second at 3000 Hz. Typically, marker-beacon transmitters are mounted off airport property on masts 2 to 10 ft high and are unattended.

The use of marker beacons is decreasing. Real estate for installation is a major problem. The increased use of DME in conjunction with the ILS has diminished the pilot's dependence on the markers, as will the increasing availability of Loran-C and GNSS for IFR navigation.

13.5.5   Receivers

The typical aircraft receiver is a double-conversion, superheterodyne design with an AM detector. Filters after the detector separate the 90- and 150-Hz tones which, in the most basic circuit, are rectified and fed to a dc microammeter. The meter needle deflects in a direction corresponding to the stronger of the two tones. A DDM of 0.175, with a larger 90-Hz modulation (left of course), produces 150 μ of current and deflects the course needle to the full-scale limit on the right ("steer right"). On the right side of the course, the 150-Hz modulation dominates with a corresponding needle deflection to the left. The arrangement for the glide path vertical needle is similar. Glass displays emulate needles. When approaching in the opposite direction on the back course, the course needle deflections are reversed (a right deflection means "steer left"). No glide-path signals are present in the back course.

The marker-beacon receiver is a dedicated crystal controlled receiver, fixed-tuned to 75 MHz. Most have audio filters that control colored lights in a cockpit indicator. Thus, in addition to the audible tones, the outer marker lights a purple lamp on the instrument panel; the middle marker, an amber lamp; and the inner marker, a white lamp.

Closely harmonized airborne equipment standards and specifications are produced by RTCA [18] and ARINC in the United States [32b] and by EUROCAE [31] in Europe.

13.5.6   ILS Limitations

The major limitation of ILS is its sensitivity to the environment. At the ILS operating frequencies, the very narrow beam widths, necessary to avoid significant illumination of the environment surrounding the approach course, require array structures which are too large to be practical. Accuracy degradations (beam bends) due to reflections from buildings, terrain, airborne aircraft, taxiing aircraft, and ground vehicles are the result. The problem is particularly stressful for the glide-slope array, where the shallow elevation angle of the glide path requires large apertures in order to reduce ground reflections sufficiently to meet the stringent vertical error limits close to runway threshold. These reflected (*multipath*) signals mix with the direct signal at the aircraft and cause bends in the glide path (the received signal is the instantaneous sum of all energy arriving at the aircraft's antenna, including reflections, Figure 13.12). The use of image arrays helps to reduce the glide slope's physical size but requires an extensive area of terrain along the direction of the approach to be flat within $\frac{1}{4}$ wavelength (about 1 ft). To achieve Category II and III operations, the glide path must be free of sharp bends (though not necessarily straight) down to a height of 50 ft (threshold), where the allowed 95% error of ±4 ft is evenly divided between bias and bend components (Table 13.3).

The small vertical aperture of typical localizer antenna elements creates an omni-directional vertical pattern. Reflections from aircraft taking off (or executing a missed approach) can cause bends of ±20 μamp (±0.3 deg = ±60 ft laterally 2 nmi from the localizer) that persist for as long as 10 sec [13]. Such bends combined with the localizer error estimate of ±22ft at the threshold of a long runway (Table 13.3) could move the course outside the runway edges. The overflight errors are controlled during instrument operations by increasing the spacing between aircraft approaching and departing.

Aircraft and vehicles on the airport surface, especially on taxiways near the antennas, can reflect ILS signals and cause significant errors along the approach

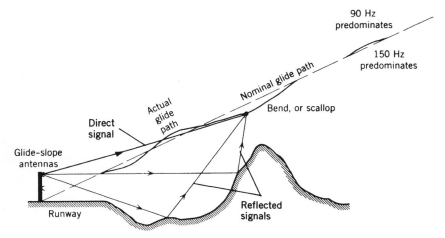

Figure 13.12   Formation of bends in the glide path.

path. Experience has shown that reflection errors can be eliminated only by prohibiting traffic in sensitive regions of the airport during periods of low-visibility operations. Activation of these "critical and sensitive areas" reduces the airport capacity [5g].

The localizer is subject to interference from strong FM stations. This is a particular problem in congested Northern Europe. The ICAO has issued standards for ILS receivers with improved FM rejection characteristics, effective in 1998.

13.6   THE MICROWAVE-LANDING SYSTEM

During the late 1960s, the requirements of civil aviation were forecast to exceed the capabilities of the ILS. The U.S. military services, for more than a decade, had been developing microwave approach guidance systems to support tactical deployments and aircraft-carrier operations (Section 13.8). In 1974, the ICAO solicited proposals from member states for a new guidance system to replace the ILS as the international standard for civil aviation [4d].

Designs were sought that retained the desirable features of the ILS while mitigating its weaknesses. The runway-resident architecture of ILS was carried forward to MLS, because, as the landing aircraft approaches the runway, linear offsets (due to errors in the *angular* guidance) continually decrease, while the signal-to-noise ratio generally increases. Thus, in the most demanding phase of flight close to the ground, the positional accuracy is constantly improving and the noise content is generally decreasing.

The main weakness of the ILS, its sensitivity to the environment (Section 13.5.6), would be essentially eliminated by narrow beam-width antennas that are physically small at microwave frequencies. As a result, a single-accuracy standard, equivalent to Category III ILS, was chosen for MLS. Also, at microwave frequencies, the lack of available channels, which limits multiple ILS deployments in metropolitan areas, would no longer be a problem.

In 1978, ICAO recommended the Time-Reference Scanning Beam MLS proposed by Australia and the United States. In 1985, ICAO adopted the MLS standard and endorsed a transition from the ILS to the MLS beginning in 1998. However, in the spring of 1995, recognizing satellite navigation applications to precision landing, ICAO recommended that standards for both ILS and MLS should stand for the next 15 to 20 years, along with anticipated standards for satellite-based landing systems. In 1987, ARINC issued a specification for an MLS receiver [32c], and, in 1995, EUROCAE issued a specification for a combined ILS-MLS receiver [31].

13.6.1 Signal Format

A basic MLS consists of azimuth and elevation ground stations and a conventional DME (Section 4.4.6) for 3D positioning on approach courses to 40 deg on either side of center line and to 15 deg elevation above the runway. An expanded MLS may include a *back-azimuth* station for departure/missed-approach lateral guidance to ±40 deg of center line, additional approach coverage to ±60 deg of center line, and a more accurate DME (DME/P, Section 13.6.6).

The MLS ground stations transmit both angle and data functions (messages) on one of 200 frequencies between 5031.0 and 5190.7 MHz, Figure 13.13 [4a]. Each transmitted function begins with a differentially encoded binary preamble, modulated by 2-state phase-shift keying. The preamble establishes the carrier phase in the airborne receiver for decoding the preamble message, provides a 5-bit correlation (Barker) code as a timing reference, and identifies the function (e.g., elevation angle) being transmitted. Preambles are radiated from low-gain, fixed-pattern antennas at the azimuth (Az) and elevation (El) stations that fill the approach coverage volume. Therefore, the acceptable bit-error rate of the differential phase-shift key (DPSK) transmissions at the 20-nmi limit determines the transmitter power (about 20 w) needed in the ground stations.

13.6.2 The Angle Functions

The requirement for proportional angle encoding throughout broad coverage sectors is achieved by scanning a narrow antenna pattern. For Az antennas, the antenna beam width may be chosen so that, when the aircraft is near the extended center line, undesirable (multipath) reflections from large hangars along the runway will arrive at the airborne antenna earlier or later than the desired (direct-path) signal. Similarly, for elevation antennas scanning vertically, the antenna beamwidth is chosen to avoid interference from signals reflected from the ground when scanning to the lowest required elevation angle, Figure 13.14. Note that reflected energy can arrive before direct energy if the

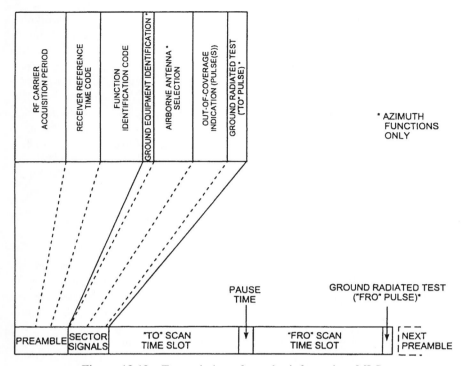

Figure 13.13   Transmission of angular information, MLS.

scanning beam illuminates a nearby object before it illuminates the landing air-craft.

Early military MLS with mechanically scanned antennas used a varying audio tone (or a pulse train) to encode the pointing angle on the scanning-beam pattern. The particular audiofrequency (or pulse code) modulation on the RF carrier, which represented the instantaneous pointing angle, was detected when the scanning pattern illuminated the aircraft's MLS antenna (Section 13.8.3). The FAA adopted time interval between successive passages of the unmod-ulated beam as an efficient method of angle encoding. This reduced spectral requirements and was compatible with high (electronic) scanning rates. In this system, the angle coding is a linear function of time, as follows:

$$\Theta = \frac{V(T_0 - t)}{2}$$

where Θ is the azimuth or elevation guidance angle in degrees, V is the scan velocity (typically 20 deg/msec), T_0 is the value of the time difference at the vertical center line plane (Az functions) or at the horizontal plane through the phase center (El functions) in milliseconds, and t is the time interval in mil-

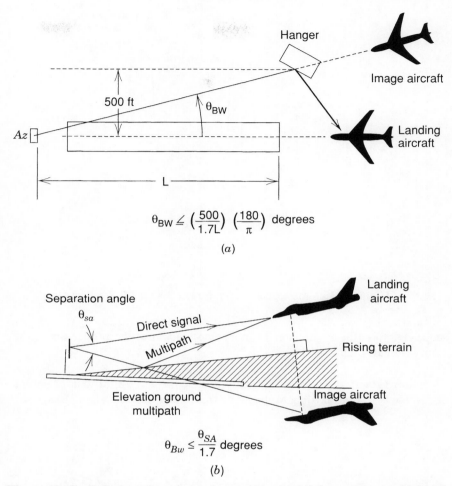

$$\theta_{BW} \leq \left(\frac{500}{1.7L}\right) \left(\frac{180}{\pi}\right) \text{ degrees}$$

(a)

$$\theta_{Bw} \leq \frac{\theta_{SA}}{1.7} \text{ degrees}$$

(b)

Figure 13.14   Selection of maximum beamwidths for MLS antennas: *(a)* Multipath geometry for azimuth antennas; *(b)* Multipath geometry for elevation antennas.

liseconds between TO and FRO passages of the beam centroid at the aircraft. A complete angle-transmission format is shown in Figure 13.15 [4a].

The high scanning rate (20,000 deg/sec) provides about 40 samples per second of the angle data, a rate ten times higher than needed to control the aircraft. Filtering this high rate at the receiver output can significantly reduce the angular errors due to multipath interference and other "noise" sources.

An electronically scanned antenna is an array of radiating elements with a feed network incorporating variable propagation delays (i.e., a "phased array"). These arrays cause the antenna pattern to rotate by "phase shifting" (delaying) the RF signal provided to each radiating element according to a predetermined

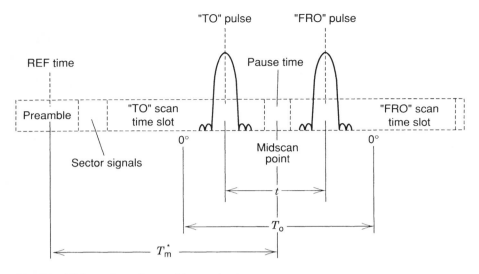

*Used by MLS receivers for confidence check

Figure 13.15   Angle scan-timing parameters for MLS antenna.

sequence stored in a computer memory and readout at the proper time in the signal format, Figure 13.16 [9].

Elevation Function   Quality vertical guidance is the most difficult to provide, due to the ever-present signal reflections from the ground. An antenna vertical-

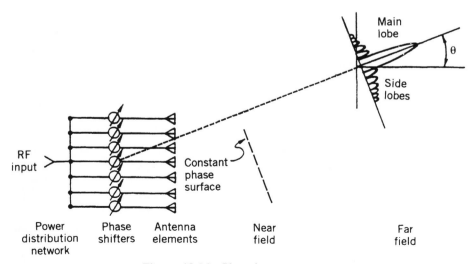

Figure 13.16   Phased array antenna.

pattern beam width of 2 deg or less is needed, in the absence of rising terrain under the approach path, to avoid interference from the ground reflection from the major lobe at lower glidepath angles (e.g., 2.5 deg), Figure 13.14*b*. The MLS elevation antennas that support Category II and III operations have vertical beam widths from 1.5 deg to 1.0 deg, created at 5 GHz by phased arrays from 8 to 12 ft high. Because the vertical pattern's side lobes are reflected from the ground in front of the elevation array, the antennas are designed to suppress side lobes 30 dB or more below the major-lobe peak at critical reflection angles. The lateral pattern of the elevation array is a fan that is shaped to reduce the effect of reflections from aircraft awaiting takeoff or from buildings.

After transmission of the preamble, the elevation array is scanned from its lowest to highest angle. This "TO-SCAN" begins slightly below the horizontal and may end as high as 32 deg, although an upper limit of 15 deg is typical of 1996 equipment designs. The scan pauses briefly before beginning the "FRO-SCAN", back to the lower elevation angle limit [4a].

Azimuth Functions   An azimuth function provides lateral guidance in the approach (forward) sector, and a back-azimuth function provides lateral guidance in the departure/missed-approach sector. While the signal formats in the two sectors differ, they are equivalent in principle and are transmitted by similar ground stations. The challenges to accurate azimuth guidance are the low coverage required over the runway and the suppression of interfering reflections from lateral structures (e.g., large hangars). A lateral-pattern beam width of 2 deg is generally satisfactory for runways less than 8,000 ft long; longer runways typically require a 1-deg major lobe, Figure 13.14*a*. The electrical and physical size relationships of the aperture are the same as those for elevation arrays because the transmission frequency is the same. Because the lateral pattern's side lobes are reflected from structures at wide angles, suppression of the side lobes is less stringent than for the elevation array. Thus, typical side lobes near the major lobe are attenuated 25 dB or more relative to the major lobe. The vertical pattern of the azimuth array is a fan, shaped at lower angles to reduce the amplitude of reflections from the ground and at higher angles to suppress reflections from overflying aircraft.

After transmission of an approach-azimuth or back-azimuth preamble, the azimuth-phased array is scanned clockwise (seen from above) from minimum to maximum pointing angle (e.g., from −40 deg through zero (on center line) to +40 deg). This 80-deg sector is the TO-SCAN limit for back-azimuth, although one version of the approach azimuth can provide a sector of 120 deg [4a]. The scan pauses briefly before beginning the FRO-SCAN back to the minimum azimuth angle.

13.6.3   The Data Functions

Like the angle-function transmissions, the data-word transmissions begin with a preamble that contains an identification code for a "basic" or an "auxiliary" data word. Following the preamble, predefined items of data are transmitted from the azimuth preamble antenna for the allowable time period, using the same DPSK coding scheme as the preamble. The differential encoding provides good resistance to multipath interference, since decoding either the direct or reflected signal yields the same information. (i.e., change = 1, no change = 0).

Basic-Data Words   The eight basic-data words contain information about the MLS ground station that is needed for proper operation of the MLS receiver [4a]. Each word contains 26 bits of information plus 2 parity bits. Included in these words are the facility identifier, proportional-guidance limits, minimum glide-path angle, runway length, and so on. All MLS installations must transmit all eight basic-data words.

Auxiliary-Data Words   In contrast, the number of auxiliary-data words is expandable to fit local needs, and the amount of data transmitted is limited only by the excess time available in the signal format [4a]. Each word contains 57 information bits plus 7 parity bits. The auxiliary-data words are demodulated within the MLS receiver and output directly to an avionic unit (e.g., a display) or to a data bus. If the addresses of on-board avionics boxes were standardized, the first eight information bits of the word could be the address of the receiving avionics. The remaining bits may include information ranging from weather data to waypoint coordinates for a particular area navigation (RNAV) procedure associated with this runway [18b].

13.6.4   Aircraft Antennas and Receivers

A large aircraft (e.g., B-747) may have one antenna above the cockpit area for forward coverage and one on the fuselage below the tail for coverage to the rear. The azimuth signal format provides a time period for the receiver to select the antenna with the stronger signal. Also, microwave horns are provided behind the nose radome in some installations (e.g., on the USAF C-130) for the final approach. Smaller aircraft may achieve the required ± 70 deg forward lateral coverage with only a single antenna typically above the cockpit. The body-mounted antennas are stubs about 0.6-in. high or, on high-performance aircraft, a flush-mounted slot. Generally, the MLS receiver has no RF amplification; signals enter the mixer through a fixed filter that covers the MLS band. After the IF, the preambles and data words are demodulated in a phase-locked loop that has dynamics selected in consideration of the desired bit-error rate and signal-acquisition time. The receiver decodes the basic-data words for internal use and passes the auxiliary-data words through to the avionics data bus. The video envelopes resulting from the TO-FRO passages of the scanning beams are processed in another path to measure the elapsed time between beam passages and to validate envelope width and amplitude. The consistency of these continuous validation checks increments a confidence counter so that the receiver can maintain track in the presence of momentary noise spikes or multipath signals.

However, a persistent signal of greater amplitude than the signal being tracked will decrement the confidence counter and cause the receiver to drop the current track and acquire the larger signal after the confidence is zero. In general, MLS receiver designs are centered on a microprocessor. For Category II and III operations, dual or triplicated receiver installations are required (Section 13.2).

13.6.5   Mobile MLS

The U.S. military have devloped a mobile MLS (MMLS) that is transported by a three-man team and installed within 30 minutes each for the azimuth-DME/P or elevation station on runways up to 12,000 ft long [19]. Favorable initial operational test results were achieved in 1993, and production systems were tested in 1995 (Figure 13.17).

The accuracy specifications for the MMLS are based on Category I and II ILS requirements rather than on the MLS accuracy (i.e., Category III) standard; however, the radiated signals conform to the MLS signal format. These relaxed accuracy requirements were chosen to allow smaller scanning antennas (with wider beam widths) than would be typical of permanent installations. Although antenna aperture has been reduced for lighter weight, the MMLS performance meets MLS standards except at heights below 100 ft.

13.6.6   Precision DME (DME/P)

A precise L-band distance measuring equipment (DME) (±100 ft at runway threshold, 95%) was developed in parallel with the MLS to improve range accu-

Figure 13.17   View up the glide path of a Mobile MLS at the Farnborough Air Show, 1994 (courtesy, Textron Defense Systems).

racy when using MLS to calculate the three-dimensional position of the aircraft with respect to the runway (RNAV mode) [4a]. (For a description of the conventional DME, see Section 4.4.6.) The transponder and interrogator designs for DME/P are fully interoperable with conventional DME (DME/N) such that reception of the DME/P signal with a DME/N receiver improves accuracy but does not realize full performance.

Obtaining this accuracy improvement requires a faster rise-time pulse to increase accuracy which causes a wider bandwidth. It reduces the radiated transponder power and the power introduced in adjacent L-band channels [4a]. These trade-offs resulted in specifying two modes of operation—an initial approach (IA) mode using the Gaussian pulse shape of DME/N and a lower-power final approach (FA) mode using a composite pulse with a shorter rise-time, an earlier thresholding point, and the standard Gaussian decay time [20]. Thus, the DME/P operates in the IA mode from the MLS outer coverage boundary to 8 nmi from the DME/P antenna, where the interrogator begins changing the pulse-spacing code in order to complete the transition to FA mode by the 7-nmi point.

Typically, the DME/P transponder is located with the MLS approach azimuth station but operates autonomously on the paired frequencies also used with the ILS (Section 13.5.2). Civil DME receivers are standardized by ARINC [32a].

13.7   SATELLITE LANDING SYSTEMS

Even before the global positioning system (GPS) (Chapter 5) was declared operational by the United States in 1993, efforts had been underway to use it for approach and landing. Later efforts involved the Russian GLONASS (Section 5.6) and may result in the combination of GPS Glonass, and other satellites into a Global Navigation Satellite System (GNSS, Section 5.7).

An operational concept called *Special Category I Precision Approach Operations Using DGPS*, based on the differential GPS (DGPS) technique (Chapter 5.5.9), was developed, tested, and certified for specific airports (Sections 5.5.9 and 5.5.10, and Table 5.7). The test results of the DGPS have been very promising (Refs. [66–68] of Chapter 5). However, the unique requirements of the approach and landing phase with regard to *accuracy, integrity, and availability* (Section 5.7.1) have led to the development of additional GPS augmentation methods, which are discussed in the next section.

13.7.1   Augmentation Concepts

The basic GPS, without differential corrections, cannot be the primary means of navigation for precision approach and landing operations because of the following limitations:

1. *Accuracy.* The nominal vertical error is ±150 meters, compared to require-

ments of about ±8 meters for Category I, ±4 meters for Category II, and ±1.3 meters for Category III (Section 13.5.2, Table 13.3).

2. *Integrity.* The GPS design lacks a monitoring system which can provide timely warnings of guidance-data faults within 10 sec for Category I, or less than 2 sec for Category III.

3. *Availability.* The number of satellites in view in certain time periods may not be adequate.

For nonprecision and Category I precision approaches, the above limitations are expected to be mitigated by the FAA's Wide Area Augmentation System (WAAS, Section 5.7.3), which was under development in 1996. WAAS may be supplemented with local differential GPS (LDGPS) stations located near runways. Specifically, the WAAS will address the above three limitations by providing (1) wide area differential GPS (WADGPS) error corrections, (2) a ground integrity broadcast (GIB), and (3) a GPS-ranging function from three or four geostationary satellites providing additional availability or reliability (Ref. [69] of Chapter 5 and [5f] of Chapter 13]. The details of the WAAS are described in Chapter 5.7.3.

To support Category II and III landing operations, the additional requirement for accuracy and monitor-response times can be achieved with a differential GPS reference station and a high-integrity ground-to-air data-link located near the landing runway, constituting a local differential GPS (LDGPS) configuration (Chapter 5.5.9). LDGPS operation leads to nearly complete cancellation of errors due to satellite ephemerides, clock offset, selective availability, and tropospheric/ionospheric propagation, leaving mainly those errors due to multipath-signal interference and receiver noise. The runway reference station (with appropriate monitoring) can supply integrity warnings within the required response time of 1 to 2 sec. Additional availability augmentation may be provided by the WAAS, by a precise time standard on-board the aircraft, and by pseudolites (Chapter 5.7.1) located along the final-approach path or on the airport. In 1996, the U.S. FAA was investigating the feasibility of Category II and III DGPS operations [5f].

13.7.2 Position Solutions

In 1996, GPS receivers were capable of about ten solutions per second, which is sufficient for coupled flight. The accuracy of the SPS code solution without augmentation is satisfactory for nonprecision approaches. The code solution is improved by integrating the Doppler shift of the carrier to obtain a velocity component for smoothing the position measurements and for removing latency effects. Satisfactory auto-land performance has been demonstrated using a velocity-smoothed, code-tracking, differential solution [21]. Carrier phase-tracking differential techniques (Chapter 5.5.9) have also been demonstrated in real time to provide even higher accuracies [22]. This *kinematic technique* results in a position solution with much lower noise content than code-tracking solutions. Also, good error-canceling properties contribute to the attractiveness of this method. One disadvantage of carrier-phase tracking is that the receiver must accurately determine the range to the satellite in terms of the specific cycle of the carrier frequency. Resolving this integer-wavelength ambiguity for each satellite and guarding against *cycle slips* requires special techniques (Section 5.5.9 and [23]). Some *pseudolite* architectures (Section 5.7.4) favor this type of solution, since the high pseudorange rates generated by the aircraft on final approach are a significant aid to cycle ambiguity resolution [24].

13.7.3   Research Issues

Data Latency   Delays of two seconds are intolerable during Category III landings. The need for differential corrections to be radioed from a ground station and the on-board processing may delay the position solution for several seconds before it is available to the pilot and flight-control system. Delays can be reduced in the ground computations, in the on-board computations, in the receiver design (e.g., the Doppler shift of the carrier can measure the velocity vector of the aircraft during the latency period) or by mixing satellite and inertial data. A state-estimator (e.g., a Kalman filter) can combine the aged position data with the velocity history of the aircraft to provide an accurate estimate of the aircraft's current position.

RF Interference   The low level of power received from the spacecraft makes the satellite-based landing system more susceptible to RF interference and receiver noise than has been experienced with the ILS and MLS. In 1996, these effects were under intensive investigation.

Multipath Effects   Satellite-signal reflections from terrain or ground objects to the aircraft antenna are small, for air-carrier aircraft, but reflected-signals at the differential stations are larger. Siting criteria, analogous to the airport *critical areas* developed for ILS and MLS, are likely to be required to ensure integrity.

13.8   CARRIER-LANDING SYSTEMS

13.8.1   Description of the Problem

Aircraft are operated from large aircraft carriers, of which there are about 20 worldwide. Each carrier stores fewer than 100 aircraft and helicopters. Helicopters and vertical-takeoff-and-landing (VTOL) aircraft are operated from smaller ships, of which there are about 300 worldwide. There are approximately 1000 carrier-compatible aircraft in the world, about 2000 naval helicopters and about 250 VTOL aircraft (in 1996, AV8 "jump-jets") [30].

The typical aircraft carrier landing deck is 600 ft long by 100 ft wide (Figure

13.18). The mean touchdown point is 180 ft forward of the runway threshold at the center of the arresting cable area, which is about 120 ft long by 75 ft wide. Two or three cables are stretched across the deck forward of this point and two or three aft; they are spaced about 40 ft apart. Jet aircraft approach the carrier deck at airspeeds up to 125 knots; their speed relative to the deck is lower due to the "wind-over-deck," which is normally about 30 knots. The approach path, often called the *glide slope*, is in the vertical plane containing the runway centerline and is inclined 3.5 deg relative to the sea at the touchdown point (4 deg relative to the moving flight deck) [25]. Aircraft cross the threshold at a mean hook-to-ramp clearance of 11 ft and touch down about 1 sec later without flaring out, at descent rates of 10 to 13 ft/sec. One aircraft may be landed per minute by an automatic landing system; as many as two per minute are landed in VFR weather. The ship's deck "heaves" vertically with a standard deviation of 4 to 5 ft, rolls with a standard deviation of 2.3 deg to 5 deg, pitches with a standard deviation of 1 deg, and yaws with a standard deviation of 1.8 deg during an aircraft approach [25]. The frequency spectra of ship motions are discussed and their relative phasing is shown in [25]. A 1-ft heave causes a 1-ft change in threshold clearance and a 14-ft change in the touchdown point. A 1 deg-pitch causes a 9-ft change in threshold clearance and an 80-ft change in the touchdown point. Ship motions affect impact velocity at touchdown. A night landing on a carrier deck is the most difficult and dangerous task in aviation.

The prevailing wind and forward motion cause a wind-over-deck between 20 and 40 knots; the ship is steered to place the relative wind along the canted flight deck. Forward motion of the ship aids the pilot by reducing landing speed relative to the deck, but it causes a turbulent-air wake which interferes with the landing. The air wake may be divided into the following components [25]:

1. Steady-state wake caused by the relative wind over the ship's deck and around the superstructure ("island"), which causes a "burble" at the threshold and along the glide path. The vertical and horizontal components of the steady-state wake for a Forrestal-class carrier are shown in [25].

2. Wake perturbations, due to the ship's angular motions and heave, causing fluctuations with standard deviations of 2 ft/sec (horizontal) and 4 ft/sec (vertical) about the steady state wake. These become negligible 2000 ft aft of the threshold.

Aircraft are stacked in holding patterns 8 to 12 nmi astern at altitudes from 4000 to 40,000 ft. They are vectored onto the glide path by carrier air traffic control (CATC) using voice radio and the approach control radar until they are within the coverage of the SPN-41 and the SPN-46 instrument landing systems. Pilots are released from the holding pattern to the glide path and may follow

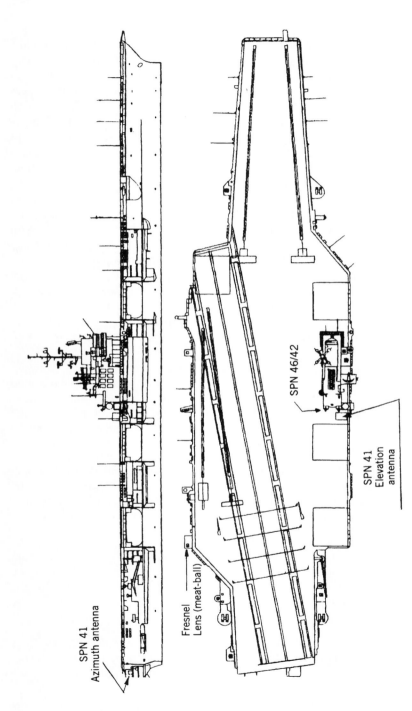

SPN 41
Azimuth antenna

Fresnel
Lens (meat-ball)

SPN 46/42

SPN 41
Elevation
antenna

Figure 13.18 Canted-deck aircraft carrier showing installation of SPN-46/42, SPN-41, and FLOLS (courtesy of U.S. Naval Air Engineering Center).

Figure 13.19   Fresnel-lens optical landing aid (courtesy, U.S. Navel Air Engineering Center).

cockpit display "needles" or may couple the autopilot to the Automatic Carrier Landing System (ACLS). Carrier pilots are also provided cues by a Fresnel-lens optical system nicknamed the "meatball," Figure 13.19.

An aircraft executing a missed approach from low altitude is called a *bolter*. Bolters fly to a Tacan-defined holding point that moves with the aircraft carrier.

The ACLS is capable of landing an aircraft "hands off," giving the U.S. Navy an operational capability in zero-zero visibility conditions. In 1996, ACLS could fly an F-18 down the glide path more accurately than an average F-18 pilot [28]. However, peacetime air operations were routinely discontinued when visibility fell below $\frac{1}{2}$ to $\frac{3}{4}$ nmi or ceilings dropped below 200 ft. Carrier commanders were reluctant to depend on the ACLS in zero-zero visibility, except in emergencies or during hostilities.

13.8.2 Optical Landing Aids

In the early days of aircraft carriers, the landing signal officer (LSO) guided the pilot using a paddle in each hand. The LSO signalled with colored lights and voice radio. Mirrors fixed to the deck were used during the 1950s to assist pilots in following the glide path by projecting an image of a light along the path.

Because of deck rotation, the aid was gyrostabilized. In 1968, mirrors began to be replaced by a vertical array of five Fresnel lenses (Figure 13.19) showing the pilot a yellow spot ("meatball") that moves vertically relative to a lighted green horizontal reference bar. The array is called the "Fresnel lens optical landing system (FLOLS) [29]. The beamwidth of each lens is about 5 mrad. Because the five lenses are spaced about 3 ft apart, adjacent beams overlap beyond 1000 ft, so pilots cannot easily estimate their displacement above or below the glide path beyond 2000 to 4000 ft. The meatball gives only vertical cues.

The optical axis of the system is servoed to remove ship's pitch and roll but not heave, although "compensated meatball" methods have been proposed which correct for all three [25]. Lens pointing angles would be adjusted according to a quadratic function of heave in much the same way as the angle data from the radar-controlled ACLS is compensated (Section 13.8.3). This compensation has not been implemented because the "meatball" is often shared by two aircraft on final approach simultaneously.

13.8.3 Electronic Landing Aids

Two electronic instrument landing systems are found on all large U.S. aircraft carriers. The ACLS involves two identical conically scanned K_a-band radars (SPN-46 or SPN-42) set aft of the island about 125 and 135 ft above the water line. They can track two aircraft simultaneously and, if desired, can uplink guidance signals to aircraft autopilots. An independent sensor (the SPN-41) monitors the approach. SPN-41 consists of a Ku-band transmitter and an airborne receiver/processor. The receiver measures angular displacements from the glide path.

Automatic Carrier Landing System   Both SPN-42 and its successor, SPN-46, are tracking-radar systems differing primarily in the antenna. Figure 13.20 shows one of the dual-frequency SPN-46 antennas. The SPN-42 was retrofitted with a small X-band parabola when aircraft beacons were added to eliminate errors caused by spatial "wandering" of the skin echo. Both radars radiate short pulses at K_a-band and receive replies from an X-band beacon (APN-202 or APN-154) as well as K_a-band skin-echoes. Radar pulses, modulated by the conically-scanning beam, are detected by the aircraft beacon receiver and data are returned to the radar in the form of the arrival time and power of a pulse from the X-band beacon's magnetron. Range and angle estimates are formed aboard ship by conventional processing of the X-band signals. Angular accuracy is primarily a function of the conically scanned K_a-band beam pattern (the X-band receiving beam is relatively broad). Errors in transferring uplink pulse height to the X-band beacon tend to cancel in the radar tracking processor.

Data measured by the radar/beacon system are digitized and delivered to a central ACLS computer which computes commands intended for the autopilot of the aircraft in track suitable for flying the aircraft down the glide path. These data are communicated to the aircraft via a UHF "Link 4" data channel

Figure 13.20   SPN-46 antenna (courtesy, U.S. Navy).

[26]. The pilot has the option of coupling the autopilot to the ACLS, flying the aircraft from "needles" controlled by ACLS data, or being "talked down" by a shipboard controller. A manual carrier landing with ACLS resembles a land-based ground-controlled approach.

SPN-41 Independent Landing Monitor   This system consists of a single K_u-band transmitter switched between separate azimuth and elevation antennas. Although driven synchronously, the horizontally and vertically scanning fan-beam antennas are separate units. A small rotating parabolic reflector provides a shaped vertical fan beam that scans the azimuth sector ±20 deg; at the end of the scan the transmitter is switched to the vertically scanning fan-beam formed by a horn antenna that mechanically oscillates at the end of a short torsion bar. The aircraft receiver converts pairs of short K_u-band transmitter pulses to IF and video. The pair spacing identifies the azimuth or elevation beam, and the spacing between pairs indicates its pointing angle. Airborne processing uses amplitude information in estimating aircraft azimuth and elevation angles relative to the desired glide path, which then drives localizer and glide-slope needles in the aircraft. Land-based SPN-41 was used for landing Space Shuttles in the 1980s and 1990s.

Figure 13.18 shows the locations of the SPN-41 azimuth and elevation antennas aboard ship. The SPN-41 is shared by all aircraft in the landing sequence. Antennas are stabilized in roll and pitch but are not compensated for heave. Since it conveys no range information and has no data link, the SPN-41 cannot be used as an ACLS. Common practice is to use SPN-41 signals at relatively long range (20 to 30 nmi) for ACLS acquisition guidance. It also serves as a monitor of SPN-42/46 angle information until the aircraft is within 12 sec of touchdown.

SPN-35 Precision Approach Radar (PAR)   A few fixed-wing aircraft carriers carry the SPN-35 as a backup to the SPN-46 or SPN-42. However, just as in commercial avaiation, PARs are seldom used because they depend on a trained operator.

In 1996, there were more than 20 LHA, LHD, and LPH assault ships that carried only SPN-43 (approach surveillance) and SPN-35 PAR for low-visibility landings of helicopters and VTOL aircraft such as the AV-8 *Harrier*. Nonprecision low-visibility landings may be accomplished with Tacan (Chapter 4).

13.9   FUTURE TRENDS

13.9.1   Pilot Aids

In 1996, several technologies were being pursued to reduce pilot work load during approach and landing or to improve the pilot's ability to monitor an automatic landing.

Enhanced vision systems, such as active millimeter-wave radar (MMWR) and passive forward-looking infrared (FLIR) detectors, can create an artificial image of the runway and project it on a head-up display. MMWR developments have focussed on 35 and 94 GHz where RF transmission "windows" exist in the atmosphere. The higher frequency (3-mm wavelength) gives better resolution, but the lower frequency (9-mm wavelength) requires less power to penetrate rain. Even at these short wavelengths, physical constraints limit the beam width to about 0.25 deg, which does not provide adequate runway resolution at the shallow elevation angles encountered during the landing maneuver [4f]. Passive infrared systems with wavelengths near 10 nm (10^{-5} mm) have adequate resolution and reasonable fog penetration characteristics, but heat sources may be required along the runway to improve the contrast of the runway against background terrain. The minimum range of a laser radar (200 ft in 1996) may deny its use for taxi guidance.

Infrared cameras (FLIR) combined with head-up displays (HUD, Chapter 15) are found in some navy aircraft. FLIRs have been tested on aircraft carriers for monitoring operations in clear weather at night but IR energy suffers almost the same absorption as visible light in fog and rain.

Figure 13.21 shows a 4 × 4 element 94 GHz focal-plane array of antenna/detector elements that was fabricated photolithographically on a wafer of GaAs (about 1-in. square) [28]. If replication of such a tile were inexpensive, 94 visual GHz cameras and focal plane array radars might enter service as landing aids.

The HUD, which was developed by the military services to aid a single pilot in transitioning to visual cues, is finding applications on air-carrier aircraft (Chapter 15.3). It projects sensor information (navigation, attitude, and other sensors) and stored runway data onto the windscreen at infinite focus, so that the pilot can see the data and images superimposed on the view through the windshield. In 1996, European and American authorities had certified many aircraft for Category IIIA operations and most glass-cockpit models of Boeing, Douglas, and Airbus aircraft for Category IIIB operations. These authorities have allowed a HUD and a fail-passive autopilot to substitute for a fail-operational autopilot.

Figure 13.21 A 4 × 4 element 94 GHz focal plane array (courtesy, U.S. Naval Research Laboratory and Westinghouse Corporation).

13.9.2   Satellite Landing Aids

There is a need for low-cost, nonprecision and Category I procedures at low-density airports in the industrialized countries and in underdeveloped countries that have poor highway and railroad systems and are therefore dependent on air travel. Satellite-navigation systems are likely to be the solution. Similarly military needs for tactical approach guidance systems (including landing on ships) will probably be satisfied by differential or relative satellite techniques (Section 13.7). For lower-visibility landing, an independent monitor is required.

During the period of transition from ground-based to satellite-based guidance, multi-function airborne receivers, capable of processing ILS, MLS, and GPS guidance signals, will be used in civil and military aviation. An ARINC characteristic for such a *multimode receiver* was in preparation in 1996. Two technologies that will aid the application of satellite navigation to precision approach operations are lower-cost inertial devices and lower-cost precision (atomic) clocks.

13.9.3   Airport Surface Navigation

The use of differential satellite-based systems for guidance and surveillance of rollout, taxi, and departure operations under low-visibility conditions will be widespread and will extend to vehicles other than aircraft (e.g., fuel trucks, police and emergency vehicles). The combination of LDGPS (Section 5.5.9) and pseudolites (Section 5.7.5) is likely to offer a solution to the airport surface navigation and surveillance problem.

13.9.4 Carrier Landing

The challenging problem of all-weather recovery of large (up to 65,000 lb) aircraft aboard a floating platform less than 1000 ft long will lead to new guidance systems. The combination of relative GPS (Chapter 5), a data link, and aircraft and ship attitude sensors could provide the basis of a new ACLS. In the difficult environment of carrier landings, an independent monitor will be needed, perhaps a descendant of SPN-41 or SPN-46.

PROBLEMS

13.1. A particular aircraft can deviate ±85ft laterally from the runway center line without leaving the hard surface. Assuming the errors are normally distributed, calculate the allowable lateral deviation (1-sigma) if the lateral performance requirement for the auto-land system states that the risk (probability) of touchdown off the runway is not to exceed 3×10^{-8}.

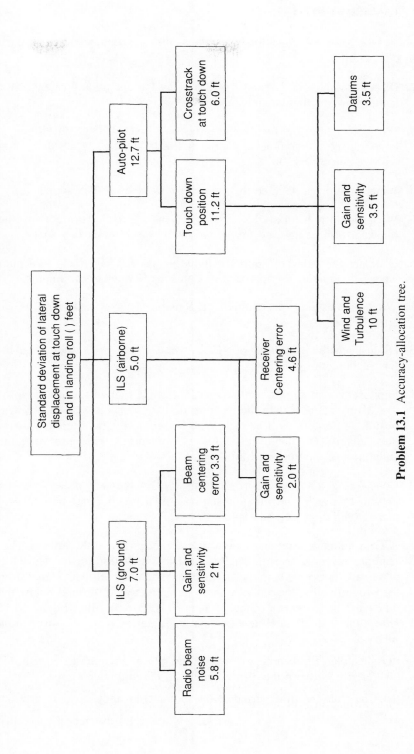

Problem 13.1 Accuracy-allocation tree.

The accuracy-allocation tree below allocates this accuracy to each error source.

13.2. Discuss the factors associated with raising the operational category for landing on a particular runway.

13.3. What precision in terms of linear measure is needed for positioning a large transport aircraft at the approach threshold for a successful landing with visibility of 1200 ft?

13.4. What factors can degrade the guidance quality of an ILS signal?

13.5. What would prevent a DME, associated with an ILS serving a specific runway, from being used for an approach to a different runway?

13.6. Compare the value of approach lights versus runway lights to a pilot who is conducting an approach to Category II weather minimums.

13.7. What justifies the expectation that ILS marker beacons will find less use after 1997?

13.8. Discuss whether a building of a given size will be a more efficient reflector (scatterer) of glide-slope or localizer signals.

13.9. Suppose that an airport is adding a new runway that will be much closer to existing hangers than the existing runway. Assume that all physical obstruction criteria will be met. With respect to the radio guidance signals, identify the factors that must be considered when assessing the suitability of the new runway.

13.10. Considering all GPS error sources, what is the key issue in siting ground reference and monitor stations for a pseudorange-based, local area, DGPS approach and landing system?

13.11. A GPS antenna is mounted on a pole 2 meters above the ground. Signals are received from a satellite at a low-elevation angle. The GPS antenna will receive both the direct signal and one multipath signal from the ground below the antenna. Assume perfectly flat ground below the antenna. The frequency of the GPS signal is 1575.42 MHz. Figure 13.11 shows the (actual) satellite elevation angle during a 10-sec observation interval.

(a) Calculate (in meters per second) the rate of change of the path difference between the direct signal and the multipath signal.

(b) Calculate the time period of the fading frequency.

Ans.: 3.78×10^{-4} m/sec; 503 seconds.

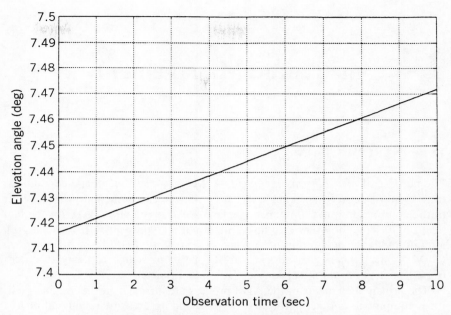

Problem 13.11   GPS Satellite elevation angle.

14 Air Traffic Management

14.1 INTRODUCTION

The purpose of *air traffic management* (ATM) is safe, efficient, and expeditious movement of aircraft in the airspace. It comprises two principal processes: air traffic control and traffic flow management. *Air traffic control* (ATC) is the tactical safety separation service whose function it is to prevent collisions between aircraft and between aircraft, terrain and obstructions. *Traffic flow management* (TFM) is the process that allocates traffic flows to scarce capacity resources (e.g., it meters arrival at capacity constrained airports).

The principal elements of the ATM process are airspace; air navigation facilities, equipment, and services; airports and landing areas; aeronautical charts, information, and services; rules, regulations, and procedures; technical information; and work force, including flight crews, air traffic controllers, traffic managers, and facilities technicians [37, 18]. This chapter describes these elements of the ATM process and explains how they interact to fulfill its purpose.

14.1.1 Services Provided to Aircraft Operators

A principal service provided by ATM is separation assurance for the prevention of collisions between aircraft and to prevent aircraft collisions with terrain and obstructions. Traffic flow management services are designed to meter traffic to taxed capacity resources, both to assure that unsafe levels of traffic congestion do not develop and to distribute the associated movement delays equitably among system users.

Relevant aeronautical information is provided both in the form of charts and publications and as current (real-time) information communicated to the user prior to and during flight. The real-time information provided includes weather observations and forecasts, traffic congestion conditions and delays, status of air navigation facilities and airports, and the positions and movement intentions of other aircraft in the vicinity. The final principal service is search and rescue, that is, notification of appropriate organizations regarding aircraft in need of search and rescue assistance and support of these organizations during the ensuing operations.

Navigation generally is the responsibility of the aircraft operator. Nonetheless, air traffic controllers assist lost and distressed aircraft in determining their positions and navigating to their destinations.

14.1.2 Government Responsibilities

Within the United States, the Federal Aviation Administration (FAA) is responsible for establishing rules and regulations allocating airspace to various uses and regulating those uses; for establishing and operating air navigation facilities; for ensuring the publication of aeronautical charts and information necessary for the safe and efficient movement of aircraft; and for establishing the rules, regulations, and procedures governing air traffic control and traffic flow management. The agency is also responsible for establishing minimum standards regarding the design, construction, certification, and maintenance of aircraft and airports as well as for certifying flight crews [4]. The United States has a *common system* for ATM in the sense that the same practices apply to civil and military users and both the FAA and Department of Defense (DOD) provide ATM services to both civil and military users. In many areas of the world, most of the airspace is devoted to military operations for which military authorities have jurisdiction; civil aviation authorities provide services to civil users in limited corridors set aside for that purpose.

Standardization of air traffic management practices internationally is the responsibility of the International Civil Aviation Organization (ICAO), which is affiliated with the United Nations and located in Montreal, Quebec, Canada. Procedures for aircraft operation and the provision of ATM services are standardized to the extent necessary to ensure regularity worldwide. In addition, appropriate technical standards for aircraft, air navigation facilities, aeronautical information and charts, and airports are established. ICAO does not provide ATM services directly to operators.

Countries align their procedures and equipment with ICAO standards so that there is safe and efficient operation of international aircraft within their borders. As a result, ATM practices and procedures around the world have few differences. Some regions have less developed ground facility infrastructures than others and the efficiency of aircraft operations suffers as a result (Section 1.7). Increased use of satellite-based services for communication, navigation and surveillance will foster better ATM services worldwide.

14.2 FLIGHT RULES AND AIRSPACE ORGANIZATION

14.2.1 Visual and Instrument Flight Rules

Aircraft in flight operate in accordance with one of two sets of rules, visual flight rules or instrument flight rules. *Visual flight rules* (VFR) operators are responsible for avoiding collisions with obstacles, terrain and other aircraft by visually observing these hazards and maintaining visual separations. Therefore, visibility conditions must meet prescribed minimums and the aircraft must remain clear of clouds. In much of U.S. airspace, VFR operators are not required to contact ATM authorities and are not subject to their directions. VFR flights usually involve small aircraft operated for personal or business purposes. Navigation may be by visual reference to the ground and aircraft attitude control may be based solely on the natural horizon visible through the windscreen of the aircraft.

Flight crews following *instrument flight rules* (IFR) must be capable of navigating and controlling the aircraft without reference to an outside visual scene. In *controlled airspace*, the crew must operate in accordance with instructions received from ATM authorities, and these authorities are responsible for separating the aircraft from all other IFR aircraft. Aircraft operating IFR must be equipped with appropriate attitude indicators, radios and other navigation equipment not required on VFR aircraft. Further, the flight crew requires additional training and experience to qualify for the necessary instrument rating. All aircraft operators holding Air Carrier Operating Certificates are required to follow instrument flight rules [6].

14.2.2   Altimetry

In order to understand airspace organization and navigation practices in the ATM community, it is necessary to know how altitude measurements are made in aircraft [7]. The barometric altimeter (Chapter 8) measures the static air pressure outside the aircraft and converts this pressure to an altitude, in feet, based on a standard model of the atmosphere adopted by ICAO [60]. According to the model, atmospheric pressure and temperature at mean sea level are 29.92 in. of mercury (1013.2 hectopascals) and 15.0°C, respectively. The temperature lapse rate in the model is 2.04°C per 1000 ft from sea level to 35,000 ft above mean sea level, after which the temperature is constant at −56.5°C.

The aircraft barometric altimeter permits the flight crew to set the reference pressure against which the outside atmospheric pressure is measured. If the reference pressure were adjusted to the actual pressure at mean sea level, and the temperature lapse rate matched the ICAO standard model, the altimeter would indicate true altitude above mean sea level. Unfortunately, due to the varying characteristics and movements of air masses, the pressure at mean sea level varies around the world, and these pressures may not be representative of air masses over land areas. Hence, altimeters are installed in ground stations (e.g., airports and flight service stations) and in the USA the setting is adjusted so that the altimeter indicates the true altitude above mean sea level of the station. A flight crew operating below 18,000 ft above mean sea level uses the setting provided by a station located along the route of flight within 100 nmi of its position. The resulting altimeter reading displayed to the crew is termed the *indicated altitude* and is a good estimate of the aircraft altitude above mean sea level. Hence, indicated altitude is expressed in feet above mean sea level, which is abbreviated "ft MSL." Above 18,000 ft MSL, flight crews set their altimeters to 29.92 in. of mercury. The resulting altimeter reading is referred to as the *pressure altitude*. Pressure altitudes are expressed as *flight levels* (FLs) in hundreds of feet using three digits. Hence, FL290 represents a pressure altitude of 29,000 ft.

14.2.3   Controlled Airspace

A fundamental construct of the ATM system is the organization of airspace into various categories and the establishment of rules governing aircraft operations within each category. The objective is to provide maximum freedom in public use of the airspace while ensuring the safety and efficiency of flight operations where airspace is heavily used or is required to support special purposes such as military training.

Within airspace regions categorized as *controlled*, aircraft operating IFR are required to receive separation and other services from ATM authorities. VFR operators are required to accept ATM separation services in some types of controlled airspace. Ground-to-air communications and navigation aids support these services. The following paragraphs describe a few examples of controlled airspace. Additional details are given in the *Airman's Information Manual* [8].

Class A—Positive Control Area   Virtually all of the airspace above the 50 U.S. states from 18,000 to 60,000 ft MSL is designated Class A. All aircraft within Class A airspace must follow instrument flight rules, that is, must file IFR flight plans and operate in accordance with clearances. A *clearance* is authorization for an aircraft to proceed under conditions specified by the responsible ATM authority. The purpose of Class A is to ensure separation for all aircraft operating in this regime wherein aircraft speeds are high and VFR see-and-avoid techniques are ineffective.

Class B—Terminal Control Areas   Class B airspace exists around the nation's busiest airports. The general shape of Class B airspace resembles an upside-down wedding cake centered on the primary airport. The radius increases with altitude so that aircraft remain within the airspace as they descend into or climb away from the airport. All aircraft, whether operating by IFR or VFR, must receive a clearance to enter the airspace and the associated ATM authority is responsible for separating all aircraft. Aircraft at the same altitude are kept at least 3 nmi apart. Aircraft within 3 nmi of one another are separated vertically by at least 1000 ft if both aircraft are IFR, or by at least 500 ft if one aircraft is IFR and the other is VFR. VFR aircraft are required to operate in accordance with visual flight rules including maintaining prescribed distances from clouds. IFR operators have priority over VFR aircraft, and access by VFR operators can be denied if the ATM authority believes VFR operations would compromise safety.

14.2.4   Uncontrolled Airspace

Uncontrolled airspace is airspace that is not designated as controlled. In the United States, most uncontrolled airspace is within 1200 ft of the ground in sparsely traveled areas. ATM authorities will not provide separation services in uncontrolled airspace. IFR operators need not file flight plans and may operate in any visibility conditions. VFR operators below 10,000 ft MSL must have a minimum visibility of one statute mile.

14.2.5   Special Use Airspace

Special use airspace is set aside for unique purposes, notably national security operations. A *prohibited area* such as the airspace surrounding the White House is closed to aircraft operations at all times. *Restricted areas* such as military missile ranges are closed to civil traffic when operating and are open at other times. *Military operations areas* (e.g., for training) will accept IFR traffic if the responsible ATM authority can safely provide separation services. VFR aircraft must exercise extreme caution when operating in active military operations areas.

14.3   AIRWAYS AND PROCEDURES

14.3.1   Victor Airways and Jet Routes

A system of *federal airways* (highways in the sky) is defined by Parts 71 and 93 of the US Federal Aviation Regulations. As illustrated in Figure 14.1, most of these airways originate and end at VOR stations (Section 4.4.4). A few airways based on nondirectional beacons (Section 4.4.2) are still in use.

From 1200 ft above the surface to 18,000 ft MSL, VOR-based airways are called *victor airways* and use "V" as a prefix in their identifiers (e.g. V51). From 18,000 ft MSL to 45,000 ft MSL, all federal airways are based on VORs and are referred to as *jet routes*. The prefix of the identifiers is "J" (e.g. J64).

Changeover points (COPs) are defined along airways to indicate where the pilot should adjust the navigation receiver to obtain course guidance from the facility (e.g. VOR) ahead instead of the one behind. The COP is normally at the midpoint between facilities but may be elsewhere if required due to radio frequency interference, facility siting limitations, or other effects. If a COP is not at the midpoint of the airway, its position is shown on aeronautical charts.

Every VOR is designated either as a *terminal VOR* or as a *low-* or *high-altitude en-route VOR*. Terminal VORs are intended for use in the local area, for example, to support instrument approach and departure procedures. They are usable only at ranges less than 25 nmi and may not be used for en-route navigation. Low-altitude en-route VORs have service volumes extending up to 40 nmi from the facility and are used at and below 18,000 ft MSL for navigation on victor airways. High-altitude VORs support navigation on jet routes; their service volumes may extend 200 nmi from the facility. Service volumes of VORs are controlled as described above in order to limit frequency interference effects among adjacent ground stations.

Airway widths are determined by navigation system performance, taking into account errors in the ground station equipment, errors in the aircraft receiver

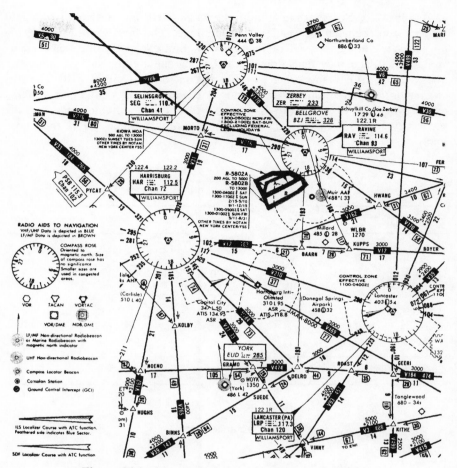

Figure 14.1. Aeronautical chart showing victor airways.

and display system, and flight technical error. *Flight technical error* is the lateral displacement error of the aircraft from the intended track given the course deviation error signal generated by the navigation system. For the manually piloted aircraft, it represents the pilot's ability to stay on course based on the displayed cross-track error. Typical 2-standard-deviation allowances for VOR/DME in the en-route environment are 1.9 deg for the VOR ground equipment and 3.0 deg for the airborne receiver. DME ground station error allowance is 0.1 nmi with the avionics allowed 0.5 nmi or 3% of range, whichever is greater. Under the assumption that the aircraft is piloted manually, the cross-track component of the flight technical error allowance is 2.0 nmi. The total error in positioning the aircraft on the intended route is derived by root-sum-squaring these errors, treated as independent [9].

As illustrated by the example in Figure 14.2*a*, VOR airways are divided into primary and secondary obstacle clearance areas. The *primary obstacle clearance area* is 4 nmi either side of the route centerline except that the area expands along system accuracy lines drawn at a 4.5-deg angle either side of the center line when the COP is more than 51 nmi from one of the facilities [10]. The *secondary obstacle clearance area* extends 2 nmi either side of the primary area except where the COP is more than 51 nmi from the facility, in which case the secondary area is bounded by a 6.7-deg angle drawn either side of the center line.

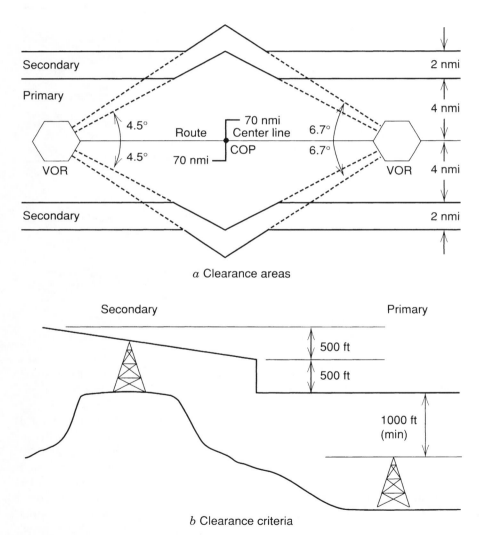

a Clearance areas

b Clearance criteria

Figure 14.2. Primary and secondary obstacle clearance areas for an airway.

Along each airway segment between facilities, a *minimum en-route altitude* (MEA) is established to ensure adequate clearance of obstacles and terrain in both the primary and secondary obstacle clearance areas. In nonmountainous areas, obstacle clearance in the primary area must be at least 1000 ft. The secondary area is raised above and slopes upward and outward from the primary area, as shown in Figure 14.2*b*. No obstacle may protrude above the secondary area. The MEA is set high enough to ensure adequate navigation signal reception and air-to-ground communications coverage. Aircraft operating IFR are assigned altitudes at or above the MEA.

14.3.2 Random Routes

Many aircraft have *area navigation* (RNAV) capability, also known as *random route navigation*. The on-board navigation system determines the position of the aircraft and a navigation computer, often embedded in the flight management system (see Section 14.5.7), carries out the necessary course computations for reaching the next waypoint according to the principles described in Chapter 2. (A *waypoint* is simply a geographically fixed position identified by its latitude and longitude.) Aircraft with RNAV capability are not constrained to travel directly toward or away from ground-based VORs as are aircraft flying along airways. Position determination may be based on station-referenced navigation signals such as VOR, DME, Loran-C, Omega, and GNSS (Chapters 4 and 5) or a self-contained capability, such as an inertial reference unit (Chapter 7). Random routes generally reduce flight distances between origins and destinations as compared to operation along the associated victor airways and jet routes. In addition, they serve to disperse traffic geographically, thereby reducing traffic congestion and increasing the traffic capacity of the airspace.

Procedures for constructing random routes that meet obstacle clearance requirements for IFR operations are provided in FAA Advisory Circular 90-45 [9]. Routes are defined based on charted waypoints, which are assigned five-letter pronounceable names. Primary and secondary obstacle clearance areas are established around the route center line. Then minimum en-route altitudes (MEAs) are established using the obstacle clearance criteria applied to airways. The lateral dimensions of the obstacle clearance areas depend upon the accuracy of the navigation system used. In the case of VOR/DME, dimensions depend upon the perpendicular distance of the facility from the route center line and the distance along the route from that tangent point (see Figure 14.3).

14.3.3 Separation Standards

There are three dimensions of separation—lateral, longitudinal, and vertical—and three sets of rules for ensuring that adequate separation exists—procedural, radar, and visual. *Procedural separation* is based on aircraft following pre-planned flight trajectories so that all aircraft in the region will be adequately separated. Flight progress and adherence to flight plans are monitored by con-

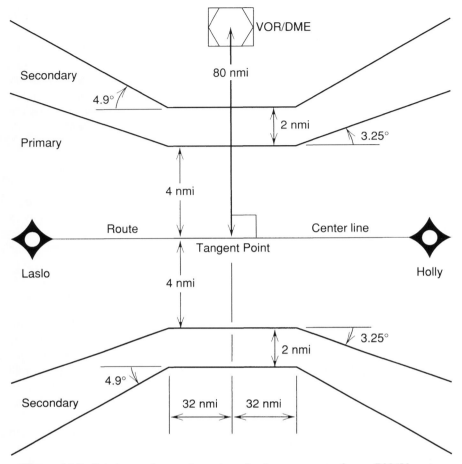

Figure 14.3. Primary and secondary obstacle clearance areas for an RNAV route.

trollers based on position reports received from flight crews. In the future, these position reports may be provided by the automatic dependent surveillance function (Section 14.5.3). Lateral and longitudinal separation standards (minimums) are based on the accuracies of the navigation systems in the aircraft. The minimums are lower along routes based on VOR/DME than over oceans, where less accurate inertial reference units (IRUs) and Omega have been the only means for navigation until the introduction of GPS (Section 5.5) on some aircraft. For example, aircraft procedurally separated at the same altitude and speed along the same airway or RNAV route using VOR/DME must be spaced at least 20 nmi apart [11 (Chapter 6)]. In oceanic airspace, aircraft at the same altitude on the same route are commonly separated by 10-min intervals that correspond to about 80 nmi for aircraft traveling at M 0.85 at FL350. Similar results apply to separation minimums for aircraft with crossing flight trajectories and for lateral separation of routes and flight trajectories.

Radar separation standards depend upon the accuracy of the radar position report displayed to the controller and hence the distance of the aircraft from the radar ground station. In the United States, the radar separation standard is 3 nmi in terminal areas when the traffic is within 40 nmi of the radar site and 5 nmi for traffic beyond 40 nmi. In the en-route environment, it is 5 nm below FL600 and 10 nmi above [11, Chapter 5].

Vertical separation is applied by fixed rules for IFR aircraft operating in uncontrolled airspace and for VFR aircraft [8, Chapter 3]. For example, VFR aircraft cruising on magnetic courses between zero and 179 deg, more than 3000 ft above the surface but below 18,000 ft MSL, fly at odd thousands (MSL) plus 500 ft (e.g., at 3500 ft, 5500 ft, and 7500 ft MSL). Opposite direction VFR traffic cruises 1000 ft away (e.g., at 4500 ft, 6500 ft, and 8500 ft MSL). IFR traffic in controlled airspace operates at altitudes or flight levels assigned by ATM. These altitudes are integer thousands of feet (e.g., 4000 ft, 12,000 ft, FL310) with a 1000-ft separation minimum up to and including FL290 and a 2000-ft separation standard above FL290. In 1988, ICAO concluded that a 1000-ft vertical separation standard can be established worldwide between FL290 and FL410 without imposing unreasonable technical requirements on aircraft. Planning was initiated at that time to implement the 1000-ft standard in heavily traveled airspace over the North Atlantic by 1996 [21].

Visual separation may be applied to IFR aircraft by a controller who is in communication with at least one of the aircraft and who can see both aircraft and maintain visual separation or at least is advised by one of the pilots that s/he can see the other aircraft and accepts responsibility for maintaining visual separation. Because visual procedures effectively reduce separation standards, they can expedite the flow of traffic and increase both the capacity of the airspace and the flexibility of flight operations. They are commonly used in terminal airspace, for example, in sequencing and spacing arrivals. They have less application in the en-route environment where aircraft relative speeds are high.

14.3.4 Terminal Instrument Procedures

When meteorological conditions permit VFR operations (i.e., *visual meteorological conditions*, VMC, prevail), flight crews may be able to make *transitions* from en-route airspace to terminal areas and runways visually. Alternatively, a crew may conduct an instrument approach procedure to guide the aircraft to the runway environment while ensuring safe clearances above terrain and obstacles. The *runway environment* consists of the runway threshold and approved lighting aids and markings identifiable with the runway (Chapter 13). Similarly, departures in VMC may be conducted visually. When meteorological conditions do not permit the visibility required for VFR, *instrument meteorological conditions* (IMC) prevail. In IMC, or when elected by the flight crew, departures are conducted in accordance with prescribed instrument departure procedures.

Instrument procedures are designed specifically for the associated runway end, and then charted and distributed to the user community [5, 12, 28, 36]. The criteria for designing instrument procedures are provided by TERPS [10].

An *instrument approach procedure* describes the route the aircraft should follow (the procedure course) and the altitudes to be flown. The route and altitudes are referenced to the navigation aids supporting the procedure, which may include VOR, VOR/DME, TACAN, VORTAC, NDB, Loran-C, IRUs, ILS, the localizer-only portion of ILS, MLS, GNSS, precision approach radar, and airport surveillance radar. Radar serves as a navigation aid when a controller uses the system to monitor the position of an aircraft relative to obstacles, terrain, and the runway and issues heading, speed, and/or altitude instructions to guide aircraft to the runway environment. Instrument approach procedures also state *landing minimums*, which have two parts—the minimum altitude to which the aircraft can descend without sighting elements of the runway environment and the minimum visibility on the runway surface permitted when conducting the procedure, e.g., visibility required to be at least 1 smi (Chapter 13).

The following description of a simple instrument approach procedure illustrates the essential considerations in their construction [5, 28]. As shown in Figures 14.4 and 14.5, the procedure has four segments. The *initial approach*

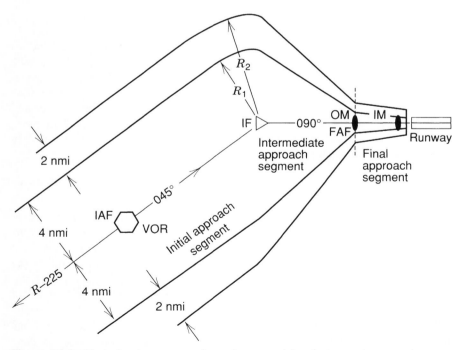

Figure 14.4. Obstacle clearance surfaces for a precision instrument approach procedure.

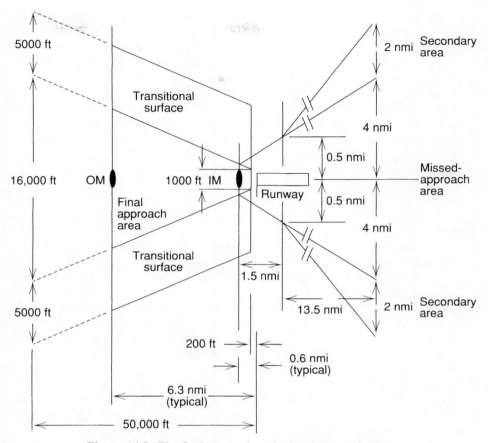

Figure 14.5. The final approach and missed approach areas.

segment commences at the *initial approach fix* (IAF). In Figure 14.4, this segment derives positive course guidance from a VOR ground station. VOR system inaccuracies, including flight technical error, require a primary obstacle clearance area extending 4 nmi on each side of the course and a secondary obstacle clearance area extending laterally 2 nmi on each side of the primary area. If any portion of the segment is more than 51 nmi from the VOR, the obstacle clearance areas must be expanded in accordance with the en-route VOR airway criteria described earlier (Figure 14.2). Along the initial approach segment, the obstacle clearance in the primary area must be at least 1000 ft. In the secondary area, the clearance must be at least 500 ft at the inner edge tapering uniformily to zero at the outer edge.

The *intermediate segment* begins at the *intermediate fix* (IF) and extends to the *final approach fix* (FAF). For an ILS approach procedure, the IF is defined by the intersection of the initial approach course and the localizer course.

Along the intermediate approach segment, aircraft make speed and configuration changes in preparation for the final approach. The length of the segment is between 5 and 15 nmi, and the obstacle clearance areas have the same widths as the areas of the adjoining segments. The minimum obstacle clearance in the primary area is 500 ft. In the secondary area, 500 ft of clearance must be provided at the inner edge tapering to zero at the outer edge.

The *final approach segment* shown in Figures 14.4 and 14.5 assumes a Category I ILS installation consisting of localizer and glide-slope stations of Category I quality and outer and inner marker beacons. (See Chapter 13 for detailed descriptions of the ILS equipment.) The outer marker (OM) serves as the FAF and is located at the point where an aircraft at the prescribed altitude at the end of the intermediate approach segment intercepts the glide slope. The length of the final approach course is dictated by the altitude prescribed at the FAF and the glide-slope angle, which is usually 3 deg. The course generally is less than 50,000 ft long.

As shown in Figure 14.5, the width of the final approach area is 1000 ft at a point 200 ft from the threshold and expands uniformly to a width of 16,000 ft at a point 50,000 ft from the beginning of the area. The final approach obstacle clearance surface is an inclined plane which originates at the runway threshold elevation 975 ft down the runway from the glide-slope ground point of intercept and overlies the final approach area. (The *glide-slope ground point of intercept* is the point on the runway center line at which the straight-line extension of the glide slope intercepts the runway surface.) The slope of the final approach obstacle clearance surface is a function of the glide-slope angle. Transitional surfaces for ILS Category I are inclined planes with a slope of 7 : 1 extending outward and upward from the final approach obstacle clearance surface. The final approach obstacle clearance criterion increases with distance from the runway threshold. (Chapter 13 covers categories II and III).

Aircraft approaching the runway using an ILS procedure may descend to the *decision height* without visually acquiring the runway environment. If the runway is not acquired at the decision height, the pilot must conduct a *missed approach procedure*. For Category I ILS approaches, the decision height may be no lower than 200 ft above the touchdown zone. If obstacles protrude through the final approach obstacle clearance surface, either the glide-slope ground point of intercept must be moved farther down the runway until the protrusions disappear, which reduces the usable length of the runway in IMC, or the decision height must be raised for adequate clearance. Raising the decision height reduces the utility of the runway in IMC.

The final portion of the approach procedure is the *missed approach segment*. It begins at the *missed approach point* (MAP), which is the point on the final approach course where the height of the glide slope above the touchdown zone equals the decision height. The inner marker (IM) generally is located near the MAP. The missed approach segment extends to a point or fix where either initial approach or en-route obstacle clearance is achieved. A straight missed approach course is 15 nmi long, and the missed approach area is 8 nmi wide at the end.

Obstacle clearance criteria ensure that an aircraft in the missed approach area climbing with a gradient of at least 150 ft/nmi from the MAP will not collide with an obstacle. In addition, obstacles may not penetrate the secondary areas that slope upward at 12 : 1 from the missed approach area.

14.3.5   Standard Instrument Departures and Arrivals

Standard instrument departures (SIDs) and *standard terminal arrival routes* (STARs) are published IFR procedures for transitioning between terminal and en-route airspace. Their purpose is to assure obstacle clearance for departing and arriving aircraft. The tracks prescribed may extend 100 to 200 nmi from the airport. A typical SID will show the track to be followed from the airport referenced to appropriate radio navigation facilities and the minimum altitudes at which the various segments may be flown [28]. The outlying track segments may provide more than one transition path to the en-route airway structure. In some cases, STARs terminate at initial approach fixes and approaches can be made without vectoring. (A *vector* is a heading instruction issued by a controller to provide navigation guidance, e.g., "Turn right to heading 090 degrees.")

14.4   PHASES OF FLIGHT

An aircraft transitions through a number of flight phases as it travels from the airport surface through the terminal airspace to the en-route environment and, finally, the destination terminal airspace and airport. *Terminal airspace* is the volume of airspace generally extending from the ground to 10,000 ft MSL to a distance of 30 or 40 nmi from the airport. Responsibility for air traffic management in this airspace is assigned to an *approach control facility*. There are five principal areas of control responsibility within the approach control facility. The *clearance delivery controller* provides the predeparture clearance to the pilot prior to the aircraft leaving the gate or parking area. The *ground controller* ensures the safe and expeditious movement of the aircraft from the gate to the runway for takeoff. Takeoff and the first few miles of the flight are under the jurisdiction of the *local controller*, after which the *departure controller* is responsible for the remainder of the flight leaving the terminal area. As the aircraft approaches the destination terminal area, an *approach controller* assumes responsibility for the flight. The clearance delivery, ground, and local controllers are in the tower cab from which they visually monitor aircraft movements on and near the airport. At major airports serviced by radar, the departure and approach controllers are in a radar approach control facility (TRACON, formerly RAPCON), which is often located inside the tower building below the cab. In 1996 in the U.S., regional TRACONS were being established to serve many airports.

Control responsibilities are assigned geographically, and each area, or *sector*, is assigned a unique VHF frequency for controller-pilot radio voice communications. The ground controller is responsible for the entire movement area of the airport except for the runways actively in use for takeoffs and landings. These *active runways* are the responsibility of the local controller, who also has control over small arrival and departure areas off the ends of the active runways under his/her jurisdiction. The remainder of the terminal airspace is divided into *departure and arrival sectors* with a controller or controller team assigned to each. Terminal areas are embedded in *en-route airspace*, also subdivided into sectors. In the en-route environment, *low-altitude sectors* extend from the floor of controlled airspace to 18,000 ft MSL, *high-altitude sectors* extend from FL180 to FL350, and *super-high-altitude sectors* are established from FL350 to FL600 [37]. In airspace surrounding major terminals, the en-route airspace may include arrival and departure sectors that interface with terminal arrival and departure sectors respectively. In the United States, en-route airspace is managed by controllers in air route traffic control centers (Sections 14.4.4 and 14.6.2).

14.4.1   Pre-flight Planning

Prior to departure, the flight crew requires a weather briefing describing current and forecast weather along the intended route of flight as well as at the departure, destination and alternate airports. Weather information of interest includes ceilings, visibilities, surface winds, winds aloft, turbulence, icing conditions, and storm activities. Information is also required regarding relevant air navigation facility outages, status of relevant special use airspace, and anticipated traffic congestion and delays. Private pilots and flight crews of business aircraft obtain this preflight briefing information from FAA *flight service stations* (FSSs) or from commercial flight planning services. Major airlines brief their crews in dedicated flight operations centers at the origin airport. The flight crew then develops the *flight plan* that states whether the operation will be IFR or VFR and provides the aircraft identification number (e.g., N446L or Central Airlines Flight 242); the aircraft make and model (e.g., Boeing 737-400); the planned true airspeed and cruising altitude; the origin, destination, and alternate airports; the planned departure time and estimated time en route; the planned route of flight including SIDs and STARs; the fuel on board (e.g., 5 hr, 40 min); the number of people on board; the aircraft color; and the pilot's name. The navigation equipment installed on the aircraft and the transponder capability (Section 14.5.2) are shown as well. The *aircraft identification number* is used by pilots and controllers as the call sign in radio communications. The description of the intended route of flight must be complete and unambiguous [8].

IFR flight plans are filed with FAA through FSSs, air route traffic control centers, towers, and commercial flight planning service providers. The FAA reviews each of the flight plans, amends it if necessary to incorporate ATM constraints (e.g., to route the aircraft around special use airspace), and stores the amended flight plan until 30 minutes prior to the planned departure time. At that time, the amended flight plan in the form of the predeparture clearance is sent to the clearance delivery controller in the tower at the airport from which the flight will originate. While flight plans for VFR flights are not required, they are strongly recommended. Their principal purpose is to notify the FAA of the flight so that search-and-rescue procedures can be initiated in the event the aircraft is overdue at its destination.

14.4.2   Departure

Flight crews operating IFR call the clearance delivery controller on the radio approximately 10 minutes prior to their planned taxi times. The controller issues the *predeparture clearance* including the departure procedure or SID, the route of flight and assigned altitude, the radio frequency for contacting the departure controller, and the assigned aircraft transponder code (Section 14.5.2). The flight crew next tunes a VHF receiver to the Automatic Terminal Information Service (ATIS) broadcast frequency to obtain noncontrol information, including ceiling and visibility, temperature, dew point, wind direction and velocity, the altimeter setting, the runways in use, and any facility outages affecting airport operations. The crew then contacts the ground controller for a taxi clearance. The local controller clears the flight onto the departure runway, clears the aircraft for takeoff, and, before the aircraft enters departure airspace, instructs the crew to contact the departure controller.

FAA standard takeoff visibility minimums are 1 smi for aircraft having two engines or less and $\frac{1}{2}$ smi for aircraft having more than two engines. Site-specific *instrument departure procedures* and ceiling/visibility minimums are established at some airports to assist pilots in avoiding terrain and obstacles during climb to the minimum en-route altitude (MEA). These minimums and procedures are published in *U.S. Terminal Procedures* [36]. During an instrument departure, aircraft are required to climb at a rate of at least 200 ft/nmi, cross the departure end of the runway at least 35 ft above ground level, and climb at least 400 ft above the airport elevation before turning. Based on this performance, obstacles to be overflown must be below a clearance plane sloping upward at 152 ft/nmi starting no higher than 35 ft above the departure end of the runway. This assures obstacle clearance not less than 48 ft/nmi of flight. Where obstacles meet the above criteria, specific IFR departure procedures are not required. Otherwise, site-specific procedures are developed and may include ceiling/visibility requirements to allow obstacles to be seen and avoided, climb gradients greater than 200 ft/nmi, and/or specified turning and/or climbing maneuvers. In extreme cases, takeoffs in IMC may not be authorized.

At airports with little traffic and no obstacles, it may be possible for the flight crew to takeoff, climb to 400 ft on the runway heading, and then turn to the cleared route of flight while climbing to the assigned en-route altitude. In busy terminal areas, the parsing of airspace into sectors with separate controller teams assigned to each and the need to segregate departure and arrival traffic flows impose restrictions on the flight paths available to departing aircraft.

14.4.3   En Route

An aircraft is under the control of one and only one controller at any one time. As the aircraft moves from one sector to another, control responsibility is handed off from the transferring controller to the receiving controller. The *hand-off process* involves three principal functions, namely, identification of the aircraft to be handed off, transfer of pilot communications and control responsibility to the receiving controller and verification of the aircraft's automatic pressure altitude report. The aircraft's transponder provides altitude reports in response to interrogations from the ground-based surveillance radar system (Section 14.5.2), and this information is displayed to controllers (Section 14.5.6). If the aircraft's altitude shown on the receiving controller's display differs from the altitude reported by the pilot over the radio (based on the pilot's altimeter reading) by more than 300 ft, the controller cannot use the data for controlling vertical separations relative to that aircraft and will instruct the pilot to switch off the altitude reporting capability of the transponder.

As the aircraft proceeds through en-route airspace, the responsible controller monitors the flight to assure separation from other aircraft and from special use airspace and to assure that the flight conforms to its flight plan as well as applicable flow instructions. *Flow instructions* place limitations on the rate at which traffic can enter saturated fixes, downstream sectors and airports. Traffic managers issue these instructions to controllers who then adjust aircraft trajectories by using vectors and speed instructions, by holding aircraft and, in some cases, by diverting (rerouting) aircraft to ensure that the instructions are followed. *Holding* is a procedure that keeps an aircraft within a specified airspace (generally in a pattern shaped like a racetrack and referenced to a fix) while awaiting further clearance from ATM.

Pilot-controller communications are restricted to matters related to management and control of aircraft including controller instructions (e.g., vectors for maintaining separations), pilot requests for local flight path modifications (e.g., around severe weather), and traffic advisories. *Traffic advisories* are issued to flight crews to alert them to other aircraft in their immediate vicinity. The advisory includes the direction in which the pilot should look to see the other aircraft, the distance to the aircraft, the direction in which it is flying, and the aircraft altitude and type. The flight crew is responsible to visually search for the aircraft, to notify the controller when the traffic is in sight, and to maintain visual separation.

Arrival metering programs have been established for some major airports. The objective is to match the arriving traffic flow to the airport's acceptance rate. If aircraft must be delayed, the goal is to delay them en route with speed reductions, a strategy that is more fuel efficient than holding. Metering uses a software program to predict the arrival flow rate and, when the rate is predicted to exceed the acceptance rate, to assign metering fix crossing times to individual aircraft to reduce the rate. *Metering fixes* are inbound fixes in en-route airspace close to the boundary of terminal airspace. Metering fix crossing times are displayed to the responsible en-route controller who then issues instructions to adjust the flight paths and speeds of individual aircraft to meet their assigned times, usually within ±1 minute.

14.4.4 Approach and Landing

In 1996 terminal airspace is frequently organized around *arrival fixes* (*corner posts*), which may serve as metering fixes, at the vertices of a square (more generally, a quadrilateral) with the airport near the center. Typically, the fixes are 30 nmi or so from the airport. Arriving IFR aircraft enter the terminal airspace at the corner posts, while departures are routed out of the area through the sides of the quadrilateral. Arriving and departing aircraft follow prescribed routes and altitude profiles to and from the airport. Different routes are established for propeller and turbojet aircraft because they operate at significantly different speeds. This arrangement procedurally separates the traffic, thereby reducing the likelihood of traffic conflicts and simplifying the controllers' tasks. However, the process does not necessarily provide direct routes to and from the airport or fuel-efficient altitudes.

The approach controller's principal task is to sequence, merge, and space the aircraft for landing. When the leading aircraft crosses the approach end of the runway, the following aircraft must be spaced a prescribed minimum distance behind. The separation minima are established to protect the following aircraft from upset due to *wake turbulence* generated by the leading aircraft and depend upon the maximum takeoff weights of the aircraft involved (Table 14.1).

The task of establishing properly spaced landing sequences without encumbering aircraft operations with unnecessary vectors and speed adjustments is demanding in heavy traffic conditions. In vectoring an aircraft off its route of flight, a controller assumes responsibility for its navigation including clearance from obstacles and terrain. To ensure adequate clearances, *minimum vectoring altitudes* are established within terminal areas, and IFR aircraft are maintained above these altitudes as they approach the airport.

Aircraft noise is the principal factor limiting airport capacity. At some air-

TABLE 14.1 Wake turbulence separations for landing

Following aircraft Category	Leading Aircraft Category		
	Small	Large	Heavy
Small	3 nmi	4 nmi	6 nmi
Large	3	3	5
Heavy	3	3	4

Note. Maximum takeoff weight, *small:* <12,500 lb; *large:* 12,500–300,000 lb; *heavy:* >300,000 lb.

ports, specific noise abatement routes have been defined (e.g., along rivers) to channel the noise where relatively few people will be affected. These routes can be complex (require a number of turns within restricted geographical areas), and it may not be feasible to site navigation aids to support them. Hence, some of these routes are flown with visual reference to the ground and published as *charted visual procedures*.

The approach controller generally hands off traffic to the local controller when it has been established on the final approach course. The local controller hands off to the ground controller as the aircraft taxis off the active runway.

The flight crew's principal responsibilities during approach and landing are to navigate the cleared route of flight, respond to controller instructions, maintain visual separation from other aircraft, and manage the aircraft configuration (flaps and gear) and dynamics (speed, altitude, and rate of descent) for landing. Entry into the terminal area may be along the STAR requested in the filed flight plan, and the aircraft may then be cleared for an instrument, visual, or charted visual approach, with or without vectoring for the transition. Prior to entering the terminal airspace, the crew listens to ATIS communications for essential information on airport and navigation aid conditions.

14.4.5 Oceanic

The ATM processes described in the preceding sections are based on controllers having reliable, accurate radar surveillance information and nearly instantaneous voice communications with flight crews. With some exceptions, neither of these conditions prevailed in oceanic airspace in 1996. Flight crews generated *position reports* by determining their positions using inertial reference units having drift rates of 1 to 2 nmi/hr and communicated these reports to controllers when over mandatory *reporting points* spaced every 500 nmi along the route of flight. Air-to-ground communication for periodic position reporting and control instructions in 1996 was dependent upon relatively unreliable high-frequency (HF) voice links with contacts sometimes requiring many minutes to establish. The consequence of the relatively inaccurate navigation, cumbersome communication, and little real-time surveillance was large separation standards that poorly utilized the available airspace. By 1995, successful experiments had been accomplished using air-to-ground data communication via geostationary satellites for the transmission of aircraft derived position and velocity data (either from inertial navigation systems or GPS receivers) to achieve satellite-based automatic dependent surveillance (ADS) (Section 14.5.3). A few transoceanic airlines had implemented the required equipment (called FANS 1) to use this technique on a regular basis on commercial flights in 1996.

To organize traffic so that controllers can effectively ensure separation, *track systems* are established in heavily traveled oceanic areas (e.g., over the North Atlantic). A *track system* is a set of nearly parallel routes, each joining oceanic gateways (fixes) on the two sides of the ocean. Some of the track systems are *flexible* in that they are adjusted twice daily to provide fuel-efficient routing matched to prevailing wind and weather conditions (*pressure pattern naviga-tion*). Overland en-route airways carrying traffic to and from the ocean connect with the oceanic routes at the gateways. The overland route segments are within radar coverage and are used for sequencing, merging, and spacing the traffic into oceanic airspace. Track systems, even with flexible tracks, are inefficient in terms of aircraft operations. The ultimate objective is to achieve RNAV routing, with each aircraft following the route, speed and altitude profiles best adapted to its operating requirements. It is expected that GNSS (Section 14.5.1 and Chapter 5), automatic dependent surveillance (Section 14.5.3) and satellite communications (Section 14.5.4) will facilitate widespread use of RNAV routing in oceanic airspace.

14.5 SUBSYSTEMS

The ATM system is a complex interconnection of communications, navigation, surveillance, aviation weather, and automation subsystems [13, 39]. They interact with corresponding aircraft subsystems, in some cases automatically via RF data links and in others manually based on pilot-controller voice communications and manual data entry. In the future, automatic data link communications will increasingly be used to reduce the work load and the potential for errors inherent in voice communications.

The voice and data communications network that interconnects U.S. traffic control centers, radars, etc. is operated by the FAA is second in size only to that operated by the U.S. Department of Defense. The network carries weather data, flight plans, facility control and maintenance information, traffic flow management and control instructions, surveillance data, and *notices to airmen* (NOTAMs) [13, 39]. A NOTAM provides current information for flight operators concerning the establishment, condition, or change in any component (facility, service, procedure, or hazard) in the system. The interfacility communications network interfaces with international ATM networks (e.g., to exchange international flight plans) using protocols established by ICAO. It interconnects with other U.S. government agencies to coordinate aircraft operations and exchange data (e.g., weather information).

14.5.1 Navigation

In 1996, only VOR/DME and VORTAC were accepted as sole means systems for en-route and terminal navigation under instrument flight rules. A *sole means air navigation system* is a system approved for specific phases of air navigation without the need for any other navigation system in the aircraft. A *supplemental air navigation system* can be used to enhance navigation performance, but the aircraft must also carry and use the sole means system approved for the particular phase of flight. In 1996, IRUs (Chapter 7) and Omega (Chapter 4) were accepted internationally as sole means navigation systems over oceans and GPS was accepted as a sole means navigation system in U.S. oceanic flight information regions (Section 14.6.6).

In 1996, guidance for precision approaches and landings was primarily provided by ILS and MLS (Chapter 13). NDBs, localizers, VOR/DME, GPS, VORTAC, Loran-C (Chapter 4), and IRUs (Chapter 7) are approved for nonprecision approaches to appropriate minimums. Precision and nonprecision *radar approach procedures* had been established. During radar approaches, flight path guidance instructions are transmitted to pilots by controllers monitoring the aircraft positions relative to prescribed approach paths. A *precision approach procedure* provides the pilot both route and height (glidepath) guidance to the decision height. A *nonprecision approach procedure* provides only route guidance; the pilot controls the altitude of the aircraft based on the altimeter reading and the charted procedure. Decision heights are lower for precision approaches because the height of the aircraft relative to terrain and obstructions can be controlled more accurately.

The development of navigation systems to meet civil and military requirements is a subject of intensive planning among users and providers of navigation services. In the United States, a principal product is the comprehensive Federal Radionavigation Plan (FRP) published biannually [59]. In 1996, two major development efforts were underway, one focused on establishing and implementing the concept of required navigation performance (RNP) and the second focused on the global navigation satellite system (GNSS). In contrast, the method most commonly used at that time to prescribe the minimum navigation performance required of aircraft operating in a given airspace was to require the installation of specific equipment. These requirements were often subject to international agreement and hence cumbersome to change. This situation constrained the effective application of new navigation technology and imposed heavy work loads on standards organizations, especially ICAO, trying to keep pace with technology improvements. Initially, the GNSS was represented by the U.S. Global Positioning System (GPS, Section 5.5). In the future, the Russian Global Orbiting Satellite System (GLONASS, Section 5.6) is expected to be part of the GNSS. The emergence of the GNSS has lead to the RNP approach.

RNP is a statement of the navigation accuracy required for operation within a defined airspace [22]. The navigation accuracy prescribed is the total *system use accuracy* (Section 2.7), including ground-station or space-station equipment errors, aircraft receiver and display errors, and flight technical errors. The RNP concept will be applied to all phases of flight, including approach and landing. Providers of air traffic services, in collaboration with airspace users, will determine the level of navigation performance required to ensure the safety and efficiency of flight operations in a given airspace and will specify the corresponding RNP. To simplify this process for airspace planners, manufacturers, and operators, ICAO is standardizing the RNP types. In 1996, four types were proposed, RNP 1, RNP 4, RNP 12.6, and RNP 20. The figure in each RNP type designation is the accuracy required on a 95% probability basis. For example, aircraft in RNP 4 airspace are required to be within 4 nmi of their intended positions in the horizontal plane 95% of the time. RNP 1 is expected to support operations along routes where the most accurate positioning of aircraft is required, including transitions to and from airports and operations along parallel offsets. *Parallel offsets* are routes offset laterally from the planned route center line by a specified distance. Operations along such offsets may be advantageous for passing slower traffic, for increasing airspace capacity and/or for facilitating uninterrupted climbs and descents where there is conflicting traffic on the planned route. RNP 4 corresponds to VOR navigation practices (Section 14.3.1), and RNP 12.6 corresponds to the nominal capabilities of inertial reference units on long oceanic flights. RNP 20 is the minimum capability considered acceptable to support en route traffic operations. It is the responsibility of the aircraft operator and the associated national airworthiness authority (the FAA in the United States) to ensure that an aircraft operating in a specified RNP airspace is equipped to achieve the required navigation performance. It is expected that the RNP concept will foster the effective application of new navigation technologies as they become available.

In 1996, the civil aviation community is moving rapidly to establish and exploit the two global navigation satellite systems, namely, GPS and GLONASS. After GPS achieved initial operational capability in late 1993, aircraft in the United States equipped with the approved avionics [15] were authorized to use GPS on a supplemental basis for some phases of flight. The principal issues in relying solely on GPS for route navigation are the *availability* of the system (i.e., the percentage of the time GPS will provide a sufficiently accurate navigation service) and the *integrity* of the system to provide timely warnings of satellite or system failures. In 1996, the development of the FAA's Wide Area Augmentation System (WAAS) (Section 5.7.3) was under way, with the aim of providing the required availability, integrity and accuracy for certain flight operations, including nonprecision and Category I precision approach (Section 13.8). In 1995, GPS augmented by local differential GPS (DGPS) corrections satisfied the requirements for so-called DGPS Special (private use) Category I approaches (Section 5.5.9). A number of issues pertain to the use of differential GPS for Category II and III approaches, including the practicality and cost effectiveness of augmentations to provide the necessary accuracy and integrity (Section 13.8); however, experimental test results obtained in 1995 were extremely promising.

Separation Standards   Separation standards are determined by the density and complexity of traffic movements, the aircraft's navigation system performance and the communication, surveillance and automation capabilities that permit controllers to intervene in order to modify flight paths. Separation standards, in turn, determine the amount of airspace that must be protected for each route and aircraft. Efficient, flexible use of airspace is fostered by reduced separation standards and application of RNAV capabilities that permit operators to fly point to point as opposed to constraining their operations to fixed routes and airways. GNSS is expected to provide the accurate, ubiquitous RNAV capability required. The international airline community has estimated that a fully implemented GNSS will reduce their costs by $5 billion annually.

14.5.2   Radar Surveillance

The radar surveillance subsystem provides controllers information on the three-dimensional positions of aircraft. In ATM, two types of radar systems are used: *Primary radar* (or *search radar*) determines target range, azimuth, and, in some cases, elevation from the signal reflected back by the aircraft skin. The main beam of the ground station antenna pattern is broad in elevation (e.g., 40 deg) and narrow in azimuth (e.g., 2 deg). As the antenna rotates (scans) about its vertical axis, pulses of radio frequency (RF) energy are transmitted, and the signals scattered back by the surface of the aircraft are received. The target range is determined by the round-trip time between transmission and reception (the energy travels at the speed of light) and the azimuth is determined by the pointing angle of the antenna main beam (Figure 14.6). Target elevation can be determined by a second antenna pointed along the target azimuth which scans in elevation (i.e., scans around a horizontal axis). Such *height-finding radars* are not commonly used for civil aviation purposes but do find application in the military (e.g., radars used for controlling precision approaches).

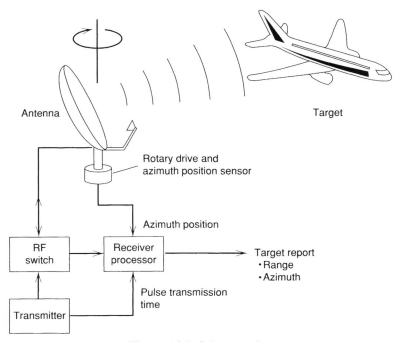

Figure 14.6. Primary radar.

The advantage of primary radar is that accurate target position reports can be generated without any participation on the part of the target aircraft. The disadvantages are that the ground-station equipment is expensive, performance is limited by the target size and the range, and it degrades when the transmitted signal is reflected back by terrain or weather (clutter), obscuring returns from aircraft, and the aircraft cannot be uniquely identified (e.g., as "Central Airlines Flight 34"). Nonetheless, primary radar is a mature technology widely used for ATM [57].

In the *secondary radar* (or *beacon radar*) system, the signal transmitted from the ground-station antenna (commonly mounted on top of a primary radar antenna) initiates the transmission of a reply signal from the *transponder* (receiver/transmitter) in the aircraft. The system essentially provides two-way air-to-ground data communications and operates in several modes [23, 38, 45, 56]. As shown in Figure 14.7, Modes A and C use simple three-pulse, pulse-position modulated (PPM) interrogations with P1 and P3 having equal amplitudes. When the P_1–P_3 interval is 8 μsec (Mode A interrogation), the transponder replies with the *aircraft identification* (ACID) *code*; if the interval is 21 μsec (Mode C), the reply will contain the encoded altitude of the aircraft. At short ranges, the signal strength may be sufficient to interrogate transponders via the antenna side lobes. To avoid this situation, aircraft in the antenna side lobes are prevented from replying through a technique called *transmit side lobe suppression*. The P_2 pulse of the interrogation (Figure 14.7) is transmitted on an omnidirectional antenna at a slightly higher power density than that produced by the antenna side lobes. Transponders are designed to reply only if the received P_1 pulse is greater than the received P_2 pulse. This condition is not satisfied in the side lobes of the antenna. It is essential to suppress transponders in the side lobe regions because their replies would be erroneously displayed at the azimuth corresponding to pointing direction of the mainbeam.

The replies are also PPM signals with 13 data pulses uniformly spaced between 2 "framing pulses" separated by 20.3 μsec. The *special identification* (SPI) *pulse*, when present, follows the last framing pulse in Mode A replies only. In Mode A replies, 12 of the 13 data pulses are used to communicate the ACID code as one of 4096 possible codes. This code is assigned to each IFR aircraft in the route clearance and is set into the transponder by the flight crew. (VFR aircraft transmit the code "1200" to identify the aircraft as not under ATM control.) The controller writes the assigned ACID code into the flight plan. Ground automation equipment correlates ACID codes with flight plans as Mode A replies are received from the surveillance site and determines the aircraft identification (e.g., Central Airlines Flight 34) corresponding to each. Through this process, the aircraft identification can be displayed in the data tag associated with the aircraft position on the controller's display (Figure 14.12 and Section 14.5.6). To contact the aircraft over VHF, the controller can read the identification directly from the display.

The SPI pulse is inserted into a Mode A reply as a result of the pilot depressing the *IDENT* button on the transponder control panel. Ground-based automa-

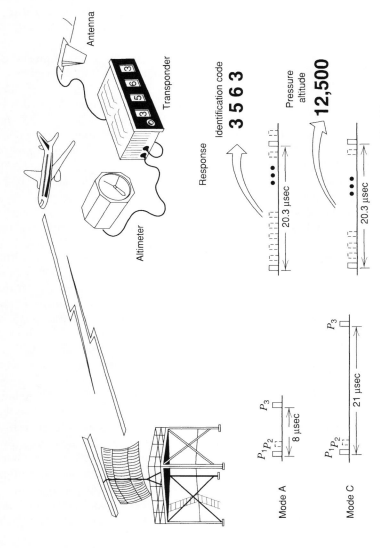

Figure 14.7. Secondary radar Mode A and Mode C (interrogation at 1030 MHz; response at 1090 MHz).

tion display equipment will blink the data block of an aircraft from which the SPI pulse is being received. Controllers will ask pilots to *IDENT* during hand-offs and at other times when it is necessary to quickly and unambiguously locate aircraft on their displays.

In Mode C replies, the 12 data pulses provide the pressure altitude of the aircraft in 100-ft increments from -1000 ft to 126,700 ft. All Mode C altitude reports are referenced to 29.92 in. of mercury.

Mode S has been added to reduce or eliminate some of the performance lim-itations of Modes A and C. For example, with Mode A, at some major airports two or more aircraft with the same ACID code may be operating in the area at the same time and the ground automation will not be able to identify these aircraft unambiguously. In addition, when two aircraft are in the antenna main beam simultaneously and close in range (e.g., one above the other), their replies will overlap and obscure one another. Mode S uses more sophisticated inter-rogation and reply modulation wave forms (Figure 14.8 and Section 14.5.4), permitting discrete addressing of aircraft. Data blocks have 56 bits in the short format and 112 bits in the long format. Error detection coding ensures unde-tected message error rates of less than 1 in 10^7. A unique 24-bit Mode S address is assigned to each aircraft so that aircraft can be unambiguously identified and addressed worldwide. There are more than 16 million available Mode S addresses with specific blocks assigned to every country by ICAO. A ground station with Mode S capability addresses each Mode S capable transponder by its unique address thereby establishing a *private-line* communications service. By managing interrogations properly in time, the ground station can ensure that replies do not overlap. The *private-line* capability of Mode S supports general-purpose data link communications capability for which a number of applications were under development in 1996 (Section 14.5.4). All secondary radar ground stations and aircraft transponders are capable of operating in Modes A and C. Where Mode S is added, Modes A and C are retained in the equipment to ensure that the equipment can operate compatibly with equipment having only Modes A and C.

Given a series of surveillance reports from a target aircraft, ground automa-tion equipment estimates the target velocity over the ground by tracking the reports. In the most elementary form, the estimated velocity vector is in the direction joining the two most recent reports with a magnitude equal to the distance between the reports divided by the time interval between them. U.S. Federal Aviation Regulations require the installation of transponders with spec-ified capabilities in various airspace regimes in order to support the ATM ser-vices provided and to ensure safety.

14.5.3   Automatic Dependent Surveillance

As shown in Figure 14.9, *automatic dependent surveillance* (ADS) is a func-tion that automatically transmits, via a data link, position data derived from the on-board navigation system [2, 46]. The purpose of ADS is to provide real-time

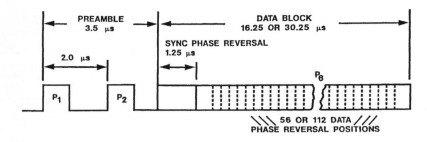

DIFFERENTIAL PHASE SHIFT KEYING (DPSK) MODULATION
DATA RATE 4 Mb/s

INTERROGATION WAVEFORM

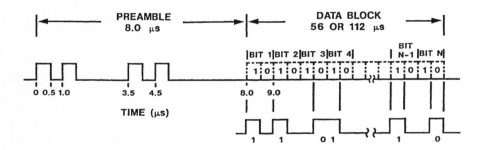

PULSE POSITION MODULATION (PPM)
DATA RATE 1 Mb/s

REPLY WAVEFORM

Figure 14.8. Mode S waveforms.

surveillance information to ATM authorities and aircraft operator dispatch facilities. Surveillance is *dependent* in this technique because its operation and quality depend upon the performance of the aircraft's navigation system. According to the ICAO definition [24, 30], as a minimum, the ADS data includes time of day, aircraft identification, and three-dimensional position information. Additional information reported may include aircraft speed and direction, the waypoints stored in the navigation system and meteorological data (static air temperature and velocity of winds aloft). The route data, including the next waypoint and the (next +1) waypoint with the corresponding planned altitudes,

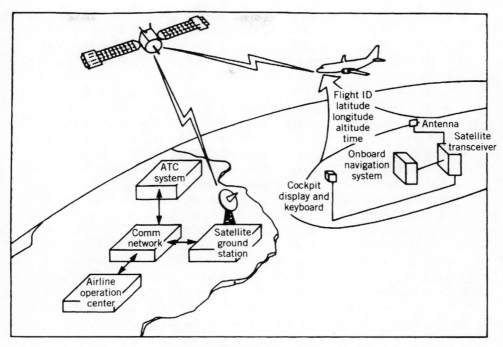

Figure 14.9. ADS concept.

enable controllers to check conformance with their flight plans for the aircraft. The interval between ADS reports can be controlled by the controller responsible for the flight. Figure 14.9 depicts the use of a geostationary communication satellite as a relay data link medium, which is typical for overocean operations. Alternatively, over land areas, a VHF or L-band direct data link from the aircraft can be used. In 1996, the initial application of ADS was in oceanic airspace, with reporting intervals ranging between 30 sec and 5 min. Over land areas, the reporting intervals are expected to range from 10 sec (or more) in en-route airspace to 4 sec in terminal areas.

14.5.4   Air-to-Ground Data Link Communications

The performance of the ATM system is critically dependent upon the quality of the information available within cockpits and within ATM and aircraft dispatch and flight planning facilities. Much of the information required at any of these sites is collected and processed elsewhere. Therefore, high capacity, reliable air-to-ground communications capabilities are required.

In 1996, air-to-ground information exchange was accomplished largely using VHF voice communications, but it was realized that digital data communications (*data link*) capability is needed to support a number of applications. The development of this data link capability was well under way in 1996 and in use for some applications; for example, for providing predeparture clearances to aircraft. In the future, aeronautical information such as weather data, ATIS, and NOTAMs will be provided on pilot displays and as printed copy in the cockpit to reduce transmission errors and work load. As more automated ATM processes are developed, data link will be required for exchanging flight plans and route clearances as well as to transmit ATC instructions to the cockpit, including altitude and VHF frequency assignments. In addition, data link will be used to provide automatically to flight crews traffic advisories derived from surveillance data resident in ground-based ATC centers, as well as to alert crews when they are about to enter special use airspace or airspace such as Class B (terminal control areas) where operations are closely regulated.

Efforts by ATM authorities to establish data link capability have been bolstered by complementary initiatives among the airlines focused on more efficient fleet scheduling and flight and maintenance operations. Engine performance is monitored automatically on board many aircraft with some parameters automatically transmitted to the ground for fault detection and trend analysis. Flight plans are sent to aircraft and stored automatically in the flight management system (FMS) computer (Section 14.5.7), with position reports, fuel status, estimated arrival times at waypoints, and wind and static air temperature data sent to the ground automatically on request. Data link is a principal tool for fostering improved safety and efficiency of flight operations.

Three data link media have been developed. *VHF data link* uses 25 kHz-wide channels in the frequency band 118–136 MHz [41]. Minimum shift keying (MSK) modulation is used to achieve data rates of 2400 bits/sec. In the future, offset quaternary phase shift keying (OQPSK) modulation techniques may be employed to increase the data rate to 21 kbits/sec. A principal advantage of VHF data link is the low cost of the ground stations and aircraft equipment. While a large number of data channels could be accommodated within the 18 MHz assigned for air-to-ground VHF aeronautical communications, the number available in practice is limited by heavy competition from voice traffic. In 1996, the airlines employed VHF data link for company purposes worldwide, and the FAA is using it to transmit predeparture clearances to appropriately equipped aircraft in order to reduce the work load of tower controllers and expedite departure operations.

By virtue of its discrete address feature, Mode S provides a wideband, general-purpose data link capability [47]. Ground-station interrogations are transmitted in a single 8 MHz-wide channel at 1030 MHz with transponder replies provided in a 2.6 MHz-wide channel at 1090 MHz. Differential phase shift keying (DPSK) modulation is used within interrogations to provide a 4-Mbit/sec data rate. Replies use pulse position modulation (PPM) with a 1-Mbit/sec data rate (Figure 14.8). A principal advantage of the Mode S data link is that it is integral with the surveillance system, which permits ground stations and transponders to efficiently support both functions. In addition, a number of data link applications (e.g., automatic traffic advisories and airspace alerts) are closely related to surveillance functions, and their implementation is therefore simplified when Mode S is used as the data link medium.

Digital data and digitized voice communications capability via *satellite data link* have been implemented to service oceanic regions and continental land areas where the establishment of ground stations is impractical [3, 25, 48, 49]. Satellite-to-aircraft communications operate in the 1.5 to 1.6 GHz frequency band (L-band) with ground-to-satellite links operating in Ku- or C-bands. In 1996, four channel types were used, three operating in a packet-switching mode for data and one in a circuit-switching mode for voice. For data rates between 600 and 2400 bits/sec, binary phase shift keying (BPSK) is used on the satellite-to-aircraft link with quaternary phase shift keying (QPSK) used for data rates from 4800 to 21,000 bits/sec. Voice transmissions are encoded at 9600 bits/sec. The principal advantage of satellite-based data link is that ground stations within line of sight of the aircraft are not required. Worldwide communications with aircraft can be assured between 80-deg North and 80-deg South latitudes using only three or four geostationary satellites and a like number of ground stations servicing those satellites. In practice, a larger number of ground stations (about 10 in 1996) is used to ensure reliability and competition in the provision of services. The principal disadvantage is that service is relatively expensive due to the high costs of satellites in orbit and the associated ground stations and aircraft equipment. In 1996, the principal application of satellite-based data link was ADS and associated oceanic ATM improvements.

A *communications protocol* is a set of rules governing the exchange of data between two entities (e.g., a computer and a display system or two computers). Typically, protocols specify data formats and coding; processes for coordination, acknowledgment, and error handling; and how data are to be sequenced and timed [55]. For air-to-ground data links, there could be a protocol developed for every cockpit entity to communicate with every ground-based entity over each of the three available media. This approach is impractical, because each entity would be required to implement a large number of protocols, and every time an entity was modified or added, every connecting entity would need to be modified to add a new protocol. One alternative is the approach selected for packet data communications within the *Aeronautical Telecommunications Network* (ATN). (See Figure 14.10). Avionics entities and ground-based entities are grouped into subnetworks connected to the three data link subnetworks through *routers*, subnetwork nodes that serve as gateways. A *network* (or *subnetwork*) is a collection of nodes to which communications entities connect, the links joining those nodes, and the protocol governing exchange of data among the nodes (Figure 14.11). Within the ATN, the avionics subnetwork connects the avionics entities and may operate according to a protocol peculiar to the aircraft type or the avionics equipment (Figure 14.10). The data link subnetworks connect any number of aircraft through avionics routers in each to any number of ground subnetworks through ground routers. The ground subnetworks operate in accordance with whatever protocols are best suited to their requirements.

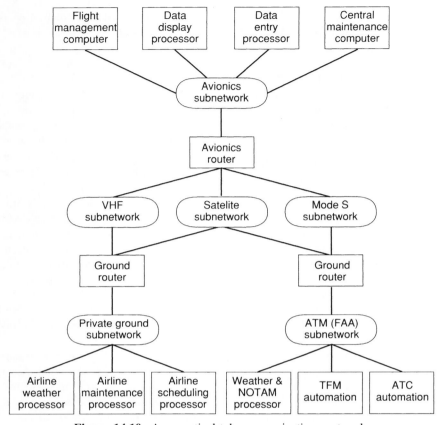

Figure 14.10. Aeronautical telecommunications network.

The ATN has adopted the International Organization for Standardization (ISO) open-system interconnection (OSI) 7-layer protocol architecture [50, 55].

14.5.5 Aviation Weather

Flight crews require reliable weather information both pre-flight and in flight for planning and decision making, and air traffic controllers need timely, accurate weather data in order to aid flight crews in avoiding hazardous weather areas. Weather products unique to aviation are needed to support these requirements including forecasts of icing conditions, turbulence, ceilings and visibilities, winds aloft, and wind shear.

The production of weather products requires observations of existing weather conditions, modeling to forecast future conditions, and timely distribution of products tailored to user needs. In the United States, this is a large-scale undertaking that involves the FAA, the National Weather Service (NWS), and

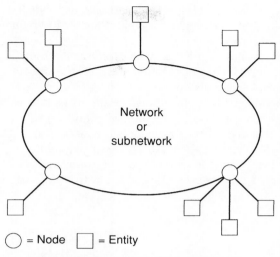

Figure 14.11. Communications network.

airspace users [29, 31]. Observations are made by ground-based sensors, aircraft sensors, flight crews in flight and space-based systems. These observations are assimilated into FAA, NWS, and airline data bases from which products are derived by forecasting models and filters that screen out the relevant information for users. The long-range objective is to ensure that all aviation weather product users are provided forecasts that are both reliable and consistent. Consistency is essential for ensuring that all concerned planners and operators are working with the same assessment of the weather situation so that their decisions and plans will be consistent as well.

14.5.6   Automation and Display Subsystem

The automation and display subsystem interfaces with the air traffic controller and with surveillance, weather, and communication subsystems [57]. The surveillance subsystem provides primary and secondary radar target reports in both analog video and digital message formats. The analog formats are used only with the older radars that are gradually being replaced. The surveillance subsystem also provides radar reflectivity data from weather cells. The weather subsystem provides weather information derived from weather radars, such as the next generation weather radar (NEXRAD) and the terminal Doppler weather radar (TDWR), and from weather sensors, such as the low-level wind shear alert system (LLWAS). The communication subsystem provides the voice and data communications circuits necessary for controllers to communicate with aircraft and with other controllers both within the local facility and at other ATM facilities. Although the controller workstation configuration is somewhat different in the en-route and terminal environments, the basic components are similar.

It consists of a situation display, a keyboard, and a trackball as well as means for displaying flight plan data and for controlling air-to-ground and ground-to-ground communications.

The situation display is oriented with North at the top and depicts the air traffic situation in relation to navigation aids, fixes, obstructions, airports, and runways (Figure 14.12). It also presents automation alerts to the controller. The information provided includes the aircraft position symbols with the associated flight data blocks, a list of safety alerts (i.e., minimum safe altitude warnings and conflict alerts), and a list of arriving and departing aircraft.

In Figure 14.12, there is an aircraft on the display in the 9 o'clock position. The data block shows that the aircraft identification (radio call sign) is N24F, its altitude is 5000 ft, and its ground speed is 210 Kn. The position of the aircraft

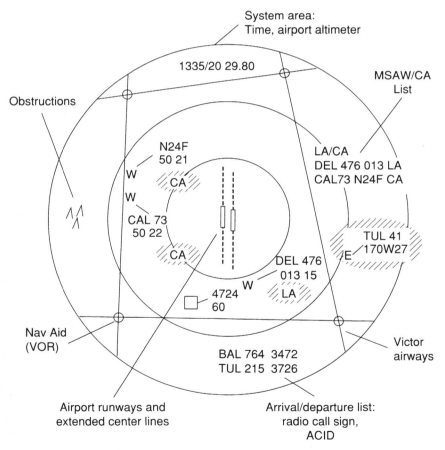

Figure 14.12. Radar situation display data presentation.

is at the symbol "W" at the end of the leader line from the data block. The "W" designates the controller responsible for the aircraft, in this case, the west controller. The blinking "CA" displayed below the data block is a conflict alert to the controller based on the automation system's assessment that N24F and CAL 73 may lose standard ATC separation if the controller does not intervene. Aircraft DEL 476 in the 5 o'clock position is below the minimum safe altitude at its location and the automation therefore has generated a low altitude alert (a blinking "LA"). The aircraft to the southwest has only a limited data block because the automation has no flight plan for the aircraft. The data block shows that the aircraft's identification code (ACID) is 4724 and its altitude is 6000 ft. Its radio call sign is unknown. The data block for TUL41 at 4 o'clock is blinking because the aircraft is in the process of being handed off from the east controller to the west controller.

14.5.7 Airborne ATM Subsystems

Aircraft are equipped with a number of subsystems that interoperate with the ground-based system, including voice and data link transceivers, navigation receivers and computers, secondary surveillance radar transponders, and, in some cases, instrumentation for sensing wind velocity, air temperature, and turbulence. Increasingly, flight management systems (FMSs) are employed routinely to guide aircraft along their flight trajectories, and their full integration with the ATM system has been a subject of intense development [19, 32, 54]. The FMS has three specific capabilities relevant to ATM [54]. It can guide the aircraft accurately along a predetermined path defined by a sequence of two-dimensional (latitude and longitude) or three-dimensional (adding altitude) waypoints. The FMS also can minimize the cost of the flight by selecting optimum speeds and/or altitudes along the predetermined path. The optimization is based on a pilot-selected cost index that weights fuel cost relative to flight time cost. When the cost of flight time is weighted heavily, the FMS will complete the trip more quickly, burning more fuel in the process. Finally, the FMS has a required time of arrival (or 4D navigation) capability that assures the aircraft will arrive at selected waypoints within small time windows (e.g., within ten seconds of the prescribed times).

Extensive data bases are resident in the FMS, as well as the current flight plan and the wind velocities and air temperatures expected along the route of flight. The FMS provides pitch, steering, and autothrottle commands to the aircraft guidance systems, as well as essential flight data, including estimated times of arrival at waypoints, fuel remaining and fuel flow, current position and speed, and current winds. The FMS integrates the various avionic subsystems, including the navigations sensors (Figure 14.13). It will automatically tune VOR/DME receivers according to the position of the aircraft and the navigation aid frequencies stored in the data base.

Data link interfaces with the FMS have been developed to allow flight data, flight plans, ADS position reports, and wind data to be transmitted to the ground

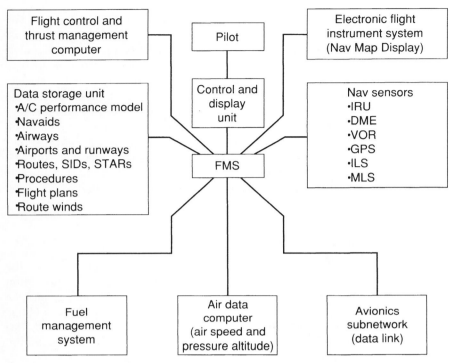

Figure 14.13. Flight management system.

and new flight plans and information requests to be sent to the aircraft. The FMS is the airborne automation system controlling the flight path of the aircraft. Linking its operation to ground-based ATM and airline dispatch operations has obvious advantages. An early application was FMS-guided terminal procedures that allowed aircraft to fly complex procedures (e.g., charted visual approaches) in IMC, thereby providing access to runways otherwise not available. An early application of 4D navigation capability is expected to be assigning aircraft specific times for crossing oceanic gateways into track systems so that the tracks with the most favorable winds can be fully utilized. Future applications will involve dynamic reclearance of aircraft on long oceanic flights to adjust flight paths to changing wind and weather patterns. The scenario may involve the airline dispatch automation system computing a new plan and data linking it to the FMS where it would be checked by the flight crew. The crew would then send the plan to ATM via the FMS. When ATM is satisfied that the plan meets separation requirements, it would be transmitted back to the FMS where the crew would select it to actively guide the aircraft. Similar applications are expected in terminal airspace for metering traffic to airports and negotiating fuel-optimal descent and approach trajectories [19].

14.6   FACILITIES AND OPERATIONS

The air traffic management services provided in the United States include traffic flow management, air traffic control, and flight services. The traffic flow management services are provided by the air traffic control system command center (ATCSCC) in conjunction with regional (en-route) and terminal facilities. ATM services are provided by air route traffic control centers (ARTCCs) at the regional level, and by approach control facilities and airport-traffic control towers (ATCTs) at the terminal and local levels. Flight services are provided principally by flight service stations (FSSs). Some flight services are also provided by ARTCCs and terminal facilities.

14.6.1   National Traffic Management

In the (ATCSCC), traffic managers monitor the national flow of air traffic to identify situations where demand may exceed capacity. This capacity/demand imbalance is often caused by a surge in demand at peak traffic hours or by a reduction in the capacity of a resource (e.g., an airport) due to weather or loss of essential ATM equipment. The traffic managers are provided graphical displays of traffic as well as automation tools to predict capacity/demand imbalances and to aid in formulating and implementing strategies to cope with excess demand. An automation tool periodically checks the estimate of demand on each resource (airport, sector, and fix), monitored against the capacity of the resource to determine if any will be saturated in the near future. As soon as a capacity/demand imbalance is predicted, the traffic manager develops a strategy to prevent its occurrence or to mitigate its effects. The strategy is often developed in coordination with the affected airlines and with the traffic management units (Section 14.6.2) and terminal facilities responsible for the aircraft involved. The strategies commonly considered include reroutes, delaying aircraft en route using speed adjustments or holding procedures, and holding aircraft on the ground at the origin airports.

14.6.2   En-route Facilities

The ARTCC provides ATM services to aircraft operating under instrument flight rules (IFR) in the en-route airspace. In 1996, there were 20 ARTCCs in the continental United States. They received surveillance data from long-range air route surveillance radars (ARSRs), and communicated with aircraft through ground-based transmitter/receivers located at remote communication air-to-ground (RCAG) sites. The ATM services provided at the ARTCC include aircraft separation and traffic flow management. The flow management services are provided by traffic managers in the *traffic management unit* (TMU) who work closely with sector controllers and traffic managers at the ATCSCC to formulate and implement traffic flow instructions. These instructions are designed to minimize delays while ensuring that excessive levels of traffic congestion are prevented.

The principal automation functions performed at the ARTCC include flight plan processing, metering, weather data processing, target tracking, tactical separation assurance based on target tracks, and strategic separation assurance based upon aircraft flight plans.

Flight Plan Processing   The flight plan processing function accepts flight plans filed by pilots through flight service stations, airline operations offices, military base operations, or sector controllers, as well as flight plans filed in bulk form by airlines. The flight plans are processed to produce flight progress data for posting to all controllers (en route and terminal) within the ARTCC airspace who will control the flight. The flight plan data are then passed to the next ARTCC along the route of flight. In 1996, flight progress data were presented to controllers on paper medium (i.e., as flight strips).

Metering   The metering function assists traffic managers and sector controllers in avoiding unacceptable levels of traffic congestion. Traffic managers in the TMU run the metering program during periods of peak traffic demand or may issue metering instructions based on their judgment and experience. The metering instructions provided to controllers may include meter fix crossing times for specific aircraft arriving at a terminal area or in-trail spacing restrictions (e.g., all aircraft being handed off to a congested adjacent sector must be separated longitudinally by 20 nmi).

Weather Processing   The weather-processing function accepts real-time weather radar reflectivity data from FAA and military surveillance and weather radars as well as processed weather data from a number of sources. These data are displayed on the sector controllers' plan-view situation displays so that the controllers can route aircraft around severe weather and provide weather advisories to pilots as required.

Track Processing and Tactical Automation Functions   The en-route automation system tracks all aircraft from which either primary or secondary target reports are received. The tracking process establishes a velocity vector for each tracked aircraft. The aircraft position and velocity data are used by the tactical separation assurance functions, namely, conflict alert, Mode C intruder, and minimum safe altitude warning (MSAW). The conflict alert function projects aircraft positions approximately two minutes into the future and searches for violations of separation minima between IFR aircraft (i.e., between aircraft for which the automation system has flight plans). The Mode C intruder function provides a similar service for IFR aircraft in relation to VFR aircraft that are providing automatic pressure altitude reports in the Mode C replies from their transponders. MSAW provides the controller a warning when an aircraft is projected to drop below the minimum safe altitude for the occupied airspace. Aural and visual alerts are provided to the responsible controller when separation problems are predicted.

Strategic Separation Assurance Functions   Automated problem detection (APD) and automated problem resolution (APR) are automated en-route ATC (AERA) functions planned for implementation in the late 1990s. They generate a four-dimensional (x, y, z, and t) trajectory for each aircraft derived from its current position, flight plan, and aerodynamic performance characteristics. As long as the aircraft operates within a threshold distance of this trajectory, the trajectory reliably represents the future path of the aircraft and therefore can be used as the basis for detecting conflicts. Based on 15- to 20-min projections of aircraft along these trajectories, APD detects conflicts with other aircraft, with restricted airspace and with flow instructions. APR will generate recommended control instructions for resolving each conflict, taking into account all aircraft operating in the vicinity. The responsible controller may elect to use these instructions or to resolve the conflict in another way.

14.6.3   Terminal Facilities

Approach control facilities provide services to aircraft within the terminal airspace delegated to the terminal facility by the parent ARTCC. The terminal airspace is sectorized into arrival and departure sectors to facilitate the flow of aircraft to and from airports.

In 1996, there were approximately 240 facilities in the United States providing approach control services. These include combined center/radar approach controls (CERAPs), terminal radar approach controls (TRACONs), terminal radar approach controls collocated with tower cabs (TRACABs), and military radar approach controls (RAPCONs). Of these terminal facilities, the FAA operated 133 small-to-medium TRACONs, 63 medium-to-large TRACONs, and 1 very large TRACON (the New York TRACON). During the mid-to-late 1990s, several large consolidated terminal facilities called *Metroplex Control Facilities* (MCFs) will be created by combining several TRACONs into a single facility. Terminal area controllers communicate with aircraft through transmitter/receiver sites known as remote transmitter/receivers (RTRs) and are provided short-range surveillance data by airport surveillance radars (ASRs). The principal automation functions performed at the terminal facilities include weather processing, target tracking, and tactical separation assurance based on target tracks (Section 14.5.6).

14.6.4   Airport Facilities

Airport ATM facilities are designated airport traffic control towers (ATCTs) or *towers* for short. In 1996, there were approximately 692 ATCTs in the United States, 419 operated by the FAA, 208 operated by the military and 65 operated by nonfederal entities.

Towers normally have four principal positions of operation (see also Section 14.4). These are the flight data, clearance delivery, ground control, and local control positions. The flight data position is responsible for the maintenance of flight data (flight strips) and for updating miscellaneous tower data, including the automatic terminal information service (ATIS). The clearance delivery position is responsible for issuing predeparture clearances to IFR aircraft. The clearance delivery and flight data positions are commonly combined and operated as a single position.

The ground controller is responsible for the safe movement of vehicles and aircraft within the airport movement area, with the exception of those operating on active runways. The local controller is responsible for movements on active runways. In 1996, local control and ground control operations were conducted principally through controller visual contact with the aircraft and ground vehicles being controlled. High-resolution primary radar is used at some major airports for detecting aircraft and ground vehicles on the airport surface and displaying this information to tower controllers. Development and testing efforts were under way in 1996 to improve surveillance and control of the surface using automatic dependent surveillance techniques (Section 14.5.3), based on the use of differential GPS (Section 5.5.9) and digital data links.

The level of automation support provided in ATCTs is dependent upon the activity level of the tower and the complexity of the air traffic operation. The smallest towers are provided with the capability to enter flight data, print flight strips and display environmental data. The largest towers have plan-view radar situation displays to support approach control operations and ground surveillance radars. During the late 1990s, the more complex towers will be provided with automation tools to assist in metering departures and to reduce the risk of aircraft collisions on the ground.

14.6.5   Flight Service Facilities

In 1996, flight services were provided in the United States by 131 flight service stations (FSSs). These facilities receive notices to airmen (NOTAMs) from the U.S. NOTAM system and weather products from the National Weather Service (NWS), including weather observations at airports, winds aloft, terminal and area forecasts, and weather warnings that describe significant meteorological events of importance to airmen. Section 14.4 describes the principal functions of flight service stations.

14.6.6   Oceanic Facilities

The oceanic airspace is divided into *flight information regions* (FIRs) under the auspices of the International Civil Aviation Organization (ICAO). The United States has responsibility for providing ATM services in several FIRs in the Atlantic and Pacific oceanic regions as well as one FIR over the Gulf of Mexico.

The associated oceanic control centers are colocated with en-route facilities in New York, Oakland, Houston, Anchorage, and Honolulu.

14.7   SYSTEM CAPACITY

The demand on the ATM system continues to rise. For example, the number of aircraft handled by U.S. air route traffic control centers (ARTCCs) is forecasted to rise from 36.6 million operations in 1989 to 47.8 million operations by the year 2000. This growth in air traffic will put increased pressure on all system resources, especially the major airports.

The FAA's Aviation System Capacity Plan [16] defines the aviation capacity problem in terms of flight delays determined by comparing actual with optimal arrival times. In 1990, 23 U.S. major airports experienced more than 20,000 aircraft hours of flight delay. That number is expected to grow to 40 major airports by the year 2000 if capacity improvements are not implemented.

During the course of a flight, a number of system resources are utilized, including the departure airport taxiways and runways; terminal fixes and sectors; en-route fixes, routes, and sectors; and the arrival airport. Although any of these resources can become saturated and contribute to flight delay, historically, the departure and arrival airports are the principal contributors. Table 14.2 shows the sources of flight delays greater than 15 minutes as reported in the 1991–92 Aviation System Capacity Plan.

The following sections discuss techniques for reducing flight delays by reducing peak demand and increasing system capacity.

14.7.1   Reducing Peak Demand

There are a number of techniques to reduce peak demand, thereby reducing the severity of demand/capacity imbalances. ATM authorities can delay departures and slow down en-route traffic in order to reduce airport arrival demand during the peak hour. Another demand management technique is to encourage smaller aircraft to use reliever airports. In the United States, of the 40 airports fore-

TABLE 14.2   Sources of flight delays greater than 15 minutes

	1988	1989	1990
Weather	70	57	53
Terminal air traffic volume	9	29	36
Center air traffic volume	12	8	2
Closed runways/taxiways	5	3	4
Equipment outages	3	2	2
Other	1	1	3

casted to experience significant delay problems, approximately one-third have 25% or more general aviation operations. For those airports, shifting a portion of the general aviation operations to a reliever airport would reduce delays. A related technique is to shift demand to underutilized airports that are equipped to handle air carrier aircraft and are within 50 mi of saturated airports. Finally, underutilized airports can be used as connecting hub airports instead of congesting major metropolitan airports with passengers who are only transferring to other flights.

14.7.2   Increasing System Capacity

The largest capacity gainers at airports are new runways. In 1996, 62 new runways or major runway extensions were planned or proposed at the top 100 U.S. airports. The next most productive capacity enhancement is improved terminal airspace procedures. In 1989, 57% of all flight delays were attributed to adverse weather conditions. These delays are a consequence of the fact that instrument approach procedures are more restrictive than the visual procedures in effect during better weather conditions. Much delay could be eliminated if the approach procedures used during IMC were more flexible.

Dependent Parallel IFR Approaches   In 1996, simultaneous parallel approaches could be made in IMC to parallel runways spaced as closely as 2500 ft provided a 2 nmi diagonal separation is maintained between aircraft (Figure 14.14*a*). Parallel runways with dependent approach streams can accommodate 45% to 65% more arrivals per hour than a single runway (i.e., 44 to 50 arrivals per hour as opposed to 30 arrivals per hour for a single runway).

The diagonal separation requirement makes one arrival stream dependent on the other because it prevents a faster aircraft on one approach from passing a slower aircraft on the adjacent approach. This dependency limits the capacity benefit of the parallel runways as compared with two independent arrival streams. In the late 1990s, technology advances and procedural improvements may permit the reduction of the diagonal separation requirement to 1.5 nmi, which would increase the capacity of the dependent parallels by approximately four aircraft per hour.

Independent Parallel IFR Approaches   During simultaneous independent approaches to parallel runways, each approach course is monitored by a dedicated monitor controller using a plan-view radar situation display. An aircraft on an ILS approach is established on the final approach course prior to the point where it intercepts the glide slope. Aircraft on parallel approaches (e.g., to 25L and 25R in Figure 14.14*b*) are separated vertically by 1000 ft prior to the point where the higher approach path intercepts the glide slope. From this point forward until the aircraft land, the monitor controllers ensure that the aircraft maintain lateral separation. In particular, if an aircraft on one approach course enters the no transgression zone shown in Figure 14.14*b*, the monitor controller

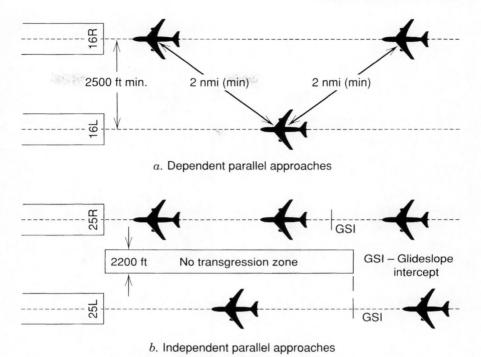

a. Dependent parallel approaches

b. Independent parallel approaches

Figure 14.14. Parallel approaches.

responsible for the parallel approach course will vector away any aircraft that is threatened by the "blundering" aircraft.

For parallel runways with center lines separated by 4300 ft or more, the standard airport surveillance radar, which has a 4.8-sec antenna scan period and uses a conventional controller display, has been used for monitoring simultaneous independent approaches. In 1995, a quick-scan radar called the *precision runway monitor* (PRM) was developed. Its 0.5-second scan period and its high resolution controller display system permit independent parallel approaches where runway center lines are separated by 3400 ft or more.

Independent approaches provide an increase in arrival capacity of approximately 30% as compared to dependent parallel approach procedures. A decrease in the runway center line spacing requirement through the use of new technology and procedures provides capacity gains for airports with limited real estate resources.

Dependent Converging IFR Approaches   Independent converging approaches are generally permitted only when the ceiling is greater than 700 ft and the visibility is 2 smi or more. The principal aircraft separation problem arises when both aircraft execute missed approaches. If the ceiling is above 700 ft the pilots and tower controllers can ensure separation visually. At lower ceilings, either

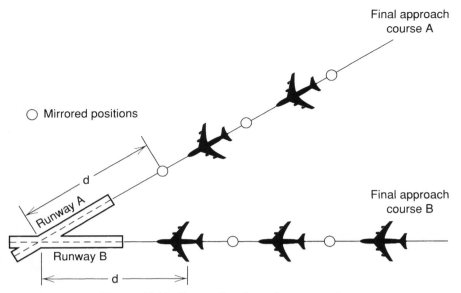

Figure 14.15. Converging dependent approaches.

the approaches must be confined to a single runway or dependent approach streams must be established. In dependent converging approach streams, the aircraft are spaced longitudinally such that, if any two aircraft execute missed approaches and fly over the point where the runway center lines intersect, the aircraft will be separated longitudinally.

To aid controllers in establishing dependent converging approach streams, the *converging runway display aid* (CRDA) was installed at a number of airports in 1995 (Figure 14.15). CRDA displays a mirror image of each approaching aircraft. As shown in Figure 14.15, the mirror image is drawn on the adjacent approach course at a distance from the runway intersection equal to that of the approaching aircraft. Controllers then space aircraft from the mirror images of aircraft on the converging course as well as from real aircraft on the same approach course. It is estimated that capacity increases of approximately ten arrivals per hour over single-runway arrival capacities are achievable with this procedure.

14.8   AIRBORNE COLLISION AVOIDANCE SYSTEMS

Airborne collision avoidance systems are installed in aircraft to provide ground-independent protection from midair collisions as a backup to the conventional ATC system. Within the United States, a two-member family of equipment called the *traffic alert and collision avoidance system* (TCAS) has been developed [17, 20, 26, 33, 51, 52, 61]. TCAS interrogates the secondary surveillance radar transponders installed in proximate aircraft and processes the replies to alert flight crews to potential conflicts. The protection provided is independent of ground-based equipment. Hence, a TCAS-equipped aircraft receives backup separation assurance service wherever it flies.

Figure 14.16 illustrates the conflict geometry when the aircraft velocities V_1 and V_2 are constant. V_R is the relative velocity, R is the distance between aircraft 2 and aircraft 1, and m is the miss distance, that is, the distance between the aircraft at the closest point of approach (CPA). V_θ is the tangential velocity of aircraft 2 relative to aircraft 1. The time that it will take aircraft 2 to reach CPA is T. From Figure 14.16,

$$T = \frac{[R^2 - m^2]^{1/2}}{V_R} \quad \text{sec} \tag{14.1}$$

$$\dot{\theta} = \frac{V_\theta}{R} \quad \text{rad/sec} \tag{14.2}$$

and

$$\sin \alpha = \frac{m}{R} = \frac{V_\theta}{V_R} \tag{14.3}$$

Combining Equations 14.2 and 14.3, and then using Equation 14.1, one obtains

$$\dot{\theta} = \frac{m V_R}{R^2} = \frac{m/V_R}{(m^2/V_R^2) + T^2} \quad \text{rad/sec} \tag{14.4}$$

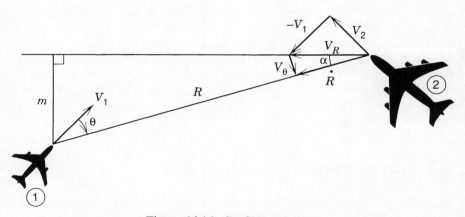

Figure 14.16. Conflict geometry.

where, from Figure 14.16,

$$V_R = [\dot{R}^2 + (R\dot{\theta})^2]^{1/2} \tag{14.5}$$

Equation 14.4 illustrates the fact that the bearing rate $\dot{\theta}$ is zero if and only if the aircraft are on a true collision course, $m = 0$.

The most widely implemented version of the TCAS family, TCAS II, interrogates the transponders on nearby aircraft, tracks the replies in range and altitude, and generates crew alerts if both range and relative altitude are predicted to be small at some time within the next 40sec or so. Two types of alerts are displayed: *traffic advisories* (TAs) and *resolution advisories* (RAs). A TA is displayed approximately 40sec prior to CPA and shows the range, bearing, and relative altitude of the intruder on a cockpit plan view display. If the intruder continues to close, an RA will be displayed approximately 25sec prior to CPA. The RA shows a vertical escape maneuver (e.g., "climb") for avoiding a collision. Two TCAS II aircraft in conflict must exchange escape maneuver intentions to ensure that they are complementary (i.e., one aircraft climbs, the other descends). TCAS II uses Mode S air-to-air data link communications to perform the coordination function.

While TCAS II uses intruder bearing θ in displaying TAs, the angle-of-arrival-sensing antenna is not sufficiently accurate to provide bearing rate data for miss-distance, m, estimation. The time to CPA is estimated as tau, τ, where

$$\tau = \frac{R}{-\dot{R}} \quad \text{sec} \tag{14.6}$$

This estimate is accurate in the critical cases where m is small. A principal disadvantage of TCAS II is that a certain number of unnecessary alerts are generated because miss distance cannot be estimated. In 1995, all air carrier aircraft with more than 30 passenger seats operating in U.S. airspace were required to be equipped with TCAS II.

TCAS I does not display RAs. The equipment tracks proximate aircraft in range and altitude and displays TAs for intruders that satisfy tau-based collision threat criteria. The TAs aid flight crews in visually acquiring threat aircraft so that visual separation can be maintained. By the end of 1995, TCAS I was required to be installed in all air carrier aircraft with 10 to 30 passenger seats operating in U.S. airspace.

14.9 FUTURE TRENDS

Since the air traffic management system consists of a large number of subsystems, including communication, navigation, surveillance, automation, and weather service, there are a wide variety of future trends toward system improvements. They are discussed below.

One of the most profound trends will be the wide-spread use of air-to-ground digital data communication (versus voice) for interfacing automated, ground-based information sources with aircraft systems and flight crews. Use of data communication will result in more reliable and timely exchange of large quantities of data. In some airspace, voice communication will be used principally for emergency and nonroutine functions.

A second major, nearly revolutionary, trend will be the widespread use of satellite-navigation systems, initially, GPS (Section 5.5), to provide highly accurate, worldwide navigation information (position, velocity, and time) to meet RNP requirements for all phases of flight. In 1996, GPS was authorized by the U.S. FAA for use as a *sole means* system of oceanic navigation and for Special nonprecision and (private use) Category I SCAT precision approaches. In the future, in conjunction with the U.S. Wide Area Augmentation System (WAAS, Section 5.7.3), it will be authorized for use as a *sole means* system for terminal and en-route navigation, as well as public-use Category I precision approach operations. With a local area augmentation system (LAAS) (Sections 5.5.9, 13.8, and 13.9), Category II and III precision approach and landing capability is likely to be provided in the more distant future. Extensive research and flight experimentation, with very promising results, were in progress in 1996. In some highly congested areas, such as in Europe, MLS is likely to be used for some time in the future for providing precision approach and landing capability, particularly for Category II and III operations.

The use of GPS will be combined with GLONASS (Section 5.6) and possibly other satellite-navigation systems, to constitute a Global Navigation Satellite System (GNSS), thereby achieving a worldwide navigation capability based on at least 48 satellites. Geostationary satellites will also include a GNSS-like ranging function. This will lead to very high *availability, integrity, and continuity of service* characteristics and hence greater safety.

Aircraft equipped with a GNSS (initially GPS) receiver and a simple navigation computer will have RNAV capability, permitting direct routes anywhere on the globe. In the more distant future, rigid airways, based on networks of ground-based navigation aids (e.g., VOR/DME) will become obsolete. The flexibility of flight operations will be greatly enhanced, and ATM service providers will discontinue operation of some ground-based navigation aids.

Another major future trend will be the combined exploitation of rapid, automatic, air-to-ground digital communication and high-accuracy (e.g., GNSS) positioning capability to provide automatic dependent surveillance (ADS, Section 14.5.3) virtually everywhere, from oceanic airspace to the airport surface. The initial implementation of ADS will be in oceanic airspace and over remote land areas, and this will result in more cost-efficient separation standards, replacing the procedural separation methods still in use in 1996. The use of ground-based primary and secondary radars is likely to continue for terminal area surveillance for some time, but in the en-route environment it will ultimately be replaced by ADS because of its greater accuracy and cost effectiveness. Based on the successful experimentation and testing in the mid-1990s, ADS techniques (using local differential GPS and data link) are likely to be used for airport surface surveillance and control at major airports.

Ground-based automation to support national and local traffic flow management, as well as the more tactical ATC process, will be implemented on a large scale as the principal means for accommodating user-preferred flight trajectories and ensuring that system capacity resources are fully utilized to accommodate demand. ATM automation linked to FMS and flight dispatch automation by data link will ensure comprehensive planning of each flight trajectory, taking the needs of all other users into account. Four-dimensional (including time) clearances will be used where necessary to ensure aircraft separations and on-time arrivals at merge points. The FMS will guide the aircraft on fuel-optimal trajectories, meeting users' schedule requirements from wheels-up to touchdown. The traffic display capability of TCAS will be enhanced with additional information, such as the locations and orientations of runways and the identifications of proximate aircraft, to provide a comprehensive display of the ATM situation for the flight crew. The crew will then be able to participate more fully in planning and decision making, including resolution of local traffic conflicts in collaboration with the other aircraft involved.

An important future trend for air-to-air collision avoidance will be the widespread mutual exchange between aircraft of their accurately known present positions, as obtained from their on-board navigation equipment, such as a GPS receiver, to provide situation and threat information on all other aircraft in the vicinity. This technique is another means, other than basic TCAS, of achieving a ground-independent collision avoidance function for all participating aircraft. In this approach, only one-way broadcasts are used, which results in much larger communication capacities than with two-way transmissions. Also, heading and other pertinent data can be exchanged to reduce the number of unnecessary resolution advisories. This approach will improve the situational awareness performance of the collision avoidance system.

There is a strong trend toward development, evolution, and implementation of the concept of *free flight* [53]. Free flight is a flight operations capability, under instrument flight rules (IFR), in which the operators have the freedom to select their flight path and speed in real time, after interaircraft communications ensure safe separation. Air traffic restrictions are only imposed to remove a threat to safety, to preclude exceeding airport capacity and to prevent unauthorized flight through Special Use Airspace (SUA) [53]. The concept is intended to greatly increase a user's operational flexibility and "self-optimization," while achieving guaranteed safety. The major technologies that will make it possible to accomplish these objectives are (1) efficient digital data links; (2) the global navigation satellite systems (e.g., GPS); (3) the concept of automatic dependent surveillance (Section 14.5.3) of continuously broadcasting (by all aircraft) the position, velocity, and short-term intent, used in conjunction with airborne collision avoidance computation equipment; (4) improved displays in the aircraft and at ATC centers to provide good situational awareness; and (5) revised air traffic procedures that capitalize on these newly available capabilities. Incremental evolution and implementation of free flight is likely to progress in the near term and continue for the foreseeable future.

The major overall theme for the future of ATM will be information fusion, providing to each user and provider of services an accurate, comprehensive awareness of the ATM situation, wherein automation systems perform the routine command, control, and guidance functions. Benefits will include increased safety and efficiency of flight operations worldwide, full utilization of system capacity resources, and reduced costs of ATM services. These new capabilities and the associated benefits will emerge first in oceanic airspace where existing ATM services have been rudimentary, the aircraft fleet is generally well equipped, and the operators are highly motivated to achieve fuel savings.

PROBLEMS

14.1. From the material in Section 14.8, show that miss distance can be expressed in terms of intruder range, range rate, and bearing rate as

$$m = \frac{\dot{\theta} R^2}{[\dot{R}^2 + (R\dot{\theta})^2]^{1/2}}$$

14.2. It is proposed to use a TCAS antenna capable of measuring the angle of arrival (θ) of the intruder's transponder reply and to track these measurements in time to derive an estimate of bearing rate ($\dot{\theta}$). The miss distance would then be estimated from the equation derived in Problem 14.1. Show that bearing rate estimation errors would need to be less than a few tenths of a degree per second. In particular, show that $\dot{\theta} = 0.90$ deg/sec for $m = 1.0$ nmi, $V_R = 300$ kn, and $T = 25$ sec.

14.3. In Figure 14.16, it is easy to show that

$$\tau = \frac{T}{\cos^2 \alpha} \qquad \text{and} \qquad \alpha = \tan^{-1} \frac{m}{V_R T}$$

Hence τ is always greater than or equal to the true time to CPA, T, and τ increases relative to T as m/V_R increases. Plot τ versus T for $V_R = 300$ kn and $m = 0, 0.5, 1.0$, and 1.5 nmi. Show that for these parameters a TCAS II that requires a τ less than 20 sec as a condition for displaying an RA:

  (a) Will not display an RA for $m = 1.0$ or 1.5 nmi.

  (b) Will display an RA for $m = 0.5$ nm approximately 18 sec prior to CPA.

(c) Would be able to inhibit the unnecessary alert for the intruder with $m = 0.5$ nmi if it could reliably detect bearing rates of 1.0 deg/sec or so.

14.4. Queuing theory can be used to model airport capacities and arrival delays. For example, the operation of a runway can be regarded as a server that lands aircraft. The time interval between the aircraft crossing the threshold and the aircraft exiting the runway (the service time) can be modeled as an exponential probability density function (pdf) with a mean of $1/\mu$ sec per landing operation. Aircraft arrivals can also be modeled as an exponential pdf with a mean of $1/x$ sec between arrivals. If the runway (server) is busy when an aircraft arrives, the aircraft must wait for the preceding aircraft to exit the runway before it can land. From queuing theory, the average number of aircraft waiting to land would be

$$\bar{n} = \frac{\rho^2}{1 - \rho}$$

where

$$\rho = \frac{x}{\mu} < 1$$

The mean time required to service a landing ($1/\mu$) is typically 50 sec, and the ultimate capacity of a runway might therefore be taken as 72 arrivals per hour. Graph $\bar{n}$ and show that for $\rho = 0.62$, on average every aircraft must wait to land ($\bar{n} = 1$) and a practical arrival capacity might be taken as $x = 0.62\mu$ or 45 aircraft per hour.

14.5. For the conditions described in Problem 14.4, the average value and standard deviation of the delay time experienced by each aircraft is

$$\bar{d} = \frac{1}{\mu} \left(\frac{\rho}{1 - \rho} \right)$$

and

$$\sigma_d = \frac{[\rho(2 - \rho)]^{1/2}}{\mu(1 - \rho)}$$

respectively. Graph these equations. Does 45 arrivals per hour still appear to be a practical arrival capacity for a single runway?

15 Avionics Interfaces

15.1 INTRODUCTION

Interfaces are an integral part of any avionics navigation system. There is no value in a navigation system that can determine the position of its host vehicle if that position cannot be made known to the host and other avionics elements through some type of a signal interface. However, interfaces between the aircraft navigation system and the rest of the vehicle extend beyond the signal interface, commonly a data bus, to include displays for the flight crew and the maintenance crew, the power system, packaging and mounting, and a suitable operating environment.

15.2 DATA BUSES

Data buses are the cornerstone of modern integrated avionics systems. They are the principal means by which information is exchanged among the navigation and other systems. Data buses are the paths by which the necessary inputs are received by a system or subsystem to perform its process(es) and the paths by which the outputs from the process(es) are sent to users.

Data buses fall into one of two broad categories: simplex (one way) with a single transmitter and multiple receivers, and duplex (two way) with multiple transmitters and receivers. Also, there are at least two possible transmission media, wire or optical cable. Other important features of data buses are bus control, bit rate, and word and message structure.

For military aircraft the principal data bus is the MIL-STD-1553, Digital Time Division Command/Response Multiplex Data Bus. This bus is used on many types of military aircraft including the F-15, F-22, and C-17. It is also being installed in older aircraft as part of retrofitting upgraded avionics, and even in some ground military equipment such as the U.S. Bradley Fighting Vehicle.

There can be three types of terminals on a MIL-STD-1553 bus: the bus controller, a remote terminal, and the bus monitor. As the name implies, the bus controller manages all activity on the bus. It directs a designated remote terminal to transmit a message back to the bus controller or to another designated remote terminal. A remote terminal is the interface between the bus and the remote terminal's host system or subsystem. It can be a separate line replaceable unit (LRU) or embedded in its host. The bus monitor, if used, records all (or a designated subset of) messages on the bus. However, in most production systems, the bus monitor is not used. Figure 15.1 shows a quadruple-redundant MIL-STD-1553 data bus.

The smallest unit of exchange on a data bus is a word. In MIL-STD-1553 a word is 20 bit times long. MIL-STD-1553 operates at one megabit per second which means each bit is one microsecond long. The first three bit times are used for synchronization and the last bit is used for parity check (odd parity). MIL-STD-1553 uses Manchester coding which means that every bit has a mid-bit transition. For a logical 1, the signal starts at a positive or high state and transitions at midbit to a negative or low state. For a logical 0, the signal starts at a negative or low state and transitions at midbit to a positive or high state.

Figure 15.2 shows the three types of words found on a 1553 bus, namely, command, status, and data. A command word, shown in Figure 15.2a, is always the first word in any message and can be transmitted only by the bus controller. Note that the synchronization code is a positive $1\frac{1}{2}$ bit time followed by a negative $1\frac{1}{2}$ bit time. Details on the significance of each bit can be found in MIL Standard 1553 [5].

MIL-Standard 1553 describes ten ways that words can be assembled into messages. There are six command/response formats and four broadcast formats. The appropriate message format is determined by the bus controller. Command/response formats require confirmation of receipt of a message by the designated addressee remote terminal. Broadcast message formats do not require confirmation of receipt and therefore cannot guarantee that the address(es) received the message. Because of this lack of a guarantee broadcast message formats have never been approved for use in U.S. AirForce applications.

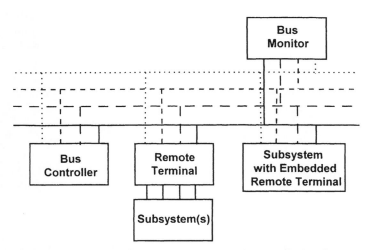

Figure 15.1   Quadruple redundant MIL-STD-1553 data bus.

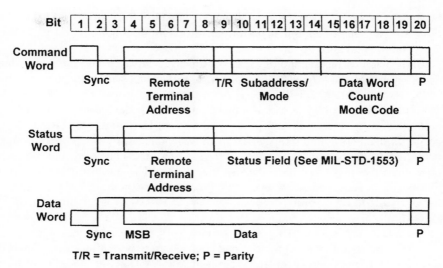

Figure 15.2   MIL-STD-1553 word types.

MIL-STD-1553, as the pioneer avionics data bus, demonstrated the advantages of data buses and whetted the appetite of the avionics designers for even higher capacity. Consequently, buses such as the high-speed data bus (HSDB) and the NATO STANAG 3910 are used in advanced fighters such as the F-22 and the Eurofighter 2000, respectively. The HSDB can operate at up to 50 Mbits/sec over either wire or optical media. STANAG 3910 is a hybrid bus design in which a MIL-STD-1553 bus controls a higher-speed (20 Mbit/sec) optical bus.

To counter the escalating electromagnetic interference (EMI) hazard in aircraft built primarily of composite material, MIL-STD-1773 has been developed to apply fiber-optic technology to aircraft data buses. Additionally, fiber-optic data buses intrinsically have enormous bandwidth, hundreds of times larger than that of MIL-STD-1553. MIL-STD-1773 is very similar to 1553, except for using the optical media. In fact, 1773 makes extensive reference to 1553 for details on word structure, and so on. Furthermore, because of the rapidly improving performance of optical bus components, 1773 defers to the specification for the system in which the bus is being used for detailed performance requirements for the optical components.

ARINC 429 [2] is the proven data bus for civil transport application and also has limited military application where civil products are being adapted to military use, such as the engines on the U.S. Air Force C-17. ARINC 429 is a one-way broadcast bus with a single transmitter and up to 20 receivers. When compared to MIL-STD-1553, ARINC 429 is very slow, having a capacity of only 12 to 14.5 kbps or 100 kbps, depending on the application. Despite these limitations, ARINC 429 is very popular on commercial transport aircraft because it is very easy to certify; that is, it is straightforward to establish the effects of the bus off-nominal performance or failure condition and to design the system to tolerate the condition.

Civil avionics bus technology has continued to advance and has led to the development of the ARINC 629 multi-transmitter data bus as the cornerstone of the B-777 aircraft. This bus incorporates many of the features of MIL-STD-1553 words but avoids the use of a bus controller, considered by civil aviation authorities to be the weak link in 1553 operation. In the case of ARINC 629, each terminal acts autonomously to transmit when, and only when, three timing conditions (one of which is unique to a given terminal) are met, and receives only those messages that have the specified labels. These transmit timing conditions are stored in an easily updatable "transmit personality" PROM, and the list of words to be recorded from the bus are stored in a "receive personality" PROM.

As noted in Section 15.1, every airborne system requires extensive integration to accomplish its function, and this integration is often accomplished through data buses. However, integration inexorably leads to the issue of data latency, since no information can be instantaneously sensed, transmitted, processed, and used. Data latency can be defined as the time from when a quantity or condition is sensed until the quantity or condition is available in suitable form to the ultimate user. Common examples of where data latency is an issue include data fusion from multiples sources such as digital maps and radar or infrared images, and feedback of sensor and state information in navigation, flight control and fire control systems.[1]

Precisely quantifying data latency is a very difficult problem. Estimates can be made based upon an understanding of sensor and bus characteristics, such as data transmission and frame rates. However, the only method to accurately determine the data latency is to assemble an actual system, complete with representative connectors, cable lengths, interfaces, and signal conditioners and converters and to make the appropriate measurements. Since there is no way to completely eliminate data latency, one method often used to compensate for it is to compute the probable value of a parameter at the time it is used and/or displayed, based on the most recent measured value and the trend in the measured values.

15.3 CREW DISPLAYS

The single most important interface in any manned aircraft is that between the aircraft and the crew. This interface takes the form of cockpit displays and

---

[1] As an example, FAA Advisory Circular 25-11 Transport Category Airplane Electronic Display Systems, requires that there be no more than 0.1-sec latency in displaying aircraft attitude information on cockpit electronic displays. Also it is important to note that typical subsonic large aircraft velocities are over 600 ft/sec, which can generate substantial data latency issues in precision navigation systems.

controls (and test and maintenance panels.) Since crew error has historically played a major role in most aircraft accidents, the cockpit must be designed to be user friendly and to ensure error-free operation, especially under high work load and emergency conditions.

Electronic displays dominate the cockpit of modern aircraft, although there are still many electromechanical displays in use, in some cases as backup to the electronic displays. Electronic displays offer substantial increases in reliability and virtually unconstrained flexibility relative to electromechanical displays.[2] The principal electronic display device is a cathode-ray tube (CRT). However, flat panel liquid crystal display (LCD) devices are rapidly capturing more of the market since, when compared to CRTs, they require less power and depth behind the instrument panel, have improved sun light readability, and are intrinsically digitally compatible.

Table 15.1 compares the performance of CRT displays on the Douglas MD-11 to LCD displays on the Boeing B-777. Note the reductions in power and weight for the LCD displays and the improved reliability, measured in mean time between failures (MTBF).

Figure 15.3a is a picture of the Boeing B-777 cockpit. From left to right on the instrument panel are the Captain's primary flight display (PFD) and navigation display (ND), the engine indicator, the crew alerting system (EICAS), and the first officer's ND and PFD. Below the EICAS display is the multi-function display (MFD). Because of the intrinsic flexibility of electronic displays, the B-777 can be dispatched with one display inoperative since the MFD is not required and the information normally displayed on the other five displays can be redistributed automatically by the symbol generators to the five remaining operating displays. Furthermore, all six display units are identical. Any display unit will operate in any position on the instrument panel, since the information displayed on it is determined by the display generators and the display unit connector pin arrangement.

Figures 15.3b and c are close-up views of the PFD and ND, respectively.

TABLE 15.1   Comparison of MD-11 CRT displays to B-777 LCD displays

Characteristic	MD-11 CRT	B-777 LCD
Viewable area (in.)	6.5×6.5	6.7×6.7
Power (w)	160	90
Weight (lb)	35	19.5
Volume (in.^{3})	896 ($8 \times 8 \times 14$)	624 ($8 \times 8 \times 9.75$)
Reliability (MTBF, hr)	8500	11765

[2]One avionics manufacturer has developed a drop-in replaceable LCD horizontal situation indicator (HSI) for the United States Air Force F-15 that is lighter and 100 times more reliable than the original electromechanical design.

(a)

Figure 15.3a   Boeing B-777 cockpit (courtesy of the Boeing Company).

The PFD contains five basic display formats. In the center is the attitude display that is dominated by the aircraft symbol, the pitch ladder, and the roll scale. ILS localizer and glide-slope guidance dots are below and to the right of the attitude display, respectively. On the left is the airspeed tape with a digital readout of the airspeed visible in the window. Across the top in the three windows are the navigation modes currently engaged. The abbreviations in the three upper windows stand for SPD = speed; LNAV = lateral navigation, and VNAV PTH = vertical navigation path following. On the right is the altitude tape with a digital readout of the altitude visible in the window. To the right of the altitude display is an instantaneous vertical speed indicator. At the bottom is a segment of a compass rose showing the current magnetic heading. The ND is dominated by the compass arc across the top. Also near the top are arrival time and distance to go to the next waypoint. (In the case shown the aircraft is over the LACRE waypoint (0 nmi) at 1434 Z.) Near the bottom of the ND is a triangle that represents the aircraft ("ownship").

In addition to the conventional electronic displays described above that are mounted in the instrument panel, two other types of cockpit electronic displays are the head-up display (HUD) and the helmet-mounted display (HMD), frequently used on military aircraft. A HUD, which is essentially a transparent display, is mounted in the cockpit above the instrument panel in the pilot's (or

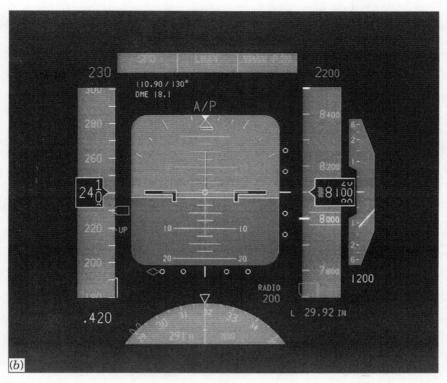

Figure 15.3b Closeup view of the Boeing B-777 primary flight display (courtesy of the Boeing Company).

first officer's) field of view when looking outside straight ahead. HUDs display a limited subset of aircraft, system, and/or target information or a synthetic or enhanced image of the outside scene. Obviously, the amount and format of information displayed on a HUD must be very carefully selected to avoid obscuring the outside scene and confusing the crew members. HUDs are relatively difficult to install, since they consume a lot of valuable space in the cockpit and must be precisely aligned to ensure the displayed information is true relative to the outside scene. A typical modern HUD will have a 24° high × 30° wide field of view and approximately 90% transmittance of the outside scene. A helmet-mounted display (HMD) is a compact, lightweight display installed on a helmet that projects into the wearer's field of view critical aircraft, system, and target information regardless of the direction in which the wearer is looking. Thus, it has a significant advantage over a HUD in military high-performance aircraft where the pilot must often look other than straight ahead while still maintaining control of the aircraft. It is essential that HMDs be very lightweight and carefully mounted for wearer comfort and to avoid neck injury in the case of high-acceleration maneuvers or ejection. To correlate the display with the

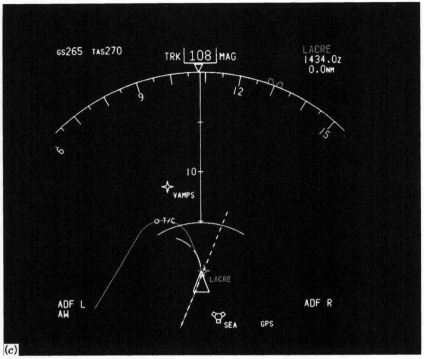

(c)

Figure 15.3c Closeup view of the Boeing B-777 navigation display (courtesy of the Boeing Company).

outside scene, the HMD host aircraft must have a helmet tracker to determine the direction the helmet is pointing at any time.

Modern civil transport aircraft are often equipped with a flight-management computer (FMC), typically part of the flight-management system (FMS), which optimizes the performance and/or flight path of the aircraft in terms of some parameter such as flight time or fuel cost. (Chapter 14.5.7). The most common interface between the crew and the FMC is a control display unit (CDU) like the one for a MD-11 shown in Figure 15.4. In addition to the alphanumeric keyboard, there are special-purpose keys that support the flight-planning and management processes. There are typically three CDUs in the aisle stand of the cockpit, one each for the captain and first officer in the front of the aisle stand, and one at the rear of the aisle stand intended primarily for use by maintenance personnel to operate the on-board maintenance system. This third CDU also is a backup to the other two.

The B-777 aircraft uses a touchpad cursor control for interacting with some of the displays. Immediately below each flight crew CDU is a touchpad that can be used to control a cursor on either the CDU display screen or the electronic library system if installed.

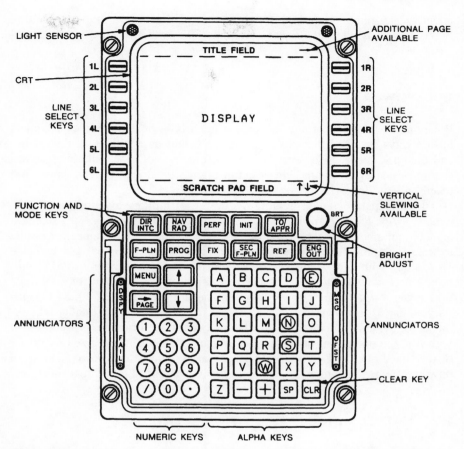

Figure 15.4   MD-11 Control display unit (courtesy of the McDonnell-Douglas Corporation).

Another medium for cockpit/crew interfaces is speech. Speech-interactive systems are of two basic types: synthetic and recognition. Synthetic speech systems have been used for many years to provide aural warnings to the crew such as "Glide slope, glide slope," and more recently for the traffic alerting and collision avoidance system (TCAS), "Traffic, traffic" (Section 14.8).

Speech is another option for the crew to control the aircraft, particularly under high work load conditions; however, so far it has found only limited application. The major challenge in using speech as a crew/aircraft interface is the same as that in many other speech-interface applications, which is *word recognition*. Speech systems with a large vocabulary capability are *user dependent*, which means that the system must be calibrated for each individual user through the generation and use of a pronunciation template. Where speed recognition systems have been tested in aircraft the correct recognition rate is only about in the mid-90% during straight and level flight. In a high-acceleration flight conditions, the rate drops to less than 80%, which is clearly not acceptable.

15.4 POWER

Power is the lifeblood of any avionics system. Historically, avionics designers have always assumed that electrical power would be available in the quality and quantity needed. However, as the performance of electronic devices has improved and avionics are being used in flight critical applications, power has taken on new importance.[3]

The principal document guiding the design of military aviation power systems is *MIL-STD-704 Aircraft Electrical Power Characteristics*. This standard levies requirements on the power quality as delivered to the connectors on the avionics equipment. A portion of these requirements is summarized in Table 15.2, along with equivalent requirements from DO-160, *Environmental Conditions and Test Procedures for Airborne Equipment*, the standard for civil aviation power. Civil avionics power requirements are very similar to those for military avionics, as shown in Table 15.2. DO-160 presents power quality requirements for avionics and describes the laboratory test facilities and procedures for verifying the operation (or survival) of the avionics when energized by power of the stated quality. Like the military power systems, a major driver for civil power systems is continuous availability of power for flight critical functions.

15.5 MAINTENANCE

Maintenance is a major factor in avionics life cycle costs, since much of the life-cycle cost is incurred as maintenance cost after the avionics has entered service. Maintenance costs are driven in large part by the amount of attention maintenance receives early in the design stage. The key drivers in maintenance are well designed built-in test (BIT) of the line replaceable unit (LRU, commonly called a *black box*) or line replaceable modules (LRMs); clearly written maintenance manuals; easy-to-use maintenance aids, and accessibility, both to replace the LRU and internal to it.

BIT is always a major concern in avionics, since approximately one-half of the avionics LRUs removed from an aircraft that are thought to be faulty are in fact not faulty. A common way of stating this situation is that the mean time between unscheduled removals (MTBUR) is approximately 50% of the mean time between failure (MTBF). One avionics manufacturer has stated a goal of increasing MTBUR to 90% of MTBF for some of the B-777 avionics.

---

[3]In the X-29 aircraft a loss of power for more than 0.2 sec will result in an unrecoverable attitude requiring immediate ejection by the pilot.

TABLE 15.2   MIL-STD-704 and DO-160 aircraft power characteristics

Voltage	MIL-STD-704	DO-160
ac voltage (volts)		
Normal	108.0–118.0	104.0–122.0 A, E, Z^a
Over- and undervoltage	100.0–125.0	97.0–134.0
Emergency	108.0–118.0 (steady state)	NA
ac frequency (Hertz)		
Normal	393–407	380–420 A, E, Z
Abnormal	375–425	(Same as normal)
Emergency	360–440 (steady state)	350–440 A, E, Z
dc voltage (volts)		
Normal	22.0–29.0 (28.0 normal)	22.0–29.5 (27.5 nominal) A, Z
		24.8–30.3 (27.5 nominal) B
Over- and undervoltage	20.0–31.5	20.5–32.2 A, Z
		22.0–32.2 B
Emergency	18.0–29.0 (steady state)	18.0 A, Z; 20.0 B
Starting	12.0–29.0 (steady state)	10.0 for 15

[a]Refers to equipment categories described in DO-160 as follows: *category A* = mainly ac power with dc from transformer/rectifier units, *category B* = solely dc power with a large battery on the bus, *category E* = soley ac power, *category Z* = mainly ac power with dc from variable speed generators.

Maintenance costs are nearly the same for a LRU removal from service that later checks out to be operational, as for a LRU removal from service that is truly faulty. This is so because of the additional testing that a good LRU must undergo to determine that it is, in fact, good. As a typical design goal, BIT should be able to detect 98% of all possible faults in an LRU and identify 95% of them. Also, the BIT function should be able to test itself.

Another major consideration is who will be performing the maintenance and where it will be performed. The general lack of well-experienced maintenance personnel, coupled with the frequent need to get the aircraft back into the air as quickly as possible, dictate that maintenance procedures and equipment be easy to use. The availability of *notebook computers* with large memory capacity and reasonably large displays has led to advanced portable maintenance aids that significantly reduce the troubleshooting time and the unscheduled removals discussed earlier.

15.6   PHYSICAL INTERFACE

Packaging of the avionics is another type of "interface" with the host aircraft. Most modern avionics is packaged in accordance with widely used packaging standards. In commercial transport aircraft and some business jets, the avionics LRUs are designed in compliance with the *ARINC 600 Air Transport Avionics*

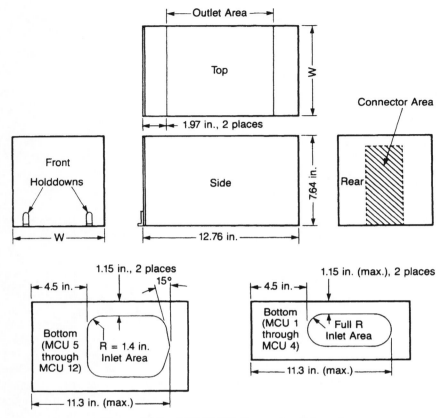

Figure 15.5 ARINC 600 line replaceable unit.

Equipment Interface. ARINC 600 establishes standards for the LRU dimensions, cooling air, and connector and tiedown placement. Figure 15.5 shows an ARINC 600 LRU. For any ARINC 600 LRU, the height and depth are constant, 7.64 and 12.75 in., respectively, but the width can vary from 25 to 256 mm in increments of 33.0 mm, depending on the number of printed circuit boards inside.

Military avionics is packaged in a much larger variety of nonstandard LRU designs. The principal approach to standardized packaging for military avionics is the standard electronics module specified in *MIL-M-28787 Modules, Standard Electronic, General Specification for.* There are several optional sizes described in that specification; however, the one most often selected is size E. Hence, there are frequent references to SEM-E modules in military avionics publications. SEM-E modules have a standard height of 6.68 in. and span (depth when viewed from the front) of 5.88 in. The width (thickness) can range from 0.38 to 0.58 in., in 0.1-in. increments.

Cooling the avionics is often a major issue, since conventional wisdom holds that the lower the operating temperature of the avionics, the lower is the failure rate, or, conversely, the longer is the MTBF.[4] For a cooling air inlet temperature of 30°, ARINC 600 requires a flow rate of 136 kg/hr/kw; for 40°C inlet air, the flow rate climbs to 200 kg/hr/kw, where kw is the amount of power being dissipated by the LRU. Another requirement in ARINC 600 is that the equipment must operate for up to 90 minutes *without* cooling air.[5] As a possible indicator of future practice, the B-777 aircraft will use passive cooling, in which the entire avionics bay is cooled and the ambient bay air cools the modules using natural convection flow through air gaps between the modules.

Military avionics is often cooled by the metal rail that grips the printed circuit board card when it is installed in the module. The rail is actively cooled by circulating fluid through it. Also under consideration for advanced military avionics high-power-dissipation cards is the immersion of the entire card directly into a liquid coolant.

Another critical interface with the host aircraft is the mounting alignment of certain types of avionics such as navigation sensors, radio-navigation antennas, radar antennas, and other sensor apertures. Alignment is also critical for HUDs and HMDs, which were discussed earlier in this chapter.

One concern in the realm of physical interfaces is electromagnetic interference (EMI). Modern digital avionics is especially susceptible to EMI because of the higher processor speeds, smaller element sizes on the electronic chips, and the increasing loss of shielding as composite materials replace metals. Because of the complex interaction of electromagnetic fields with the aircraft metallic structure and wiring, EMI effects are difficult to model and predict, although progress is being made using finite element methods in programs operating on high-speed computers. Consequently, many adverse EMI effects are not discovered until the avionics is installed on the host aircraft.

15.7 FUTURE TRENDS

The present trend toward increased integrated avionics and automated cockpits is expected to continue, driven by even higher-speed microprocessors and data buses. Over the next decade, more aircraft functions will be allocated to the avionics, including additional flight critical ones. The avionics will be packaged in modules that fit into a cabinet that provides cooling air, a backplane data bus,

---

[4]Some reliability experts doubt the validity of the assumption that higher temperatures, at least within the operating temperature range of most avionics, lead to higher failure rates. They believe instead that the failure rate is coupled to the number of on/off cycles an electronic component or device experiences.

[5]This requirement is driven by the typical airline route structure, which dictates that an aircraft should be capable of being dispatched with an *inoperative* cooling system for up to a $1\frac{1}{2}$-hr flight from a remote "spoke" or feeder airport with limited or no-maintenance capability to an airport where the cooling system can be repaired.

power, and EMI protection. Fault tolerance will become a major design feature of avionics to ensure error-free operation for long periods. Enhanced vision systems will be the cornerstone of enhanced situation awareness, leading to safer operations in low-visibility and crepuscular (e.g., twilight) conditions.

In the longer term, distributed avionics architectures may develop in which there will be multiple identical avionics cabinets distributed throughout the aircraft, each of which will be capable of performing all functions, including centralized ones such as maintenance monitoring and recording.

PROBLEMS

15.1. Discuss the advantages of fiber-optic data buses over wire data buses.

15.2. Why is it necessary to determine the data latency for a given data bus configuration?

15.3. State the advantages of liquid crystal displays over cathode-ray tube displays.

References

CHAPTER 1

[1] Aeronautical Radio, Inc. *ARINC Characteristics and Specifications for Navigation Equipment.* Airlines Electronic Engineering Committee. Annapolis, MD.

[2] Cohen, C. E., et al. Real-time flight testing using integrity beacons for GPS Category III precision landing. *Navigation* 41:2 (1994): 145–157, Institute of Navigation, Alexandria, VA.

[3] International Civil Aviation Organization. Reports of the Future Air Navigation System (FANS) Committee. Montreal, Canada.

[4] Kelly, R. J., and J. M. Davis. Required navigation performance for precision approach and landing with GNSS application. *Navigation* 41:1 (1994): 1–30, Institute of Navigation, Alexandria, VA.

[5] RTCA, Inc. Produces Minimum Aviation System Performance Specifications (MASPS) and Minimum Operational Performance Standards (MOPS) for airborne and ground avionics using DO-xxx report numbers. Washington, D.C.
(a) "Environmental Conditions and Test Procedures for Airborne Equipment," DO-160D (1997).
(b) "Minimum Operational Performance Standards for Airborne Automatic Dependent Surveillance (ADS) Equipment," DO-212, 1992.

[6] U.S. Departments of Transportation and Defense. *Federal Radionavigation Plan.* Published in even-numbered years. Washington, DC, 175 pp.

[7] U.S. Federal Aviation Administration. *Advisory Circulars* Especially the AC-20-xx and AC-90-xx series on navigation. Washington, DC.

[8] U.S. Federal Aviation Administration. *Technical Standard Orders TSO-C91* (1985) and *TSO C126* (1992). *Emergency Locator Transmitter* (Most recent issue). Washington, DC.

[9] U.S. Federal Aviation Administration. *Federal Aviation Regulations.* Part 25: Instruments, navigation, lightning protection; Part 91: Rules of flight; Part 97: Standard instrument approach procedures. Updated at irregular intervals. Washington, DC.

[10] U.S. Federal Aviation Administration. *Airman's Information Manual.* Updated quarterly. Washington, DC.

[11] EUROCAE (European Organisation for Civil Aviation Equipment). Produces Minimum Performance Specifications (MPS) and Minimum Operational Performance Specifications (MOPS) for airborne and ground avionics using ED-xxx report numbers. Some are issued jointly with RTCA. Paris, France.

[12] Bellamy, J. C. History of pressure pattern navigation. *Navigation* 43:1 (1996): 1–7, Institute of Navigation, Alexandria, VA.

[13] Bussert, J. C. China's great march toward modernization. *Avionics* 20:6 (1996): 34–39, Phillips Publishing, Rockville, MD.

CHAPTER 2

[1] Boozer, D. D., and J. R. Fellerhoff. Terrain aided navigation test results in the AFTI/F-16 aircraft. *Navigation* (Summer 1988): 161–175.

[2] Bowditch, N. *The American Practical Navigator.* United States Navy, Hydrographic Office, No. 9. Washington, DC. 1995.

[3] Forssell, B. *Radio Navigation Systems.* Englewood Cliffs, NJ: Prentice-Hall, 1991.

[4] Goehler, D. J. Future electronic charts and aeronautical databases. *Proceedings of the Annual Meeting of the Institute of Navigation*, June 1991, Institute of Navigation, Alexandria, VA.

[5] Golden, J. P. Terrain contour matching (TERCOM): A cruise missile guidance aid. *Society of Photo-optical Instrumentation Engineers* 238 (1980): 10–18.

[6] Goodchild, M., D. Maguire, and D. Rhind (eds.). *Geographical Information Systems: Principles and Applications.* London: Longmans, 1991.

[7] Hald, A., *Statistical Theory with Engineering Applications.* New York: Wiley, 1952.

[8] Hinrichs, P. R., Advanced terrain correlation techniques. *IEEE Plans Conference*, 1976, San Diego, CA. pp. 89–96.

[9] Hsu, D. Y. Closed-form solution for geodetic coordinate transformations. *Proceedings of the National Technical Meeting of the ION*, January 1992, pp. 397–400.

[10] Kayton, M. *Coordinate frames in inertial navigation*, 2 vols. MIT Draper Laboratory Report T-260. Cambridge, MA, 1960.

[11] Kovalesky, J., I. Mueller, and I. Kolaczek. *Reference Frames in Astronomy and Geophysics.* Dordrecht: Kluwer, 1989.

[12] Massatt, P., and K. Rudnick. Geometric formulas for dilution of precision calculations. *Navigation:* 37, 4 (Winter 1990–91): 379–391.

[13] McDonnell, P. W. *Introduction to Map Projections*, 2d ed. New York, M. Dekker, 1991, 174 pp.

[14] Pierce, B. O. *A Short Table of Integrals.* Boston: Ginn, 1929.

[15] Schwartz, C. (ed). *North American Datum of 1983.* Reston, VA: U.S. Geological Survey, 1990.

[16] Torge, W. *Geodesy*, 2d ed. Berlin, deGruyter, 1991, 264 pp.

[17] U.S. Departments of Transportation and Defense. *Federal Radionavigation Plan.* Published in even years. Washington, DC.

[18] U.S. Defense Mapping Agency. *Geodetic Distance and Azimuth Computations for Lines over 500 Miles.* ACIC Report 80. Washington, DC, 1959.

[19] U.S. Defense Mapping Agency. *Product Specification for ARC Digitized Raster Graphics.* ACIC Report PS/2DJ/100, Washington, DC, 1989.

[20] U.S. Defense Mapping Agency. *World Geodetic System 1984 (WGS 84)—Its Definition and Relationship with Local Geodetic Systems.* Maryfield, VA: Washington, DC. 1991.

[21] *Global Positioning System Standard Positioning Service: Signal Specification*, 2d ed. Los Angeles, CA: U.S. Air Force Space Division. March 1995.

[22] White, M. Technical requirements and standards for a multipurpose geographic data system. *The American Cartographer* 11(1984): 15–26.

[23] Zhu, J. Conversion of Earth-centered Earth-fixed to geodetic coordinates. *IEEE Transactions on Aerospace and Electronic Systems* 30, 3 (July 1994): 957–961.

[24] RTCA, Inc., Washington, DC.
(a) "Preparation, Verification, and Distribution of User-Selectable Navigation Data Bases," DO-200, 1988.
(b) "Minimum Operational Performance Standards for Airborne Area Navigation Equipment Using Multi-Sensor Inputs," DO-187, 1984.
(c) "Minimum Operational Performance Standards for Airborne Area Navigation Equipment Using a Single Colocated VOR/DME Sensor Input," DO-180A, 1990.
(d) "Software Considerations in Airborne Systems and Equipment Certification, DO-178D 1997.

[25] EUROCAE Paris, France.
(a) "Minimum Performance Specification for Airborne Area Navigation Computing Equipment Based on VOR and DME as Sensors," ED-28, 1982.
(b) "Minimum Operational Performance Requirement for Airborne Area Navigation Systems Based on Two DME as Sensors." ED-39, 1984.
(c) "Minimum Operational Performance Requirement for Area Navigation Equipment Using Multi-Sensor Inputs," ED-58, 1988.

[26] Aeronautical Radio, Inc. (ARINC), Annapolis, MD.
(a) ARINC Characteristic 424-13 "Navigation System Data Base," November 1998
(b) ARINC Characteristic 652 "Guidance for Avionics Software Management," January 1993.

CHAPTER 3

[1] Kalman, R. E. A new approach to linear filtering and prediction problems. *Transactions of the ASME, Journal of Basic Engineering* (March 1960).

[2] Huddle, J. R. Applications of Kalman filtering theory to augmented inertial navigation systems. *NATO-AGARDograph 139* (February 1970): ch. 11.

[3] Brown, R. G., and D. T. Friest. Optimization of a hybrid inertial solar-tracker navigation system. *1964 IEEE International Convention Record*, pt. 7.

[4] Bona, B. E., and R. J. Smay. Optimum reset of ship's inertial navigation system. *IEEE Transactions on Aerospace and Electronic Systems*, AES-2, no. 4 (July 1966).

[5] Meditch, J. S. *Stochastic Optimal Linear Estimation and Control*. New York: McGraw-Hill, 1969.

[6] Brown, R. G., and P. Y. C. Hwang. *Introduction to Random Signals and Applied Kalman Filtering*, 2d ed. New York: Wiley, 1992.

[7] Maybeck, P. S. *Stochastic Models, Estimation and Control*, vol. 1. New York: Academic Press, 1979.

[8] Van Dierendonck, A. J., J. B. McGraw, and R. G. Brown. Relationship between Allan variances and Kalman filter parameters. *Proceedings of the 16th Annual*

Precises Time and Time Interval (PTTI) Applications and Planning Meeting, NASA Goddard Space Flight Center, November 27–19, 1984.

[9] Ausman, J. S. A Kalman filter mechanization for the baro-inertial vertical channel. *Proceedings of the Institute of Navigation Forty-Seventh Annual Meeting,* Williamsburg, VA. June 1991, pp. 153–159.

[10] Bierman, G. J. *Factorization Methods for Discrete Sequential Estimation.* New York: Academic Press, 1977.

[11] Anoll, R. K. Integrated GPS/Loran-C receivers: A perspective on user acceptance, integration schemes, and certification status. *Proceedings of the 1994 National Technical Meeting,* Institute of Navigation, January 24–26, 1994.

[12] Buell, H., and A. Hunton. Development of an advanced system for helicopter applications. *Proceedings of the 1994 National Technical Meeting,* Institute of Navigation, San Diego, CA, January 24–26, 1994.

[13] Carlson, N. A., and M. P. Berarducci. Federated Kalman filter simulation results. *Institute of Navigation* 41, 3(Fall 1994): 297–321.

[14] Enge, P. K., and McCullough, J. R. Aiding GPS with calibrated Loran-C. *Institute of Navigation* 35, 4 (Winter 1988–89): 469–482.

[15] Per Enge, et al. Combining pseudoranges from GPS and Loran-C for air navigation. *Navigation, Journal of the Institute of Navigation* 37, 1 (Spring 1990).

CHAPTER 4

[1] Alford, A. Variable frequency radio beacon. U.S. Patent 2,241,897. May 13, 1941.

[2] Alford, A., and A. G. Kandoian. Ultra-high-frequency loop antennae. *Electrical Communication* 18, 5 (April 1940).

[3] Anderson, S. R., and R. B. Flint. The CAA Doppler omnirange. *Proceedings of the IRE* 47, 5 (May 1959).

[4] Britting, K. R. *Inertial Navigation Systems Analysis.* New York: Wiley-Interscience, 1971.

[5] Bose, K. W. *Aviation Electronics.* Casper, WY: IAP, Inc., 1983. Wild Goose Association. Loran-C system description. *Radionavigation Journal,* 1975 and 1976.

[6] Bourasseau, S. Differential Omega in Indonesia. *Proceedings of the Fifteenth Annual Meeting of the International Omega Association,* Sanur, Bali, Indonesia, September 1990.

[7] Colin, R. I., and S. H. Dodington. Principles of Tacan. *Electrical Communication* 33, 1 (March 1956).

[8] DeGroot, L. E. Navigation and control from Loran-C. *Navigation, Journal of the Institute of Navigation* 11, 3 (Autumn 1964).

[9] Doherty, R. H., G. Heffley, and R. F. Linfield. Timing potentials of Loran-C. *Proceedings of the IRE* 49, 11 (November 1961).

[10] Durbin, E. Current developments in the Loran-C system. *Navigation, Journal of the Institute of Navigation* (U.S.) 9, 2 (Summer 1962).

[11] Frank, R. L., and A. H. Phillips. Digital Loran-C receiver uses microcircuits. *Electronics* 37 (January 31, 1964).

[12] Gupta, R., S. Donnelly, P. Morris, and R. Vence, Jr. Omega system 10.2 kHz signal coverage diagrams. *Proceedings of the Fifth Annual Meeting of the Internal Omega Association,* Bergen, Norway, August 1980.

[13] Gupta, R., and P. Morris. Overview of Omega signal coverage. In *AGARDograph No. 314, Analysis, Design, and Synthesis Methods for Guidance and Control Systems*, ed. C. Leondes, Advisory Group for Aerospace Research and Development, North Atlantic Treaty Organization, June 1990.

[14] Hawkins, H. E., and O. LaPlant. Radar performance degradation in fog and rain. *IRE Transactions on Aerospace and Navigational Electronics* 6, 1 (March 1959).

[15] Henney, K. (ed.). *Radio Engineering Handbook* (section on Aviation Electronics by A. Casabona), New York: McGraw-Hill, 1959.

[16] Hildebrand, V. Omega validation highlights. *Proceedings of the Fourteenth Annual Meeting of the International Omega Association*, Long Beach, CA, October 1989.

[17] Hurley, H. C., S. R. Anderson, and H. F. Keany. The CAA VHF Omnirange. *Proceedings of the IRE* 39, 12 (December 1951).

[18] Johler, J. R. The Propagation Time of a Radio Pulse. *IEEE Transactions on Antennas and Propagation* 11, 4 (November 1963).

[19] Kraus, J. D. *Antennas*. New York: McGraw-Hill, 1950.

[20] Litchford, G., and J. Saganowich. An Omega/transponder display system. *Proceedings of the Eleventh Annual Meeting of the International Omega Association*, Quebec City, Canada, August 1986.

[21] Metz, H. I. International short distance navigation—after the ICAO February Meeting. *Proceedings of the National Aeronautical Electronics Conference*, Dayton, OH, 1959.

[22] Morris, P., and M. Cha. Omega propagation corrections: Background and computational algorithm. Report ONSOD 01-74. U.S. Coast Guard, Washington, DC. December 1974.

[23] Morris, P., and E. Swanson. New coefficients for the Swanson propagation correction model. *Proceedings of the Fifth Annual Meeting of the International Omega Association*, Bergen, Norway, August 1980.

[24] Morris, P. Omega system availability as a global measure of navigation accuracy. Report no. CG-ONSCEN-05-90. National Technical Information Service, no. AD-A229492, 1990.

[25] Morris, P., and R. Gupta. New approach to Omega PPCs. Report no. CG-ONSCEN-03-92. U.S. Coast Guard, Washington, DC. February 1992.

[26] Nard, G. P., 1980, Differential Omega navigation and equipments. *Proceedings of the Fourth Annual Meeting of the International Omega Association*, San Diego, CA, September 1979.

[27] Panter, R. *Modulation, Noise and Spectro-analysis Applied to Information Transmission*. New York: McGraw-Hill, 1965.

[28] Peterson, B., K. Gross, E. Chamberlin, and T. Montague. Integrated CIS VLF/Omega receiver design. *IEEE Aerospace and Electronic Systems Magazine* 8, 1 (January 1993).

[29] Pierce, J. The use of composite signals at very low radio frequencies. Harvard University Engineering and Applied Physics Division Technical Report 552, February 1968.

[30] Poritsky, P. (ed.). Special Issue on VOR/DME. *IEEE Transactions on Aerospace and Navigational Electronics* 12, 1 (March 1965).

[31] Powell, C. The Decca navigator system for ship and aircraft use. *Proceedings of the IEE*, England, 105, suppl., March 1968.

[32] Reynolds, P. Pan American World Airways Omega experience. *Proceedings of*

the Fourth Annual Meeting of the International Omega Association, San Diego, CA, September 1979.

[33] Sakran, C. U.S. Navy flight test results with the LTN-211 ONS. *Proceedings of the Sixth Annual Meeting of the International Omega Association,* Montreal, Canada, August 1981.

[34] Samaddar, S. N. The Theory of Loran-C ground wave propagation—A review. *Journal of the Institute of Navigation* 26, 3 (Fall 1979).

[35] *Reference Data for Engineers: Radio, Electronics, Computer and Communications,* 7th ed. Indianapolis, IN: Sams, Howard W. & Co., 1985.

[36] Skolnik, M. I. *Introduction to Radar Systems.* New York: McGraw-Hill, 1962.

[37] Swanson, E. A new approach to Omega coverage diagrams. *Proceedings of the Eighth Annual Meeting of the International Omega Association,* Lisbon, Portugal, July 1983.

[38] Terman, F. E. *Radio Engineering.* New York: McGraw-Hill, 1947.

[39] Warren, R., K. Tench, R. Gupta, and P. Morris. Omega ACCESS: A microcomputer display of Omega signal coverage diagrams. *Proceedings of the Eleventh Annual Meeting of the International Omega Association,* Quebec City, Canada, August 1986.

[40] Watt, A. *VLF Radio Engineering.* International Series of Monographs in Electromagnetic Waves, vol. 14. London: Pergamon Press, 1967.

[41] Westling, G. R. Joint Soviet/American Loran operations, The Bering Sea chain. IEEE Position, Location and Navigation and Symposium, 1988.

[42] Westman, H. P. (ed.). *Reference Data for Radio Engineers.* New York: International Telephone and Telegraph Corp., 1956.

[43] U.S. Department of Transportation, U.S. Coast Guard. *Omega Navigation System User's Guide.* COMDTPUB P1656.3. Washington, DC, 3 July 1990.

[44] U.S. Department of Transportation, U.S. Coast Guard, *Specification of the Transmitted Loran-C Signal.* Washington, DC, 29 May 1990.

[45] Per Enge, et al. Combining pseudoranges from GPS and Loran-C for air navigation. *Navigation, Journal of the Institute of Navigation* 37, 1 (Spring 1990).

[46] Anoll, R. K. Integrated GPS/Loran-C receivers: A perspective on user acceptance, integration schemes, and certification, status. *Proceedings of the 1994 National Technical Meeting, Institute of Navigation,* January 24–26, 1994, San Diego, CA.

[47] Enge, P. K., and J. R. McCullough. Aiding GPS with calibrated Loran-C. *Navigation, Journal of the Institute of Navigation* 34, 4 (Winter 1988–89): 469–482.

[48] U.S. Departments of Defense and Transportation. *1994 Federal Radionavigation Plan,* DOT-VNTSC-RSPA-95-1/DOD-4650.5. Washington, DC, 1994.

[49] RTCA, Inc. Washington, DC.
(a) "Minimum Operational Performance Standards for Airborne VOR Receiving Equipment," DO-196, 1986.
(b) "Minimum Operational Performance Standards for Airborne Area Navigation Equipment Using Loran-C Inputs," DO-194, 1986.
(c) "Minimum Operational Performance Standards for Airborne DME," DO-189, 1985.
(d) "Minimum Operational Performance Standards for Airborne ADF Equipment," DO-179, 1982.
(e) "Minimum Performance Standards for Airborne Omega Receiving Equipment," DO-164A, 1979.

(f) "Minimum Performance Standards for Airborne Radio Marker Receiving Equipment," DO-143, 1970.
[50] EUROCAE, Paris, France.
(a) "Minimum Performance Specification for Airborne VOR Receiving Equipment," ED-22B, 1988.
(b) "Minimum Performance Specification for Airborne Omega Navigation Equipment," ED-29, 1977.
(c) "Minimum Performance Specification for Airborne ILS Receiving Equipment (Localiser)," ED-46A, 1988.
(d) "Minimum Performance Specification for Airborne ILS Receiving Equipment (Glide Path)," ED-47A, 1988.
(e) "Minimum Performance Specification for Conventional and Doppler VHF Omnirange, Ground Equipment." ED-52, 1984.
(f) "Minimum Operational Performance Requirements for DME Interrogators" ED-54, 1987.
(g) "Minimum Performance Specification for DME Ground Equipment," ED-57, 1986.

CHAPTER 5

[1] International Civil Aviation Organization. *Special Committee on Future Air Navigation Systems, Fourth Meeting*, Doc. 9524, FANS/4, Montreal, May 2–20, 1988.

[2] RTCA, Inc. *RTCA Task Force Report on the Global Navigation Satellite System (GNSS), Transition and Implementation Strategy*, Washington, DC, September 1992.

[3] Stansell, T. A. *The Transit Navigation Satellite System—Status, Theory, Performance, Applications.* Magnavox Government and Industrial Electronics Company, 1978.

[4] Danchik, R. J. Navy navigation satellite system status. *Record of the Position Location and Navigation Symposium*, PLANS '88, Orlando, FL, November 29–December 2, 1988, pp. 21–24.

[5] Chobotov, V. A. (ed.). *Orbital Mechanics.* AIAA Education Series, Washington, DC: American Institute of Aeronautics and Astronautics, 1991.

[6] Kopitzke, E. *NAVSTAR Global Positioning System Satellite Navigation Ephemeris Algorithm.* Magnavox Government and Industrial Electronics Company Report R-5226, December 19, 1975.

[7] ARINC Research Corporation. *ICD-GPS-200B-PR, NAVSTAR GPS Space Segment/Navigation User Interfaces (Public Release Version).* Fountain Valley, CA, July 3, 1991.

[8] U. S. Department of Defense. *Global Positioning System Standard Positioning Service Signal Specification.* Washington, DC, June 2, 1995.

[9] U.S. Department of the Air Force, Headquarters Space Division (AFSC) SD/YED. *NAVSTAR Global Positioning System User Equipment Relevant World Geodetic System 1984 Technical Data Package*, Los Angeles, December 2, 1986.

[10] Blair, B. E. (ed.). *Time and Frequency, Theory and Fundamentals.* National Bureau of Standards Monograph 140. Boulder, CO, May 1974.

[11] Barnes, J. A. *Models for the Interpretation of Frequency Stability Parameters.* National Bureau of Standards Technical Note 683. Boulder, CO, August 1976.

[12] Van Dierendonck, A. J., J. B. McGraw, and R. G. Brown. Relationship between Allan variances and Kalman filter parameters. *Proceedings of the Sixteenth Annual Precise Time and Time Interval (PTTI) Applications and Planning Meeting.* NASA Goddard Space Flight Center, Greenbelt, MD, November 27–29, 1984, pp. 273–293.

[13] Klobuchar, J. A. Ionospheric effects on GPS. *GPS World.* Eugene, OR, April 1991, pp. 48–51.

[14] Boithias, L. *Radio Wave Propagation*, transl. by D. Beeson. New York: McGraw-Hill, 1987.

[15] Chao, C. C. *The Tropospheric Calibration Model for Mariner Mars 1971.* JPL Technical Report 32-1587, vol. 14. 1974.

[16] Parkinson, B. W., and S. W. Gilbert. NAVSTAR: Global positioning system—Ten years later. *Proceedings of the IEEE*, vol. 71, no. 10, October 1983, New York, NY, pp. 1177–1186.

[17] Van Melle, M. Cesium and rubidium frequency standard status and performance on the GPS program. *Proceedings of ION GPS-90, Third International Technical Meeting of the Satellite Division of the Institute of Navigation*, Colorado Springs, CO, September 19–21, 1990, pp. 123–137.

[18] Spilker, J. J., Jr. GPS signal structure and performance characteristics. *Global Positioning System*, vol. 1. Washington, DC: Institute of Navigation, 1980, pp. 29–54.

[19] Campbell, S. D., and R. R. LaFrey. Flight test results for an experimental GPS C/A-code receiver in a general aviation aircraft. *Global Positioning System*, vol. 2. Washington, DC: Institute of Navigation, 1984, pp. 239–257.

[20] Kalafus, R. M., J. Vilcans, and N. Knable. Differential Operation of NAVSTAR GPS. *Global Positioning System*, vol. 2. Washington, DC: Institute of Navigation, 1984, pp. 197–214.

[21] McNeff, J. G. GPS signal policy. *Proceedings of ION GPS-91, Fourth International Technical Meeting of the Satellite Division of the Institute of Navigation*, Albuquerque, September 11–13, 1991, pp. 33–37.

[22] Leach, M. P., M. A. Cardoza, and K. A. Duff. Assessment of GPS receiver performance in the presence of A-S. *Proceedings of the 1993 National Technical Meeting*, the Institute of Navigation, January 20–22, 1993, pp. 613–622.

[23] Van Dierendonck, A. J., Understanding GPS receiver terminology: A tutorial. *GPS World*, Eugene, OR, January 1995, pp. 34–44.

[24] U.S. Department of Transportation and Department of Defense. *1992 Federal Radio Navigation Plan*, Washington, DC, January 1993.

[25] Martinez, R. *Global Positioning System (GPS) Status Briefing.* RTCA Paper 316-93/SC159-453. Washington, DC, July 16, 1993.

[26] Milliken, R. J., and C. J. Zoller. Principle of operation of NAVSTAR and system characteristics. *Global Positioning System*, vol. 1. Washington, DC: Institute of Navigation, 1980, pp. 3–14.

[27] Durand, J. M., et al. GPS availability. Part I: Availability of service achievable for different categories of civil users. *Navigation, Journal of the Institute of Navigation*, 37, 2 (Summer 1990): 123–139.

[28] Durand, J. M., and A. Caseau. GPS availability. Part II: Evaluation of state prob-

abilities for 21 and 24 satellite constellations. *Navigation, Journal of the Institute of Navigation* 37, 3 (Fall 1990): 285–296.

[29] Phlong, W. S., and B. D. Elrod. Availability characteristics of GPS and augmention alternatives. *Navigation, Journal of the Institute of Navigation* 40, 4 (Winter 1993–94): pp. 409–428.

[30] Sams, M., A. J. Van Dierendonck, and Q. Hua. Satellite navigation accuracy and availability modeling as an air traffic management tool. *Proceedings of the 1995 National Technical Meeting*, the Institute of Navigation, Anaheim, CA, January 18–20, 1995.

[31] Francisco, S. G. Operational control segment of the global positioning system. *Record of Position Location and Navigation Symposium PLANS '84*, San Diego, November 26–29, 1984, pp. 51–58.

[32] Nagle, J. R., A. J. Van Dierendonck, and Q. D. Hua. Inmarsat-3 navigation signal C/A-code selection and interference analysis, *Navigation, Journal of the Institute of Navigation* 39, 4 (Winter 1992–93): 445–461.

[33] Dixon, R. C. *Spread Spectrum Systems*, 2d ed. New York: Wiley 1984.

[34] Van Dierendonck, A. J., S. S. Russell, E. R. Kopitzke, and M. Birnbaum. The GPS navigation message. *Global Positioning System*, vol. 1, Washington, DC: Institute of Navigation, 1980, pp. 55–73.

[35] Jorgensen, P. S. Relativity correction in GPS user equipment. *Record of the Position Location and Navigation Symposium*, PLANS '86, Las Vegas, November 4–7, 1986, pp. 177–183.

[36] Klobuchar, J. A. Design and characteristics of the GPS ionospheric time delay algorithm for single frequency users. *Record of the Position Location and Navigation Symposium*, PLANS '86, Las Vegas, November 4–7, 1986, pp. 280–286.

[37] Feess, W. W., and S. G. Stephens. Evaluation of GPS ionospheric time delay algorithm for single frequency users. *Record of the Position Location and Navigation Symposium*, PLANS 86, Las Vegas, November 4–7, 1986, pp. 206–213.

[38] Bartholomew, R. G., K. L. Nelson, L. S. Snow-Sollum, and K. L. Therkelsen. Software architecture of the family of DoD standard GPS receivers. *Proceedings of the Satellite Division First Technical Meeting*, the Institute of Navigation Satellite Division, Colorado Springs, CO, September 21–25, 1987, pp. 23–35.

[39] U.S. Department of Transportation, Federal Aviation Administration (FAA). TSO-C129, *Technical Standing Order, Airborne Supplemental Navigation Equipment Using the Global Positioning System (GPS)*. Washington, DC, 1992.

[40] RTCA, Inc. *Minimum Operational Performance Standards for Airborne Supplemental Navigation Equipment Using Global Positioning System (GPS)*. Document No. RTCA/DO-208, Prepared by SC-159, July 1991, including change 1, 1993. Washington, DC.

[41] RTCA, Inc. *Minimum Aviation System Performance Standards, DGNSS Instrument Approach System: Special Category 1 (SCAT-1)*. Document No. RTCA/DO-217. Prepared by SC-159, August 27, 1993. Washington, DC.

[42] Aeronautical Radio, Inc. *ARINC Characteristic 743A-1 GNSS Sensor*. Prepared by the Airlines Electronic Engineering committee, November 8, 1993. Annapolis, MD.

[43] Fenton, P., B. Falkenberg, T. Ford, K. Ng, and A. J. Van Dierendonck. NovAtel's GPS receiver—The high performance OEM sensor of the future. *Proceedings of ION GPS-91, Fourth International Technical Meeting of the Satellite Division of the Institute of Navigation*, Albuquerque, September 11–13, 1991, pp. 49–58.

[44] Van Dierendonck, A. J., P. Fenton, and T. Ford. Theory and performance of narrow correlator spacing in a GPS receiver. *Navigation, Journal of the Institute of Navigation*, 39, 3 (Fall 1992): 265–283.

[45] Chang, H. Presampling filtering, sampling and quantization effects on digital matched filter performance. *Proceedings of the International Telemetering Conference*, San Diego, 1982, pp. 889–915.

[46] Turin, G. L. An introduction to digital matched filters. *Proceedings of the IEEE*, vol. 64, no. 7, July 1976, pp. 1092–1112.

[47] Lim, T. L. Noncoherent digital matched filters: Multi-bit quantization. *IEEE Transactions on Communications*, vol. COM-26, no. 4, April 1978, pp. 409–419.

[48] Amoroso, F. Adaptive A/D converter to suppress CW interference in spread spectrum communications. *IEEE Transactions on Communications* 31, 10 (October 1983): 1117–1123.

[49] Amoroso, F., and J. L. Bricker. Performance of the adaptive A/D converter in combined CW and Gaussian interference. *IEEE Transactions on Communications*, vol. COM-34, no. 3, March 1986, pp. 209–213.

[50] Bricker, J. L. Mathematical methodology for performance analysis in combined CW and Gaussian interference. *MILCOM '84 Conference Record*, October 1984, pp. 39.2.1–39.2.7.

[51] Litton Guidance and Control Systems. Data Sheet for the Global Positioning Guidance Package, Woodland Hills, CA, 1992.

[52] NAVSTAR GPS Joint Program Office. *Specification for NAVSTAR Global Positioning System (GPS) Miniaturized Airborne GPS Receiver (MAGR)*. Final draft. Specification No. CI-MAGR-300, Code Identification: 07868, March 30, 1990.

[53] NAVSTAR GPS Joint Program Office. *Guidelines for NAVSTAR Global Positioning System (GPS) Embedded GPS Receiver (EGR) Applications*. GPS-EGR-600. Code Identification: 07868, April 30, 1992.

[54] Latterman, D., E. Emile, and C. A. Wu. Guidelines for NAVSTAR GPS embedded receiver applications. *Proceedings of the National Technical Meeting*, the Institute of Navigation, San Diego, January 27–29, 1992, pp. 81–88.

[55] Frank, G. B., and M. D. Yakos. Collins next generation digital GPS receiver. *Record of IEEE PLANS '90 Position Location and Navigation Symposium*, Las Vegas, March 21–23, 1990, pp. 286–292.

[56] Private communication with Ken Plate of Dorne & Margolin, 1992.

[57] Gray, D. E., and D. C. Forseth. Rockwell International's miniature high performance GPS receiver. *Proceedings of ION GPS-89, Second International Technical Meeting of the Satellite Division of the Institute of Navigation*, Colorado Springs, CO, September 27–29, 1989, pp. 207–216.

[58] Rambo, J. C. Receiver processing software design of the Rockwell International DoD standard GPS receivers. *Proceedings of ION GPS-89, Second International Technical Meeting of the Satellite Division of the Institute of Navigation*, Colorado Springs, CO, September 27–29, 1989, pp. 217–225.

[59] Moen, V., S. Rankin, J. Kacirek, and R. Bartholomew. Design, capabilities and performance of the miniaturized airborne GPS receiver. *Record of IEEE PLANS '92 Position Location and Navigation Symposium*, Monterey, CA, March 25–27, 1992, pp. 290–297.

[60] Karicek, J. L., and R. G. Bartholomew. Signal acquisition and tracking in the

DoD standard miniaturized airborne GPS receiver. *Proceedings of the Forty-Eight Annual Meeting*, the Institute of Navigation, Washington, DC, June 29–July 1, 1992, pp. 53–61.

[61] Johnson, C. R., P. W. Ward, M. D. Turner, and S. D. Roemerman. Applications of a multiplexed GPS user set. *Global Positioning System*, vol. 2. Washington, DC: Institute of Navigation, 1984, pp. 61–77.

[62] Beser, J., and B. W. Parkinson. The application of NAVSTAR differential GPS in the civilian community. *Global Positioning System*, vol. 2. Washington, DC: Institute of Navigation, 1984, pp. 167–196.

[63] RTCM Special Committee No. 104. *RTCM Recommended Standards for Differential NAVSTAR GPS Service*. Version 2.0. Washington, DC: Radio Technical Commission for Maritime Services, January 1, 1990.

[64] Kalafus, R. M., A. J. Van Dierendonck, and N. A. Pealer. Special committee 104 recommendations for differential GPS service. *Global Positioning System*, vol. 3. Washington, DC: Institute of Navigation, 1986, pp. 101–116.

[65] Gloeckler, F., A. J. Van Dierendonck, and R. R. Hatch. Proposed revisions to RTCM SC-104 recommended standards for differential NAVSTAR GPS service for carrier phase applications. *Proceedings of ION GPS-92, Fifth International Technical Meeting of the Satellite Division of the Institute of Navigation*, Albuquerque, September 16–18, 1992, pp. 625–634.

[66] Hundley, W., S. Rowson, G. Courtney, V. Wullschleger, R. Velez, and P. O'Donnell. Flight evaluation of a basic C/A code differential GPS landing system for Category I precision approach. *Navigation, Journal of the Institute of Navigation* 40, 2 (Summer 1993): 161–178.

[67] Wullschleger, V., R. Velez, W. Hundley, S. Rowson, G. Courtney, and P. O'Donnell. FAA/Wilcox flight test results of DGPS system for precision approach. *Proceedings of the Forty-Ninth Annual Meeting*, the Institute of Navigation, Cambridge, MA, June 21–23, 1993, pp. 111–118.

[68] Rowson, S. V., G. R. Courtney, and R. M. Hueschen. Performance of Category IIIB automatic landings using C/A-code tracking differential GPS. *Navigation, Journal of the Institute of Navigation* 41, 2 (Summer 1994): 127–144.

[69] Loh, R. Seamless aviation: FAA's wide area augmentation system. *GPS World*, Eugene, OR, April 1995, pp. 20–30.

[70] Cannon, M. E. High-accuracy GPS semikinematic positioning: modeling and results. *Navigation, Journal of the Institute of Navigation* 37, 1 (Spring 1990): 53–64.

[71] Hatch, R. R. Ambiguity resolution while moving—Experimental results. *Proceedings of ION GPS-91, Fourth International Technical Meeting of the Satellite Division of the Institute of Navigation*, Albuquerque, September 11–13, 1991, pp. 707–713.

[72] Lachapelle, G., M. E. Cannon, and G. Lu. Ambiguity resolution on the fly—A comparison of P code and high performance C/A code receiver technologies. *Proceedings of ION GPS-92, Fifth International Technical Meeting of the Satellite Division of the Institute of Navigation*, Albuquerque, September 16–18, 1992, pp. 1025–1032.

[73] Cohen, C. E., B. Pervan, H. S. Cobb, D. Lawrence, J. D. Powell, and B. W. Parkinson. Cycle ambiguity resolution using a pseudolite for precision landing of aircraft with GPS. *Proceedings of the Second International Symposium on Differential Satellite Navigation Systems*, Amsterdam, March 30–April 2, 1993.

[74] Patton, S. L., D. L. Van Dusseldorp, and R. G. Bartholomew. Performance analysis of a miniaturized airborne GPS receiver. *Proceedings of ION GPS-91, Fourth International Technical Meeting of the Satellite Division of the Institute of Navigation*, Albuquerque, September 11–13, 1991, pp. 413–422.

[75] Conley, R. GPS performance: What is normal? *Navigation, Journal of the Institute of Navigation* 40, 3 (Fall 1993): 261–282.

[76] Kazantsev, V. N., M. F. Reshetnev, A. G. Kozlov, and V. F. Cheremisin. Current status, development program and performance of the GLONASS System. *Proceedings of ION GPS-92, Fifth International Technical Meeting of the Satellite Division of the Institute of Navigation*, Albuquerque, September 16–18, 1992, pp. 139–144.

[77] Dale, S. A., and P. Daly. Recent observations with the Soviet Union's GLONASS navigation satellites. *Record of the Position Location and Navigation Symposium*, PLANS '86, Las Vegas, November 4–7, 1986, pp. 20–25.

[78] Daly, P. Review of GLONASS system characteristics. *Proceedings of ION GPS-90, Third International Technical Meeting of the Satellite Division of the Institute of Navigation*, Colorado Springs, CO, September 19–21, 1990, pp. 267–275.

[79] *Global Orbiting Navigation Satellite System (GLONASS) Interface Control Document.* International Civil Aviation Organization. GNSSP/2-WP/66, Montreal, Quebec, Canada, November 14, 1995.

[80] Lennen, G. R. The USSR's GLONASS P-code—Determination and initial results. *Proceedings of ION GPS-89, Second International Technical Meeting of the Satellite Division of the Institute of Navigation*, Colorado Springs, CO, September 27–29, 1989, pp. 77–83.

[81] International Frequency Registration Board (IFRB). *Circular 1522.* Special Section No. AR 11/A/3, ITU, Geneva, Switzerland, 1985.

[82] International Telecommunication Union Radiocommunication Study Groups. *Technical Description and Characteristics of Global Space Navigation System GLONASS-M*, Document 8D/46-E, November 22, 1994.

[83] International Telecommunication Union Radiocommunication Study Groups. *Technical Description and Characteristics of Global Space Navigation System GLONASS-M, Addendum 1.* Document 8D/46(Add. 1)-E, December 6, 1994.

[84] Kitching, I. D., S. A. Dale, and P. Daly. Time transfer with GLONASS navigation satellites. *Proceedings of the Satellite Division's International Technical Meeting, the Institute of Navigation*, Colorado Springs, CO, September 19–23, 1988, pp. 487–493.

[85] Misra, P., R. I. Abbot, and E. M. Gaposchkin. Transformation between WGS84 and PZ-90. *Proceedings of ION GPS-96, Ninth International Technical Meeting of the Satellite Division of the Institute of Navigation*, Kansas City, MO, September 17–20, 1996.

[86] Misra, P., M. Pratt, R. Muchnik, B. Burke, and T. Hall. GLONASS Performance: Measurement Data Quality and System Upkeep. *Proceedings of ION GPS-96, Ninth International Technical Meeting of the Satellite Division of the Institute of Navigation*, Kansas City, MO, September 17–20, 1996.

[87] GPS World. *Global View—Newsfront.* Eugene, OR, October 1996, p. 18.

[88] Loh, R. FAA program plan for GIB/WADGPS. RTCA Paper 89-93/SC-159-415, February 17, 1993.

[89] Shaw, M., T. Simpson, and K. Sandhoo. FAA implementation planning for the use of GPS in air navigation and landing, *Proceedings of the National Technical Meeting, the Institute of Navigation,* Anaheim, CA, January 18–20, 1995, pp. 11–18.

[90] Van Graas, F., and J. Farrell. Baseline fault detection and exclusion algorithm. *Proceedings of the 49th Annual Meeting, the Institute of Navigation,* Cambridge, MA, June 21–23, 1993, pp. 413–420.

[91] Brown, R. G., J. Kraemer, and G. Nim. A partial identification RAIM algorithm for GPS sole means navigation. *Proceedings of ION GPS-94, Seventh International Technical Meeting of the Satellite Division of the Institute of Navigation,* Salt Lake City, September 20–23, 1994, pp. 557–565.

[92] Brown, R. G. A baseline RAIM scheme and a note on the equivalence of three RAIM methods. *Navigation, the Journal of the Institute of Navigation* 39, 3 (Fall 1992): 301–316.

[93] Parkinson, B. W., and P. Axelrad. Autonomous GPS integrity monitoring using the pseudorange residual. *Navigation, Journal of the Institute of Navigation* 35, 2 (Summer 1988): 255–274.

[94] Van Dyke, K. L. RAIM availability for supplemental GPS navigation. *Navigation, Journal of the Institute of Navigation,* 39, 4 (Winter 1992–93): 429–444.

[95] Brown, R. G. RAIM and GIC working together: The ultimate solution to the GPS integrity problem. *Navigation, Journal of the Institute of Navigation* 36, 2 (Summer 1989): 173–178.

[96] Daly, P. Progress towards joint civil use of GPS & GLONASS. *Record of IEEE PLANS '92 Position Location and Navigation Symposium,* Monterey, CA, March 25–27, 1992, pp. 1–6.

[97] Daly, P., S. Riley, and P. Raby. GLONASS status and initial C/A and P-code ranging tests. *Proceedings of ION GPS-92, Fifth International Technical Meeting of the Satellite Division of the Institute of Navigation,* Albuquerque, September 16–18, 1992, pp. 145–151.

[98] U.S. Department of Transportation, Federal Aviation Administration. *Wide Area Augmentation System (WAAS) Specification.* FAA-E-2892. Washington, DC, May 9, 1994.

[99] Lundberg, O. Way points for radio navigation in the 21st century. *Proceedings of ION GPS-94, Seventh International Technical Meeting of the Satellite Division of the Institute of Navigation,* Salt Lake City, September 20–23, 1994, pp. 3–15.

[100] Van Dierendonck, A. J., J. Nagle, and G. V. Kinal. Evolution to civil GNSS taking advantage of geostationary satellites. *Proceedings of the 49th Annual Meeting, the Institute of Navigation,* Cambridge, MA, June 21–23, 1993, pp. 231–240.

[101] Kinal, G. V., and A. J. Dierendonck. Susceptibility of Inmarsat navigation payloads to jamming and spoofing: Fact or (science) fiction? *Proceedings of the 1993 National Technical Meeting, the Institute of Navigation,* San Francisco, January 20–22, 1993, pp. 459–468.

[102] Van Dierendonck, A. J., and P. Enge. The Wide Area Augmentation System (WAAS) signal specification. *Proceedings of ION GPS-94, Seventh International Technical Meeting of the Satellite Division of the Institute of Navigation,* Salt Lake City, September 20–23, 1994, pp. 985–994.

[103] Enge, P., A. J. Van Dierendonck, and G. Kinal. A signal design for the GIC which includes capacity for WADGPS data. *Proceedings of ION GPS-92, Fifth International Technical Meeting of the Satellite Division of the Institute of Navigation,* Albuquerque, September 16–18, 1992, pp. 875–884.

[104] RTCA, Inc. *Minimum Operational Performance Standard for Global Positioning System/Wide Area Augmentation System Airborne Equipment.* RTCA/DO-229, January 16, 1996.

[105] Schwartz, M. *Information Transmission, Modulation, and Noise,* 4th ed. New York: McGraw-Hill, 1990.

[106] Walter, T., P. Enge, and F. Van Graas. Integrity for the Wide Area Augmentation System. *Proceedings of DSNS 95, Fourth International Conference on Differential Satellite Navigation Systems,* Bergen, Norway, April 24–28, 1995.

[107] Klein, D., and B. W. Parkinson. The use of pseudo-satellites for improving GPS performance. *Global Positioning System,* vol. 3. Washington, DC: Institute of Navigation, 1986, pp. 135–146.

[108] Stansell, T. A., Jr. RTCM SC-104 recommended pseudolite signal specification. *Global Positioning System,* vol. 3. Washington, DC: Institute of Navigation, 1986, pp. 117–134.

[109] Parkinson, B. W., and K. T. Fitzgibbon. Optimal locations of pseudolites for differential GPS. *Navigation, Journal of the Institute of Navigation* 33, 4 (Winter 1986–87): 259–283.

[110] Van Dierendonck, A. J., B. D. Elrod, and W. C. Melton. Improving the integrity, availability and accuracy of GPS using pseudolites. *Proceedings of NAV '89, Royal Institute of Navigation Conference,* October 17–19, 1989, no. 32.

[111] Schuchman, L., B. D. Elrod, and A. J. Van Dierendonck. Applicability of an augmented GPS for navigation in the national airspace system. *Proceedings of the IEEE* 77, 11 (November 1989): 1709–1727.

[112] Sams, M. Cat II/III accuracy and availability for local area differential GPS including geostationary satellite and pseudolite augmentations. *Proceedings of the 51st Annual Meeting, the Institute of Navigation,* Colorado Springs, CO, June 5–7, 1995, pp. 481–490.

[113] Elrod, B., K. Barltrop, and A. J. Van Dierendonck. Testing of GPS augmented with pseudolites for precision approach applications. *Proceedings of ION GPS-94, Seventh International Technical Meeting of the Satellite Division of the Institute of Navigation,* Salt Lake City, September 20–23, 1994, pp. 1269–1278.

[114] Van Dierendonck, A. J. The role of pseudolites in the implementation of differential GPS. *Record of IEEE PLANS '90 Position Location and Navigation Symposium,* Las Vegas, March 21–23, 1990, pp. 370–377.

[115] Elrod, B. D., and A. J. Van Dierendonck. Testing and evaluation of GPS augmented with pseudolites for precision landing applications. *Proceedings of the Second International Symposium on Differential Satellite Navigation Systems,* Amsterdam, March 30–April 2, 1993.

[116] McGraw, G. A. Analysis of pseudolite code interference effects for aircraft precision approaches. *Proceedings of the 50th Annual Meeting, the Institute of Navigation,* Colorado Springs, CO, June 6–8, 1994, pp. 433–437.

[117] Klass, P. J., Inmarsat orbit to impact aviation. *Aviation Week & Space Technology,* March 29, 1993, pp. 58–59.

[118] Time for action on global NAVSAT. Editorial, *Aviation Week & Space Technology,* August 2, 1993, p. 70.

[119] Inmarsat narrows choices for new satellite orbit. *Aviation Week & Space Technology*, August 2, 1993, p. 26.

[120] *Report of the FANS GNSS Technical Subgroup Third Meeting*, London, England, February 18–21, 1992, sec. 7.1.

[121] Parkinson, B. W. and J. J. Spilker, Jr. (eds.) *Global Positioning System: Theory and Applications*, Vols. 1 and 2. (Vols. 163 and 164 of *Progress in Astronautics and Aeronautics*), Washington, DC: American Institute of Aeronautics and Astronautics, Inc., 1996.

[122] RTCA Inc. Washington, DC.
(a) "Minimum Operational Performance Standards for Global Positioning System/Wide Area Augmentation System Airborne Equipment," DO-229, 1996.
(b) "Minimum Operational Performance Standards for GNSS Airborne Antenna Equipment," DO-228 1995.

[123] EUROCAE "Minimum Operational Performance Specification for Airborne GPS Receiving Equipment." ED-72, 1992. Paris, France.

[124] Filter Center. *Aviation Week & Space Technology*, October 14, 1996, p. 64.

[125] GPS World. *Global View—Newsfront*. Eugene, OR, June 1996, pp. 16–18.

[126] GPS World. *Washington View—GNSS/MSS Spectrum Battle*. Eugene, OR, June 1996, pp. 12–14.

[127] GPS World Newsletter. Eugene, OR, September 27, 1996.

[128] GPS World Newsletter. Eugene, OR, October 11, 1996.

[129] RTCA, Inc. *Appendix A, Wide Area Augmentation System Signal Specification*, RTCA Paper 379-96/SC159-721, October 6, 1996.

[130] Sams, M., A. J. Van Dierendonck, and Q. Hua. Availability and continuity performance modeling. *Proceedings of the 52nd Annual Meeting of the Institute of Navigation*, Cambridge, MA, June 19–21, 1996, pp. 289–298.

[131] Barltrop, K. J., J. F. Stafford, and B. D. Elrod. Local DGPS with pseudolite augmentation and implementation considerations in LAAS. *Proceedings of ION GPS-96, Ninth International Technical Meeting of the Satellite Division of the Institute of Navigation*, Kansas City, MO, September 17–20, 1996.

[132] Elrod, B., K. Barltrop, J. Stafford, and D. Brown. Test results of local area augmentation with an in-band pseudolite. *Proceedings of the National Technical Meeting of the Institute of Navigation*, Santa Monica, CA, January 22–24, 1996, pp. 145–154.

[133] Kaplan, Elliott D. et al, *Understanding GPS, Principles and Applications*, Artech House, 1996.

[134] Sandhoo, K. S. Second Civil GPS Frequency Targeted for Block IIF Satellites. *ION Newsletter*, The Institute of Navigation, Alexandria, VA, Spring/Summer 1996, p. 3.

[135] Van Dyke, K. L. Removal of SA: Benefits to GPS Integrity. *Proceedings of ION-GPS-96, Ninth International Technical Meeting of the Satellite Division of the Institute of Navigation*, Kansas City, MO, September 17–20, 1996.

[136] Enge P. and A. J. Van Dierendonck. Design of the Signal and Data Format for Wide Area Augmentation of the Global Positioning System. *Proceedings of the 1996 IEEE Position Location and Navigation Symposium (PLANS)*, Atlanta, GA, April 1996, pp. 485–495.

[137] Brown, D. et al, FAA Static and Flight Test Results for Testbed Version 2 of the Wide Area Augmentation System (WAAS). Proceedings of the 1996 IEEE Posi-

tion Location and Navigation Symposium (PLANS), Atlanta, GA, April 1996, pp. 496–502.

CHAPTER 6

[1] Altrichter, W. W. JTIDS relative navigation and data registration. *Proceedings of the 1992 IEEE Position Location and Navigation Symposium (PLANS)*, Monterey, CA, March 24–27, 1992. IEEE Catalog No. 92ch3085-8.

[2] Altrichter, W. W. Development of JTIDS geodetic and relative navigation theory. GEC-Marconi Electronic Systems Corporation, Document R256A152, April 1989.

[3] Amoroso, F., and J. Kivett. Simplified MSK signaling technique. *IEEE Transactions on Communications* (April 1977).

[4] Chadwick, J. B. JTIDS RelNav network off-line simulation. *Proceedings of the 1980 IEEE Position Location and Navigation Symposium (PLANS)*, Atlantic City, NJ, November 1980.

[5] Cook, R. E., et al. Enhanced PLRS development testing and overview and status report. *Proceedings of the 1988 IEEE Position Location and Navigation Symposium (PLANS)*, November 1988.

[6] Dell-Imagine, R. A. JTIDS—An overview of the system design and implementation. *Proceedings of the IEEE Position Location and Navigation Symposium.* IEEE Publication 76-CH1138-7 AES, pp. 212–218, November 1976.

[7] Dunn, R. A natural parameter-controller specification procedure for an integrated radio/dead reckoner navigation system. *Proceedings of the 1980 IEEE Position Location and Navigation Symposium (PLANS)*, November 1980.

[8] Dunn, R., et al. Design and development of a second generation relative navigation analytical simulator for JTIDS full scale development. *Proceedings of the 1982 IEEE Position Location and Navigation Symposium (PLANS)*, November 1982.

[9] Fried, W. R. Principles and simulation of JTIDS relative navigation. *IEEE Transactions on Aerospace and Electronic Systems* 14, 1 (January 1978).

[10] Fried, W. R., and R. Loeliger. Principles, system configuration and algorithm design of the inertially aided JTIDS relative navigation function. *Navigation, Journal of the Insitute of Navigation* 26, 3 (Fall 1979).

[11] Fried, W. R. Operational benefits and design approaches for combining JTIDS and GPS navigation. *Navigation, Journal of the Institute of Navigation* 31, 2 (Summer 1984).

[12] Fried, W. R. JTIDS relative navigation—principles, architecture and inertial mixing. *AGARDograph No. 314. Analysis, Design and Synthesis Methods for Guidance and Control Systems.* North Atlantic Treaty Organization (NATO) Advisory Group for Research and Development, June 1990.

[13] Fried, W. R. Principles, simulation results and interoperability of JTIDS relative navigation. *Proceedings of the 1976 IEEE Position Location and Navigation Symposium (PLANS).* IEEE Publication 76CH1138-7 AES, pp. 216–222.

[14] Greenberg, M. S. Post-flight assessment of JTIDS RelNav. *Proceedings of the 1980 IEEE Position Location and Navigation Symposium (PLANS)*, November 1980. Atlantic City, NJ.

[15] Kerr, H. A stable decentralized filtering implementation for JTIDS RelNav.

Proceedings of the 1980 IEEE Position Location and Navigation Symposium (PLANS), November 1980.

[16] Kivett, J. A. PLRS—A new spread spectrum position location reporting system. Proceedings of the 1976 IEEE Position Location and Navigation Symposium (PLANS), November 1976, IEEE Publication 76CH1138-7AES.

[17] Kivett, J. A., and U. S. Okawa. PLRS—A new spread spectrum position location reporting system. *AGARDograph No. 314. Analysis Design and Synthesis Methods for Guidance and Control Systems.* North Atlantic Treaty Organization (NATO) Advisory Group for Research and Development, June 1990.

[18] Kivett, J. A., and F. L. Morse. PLRS engineering development testing-early results. *Proceedings of the 1980 IEEE Position Location and Navigation Symposium (PLANS)*, November 1980.

[19] Kivett, J. A., and R. E. Cook. Enhancing PLRS with user-to-user data capability. *Proceedings of the 1986 IEEE Position Location and Navigation Symposium (PLANS)*, November 1986.

[20] Klein, D. Ship response model development and RelNav performance impact analysis. *Proceedings of the 1982 IEEE Position Location and Navigation Symposium (PLANS)*, November 1982.

[21] Kriegsman, B. A., and W. M. Stonestreet. A navigation filter for an integrated GPS-JTIDS-INS system for tactical aircraft. *Proceedings of the 1978 IEEE Position Location and Navigation Symposium (PLANS)*, November 1978.

[22] Leondes, C. T. (ed.). Principles and operational aspects of precision position determination. *NATO AGARDograph No. 245*, July 1979, pp. 25-1–40-8.

[23] Lull, J. M., et al. Coordinate transformation in PLRS. *Proceedings of the 1980 IEEE Position Location and Navigation Symposium (PLANS)*, November 1980.

[24] Okawa, U. S., and R. Fincher. PLRS development testing—An update. *Proceedings of the 1986 IEEE Position Location and Navigation Symposium (PLANS)*, November 1986.

[25] Rome, J. H., et al. Enhanced noise immunity and error control in a fully integrated JTIDS/GPS receiver. *Proceedings of the 1980 IEEE Position Location and Navigation Symposium (PLANS)*, November 1980.

[26] Rubin, J., and S. Welt. JTIDS distributed TDMA (DTDMA) terminal development—Results with emphasis on relative navigation performance. *Proceedings of the 1980 IEEE Position Location and Navigation Symposium (PLANS)*, November 1980.

[27] Sacks, J. Grid merging approaches for JTIDS stage one operation. *Proceedings of the 1982 IEEE Position Location and Navigation Symposium (PLANS)*, November 1982. Atlantic City, NJ.

[28] Schneider, A. M. Observability of relative navigation using range-only measurements. *Proceedings of the 1982 IEEE Position Location and Navigation Symposium (PLANS)*, November 1982.

[29] Westbrook, E. A., and R. C. Snodgrass. Relative navigation by passive ranging in a synchronous time division multiple access data net. The MITRE Corporation MTR-2996. Prepared for Electronic Systems Division, U.S. Air Force, March 1975.

[30] Widnall, W. S., and J. F. Kelley. JTIDS relative navigation with measurement sharing: Design and performance. *Proceedings of the 1984 IEEE Position Location and Navigation Symposium (PLANS)*, November 1984.

CHAPTER 7

[1] Anderson, D. Z. Optical gyroscopes. *Scientific American* 254 (April 1986): 86–91.

[2] Ausman, J. S. A Kalman filter mechanization for the baro-inertial vertical channel. *Proceedings of the Institute of Navigation Forty-Seventh Annual Meeting,* Williamsburg, VA. June 1991, pp. 153–159.

[3] Ausman, J. S. Baro-inertial loop for the USAF standard RLG INU. *Navigation, Journal of the Institute of Navigation* 38, 2 (Summer 1991): 205–220.

[4] Bortz, J. E. A new mathematical formulation for strapdown inertial navigation. *IEEE Transactions on Aerospace and Electronics Systems* Vol. AES-7, 1 (January 1971), pp. 61–66.

[5] Brown, A., R. Ebner, and J. Mark. A calibration technique for a laser gyro strapdown inertial navigation system. *Proceedings Symposium Gyro Technology,* Stuttgart, Germany, September 1982.

[6] Caliguiri, J. F. SGN-10, first commercial inertial navigator. *Journal of the Institute of Navigation* (Spring 1967): 85–92.

[7] Carpentier, J., J. C. Radix, J. Bouvet, and G. Bonnevalle. *Navigation par Inertie.* Paris: Dunod, 1962, 287 pp.

[8] Chow, W. W., J. Gea-Banacloche, L. M. Pedrotti, V. E. Sanders, W. Schlech, and M. O. Scully. The ring laser gyro. *Review of Modern Physics* 57, 1 (January 1985).

[9] Craig, R. J. Theory of errors of a multigimbal, elastically supported, tuned gyroscope. *IEEE Transactions on Aerospace and Electronic Systems* 8, 3 (May 1972), 289–297.

[10] Craig, R. J. Theory of operation of elastically supported tuned gyroscope. *IEEE Transactions on Aerospace and Electronic Systems* 8, 3 (May 1972), pp. 280–288.

[11] Ezekiel, S., and H. Arditty (eds.). *Fiber Optic Rotation Sensors and Related Technologies.* New York: Springer-Verlag, 1982.

[12] Fernandez, M., and G. R. Macomber. *Inertial Guidance Engineering,* Englewood Cliffs, NJ: Prentice-Hall, 1962, 530 pp.

[13] Friedman, A. L. Use of speed and position fix information in inertial navigators, AIAA Paper 1957–61, 1961, 13 pp.

[14] Gelb, A. (ed.). The analytic sciences corporation. *Applied Optimal Estimation,* Cambridge: MIT Press, 1984.

[15] Gilster, G. High accuracy performance capabilities of the military standard ring laser gyro inertial navigation unit. *Proceedings of IEEE PLANS,* April 11–15, 1994, Las Vegas, NV, pp. 464–473.

[16] Goodman, L. E., and A. R. Robinson. Effect of finite rotations on gyroscopic sensing devices. ASME Paper 57-A-30, 1957, 4 pp.

[17] Hadfield, M. J., and R. E. Wheeler. Update 89: Additional results with the multifunction RLG system. *AIAA Guidance, Navigation and Control Conference,* Boston, MA, August 14–16, 1989, Paper AIAA-89-3583.

[18] Huddle, J. R. Historical perspective on position and gravity survey with inertial systems. *AIAA Journal of Guidance, Control, and Dynamics,* May 1986, 11 pp.

[19] Hulsing, R. H. Single Coriolis inertial rate and acceleration sensor. *Journal of the Institute of Navigation* 35, 3 (Fall 1988): 347–359.

[20] Johnson, D. Frequency domain analysis for RLG system design. *Navigation, Journal of the Institute of Navigation* 34, 3 (Fall 1987), pp. 178–189.

[21] Jordan, J. W. An accurate strapdown direction cosine algorithm. NASA TN D-5384, September 1969.

[22] Jordan, S. K. Effects of geodetic uncertainty on a damped inertial navigation system. *IEEE Transactions on Aerospace and Electronic Systems*, September 1973, 12 pp.

[23] Kayton, M., and W. Fried (eds.). *Avionics Navigation Systems*. New York: Wiley, 1969.

[24] Kayton, M. Coordinate frames in inertial navigation. Massachusetts Institute of Technology Instrumentation Laboratory Report T-260, August 1960.

[25] Kayton, M., and R. C. Gilbert. Effect of component orthogonality on aircraft inertial navigation system errors. AIAA Paper 64-237, June 1964.

[26] Kayton, M. Fundamental limitations on inertial navigation. In *Progress in Astronautics and Rocketry: Guidance and Control*, vol. 8. New York: Academic Press, 1962, pp. 367–394.

[27] Kayton, M. Gyro torquing signals at an arbitrary azimuth on an ellipsoidal earth. *AIAA Journal* (December 1963).

[28] Lee, J. G., Y. J. Yoon, J. G. Mark, and D. A. Tazartes. Extension of strapdown attitude algorithm for high-frequency base motion. *Journal of Guidance, Control, and Dynamics* 13, 4 (July–August 1990), 738–743.

[29] Levine, S., R. Dennis, and K. Bachman. Strapdown astro-inertial navigation utilizing the optical wide-angle lens startracker. *Navigation, Journal of the Institute of Navigation* 37, 4 (Winter 1990–91), 347–362.

[30] Lottman, D., K. Homb, and A. Brown. Kalman filter implementation in the Litton LR-80 AHRS. *IEEE NAECON Proceedings*, 1984, 8 pp., Dayton, OH.

[31] Mansour, W., and C. Lacchini. Two axis dry tuned-rotor gyroscopes design and technology. *Journal of Guidance, Control, and Dynamics* 16, 3 (May–June 1993).

[32] Mark, J. G., D. A. Tazartes, B. Fidric, and A. Cordova. A rate integrating fiber optic gyro. *Navigation, Journal of the Institute of Navigation* 38, 4 (Winter 1991–92), 341–353.

[33] Mark, J. G., R. E. Ebner, and A. K. Brown. Design of RLG inertial systems for high vibration. *Proceedings of IEEE PLANS Conference*, December 1982, Atlantic City, NJ.

[34] Mark J. G., D. A. Tazartes, and T. Hilby. Fast orthogonal calibration of a strapdown inertial system. *Proceedings, Symposium Gyro Technology*, Stuttgart, Germany, September 1986.

[35] Matthews, A., and H. Welter. Cost effective, high-accuracy inertial navigation. *Navigation, Journal of the Institute of Navigation* 36, 2 (Summer 1989), 157–172.

[36] Matthews, A. Utilization of fiber optic gyros in inertial measurement units. *Navigation, Journal of the Institute of Navigation* 37, 1 (Spring 1990), 17–38.

[37] McKern, R. A. A study of transformation algorithms for use in a digital computer. Master's thesis. Department of Aeronautics and Astronautics, Massachusetts Institute of Technology, Cambridge. January 1968.

[38] Norling, B. Emergence of miniature quartz vibrating beam accelerometer technology for tactical navigation and flight control. *Proceedings 19th Annual Joint Services Data Exchange for Guidance, Navigation, and Control*, Cambridge, MA. November 14–17, 1988.

[39] Norling, B. L. Superflex: A synergistic combination of vibrating beam and quartz flexure accelerometer technology. *Navigation, Journal of the Institute of Navigation* 34, 4 (Winter 1987–88), 337–353.

[40] O'Donnell, C. F. (ed.). *Inertial Navigation Analysis and Design*. New York: McGraw-Hill, 1964.

[41] Pitman, G. R. (ed.). *Inertial Guidance*. New York: Wiley, 1962.

[42] Poor, W. A. A geometric description of wander azimuth frames. *Navigation, Journal of the Institute of Navigation* Fall 1989, vol. 36, No. 3, 303–318.

[43] Post, E. J. Sagnac effect. *Review of Modern Physics* 39, 2 (April 1967).

[44] Savage, P. G. Advances in strapdown sensors. *AGARD Lecture Series* 133, May 1984.

[45] Savage, P. G. Laser gyros in strapdown inertial navigation systems. *Proceedings of IEEE PLANS*, November 1976, San Diego, CA.

[46] Savage, P. G. Strapdown system algorithms. *AGARD Lecture Series* 133, May 1984.

[47] Savet, P. H. (ed.). *Gyroscopes: Theory and Design*. New York: McGraw-Hill, 1961.

[48] Scarborough, J. B. *The Gyroscope—Theory and Applications*. New York: Interscience Publishers, 1958.

[49] Statz, H., T. Dorschner, M. Holtz, and I. Smith. The multioscillator ring laser gyroscope. *Laser Handbook*, vol. 4. Amsterdam: North Holland Physics, Division of Elsevier 1985, pp. 229–332.

[50] Stripling, W. W., and J. R. Baskett. Hemispherical resonator gyro: Principle, design, and performance. *AGARD Conference Proceedings 525, Integrated and Multi-Function Navigation*, May 1992.

[51] Synge, J. L., and B. A. Griffith. *Principles of Mechanics*. New York: McGraw-Hill, 1949.

[52] Tazartes, D. A. Fiber optics based IMU. *SAE Technical Paper Series 901826*. October 1990.

[53] Tazartes, D. A., R. J. Buchler, H. J. Tipton, and R. Grethel. Synergistic interferometric GPS-INS. *Proceedings, Institute of Navigation 1995 National Technical Meeting*, January 18–20, 1995, Anaheim, CA, pp. 657–671.

[54] Tazartes, D. A., and J. G. Mark. Integration of GPS receivers into existing inertial navigation systems. *Navigation, Journal of the Institute of Navigation* 35, 1 (Spring 1988), 105–119.

[55] Warren, K. Electrostatically force-balanced silicon accelerometer. *Navigation, Journal of the Institute of Navigation* 38, 1 (Spring 1991): 91–99.

[56] Warzynski, R., and R. Ringo. The evolution of ESG technology. *Proceedings of AGARD Symposium*, June 1971.

[57] Weber, D. A three axis monolithic ring laser gyro. *Navigation, Journal of the Institute of Navigation* 35, 1 (Spring 1988): 15–22.

[58] Wei, M., and K. Schwarz. A strapdown inertial algorithm using an Earth-fixed cartesian frame. *Navigation, Journal of the Institute of Navigation* 37, 2 (Summer 1990), 153–167.

[59] Wilkinson, J. R. Ring lasers. *Progress Quantum Electronics* 11 (1987): 1–103.

[60] ARINC Airlines Electronic Engineering Committee, Aeronautical Radio, Inc., Annapolis, MD.

(a) ARINC Characteristic 561-11 "Air Transport Inertial Navigation System," January 17, 1975.

(b) ARINC Characteristic 704-6 "Inertial Reference System," May 25, 1990.

(c) ARINC Characteristic 738-1 "Air Data and Inertial Reference System," November 18, 1994.

[61] Institute of Electrical and Electronics Engineers, New York, NY. IEEE Standards.
(a) 337-1972 "IEEE Standard Specification Format Guide and Test Procedures for Linear Single-axis Pendulous Analog Torque Balance Accelerometer." 48 pp.
(b) 517-1974 "IEEE Standard Specification Format Guide and Test Procedure for Single-Degree-of-Freedom Rate-Integrating Gyro." 60 pp.
(c) 529-1980 "IEEE Supplement for Strapdown Applications to IEEE Standard 517." 24 pp.
(d) 528-1994 "IEEE Standard for Inertial Sensor Terminology." 20 pp.
(e) 530-1978 "IEEE Standard Specification Format Guide and Test Procedure for Linear Single-Axis, Digital Torque-Balance Accelerometer." 44 pp.
(f) 671-1985 "IEEE Standard Specification Format Guide and Test Procedure for Non-Gyroscopic Inertial Angular Sensors: Jerk, Acceleration, Velocity, and Displacement." 52 pp.
(g) 813-1988 "IEEE Standard Specification Format Guide and Test Procedure for Two-Degree-of-Freedom Dynamically-Tuned Gyros." 64 pp.
(h) 836-1991 "IEEE Recommended Practice for Precision Centrifuge Testing of Linear Accelerometers." 80 pp.
(i) 647 "IEEE Specification Format Guide and Test Procedure for Single Axis Laser Gyros," under preparation.
[62] U.S. Air Force, "Specification for USAF Form, Fit and Function (F^3) Medium Accuracy Navigation Unit," SNU-84-1, Revision D. Wright Patterson Air Force Base, OH: Aeronautical Systems Division, Air Force Systems Command, September 21, 1992.

CHAPTER 8

[1] Abbott, W. Y., S. C. Spring, R. J. Stewart. Flight evaluation of J-Tec VT-100 vector airspeed sensing system, final report. USAAEFA Proj. No. 75-17-2, May 1977.
[2] Aiken, W. S., Jr. Standard nomenclature for airspeeds with tables and charts for use in calculation of airspeed. NACA Report No. 837, 1946, Washington, DC.
[3] Aeronautical Radio, Inc., Annapolis, MD. ARINC characteristic 706-4, Mark 5 subsonic air-data system. January 1988.
[4] Ames Research Staff, Equations, tables and charts for compressible flow. NACA Report 1135, Ames Aeronautical Laboratories, Moffett Field, CA, 1953.
[5] Ames Research Staff, US standard atmosphere. 1962 and 1976 Revision. Langley, VA: NASA and U.S. Committee on Extension to Standard Atmosphere (COESA).
[6] Baumker, M., and W. Hassenpflug. Analytical evaluation of helicopter true airspeed and associated flight tests, papers no. 112, 119. *14th European Helicopter Forum*, Milan, Italy, Sept. 1988.
[7] Bogue, R. K. Recent flight-test results of optical air-data techniques. NASA Technical Memorandum 4504, 1993.
[8] Colton, R. F. Vortex anemometry—new applications of an old principle. *20th International Instrumentation Symposium*, May 21–23, 1974, Instrument Society of America, Albuquerque, NM.

[9] Colton, R. F. Vortex anemometers—second generation. Instrument Society of America (ISA) Report 75-757, Industry Oriented Conference and Exhibit, Milwaukee, WI, October 6–9, 1975.

[10] DeLeo, R. V., and F. W. Hagen. Flight calibration of aircraft static pressure systems. Federal Aviation Agency, Report SRDS RD-66-3, February 1966.

[11] Diehl, W. S. Standard atmosphere tables and data. NACA Report No. 218, 1948.

[12] Emrich, R. J. Methods of experimental physics, vol. 18, Fluid Dynamics. Academic Press, 1981.

[13] Erickson, R. A. Accuracy of in-flight computation of altitude from air-data inputs. NAVWEPS Report No. 7784, NOTS TP 2771, U.S. Naval Ordnance Test Station, China Lake, CA, December 1961, 28 pp.

[14] U.S. Army. Final Report III, USAASTA, Project No. 71-30, Flight Evaluation Pacer Systems, Inc. LORAS II low airspeed system. March 1974.

[15] U.S. Army. Final Report I, USAASTA, Project No. 71-30, Flight evaluation, Elliott dual-axis low airspeed system, LASSIE II, low airspeed sensor. September 1975.

[16] U.S. Army. Final Report VI, USAAEFA, Project No. 71-30, Flight evaluation, Elliott dual-axis low airspeed system, LASSIE II, low airspeed sensor. September 1975.

[17] U.S. Army. Final Report IV, U.S. Army Aviation Systems Test Activity USAAST SUP Project No. 71-30, Flight evaluation, J-TEC airspeed system, low airspeed sensor, April 1974.

[18] Gracey, W. Recent developments in pressure altimetry. *AIAA Journal of Aircraft*, May–June 1965.

[19] Goldstein, S. (Ed.). Modern developments in fluid dynamics. Volumes I and II, Oxford Engineering Science Series, Oxford University Press, 1952.

[20] Hansman, R. J., and B. H. Kang. Preliminary definition of pressure sensing requirements for hypersonic vehicles. *AIAA-88-4652; Proceedings of AIAA/NASA/AFWAL Conference on Sensors and Measurement Techniques for Aeronautical Applications*, September 1988.

[21] Hess, J. L., and A. M. O. Smith. Static pressure probes derived from supersonic slender-body theory. *AIAA Journal of Aircraft*, September–October 1967, pp. 409–415.

[22] Hillje, E. R. The orbiter air-data system. NASA CP-2342-PT-1, June 1983.

[23] Hillje, E. R., and D. E. Tymms. The ascent air-data system for the space shuttle. *Proceedings of 11th Aerodynamic Testing Conference*, March 1980, Colorado Springs.

[24] Hilton, W. F. High speed aerodynamics. New York: Longmans, Green and Co., 1951.

[25] Hoad, D. R., and D. B. Rhodes. Preliminary rotor wake measurements with a laser velocimeter. NASA Tech. Memorandum 83246, 1986.

[26] Kaletka, J. Evaluation of the helicopter low airspeed system LASSIE. *DGLR Seventh European Rotorcraft and Powered Lift Aircraft Forum*, Garmisch Parten Kirchen, Germany. 1981.

[27] Kayton, M., and W. R. Fried (Eds.). Avionics navigation systems. John Wiley & Sons, N.Y. 1969.

[28] Ladenburg, R. W., B. Lewis, R. N. Pease, and H. S. Taylor (Eds.). Physical measurements in gas dynamics and combustion. Princeton University Press, 1954.

[29] Lion, K. S. Instrumentation in scientific research. McGraw-Hill Book Co., Inc., New York, 1959.

[30] Livingston, S. P., and W. Gracey, Tables of airspeeds, altitutde and Mach number based on latest values for atmospheric properties and physical constants, NASA TN D-822, August 1961.

[31] Mandle, J. A promising low speed air-data system for helicopters. Twelfth European Rotorcraft Forum, Paper No. 51, Garmisch Parten Kirchen. 1986.

[32] Minzner, R. A., K. S. W. Champion, and H. L. Pond. The ARDC model atmosphere. 1959, Air Force Surveys in Geophysics No. 115 (AFCRC-TR-59-267), Air Force Cambridge Research Center, Aug. 1959.

[33] Onksen, P. J. Helicopter omnidirectional air-data systems. *Proceedings of IEEE NAECON Conference*, Dayton, OH, May 1983. IEEE, New York, NY.

[34] Pruett, C. D., J. Wolf, M. L. Heck, and P. M. Siemers, Innovative air-data system for the space shuttle orbiter. *Journal of Spacecraft and Rockets*, 20, 1 (January–February 1983).

[35] Roshko, A. On the development of turbulent wakes from vortex streets. NACA TN 2913 (1953).

[36] Sebring, D., and M. McIntyre. An air-data inertial reference system for future commercial airplanes. Paper No. 88-3918, *AIAA/IEEE 8th Digital Avionics Systems Conference Proceedings*, October 1988.

[37] Shapiro, A. H. The dynamics and thermodynamics of compressible flow. Vols. 1 and 2. Ronald Press, New York, 1953.

[38] Siemers, P. M. et al. Shuttle flight pressure instrumentation: experience and lessons for the future. NASA CP-2283, March 1983.

[39] Society of Automotive Engineers. Aerospace Recommended Practices (ARP):
(a) ARP-920: "Design and Installation of Pitot Static Systems for Transport Aircraft," 1967.
(b) ARP-942: "Pressure Altimeter Systems," 1967. Society of Automotive Engineers, New York 17, N.Y.

[40] Stephan, S. C., Jr., Study and experimental research into flight instrumentation for vehicle operation in the fringe or outside of the atmosphere. ASD Tech Report 61-142, Volumes I and II, Nov. 1961.

[41] Von Karman, Mechanism of drag. Physikalische Zeitschraft, 1912.

[42] Webb, L. Characteristics and use of X-15 air-data sensors. NASA TND-4597, November 1967.

[43] Wolf, H., M. Henry, and P. M. Siemers, Shuttle entry air-data system (SEADS)—flight verification of an advanced air-data concept, AIAA Paper 88-2104, May 1988.

[44] Wolowicz, C., and T. Gossett, Operational and performance characteristic of the X-15 spherical hypersonic flow direction sensor, NASA TND-3070, August 1965.

[45] Winn, A. L., and J. S. Kishi, Flight evaluation Elliott low airspeed system, Final Report 21 June–15 November, 1971, Report No. AD-753343, USAASTA-71-30, September 1972.

[46] Young, W. L. Test evaluation of J-Tec true airspeed sensor, AFFDL-TM-70-1-FGS, December 1970.

[47] EUROCAE, Minimum performance standards for airborne altitude measurements and coding systems. ED-26, Paris, France, 1979.

[48] RTCA, Inc. Altimetry, DO-88, Washington, DC, 1958.

CHAPTER 9

[1] Archibald, J. B. New technology alternatives to the existing aircraft magnetic azimuth detector. *IEEE NAECON Conference*, Dayton, OH 1993, pp. 333–344.

[2] Barton, C. E. International geomagnetic reference field: The Seventh generation 1995. *Journal of Geomagnetism and Geoelectricity*, special issue on IGRF 1996.

[3] Bickel, S. H. Error analysis of an algorithm for magnetic compensation. *IEEE Transactions on Aerospace and Electronic Systems* (September 1979).

[4] Hine, A. *Magnetic Compasses and Magnetometers.* Toronto: University of Toronto Press, 1968.

[5] Jacobs, J. A. (ed.). *Geomagnetism*, 4 vols. San Diego: Academic Press, 1987–91.

[6] Journal of Geomagnetism and Geoelectricity, special issue on IGRF, 44(1992): 679–707. Articles by R. A. Langel, J. Quinn, and others.

[7] Quinn, J., et al. 1995 revision of joint US/UK geomagnetic field models; Main field. *Journal of Geomagnetism and Geoelectricity* (Fall 1996).

[8] Ramsden, E. Measuring magnetic fields with fluxgate sensors. *Sensors* (September 1994): 87–90.

[9] Savet, P. H. *Gyroscopes: Theory and Design.* New York: McGraw-Hill, 1961.

[10] Shapiro, A. H. Precision azimuth reference systems. *Proceedings of the IRE (IEEE) National Conference on Aeronaturical Electronics*, Dayton, OH, May 1957.

[11] Spencer, H. S., and G. F. Kucera. *Handbook of Magnetic Compass Adjustment and Compensation.* Washington, D.C: U.S. Hydrographic Office Pub. 226. latest issue.

[12] Truxall, J. C. *Control System Synthesis.* New York: McGraw-Hill, 1955.

[13] U.S. Federal Aviation Administration. Technical standard order for magnetic non-stabilized type direction instrument. TSO-C7d. Washington, DC, 1989.

[14] U.S. Federal Aviation Agency. Technical standard order for direction instruments (magnetic). Direct and remote reading. Report TSO-C7c. Washington, DC, 1958.

[15] *U.S. Federal Aviation Regulations.* Part 25.1303. Washington, DC.

[16] Washburn, J., J. Galloway, and H. Kent. Standard compass/attitude and heading reference system (C/AHRS) utilizing fiber optic gyroscopes. *IEEE NAECON Conference*, Dayton, OH, May 1993. IEEE, New York, pp. 362–369.

CHAPTER 10

[1] Barton, D. K. *Modern Radar Systems.* Boston: Artech House, 1988.

[2] Benjamin, S. K. The AN/APN-153(V) Doppler navigation equipment. *IEEE Transactions* AS-1, 2 (August 1963).

[3] Berger, F. B. The design of airborne Doppler velocity measuring systems. *IRE Transactions* ANE-4 (December 1957): 176–196.

[4] Berger, F. B. The nature of Doppler velocity measurement. *IRE Transactions* ANE-4 (September 1957): 157–175.

[5] Buell, H. Doppler radars for low cost, medium accuracy navigation. *AGARD Conference Proceedings: Medium Accuracy Low Cost Navigation*, no. 176. Sandefiord, Norway, September 1975.

[6] Buell, H. Performance of the AN/ASN-128 Doppler navigation system for helicopters. *1982 IEEE Position Location and Navigation Symposium*, Atlantic City, October 1982.

[7] Buell, H., and J. Fiore. Accuracy certification flight test report for Doppler velocity sensor (DVS) for C-130 SCNS. GEC Marconi Electronic Systems, April 18, 1985.

[8] Buell, H., and D. Doremus. The AN/ASN-157; A single LRU Doppler navigation system for helicopters. 12th Digital Avionics Systems Conference, Ft. Worth, TX, October 1993.

[9] Buell, H., and A. J. Hunton. Synergistic effects of Doppler radar/GPS navigation integration and the development of an advanced navigation system for helicopter applications. *Proceedings of the Institute of Navigation National Technical Meeting*, San Diego, January 1994.

[10] Buell, H. The AN/ASN-137 advanced Doppler navigation system for helicopters. *IEEE Position Location and Navigation Symposium* (PLANS), Atlantic City, NJ, 1980.

[11] Bussey, H. C., and C. A. Zielinski. The AN/APN-79 navigation system CW Doppler radar groundspeed sensor. *Proceedings of the 1958 National Conference on Aeronautical Electronics*, IRE, Dayton, OH, May 1958, pp. 128–132. IEEE, New York, NY.

[12] Campbell, J. P. Back-scattering characteristics of land and sea at X-band. *Proceedings of the 1958 National Conference on Aeronautical Electronics*, IRE, Dayton, OH, May 1958.

[13] Condie, M. A. Basic design considerations—Automatic navigator AN/APN-67. *IRE Transactions* ANE-4 (December 1957): 197–201.

[14] Durst, C. J. The sea surface and Doppler. *Journal of the Institute of Navigation* (England) 11, 2 (April 1958): 143–149.

[15] Eaves, J. L., and E. K. Reedy. *Principles of Modern Radar.* New York: Van Nostrand Reinhold, 1986.

[16] Fried, W. R. An FM-CW radar for simultaneous three dimensional velocity and altitude measurement. *IEEE Transactions* ANE-11, 1 (March 1964).

[17] Fried, W. R. The application of Doppler navigation equipment in the air traffic control environment. *Proceedings of the 1949 National Aeronautical Electronics Conference*, IRE, Dayton, OH, May 1959, pp. 502–517.

[18] Fried, W. R. Principles and performance analysis of Doppler navigation systems. *IRE Transactions* ANE-4, 4 (December 1957): 176–196.

[19] Fried, W. R., and J. A. Losier. The nature and correction of Doppler radar errors over water. *Proceedings of the 1964 National Conference on Aerospace Electronics*, IEEE, Dayton, OH, May 1964.

[20] Glegg, K. C. M. A low noise CW Doppler technique. *Proceedings of the 1958 National Conference on Aeronautical Electronics*, IRE, Dayton, OH, May 1958, pp. 133–144.

[21] Grant, C. R., and B. S. Yaplee. Back-scattering from water and land at centimeter wavelengths. *Proceedings of the IRE*, vol. 45, July 1957, pp. 976–982.

[22] Gray, T., and J. Moran. Decca Doppler and airborne navigation. *British Communications and Electronics* 5 (October 1948): 764–771.

[23] Grocott, D. F. H. Doppler correction for surface movement. *Journal of the Institute of Navigation* (England) 16, 1 (January 1963).

[24] Gustin, R. M. Flight test results of navigation set radar, AN/APN-66 and AN/APN-82. Wright Air Development Center Technical Note TN-55-746. Wright Patterson Air Force Base, Dayton, OH, 1955.

[25] Laschever, N. L. AN/APN-78, A Doppler navigator for helicopters. *Proceedings of the 1958 National Conference on Aeronautical Electronics*, IRE, Dayton, OH, May 1958, pp. 117–122.

[26] McKay, M. W. The AN/APN-96 Doppler radar set. *1958 IRE National Convention Record*, pt. 5, pp. 71–77.

[27] McMahon, F. A. The AN/APN-81 Doppler navigation system. *IRE Transactions* ANE-4 (December 1957): 202–211.

[28] Miller, G. M. Design considerations for the APN-79 airborne Doppler navigation system. *Navigation* 6, 3 (Autumn 1958): 147–156.

[29] Moore, R. K., and C. S. Williams. Radar terrain return at near vertical incidence. *Proceedings of the IRE* 45, 2 (February 1957): 228–238.

[30] Oleinik, L., et al. Doppler/GPS navigation set (DGNS) flight test report. U.S. Army Electronics Systems Division, Command/Control and Systems Integration Directorate, Cecom Research, Development and Engineering Center, Fort Monmouth, NJ, February 27, 1995.

[31] Saltzman, H., and G. Stavis. A dual beam planar array antenna for Janus type Doppler navigation systems. *IRE Convention Record 1958*, pt. 1, p. 241.

[32] Schultheiss, P. M., C. A. Worgin, and F. Zweig, Short-time frequency measurement of narrow band random signals in the presence of wide-band noise. *Journal of Applied Physics*, 25, 8 (1954): 1025–1036.

[33] Smith, P. G. Leakage rejection in beam-switched CW radars. *IRE Transactions* ANE-10, 1 (March 1963).

[34] Smith, P. G. Null tracking Doppler navigation radar. *IEEE Transactions*, ANE-10, 1 (March 1963).

[35] Willis, D. C. Operational experience with Doppler. *Navigation* 9, 3 (Autumn 1962).

[36] Wiltse, J. C., S. P. Schlesinger, and C. M. Johnson. Backscattering characteristics of the sea in the region from 10 to 50 KMC. *Proceedings of the IRE* 45, 2 (February 1957): 220–228.

[37] Wulfsberg, P. G. A coherent high performance FM-CW Doppler radar navigation system. *Proceedings of the 1959 National Aeronautical Electronics Conference*, Dayton, OH, May 1959, pp. 343–347.

[38] ARINC Characteristic No. 540. Airborne Doppler radar, Annapolis, MD: Aeronautical Radio Inc., 1965.

[39] RTCA, Inc. Minimum operational performance standards for airborne Doppler radar navigation equipment, DO-158, 1975, Washington, D.C.: RTCA.

[40] Fried, W. R., History of Doppler radar navigation, *Navigation* 40:2 (1993) 121–136, Institute of Navigation, Alexandria, VA.

[41] Tull, W. J., The early history of airborne Doppler systems, *Navigation*, 43:1 (1996), 9–24, Institute of Navigation, Alexandria, VA.

CHAPTER 11

[1] Kovaly, J. J. *Synthetic Aperture Radar.* Boston: Artech House, 1976.

[2] Hovanessian, S. A. *Introduction to Synthetic Array and Imaging Radars.* Boston: Artech House, 1980.

[3] Rihaczek, A. W. *Principles of High Resolution Radar.* New York: McGraw-Hill, 1969, ch. 13.

[4] Wehner, R. R. *High Resolution Radar.* Boston: Artech House, 1987, ch. 6.

[5] Morris, G. V. *Airborne Pulse Doppler Radar.* Boston: Artech House, 1988, ch. 9.

[6] Stimson, G. W. *Introduction to Airborne Radar.* Hughes Aircraft Company, El Segundo, CA, 1983, pt. VIII.

[7] Oppenheim, A. V. (ed.). *Applications of Digital Signal Processing.* Englewood Cliffs, NJ: Prentice-Hall, 1978, ch. 5.

[8] Farrell, J. L., J. H. Mims, and A. Sorrell. Effects of navigation errors in maneuvering SAR. *IEEE Transactions on Aerospace and Electronic Systems* 9, 5 (September 1973): 758–775.

[9] Robinson, P. N. Depth of field for SAR with aircraft acceleration. *IEEE Transactions on Aerospace and Electronic Systems* 20, 5 (September 1984): 603–615.

[10] Ausherman, D. A., A. Kozma, J. L. Walker, H. M. Jones, and E. C. Poggio. Developments in radar imaging. *IEEE Transactions on Aerospace and Electronic Systems* 20, 4 (July 1984): 363–400.

[11] Cafforio, C., C. Prati, and F. Rocca. SAR data focusing using seismic migration techniques. *IEEE Transactions on Aerospace and Electronic Systems* 27, 2 (March 1991): 194–207.

[12] Perkins, L. C., H. B. Smith, and D. H. Mooney. The development of airborne pulse Doppler radar. *IEEE Transactions of Aerospace and Electronic Systems* 20, 3 (May 1984): 292–303.

[13] Scott, W. B. ITT Gilfillan focuses on advanced phased array and bistatic radars. *Aviation Week and Space Technology*, June 1988, pp. 93–97.

[14] Scott, W. B. First model of ATF common integrated processor delivered to Lockheed. *Aviation Week and Space Technology*, August 29, 1988, pp. 65–67.

[15] New circuits expected to exceed projections. *Aviation Week and Space Technology*, July 30, 1984, pp. 46–51.

[16] Zorpette, G. The beauty of 32 bits. *IEEE Spectrum* (September 1985): 65–71.

[17] Hutcheson, L. D., P. Haugen, and A. Husain. Optical interconnects replace hardware. *IEEE Spectrum* (March 1987): 30–35.

[18] Skolnik, M. I. *Introduction to Radar Systems*, 2d ed. New York: McGraw-Hill, 1980.

[19] Barton, D. K. *Radar Systems Analysis.* Englewood Cliffs, NJ: Prentice-Hall, 1964.

CHAPTER 12

[1] Kayton, M., and W. R. Fried. *Avionics Navigation Systems.* New York: Wiley, 1969.

[2] Pitman, G. R. *Inertial Guidance.* New York: Wiley, 1962.

[3] Britting, K. R. *Inertial Navigation Systems Analysis.* New York: Wiley-Interscience, 1971.

[4] Jazwinski, A. H. *Stochastic Processes and Filtering Theory.* New York: Academic Press, 1970.

[5] Nahi, N. E. *Estimation Theory and Applications.* New York: Wiley, 1969.

[6] Brockstein, A., and J. Kouba. Derivation of free inertial, general wander azimuth error model equations. Litton Systems, Inc., Woodland Hills, CA, Technical Memo 69-72, Pub. No. 9325A.

[7] Ogata, K. *State Space Analysis of Control Systems.* Englewood Cliffs, NJ: Prentice-Hall, 1967.

[8] Gelb, A. *Applied Optimal Estimation.* Cambridge: MIT Press, 1974.

[9] Knobbe, E. J. Application of the information matrix to inertial component/system test. *Proceedings Seventh Biennal Guidance Test Symposium,* Holloman AFB, NM, May 1975.

[10] Van Trees, H. L. *Detection, Estimation, and Modulation Theory.* Part I. New York: Wiley, 1968.

[11] Wiley, R. R. Cassegrain-type telescopes. *Sky and Telescope,* April 1962.

[12] Herzberger, M. *Modern Geometrical Optics.* New York: Interscience, 1958.

[13] Valasek, J. *Introduction to Theoretical and Experimental Optics.* New York: Wiley, 1949.

[14] Sherwin, C. *Introduction to Quantum Mechanics.* New York: Henry Holt, 1959.

[15] Bell, E. Radiometric quantities, symbols, and units. *Proceedings of the IRE,* September 1959.

[16] Celestial trackers, theory, design, applications. *IEEE Transactions on Aerospace and Navigational Electronics* (special issue) 10, 3 (September 1963).

[17] Quasius, G., and F. McCanless. *Star Trackers and Systems Design,* Spartan Books, 1966.

[18] The brightness and polarization of the daylight sky at altitudes of 18,000 to 38,000 feet. *Journal of the Optical Society of America* 41, 7 (July 1951).

[19] Laverty, N. P., and W. M. Clark. High altitude daytime sky radiance measurement. Nortronics Report 64-219. June 1964.

[20] U.S. Air Force. *The Handbook of Geophysics,* rev. ed. New York: Macmillan, 1960.

[21] Deters, R. A., and R. L. Gutshall. Charge transfer device star tracker applications. *AIAA Journal of Guidance* 10, 1 (January–February 1987).

[22] Dereniak, E., and D. Crowe. *Optical Radiation Detectors.* New York: Wiley, 1984.

[23] Duncan, T. M. A daylight stellar sensor using a charge coupled device. *Proceedings, SPIE* 111 (1989).

[24] Cox, J. A. Evaluation of peak location algorithms with subpixel accuracy for mosaic focal planes. *Proceedings SPIE* (August 1981).

[25] Boxenhorn, B. Covariance analysis of a charge carrier device processing algorithm for stellar sensors. *AIAA Journal of Guidance* 6 (July–August 1983).

[26] Dennison, E. W., and R. H. Stanton. Ultra-precise star tracking using charge coupled devices (CCDs). *Proceedings SPIE* 252 (1980).

[27] Norville, W. *Celestial Navigation Step by Step,* 2d ed. Camden, ME: International Marine Publishing, 1984.

[28] Meeus, J. *Astronomical Algorithms.* Richmond, VA: Willman-Bell, 1991.

[29] Stoll, R., P. F. Forman, and J. Edelman (Perkin-Elmer Corp.). The effect of dif-

ferent grinding procedures on the strength of scratched and unscratched fused silica. Symposium on the Strength of Glass and Ways to Improve it, Perkin-Elmer Corp., September 1961.

[30] Thomas, C. O., and M. F. Popelka, Jr., Viewing condition during supersonic flight. Autonetics Division, North American Rockwell Corporation. Report EM-1314. September 16, 1957.

[31] Johnson, F. S. Atmospheric structure. *Astronautics* (August 1962): 54.

[32] Hosfeld, R. Comparison of stellar scintillation with image motion. *Journal of the Optical Society of America* 44, 4 (April 1954).

[33] National Advisory Committee for Aeronautics. Light diffusion through high speed turbulent boundary layers, NACA RM A56B21, May 25, 1956.

[34] Ellison, M. H. The effects of scintillation on telescopic images. *Symposium Proceedings on Astronomical Optics.* Amsterdam: North Holland, 1956.

[35] Deflection and diffusion of a light ray passing through a boundary layer. Douglas Aircraft Co., DDC AD264. May 16, 1952.

[36] Gottlieb, D. M. SKYMAP, A New Catalog of Stellar Data. *Astrophysical Journal,* suppl. series, 38 (November 1978): 287–308.

[37] Glasby, J. S. *Variable Stars.* London: Constable Ltd., 1968.

[38] Burnham, R., Jr. *Burnham's Celestial Handbook*, vols. 1–3. New York: Dover, 1978.

[39] Green, R. *Spherical Astronomy.* Cambridge: Cambridge University Press, 1985.

[40] Kennel, J., S. Havstad, and D. Hood. Star sensor simulation for astroinertial guidance and navigation. *SPIE Proceedings* No. 1694 (April 1992).

[41] Bowditch, N. *The American Practical Navigator.* U.S. Navy Hydrographic Office Publication 9, Washington, DC, 1995.

CHAPTER 13

[1] Kelly, R. J., and J. M. Davis. Required navigation performance (RNP) for precision approach and landing with GNSS application. *Navigation* 41, 1 (Spring 1994).

[2] Directorate of All Weather Operations. National Air Traffic Service (U.K.). Internal memos of various dates in 1982.

[3] Correspondence titled "RVR Statistiek" from A. J. G. J. Brakke, Aeronautical Inspection Directorate, Hoofdorp, The Netherlands, November 1994.

[4] The International Civil Aviation Organization (ICAO), headquartered in Montreal, an entity of the United Nations, maintains various Annexes to the (Chicago) Convention on International Civil Aviation that promulgate standards and recommended practices (SARPs) to ensure ground/air avionics interoperability. In addition a large number of manuals and other documents provide guidance on implementing and operating most aspects of the civil aviation infrastructure:

(a) Annex 10, *Aeronautical Telecommunications*, vol. 1, 4th ed., April 1985

(b) Annex 14, *Aerodromes*, 8th ed., March 1983

(c) Doc 9365, *Manual of All Weather Operations*, Second. Ed., 1991

(d) Operational requirements for a new non-visual precision approach and landing guidance system for international civil aviation, *Report of the 7th Air Navigation Conference*, 1974

(e) *Report of the All Weather Operations Divisional Meeting (AWO 78)*, Doc 9242, 1978

(f) *Report of the Fifteenth Meeting of the All Weather Operations Panel*, October 1994

(g) *Proposed Operational Requirements for Advanced Surface Movement Guidance & Control Systems (A-SMGCS)*, European Air Navigation Planning Group, Draft 6, October 1994.

[5] Federal Aviation Administration: Washington, D.C.

(a) Advisory Circular 120-29, *Criteria for Approving Category I and Category II Landing Minima for FAR121 Operations*, September 1970

(b) Advisory Circular 120-28c, *Criteria for Approval of Category III Landing Weather Minima*, March 1984

(c) Advisory Circular 20-57A, *Automatic Landing Systems*, January 1971

(d) Handbook 8260.3B, *United States Standard for Terminal Instrument Procedures (TERPS)*, 3d ed., July 1976, reprinted August 1993.

(e) *Federal Aviation Regulations (FAR)*, current edition

(f) *GPS Implementation Plan for Air Navigation and Landing*, August 1994.

(g) Order 6750.54, Electronic Installation and Instructions for Instrument Landing System (ILS) Facilities. Dec. 17, 1993. pp. 20. Washington, DC.

[6] Instrument Approach Procedure Charts. Published by the U.S. Coast and Geodetic Survey and by the USAF Aeronautical Chart and Information Center at two-week intervals.

[7] Litchford, G. B. The 100 ft barrier. *Aeronautics and Astronautics, AIAA* (July 1964): 58–65.

[8] *AIAA/ION Guidance and Control Conference Proceedings*, August 1965:

(a) G. L. Teper and R. L. Stapleford. An assessment of the lateral-directional handling qualities of a large aircraft in the landing approach, pp. 351–359

(b) J. R. Utterstrom and R. E. Kestek. The Boeing-Bendix precision approach and landing system, pp. 319–324

American Institute of Aeronautics and Astronautics, New York, 1965.

[9] Redlien, H. W., and R. J. Kelly. Microwave landing system: The new international standard. *Advances in Electron Physics* 57 (1981).

[10] Cole, D. *A Brief Early History of Automatic Landing in the UK*. London: CAA House, undated.

[11] McGrath, J. L. *Trident All Weather Operations*. Islip, Middlesex: British European Airways, June 1973.

[12] Lloyd, E., and W. Tye. *Systematic Safety.* London: CAA House, July 1982.

[13] Metz, H. I. A survey of instrument landing systems in the United States. *IRE Transactions on Aeronautical and Navigational Electronics* (January 1959): 78–84.

[14] Sandretto, P. C. *Electronic Avigation Engineering*, ITT Corp., New York, 1958.

[15] Brockway, A. L. *ILS accuracy requirements per ICAO Annex 10*. Bendix Corporation, Baltimore, MD, January 1974.

[16] *Instrument Landing System Concepts*, 7th ed. Federal Aviation Administration Academy, Catalog No. 40233, March 1986.

[17] Kayton, M. The near field of the instrument landing system glide slope. *IEEE Transactions on Aerospace and Electrical Systems* (July 1968): 237.

[18] RTCA, Inc. Washington, DC.

(a) A new guidance system for approach and landing. Document DO-148. December 18, 1970

(b) Minimum operational performance standards for MLS area navigation equipment. Document DO-198, 1995

(c) Minimum aviation system performance standards DGNSS instrument approach system, DO-217 including Change 1, 1994.

(d) Minimum operational performance standards for airborne ILS localizer receiving equipment, DO-195, 1986.

(e) Minimum operational performance standards for airborne ILS glide slope receiving equipment, DO-192, 1986.

(f) Minimum operational performance standards for MLS airborne receiving equipment, DO-177, 1992.

[19] U.S. Air Force, Electronic Systems Center. *Specification for the Mobile Microwave Landing System. Revision A.* Hanscom Air Force Base, MA, September 1993.

[20] Kelly, R. J., and D. R. Cusick, Distance measuring equipment and its evolving role in aviation. *Advances in Electronics and Electron Physics* 68 (1986).

[21] van Graas, F., et al. FAA/Ohio University/UPS Autoland flight test results. *1995 National Technical Meeting*, Institute of Navigation, Washington, DC, January 1995.

[22] van Graas, F., D. Diggle, and R. Hueschen. Interferometric GPS Flight Reference/Autoland System: Flight test results. *Satellite Division Conference GPS-93*, Institute of Navigation, Washington, DC, September 1993.

[23] Diggle, D. *An Investigation Into the Use of Satellite-Based Positioning Systems for Flight Reference/Autoland Operations*, Ph.D. dissertation Ohio University, March 1994.

[24] Institute of Navigation, *Satellite Division Conference GPS-95*, September 1995:
(a) W. Hundley, et al. FAA-Wilcox Electric Category IIIb Feasibility Demonstration Program—Flight test results
(b) F. van Graas, et al. FAA/Ohio University/United Parcel Service DGPS Autoland flight test demonstration
(c) B. Elrod, et al. Stanford Telecommunications, Test results of local area augmentation of GPS with an in-band pseudolite.

[25] Durand, T. S., and R. J. Wasicko, Factors influencing glide path control in carrier landing. *AIAA J. Aircraft* 4:2 pp. 146–158. 1967. Washington, DC: American Institute of Aeronautics and Astronautics.

[26] Schoppe, W. J. *The Navy's Use of Digital Radio*, IEEE Transactions on Communication, 1979, pp. 1938–1945.

[27] Dornheim, M. A. *MM-Radar Shows Commercial Utility*, Aviation Week, November 2, 1994, pp. 55–59.

[28] Henry, H. G., R. C. Frietag, R. R. Shaller, and M. Cohn. *1989 IEEE Electron Devices Symposium (IEDM Convention Record)*, July 1989, pp. 7.7.1–7.7.4.

[29] Gold, T. Visual perception of pilots in carrier landing. *AIAA J. Aircraft* 11:12 pp. 723–729. 1974. Washington, DC: American Institute of Aeronautics and Astronautics.

[30] Sharp, R. (ed). JANES FIGHTING SHIPS 1995–96. Surrey, UK: Janes Information Group.

[31] EUROCAE, Minimum operational performance specification for ILS/MLS airborne receiving equipment. ED-74. Paris, France.

[32] Airlines Electronic Engineering Committee, Aeronautical Radio, Inc. Annapolis, MD.
(a) ARINC Characteristic 709-8 "Precision Airborne Distance Measuring Equipment. 8/94.
(b) ARINC Characteristic 710-10 "Mark 2 Airborne ILS Receiver." 7/96.
(c) ARINC Characteristic 727-1 "Airborne Microwave Landing System" 8/87.

CHAPTER 14

[1] ARINC, Inc. *Airborne Global Positioning System Receiver.* ARINC Characteristic 743. Aeronautical Radio, Inc., Annapolis, MD 21401. Current edition.

[2] ARINC, Inc. *Automatic Dependent Surveillance.* ARINC Characteristic 745. Aeronautical Radio, Inc., Annapolis, MD 21401. Current edition.

[3] ARINC, Inc. *Aviation Satelllite Communications System.* ARINC Characteristic 741. Aeronautical Radio, Inc., Annapolis, MD 21404. Current edition.

[4] Congress of the United States of America, *Federal Aviation Act of 1958.* Public Law 85-726, as amended. U.S. Congress, Washington, DC, August 23, 1958.

[5] Department of the Air Force, *Instrument Flying.* AFM 5137. Government Printing Office, Washington, DC 20402. Current edition.

[6] Federal Aviation Administration, *Federal Aviation Regulations, Part 121.* Government Printing Office, Washington, DC 20402. Current edition.

[7] Federal Aviation Administration, *Pilot's Handbook of Aeronautical Knowledge.* Advisory Circular 61-23. Government Printing Office, Washington, DC 20402. Current edition.

[8] Federal Aviation Administration, *Airman's Information Manual.* Government Printing Office, Washington, DC 20402. Current edition.

[9] Federal Aviation Administration, *Approval of Area Navigation Systems for Use in the U.S. National Airspace System.* Advisory Circular 90-45. Government Printing Office, Washington, DC 20402. Current edition.

[10] Federal Aviation Administration. *United States Standard for Terminal Instrument Procedures (TERPS).* FAA Handbook 8260.3. Government Printing Office, Washington, DC 20402. Current edition.

[11] Federal Aviation Administration, *Air Traffic Control.* FAA Order 7110.65. Government Printing Office, Washington, DC 20402. Current edition.

[12] Federal Aviation Administration, *Instrument Flying Handbook.* Advisory Circular 61-27. Government Printing Office, Washington, DC 20402. Current edition.

[13] Federal Aviation Administration. *National Airspace System Operational Concept.* NAS-SR-130, DOT/FAA/SE-93/1. National Technical Information Service, Springfield, VA 22161. June 1993.

[14] Federal Aviation Administration. *U.S. National Aviation Standard for the Global Positioning System Standard Positioning Service.* FAA Order 6880.1. Federal Aviation Administration, Washington, DC 20591. Current edition.

[15] Federal Aviation Administration. *Airborne Supplemental Navigation Equipment Using the Global Positioning System (GPS).* Technical Standard Order C129. Federal Aviation Administration, Washington, DC 20591. Current edition.

[16] Federal Aviation Administration. *Aviation System Capacity Plan.* Federal Aviation Administration, Washington, DC 20591. Current edition.

[17] Federal Aviation Administration. *Introduction to TCAS II*. Federal Aviation Administration, Washington, DC 20591. March 1990.

[18] Field, A. *International Air Traffic Control*. New York: Pergamon Press, 1985.

[19] Green, S. M., et al. Profile negotiation: A concept for integrating airborne and ground-based automation for merging arrival traffic. *Proceedings of the 1991 RTCA Technical Symposium*. RTCA, Inc., Washington, DC 20036. November 1991.

[20] Harman, W. H. TCAS: A system for preventing midair collisions. *Lincoln Laboratory Journal*, vol. 2, no. 3 (special issue on ATC). Lincoln Laboratory, Massachusetts Institute of Technology, Lexington, MA 02173-91108. Fall 1989.

[21] International Civil Aviation Organzation. *Manual on Implementation of a 1000 ft Vertical Separation Minimum (VSM) Between FL290 and FL410*. International Civil Aviation Organization, Montreal, Canada H3A 2R2. Current edition.

[22] International Civil Aviation Organization. *Guidance Material for Required Navigation Performance (RNP)*. International Civil Aviation Organization, Montreal, Canada H3A 2R2. To be published.

[23] International Civil Aviation Organization. *Secondary Surveillance Radar Mode S Advisory Circular*. Circular 174-AN/110. International Civil Aviation Organization, Montreal, Canada H3A 2R2. 1983.

[24] International Civil Aviation Organization. *Automatic Dependent Surveillance*. Circular 226-AN/135. International Civil Aviation Organization, Montreal, Canada H3A 2R2. 1983.

[25] International Civil Aviation Organization. *Aeronautical Mobile Satellite Service Advisory Circular*. International Civil Aviation Organization, Montreal, Canada H3A 2R2. To be published.

[26] International Civil Aviation Organization. *Airborne Collision Avoidance Systems*. Circular 195-AN/118. International Civil Aviation Organization, Montreal, Canada H3A 2R2. 1985.

[27] International Civil Aviation Organization. *Report of Special Committee on Future Air Navigation Systems Fourth Meeting*. International Civil Aviation Organization, Montreal, Canada H3A 2R2. May 1988.

[28] Jeppesen Sanderson, *Instrument Rating Manual*. Jeppesen Sanderson, Inc., Englewood, CO. Current edition.

[29] Mahapatra, P. R., and D. S. Zrnic. Sensors and systems to enhance aviation safety against weather hazards. *Proceedings of the IEEE*, vol. 79, no. 9. September 1991.

[30] Massoglia, P. L., et al. The use of satellite technology for oceanic air traffice control. *Proceedings of the IEEE*, vol. 77, no. 11 (special issue on ATC). November 1989.

[31] McCarthy, J. Advances in Weather Technology for the Aviation System. *Proceedings of the IEEE*, vol. 77, no. 11 (special issue on ATC). November 1989.

[32] Miller, C. A. et al. *A View of Air Traffic Management as a Challenge in Information Management*. Paper presented at GLOBAL NAVCOM 93, Seattle, WA, June 28–July 1, 1993. (Available from Secretariat of International Civil Aviation Organization.)

[33] Miller, C. A., et al. *Results of the Traffic Alert and Collision Avoidance System Transition Program*. Paper presented at 46th Annual International Air Safety Seminar, Kuala Lumpur, Malaysia. November 8–11, 1993. (Available from Flight Safety Foundation, Arlington, VA.)

[34] Miller, C. A., et al. *Report of the Future System Design Working Group.* Federal Aviation Administration, Washington, DC 20591. November 1990.

[35] Miller, C. A., and J. J. Fee. Future management of oceanic air traffic. *Avionics.* August 1991.

[36] National Oceanic and Atmospheric Administration. *U.S. Terminal Procedures.* National Ocean Service, NOAA, N/CG33, Rieverdale, MD 20737.

[37] Nolan, M. S. *Fundamentals of Air Traffic Control.* Belmont, CA: Wadsworth, 1990.

[38] Orlando, V. A. The Mode S beacon radar system. *Lincoln Laboratory Journal* vol. 2, no. 3 (special issue on ATC). Lincoln Laboratory, Massachusetts Institute of Technology, Lexington, MA 02173-91108. Fall 1989.

[39] Pozesky, M. T., and M. K. Mann. The US air traffic control system architecture. *Proceedings of the IEEE*, vol. 77, no. 11 (special issue on ATC). November 1989.

[40] Pozesky, M. T. *Concepts and Description of the Future Air Traffic Management System for the United States.* Federal Aviation Administration, Washington, DC 20591. April 1991.

[41] Roy, A., and A. D. Martelli. Aeronautical VHF datalink: Present and future. *Proceedings of Third Annual International Aeronautical Telecommunications Symposium on Data Link Integration.* Federal Aviation Administration and Aeronautical Radio, Inc., Annapolis, MD, May 1991.

[42] RTCA, Inc. *Report of Special Committee 159 on Minimum Aviation System Performance Standards (MASPs) for Global Positioning System (GPS).* DO-202. RTCA, Inc., Washington, DC 20036. Current edition.

[43] RTCA, Inc. *Minimum Operational Performance Standards for Airborne Supplemental Navigation Equipment Using Global Positioning System (GPS).* DO-208. RTCA, Inc., Washington, DC 20036. Current edition.

[44] RTCA, Inc. *Task Force Report on the Global Navigation Satellite System (GNSS) Transition and Implementation Strategy.* RTCA, Inc., Washington, DC 20036. September 1992.

[45] RTCA, Inc. *Minimum Operational Performance Standards for Air Traffic Control Radar Beacon System/Mode Select (ATCRBS/Mode S) Airborne Equipment.* DO-181A. RTCA, Inc., Washington, DC 20036. Current edition.

[46] RTCA, Inc. *Minimum Operational Performance Standard for Airborne Automatic Dependent Surveillance (ADS).* DO-212. RTCA, Inc., Washington, DC 20036. Current edition.

[47] RTCA, Inc. *Minimum Operational Performance Standard for Mode S Airborne Data Link Processor.* DO-203. RTCA, Inc., Washington, DC 20036. Current edition.

[48] RTCA, Inc. *Minimum Operational Performance Standards for Aeronautical Mobile Satellite Services (AMSS).* DO-210. RTCA, Inc., Washington, DC 20036. Current edition.

[49] RTCA, Inc. *Guidance on Aeronautical Mobile Satellite Service (AMSS) End-to-End System Performance.* RTCA, Inc., Washington, DC 20036. To be published.

[50] RTCA, Inc. *Design Guidelines and Recommended Standards to Support Open Systems Interconnection for Aeronautical Mobile Digital Communications.* DO-205. RTCA, Inc., Washington, DC 20036. Current edition.

[51] RTCA, Inc. *Minimum Operational Performance Standards for Traffic Alert and Collision Avoidance System (TCAS II) Airborne Equipment.* DO-185, Vol I and II. RTCA, Inc., Washington, DC 20036. Current edition.

[52] RTCA, Inc. *Minimum Operational Performance Standards for an Active Traffic Alert and Collision Avoidance System I (Active TCAS I) for Revenue Passenger Operations.* DO-197A. RTCA, Inc., Washington, DC 20036. Current edition.

[53] RTCA, Inc., Final report of RTCA Task Force 3 *Free flight implementation*, October 26, 1995.

[54] Sorensen, J., et al. Opportunities for integrating the aircraft FMS with the future air traffic management system. *37th Annual Air Traffic Control Association Conference Proceedings.* Air Traffic Control Association, Inc., Arlington, VA 22201. November 1992.

[55] Stallings, W. *Data and Computer Communications.* New York: Macmillian, 1988.

[56] Stevens, M. C. *Secondary Surveillance Radar.* Boston: Artech House, 1988.

[57] Stone, M. L., and J. R. Anderson. Advances in Primary Radar. *Lincoln Laboratory Journal* vol. 2, no. 3 (special issue on ATC). Lincoln Laboratory, Massachusetts Institute of Technology, Lexington, MA 02173-91108. Fall 1989.

[58] Unisys Corporation. *Automated Radar Terminal System (ARTS IIA) System Description.* Unisys Corporation, Paoli, PA 19301. May 1987.

[59] U.S. Department of Transportation and U.S. Department of Defense. *Federal Radionavigation Plan.* National Technical Information Service, Springfield, VA 22161. Published in even-numbered years.

[60] Van Sickle, N. D. *Modern Airmanship.* Tab Books, Inc., Blue Ridge Summit, PA. Current edition.

[61] Williamson, T., and N. Spencer. Development and operation of the Traffic Alert and Collision Avoidance System (TCAS). *Proceedings of the IEEE*, vol. 77, no. 11 (special issue on ATC). November 1989.

CHAPTER 15

[1] U.S. Department of the Navy. Advanced avionics architecture and technology review. Final report. U.S. Navy Naval Air Systems Command, August 1993.

[2] Aeronautical Radio, Inc. ARINC 429 *Digital Information Transfer System.* Annapolis, MD. 1993.

[3] Spitzer, C. R. *Digital Avionics Systems*, 2d ed. New York: McGraw-Hill, 1993.

[4] RTCA, Inc. DO-160 *Environmental Conditions and Test Procedures for Airborne Equipment.* Washington, DC, 1989.

[5] MIL-STD-1553 Digital Time Division Command/Response Multiplex Bus, September 1986.

[6] MIL-STD-704 Aircraft Electrical Power Characteristics, May 1991.

INDEX

Use this index in conjunction with the Table of Contents. Acronyms are defined in italics. Listings are under the noun except where an expression is commonly used, in which case the listing is under the adjective (e.g., *Kalman filter* or *multisensor navigation*). When *navigation* is the noun, the listing is under the adjective or the navaid. Some expressions are listed only under the acronym (e.g., FIR, FANS, AGC). Page numbers lead the reader to the vicinity of the topic in the book but not necessarily to all pages on which the expression occurs.

Printed and bound by CPI Group (UK) Ltd, Croydon, CR0 4YY